To Our Children

The Distance Formula

33. The distance between two points $P(x_1, y_1)$ ans $Q(x_2, y_2)$ is

$$d = \sqrt{(x_2 - x_1)^2 + (y_2 - y_1)^2}$$

Analytic Geometry

34. Slope of the line, $m = \dfrac{y_2 - y_1}{x_2 - x_1}$, $x_2 \neq x_1$

35. Equation of a line in slope-intercept form: $y = mx + b$

36. Circle: $(x - h)^2 + (y - k)^2 = r^2$, $C(h, k)$ is the center and r = radius

37. Parabolas: $y = ax^2 + bx + c$, $a \neq 0$

38. $y^2 = 4px$

39. Ellipse: $\dfrac{x^2}{a^2} + \dfrac{y^2}{b^2} = 1$

40. Hyperbola: $\dfrac{x^2}{a^2} - \dfrac{y^2}{b^2} = 1$

Properties of Logarithms

41. $\log_b (mn) = \log_b m + \log_b n$

42. $\log_b \left(\dfrac{m}{n}\right) = \log_b m - \log_b n$

43. $\log_b(m^r) = r(\log_b m)$

Sequences

44. Arithmetic sequence: $a_n = a_1 + (n - 1)d$

45. Sum of n terms: $S_n = \dfrac{n(a_1 + a_n)}{2}$

46. Geometric sequence: $a_n = a_1 r^{n-1}$

47. Sum of n terms: $S_n = \dfrac{a_1(r^n - 1)}{r - 1}$

Systematic Counting

48. $P(n,r) = n(n - 1)(n - 2) \cdots (n - r + 1)$
$$= \dfrac{n!}{(n - r)!}$$

49. $\dbinom{n}{r} = \dfrac{P(n,r)}{r!} = \dfrac{n!}{(n - r)!r!}$

50. Binomial theorem:

$$(a + b)^n = \binom{n}{0}a^n + \binom{n}{1}a^{n-1}b$$

$$+ \binom{n}{2}a^{n-2}b^2 + \cdots$$

$$+ \binom{n}{r}a^{n-r}b^r + \cdots + \binom{n}{n}b^n$$

Definitions of Trignometric Ratios

1. For (x,y), any point on the terminal sides of θ,

$$r = \sqrt{x^2 + y^2} \text{ and } r \neq 0$$

2. $\sin \theta = \dfrac{y}{r}$

3. $\cos \theta = \dfrac{x}{r}$

4. $\tan \theta = \dfrac{y}{x}$, $x \neq 0$

5. $\cot \theta = \dfrac{x}{y}$, $y \neq 0$

COLLEGE ALGEBRA AND TRIGONOMETRY WITH APPLICATIONS

SECOND EDITION

Bodh R. Gulati
Southern Connecticut State University

Helen G. Bass
Southern Connecticut State University

ωcb
Wm. C. Brown Publishers
Dubuque, Iowa

Copyright © 1986 by Bodh R. Gulati and Helen Bass.

Copyright © 1988 by Wm. C. Brown Publishers. All rights reserved

A previous edition was published under the title, ALGEBRA AND
TRIGONOMETRY: PRECALCULUS MATHEMATICS, copyright
© 1982 by Bodh R. Gulati and Helen Bass.

Library of Congress Catalog Card Number: 85-28761

ISBN 0-697-06860-9

Printed in the United States of America by Wm. C. Brown Publishers
2460 Kerper Boulevard, Dubuque, IA 52001

10 9 8 7 6 5 4 3 2

CONTENTS

8 TRIANGLES, VECTORS, AND COMPLEX NUMBERS   443

9 SYSTEMS OF LINEAR EQUATIONS AND LINEAR INEQUALITIES   491

10 SEQUENCES AND SERIES   555

11   ADDITIONAL TOPICS IN ALGEBRA   591

PREFACE

APPROACH

This book is intended to serve the needs of students with a wide range of abilities. In particular, we have in mind those students who approached mathematics with fear or considered mathematics of very little significance and so forgot what they learned in high school. We have made every effort to ease this "math anxiety" by using an intuitive, interesting, and down-to-earth approach that thoroughly relates basic concepts to realistic situations. We have presented problem solving as an integral part of the text and have emphasized practical skills and computational techniques. Applications from business, economics, management, and the social and natural sciences are used to motivate the discussion. Although the mathematical background necessary to use this text is minimal, we have not compromised theoretical accuracy.

INTENDED AUDIENCE

The traditional sequencing and choice of topics will enable an instructor to adapt the text to various course lengths. **The first two chapters constitute a careful review of intermediate algebra and can be skimmed or skipped depending on the background of the students.** It is our experience that material carefully selected from the first seven chapters provides a good foundation for calculus courses. Chapter 8 may be omitted, if necessary, without any loss of continuity. Then matrix algebra and its applications in Chapter 9, sequences and series in Chapter 10, and circles, conic sections, and counting problems in Chapter 11 can be covered.

This text can be used in a one-year sequence for students who need an intensive review of intermediate algebra and have no background in trigonometry. For this group, we recommend that Chapters 1 through 5 be covered during the first semester and Chapters 6 through 11 in the second semester.

The revision to the first edition is made in the light of valuable feedback from users of the text. Some outstanding features of this edition are:

Presentation

- There are clear, concise explanations in those areas that students ordinarily have trouble understanding.
- The underlying mathematical concepts are discussed in simple terms.
- Careful attention has been given to **readability.**
- There are noticeably detailed explanations of difficult algebraic concepts and trigonometric identities.

Learning Resources

- **Key Formulas** have been added immediately after **Key Words.**
- Over 500 carefully worked-out examples provide comprehensive explanations of concepts.
- **Exercise Sets** have been expanded. Over 3800 carefully graded exercises provide an opportunity for practice and building of skills.
- **Exercises for Review** (at the end of each chapter) have been enlarged. A **Chapter Test** has been introduced at the end of each chapter.
- Over 330 **calculator-oriented problems** are integrated in the text.

Content

- The use of a scientific calculator is introduced early in the text and is required in some chapters. More calculator-related problems have been added, where appropriate.
- A review of basic algebraic concepts is provided in one single chapter on **Fundamental Concepts.**
- The section on quadratic functions has been expanded and then moved to Chapter 3 on **Functions.** The section on variation has also been moved to an appropriate place in Chapter 3.
- The sections on circles and conic sections have been moved to Chapter 11 on **Additional Topics in Algebra.** This new chapter also contains material on systematic counting.
- The chapter on **Exponential and Logarithmic Functions** is expanded. New applications have been added. More graphs have been introduced.
- The chapter on **Trigonometric Functions** is rewritten with more applications in mind.
- Discussion of matrices and their properties has been moved to Chapter 9 on **Systems of Linear Equations and Linear Inequalities.**

- Maximum/minimum problems in Chapter 3 are solved by completing the squares.
- **The limit of a sequence,** a difficult topic frequently ignored in **Algebra and Trigonometry texts,** is discussed in Chapter 10.
- An intuitive proof of the **binomial theorem** (without mathematical induction) is provided in Chapter 11.

PROBLEM SOLVING

- Strategies for solving traditional algebra problems are introduced.
- The abundance of carefully worked-out examples enable the student to gain a firm grasp of the principles presented.
- Applications and a varied collection of interesting real-life problems depicting realistic situations in business, economics, management, natural and social sciences facilitate an intuitive understanding of basic concepts.

ACCURACY

- Every possible effort has been made to make this edition error-free. Despite our best efforts, it is possible that some errors may have escaped our attention. We will appreciate if any reader brings to our attention any error that remained unnoticed.

DESIGN

- The design features include functional use of second color, key words repeated in the margins and numerous illustrations.
- Key definitions, procedures, and propositions are highlighted.

CALCULATOR USE

Since the text continues to stress fundamental concepts and problem solving skills, we have kept the use of a calculator to a minimum. We believe that a calculator, when it is not being used as a learning device, should be used for lengthy calculations or for certain exponential and logarithmic values. For example, some problems in Chapter 4 that approximate the real zeros of a polynomial would be difficult to attempt without a calculator. The same is true for problems dealing with compound interest and exponential functions in Chapter 5. To assist the instructor and the student, we have used a calculator symbol to designate those problems for which a calculator is helpful. Although a calculator is not required for a vast majority of problems, its use may help in the learning process. The student will find it advantageous to own a hand-held scientific calculator that has common logarithms (log), natural logarithms (ln), trigonometric functions, exponential y^x and e^x keys, and a root key $\sqrt[x]{x}$ which allows easy calculations of powers and roots.

STUDY GUIDE

The study guide provides detailed solutions to representative problems in the text. This guide is designed to aid the student in problem solving and building of basic skills. Two practice chapter tests are provided at the end of each chapter to serve as self-tests. The detailed solutions to these chapter tests are also included.

INSTRUCTOR'S MANUAL

The instructor's manual provides answers to even-numbered problems and three forms of every chapter test along with their answers.

B. R. G.
H. G. B.

TO THE STUDENT

A few words of advice and perspective are in order. The mathematical concepts and techniques you will encounter in this book are relatively easy. The difficulties, if any, will be problems left over from your earlier experiences with the subject at the high school level and the amount of effort you put in. We expect that you have two years of high school algebra to draw on as you progress through the text. With less preparation, you will have to work harder.

The mathematics in this book is not difficult, but some concepts are new. To master them will require patience and perseverence. We hope that you find our style and exposition lucid; but do not expect to read this or any other mathematics book like a novel. Mathematics is not a spectator sport. You have to get involved.

Here are some suggestions that have been useful to other students who have taken similar courses:

1. Expect to read and reread the text thoughtfully.

2. Make a habit of having a paper and pencil at hand to answer questions raised in the text.

3. Work through the exercises and create examples of your own. This will reinforce the learning process.

4. Complete as many exercises as possible at the end of each section. Attempt the odd-numbered problems first and check your answers with those at the back of the text. At times your answer may be in a different form from that given in the book. If so, show that the two are equivalent.

5. Test your own understanding by working through the review exercises at the end of each chapter. Answers for these exercises appear at the back of the text.

6. Do not go on to new material until you master the previously covered material.

7. Do not hesitate to ask questions of your instructor. Remember that you are in the class to learn. If you have questions not yet answered in the class, be sure to ask for additional help.

8. Many schools offer a mathematics laboratory or some form of tutorial service. If your school does, take full advantage of it.

9. Detailed solutions to representative problems in the text have been compiled in a study guide. This study guide, complete with practice chapter tests, should be available in your college bookstore.

10. Complete the chapter test (at the end of each chapter) in a 50-minute period as though you were in class. Correct your work by checking your answers with the answers given in the text.

11. It may be to your advantage to own a scientific hand-held calculator equipped with logarithmic, exponential, and trigonometric functions. Check with your instructor for specific details.

We believe that you can gain a significant appreciation of many applications of mathematics from our text. All that we ask is that you give yourself a chance to learn what mathematics can do for you.

Have an enjoyable semester studying from this book. Good luck!

B. R. G
H. G. B

ACKNOWLEDGMENTS

We wish to express our appreciation to our colleagues at Southern Connecticut State University whose suggestions, comments, and criticism have helped us in this pleasant undertaking. We would also like to convey our sincere thanks to the reviewers for their assistance. These people reviewed the first edition:

Helen Baril, Quinnipiac College
Ignacio Bello, Hillsboro Community College
Stephen H. Brown, Auburn University
Charles Burrell, Western Carolina University
Louis J. Chatterly, Brigham Young University
Ronald Christensen, University of Wisconsin (LaCrosse Campus)
Charles Cook, Tri State University
John F. Davis, Cloud County Community College
Robert Denton, Orange Coast College
Sally Ann Folley, University of New Hampshire
Albert Giambrone, Sinclair Community College
Gary Grimes, Mt. Hood Community College
Jerry Karl, Golden West College
Anthony Lepre, Lehigh County Community College
Peter Lindstrom, Genesee Community College
Norman Locksley, Prince Georges Community College
Raymond McKellips, Southwestern Oklahoma State University
Ronald Marsh, University of Michigan (Dearborn Campus)
George Miller, Central Connecticut State University
Paul Miller, Community College of Baltimore
Allan Olinsky, Bryant College
Jerry Paul, University of Cincinnati
Adale Shapiro, Central Piedmont Community College
James Thorpe, Saddleback College

John Whitcomb, University of North Dakota
Jerry Williams, Georgia Southwestern College

Several users of the first edition have contributed significantly to the development of this version. These people reviewed the second edition:

Sister M. Alicia, Maria Regina College
Kenneth B. Black, Greenfield Community College
Joseph Cleary, Massasoit Community College
Irving H. Hart III, Oregon Institute of Technology
H. T. Hayslett, Jr., Colby College
Veronica T. McConnell, Bristol Community College
Mike Matus, Santa Monica College
Rosalyn Merzer, The College of Staten Island
Richard S. Montgomery, University of Connecticut at Avery Point
Gerald Stein, Merced College
John Tobey, Jr., North Shore Community College

In addition, we would like to single out as especially helpful the line-by-line review of this manuscript by Professor Thomas Woods of Central Connecticut State University. He carefully checked all the examples, problems, and exercises (a tedious but extremely important job).

We are grateful to the staff of Allyn and Bacon for their cooperation and valuable assistance. Notable among these are Deborah Schneck and Carl Lindholm, who worked with us on the first edition, and Carol Nolan-Fish, Louise Lindenberger, and Judy Fiske, who worked with us on this edition. The book could not have become a reality without and assistance of each of these fine people. To them we offer our sincere thanks.

We cannot afford to forget our indebtedness to our families. They endured many long separations, both physical and psychological as we concentrated on bringing this edition to a conclusion. We are profoundly grateful for their patience and understanding during this difficult period. They, good-naturedly (as always), accepted us when we were absent even while we were present.

B. R. G.
H. G. B

INDEX TO SELECTED APPLICATIONS

FUNDAMENTAL CONCEPTS

1

1.1

THE REAL NUMBERS

The numbers most familiar to you, namely,

$$1, 2, 3, 4, \ldots$$

natural numbers   are called the *natural numbers* (also known as *counting numbers*, or *positive integers*). The natural numbers are *closed* under the operations of addition and multiplication; that is, if we add or multiply two natural numbers, we again obtain a natural number. Certain basic rules of elementary algebra are so familiar that you probably apply them without even thinking. We state some of them explicitly here.

If a, b, and c are natural numbers, then

$a + b = b + a,$	commutativity for addition	**(1.1)**
$ab = ba,$	commutativity for multiplication	**(1.2)**
$a + (b + c) = (a + b) + c,$	associativity for addition	**(1.3)**
$a(bc) = (ab)c,$	associativity for multiplication	**(1.4)**
$a(b + c) = ab + ac,$	distributive property	**(1.5)**
$a \cdot 1 = 1 \cdot a = a.$	multiplicative identity	**(1.6)**

On the basis of these properties, we can justify other statements about the natural numbers. The operation of subtraction, for example, can be defined in terms of the familiar operation of addition.

Definition 1.1.1

> If a and b are natural numbers and c is another natural number such that $a + c = b$, then we write $c = b - a$.

For example, if $a = 5$ and $b = 8$, then $8 - 5 = c$ implies the existence of a natural number c such that $5 + c = 8$. In this case, $c = 3$. But what about $5 - 8 = c$? Clearly, no natural number satisfies this expression. Similarly, there exists no natural number that satisfies the equation $5 - 5 = c$. This demonstrates that the natural numbers are not closed under the operation of subtraction, since subtraction of one natural number from another does not *always* produce a natural number. Fortunately, this difficulty can be overcome by introducing the negative numbers $\ldots, -3, -2, -1$ and the number 0. Taken together,

integers   negative integers, zero, and the positive integers form the system of *integers*:

$$\ldots, -3, -2, -1, 0, 1, 2, 3, \ldots.$$

In this system every equation of the form $8 + c = 5$ always has a solution. Further, there exists a unique number 0 (zero) in this system with the property that for every integer a

$$a + 0 = 0 + a = a. \tag{1.7}$$

identity element We refer to zero as an *identity element* under the operation of addition. Also, for each integer a there exists another integer $(-a)$ such that

$$a + (-a) = (-a) + a = 0. \tag{1.8}$$

additive inverse The integer $(-a)$ is called the *additive inverse* of a. We must reinforce the fact that we have extended the system of natural numbers to the system of integers in which subtraction is always possible. Thus we have the following definition.

Definition 1.1.2

If a and b are integers and c is another integer such that $a + c = b$, then we write $c = b - a$.

Thus, given any two integers a and b, we can subtract a from b; that is, we can always find an integer c such that $b = a + c$. But can we divide a by b if a and b are integers? In other words, is there an integer d such that $a = bd$? If $a = 15$ and $b = 3$, there certainly exists an integer $d = 5$ such that $a = bd$. What happens if $a = 12$ and $b = 5$? There is no integer d satisfying the equation $12 = 5d$. This means that if we want a number system in which exact division is always possible, we must extend the system of integers so as to include
rational numbers fractions of the form $-\frac{1}{2}, \frac{2}{3}$, and $-\frac{3}{4}$. This system is called the system of *rational numbers*.

Definition 1.1.3

A *rational number* is a number that can be represented in the form a/b, where a and b are integers and $b \neq 0$.

Why do we make the restriction $b \neq 0$? Suppose that a is divisible by zero. Then what is the number d such that $\dfrac{a}{0} = d$? Note that $\dfrac{a}{0} = d$ means that $a = 0 \cdot d$. If $a = 0$, then the equation

$$a = 0 \cdot d$$

is valid no matter what value we choose for d—a situation we do not wish to permit. However, if $a \neq 0$, then the equation

$$a = 0 \cdot d$$

is *not* valid, because $0 \cdot d = 0$ no matter what number we choose for d, and we have assumed specifically that $a \neq 0$. Hence, in both cases, there is no satisfactory number d for which $\dfrac{a}{0} = d$. The expression $\dfrac{a}{0}$ remains undefined, and *division by zero is not permitted.*

For any number $a \neq 0$, there exists another number $1/a$ such that

$$a \cdot \left(\frac{1}{a}\right) = \left(\frac{1}{a}\right) \cdot a = 1. \tag{1.9}$$

multiplicative inverse   The number $1/a$, also written as a^{-1}, is called the *multiplicative inverse* of the number a.

We now state some of the basic rules that govern the operations with rational numbers.

$$\frac{a}{1} = a. \tag{1.10}$$

$$\frac{a}{b} = \frac{c}{d} \quad \text{if and only if} \quad ad = bc, \quad b \neq 0, \quad d \neq 0. \tag{1.11}$$

$$\frac{ac}{bc} = \frac{a}{b}, \quad b \neq 0, \quad c \neq 0. \tag{1.12}$$

$$\frac{1}{a} \cdot \frac{1}{b} = \frac{1}{ab}, \quad a \neq 0, \quad b \neq 0. \tag{1.13}$$

$$\frac{a}{b} \cdot \frac{c}{d} = \frac{ac}{bd}, \quad b \neq 0, \quad d \neq 0. \tag{1.14}$$

$$\frac{a}{c} + \frac{b}{c} = \frac{a+b}{c}, \quad c \neq 0. \tag{1.15}$$

$$\frac{a}{b} + \frac{c}{d} = \frac{ad+bc}{bd}, \quad b \neq 0, \quad d \neq 0. \tag{1.16}$$

$$\frac{a}{b} - \frac{c}{d} = \frac{ad-bc}{bd}, \quad b \neq 0, \quad d \neq 0. \tag{1.17}$$

$$\frac{a}{b} \div \frac{c}{d} = \frac{ad}{bc}, \quad b \neq 0, \quad c \neq 0, \quad d \neq 0. \tag{1.18}$$

Let us now illustrate some of these rules.

Example 1.1

(a) $\dfrac{5}{20} = \dfrac{5 \cdot 1}{5 \cdot 4} = \dfrac{1}{4}$ (1.12)

(b) $\dfrac{1}{2} \cdot \dfrac{1}{3} = \dfrac{1}{6}$ (1.13)

(c) $\dfrac{2}{3} \cdot \dfrac{4}{7} = \dfrac{8}{21}$ (1.14)

(d) $\dfrac{2}{7} + \dfrac{3}{7} = \dfrac{2 + 3}{7} = \dfrac{5}{7}$ 　　　　　　　　　　　　　(1.15)

(e) $\dfrac{2}{3} + \dfrac{3}{7} = \dfrac{(2 \cdot 7) + (3 \cdot 3)}{3 \cdot 7} = \dfrac{14 + 9}{21} = \dfrac{23}{21}$ 　　　(1.16)

(f) $\dfrac{3}{4} - \dfrac{2}{5} = \dfrac{(3 \cdot 5) - (4 \cdot 2)}{4 \cdot 5} = \dfrac{15 - 8}{20} = \dfrac{7}{20}$ 　　　(1.17)

(g) $\dfrac{2}{3} \div \dfrac{5}{7} = \dfrac{2 \cdot 7}{3 \cdot 5} = \dfrac{14}{15}$ 　　　　　　　　　(1.18)

The rational numbers, which include positive integers, negative integers, zero, and fractions, form a system in which the operations of addition, subtraction, multiplication, and division by nonzero numbers are always possible. The system of rational numbers we have developed seems to be all that we need in mathematics. However, this is not true, and the system of rational numbers still lacks some desirable properties. For example, what rational number a/b we can choose for x such that

$$x^2 = 2?$$

We can find rational numbers a/b for which $(a/b)(a/b)$ gets *closer and closer* to 2, but we cannot find any rational number a/b whose square is *exactly* 2. This means that we must extend the system of rational numbers. The extension of rational numbers to the system of real numbers is much more difficult than to describe the extension of the positive integers to the integers, or the integers to the rational numbers. Our discussion will, therefore, be mostly intuitive and geometric.

Every rational number can be expressed as a fraction a/b, $b \neq 0$, and every fraction can be written as a decimal expression, either repeating, such as

$$\frac{1}{3} = 0.3333333 \ldots \qquad \frac{1}{7} = 0.142857142857 \ldots$$

or terminating, such as

$$2 = 2.000000 \ldots$$

$$\frac{3}{4} = 0.750000 \ldots$$

$$\frac{1}{2} = 0.500000 \ldots$$

$$\frac{1}{5} = 0.200000 \ldots$$

Thus the *rational numbers can be expressed in either a repeating or a terminating decimal representation*. Let us now consider a nonrepeating and nonterminating decimal representation. You are familiar with the symbol π used in the area of the circle with radius r. The decimal representation of π is

$$3.141592654\ldots.$$

Another number e, which we use in Chapter 5, plays an important role in mathematics, and scientific calculators compute values of e^x at the push of a button. The decimal representation of e is

$$2.718281828459\ldots.$$

These numbers, π and e, and infinitely many more such as $\sqrt{3}$, $-\sqrt{5}$, and so forth, which cannot be represented in the form a/b, $b \neq 0$, are called **irrational numbers.** Taken together, the rational numbers and the irrational numbers form the real number system R.

real numbers   A geometric representation of the real number system is helpful in building an intuitive understanding of what the real numbers are. On a straight line, taken to be horizontal for convenience, we choose an arbitrary point and call it the origin 0. We then choose an arbitrary unit length on the line segment (see Figure 1.1). The unit point is usually placed to the right of the origin 0. With this unit of length, we represent positive and negative integers by a set of points equally spaced on the line, placing positive integers to the right and negative integers to the left of the origin 0.

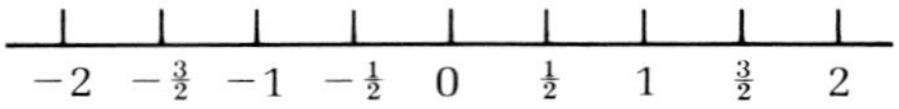

FIGURE 1.1

By dividing each of the segments into equal parts, we can associate fractions such as $-\frac{3}{2}, -\frac{1}{2}, \frac{1}{2}, \frac{3}{2}, \ldots$ with particular points. In general, we can represent the fraction m/n by dividing the unit length into n equal parts and counting off m of these parts. It may appear that this process of representing rational numbers by marking off the corresponding points would eventually cover the entire line. However, this is not the case, since there are an infinite number of points on the line which do not correspond to any rational number. For instance, there is no rational number that can express exactly the length of the hypotenuse of a right triangle that has legs of unit length.

Consider a right triangle OAB with legs OA and AB of unit length (see Figure 1.2). Then, according to the famous theorem of Pythagoras, the length of hypotenuse OB, denoted by x, is given by

$$x^2 = 1^2 + 1^2$$
$$x^2 = 2$$
$$x = \sqrt{2}.$$

It can be shown that the distance $OB = OC = \sqrt{2}$ cannot be represented in the form a/b where a and b are integers and $b \neq 0$. One can intuitively see by using

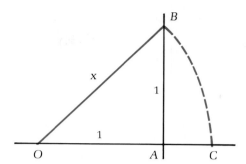

FIGURE 1.2

a scientific calculator that $\sqrt{2} = 1.414213562\ldots$, which is a nonrepeating decimal representation. However, the number $\sqrt{2}$, which is not rational, does correspond to a point on the number line.

Definition 1.1.4

> The collection of all rational numbers and irrational numbers is called the *real number system.*

Figure 1.3 will help you picture the hierarchy of the real number system *R* we have developed.

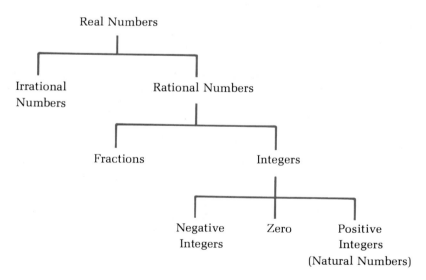

FIGURE 1.3

Observe that corresponding to every point on the number line there is one and only one real number, and conversely, to each real number there corresponds one and only one point on the number line. This correspondence (one number for one point, and vice versa) is called a *one-to-one correspondence.*

coordinate
graph

When a point on the number line is associated with a real number, the number is called the **coordinate** of the point, and the point is called the **graph** of the number.

EXERCISES

Classify the statements in Exercises 1 to 15 as true or false.

1. Every natural number is a rational number.
2. Every rational number is a natural number.
3. Every rational number is an integer.
4. Some integers are not rational.
5. Some rational numbers are not integers.
6. The sum of two rational numbers is rational.
7. The rational numbers are closed under multiplication.
8. The division of an integer by zero is not permitted.
9. The irrational numbers are closed under multiplication.
10. The integers are closed under subtraction.
11. The nonzero integers are closed under division.
12. The real numbers are closed under multiplication.
13. There exists a one-to-one correspondence between the rational numbers and the points on the real number line.
14. There exists a one-to-one correspondence between the real numbers and the points on the real number line.
15. Corresponding to every irrational number, there is a point on the real number line.
16. Specify whether the following real numbers are rational or irrational.

    (a) $\frac{4}{5}$ **(b)** -4 **(c)** 2546 **(d)** 0 **(e)** $\frac{22}{7}$ **(f)** $\sqrt{2}$

    (g) $\sqrt{9}$ **(h)** -98 **(i)** π **(j)** $-\sqrt{3}$ **(k)** $\frac{\sqrt{25}}{3}$ **(l)** $\frac{2\pi}{3}$

In Exercises 17 to 34, perform the indicated operation, and write the resulting expression as a rational number.

17. $\frac{3}{4} \cdot \frac{1}{2}$ 18. $\frac{2}{5} \cdot \frac{4}{5}$ 19. $\frac{2}{5} + \frac{1}{5}$ 20. $\frac{3}{7} + \frac{2}{7}$
21. $\frac{1}{2} + \frac{1}{3}$ 22. $\frac{1}{3} + \frac{1}{4}$ 23. $\frac{1}{3} - \frac{1}{4}$ 24. $\frac{1}{4} - \frac{1}{7}$
25. $\frac{2}{3} + \frac{3}{4}$ 26. $\frac{2}{3} - \frac{3}{4}$ 27. $\frac{2}{3} \div \frac{3}{4}$ 28. $\frac{4}{5} \div \frac{7}{8}$
29. $\frac{1}{2} + (\frac{1}{3} + \frac{1}{5})$ 30. $(\frac{1}{2} + \frac{1}{3}) + \frac{1}{5}$ 31. $\frac{2}{3}(\frac{1}{4} + \frac{1}{5})$ 32. $\frac{2}{3} \cdot \frac{1}{4} + \frac{2}{3} \cdot \frac{1}{5}$
33. $\frac{4}{5}(\frac{1}{5} - \frac{1}{6})$ 34. $\frac{4}{5} \cdot \frac{1}{5} - \frac{4}{5} \cdot \frac{1}{6}$

In Exercises 35 to 42, find the decimal representation.

35. $\frac{1}{6}$ 36. $\frac{2}{7}$ 37. $\frac{1}{16}$ 38. $\frac{1}{9}$
39. $\frac{1}{11}$ 40. $\frac{1}{33}$ 41. $\frac{10}{99}$ 42. $-\frac{4}{11}$

43. Show by means of examples that

$$\frac{1}{a} + \frac{1}{b} \neq \frac{1}{a+b}, \qquad a \neq 0, \quad b \neq 0, \quad b \neq -a.$$

44. Show by means of examples that

$$\frac{a}{b} + \frac{a}{c} \neq \frac{a}{b+c}, \qquad b \neq 0, \quad c \neq 0, \quad b + c \neq 0.$$

INEQUALITIES, INTERVALS, AND ABSOLUTE VALUE

inequalities

In elementary algebra and geometry, we study equalities almost exclusively. As we progress in mathematics, we shall see that the study of inequalities is both interesting and useful.

At present, we are concerned with inequalities among real numbers. We begin by recalling that positive numbers are associated with points to the right of 0 on the real line, while negative numbers are associated with points to the left of 0. A number a is greater than 0 if and only if a is positive. In symbols, we write

$$a > 0,$$

where the symbol $>$ is read "is greater than." Similarly, a number a is less than 0 if and only if a is negative. This statement is symbolized by

$$a < 0,$$

where the symbol $<$ is interpreted "is less than."

Definition 1.2.1

Given any two real numbers a and b, we say that a is greater than b $(a > b)$ or, equivalently, b is less than a $(b < a)$ if $a - b$ is a positive number. The symbols $<$ and $>$ are called *inequality signs*.

This definition is related to our geometric idea that larger numbers lie to the right of smaller ones on the number line. Figure 1.1 shows point 2, for example, to the right of -1, so $2 > -1$. Using Definition 1.2.1, we see that 2 is greater than -1 or, equivalently, -1 is less than 2 because $2 - (-1) = 3$, a positive number. The following propositions outline some basic properties of inequalities.

Proposition 1.2.1

The Trichotomy Law: For any real number a, one and only one of the following relations is true:

$$a < 0, \quad \text{or} \quad a = 0, \quad \text{or} \quad a > 0. \qquad \textbf{(1.19)}$$

Proposition 1.2.2

For any real numbers a and b, the following hold:

If $a > 0$ and $b > 0$, then

$$a + b > 0 \quad \text{and} \quad ab > 0. \tag{1.20}$$

If $a < 0$ and $b < 0$, then

$$a + b < 0 \quad \text{and} \quad ab > 0. \tag{1.21}$$

If $a < 0$ and $b > 0$, then

$$ab < 0. \tag{1.22}$$

Example 1.2

(a) To illustrate (1.20), let $a = 5$ and $b = 3$. Then

$$5 + 3 = 8 \quad \text{and} \quad 8 > 0$$

and

$$5 \cdot 3 = 15 \quad \text{and} \quad 15 > 0.$$

(b) To illustrate (1.21), let $a = -5$ and $b = -3$. Then

$$(-5) + (-3) = -8 \quad \text{and} \quad -8 < 0,$$

and

$$(-5) \cdot (-3) = 15 \quad \text{and} \quad 15 > 0.$$

(c) Let $a = -5$ and $b = 3$. Then

$$(-5) \cdot (3) = -15 \quad \text{and} \quad -15 < 0.$$

This illustrates property (1.22).

The following proposition concerning inequalities can be justified by using statements (1.19) to (1.22) and Definition 1.2.1.

Proposition 1.2.3

For any real numbers a, b, and c, the following are true:

If $a < b$ and $b < c$, then $a < c$. $\tag{1.23}$

If $a < b$, then $a + c < b + c$ for any number c, where c may be positive, negative, or zero. $\tag{1.24}$

If $a < b$ and c is any *positive* number, then $ac < bc$. $\tag{1.25}$

If $a < b$ and c is any *negative* number, then $ac > bc$. $\tag{1.26}$

We suggest that you create examples of your own to convince yourself that properties (1.23) to (1.26) are valid for all real numbers a, b, and c.

The symbol $a \geq b$ means that a is greater than or equal to b. Similarly, the notation $a \leq b$ means that a is less than or equal to b. Note that the properties of the relation $<$ may be extended to the relation $\leq$.

The following examples illustrate the use of some of these properties in solving sentences which involve inequalities.

Example 1.3 If $x + 5 < 9$, then by using (1.24) we add -5 to both sides of the inequality to obtain

$$x + 5 + (-5) < 9 + (-5),$$

or

$$x < 4.$$

Example 1.4 If $3x > 6$, then by using (1.25) we multiply both sides of the inequality by the positive number $\frac{1}{3}$ and obtain

$$\tfrac{1}{3}(3x) > \tfrac{1}{3}(6),$$

or

$$x > 2.$$

Remember that multiplying both sides of an inequality by a positive number does not change the direction of an inequality.

Example 1.5 If $-2x < 8$, then we multiply both sides of the inequality by the negative number $(-\frac{1}{2})$, thus obtaining

$$(-\tfrac{1}{2})(-2x) > (-\tfrac{1}{2})(8),$$

or

$$x > -4.$$

Recall from property (1.26) that multiplication of both sides of an inequality by a negative number reverses the direction of an inequality.

Sets and Subsets

A brief discussion of sets and subsets is now appropriate. In Section 1.1, we referred repeatedly to **a system, or collection,** of numbers satisfying certain properties. It is convenient to use the notation and terminology from set theory in discussing such systems of numbers. We define a **set** as a collection of objects of some kind. The objects in the collection are called the members, or **elements,** of the set.

element

Sets are designated by capital letters $A, B, C, \ldots$, while elements of sets are denoted by lowercase letters $a, b, c, \ldots$. If a set A consists of the numbers 2, 4, 6, and 8, we write

$$A = \{2, 4, 6, 8\}.$$

When a set has a large number of elements, we may abbreviate the listing of its members. For example, if set B consists of positive integers less than or equal to 1000, we may write

$$B = \{1, 2, 3, \ldots, 1000\}.$$

We interpret the dots to mean that the unlisted positive integers 4 through 999, inclusive, are also members of set B.

Essentially, a set may be specified in two different ways. One way is to list all the elements of a set; another is to enclose in braces a defining property to determine whether a given element belongs to that set. Consider, for example, the set of natural numbers less than or equal to 4. We may either write

$$A = \{1, 2, 3, 4\}$$

or denote this set by

$$\{x \,|\, x \text{ is a natural number and } x \leq 4\}.$$

We read this notation as "the set of all x such that x is a natural number and x is less than or equal to 4." The vertical line in the notation is read "such that." The letter of the alphabet in the notation (in this case, x) is called a
variable   **variable.** For the most part, we use the second method of defining a set, namely,

$$A = \{x \,|\, x \text{ has the property } T\},$$

and we interpret A to be the set of all elements x which have the property T.

To express the fact that a is the member of the set A, we write

$$a \in A,$$

where the symbol $\in$ is read "is an element of" or "belongs to." Thus if Q is the set of rational numbers, then

$$\tfrac{1}{2} \in Q.$$

To denote that $\sqrt{2}$ is *not* an element of Q, we write

$$\sqrt{2} \notin Q.$$

Definition 1.2.2 | A set with no elements is called the *null set;* it is denoted by $\{\ \}$ or $\varnothing$.

The set of all secretaries who can type 1200 words per minute, the set of sales executives in Sears and Roebuck over 400 years old, the set of all men who weigh 2000 pounds, and the set of all counting numbers greater than 5 and less than 6 are all examples of empty sets.

Let us now consider the sets

$$P = \{x \mid x \text{ is a positive integer}\}$$

and

$$Q = \{x \mid x \text{ is a rational number}\}.$$

How do these sets P and Q compare? An obvious statement is that every element of set P is also an element of the set Q.

Definition 1.2.3	A set A is a subset of a set B if and only if every element of A is also an element of B.

The notation $A \subseteq B$ or $B \supseteq A$ means that "A is a subset of B" or "B contains A." If there exists some element $x \in A$ which does not belong to B, then A is not a subset of B, and we write $A \nsubseteq B$ or $B \nsupseteq A$.

Example 1.6	In the set P of positive integers,

$$\{1, 2, 3\} \subseteq \{1, 2, 3, 4\},$$

$$\{2, 4, 6, 8\} \subseteq \{2, 4, 6, 8, 10\},$$

$$\{5, 6, 7\} \subseteq \{5, 6, 7\},$$

but

$$\{1, 2, 3\} \nsubseteq \{1, 2, 4\}.$$

Intervals

intervals Certain subsets of real numbers are so important to the study of mathematics that we give them special consideration. These subsets are called **intervals.**

Definition 1.2.4 *open intervals*	If a and b are any real numbers and $a < b$, then the *open interval* from a to b, written (a, b), is the collection of all real numbers greater than a and less than b. In symbols, $$(a, b) = \{x \mid a < x < b\}.$$

Geometrically, the open interval (a, b) may be illustrated on the real number line as shown in Figure 1.4. The open circles at the endpoints a and b represent the fact that the numbers corresponding to these points are not included in the interval.

FIGURE 1.4 a b

Definition 1.2.5
closed intervals

If a and b are any real numbers and $a < b$, then the *closed interval* from a to b, written $[a, b]$, is the collection of all real numbers greater than or equal to a and less than or equal to b. Thus.

$$[a, b] = \{x \mid a \leq x \leq b\}.$$

The closed interval $[a, b]$ is illustrated on the real number line in Figure 1.5. The closed circles at the endpoints a and b indicate that the numbers represented by the endpoints are also included in the interval. An interval which *half-open intervals* contains the endpoint a but not b is said to be *half-open on the right*. That is,

$$[a, b) = \{x \mid a \leq x < b\}.$$

FIGURE 1.5 a b

Similarly, an interval which contains the endpoint b but not a is said to be *half-open on the left*. In other words,

$$(a, b] = \{x \mid a < x \leq b\}.$$

The concept of an interval can be extended to cover some unusual cases. Suppose that the interval has no upper limit, as is the case for the set of all real numbers greater than or equal to 6, $\{x \mid x \geq 6\}$. In interval notation, this can be represented by $[6, \infty)$. Geometrically, this may be represented on the real number line as shown in Figure 1.6. The symbol ∞ is not a number and is interpreted as "no upper limit," or infinity. Similarly, $(-\infty, 3]$ represents the set $\{x \mid x \leq 3\}$ and consists of all real numbers less than or equal to 3. Graphically, this may be represented as shown in Figure 1.7. The symbol $-\infty$ means "no lower limit," or negative infinity.

FIGURE 1.6 $-1 \quad 0 \quad 1 \quad 2 \quad 3 \quad 4 \quad 5 \quad 6 \quad 7 \quad 8 \quad 9$

FIGURE 1.7 $-1 \quad 0 \quad 1 \quad 2 \quad 3 \quad 4 \quad 5 \quad 6 \quad 7 \quad 8$

Example 1.7   Sketch the graph of the following intervals:

(a) $(2, 4)$ **(b)** $[2, 4]$ **(c)** $(2, 4]$

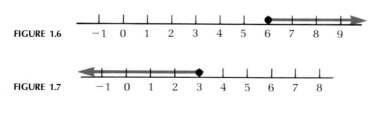

Solution:

(a) The open interval (2, 4) is the set of all real numbers greater than 2 and less than 4. The endpoints 2 and 4 are *not* included. The graph of the open interval (2, 4) is shown in Figure 1.8.

FIGURE 1.8   2   4

(b) The closed interval [2, 4] is the set of all real numbers greater than or equal to 2 and less than or equal to 4. The closed circles at the endpoints 2 and 4 indicate that the numbers 2 and 4 are included in the interval. The graph of the closed interval [2, 4] is shown in Figure 1.9.

FIGURE 1.9   2   4

(c) The half-open interval (2, 4] is the set of all real numbers greater than 2 and less than or equal to 4. Its graph is shown in Figure 1.10.

FIGURE 1.10   2   4

Example 1.8

Sketch the graph of the following intervals:

(a) $(5, \infty)$ **(b)** $[5, \infty)$ **(c)** $(-\infty, 2)$

Solution:

(a) The interval $(5, \infty)$ consists of all real numbers greater than 5. Graphically, this is shown in Figure 1.11. Note that the number 5 is *not* included in the interval.

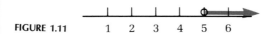

FIGURE 1.11   1   2   3   4   5   6

(b) The interval $[5, \infty)$ represents the set of all real numbers greater than or equal to 5. The graph of $[5, \infty)$ is shown in Figure 1.12. Note that the closed circle at 5 indicates that the number 5 is included in the interval.

FIGURE 1.12   1   2   3   4   5   6

(c) The interval $(-\infty, 2)$ is the set of all real numbers less than 2. The graph of the interval is shown in Figure 1.13.

FIGURE 1.13   1   2   3   4   5   6

Absolute Value

absolute value

We now introduce the geometric concept of distance in the real number system. Let a be the coordinate of a point on the number line. The number of units between this point and the origin 0 is called the *absolute value* of a and is denoted by $|a|$. Thus $|3| = 3$, $|-4| = 4$, $|4| = 4$, and so on. Thus the absolute value of a number gives only its distance from the origin and not its direction.

It is helpful to formulate a definition of absolute value that does not depend on a geometric notion. Of course, this definition gives us the same result as the geometric concept.

Definition 1.2.6

For any real number a,

$$|a| = \begin{cases} a & \text{if } a \geq 0 \\ -a & \text{if } a < 0. \end{cases}$$

From the definition, we can see that $|0| = 0$ and $|5| = 5$ because $5 > 0$, while $|-5| = -(-5) = 5$ because $-5 < 0$. These examples illustrate that the absolute value of a number is never negative.

We now state some of the important properties of absolute value.

Proposition 1.2.4

If a and b are arbitrary real numbers, then

$$|-a| = |a|, \tag{1.27}$$

$$-|a| \leq a \leq |a|, \tag{1.28}$$

$$|ab| = |a| \cdot |b| \tag{1.29}$$

$$ab \leq |a| \cdot |b|, \tag{1.30}$$

$$\left|\frac{a}{b}\right| = \frac{|a|}{|b|}. \tag{1.31}$$

We now construct some numerical examples to illustrate property (1.29) and leave the verification of other properties as an exercise to the reader.

If $a = 0$ or $b = 0$, then $ab = 0$ and (1.29) is trivial. If $ab > 0$, then $|ab| = ab$. Also, a and b must have the same sign; so either a and b are both positive, or a and b are both negative. Let $a = 2$ and $b = 3$; then $ab = 6$. By Definition

1.2.6, we see that

$$|6| = 6, \qquad |2| = 2, \qquad \text{and} \qquad |3| = 3.$$

Thus,

$$|ab| = |a| \cdot |b|.$$

If $a = -2$ and $b = -3$, then also $(-2) \cdot (-3) = 6$. By Definition 1.2.6, we see that

$$|6| = 6, \qquad |-2| = -(-2) = 2 \qquad \text{and} \qquad |-3| = -(-3) = 3,$$

and thus

$$|ab| = |a| \cdot |b|.$$

If $ab < 0$, then $|ab| = -ab$. Also, a and b must have opposite signs. Thus, if $a = 3$ and $b = -2$, then $ab = 3(-2) = -6$. By Definition 1.2.6, we have

$$|-6| = -(-6) = 6, \qquad |3| = 3, \qquad \text{and} \qquad |-2| = -(-2) = 2.$$

If $a = -3$ and $b = 2$, then $ab = (-3) \cdot (2) = -6$. Using Definition 1.2.6, we have

$$|-6| = -(-6) = 6, \qquad |-3| = -(-3) = 3, \qquad \text{and} \qquad |2| = 2.$$

Thus

$$|ab| = |a| \cdot |b|.$$

Some interesting and useful inequalities are combined into the following proposition.

Proposition 1.2.5

If a and b are arbitrary real numbers, then

$$|a + b| \le |a| + |b|, \tag{1.32}$$

$$|a| - |b| \le |a - b|. \tag{1.33}$$

Let us construct some numerical examples to illustrate the plausibility of these properties.

Example 1.9

To illustrate (1.32), let $a = 5$ and $b = -3$. Then

$$|5 + (-3)| \le |5| + |-3| = 5 + 3 = 8.$$

Since $|5 + (-3)| = |2| = 2$, clearly in this case

$$|5 + (-3)| < 8.$$

If we let $a = 5$ and $b = 3$, then

$$|5 + 3| = |5| + |3| = 5 + 3 = 8.$$

Example 1.10

To illustrate (1.33), let $a = -12$ and $b = -5$. Then

$$|a| - |b| = |-12| - |-5| = 12 - 5 = 7,$$

and

$$|a - b| = |-12 - (-5)| = |-12 + 5| = |-7| = 7.$$

Thus we have

$$|a| - |b| = |a - b|.$$

If we let $a = -12$ and $b = 5$, then

$$|a| - |b| = |-12| - |5| = 12 - 5 = 7$$

and

$$|a - b| = |-12 - 5| = |-17| = 17,$$

in which case we have

$$|a| - |b| < |a - b|.$$

You should construct other numerical examples to convince yourself that properties (1.32) and (1.33) hold for all real numbers.

EXERCISES

Express the statements in Exercises 1 to 10 by means of the appropriate symbol: $<$, $\leq$, $>$, or $\geq$.

1. 5 is greater than 2.

2. -6 is less than -4.

3. -4 is greater than -10.

4. x is between -2 and 3, inclusive.

5. x is positive.

6. x is nonpositive.

7. x is nonnegative.

8. x is negative.

9. $3x$ is less than or equal to 18.

10. $2x$ is greater than or equal to 10.

Justify the statements in Exercises 11 to 18 by citing the appropriate property (1.23) to (1.26).

11. If $x + 3 > 4$, then $x > 1$.

12. If $x - 2 < 6$, then $x < 8$.

13. If $2x < 8$, then $x < 4$.

14. If $3x > 6$, then $x > 2$.

15. If $\dfrac{x}{2} > 4$, then $x > 8$.

16. If $-3x < 15$, then $x > -5$.

17. If $-2x > -6$, then $x < 3$.

18. If $-\dfrac{1}{3}x < -4$, then $x > 12$.

Sketch the graph of the interval in Exercises 19 to 34.

19. $(-1, 3)$

20. $(-1, 4]$

21. $[1, 4)$

22. $[1, 4]$

23. $(-2, \frac{5}{2})$

24. $(-1, \frac{3}{4}]$

25. $(-\frac{1}{2}, 3)$

26. $(3, \frac{9}{2}]$

27. $(5, \infty)$ **28.** $[5, \infty)$ **29.** $(-\infty, \frac{1}{2}]$ **30.** $(-\infty, 3)$
31. $(-\infty, -2]$ **32.** $(-\infty, -2)$ **33.** $(-\infty, \frac{5}{2})$ **34.** $(-\infty, 5]$

Using interval notation, represent the sets in Exercises 35 to 46.

35. $\{x \mid 2 < x < 6\}$ **36.** $\{x \mid -1 \le x \le 1\}$ **37.** $\{x \mid -3 \le x < 5\}$ **38.** $\{x \mid -3 \le x \le 5\}$
39. $\{x \mid -4 < x \le 6\}$ **40.** $\{x \mid \frac{3}{2} \le x \le 5\}$ **41.** $\{x \mid -\frac{1}{2} < x \le 1\}$ **42.** $\{x \mid 0.5 < x < 2.5\}$
43. $\{x \mid x \le 2\}$ **44.** $\{x \mid x \ge -1\}$ **45.** $\{x \mid x < -\frac{5}{2}\}$ **46.** $\{x \mid x \ge \frac{3}{2}\}$

Compute the expressions in Exercises 47 to 58.

47. $|-7|$ **48.** $|15|$ **49.** $|6 - 8|$ **50.** $|-8 - 6|$
51. $|12 - 7|$ **52.** $|7 - 12|$ **53.** $|7| - |-12|$ **54.** $|6| - |8|$
55. $|-3| - |-12|$ **56.** $|-2| - |-\frac{1}{4}|$ **57.** $|\frac{3}{2}| - |\frac{7}{4}|$ **58.** $|\frac{3}{2}| - |-\frac{7}{4}|$

Justify the statements in Exercises 59 to 66 by citing the appropriate property, (1.29) to (1.31).

59. $|3| \cdot |-2| = |-6|$ **60.** $|5| \cdot |4| = |20|$ **61.** $|-5| \cdot |-4| = |20|$
62. $4(-3) \le |4| |-3|$ **63.** $(-5)(-6) \le |-5| |-6|$ **64.** $(-\frac{1}{2})(-\frac{3}{4}) \le |-\frac{1}{2}| |-\frac{3}{4}|$
65. $\left|\dfrac{-5}{-4}\right| = \dfrac{|-5|}{|-4|}$ **66.** $\left|\dfrac{-12}{5}\right| = \dfrac{|-12|}{|5|}$

Illustrate by numerical examples that the statements in Exercises 67 to 70 are valid for all real numbers a and b.

67. $|ab| = |a| \cdot |b|$ **68.** $ab \le |ab|$ **69.** $|a + b| \le |a| + |b|$ **70.** $|a| - |b| \le |a - b|$

1.3

INTEGRAL EXPONENTS

You are probably familiar with symbols of the form x^m, where x is any real number and m is a nonnegative integer.

Definition 1.3.1

> For any real number x and any positive integer m,
>
> $$x^m = x \cdot x \cdot x \cdots x \qquad (m \text{ factors}).$$

exponent   In the symbol x^m, m is called an **exponent** or **power,** x is called the *base*,
base   and the expression x^m is called an **exponential.**

Example 1.11

(a) $2^3 = 2 \cdot 2 \cdot 2 = 8$.

(b) $(-3)^4 = (-3)(-3)(-3)(-3) = 81$.

(c) $(-1)^3 = (-1)(-1)(-1) = -1$.

(d) $(\frac{1}{4})^3 = (\frac{1}{4})(\frac{1}{4})(\frac{1}{4}) = \frac{1}{64}$.

The use of exponents is governed by several rules, which we review here briefly.

Proposition 1.3.1

For any real number x and any positive integers m and n,

$$x^m \cdot x^n = x^{m+n}.$$

Thus to multiply two exponentials with the same base, we simply add the exponents.

Example 1.12

(a) $2^3 \cdot 2^4 = 2^{3+4} = 2^7 = 128.$

(b) $y^4 \cdot y^2 = y^{4+2} = y^6.$

(c) $b^{3r} \cdot b^{2t} = b^{3r+2t}$,   where r and t are positive integers.

(d) $z^{n+1} \cdot z^{2n+3} = z^{3n+4}$,   where n is a positive integer.

Proposition 1.3.2

For any real number $x \neq 0$ and any positive integers m and n,

$$\frac{x^m}{x^n} = \begin{cases} x^{m-n} & \text{if } m > n \\ \dfrac{1}{x^{n-m}} & \text{if } m < n \\ 1 & \text{if } m = n \end{cases}$$

Example 1.13

(a) $\dfrac{3^6}{3^2} = 3^{6-2} = 3^4 = 81.$

(b) $\dfrac{a^5}{a^3} = a^{5-3} = a^2,$ $a \neq 0.$

(c) $\dfrac{b^5 \cdot b^6}{b^9} = \dfrac{b^{5+6}}{b^9} = \dfrac{b^{11}}{b^9} = b^{11-9} = b^2,$ $b \neq 0.$

(d) $\dfrac{c^4}{c^4} = 1,$ $c \neq 0.$

Example 1.14

(a) $\dfrac{2^5}{2^9} = \dfrac{1}{2^{9-5}} = \dfrac{1}{2^4} = \dfrac{1}{16}.$

(b) $\dfrac{a^3}{a^7} = \dfrac{1}{a^{7-3}} = \dfrac{1}{a^4},$ $a \neq 0.$

(c) $\dfrac{b^3 \cdot b^4}{b^8} = \dfrac{b^{3+4}}{b^8} = \dfrac{b^7}{b^8} = \dfrac{1}{b^{8-7}} = \dfrac{1}{b},$ $b \neq 0.$

Proposition 1.3.3

If m and n are positive integers, then for any real number x
$$(x^m)^n = x^{mn}.$$

Example 1.15

(a) $(b^3)^4 = b^{3\cdot4} = b^{12}$.

(b) $(2^2)^4 = 2^{2\cdot4} = 2^8 = 256$.

(c) $(3^3)^2 = 3^{3\cdot2} = 3^6 = 729$.

Proposition 1.3.4

If x and y are real numbers, then for any positive integer m,
$$(xy)^m = x^m y^m.$$

Example 1.16

(a) $(ab)^8 = a^8 b^8$.

(b) $(a^2 b^3)^4 = (a^2)^4 (b^3)^4 = (a^{2\cdot4})(b^{3\cdot4}) = a^8 b^{12}$.

Proposition 1.3.5

Let x and y be any numbers with $y \neq 0$; then for any positive integer m,
$$\left(\frac{x}{y}\right)^m = \frac{x^m}{y^m}.$$

Example 1.17

(a) $\left(\dfrac{2}{3}\right)^4 = \dfrac{2^4}{3^4} = \dfrac{16}{81}$.

(b) $\left(\dfrac{ab^2}{c}\right)^3 = \dfrac{a^3 b^6}{c^3}, \quad c \neq 0$.

Example 1.18

Use the properties of exponents to simplify the following expressions. All answers should have positive exponents.

(a) $(2x^3 y^2)(3xy^3)$.

(b) $(3x^2 y)^4$.

(c) $\left(\dfrac{2x^2}{y}\right)^3 \left(\dfrac{3y^2}{x^3}\right)^2, \quad x \neq 0, \quad y \neq 0$.

Solution:

(a) $(2x^3 y^2)(3xy^3) = 6 \cdot x^{3+1} \cdot y^{2+3} = 6x^4 y^5$.

(b) $(3x^2 y)^4 = 3^4 \cdot (x^2)^4 \cdot y^4 = 81x^8 y^4$.

(c) $\left(\dfrac{2x^2}{y}\right)^3\left(\dfrac{3y^2}{x^3}\right)^2 = \dfrac{(2x^2)^3}{y^3}\cdot\dfrac{(3y^2)^2}{(x^3)^2}$, $x\neq 0,\quad y\neq 0$ (Proposition 1.3.5)

$= \dfrac{2^3(x^2)^3}{y^3}\cdot\dfrac{3^2(y^2)^2}{(x^3)^2}$ (Proposition 1.3.4)

$= \left(\dfrac{8x^6}{y^3}\right)\left(\dfrac{9y^4}{x^6}\right)$ (Proposition 1.3.3)

$= 72y^{4-3}$ (Proposition 1.3.2)

$= 72y.$

Thus far, we have discussed in this section only positive integers as exponents. We now extend the concept of an exponent to the set of *all integers* so that expressions of the form x^0 and x^{-5} carry useful meanings. The definitions for this type of exponent must conform to the previously established laws of exponents.

First, we consider the meaning of x^0, where x is any real number different from zero. If Proposition 1.3.1 is to hold for $n = 0$, then we must have

$$x^m \cdot x^0 = x^{m+0},$$

but since

$$m + 0 = m,$$

it follows that

$$x^{m+0} = x^m.$$

Hence

$$x^m \cdot x^0 = x^m,$$

and we have the following definition.

Definition 1.3.2

For any real number x different from zero,
$$x^0 = 1.$$

Example 1.19

(a) $7^0 = 1.$

(b) $8(x^0) = 8\cdot 1 = 8,$ $x \neq 0.$

(c) $(8x)^0 = 1,$ $x \neq 0.$

(d) $3(xy)^0 z = 3\cdot 1\cdot z = 3z,$ $x \neq 0,\quad y \neq 0.$

We now discuss the meaning of x^{-m}, where $x \neq 0$ and m is a positive integer. Considering Proposition 1.3.1 again, with $n = -m$ and $x \neq 0$, we have

$$x^m \cdot x^{-m} = x^{m+(-m)}.$$

Since $m + (-m) = (-m) + m = 0$ for all integers, it follows that

$$x^{m+(-m)} = x^{(-m)+m} = x^0 = 1.$$

Hence we conclude that $x^m \cdot x^{-m} = 1$. We therefore have the following formal definition.

Definition 1.3.3

For any real number x different from zero,

$$x^{-m} = \frac{1}{x^m}.$$

Example 1.20

(a) $6^{-2} = \dfrac{1}{6^2} = \dfrac{1}{36}.$

(b) $x^{-3} = \dfrac{1}{x^3}, \qquad x \neq 0.$

(c) $\dfrac{1}{x^{-2}} = \dfrac{1}{1/x^2} = x^2, \qquad x \neq 0.$

(d) $x^2 y^{-3} = \dfrac{x^2}{y^3}, \qquad y \neq 0.$

We now assume that the laws of exponents established earlier are valid for all integers m and n, whether positive, negative, or zero.

Example 1.21

Use the properties of exponents to simplify each of the following expressions. All answers should have positive exponents.

(a) $(x^2 y^{-3})^{-2} \qquad (x \neq 0, \quad y \neq 0).$

(b) $\dfrac{12x^4 y^{-3} z^{-2}}{6x^{-2} y} \qquad (x \neq 0, \quad y \neq 0, \quad z \neq 0).$

Solution:

(a) $(x^2 y^{-3})^{-2} = (x^2)^{-2}(y^{-3})^{-2} = x^{-4} y^6 = \dfrac{1}{x^4} \cdot y^6 = \dfrac{y^6}{x^4} \qquad (x \neq 0, \quad y \neq 0).$

(b) $\dfrac{12x^4y^{-3}z^{-2}}{6x^{-2}y} = \dfrac{12}{6}\dfrac{x^4y^{-3}z^{-2}}{x^{-2}y} = 2 \cdot \dfrac{x^4}{x^{-2}} \cdot \dfrac{y^{-3}}{y} \cdot \dfrac{z^{-2}}{1}$

$$(x \neq 0, \quad y \neq 0, \quad z \neq 0)$$

$$= 2 \cdot x^{4-(-2)} \dfrac{1}{y^{1-(-3)}} \dfrac{1}{z^2}$$

$$= 2x^6 \dfrac{1}{y^4} \dfrac{1}{z^2}$$

$$= \dfrac{2x^6}{y^4z^2} \qquad (x \neq 0, \quad y \neq 0, \quad z \neq 0).$$

EXERCISES

Evaluate each expression in Exercises 1 to 20.

1. $2^3 \cdot 2^2$ **2.** $3^2 \cdot 3^3$ **3.** $(2^2)^3$ **4.** $(3^3)^2$

5. $(2^3 \cdot 3^2)^4$ **6.** $(5^2 \cdot 3^4)^3$ **7.** $\dfrac{2^7}{2^3}$ **8.** $\dfrac{5^4}{5^2}$

9. $\dfrac{3^6 \cdot 3^4}{3^8}$ **10.** $\dfrac{5^4 \cdot 5^6}{5^{10}}$ **11.** $\left(\dfrac{2}{3}\right)^3\left(\dfrac{3}{4}\right)^4$ **12.** $\left(\dfrac{2}{3}\right)^4\left(\dfrac{3}{2}\right)^5$

13. $\left(\dfrac{4}{5}\right)^2\left(\dfrac{5}{4}\right)^3$ **14.** $\dfrac{(3^4)^2 \cdot 2^4}{(2^3)^2 \cdot (3^2)^3}$ **15.** $\dfrac{3^{-2} \cdot 4^{-3}}{3^4 \cdot 2^{-4}}$ **16.** $\dfrac{4^{-3} \cdot 5^{-2}}{4^{-7} \cdot 2^3}$

17. $\dfrac{2^{-3} \cdot 4^{-2}}{5^{-2} \cdot 4^{-5}}$ **18.** $\dfrac{3^4 \cdot 4^{-3}}{2^{-5} \cdot 3^5}$ **19.** $\left(\dfrac{2}{3}\right)^2\left(\dfrac{3}{4}\right)^4\left(\dfrac{1}{2}\right)^{-5}$ **20.** $\left(\dfrac{4}{5}\right)^3\left(\dfrac{4}{7}\right)^{-3}\left(\dfrac{2}{5}\right)^{-6}$

Use the properties of exponents to simplify each expression in Exercises 21 to 50. All answers should have positive exponents.

21. $(2a^2)(3a^5)$ **22.** $(3a^m)(4a^{n+2})$ **23.** $(4a^2b^3)(5a^3b)$

24. $(2a^2b^3c^4)(5ab^2c^5)$ **25.** $(a^2b)^4(a^3b)^2$ **26.** $(a^2b^3c^4)^3(2ab^2c^2)^2$

27. $\dfrac{(ab^2)^4}{a^2b^3}$ **28.** $\dfrac{(3a^2b)^3}{4a^3b^2}$ **29.** $\dfrac{(c^2d^3)^4}{(c^3d^2)^3}$

30. $\dfrac{(3ab)^2(2bc)^3}{(a^2)^4(b^3)^3c^2}$ **31.** $\left(\dfrac{2a}{b}\right)^3\dfrac{b^5}{3a}$ **32.** $\left(\dfrac{a}{b}\right)^4\left(\dfrac{a^2}{b^3}\right)^2\left(\dfrac{c^4}{a}\right)^3$

33. $(3x)^{-2}$ **34.** $(2x^{-3})^2$ **35.** $(xy^{-1}z^{-2})^3$

36. $(x^2y^{-1})^{-2}$ **37.** $(x^3y^{-2}z^{-4})^{-1}$ **38.** $(3x^{-1}y)^{-4}$

39. $(x^{-2}y^3z)^3(x^2y^3z^6)^{-2}$ **40.** $(2xy^{-1})^{-2}(5x^{-2}y^2)^3$ **41.** $\dfrac{3x^{-2}y^4}{4x^4y^{-2}}$

42. $\left(\dfrac{x^{-2}y^3}{x^{-5}y^{-2}}\right)^{-2}$.

43. $\left(\dfrac{x^2}{2y^4}\right)^{-3}$

44. $\left(\dfrac{xy^{-2}z^{-3}}{x^2y^3z^4}\right)^{-4}$

45. $\left(\dfrac{3x^2y^{-3}z^{-5}}{2x^3y^2z^6}\right)^{-2}$

46. $\dfrac{(x^3y^{-4}z)^{-3}}{(x^4y^2)^{-4}}$

47. $\left(\dfrac{xy^2z^3d^{-1}}{x^2yz^5}\right)^0$

48. $\dfrac{(4x^{-3}y^2)^{-2}}{5x^{-2}y^{-3}}$

49. $\dfrac{(x^2y^{-3}z)^{-4}}{(x^{-3}y^2z)^{-2}}$

50. $\dfrac{(5x^2y^{-1}z^{-4})^{-2}}{(3x^3y^{-2}z^{-5})^{-4}}$

1.4 ROOTS AND RADICALS

We now assign meanings to expressions of the form $\sqrt[n]{x}$ so as to be consistent with the laws of exponents established in Section 1.3. In particular, for $n = 2$, if x and y are positive real numbers, then we write

$$y = \sqrt{x} \qquad \text{if and only if} \qquad y^2 = x.$$

principal square root This positive number y is called the **principal square root** of the positive number x.

Example 1.22

(a) $\sqrt{25} = 5$ because $5^2 = 25$.

(b) $\sqrt{\frac{4}{9}} = \frac{2}{3}$ because $\left(\frac{2}{3}\right)^2 = \frac{4}{9}$.

Actually, if x is a positive real number and $y^2 = x$, then so is $(-y)^2 = x$. Thus there are two real numbers, one positive and the other negative, whose square is x. However, we follow the convention that the symbol $\sqrt{x}$ always represents the positive square root of x.

Next, we consider expressions of the form $\sqrt[3]{x}$. We write

$$y = \sqrt[3]{x} \qquad \text{if and only if} \qquad y^3 = x.$$

A little reflection will show that if $x > 0$, then $y > 0$, but if $x < 0$, then $y < 0$.
principal cube root This number y is called the **principal cube root** of x, or simply the cube root of x.

Example 1.23

(a) $\sqrt[3]{27} = 3$ since $3^3 = 27$.

(b) $\sqrt[3]{-27} = -3$ since $(-3)^3 = -27$.

(c) $\sqrt[3]{\dfrac{-64}{125}} = -\dfrac{4}{5}$ since $\left(-\dfrac{4}{5}\right)^3 = -\dfrac{64}{125}$.

In general, if n is a positive integer, then we write

$$y = \sqrt[n]{x} \qquad \text{if and only if} \qquad y^n = x$$

radical
radicand
index

and call $\sqrt[n]{x}$ the nth root of x. The symbol $\sqrt[n]{x}$ is called a **radical,** x is called the **radicand,** and n is the **index of the radical.** Further, if $x > 0$, then there exists one and only one positive number y satisfying the condition

$$y = \sqrt[n]{x}.$$

principal nth root

This positive number is called the **principal nth root** of x. Thus 5 is the principal square root of 25, 2 is the principal cube root of 8, 3 is the principal fourth root of 81, and so on.

If $x < 0$ and n is odd, then there exists one negative nth root of x. This negative number is the **principal nth root** of x. For example, -2 is the principal cube root of -8, since $(-2)^3 = -8$; -3 is the principal fifth root of -243; and so on.

Example 1.24

(a) $\sqrt[3]{64} = 4$ since $4^3 = 64$.
(b) $\sqrt[5]{-32} = -2$ since $(-2)^5 = -32$.
(c) $\sqrt[6]{729} = 3$ since $3^6 = 729$.

If $x < 0$ and n is even, then the nth root of x is *not* a real number. For example, $\sqrt{-16}$ is not a real number because there is no real number whose square is -16.

We must reinforce the fact that if $x < 0$ and n is even, then $\sqrt[n]{x}$ is *not* a real number. If x is any real number and n is odd, then $\sqrt[n]{x}$ represents one and only one real number.

Let us look at some examples.

Example 1.25

(a) $\sqrt{16} = 4$.
(b) $\sqrt[3]{-125} = -5$.

(c) $\sqrt[3]{\dfrac{1}{216}} = \dfrac{1}{6}$.

(d) $\sqrt{-25}$ is not a real number.

We observed earlier that if x and y are positive real numbers and $y = \sqrt{x}$, then $y^2 = x$. This statement implies that

$$\sqrt{x} \cdot \sqrt{x} = x,$$

or

$$(\sqrt{x})^2 = x.$$

Similarly, if $y = \sqrt[3]{x}$, then $y^3 = x$; that is,

$$(\sqrt[3]{x})^3 = x.$$

In general, if n is any positive integer, then

$$(\sqrt[n]{x})^n = x.$$

Further, if $x < 0$ and n is an odd positive integer or if $x > 0$, then

$$\sqrt[n]{x^n} = x.$$

If $x < 0$ and n is an even positive integer, then

$$\sqrt[n]{x^n} = |x|.$$

You may wonder why absolute value signs are necessary when $x < 0$ and n is an even positive integer. Consider, for example, $x = -3$ and $n = 2$; then

$$\sqrt{(-3)^2} = \sqrt{9} = 3,$$

which is the absolute value of -3. We must emphasize that if $x < 0$, then $\sqrt{x^2} \neq x$.

Example 1.26

(a) $\sqrt{7^2} = 7.$

(b) $\sqrt[3]{5^3} = 5.$

(c) $\sqrt[5]{(-2)^5} = -2.$

(d) $\sqrt{(-5)^2} = |-5| = 5.$

Radicals obey some simple rules, which we now state.

Proposition 1.4.1

If x and y are real numbers, then

(a) $\sqrt[n]{x^n} = x,$ where n is an odd positive integer.

(b) $\sqrt[n]{x^n} = |x|,$ where n is an even positive integer.

(c) $\sqrt[n]{x} \cdot \sqrt[n]{y} = \sqrt[n]{xy}$ if $\sqrt[n]{x}$ and $\sqrt[n]{y}$ exist.

(d) $\sqrt[n]{\dfrac{x}{y}} = \dfrac{\sqrt[n]{x}}{\sqrt[n]{y}},$ $y \neq 0,$ $\sqrt[n]{x}$ and $\sqrt[n]{y}$ exist.

Parts (a) and (b) follow directly from the foregoing discussion. To prove (c), let

$$x = a^n, \qquad y = b^n.$$

Then

$$xy = a^n \cdot b^n = (ab)^n,$$

or

$$\sqrt[n]{xy} = \sqrt[n]{(ab)^n} = ab = \sqrt[n]{x} \cdot \sqrt[n]{y}.$$

The proof of (d) is similar and is left as an exercise for the reader.

Definition 1.4.1

A radical expression is in *simplified form* if none of the factors of the radicand (the quantity under the radical sign) can be expressed as powers greater than or equal to the index.

This means that satisfying the condition for simplified form amounts to taking as much out from under the radical sign as possible; that is, no perfect squares can be factors of the quantity under a square root sign, no perfect cubes can be factors of what is under a cube root sign, and so on.

However, writing a radical expression in simplified form does not always yield a simpler-looking expression. Simplified form for radicals is simply a way of writing radicals so that they are easiest to work with.

Example 1.27

Write the following radical expressions in simplified form.

(a) $\sqrt{18}$ (b) $\sqrt[3]{108}$ (c) $\sqrt[3]{\dfrac{-8}{27}}$ (d) $\sqrt[4]{\dfrac{81}{256}}$

Solution:

(a) The largest perfect square that divides 18 is 9. We write

$$18 = (9)(2)$$

and then apply Proposition 1.4.1(c). Thus,

$$\sqrt{18} = \sqrt{9 \cdot 2}$$
$$= \sqrt{9} \cdot \sqrt{2}$$
$$= 3\sqrt{2}.$$

(b) The largest perfect cube that divides 108 is 27. We write

$$108 = 27 \cdot 4$$

and then apply Proposition 1.4.1(c). We have

$$\sqrt[3]{108} = \sqrt[3]{27 \cdot 4}$$
$$= \sqrt[3]{27}(\sqrt[3]{4})$$
$$= 3\sqrt[3]{4}.$$

(c) Here, both -8 in the numerator and 27 in the denominator are perfect cubes. We use Proposition 1.4.1(d) and write the numerator and denominator as two separate radicals:

$$\sqrt[3]{\frac{-8}{27}} = \frac{\sqrt[3]{-8}}{\sqrt[3]{27}} = -\frac{2}{3}.$$

(d) Using Proposition 1.4.1(d), we first write the numerator and denominator as two separate radicals:

$$\sqrt[4]{\frac{81}{256}} = \frac{\sqrt[4]{81}}{\sqrt[4]{256}}.$$

Observe that $81 = 3^4$ and $256 = 4^4$, and we can express both 81 in the numerator and 256 in the denominator as fourth power of 3 and 4, respectively. Hence,

$$\sqrt[4]{\frac{81}{256}} = \frac{3}{4}.$$

Example 1.28

Write each radical expression in simplified form.

(a) $\sqrt{9x^2}$ **(b)** $\sqrt{x^3}$ **(c)** $\sqrt[3]{x^6 y^9}$ **(d)** $\sqrt[4]{x^5 y^7}$

Solution:

(a) $\sqrt{9x^2} = \sqrt{9} \cdot \sqrt{x^2} = 3|x|$
(b) $\sqrt{x^3} = \sqrt{x^2} \cdot \sqrt{x} = |x| \cdot \sqrt{x}$
(c) $\sqrt[3]{x^6 y^9} = \sqrt[3]{(x^2 y^3)^3} = x^2 y^3$
(d) We can express the radicand as follows:

$$x^5 y^7 = (x^4 y^4)(xy^3) = (xy)^4 (xy^3).$$

Thus,

$$\begin{aligned}
\sqrt[4]{x^5 y^7} &= \sqrt[4]{(xy)^4 (xy^3)} \\
&= \sqrt[4]{(xy)^4} \sqrt[4]{xy^3} \\
&= |xy| \sqrt[4]{xy^3}.
\end{aligned}$$

Example 1.29

Simplify the following expressions:

(a) $\sqrt{\dfrac{16a^4}{b^2}}$ **(b)** $\sqrt{\dfrac{9x^5}{25y^4}}$ **(c)** $\sqrt[3]{\dfrac{216x^3}{343y^6}}$

Solution:

(a) $\sqrt{\dfrac{16a^4}{b^2}} = \dfrac{\sqrt{16a^4}}{\sqrt{b^2}} = \dfrac{\sqrt{(4a^2)^2}}{\sqrt{b^2}} = \dfrac{4a^2}{|b|}, \qquad b \neq 0.$

(b) $\sqrt{\dfrac{9x^5}{25y^4}} = \dfrac{\sqrt{9x^5}}{\sqrt{25y^4}} = \dfrac{\sqrt{(3x^2)^2 \cdot x}}{\sqrt{(5y^2)^2}} = \dfrac{\sqrt{(3x^2)^2} \cdot \sqrt{x}}{\sqrt{(5y^2)^2}} = \dfrac{3x^2}{5y^2} \cdot \sqrt{x},\qquad y \neq 0.$

(c) $\sqrt[3]{\dfrac{216x^3}{343y^6}} = \dfrac{\sqrt[3]{216x^3}}{\sqrt[3]{343y^6}} = \dfrac{\sqrt[3]{(6x)^3}}{\sqrt[3]{(7y^2)^3}} = \dfrac{6x}{7y^2},\qquad y \neq 0.$

Sometimes it is convenient to eliminate radicals from the denominator when you are working with fractions. *You multiply the numerator and the denominator by some fraction so that the resulting denominator is free of radicals.*

Example 1.30

(a) $\dfrac{1}{\sqrt{2}} = \dfrac{1}{\sqrt{2}} \cdot \dfrac{\sqrt{2}}{\sqrt{2}} = \dfrac{\sqrt{2}}{2}.$

(b) $\dfrac{4}{\sqrt{3}} = \dfrac{4}{\sqrt{3}} \cdot \dfrac{\sqrt{3}}{\sqrt{3}} = \dfrac{4\sqrt{3}}{3}.$

(c) $\sqrt{\dfrac{x}{2y}} = \dfrac{\sqrt{x}}{\sqrt{2y}} \cdot \dfrac{\sqrt{2y}}{\sqrt{2y}} = \dfrac{\sqrt{2xy}}{2y},\qquad y \neq 0.$

(d) $\sqrt[3]{\dfrac{x}{y}} = \sqrt[3]{\dfrac{x}{y} \cdot \dfrac{y^2}{y^2}} = \dfrac{\sqrt[3]{xy^2}}{\sqrt[3]{y^3}} = \dfrac{\sqrt[3]{xy^2}}{y},\qquad y \neq 0.$

This example illustrates a technique called *rationalizing* the denominator.

EXERCISES

Simplify each of the expressions in Exercises 1 to 32.

1. $\sqrt{36}$ **2.** $\sqrt{64}$ **3.** $\sqrt{81}$ **4.** $\sqrt{121}$

5. $\sqrt{32}$ **6.** $\sqrt{75}$ **7.** $\sqrt[3]{64}$ **8.** $\sqrt[3]{-216}$

9. $\sqrt[3]{54}$ **10.** $\sqrt[3]{192}$ **11.** $\sqrt[3]{\dfrac{27}{125}}$ **12.** $\sqrt[3]{\dfrac{-64}{343}}$

13. $\sqrt[4]{32}$ **14.** $\sqrt[4]{162}$ **15.** $\sqrt[4]{\dfrac{16}{81}}$ **16.** $\sqrt{\dfrac{256}{625}}$

17. $\sqrt{4x^2}$ **18.** $\sqrt{8x^4y^2}$ **19.** $\sqrt{\dfrac{16x^4}{y^2}}$ **20.** $\sqrt{\dfrac{4x^3}{y^4}}$

21. $\sqrt{\dfrac{18a^4b}{c^6}}$ **22.** $\sqrt{\dfrac{64a^5}{b^8}}$ **23.** $\sqrt[3]{x^3y^6}$ **24.** $\sqrt[3]{x^7y^9}$

25. $\sqrt[3]{\dfrac{8x^6}{y^3}}$ **26.** $\sqrt[3]{\dfrac{27x^4}{y^9}}$ **27.** $\sqrt[4]{16x^4y^4}$ **28.** $\sqrt[4]{\dfrac{81x^8}{y^{12}}}$

29. $\sqrt[5]{\dfrac{-32x^5}{y^5}}$ **30.** $\sqrt[5]{\dfrac{243x^5y^6}{z^{10}}}$ **31.** $\sqrt[6]{x^6y^{12}}$ **32.** $\sqrt[6]{\dfrac{64a^6}{b^6c^6}}$

Rationalize the denominator in each fraction in Exercises 33 to 40.

33. $\dfrac{1}{\sqrt{5}}$ **34.** $\dfrac{3}{4\sqrt{2}}$ **35.** $\dfrac{1}{\sqrt{y}}$ **36.** $\sqrt{\dfrac{2x}{3y}}$

37. $\sqrt{\dfrac{x^2}{y^3}}$ **38.** $\dfrac{1}{\sqrt[3]{y}}$ **39.** $\sqrt[3]{\dfrac{2y}{x}}$ **40.** $\sqrt[3]{\dfrac{8x}{y^2}}$

1.5 RATIONAL EXPONENTS

We use radicals to assign meanings to expressions of the form $x^{m/n}$, where x is any real number and m/n is a rational number but not necessarily an integer. Recall from Proposition 1.3.3 that if x is a real number, then for any positive integers m and n,

$$(x^m)^n = x^{mn}.$$

If this rule of exponents is to hold for $m = 1/n$ and $x^{1/n}$ a real number, then we must have

$$(x^{1/n})^n = x^{(1/n)n} = x^1 = x.$$

This means that $x^{1/n}$ is the nth root of x.

Definition 1.5.1

> For any real number x and any positive integer n, if $\sqrt[n]{x}$ exists, then
> $$x^{1/n} = \sqrt[n]{x}.$$

Example 1.31

(a) $(16)^{1/4} = (2^4)^{1/4} = 2^{4(1/4)} = 2^1 = 2.$
(b) $(-243)^{1/5} = [(-3)^5]^{1/5} = (-3)^{5(1/5)} = -3.$
(c) $(729)^{1/6} = (3^6)^{1/6} = 3^{6(1/6)} = 3.$

We now generalize from rational exponents of the form $1/n$ to rational exponents of the form m/n. If the rules of exponents are to hold, then

$$(x^{m/n})^n = x^{(m/n)n} = x^m,$$

which implies that $x^{m/n}$ is the nth root of x^m. In other words,

$$x^{m/n} = \sqrt[n]{x^m}.$$

Example 1.32

(a) $(8)^{2/3} = \sqrt[3]{8^2} = \sqrt[3]{64} = 4.$
(b) $(27)^{2/3} = \sqrt[3]{27^2} = \sqrt[3]{729} = 9.$

In computing $x^{m/n}$ in Example 1.32, we first computed the mth power of x and then extracted its nth root. This method may be time-consuming and difficult, particularly if x is large. To circumvent this, we use the following result.

Proposition 1.5.1

If m/n is a rational number with $n > 0$ and if $x > 0$, then

$$x^{m/n} = \sqrt[n]{x^m} = (\sqrt[n]{x})^m.$$

Observe that the first equality is the definition of $x^{m/n}$. To prove that

$$\sqrt[n]{x^m} = (\sqrt[n]{x})^m,$$

we note that if $m > 0$, then the right-hand side is as follows:

$$(\sqrt[n]{x})^m = \underbrace{(\sqrt[n]{x})(\sqrt[n]{x})(\sqrt[n]{x}) \cdots (\sqrt[n]{x})}_{m \text{ times}}$$

$$= \sqrt[n]{\underbrace{x \cdot x \cdot x \cdots x}_{(m \text{ times})}}$$

$$= \sqrt[n]{x^m}.$$

The case $m < 0$ is similar and is left as an exercise for the reader.

Note that in the expression $\sqrt[n]{x^m}$, we first take the mth power of x and then extract its nth root. In the expression $(\sqrt[n]{x})^m$, we extract the nth root of x first and then take the mth power of the number obtained. The result is the same in both cases.

Example 1.33

Compute $(32)^{4/5}$.

Solution:

$$(32)^{4/5} = [(32)^{1/5}]^4 = [(2^5)^{1/5}]^4 = [(2)^{5(1/5)}]^4 = 2^4 = 16.$$

Alternatively, we could proceed as follows:

$$(32)^{4/5} = [(32)^4]^{1/5} = (1{,}048{,}576)^{1/5} = [(16)^5]^{1/5} = 16.$$

Clearly, the second approach is more difficult and assumes the use of a calculator. If you were to use the calculator, then a simpler approach would be to recognize that

$$\tfrac{4}{5} = 0.8$$

and then use the calculator as follows:

Enter	Press	Display
32	y^x	32.
0.8	=	16.

Example 1.34

Compute the following:

(a) $(-125)^{2/3}$ **(b)** $(64)^{5/6}$ **(c)** $(81)^{5/4}$

Solution:

(a) $(-125)^{2/3} = [(-5)^3]^{2/3} = (-5)^{3(2/3)} = (-5)^2 = 25.$
(b) $(64)^{5/6} = (2^6)^{5/6} = 2^{6(5/6)} = 2^5 = 32.$
(c) $(81)^{5/4} = (3^4)^{5/4} = (3)^{4(5/4)} = 3^5 = 243.$

Using a calculator, you can see that $\frac{5}{4} = 1.25$ in decimal form and then proceed as follows:

Enter	Press	Display
81	y^x	81.
1.25	=	243.

Applying Definition 1.3.3 to rational exponents, we can express

$$x^{-m/n} = \frac{1}{x^{m/n}}, \qquad x \neq 0,$$

where m/n is any rational number.

Example 1.35

(a) $(16)^{-5/4} = \dfrac{1}{(16)^{5/4}} = \dfrac{1}{(2^4)^{5/4}} = \dfrac{1}{2^5} = \dfrac{1}{32}.$

(b) $(216)^{-2/3} = \dfrac{1}{(216)^{2/3}} = \dfrac{1}{(6^3)^{2/3}} = \dfrac{1}{6^2} = \dfrac{1}{36}.$

(c) $(-128)^{-3/7} = \dfrac{1}{(-128)^{3/7}} = \dfrac{1}{[(-2)^7]^{3/7}} = \dfrac{1}{(-2)^3} = -\dfrac{1}{8}.$

As another consequence of

$$x^{-m} = \frac{1}{x^m}, \qquad x \neq 0,$$

we note that

$$\left(\frac{a}{b}\right)^{-m} = \left(\frac{b}{a}\right)^m, \quad a \neq 0, b \neq 0$$

holds whenever the letters are replaced by numbers for which the expression involved has a meaning. Thus

$$\left(\tfrac{3}{4}\right)^{-2} = \left(\tfrac{4}{3}\right)^2 = \tfrac{16}{9}.$$

Example 1.36

Using rules of exponents, simplify the following:

(a) $x^{1/3} \cdot x^{1/6}$.

(b) $\dfrac{y^{1/2}}{y^{1/3}}$, $y \neq 0$.

(c) $(8x^6)^{2/3}$.

Solution:

(a) $x^{\frac{1}{3}} \cdot x^{\frac{1}{6}} = x^{\frac{1}{3}+\frac{1}{6}} = x^{\frac{1}{2}} = \sqrt{x}$.

(b) $\dfrac{y^{1/2}}{y^{1/3}} = y^{\frac{1}{2}-\frac{1}{3}} = y^{\frac{1}{6}} = \sqrt[6]{y}$, $y \neq 0$.

(c) $(8x^6)^{2/3} = [(2x^2)^3]^{2/3} = [(2x^2)]^{3(2/3)} = (2x^2)^2 = 4x^4$.

Example 1.37

Using rules of exponents, simplify each expression. All answers should have positive exponents.

(a) $(25x^4)^{-3/2}$, $x \neq 0$.

(b) $\left(\dfrac{27x^3}{y^6}\right)^{1/3}$, $y \neq 0$.

Solution:

(a) $(25x^4)^{-3/2} = [(5x^2)^2]^{-3/2}$

$\qquad\qquad\quad = (5x^2)^{2(-3/2)}$

$\qquad\qquad\quad = (5x^2)^{-3}$

$\qquad\qquad\quad = \dfrac{1}{(5x^2)^3}$

$\qquad\qquad\quad = \dfrac{1}{125x^6}$, $x \neq 0$.

(b) $\left(\dfrac{27x^3}{y^6}\right)^{1/3} = \dfrac{(27x^3)^{1/3}}{(y^6)^{1/3}} = \dfrac{[(3x)^3]^{1/3}}{[(y^2)^3]^{1/3}} = \dfrac{(3x)^{3(1/3)}}{(y^2)^{3(1/3)}}$

$\qquad\qquad\qquad\qquad = \dfrac{3x}{y^2}$, $y \neq 0$.

EXERCISES

Compute each expression in Exercises 1 to 18.

1. $(9)^{3/2}$ **2.** $(16)^{3/4}$ **3.** $(125)^{1/3}$ **4.** $(81)^{3/4}$

5. $(64)^{2/3}$ **6.** $(-32)^{3/5}$ **7.** $(1000)^{2/3}$ **8.** $(729)^{5/6}$

9. $(4)^{-3/2}$ **10.** $(25)^{-1/2}$ **11.** $(125)^{-2/3}$ **12.** $(64)^{-5/6}$

13. $\left(\dfrac{4}{9}\right)^{3/2}$ **14.** $\left(\dfrac{9}{49}\right)^{-3/2}$ **15.** $\left(\dfrac{8}{125}\right)^{-2/3}$ **16.** $\left(\dfrac{32}{243}\right)^{-2/5}$

17. $\left(\dfrac{16}{81}\right)^{-3/4}$ **18.** $\left(-\dfrac{1}{128}\right)^{-2/7}$

Using rules of exponents, simplify each expression in Exercises 19 to 60. All answers should have positive exponents.

19. $x^{2/3} \cdot x^{1/3}$ **20.** $(4x^{1/2})(3x^{1/4})$ **21.** $(2y^{1/3})(3y^{1/2})$

22. $x^{1/3} \cdot x^{1/4}$ **23.** $\dfrac{y^{3/4}}{y^{1/2}}$ **24.** $\dfrac{y^{2/3}}{y^{1/2}}$

25. $(16x^4)^{3/2}$ **26.** $(8x^3)^{2/3}$ **27.** $(8x^{3/2})^{5/3}$

28. $(16y^4)^{1/4}$ **29.** $(16y^4)^{1/8}$ **30.** $(x^{2/3}y^{1/3})^2$

31. $(x^{1/2} \cdot y^{1/5})^{10}$ **32.** $(x^{1/4} \cdot y^{1/6})^{12}$ **33.** $(x^6y^3)^{2/3}$

34. $(a^4b^8)^{3/4}$ **35.** $(32x^5y^{10})^{2/5}$ **36.** $(125a^3b^9)^{2/3}$

37. $\left(\dfrac{x^2}{y}\right)^{-2}$ **38.** $\left(\dfrac{x^3}{y^2}\right)^{-3}$ **39.** $\left(\dfrac{x^6}{y^3}\right)^{1/3}$

40. $\left(\dfrac{x^4}{y^8}\right)^{3/4}$ **41.** $\left(\dfrac{x^4}{y^8}\right)^{-1/2}$ **42.** $\left(\dfrac{x^4}{y^8}\right)^{-1/4}$

43. $(x^4y^8)^{-1/8}$ **44.** $(x^3y^{-6})^{-2/3}$ **45.** $(x^3y^6)^{-1/3}$

46. $(a^5b^{-10}c^5)^{2/5}$ **47.** $(a^5b^{10}c^{-5})^{-2/5}$ **48.** $(x^3y^{-3}z^{-6})^{-4/3}$

49. $\left(\dfrac{243x^5}{32y^5}\right)^{-3/5}$ **50.** $\left(\dfrac{x^{4/3}}{y^{2/3}}\right)^{-3/2}$ **51.** $\left(\dfrac{16x^4y^8}{z^{12}}\right)^{-3/4}$

52. $\left(\dfrac{27x^3}{y^3z^3}\right)^{-4/3}$ **53.** $\left(\dfrac{64x^6}{y^6}\right)^{-1/6}$ **54.** $\left(\dfrac{128a^{-7}}{b^7}\right)^{-2/7}$

55. $x^{1/2}(x^{1/2} + x + 1)$ **56.** $x^{1/2}(x^{1/2} + x^{-1/2})$ **57.** $x^{1/3}(x^{1/2} + x^{1/3})$

58. $x^{1/3}(2x^{2/3} + x^{4/3})$ **59.** $x^{2/3}(x + x^{1/2} + 1)$ **60.** $x^{1/4}(x + x^{1/2} + x^{1/3} + 2)$

1.6

ALGEBRAIC EXPRESSIONS AND POLYNOMIALS

variable
domain

Any letter of the alphabet, say x, that may represent any element of a given nonempty set is called a **variable**. The **domain** of the variable is the set of all possible values for which the defining equation makes any sense. For example, in the expression

$$y = \sqrt{x}$$

we recall that if x is negative, then y is not a real number. Therefore, the values of x must be restricted to the set of nonnegative real numbers. Similarly, the expression

$$y = \frac{1}{x}$$

is not defined for $x = 0$. Hence the domain of the variable x is the set of non-zero real numbers.

A **constant** is a specific real number. It may be interpreted as a variable whose domain consists of a fixed real number.

Example 1.38

For the circumference of a circle of radius r, the formula is

$$C = 2\pi r.$$

If all circles are considered, then C and r are variables and 2 and π are constants.

Definition 1.6.1

Any grouping of constants and variables formed by the use of elementary operations of addition, subtraction, multiplication, division (except by zero), and extraction of roots is an **algebraic expression.**

Some examples of algebraic expressions are

$$3x^2 + \frac{4}{x}, \qquad \sqrt{x - 3}, \qquad \sqrt{x} + \sqrt{y}, \qquad \text{and} \qquad \frac{1}{x - y},$$

where x and y are variables. However, exponential and logarithmic expressions, such as

$$2^x \qquad \text{and} \qquad \log x$$

are not algebraic expressions.

Definition 1.6.2

An algebraic expression of the form ax^n, where a is some constant and n is a non-negative integer, is called a **monomial** in x. The constant a is called the **coefficient** of x^n.

Definition 1.6.3

Any finite sum of monomials in x is called a *polynomial in x.*

Some examples of polynomials in x are

$$2x + 3, \qquad x^2 + 2x + 4, \qquad \tfrac{1}{3}x^3 + \tfrac{1}{2}x^2 + 1, \qquad \text{and} \qquad 5x^4 + \tfrac{1}{4}.$$

However, the algebraic expressions such as

$$\frac{1}{x}, \qquad \sqrt{x}, \qquad \frac{x^2 + 1}{2x}, \qquad \text{and} \qquad \frac{3x + 2}{\sqrt{x}}$$

are not polynomials, since they involve division by variables or roots of variables.

The highest exponent in a polynomial of one variable is called the *degree* of the polynomial. Thus a constant is said to have degree zero, $2x + 3$ is a polynomial of first degree, $x^2 + x + 1$ is a polynomial of second degree, $\frac{1}{3}x^3 + \frac{1}{2}x^2 + 1$ is a polynomial of third degree, and $5x^4 + \frac{1}{4}$ is a polynomial of fourth degree.

If a monomial contains more than one variable, its degree is determined by the sum of the exponents on the variables. The monomial $2x^3y^4$, for example, is of degree $3 + 4 = 7$. The degree of a polynomial in more than one variable is the highest degree of any monomial appearing in a polynomial. For example, the polynomial

$$4x^6y^2 + 5x^4y^3 + 3x^3y^6$$

is of degree 9. (Why?)

Since we are often concerned with polynomials in one variable, we define this concept formally.

Definition 1.6.4

A polynomial in x of degree n is an algebraic expression

$$a_nx^n + a_{n-1}x^{n-1} + \cdots + a_2x^2 + a_1x + a_0 \qquad (a_n \neq 0),$$

where n is a nonnegative integer and $a_0, a_1, \ldots, a_n$ are real numbers. The coefficient a_n is called the **leading coefficient,** and the term a_nx^n is called the **leading term.**

Definition 1.6.5

A polynomial is said to be linear, quadratic, or cubic if its degree is 1, 2, or 3, respectively.

Thus $a_1x + a_0$ is a linear polynomial if $a_1 \neq 0$; $a_2x^2 + a_1x + a_0$ is a quadratic polynomial if $a_2 \neq 0$; and $a_3x^3 + a_2x^2 + a_1x + a_0$ $(a_3 \neq 0)$ is a cubic polynomial in x.

We denote polynomials by the capital letters $P, Q, R, S, T, \ldots$. The symbol $P(x)$, for example, is read "P of x," "P evaluated at x," or "the value of P at point x." Thus $P(3)$ represents the value of the polynomial P evaluated at $x = 3$.

Example 1.39

Let P be a polynomial in x defined by

$$P(x) = x^2 + 3x - 4.$$

Then

(a) $P(1) = 1^2 + 3(1) - 4 = 0,$
(b) $P(2) = 2^2 + 3(2) - 4 = 6,$
(c) $P(-2) = (-2)^2 + 3(-2) - 4 = -6,$
(d) $P(0.5) = (0.5)^2 + 3(0.5) - 4 = -2.25,$
(e)* $P(0.75) = (0.75)^2 + 3(0.75) - 4 = -1.1875.$

Example 1.40

Let P and Q be two polynomials in x defined by

$$P(x) = 2x + 1, \qquad Q(x) = x^2 + x + 1.$$

Determine

(a) $P(Q(2)),$
(b) $Q(P(2)).$

Solution:

(a) To determine $P(Q(2))$, we first evaluate $Q(2)$ and then evaluate P at the value obtained.

$$Q(2) = 2^2 + 2 + 1 = 7.$$

Thus

$$P(Q(2)) = P(7) = 2(7) + 1 = 15.$$

(b) Similarly, to compute $Q(P(2))$, we evaluate $P(2)$ first and then evaluate Q at the value thus obtained.

$$P(2) = 2(2) + 1 = 5$$

and

$$Q(P(2)) = Q(5) = 5^2 + 5 + 1 = 31.$$

Observe that

$$P(Q(2)) \neq Q(P(2)).$$

Sum and Difference of Polynomials

To add or subtract polynomials, we rearrange and regroup like terms, add or subtract the coefficients of like terms, and then simplify the results, using the properties of the real number system. Consider, for example, the following examples.

---

* A calculator is needed.

Example 1.41

(a) $3x^2 + 5x^2 = (3 + 5)x^2 = 8x^2.$

(b) $3x^2 - 5x^2 = (3 - 5)x^2 = -2x^2.$

Example 1.42

(a) $(2x^3 + 3x^2 - x + 4) + (x^3 + 2x^2 + 3x + 6)$
$$= (2x^3 + x^3) + (3x^2 + 2x^2) + (-x + 3x) + (4 + 6)$$
$$= (2 + 1)x^3 + (3 + 2)x^2 + (-1 + 3)x + (4 + 6)$$
$$= 3x^3 + 5x^2 + 2x + 10.$$

(b) $(2x^3 + 3x^2 - x + 4) - (x^3 + 2x^2 + 3x + 6)$
$$= (2x^3 - x^3) + (3x^2 - 2x^2) + (-x - 3x) + (4 - 6)$$
$$= (2 - 1)x^3 + (3 - 2)x^2 + (-1 - 3)x + (-2)$$
$$= x^3 + x^2 - 4x - 2.$$

Example 1.43

(a) $(x^3 + x^2y + 2xy^2 + 2y^3) + (3x^2y + 5xy^2 - y^3)$
$$= x^3 + (x^2y + 3x^2y) + (2xy^2 + 5xy^2) + (2y^3 - y^3)$$
$$= x^3 + (1 + 3)x^2y + (2 + 5)xy^2 + (2 - 1)y^3$$
$$= x^3 + 4x^2y + 7xy^2 + y^3.$$

(b) $(x^3 + x^2y + 2xy^2 + 2y^3) - (3x^2y + 5xy^2 - y^3)$
$$= x^3 + (x^2y - 3x^2y) + (2xy^2 - 5xy^2) + [2y^3 - (-y^3)]$$
$$= x^3 + (1 - 3)x^2y + (2 - 5)xy^2 + [2 - (-1)]y^3$$
$$= x^3 - 2x^2y - 3xy^2 + 3y^3.$$

EXERCISES

1. Let P be a polynomial in x defined by

$$P(x) = x^2 + 3x + 5.$$

Find the following:
(a) $P(1)$ (b) $P(3)$ (c) $P(\frac{1}{2})$

2. Let Q be a polynomial in x defined by

$$Q(x) = 2x^2 - 5x + 7.$$

Determine the following:
(a) $Q(0)$ (b) $Q(2)$ (c) $Q(-1)$ (d) $Q(-2)$ (e) $Q(\frac{1}{2})$ (f) $Q(\frac{1}{4})$

3. Let P be a polynomial in x defined by

$$P(x) = x^3 + x^2 + x + 1.$$

Determine the following:
(a) $P(-2)$ (b) $P(-3)$ (c) $P(-4)$ (d) $P(0)$ (e) $P(2)$ (f) $P(3)$

4. Let P be a polynomial in x defined by

$$P(x) = x^4 + x^2 + 3.$$

Determine the following:
   (a) $P(-3.1)$ **(b)** $P(-2.3)$ **(c)** $P(-1.1)$ **(d)** $P(1.7)$ **(e)** $P(3.2)$ **(f)** $P(2.4)$

5. Let P and Q be two polynomials in x defined by

$$P(x) = 3x + 2, \qquad Q(x) = 4x + 1.$$

Determine the following:
   (a) $P(2)$ **(b)** $Q(2)$ **(c)** $P(P(2))$ **(d)** $Q(P(2))$ **(e)** $P(Q(2))$ **(f)** Is $P(Q(2)) = Q(P(2))$?

6. Let P and Q be two polynomials in x defined by

$$P(x) = x^2 + x + 3, \qquad Q(x) = 2x + 1.$$

Determine the following:
   (a) $P(3)$ **(b)** $Q(3)$ **(c)** $P(Q(3))$ **(d)** $Q(P(3))$ **(e)** $P(Q(4))$ **(f)** $Q(P(4))$
   (g) Is $P(Q(4)) = Q(P(4))$?   **(h)** Is $P(Q(3)) = Q(P(3))$?

Perform the indicated operations in Exercises 7 to 20.

7. $(x^2 - 3x + 1) + (2x^2 + x + 3)$

8. $(3x^2 + 4x - 7) + (x^2 - x - 1)$

9. $(x^3 + x^2 + 3x + 2) + (3x^2 + 2x)$

10. $(3x^3 + 2x^2 - x + 7) - (x^3 + 4x^2 - 3x + 1)$

11. $(5x^3 - 3x^2 + 4x + 3) - (3x^3 + 5x^2 - 2x - 1)$

12. $(y^3 + 3y^2 + 3y + 6) + (y^4 + 2y^3 - 3y^2 + 3y + 1)$

13. $(3m^3 + 4m^2 + 7m - 2) - (2m^3 - 2m^2 - 4m + 12)$

14. $(8m^3 + 6m^2 + 11m - 10) - (12m^3 - 7m^2 + 4m - 12)$

15. $(x^3 + 3x^2y + 6xy^2 + 2y^3) + (2x^2y - xy^2 - y^3)$

16. $(x^3y^2 + 3x^2y + 6xy^2 + 2y^3) + (4x^2y + 10xy^2 + 5y^3)$

17. $(x^4 + 4x^3y + 6x^2y^2 + 4xy^3 + y^4) - (2x^3y - 7x^2y^2 + xy^3 + 3y^4)$

18. $(x^3 + 3y^3 + xy^2 + 2x + 5) + (4x^3 - y^3 + 2xy^2 + 3)$

19. $(3x^4 - 6x^2y^2 + 4xy^3 + y^4) - (5x^4 - 4x^2y^2 + xy^3 + 2y^4)$

20. $(6m^2 + 3mn + 8n^2) - (4m^2 - 2mn - n^2)$

1.7

PRODUCTS OF POLYNOMIALS

The product of polynomials is determined by first using the distributive property of the real number system. Then the rule of exponents

$$x^m \cdot x^n = x^{m+n}$$

is applied, and like terms are combined.

Example 1.44 Expand the following:

   (a) $2x(3x + 4y)$
   (b) $(3x + y)(2x + 3y)$

Solution:

(a) Using the distributive property of the real number system, we have

$$2x(3x + 4y) = (2x)(3x) + (2x)(4y) = 6x^2 + 8xy.$$

(b) $(3x + y)(2x + 3y) = 3x(2x + 3y) + y(2x + 3y)$
$$= (3x)(2x) + (3x)(3y) + y(2x) + y(3y)$$
$$= 6x^2 + 9xy + 2xy + 3y^2$$
$$= 6x^2 + 11xy + 3y^2.$$

Example 1.45

Expand the following:

$$(2x + 3)(x^2 + 3x + 6).$$

Solution: Using the distributive property, we have

$(2x + 3)(x^2 + 3x + 6) = 2x(x^2 + 3x + 6) + 3(x^2 + 3x + 6)$
$$= (2x)(x^2) + (2x)(3x) + (2x)(6) + 3x^2 + 9x + 18$$
$$= 2x^3 + 6x^2 + 12x + 3x^2 + 9x + 18$$
$$= 2x^3 + (6x^2 + 3x^2) + (12x + 9x) + 18$$
$$= 2x^3 + 9x^2 + 21x + 18.$$

Sometimes it is convenient to arrange this product as follows:

$$
\begin{array}{lll}
x^2 + 3x + 6 & & \\
2x + 3 & & \\
\hline
2x^3 + 6x^2 + 12x & = 2x(x^2 + 3x + 6) & \\
  3x^2 + 9x + 18 & = 3(x^2 + 3x + 6) & \\
\hline
2x^3 + 9x^2 + 21x + 18 & \text{sum} & \\
\end{array}
$$

We now consider products of polynomials that occur frequently in algebra. If x, a, b, c, and d are real numbers, then by the distributive property, we have

$(ax + b)(cx + d) = ax(cx + d) + b(cx + d)$
$$= (ax)(cx) + (ax)(d) + (b)(cx) + (b)(d)$$
$$= acx^2 + adx + bcx + bd$$
$$= acx^2 + (ad + bc)x + bd.$$

Thus

$$(ax + b)(cx + d) = acx^2 + (ad + bc)x + bd. \qquad \textbf{(1.34)}$$

If $a = 1$ and $c = 1$, then (1.34) reduces to

$$(x + b)(x + d) = x^2 + (b + d)x + bd. \tag{1.35}$$

Setting $d = b$ in (1.35), we obtain

$$(x + b)^2 = x^2 + 2bx + b^2. \tag{1.36}$$

If we replace b by $-b$ in (1.36), we get

$$(x - b)^2 = x^2 - 2bx + b^2. \tag{1.37}$$

If we set $d = -b$ in (1.35), then we obtain

$$(x + b)(x - b) = x^2 - b^2. \tag{1.38}$$

Let us now look at some examples.

Example 1.46   Form the following products:

(a) $(2x + 3)(4x + 5)$
(b) $(x + 4)(x + 6)$
(c) $(x - 2)(x - 3)$
(d) $(x + 6)(x - 1)$

Solution:

(a) Setting $a = 2$, $b = 3$, $c = 4$, and $d = 5$ in (1.34), we get
$$\begin{aligned}
(2x + 3)(4x + 5) &= 8x^2 + [(2)(5) + (3)(4)]x + (3)(5) \\
&= 8x^2 + (10 + 12)x + 15 \\
&= 8x^2 + 22x + 15.
\end{aligned}$$

(b) If we set $b = 4$ and $d = 6$ in (1.35), we obtain
$$\begin{aligned}
(x + 4)(x + 6) &= x^2 + (4 + 6)x + (4)(6) \\
&= x^2 + 10x + 24.
\end{aligned}$$

(c) With $b = -2$ and $d = -3$ in (1.35), we have
$$\begin{aligned}
(x - 2)(x - 3) &= x^2 + [(-2) + (-3)]x + (-2)(-3) \\
&= x^2 - 5x + 6.
\end{aligned}$$

(d) Using (1.35) again, we obtain
$$\begin{aligned}
(x + 6)(x - 1) &= x^2 + [6 + (-1)]x + (6)(-1) \\
&= x^2 + 5x - 6.
\end{aligned}$$

Example 1.47

Form the following products:

(a) $(2x + 5)^2$

(b) $(3x - 4y)^2$

(c) $(4x + 5y)(4x - 5y)$

Solution:

(a) This is an application of (1.36). Note that the first term in the parentheses is $2x$ and not x. Accordingly,

$$(2x + 5)^2 = (2x)^2 + 2(2x)(5) + 5^2$$
$$= 4x^2 + 20x + 25.$$

(b) Using (1.37) with $b = 4y$, we have

$$(3x - 4y)^2 = (3x)^2 - 2(3x)(4y) + (4y)^2$$
$$= 9x^2 - 24xy + 16y^2.$$

(c) Applying (1.38), we obtain

$$(4x + 5y)(4x - 5y) = (4x)^2 - (5y)^2$$
$$= 16x^2 - 25y^2.$$

We use the same rules to multiply radical expressions as we have to multiply polynomials.

Example 1.48

Form the following products:

(a) $(\sqrt{5} + \sqrt{3})^2$

(b) $(\sqrt{7} - \sqrt{5})^2$

(c) $(\sqrt{5} + \sqrt{3})(\sqrt{5} - \sqrt{3})$

Solution:

(a) This is an application of (1.36), which is the formula for the square of the sum.

$$(\sqrt{5} + \sqrt{3})^2 = (\sqrt{5})^2 + 2(\sqrt{5})(\sqrt{3}) + (\sqrt{3})^2$$
$$= 5 + 2\sqrt{15} + 3$$
$$= 8 + 2\sqrt{15}.$$

(b) We apply (1.37), which is the formula for the square of the difference.

$$(\sqrt{7} - \sqrt{5})^2 = (\sqrt{7})^2 - 2(\sqrt{7})(\sqrt{5}) + (\sqrt{5})^2$$
$$= 7 - 2\sqrt{35} + 5$$
$$= 12 - 2\sqrt{35}.$$

(c) Using the difference-of-two-squares formula (1.38), we have

$$(\sqrt{5} + \sqrt{3})(\sqrt{5} - \sqrt{3}) = (\sqrt{5})^2 - (\sqrt{3})^2$$
$$= 5 - 3$$
$$= 2.$$

In Example 1.48(c), the two expressions $(\sqrt{5} + \sqrt{3})$ and $(\sqrt{5} - \sqrt{3})$ are called **conjugates.** In general, $(\sqrt{a} + \sqrt{b})$ and $(\sqrt{a} - \sqrt{b})$ are conjugates of each other. Multiplying conjugates of this form always yields a real number.

If we want to rationalize the denominator in the expression

$$\frac{1}{\sqrt{a} + \sqrt{b}}$$

we simply multiply and divide by $\sqrt{a} - \sqrt{b}$, the conjugate of $\sqrt{a} + \sqrt{b}$, and write

$$\frac{1}{\sqrt{a} + \sqrt{b}} = \frac{1}{\sqrt{a} + \sqrt{b}}\left(\frac{\sqrt{a} - \sqrt{b}}{\sqrt{a} - \sqrt{b}}\right)$$
$$= \frac{\sqrt{a} - \sqrt{b}}{a - b}, \qquad a \neq b.$$

Example 1.49

Rationalize the denominator in

$$\frac{1}{\sqrt{5} + \sqrt{2}}.$$

Solution:

$$\frac{1}{\sqrt{5} + \sqrt{2}} = \frac{1}{\sqrt{5} + \sqrt{2}}\left(\frac{\sqrt{5} - \sqrt{2}}{\sqrt{5} - \sqrt{2}}\right)$$
$$= \frac{\sqrt{5} - \sqrt{2}}{5 - 2}$$
$$= \frac{\sqrt{5} - \sqrt{2}}{3}.$$

Example 1.50

Rationalize the denominator

$$\frac{1}{\sqrt{x + y} - \sqrt{x}}.$$

Solution:

$$\frac{1}{\sqrt{x+y}-\sqrt{x}}\left(\frac{\sqrt{x+y}+\sqrt{x}}{\sqrt{x+y}+\sqrt{x}}\right) = \frac{\sqrt{x+y}+\sqrt{x}}{(x+y)-x}$$

$$= \frac{\sqrt{x+y}+\sqrt{x}}{y} \qquad (y \neq 0).$$

If we want to rationalize a numerator of the form $\sqrt{a}+\sqrt{b}$, we multiply and divide by its conjugate, namely $\sqrt{a}-\sqrt{b}$. Thus we may write

$$\sqrt{a}+\sqrt{b} = (\sqrt{a}+\sqrt{b})\left(\frac{\sqrt{a}-\sqrt{b}}{\sqrt{a}-\sqrt{b}}\right)$$

$$= \frac{a-b}{\sqrt{a}-\sqrt{b}} \qquad (a \neq b).$$

Similarly,

$$\sqrt{a}-\sqrt{b} = (\sqrt{a}-\sqrt{b})\left(\frac{\sqrt{a}+\sqrt{b}}{\sqrt{a}+\sqrt{b}}\right)$$

$$= \frac{a-b}{\sqrt{a}+\sqrt{b}}.$$

Example 1.51

Rationalize the numerator in

$$\frac{\sqrt{x+3y}-\sqrt{x+y}}{2}.$$

Solution:

$$\frac{(\sqrt{x+3y}-\sqrt{x+y})}{2} = \frac{(\sqrt{x+3y}-\sqrt{x+y})}{2}\frac{(\sqrt{x+3y}+\sqrt{x+y})}{(\sqrt{x+3y}+\sqrt{x+y})}$$

$$= \frac{1}{2}\frac{(x+3y)-(x+y)}{\sqrt{x+3y}+\sqrt{x+y}}$$

$$= \frac{1}{2}\frac{2y}{\sqrt{x+3y}+\sqrt{x+y}}$$

$$= \frac{y}{\sqrt{x+3y}+\sqrt{x+y}}.$$

We now observe that

$$(x + a)(x^2 - ax + a^2) = x(x^2 - ax + a^2) + a(x^2 - ax + a^2)$$
$$= (x)(x^2) + (x)(-ax) + (x)(a^2)$$
$$+ (a)(x^2) + (a)(-ax) + (a)(a^2)$$
$$= x^3 - ax^2 + xa^2 + ax^2 - a^2x + a^3$$
$$= x^3 + [(-a) + a]x^2 + [a^2 + (-a^2)]x + a^3$$
$$= x^3 + a^3.$$

Thus

$$(x + a)(x^2 - ax + a^2) = x^3 + a^3. \qquad \textbf{(1.39)}$$

Replacing a by $(-a)$ in (1.39); we get

$$(x - a)(x^2 + ax + a^2) = x^3 - a^3. \qquad \textbf{(1.40)}$$

Example 1.52   Form the following products:

(a) $(3x + 2y)(9x^2 - 6xy + 4y^2)$
(b) $(4x - 5y)(16x^2 + 20xy + 25y^2)$

Solution:

(a) This is a direct application of (1.39). Accordingly,

$$(3x + 2y)(9x^2 - 6xy + 4y^2) = (3x)^3 + (2y)^3$$
$$= 27x^3 + 8y^3.$$

(b) Using (1.40), we obtain

$$(4x - 5y)(16x^2 + 20xy + 25y^2) = (4x)^3 - (5y)^3$$
$$= 64x^3 - 125y^3.$$

EXERCISES

Find the products in Exercises 1 to 58.

1. $3(2x - 3y + 1)$
2. $x(2x - 3y)$
3. $2x(x - 2y + 5)$
4. $5x^2(7x - 2y + 3)$
5. $xy(3x^2 + 4xy)$
6. $3xy(x^2 - xy + y^2)$
7. $x^2y^2(x^3 + 3xy + y^2)$
8. $xyz(xy + yz + zx)$
9. $abc(a^2 + b^2 + c^2)$
10. $a^2b^2c^2(a + b + c)$
11. $(2x + 1)(3x - 1)$
12. $(3x + 2)(2x + 3)$
13. $(4x - 5)(7x + 3)$
14. $(4x - 3)(3x + 7)$
15. $(6x + 7)(3x - 8)$

16. $(3x + 1)(4x - 3)$ **17.** $(4x + 3y)(x - 2y)$ **18.** $(3x + 2y)(x + y)$
19. $(x + 5y)(2x - 3y)$ **20.** $(4x - 3y)(4x + 3y)$ **21.** $(a + 3b)(a - 2b)$
22. $(a - 4b)(4a + b)$ **23.** $(2m + 3n)(2m + 5n)$ **24.** $(3m - 7n)(4m + n)$
25. $(x + 4)(x + 3)$ **26.** $(x + 3)(x + 6)$ **27.** $(x + 5)(x - 3)$
28. $(x - 2)(x + 7)$ **29.** $(x - 8)(x + 9)$ **30.** $(x - 4)(x - 7)$
31. $(x - 6)(x - 11)$ **32.** $(x - 1)(x + 8)$ **33.** $(x + 3y)^2$
34. $(2x - y)^2$ **35.** $(4x - 5y)^2$ **36.** $(3x + 4y)^2$
37. $(2x + 7)^2$ **38.** $(4x - 5)^2$ **39.** $(3x - 5)^2$
40. $(6x + 1)^2$ **41.** $(x - 3)(x^2 + 3x + 9)$ **42.** $(2x + 1)(4x^2 - 2x + 1)$
43. $(y + 2)(y^2 - 2y + 4)$ **44.** $(x + 2y)(x^2 - 2xy + 4y^2)$
45. $(3x - 4y)(9x^2 + 12xy + 16y^2)$ **46.** $(2x + 3y)(4x^2 - 6xy + 9y^2)$
47. $(1 - x)(1 + x)(1 + x^2)$ **48.** $(1 - x)(1 + x + x^2)$
49. $(1 - x)(1 + x + x^2 + x^3)$ **50.** $(1 - x)(1 + x + x^2 + x^3 + x^4)$
51. $(x + y + z)(x + y - z)$ **52.** $(3x + 2y + 4z)(3x + 2y - 4z)$
53. $(x^m + y^m)(x^m - y^m)$ **54.** $(x^m + 3)(x^m - 2)$
55. $(x + a)(x + b)(x + c)$ **56.** $(x - a)(x - b)(x - c)$
57. $(x + a)^3$ **58.** $(x - a)^3$

Simplify each expression in Exercises 59 to 70.
59. $(a + b)^2 + (a - b)^2$ **60.** $(3x + 4y)^2 + (3x - 4y)^2$
61. $(2x + 3y)^2 - (2x - 3y)^2$ **62.** $(5x + 4y)^2 - (5x - 4y)^2$
63. $(a + b + c)^2 - (a + b - c)^2$ **64.** $(2a + b + 3c)^2 + (2a - b + 3c)^2$
65. $(x + y)^3 + (x - y)^3$ **66.** $(x + y)^3 - (x - y)^3$
67. $(6x + y)(x + 2y) - (2x + 3y)(3x - y)$ **68.** $(5x - 2y)(3x + y) - (3x + 4y)(x - y)$
69. $(4x^2 + 12xy + 9y^2) - (2x - 3y)^2$ **70.** $(x + y + z)^2 - (x + y - z)^2$

Rationalize the denominator in each fraction in Exercises 71 to 80.
71. $\dfrac{1}{\sqrt{3} + \sqrt{2}}$ **72.** $\dfrac{1}{\sqrt{2} - 1}$ **73.** $\dfrac{3}{\sqrt{6} - \sqrt{3}}$ **74.** $\dfrac{2}{\sqrt{7} + \sqrt{3}}$
75. $\dfrac{1}{\sqrt{x} - \sqrt{y}}$ **76.** $\dfrac{1}{\sqrt{x + y} + \sqrt{x}}$ **77.** $\dfrac{1}{\sqrt{x + 4y} - \sqrt{x}}$ **78.** $\dfrac{1}{\sqrt{x + y} - \sqrt{x - y}}$
79. $\dfrac{2}{\sqrt{x + 2y} + \sqrt{x - 2y}}$ **80.** $\dfrac{3}{\sqrt{x + 3y} - \sqrt{x - 3y}}$

Rationalize the numerator in each fraction in Exercises 81 to 90.
81. $\dfrac{\sqrt{6} - \sqrt{5}}{2}$ **82.** $\dfrac{\sqrt{6} + \sqrt{5}}{2}$ **83.** $\dfrac{\sqrt{7} + \sqrt{3}}{3}$ **84.** $\dfrac{\sqrt{3} + \sqrt{2}}{2}$
85. $\dfrac{4\sqrt{3} + 3\sqrt{5}}{3}$ **86.** $\dfrac{2\sqrt{5} - 3\sqrt{2}}{4}$ **87.** $\dfrac{\sqrt{x + y} + \sqrt{x}}{2}$ **88.** $\dfrac{\sqrt{x + 2y} - \sqrt{x}}{3}$
89. $\dfrac{\sqrt{x + h} - \sqrt{x}}{h}$ **90.** $\dfrac{\sqrt{2(x + h)} - \sqrt{2x}}{h}$

1.8

FACTORING POLYNOMIALS

We know that if x, y, and z are real numbers, then by the distributive property, we have

$$x(y + z) = xy + xz.$$

The real numbers x and $(y + z)$ are *factors* of $xy + xz$. Similarly, in the expression $6a^2b + 4ab^2$, the greatest common factor of the terms $6a^2b$ and $4ab^2$ is $2ab$. Thus

$$6a^2b + 4ab^2 = 2ab(3a + 2b).$$

The terms $2ab$ and $(3a + 2b)$ are *factors* of the product $6a^2b + 4ab^2$. Factoring is thus a process of expressing a polynomial of certain degree as a product of polynomials of lower degree. When we write

$$x^2 + 7xy + 12y^2 = (x + 3y)(x + 4y),$$

we are factoring the left-hand side of the equation, namely, $x^2 + 7xy + 12y^2$. The terms $(x + 3y)$ and $(x + 4y)$ on the right-hand side are factors of the product $x^2 + 7xy + 12y^2$.

In this section, we restrict ourselves to polynomials with integral coefficients and therefore look for factors that have integral coefficients. For example,

$$x^2 + 6x + 8 = (x + 4)(x + 2),$$

$$5y^2 + 4y - 1 = (5y - 1)(y + 1),$$

$$4x^2 - 20xy + 25y^2 = (2x - 5y)^2,$$

and so on.

Formulas (1.34) to (1.40) developed in Section 1.7 are used extensively in factoring polynomials. Because these formulas hold for all real numbers x, a, b, c, and d, they hold even if these values are replaced by expressions that represent some other real values. For example, formula (1.38) is for the difference of two squares, which we restate as follows:

$$x^2 - b^2 = (x - b)(x + b).$$

Several applications follow from this formula.

Example 1.53   Factor the following:

(a) $4x^2 - 9$

(b) $5x^2 - 20y^2$

Solution:

(a) Note that

$$4x^2 = (2x)^2 \quad \text{and} \quad 9 = 3^2.$$

Thus

$$4x^2 - 9 = (2x)^2 - (3)^2 = (2x - 3)(2x + 3).$$

(b) $5x^2 - 20y^2 = 5(x^2 - 4y^2)$
$$= 5[(x)^2 - (2y)^2]$$
$$= 5(x - 2y)(x + 2y).$$

Example 1.54

Factor the following:

(a) $x^4 - 81y^4$
(b) $(3x + 2y)^2 - 49z^2$

Solution:

(a) $x^4 - 81y^4 = (x^2)^2 - (9y^2)^2$
$$= (x^2 - 9y^2)(x^2 + 9y^2).$$

Observing that

$$x^2 - 9y^2 = (x)^2 - (3y)^2 = (x - 3y)(x + 3y),$$

we conclude that

$$x^4 - 81y^4 = (x - 3y)(x + 3y)(x^2 + 9y^2).$$

(b) $(3x + 2y)^2 - 49z^2 = (3x + 2y)^2 - (7z)^2$
$$= (3x + 2y + 7z)(3x + 2y - 7z).$$

Example 1.55

Factor

$$x^2 + 8x + 12.$$

Solution: Comparing this expression with (1.35), we observe that

$$b + d = 8 \quad \text{and} \quad bd = 12.$$

The problem now reduces to finding two integers b and d such that their sum is 8 and their product is 12. Since $bd = 12$, it follows that b and d are both positive or both negative. Further, since $b + d$ is positive, we conclude that b and d are both positive. There are only three ways to express 12 as the product of two positive integers:

$$12 = 12 \cdot 1, \quad 12 = 6 \cdot 2, \quad \text{and} \quad 12 = 4 \cdot 3.$$

The second set of factors, 6 and 2, add to 8. Thus

$$x^2 + 8x + 12 = (x + 6)(x + 2).$$

Example 1.56

Factor

$$x^2 - 10x + 24.$$

Solution: If we compare this expression with (1.35), we note that

$$b + d = -10 \qquad \text{and} \qquad bd = 24.$$

The fact that bd is positive means that b and d are both positive or both negative; since $b + d$ is a negative number, b and d are both negative. We can express 24 as the product of two negative numbers as follows:

bd	b	d	$b + d$
24	-1	-24	-25
24	-2	-12	-14
24	-3	-8	-11
24	-4	-6	-10

Note that -4 and -6 are the only numbers that add to -10, and their product is 24. Hence

$$x^2 - 10x + 24 = (x - 4)(x - 6).$$

Example 1.57

Factor

$$x^2 + 8x - 33.$$

Solution: We proceed as before and compare this expression with (1.35). Observe that

$$b + d = 8 \qquad \text{and} \qquad bd = -33.$$

Since bd is negative, it follows that one factor is negative and the other positive. We can express -33 as the product of one negative and one positive number as follows:

bd	b	d	$b + d$
-33	-1	33	32
-33	1	-33	-32
-33	-3	11	8
-33	3	-11	-8

The third set of factors, $b = -3$ and $d = 11$, satisfies the given conditions that $b + d = 8$ and $bd = -33$. Thus

$$x^2 + 8x - 33 = (x - 3)(x + 11).$$

We now turn our attention to second-degree polynomials of the form

$$px^2 + qx + r, \qquad p \neq 1, \quad p \neq 0, \tag{1.41}$$

where p and q are integers. The factors of this polynomial (if they exist) must be of the form $(ax + b)$ and $(cx + d)$, where a, b, c, and d are integers. In factoring (1.41), we use the following proposition, which we state without proof.

Proposition 1.8.1

A polynomial of the form

$$px^2 + qx + r$$

is factorable over the set of integers if and only if the product pr has factors that have a sum of q.

The proposition is also applicable to polynomials of the form

$$px^2 + qxy + ry^2$$

where p, q, and r are integers with the restriction that $p \neq 0$ and $p \neq 1$. If the product pr does not have any factors that add to q, then the second-degree polynomial is said to be **prime,** or **irreducible.**

Example 1.58

Factor

$$6x^2 - 7x - 3.$$

Solution: We must find two integers whose product is $pr = (6)(-3) = -18$ and whose sum is -7. Since the product is negative, it follows that one factor is positive and the other negative. We can express -18 as the product of one positive and one negative number as follows:

Product	Factors		Sum of the Factors
-18	1	-18	-17
-18	2	-9	-7
-18	3	-6	-3
-18	-1	18	17
-18	-2	9	7
-18	-3	6	3

The second set of factors, 2 and -9, satisfies the conditions. Next we write

$$6x^2 - 7x - 3 = 6x^2 + 2x - 9x - 3.$$

Observe that $2x$ is the greatest common factor in the first two terms on the right-hand side, while 3 is the greatest common factor in the last two terms. Thus

$$6x^2 + 2x - 9x - 3 = 2x(3x + 1) - 3(3x + 1)$$

By the distributive property,

$$2x(3x + 1) - 3(3x + 1) = (2x - 3)(3x + 1).$$

Therefore,

$$6x^2 - 7x - 3 = (2x - 3)(3x + 1).$$

Example 1.59   Factor

$$7x^2 - 19x + 10.$$

Solution: Comparing this expression with (1.41), we note that $p = 7$, $q = -19$, and $r = 10$. We need to find two numbers whose product is $pr = (7)(10) = 70$ and whose sum is $q = -19$. Since the product is positive, the two numbers must be of the same sign; since their sum is negative, the numbers are both negative. We can express 70 as the product of two negative numbers as follows:

Product	Factors		Sum of the Factors
70	-1	-70	-71
70	-2	-35	-37
70	-5	-14	-19
70	-10	-7	-17

The third set of factors, -5 and -14, satisfies the conditions. Next we write

$$7x^2 - 19x + 10 = 7x^2 - 14x - 5x + 10$$
$$= 7x(x - 2) - 5(x - 2)$$
$$= (7x - 5)(x - 2).$$

Hence

$$7x^2 - 19x + 10 = (7x - 5)(x - 2).$$

Example 1.60   Factor

$$4x^2 + xy - 14y^2.$$

Solution: First we find two numbers whose product is $(4)(-14) = -56$ and whose sum is 1. The table below provides several combinations:

Product	Factors		Sum of the Factors
-56	-1	56	55
-56	-2	28	26
-56	-4	14	10
-56	-7	8	1

The last set of factors, -7 and 8, satisfies the given conditions. Next, we may express the polynomial as follows:

$$4x^2 + xy - 14y^2 = 4x^2 + 8xy - 7xy - 14y^2$$
$$= 4x(x + 2y) - 7y(x + 2y)$$
$$= (4x - 7y)(x + 2y).$$

Hence

$$4x^2 + xy - 14y^2 = (4x - 7y)(x + 2y).$$

Sum and Difference of Cubes

Recall that

$$x^2 - b^2 = (x - b)(x + b).$$

However, the sum of squares

$$x^2 + b^2$$

is irreducible. In the case of a sum or difference of two cubes, we use (1.39) or (1.40), respectively. These formulas can be restated as follows:

$$x^3 + a^3 = (x + a)(x^2 - ax + a^2)$$

and

$$x^3 - a^3 = (x - a)(x^2 + ax + a^2).$$

Let us look at some examples.

Example 1.61 Factor the following:

(a) $a^3 + 8$

(b) $27a^3 - 64b^3$

Solution:

(a) $a^3 + 8 = a^3 + 2^3 = (a + 2)(a^2 - 2a + 4)$.

(b) $27a^3 - 64b^3 = (3a)^3 - (4b)^3$.

Substituting $3a$ for x and $4b$ for a in (1.40), we get

$$(3a)^3 - (4b)^3 = (3a - 4b)(9a^2 + 12ab + 16b^2).$$

Example 1.62　　Factor

$$x^6 - y^6.$$

Solution:　First we express this polynomial as a difference of two squares. That is,

$$x^6 - y^6 = (x^3)^2 - (y^3)^2 = (x^3 - y^3)(x^3 + y^3).$$

Now we use the formulas for the difference and the sum of two cubes, and we obtain

$$\begin{aligned} x^6 - y^6 &= (x^3 - y^3)(x^3 + y^3) \\ &= (x - y)(x^2 + xy + y^2)(x + y)(x^2 - xy + y^2). \end{aligned}$$

Alternatively, we may first express the polynomial as a difference of two cubes and then apply (1.40). That is,

$$\begin{aligned} x^6 - y^6 &= (x^2)^3 - (y^2)^3 \\ &= (x^2 - y^2)(x^4 + x^2y^2 + y^4) \\ &= (x - y)(x + y)(x^4 + x^2y^2 + y^4). \end{aligned}$$

It may not be easy to recognize that the third factor, $x^4 + x^2y^2 + y^4$, can be written as a difference of two squares as follows:

$$\begin{aligned} x^4 + x^2y^2 + y^4 &= (x^4 + 2x^2y^2 + y^4) - x^2y^2 \\ &= (x^2 + y^2)^2 - (xy)^2 \\ &= (x^2 + y^2 - xy)(x^2 + y^2 + xy). \end{aligned}$$

In either case,

$$x^6 - y^6 = (x - y)(x + y)(x^2 - xy + y^2)(x^2 + xy + y^2).$$

Factor by Grouping

Frequently, we encounter polynomials that have more than three terms. These terms can be rearranged to group terms that have a common factor. This allows us to factor the polynomial easily by using the distributive property of the real number system.

Example 1.63

Factor

$$ab + a + b + b^2.$$

Solution: The first two terms have a as the common factor, while the last two terms have b as the common factor. Thus

$$ab + a + b + b^2 = a(b + 1) + b(b + 1).$$

Observe that $a(b + 1)$ and $b(b + 1)$ also have a common factor, namely, $(b + 1)$:

$$a(b + 1) + b(b + 1) = (a + b)(b + 1).$$

Thus

$$ab + a + b + b^2 = (a + b)(b + 1).$$

Example 1.64

Factor

$$x^3 - x^2y + xy^2 - y^3.$$

Solution:

$$\begin{aligned} x^3 - x^2y + xy^2 - y^3 &= (x^3 - x^2y) + (xy^2 - y^3) \\ &= x^2(x - y) + y^2(x - y) \\ &= (x^2 + y^2)(x - y). \end{aligned}$$

Example 1.65

Factor

$$x^2 + 6xy + 9y^2 - 25z^2.$$

Solution: The first three terms form a perfect square. Thus we write

$$x^2 + 6xy + 9y^2 - 25z^2 = (x + 3y)^2 - (5z)^2.$$

Applying the formula for the difference of two squares, we obtain

$$(x + 3y)^2 - (5z)^2 = (x + 3y + 5z)(x + 3y - 5z).$$

Hence

$$x^2 + 6xy + 9y^2 - 25z^2 = (x + 3y + 5z)(x + 3y - 5z).$$

EXERCISES

Factor each expression in Exercises 1 to 14.

1. $x^2 + x$

2. $3x^2 + 6x$

3. $2x^4 + 3x^2$

4. $9x^4 - 12x^3$

5. $x^3 + x^2 + 2x$

6. $x^4 + 2x^3 - x^2 + 4x$

7. $x^2y + xy^2$

8. $6x^2 - 2xy$

9. $4m^2 + 8mn$

10. $2m^2 - 4mn$

11. $a(x + y) + b(x + y)$

12. $x(x + 4) - y(x + 4)$

13. $x(x + 2y) + y(x + 2y)$

14. $p(2p + 3q) - 2q(2p + 3q)$

Factor each expression in Exercises 15 to 26, using the difference of two squares.

15. $x^2 - 4$

16. $y^2 - 9$

17. $4x^2 - 9y^2$

18. $25x^2 - 36y^2$

19. $16x^2 - 49y^2$

20. $4m^2 - 16n^2$

21. $9m^2 - 81n^2$

22. $121m^2 - 144n^2$

23. $x^2y^2 - 9z^2$

24. $(x + y)^2 - z^2$

25. $(a + b + c)^2 - (a + b - c)^2$

26. $(3x + 2y)^2 - (3x - 2y)^2$

Factor each expression in Exercises 27 to 64.

27. $x^2 + 5x + 6$

28. $x^2 + 9x + 18$

29. $x^2 - 9x + 14$

30. $x^2 - 11x + 30$

31. $y^2 - 2y - 35$

32. $y^2 - 13y - 30$

33. $y^2 + 4y - 21$

34. $y^2 + 14y - 32$

35. $m^2 + 8m - 48$

36. $m^2 + 15m - 54$

37. $m^2 + 5m - 36$

38. $m^2 + m - 6$

39. $p^2 + 4p + 4$

40. $p^2 - 6p + 9$

41. $p^2 - 14p + 49$

42. $p^2 - 16p + 64$

43. $2x^2 + 11x + 12$

44. $2x^2 + 7x + 6$

45. $3x^2 + 4x + 1$

46. $3x^2 + 5x - 2$

47. $3x^2 + 5x + 2$

48. $3x^2 - 2x - 5$

49. $4x^2 + 8x + 3$

50. $4x^2 + 11x - 3$

51. $4x^2 + 12x - 7$

52. $5m^2 + 12m + 4$

53. $6m^2 - m - 12$

54. $6m^2 + 7m - 3$

55. $6m^2 - 7mn - 5n^2$

56. $6p^2 + 11pq - 10q^2$

57. $10p^2 - 17pq + 3q^2$

58. $10p^2 - 23pq - 5q^2$

59. $10p^2 + 17pq - 48q^2$

60. $12p^2 - 7pq - 12q^2$

61. $4x^2 - 12xy + 9y^2$

62. $9y^2 + 24yz + 16z^2$

63. $4x^2 + 4xy + y^2$

64. $25x^2 - 80xy + 64y^2$

Factor each expression in Exercises 65 to 76, using the sum or difference of two cubes.

65. $a^3 + 1$

66. $a^3 + 27$

67. $a^3 - 8b^3$

68. $27a^3 - 64b^3$

69. $8x^3 + 27y^3$

70. $125y^3 - 8z^3$

71. $216x^3 - 125y^3$

72. $27x^3 - 125y^3$

73. $x^6 + y^6$

74. $27x^6 - 64y^6$

75. $64x^6 - 729y^6$

76. $64x^6 + 729y^6$

Factor each expression in Exercises 77 to 90 by grouping.

77. $x^3 + x^2 + xy + y$

78. $am^2 + bn^2 + an^2 + bm^2$

79. $ax + ay - bx - by$

80. $a^3 - a^2 + 4ab - 4b$

81. $a^2 + ab + 2a + 2b$

82. $a^2 - 3a + ab - 3b$

83. $1 + x + x^2 + x^3$

84. $x^2y + 3x - 3y - xy^2$

85. $ax^2 + x^2 - ay^2 - y^2$

86. $ax^2 + bx^2 - a - b$

87. $x^2 + 2x + 1 - y^2$

88. $x^2 + 4x + 4 - 9y^2$

89. $4x^2 - 12xy + 9y^2 - 16z^2$

90. $9x^2 + 24xy + 16y^2 - 49z^2$

1.9   **QUOTIENTS OF POLYNOMIALS
AND SYNTHETIC DIVISION**

It is frequently necessary to divide one polynomial $P(x)$ by a polynomial of the form $x - k$, where the coefficient of x is 1. We can do so by long division as shown in the following example:

$$
\begin{array}{r}
3x^2 + 8x\ \ + 6 \qquad\qquad \text{quotient} \\
x - 2\ \overline{)\,3x^3 + 2x^2 - 10x + 4} \quad \text{divisor}\,\overline{)\,\text{dividend}} \\
\underline{3x^3 - 6x^2} \\
8x^2 - 10x \\
\underline{8x^2 - 16x} \\
6x + \ 4 \\
\underline{6x - 12} \\
\underline{+ 16} \qquad \text{remainder}
\end{array}
$$

Since the like terms are aligned in columns, we can eliminate the variable and write only the coefficients. Observe that terms in the dividend are arranged in descending powers of x. *Should there be any missing term, we write a zero for its coefficient.* Further, the coefficient of x in the divisor is 1; it contributes nothing to the process and can therefore be omitted. These steps simplify the above example as follows:

$$
\begin{array}{r}
3 \quad\ 8 \quad\ \ 6 \\
-2\,\underline{)\,3 \quad\ 2 \ -10 \quad\ 4}\\
\circled{3}\ -6 \\
\underline{8 \ -10 }\\
\circled{8}\ -16 \\
\underline{6 \quad\ 4}\\
\circled{6}\ -12\\
\underline{16}
\end{array}
$$

Observe that the numbers in circles are repetitions of the numbers written immediately above. If we eliminate these repetitions, we have the following form:

$$
\begin{array}{r}
3 \quad\ 8 \quad\ \ 6 \\
-2\,\underline{)\,3 \quad\ 2 \ -10 \quad\ 4}\\
-6 \\
\underline{8 \ \boxed{-10} }\\
-16 \\
\underline{6 \quad\ \boxed{4}}\\
-12\\
\underline{16.}
\end{array}
$$

This can be condensed further by eliminating the boxed numbers, since they are only repetitions of the coefficients in the dividend. We then get

$$
\begin{array}{r}
3 \quad\ 8 \quad\ \ 6 \\
-2\,\underline{)\,3 \quad\ 2 \ -10 \quad\ 4}\\
\underline{-6 \ -16 \ -12}\\
8 \quad\ 6 \quad\ 16.
\end{array}
$$

If we now insert the leading coefficient 3 in the first position of the last row, we recognize that the first three numbers of that row are the coefficients 3, 8, and 6 of the quotient and the last number, 16, is the remainder. Since the coefficients of the quotient do not have to be repeated, we may delete the first row, thus obtaining

$$\begin{array}{r|rrrr} -2 & 3 & 2 & -10 & 4 \\ \downarrow & & -6 & -16 & -12 \\ \hline & 3 & 8 & 6 & 16. \end{array} \qquad \textbf{(1.42)}$$

Now let us pause and examine (1.42). We "bring down" the leading coefficient 3 to the third row. This is multiplied by the divisor -2, and the product -6 is written in the next position in the second row. Subtracting -6 from 2, we obtain the second number in the third row, 8. Multiplying 8 by -2, we obtain -16, which we put in the third position in the second row. Now, the product -16 is subtracted from -10, and the result, 6, is placed in the third row. This number 6 is multiplied by -2, and the product, -12, is written in the next position in the second row. This product -12 is also subtracted from 4 to yield 16 in the third row.

The first three numbers 3, 8, and 6, which are the coefficients of the variables in the quotient, and the last number 16, all appearing in the third row, have been obtained by subtracting the numbers in the second row from the corresponding elements in the first row. The subtraction at each step can be avoided if we replace -2 in the divisor by its additive inverse, 2. This step changes the signs of the numbers in the second row. If we now add instead of subtracting at each step in the above process, we obtain

$$\begin{array}{r|rrrr} 2 & 3 & 2 & -10 & 4 \\ \downarrow & & 6 & 16 & 12 \\ \hline & 3 & 8 & 6 & 16. \end{array}$$

Hence the quotient is $3x^2 + 8x + 6$ and the remainder is 16. This process is called *synthetic division*.

Let us now summarize the steps involved in using synthetic division if the polynomial $P(x)$ is divided by $x - k$.

1. Arrange the coefficients of $P(x)$ in descending powers of x. In case a term is missing, write a zero for its coefficient.
2. Place k, the additive inverse of $-k$, in the divisor.
3. Bring down the coefficient of the largest power of x to the third row. Multiply it by k, place the product in the next position under the *second* coefficient of the dividend, and add the two numbers.
4. Multiply this sum by k, place the product under the *third* coefficient of the dividend, and add.
5. Continue this process until all the coefficients in the dividend are used.

The quotient is formed by using the numbers from the third row as coefficients, reading from left to right; the last number in the third row is the remainder. Note that the degree of the quotient is 1 less than the degree of $P(x)$.

The following examples further illustrate the procedure.

Example 1.66   Use synthetic division to determine the quotient and the remainder by dividing $P(x) = 2x^4 - 3x^3 - 4x + 8$ by $x - 3$.

Solution: The divisor is $x - 3$. We therefore use 3, the additive inverse of (-3) in the divisor. The synthetic division takes the form

$$\begin{array}{r|rrrrr} 3 & 2 & -3 & 0 & -4 & 8 \\ & \downarrow & 6 & 9 & 27 & 69 \\ \hline & 2 & 3 & 9 & 23 & \boxed{77} = \text{remainder} \end{array}$$

The zero in the first row represents the coefficient of the missing x^2 term. The first four numbers in the third row are the coefficients of the quotient, and the last number is the remainder. Hence the quotient is

$$2x^3 + 3x^2 + 9x + 23$$

and the remainder is 77.

Example 1.67   Use synthetic division to determine the quotient and the remainder if $P(x) = x^3 + 8$ is divided by $x + 2$.

Solution: We write the divisor $x + 2$ in the form $x - k$ as $x + 2 = x - (-2)$ and use $k = -2$ in the divisor.

$$\begin{array}{r|rrrr} -2 & 1 & 0 & 0 & 8 \\ & \downarrow & -2 & 4 & -8 \\ \hline & 1 & -2 & 4 & 0. \end{array}$$

The quotient is

$$x^2 - 2x + 4,$$

and the remainder is 0.

EXERCISES

Using synthetic division, find the quotient and the remainder if the first expression is divided by the second in Exercises 1 to 18.

1. $x^2 - 7x + 14$; $x - 3$
2. $3x^2 - x - 12$; $x - 2$
3. $x^3 + 3x^2 + 3x + 5$; $x + 1$
4. $x^3 - 3x^2 + 3x - 1$; $x - 1$
5. $x^3 - 3x^2 + x - 4$; $x - 1$
6. $x^3 - 7x^2 - 13x + 3$; $x + 2$
7. $3x^3 - 11x^2 - 20x + 3$; $x - 5$
8. $5x^3 - 6x^2 - 24x - 2$; $x + 2$
9. $2x^4 + 5x^3 - 2x - 8$; $x + 3$
10. $3x^4 - 8x^3 + 7x + 6$; $x - 2$

11. $x^4 - 1$; $x - 1$
12. $x^5 + 1$; $x + 1$
13. $2x^5 + 3x^4 - x^3 + 7x^2 - 12x + 1$; $x + 1$
14. $x^5 - 3x^4 - 5x^3 + 2x^2 - 16x + 3$; $x + 2$
15. $x^5 + 32$; $x + 2$
16. $x^6 - 64$; $x - 2$
17. $x^6 - y^6$; $x - y$
18. $x^6 + y^6$; $x + y$

1.10

RATIONAL EXPRESSIONS

Definition 1.10.1

A fraction that includes polynomials in both the numerator and the denominator is called a *rational expression*.

The fractions

$$\frac{3x + 4}{4x + 1}, \qquad \frac{x^2 + 3x}{x^2 + 4x + 1}, \qquad \frac{xy}{x + y}, \qquad \text{and} \qquad \frac{x^3 - y^3}{x^2 + y^2}$$

are all examples of rational expressions.

The basic rules that govern operations with rational expressions are similar to those involving rational numbers. We state some of them explicitly. If P, Q, R, and S are polynomials, then

$$\frac{P}{1} = P, \tag{1.43}$$

$$\frac{P}{Q} = \frac{R}{S} \qquad \text{if and only if} \qquad PS = QR, \quad Q \neq 0, \quad S \neq 0, \tag{1.44}$$

$$\frac{PR}{QR} = \frac{RP}{RQ} = \frac{P}{Q}, \qquad Q \neq 0, \quad R \neq 0, \tag{1.45}$$

$$\frac{P}{R} + \frac{Q}{R} = \frac{P + Q}{R}, \qquad R \neq 0, \tag{1.46}$$

$$\frac{P}{Q} + \frac{R}{S} = \frac{PS + QR}{QS}, \qquad Q \neq 0, \quad S \neq 0, \tag{1.47}$$

$$\frac{P}{Q} - \frac{R}{S} = \frac{PS - QR}{QS}, \qquad Q \neq 0, \quad S \neq 0, \tag{1.48}$$

$$\frac{P}{Q} \cdot \frac{R}{S} = \frac{PR}{QS}, \qquad Q \neq 0, \quad S \neq 0, \tag{1.49}$$

$$\frac{P}{Q} \div \frac{R}{S} = \frac{PS}{QR}, \qquad Q \neq 0, \quad R \neq 0, \quad S \neq 0. \tag{1.50}$$

Let us now illustrate these operations with some examples.

Example 1.68

Reduce the rational expression

$$\frac{x^2 - 1}{x^2 + x - 2}, \qquad x \neq -2, 1,$$

to lowest terms.

Solution: The numerator $x^2 - 1$ and the denominator $x^2 + x - 2$ can be factored as follows:

$$x^2 - 1 = (x - 1)(x + 1)$$
$$x^2 + x - 2 = (x - 1)(x + 2).$$

Thus

$$\frac{x^2 - 1}{x^2 + x - 2} = \frac{(x - 1)(x + 1)}{(x - 1)(x + 2)} \qquad (x \neq -2, 1)$$

$$= \frac{x + 1}{x + 2} \qquad (x \neq -2, 1).$$

Remember that to reduce a rational expression to lowest terms, we must factor both the numerator and the denominator and then use (1.45) to cancel every nonzero common factor appearing in both the numerator and the denominator.

Example 1.69

(a)

$$\frac{x^2 + 3x}{x^2 + 7x + 12} = \frac{x(x + 3)}{(x + 3)(x + 4)} \qquad (x \neq -3, -4)$$

$$= \frac{x}{x + 4} \qquad (x \neq -3, -4).$$

(b)

$$\frac{x^2 - 25}{x^2 + 2x - 15} = \frac{(x - 5)(x + 5)}{(x - 3)(x + 5)} \qquad (x \neq -5, 3)$$

$$= \frac{x - 5}{x - 3} \qquad (x \neq -5, 3).$$

(c)

$$\frac{y^2 + 6y + 8}{y^2 + 10y + 24} = \frac{(y + 2)(y + 4)}{(y + 6)(y + 4)} \qquad (y \neq -4, -6)$$

$$= \frac{y + 2}{y + 6} \qquad (y \neq -4, -6).$$

(d)

$$\frac{2x^2 + xy - 15y^2}{x^2 - 9y^2} = \frac{(2x - 5y)(x + 3y)}{(x - 3y)(x + 3y)} \qquad (x \neq 3y, -3y)$$

$$= \frac{2x - 5y}{x - 3y} \qquad (x \neq 3y, -3y).$$

To find the sum of fractions having the same denominator, we use rule (1.46).

Example 1.70

(a) $\dfrac{2x + 1}{3} + \dfrac{3x + 4}{3} = \dfrac{(2x + 1) + (3x + 4)}{3} = \dfrac{5x + 5}{3} = \dfrac{5(x + 1)}{3}.$

(b) $\dfrac{2x}{5y} + \dfrac{x^2}{5y} + \dfrac{3}{5y} = \dfrac{2x + x^2 + 3}{5y}$ $(y \neq 0)$.

(c) $\dfrac{xy}{z} + \dfrac{4y}{z} = \dfrac{xy + 4y}{z} = \dfrac{(x + 4)y}{z}$ $(z \neq 0)$.

To find the sum (or difference) of fractions having different denominators, we first find the lowest common denominator of the fractions and then replace each of the fractions with equivalent fractions having the same denominator. We are then ready to use (1.46) to express the sum (or difference) as a single fraction. Consider, for instance, the following examples.

Example 1.71   Express

$$\frac{2}{x + 2} + \frac{3}{x - 2} \qquad (x \neq -2, 2)$$

as a single fraction.

Solution: The lowest common denominator of the fractions is $(x - 2)(x + 2)$. Now we replace each of the fractions with equivalent fractions having $(x - 2)(x + 2)$ as their common denominator and then find the sum.

$$\frac{2}{x + 2} + \frac{3}{x - 2} = \frac{2(x - 2)}{(x - 2)(x + 2)} + \frac{3(x + 2)}{(x - 2)(x + 2)}$$

$$= \frac{2(x - 2) + 3(x + 2)}{(x - 2)(x + 2)} \qquad (x \neq -2, 2)$$

$$= \frac{2x - 4 + 3x + 6}{(x - 2)(x + 2)} \qquad (x \neq -2, 2)$$

$$= \frac{5x + 2}{(x - 2)(x + 2)} \qquad (x \neq -2, 2).$$

Example 1.72   Express

$$\frac{y}{x - y} - \frac{x}{x + y} \qquad (x \neq -y, y)$$

as a single fraction.

Solution: The lowest common denominator of the fractions is $(x - y)(x + y)$. Replacing each of the fractions with equivalent fractions having $(x - y)(x + y)$ as their common denominator, we get

$$\frac{y}{x - y} - \frac{x}{x + y} = \frac{y(x + y)}{(x - y)(x + y)} - \frac{x(x - y)}{(x - y)(x + y)} \qquad (x \neq -y, y)$$

$$= \frac{y(x + y) - x(x - y)}{(x - y)(x + y)} \qquad (x \neq -y, y)$$

$$= \frac{xy + y^2 - x^2 + xy}{(x - y)(x + y)} \qquad (x \neq -y, y)$$

$$= \frac{2xy + y^2 - x^2}{(x - y)(x + y)} \qquad (x \neq -y, y).$$

Although we must exclude those values of the variable for which the denominator of the rational expression is zero, these are not stated explicitly from this point on. **We continue, however, to assume these restrictions, and the rational expressions are henceforth understood to exclude values of the variables for which the denominator is zero.**

To multiply or divide rational expressions, we factor (if possible) all numerators and denominators and then use rule (1.49) or (1.50), respectively. Then every nonzero common factor is canceled from both the numerator and the denominator to reduce the resulting rational expression to lowest form.

Example 1.73   Simplify the following:

(a) $\dfrac{x^2 - 3x + 2}{x^2 + 5x + 6} \cdot \dfrac{x + 3}{3x - 6}$

(b) $\dfrac{2x + 2}{x^2 + 2x - 8} \cdot \dfrac{x^2 - 4}{x^2 + 4x + 4}$

Solution:

(a) $\dfrac{x^2 - 3x + 2}{x^2 + 5x + 6} \cdot \dfrac{x + 3}{3x - 6} = \dfrac{(x - 1)(x - 2)}{(x + 2)(x + 3)} \cdot \dfrac{x + 3}{3(x - 2)}$

$$= \frac{(x - 1)(x - 2)(x + 3)}{3(x + 2)(x + 3)(x - 2)}$$

$$= \frac{x - 1}{3(x + 2)}.$$

(b) $\dfrac{2x + 2}{x^2 + 2x - 8} \cdot \dfrac{x^2 - 4}{x^2 + 4x + 4} = \dfrac{2(x + 1)}{(x - 2)(x + 4)} \cdot \dfrac{(x - 2)(x + 2)}{(x + 2)^2}$

$$= \dfrac{2(x + 1)(x - 2)(x + 2)}{(x - 2)(x + 4)(x + 2)^2}$$

$$= \dfrac{2(x + 1)}{(x + 4)(x + 2)}.$$

Example 1.74

Simplify the following:

(a) $\dfrac{x^2 - x - 6}{x^2 + 2x - 3} \div \dfrac{x + 2}{x + 3}$

(b) $\dfrac{x^3 + y^3}{x^3 - 8y^3} \div \dfrac{x^2 + 3xy + 2y^2}{x^2 - 4y^2}$

Solution:

(a) $\dfrac{x^2 - x - 6}{x^2 + 2x - 3} \div \dfrac{x + 2}{x + 3} = \dfrac{x^2 - x - 6}{x^2 + 2x - 3} \cdot \dfrac{x + 3}{x + 2}$

$$= \dfrac{(x - 3)(x + 2)}{(x - 1)(x + 3)} \cdot \dfrac{x + 3}{x + 2}$$

$$= \dfrac{x - 3}{x - 1}.$$

(b) $\dfrac{x^3 + y^3}{x^3 - 8y^3} \div \dfrac{x^2 + 3xy + 2y^2}{x^2 - 4y^2} = \dfrac{x^3 + y^3}{x^3 - 8y^3} \cdot \dfrac{x^2 - 4y^2}{x^2 + 3xy + 2y^2}$

$$= \dfrac{(x + y)(x^2 - xy + y^2)}{(x - 2y)(x^2 + 2xy + 4y^2)} \cdot \dfrac{(x - 2y)(x + 2y)}{(x + y)(x + 2y)}$$

$$= \dfrac{(x + y)(x^2 - xy + y^2)(x - 2y)(x + 2y)}{(x - 2y)(x^2 + 2xy + 4y^2)(x + y)(x + 2y)}$$

$$= \dfrac{x^2 - xy + y^2}{x^2 + 2xy + 4y^2}.$$

Let us now look at quotients in which the numerator and the denominator are not polynomials.

Example 1.75

Simplify the following:

$$\frac{1 - \dfrac{1}{x+1}}{1 + \dfrac{1}{x+1}}.$$

Solution: We express both the numerator and the denominator as rational expressions and then divide the two fractions.

$$1 - \frac{1}{x+1} = \frac{x+1-1}{x+1} = \frac{x}{x+1}$$

$$1 + \frac{1}{x+1} = \frac{x+1+1}{x+1} = \frac{x+2}{x+1}.$$

Thus

$$\frac{1 - \dfrac{1}{x+1}}{1 + \dfrac{1}{x+1}} = \frac{\dfrac{x}{x+1}}{\dfrac{x+2}{x+1}}$$

$$= \frac{x}{x+1} \div \frac{x+2}{x+1}$$

$$= \frac{x}{x+1} \cdot \frac{x+1}{x+2}$$

$$= \frac{x}{x+2}.$$

Another method of simplifying such expressions is to multiply both the numerator and the denominator by the least common denominator of all the denominators, as illustrated in Example 1.76.

Example 1.76

Simplify the following:

$$\frac{\dfrac{1}{x-4} + \dfrac{2}{x^2 - 16}}{3 + \dfrac{1}{x+4}}.$$

Solution: The least common denominator of all the denominators is $x^2 - 16$. Multiplying both the numerator and the denominator by $x^2 - 16$, we get

$$\frac{\dfrac{1}{x - 4} + \dfrac{2}{x^2 - 16}}{3 + \dfrac{1}{x + 4}} = \frac{(x^2 - 16)\left(\dfrac{1}{x - 4} + \dfrac{2}{x^2 - 16}\right)}{(x^2 - 16)\left(3 + \dfrac{1}{x + 4}\right)}$$

$$= \frac{(x^2 - 16)\left(\dfrac{1}{x - 4}\right) + (x^2 - 16)\left(\dfrac{2}{x^2 - 16}\right)}{(x^2 - 16)(3) + (x^2 - 16)\left(\dfrac{1}{x + 4}\right)}$$

$$= \frac{(x + 4) + 2}{3(x^2 - 16) + (x - 4)}$$

$$= \frac{x + 6}{3x^2 + x - 52}$$

$$= \frac{x + 6}{(3x + 13)(x - 4)}.$$

EXERCISES

Reduce each rational expression in Exercises 1 to 24 to lowest terms.

1. $\dfrac{x^2 - x}{x - 1}$ $(x \neq 1)$

2. $\dfrac{x + 1}{x^2 + x}$ $(x \neq 0, -1)$

3. $\dfrac{x^2 - 1}{x - 1}$ $(x \neq 1)$

4. $\dfrac{3x + 6}{x^2 - 4}$ $(x \neq -2, 2)$

5. $\dfrac{x^2 - 4}{(x - 2)^2}$ $(x \neq 2)$

6. $\dfrac{x^2 - 9}{2x + 6}$ $(x \neq -3)$

7. $\dfrac{x^2 + 5x + 6}{x + 2}$ $(x \neq -2)$

8. $\dfrac{x^2 + 5x - 14}{x - 2}$ $(x \neq 2)$

9. $\dfrac{x^2 + 7x + 12}{x(x + 4)}$ $(x \neq -4, 0)$

10. $\dfrac{x^2 + x - 2}{x^2 - 2x + 1}$ $(x \neq 1)$

11. $\dfrac{x^2 - 6x + 8}{x^2 - 9x + 20}$ $(x \neq 4, 5)$

12. $\dfrac{x^2 + 2x - 15}{x^2 - 6x + 9}$ $(x \neq 3)$

13. $\dfrac{x^2 - 8x + 15}{x^2 - 2x - 3}$ $(x \neq -1, 3)$

14. $\dfrac{y^2 - 2y - 8}{y^2 - y - 6}$ $(y \neq -3, 2)$

15. $\dfrac{x^2 - y^2}{x - y}$ $(x \neq y)$

16. $\dfrac{x^3 + y^3}{x^2 - y^2}$ $(x \neq -x, x)$

17. $\dfrac{x^3 - y^3}{x^2 - y^2}$ $(x \neq -y, y)$

18. $\dfrac{8x^3 - 27y^3}{4x^2 - 9y^2}$ $(y \neq -\frac{2}{3}x, \frac{2}{3}x)$

19. $\dfrac{x^2 - 16y^2}{x^2 - 3xy - 4y^2}$ $(x \neq -y, 4y)$

20. $\dfrac{x^2 + 5xy + 6y^2}{x^2 + 6xy + 8y^2}$ $(x \neq -2y, -4y)$

21. $\dfrac{x^2 + xy}{x^2 + 3xy + 2y^2}$ $(x \neq -y, -2y)$

22. $\dfrac{x^2 + 2xy}{x^2 + 5xy + 6y^2}$ $(x \neq -2y, -3y)$

23. $\dfrac{x^2 + 6xy + 8y^2}{x^2 - 3xy - 10y^2}$ $(x \neq -2y, 5y)$

24. $\dfrac{x^2 - 8xy + 15y^2}{x^2 + xy - 12y^2}$ $(x \neq 3y, -4y)$

Reduce each expression in Exercises 25 to 40 to a single fraction in lowest terms.

25. $\dfrac{x}{4} + \dfrac{3x + 1}{4}$

26. $\dfrac{3x}{5} - \dfrac{y - 2}{5} + \dfrac{2y}{5}$

27. $\dfrac{x}{y} + \dfrac{3x^2}{y} + \dfrac{1}{y}$ $(y \neq 0)$

28. $\dfrac{2x}{3y} + \dfrac{x^3}{3y} - \dfrac{4}{3y}$ $(y \neq 0)$

29. $\dfrac{3x - y}{y} - \dfrac{4x + 3y}{y} + \dfrac{2}{y}$ $(y \neq 0)$

30. $\dfrac{x + y}{z} + \dfrac{3x - y}{z} + \dfrac{1}{z}$ $(z \neq 0)$

31. $\dfrac{2x + y}{3y} + \dfrac{x}{y}$ $(y \neq 0)$

32. $\dfrac{3}{x} + \dfrac{4}{y} + \dfrac{1}{xy}$ $(x \neq 0, y \neq 0)$

33. $\dfrac{x}{3y} + \dfrac{2y}{x}$ $(x \neq 0, y \neq 0)$

34. $\dfrac{x + y}{2} - \dfrac{4}{x + y}$ $(y \neq -x)$

35. $\dfrac{x - y}{x + y} + \dfrac{x + y}{x - y}$ $(y \neq -x, x)$

36. $\dfrac{x + y}{x - y} - \dfrac{x - y}{x + y}$ $(y \neq -x, x)$

37. $\dfrac{2x - y}{2x + y} - \dfrac{xy}{2x - y}$ $(y \neq -2x, 2x)$

38. $\dfrac{y}{3x + 2y} + \dfrac{x - 2y}{3x + y}$ $\left(y \neq -3x, \dfrac{-3x}{2}\right)$

39. $\dfrac{x + 2y}{x + y} + \dfrac{x + y}{x + 2y}$ $(x \neq -y, -2y)$

40. $\dfrac{2x - 3y}{2x + 3y} + \dfrac{2x + 3y}{2x - 3y}$ $(y \neq \pm\frac{2}{3}x)$

Reduce each product or quotient in Exercises 41 to 70 to lowest terms.

41. $\dfrac{x}{x + 1} \cdot \dfrac{x^2 - 1}{x}$

42. $\dfrac{x}{x - 2y} \cdot \dfrac{x^2 - 4y^2}{x^2}$

43. $\dfrac{2x + 2}{5} \cdot \dfrac{10}{xy + y}$

44. $\dfrac{x^2 + x}{x^2} \cdot \dfrac{3x - 3}{x^2 - 1}$

45. $\dfrac{x + 3}{x^2 + x} \cdot \dfrac{x^2 + 2x + 1}{5x + 15}$

46. $\dfrac{4x^2 - 9}{(x + 2)^2} \cdot \dfrac{x + 2}{2x - 3}$

47. $\dfrac{x^2 - 7x + 10}{x^2 + 6x + 8} \div \dfrac{x - 5}{x + 4}$

48. $\dfrac{9x^2 - 25}{x^2 + 6x + 9} \div \dfrac{3x + 5}{x + 3}$

49. $\dfrac{m^2 - 64}{n^2 - 81} \div \dfrac{m - 8}{n - 9}$

50. $\dfrac{m^2 - 16}{m^2 - 6m + 8} \div \dfrac{m^3 + 4m^2}{m^2 - 9m + 14}$

51. $\dfrac{m^2 - 5m + 6}{m^2 - 4m + 3} \cdot \dfrac{m^2 - m}{m^2 + 2m - 8}$

52. $\dfrac{m^2 - 2m - 8}{m^2 - 3m - 10} \cdot \dfrac{m^2 - 7m + 12}{m^2 - 8m + 16}$

53. $\dfrac{x^2 + xy - 6y^2}{x^2 - 6xy + 8y^2} \cdot \dfrac{x^2 - 9xy + 20y^2}{x^2 - 4xy - 21y^2}$

54. $\dfrac{x^2 + 4xy + 4y^2}{x^2 - 4y^2} \cdot \dfrac{x^2 - xy - 2y^2}{x^2 + xy - 2y^2}$

55. $\dfrac{9x^2 - 1}{9x + 3} \cdot \dfrac{6x^2 - x - 2}{4x^2 - 1}$

56. $\dfrac{y^2 + 14y + 49}{y^2 - 7y - 30} \div \dfrac{y^3 + 7y^2}{y^2 - 100}$

57. $\dfrac{m^3 + n^3}{m^3 - n^3} \div \dfrac{(m + n)^2}{m^2 - n^2}$

58. $\dfrac{8m^3 - 125n^3}{3(2m - 5n)^2} \div \dfrac{4m^2 + 10mn + 25n^2}{4m^2 - 25n^2}$

59. $\dfrac{m^3 + n^3}{m^2 - n^2} \cdot \dfrac{m + n}{m^2 - mn + n^2}$

60. $\dfrac{m^3 - 27n^3}{m^2 - 9n^2} \cdot \dfrac{6m + 18n}{2m^2 + 6mn + 18n^2}$

61. $\dfrac{1 - \dfrac{1}{x}}{2 + \dfrac{1}{x}}$

62. $\dfrac{x - \dfrac{1}{x + 1}}{(x + 1) - \dfrac{1}{x + 1}}$

63. $\dfrac{\dfrac{x}{y} - \dfrac{y}{x}}{\dfrac{1}{x} + \dfrac{1}{y}}$

64. $\dfrac{\dfrac{x^2}{y} + \dfrac{y^2}{x}}{\dfrac{x}{y} + \dfrac{y}{x}}$

65. $\dfrac{\dfrac{1}{x - 1} - \dfrac{1}{x + 1}}{\dfrac{1}{x - 1} + \dfrac{1}{x + 1}}$

66. $\dfrac{\dfrac{2}{x - 3} - \dfrac{1}{x + 3}}{\dfrac{1}{x - 3} + \dfrac{1}{x + 3}}$

67. $\dfrac{3 - \dfrac{1}{x + 2}}{x + \dfrac{1}{x + 2}}$

68. $\dfrac{\dfrac{1}{x} - \dfrac{1}{x + 1}}{1 + \dfrac{1}{x + 1}}$

69. $\dfrac{\dfrac{1}{x + 1} + \dfrac{2x}{x + 3}}{\dfrac{3x + 1}{x + 3} - \dfrac{2}{x + 1}}$

70. $\dfrac{\dfrac{1}{x - 3} + \dfrac{4}{x^2 - 9}}{2 + \dfrac{1}{x - 3}}$

1.11

PARTIAL FRACTIONS

In Section 1.10, we reduced a group of rational expressions connected by the signs of addition and subtraction to one single fraction. Conversely, sometimes it may be necessary to break down a given rational fraction into a sum or difference of simpler fractions. Each of these simpler fractions will have as its denominator one of the factors of the denominator of the original rational *partial fraction* expression. These simpler fractions are called **partial fractions.** For example, we may verify that

$$\frac{5x - 9}{(x - 1)(x - 2)} = \frac{4}{x - 1} + \frac{1}{x - 2}.$$

The expressions

$$\frac{4}{x-1} \quad \text{and} \quad \frac{1}{x-2}$$

are partial fractions of

$$\frac{5x-9}{(x-1)(x-2)}.$$

Generally, any rational expression may be resolved into a series of partial fractions. We assume that the degree of the polynomial in the numerator is less than that of the polynomial in the denominator; otherwise, we must carry out the division process until the degree of the remainder is lower than that of the denominator. The quotient is left intact, and it is the remaining expression that needs to be resolved into partial fractions.

If $P(x)$ and $Q(x)$ are polynomials with real coefficients and $Q(x)$ factors into distinct linear factors, then to each linear factor, say, $x - a$ in the denominator $Q(x)$, there corresponds a partial fraction of the form

$$\frac{A}{x-a},$$

where A is independent of x.

Example 1.77 Resolve

$$\frac{5x+2}{x^2-4}$$

into partial fractions.

Solution: The denominator $(x^2 - 4)$ factors into two distinct linear factors, $(x - 2)$ and $(x + 2)$. We may, therefore, write

$$\frac{5x+2}{x^2-4} = \frac{A}{x-2} + \frac{B}{x+2},$$

where A and B are real numbers to be determined. Multiplying both sides of the equation by $(x^2 - 4)$, we obtain

$$5x + 2 = A(x + 2) + B(x - 2).$$

Since this equation is an identity and holds for all values of x, let us set $x = 2$ and $x = -2$ to determine the values of A and B, respectively. Thus, if $x = 2$, we get

$$12 = 4A \quad \text{or} \quad A = 3.$$

Similarly, if $x = -2$, then we have

$$-8 = -4B \quad \text{or} \quad B = 2.$$

Thus the partial-fraction decomposition is as follows:

$$\frac{5x + 2}{x^2 - 4} = \frac{3}{x - 2} + \frac{2}{x + 2}.$$

Example 1.78

Resolve

$$\frac{11x^2 - 23x}{(2x - 1)(x - 3)(x + 3)}$$

into partial fractions.

Solution: Since the denominator is composed of three distinct linear factors $(2x - 1)$, $(x - 3)$, and $(x + 3)$, we assume that

$$\frac{11x^2 - 23x}{(2x - 1)(x - 3)(x + 3)} = \frac{A}{2x - 1} + \frac{B}{x - 3} + \frac{C}{x + 3},$$

where A, B, and C are real numbers to be determined. Multiplying each member of this equation by $(2x - 1)(x - 3)(x + 3)$, we obtain

$$11x^2 - 23x = A(x - 3)(x + 3) + B(2x - 1)(x + 3) + C(2x - 1)(x - 3).$$

Setting $2x - 1 = 0$, or $x = \frac{1}{2}$, in the last equation, we have

$$11(\tfrac{1}{4}) - 23(\tfrac{1}{2}) = A(\tfrac{1}{2} - 3)(\tfrac{1}{2} + 3)$$

$$-\tfrac{35}{4} = -\tfrac{35}{4}A$$

$$A = 1.$$

Similarly, if we set $x = 3$ and $x = -3$ in succession in the above equation, we find that

$$B = 1 \quad \text{and} \quad C = 4.$$

Hence

$$\frac{11x^2 - 23x}{(2x - 1)(x - 3)(x + 3)} = \frac{1}{2x - 1} + \frac{1}{x - 3} + \frac{4}{x + 3}.$$

If the denominator $Q(x)$ contains a repeated linear factor, say, $(x - b)^2$, then there corresponds a sum of two partial fractions

$$\frac{B}{x - b} + \frac{C}{(x - b)^2}.$$

Similarly, if $Q(x)$ contains $(x - b)^3$ as its repeated linear factor, then we have a sum of three partial fractions

$$\frac{B}{x - b} + \frac{C}{(x - b)^2} + \frac{D}{(x - b)^3}$$

and so on. Let us illustrate this statement with an example.

Example 1.79

Resolve

$$\frac{x + 3}{(x + 1)(x - 1)^2}$$

into partial fractions.

Solution: Note that the denominator contains a repeated linear factor $(x - 1)^2$. We therefore assume that

$$\frac{x + 3}{(x + 1)(x - 1)^2} = \frac{A}{x + 1} + \frac{B}{x - 1} + \frac{C}{(x - 1)^2}.$$

Multiplying each member of this equation by $(x + 1)(x - 1)^2$, we have

$$x + 3 = A(x - 1)^2 + B(x - 1)(x + 1) + C(x + 1).$$

Substituting $x = 1$ in the last equation, we find that

$$4 = 2C \quad \text{or} \quad C = 2.$$

Similarly, setting $x = -1$, we have

$$2 = 4A \quad \text{or} \quad A = \tfrac{1}{2}.$$

To find B, we may substitute *any* value of x, say $x = 0$, in the equation and obtain

$$3 = A - B + C.$$

Since $A = \tfrac{1}{2}$ and $C = 2$, we find that $B = -\tfrac{1}{2}$. Thus the breakdown into partial fractions is as follows:

$$\frac{x + 3}{(x + 1)(x - 1)^2} = \frac{1}{2(x + 1)} - \frac{1}{2(x - 1)} + \frac{2}{(x - 1)^2}.$$

EXERCISES

Resolve Exercises 1 to 20 into partial fractions.

1. $\dfrac{2}{x(x + 2)}$

2. $\dfrac{x + 2}{x(x - 1)}$

3. $\dfrac{3x + 5}{(x + 1)(x + 2)}$

4. $\dfrac{x + 1}{(2 - x)(x - 3)}$

5. $\dfrac{3x - 2}{(x - 1)(x + 2)}$

6. $\dfrac{2x - 1}{(x - 2)(x + 7)}$

7. $\dfrac{1}{x^2 - 1}$

8. $\dfrac{2x - 9}{x^2 - 7x + 12}$

9. $\dfrac{3x - 2}{x^2 - 3x + 2}$

10. $\dfrac{5x - 11}{2x^2 + x - 6}$

11. $\dfrac{7x + 5}{3x^2 + 4x + 1}$

12. $\dfrac{11x - 1}{2x^2 + x - 3}$

13. $\dfrac{x+2}{(x-1)(x-2)(x-3)}$

14. $\dfrac{x^2+3x+4}{(x-1)(x-2)(x-3)}$

15. $\dfrac{1}{x^2(1-x)}$

16. $\dfrac{x-8}{x(x-2)^2}$

17. $\dfrac{8-x}{(x+1)(x-2)^2}$

18. $\dfrac{x^2}{(x+2)(x+1)^2}$

19. $\dfrac{2x+1}{(x+2)(x-3)^2}$

20. $\dfrac{x^2}{(x-2)^2(1-2x)}$

1.12

COMPLEX NUMBERS

The system of real numbers we developed earlier in the chapter does not provide solutions of the equation $x^2 = -k$, $k > 0$. The equation $x^2 = -1$, for example, has no solution in the real number system because there is no real number x whose square is -1. We introduce in this section a still broader class of number system—the complex number system C. This system is useful for solving equations with real coefficients in later chapters.

Definition 1.12.1
complex numbers

The number $z = a + bi$, a, $b \in R$, is a *complex number*. The symbol i is a number whose square is -1.

The numbers

$$3 + 4i, \qquad \tfrac{1}{2} + \tfrac{1}{3}i, \qquad \sqrt{5}i, \qquad \text{and} \qquad 2 - \sqrt{\tfrac{3}{2}}i$$

are all examples of complex numbers.

Definition 1.12.2

Two complex numbers $a + bi$ and $c + di$ are said to be equal if and only if $a = c$ and $b = d$.

For example, if $a + bi = 2 + 3i$, then $a = 2$ and $b = 3$.

Now that we know what complex numbers look like, we will define addition and multiplication in the complex number system.

Definition 1.12.3

If $z_1 = a + bi$ and $z_2 = c + di$, where a, b, c, and d are real numbers, then

$$z_1 + z_2 = (a + bi) + (c + di) = (a + c) + (b + d)i, \tag{1.51}$$

$$z_1 \cdot z_2 = (a + bi) \cdot (c + di) = (ac - bd) + (ad + bc)i. \tag{1.52}$$

Example 1.80

(a) $(2 + 3i) + (7 + 4i) = (2 + 7) + (3 + 4)i = 9 + 7i.$

(b) $(2 + 3i) \cdot (7 + 4i) = (2 \cdot 7 - 3 \cdot 4) + (2 \cdot 4 + 3 \cdot 7)i$
$$= 2 + 29i.$$

The commutative, associative, and distributive properties we studied in Section 1.1 remain valid for complex numbers. However, the inequality signs $<$, $\leq$, $>$, and $\geq$ do not have any assigned meanings in the complex number system.

The real number a and the complex number $a + 0i$ are regarded as the same. For this reason, we say that the real number system R is an essential part of the complex number system C. Thus we may write

$$
\begin{array}{ccc}
3i & \text{for} & 0 + 3i \\
0 & \text{for} & 0 + 0i \\
-4 & \text{for} & -4 + 0i \\
i & \text{for} & 0 + i
\end{array}
$$

and so on. Writing $i = 0 + 1i$ and then using rule (1.52) for multiplication of two complex numbers, we obtain

$$
\begin{aligned}
i^2 &= (0 + 1i)(0 + 1i) \\
&= (0 \cdot 0 - 1 \cdot 1) + (0 \cdot 1 + 1 \cdot 0)i \\
&= -1 + 0i \\
&= -1.
\end{aligned}
$$

Thus in the simplified notation,

$$i^2 = -1. \tag{1.53}$$

and we write

$$i = \sqrt{-1}.$$

In adding or multiplying two complex numbers $z_1 = a + bi$ and $z_2 = c + di$, we can treat all terms as if they were real numbers except that i^2 is replaced by -1. Thus

$$
\begin{aligned}
z_1 + z_2 = (a + bi) + (c + di) &= a + c + bi + di \\
&= (a + c) + (b + d)i,
\end{aligned}
$$

and

$$
\begin{aligned}
z_1 \cdot z_2 = (a + bi)(c + di) &= a(c + di) + bi(c + di) \\
&= ac + adi + bci + bdi^2 \\
&= (ac - bd) + (ad + bc)i.
\end{aligned}
$$

These results agree with (1.51) and (1.52), respectively. **Remember that we multiply complex numbers just as if we are dealing with real numbers except that i^2 is replaced by the real number -1.**

We noted that the complex number $a + bi$ with $b = 0$ is identified with the real number a. For this reason, the real number a in the complex number $a + bi$

real part
imaginary part

is called the *real part*. If $b \neq 0$, then bi in the number $a + bi$ is called the *imaginary part*.

Next we define the operation of subtraction in the complex number system.

Definition 1.12.4

For two complex numbers $z_1 = a + bi$ and $z_2 = c + di$,

$$z_1 - z_2 = (a + bi) - (c + di) = (a - c) + (b - d)i, \qquad \textbf{(1.54)}$$

where a, b, c, and d are real numbers.

Example 1.81

(a) $(5 + 8i) - (3 + 5i) = (5 - 3) + (8 - 5)i = 2 + 3i$.

(b) $(4 + 6i) - (5 + 8i) = (4 - 5) + (6 - 8)i = -1 - 2i$.

(c) $(\frac{22}{7} + 3i) - (3 + \frac{22}{7}i) = (\frac{22}{7} - 3) + (3 - \frac{22}{7})i = \frac{1}{7} - \frac{1}{7}i$.

We have observed that to add, subtract, or multiply complex numbers, we proceed as if we are dealing with real numbers, except that we replace i^2 by -1. Unfortunately, division cannot be handled this way. Therefore to determine *conjugate* the quotient of two complex numbers, we introduce the notion of the **conjugate of a complex number.**

Definition 1.12.5

The *conjugate* of the complex number $z = a + bi$ is denoted by the complex number $\bar{z} = a - bi$.

Example 1.82

(a) The conjugate of $z = 3 + 4i$ is $\bar{z} = 3 - 4i$.

(b) The conjugate of $z = 5 - 6i$ is $\bar{z} = 5 + 6i$.

The method of division in the complex number system C is based on the observation that if $z = x + yi$, then

$$z \cdot \bar{z} = (x + yi)(x - yi) = (x^2 - y^2 i^2) = x^2 + y^2, \qquad \textbf{(1.55)}$$

which is a positive number unless $z = 0$, that is, $x = 0$ and $y = 0$. To calculate an expression of the form

$$\frac{a + bi}{c + di},$$

we multiply the numerator and the denominator by the conjugate of $c + di$. The result is

$$\frac{a + bi}{c + di} = \frac{a + bi}{c + di} \cdot \frac{c - di}{c - di} = \frac{(ac + bd) + (bc - ad)i}{c^2 + d^2}$$

$$= \frac{ac + bd}{c^2 + d^2} + \frac{(bc - ad)}{c^2 + d^2} i.$$

Example 1.83

Express

$$\frac{3 + 2i}{5 + 4i}$$

in the form $a + bi$.

Solution: We multiply the numerator and denominator by $5 - 4i$ (the conjugate of $5 + 4i$) and obtain

$$\frac{3 + 2i}{5 + 4i} = \frac{3 + 2i}{5 + 4i} \cdot \frac{5 - 4i}{5 - 4i} = \frac{15 - 2i - 8i^2}{25 - 16i^2}$$

$$= \frac{15 - 2i + 8}{25 + 16}$$

$$= \frac{23 - 2i}{41}$$

$$= \frac{23}{41} - \frac{2}{41} i.$$

Example 1.84

Express

$$\frac{1}{2 - 3i}$$

in the form $a + bi$.

Solution: Multiplying the numerator and the denominator by $2 + 3i$ (the conjugate of $2 - 3i$), we get

$$\frac{1}{2 - 3i} = \frac{1}{2 - 3i} \cdot \frac{2 + 3i}{2 + 3i} = \frac{2 + 3i}{4 - 9i^2}$$

$$= \frac{2 + 3i}{4 + 9}$$

$$= \frac{2 + 3i}{13}$$

$$= \frac{2}{13} + \frac{3}{13} i.$$

Conjugates of complex numbers have interesting and useful results, which we now state and prove. An important application of these properties is made in Chapter 4.

Proposition 1.12.1   If z_1 and z_2 are complex numbers, then

$$\overline{z_1 + z_2} = \overline{z_1} + \overline{z_2}, \qquad (1.56)$$

$$\overline{z_1 \cdot z_2} = \overline{z_1} \cdot \overline{z_2}. \qquad (1.57)$$

Proof: Let $z_1 = a + bi$ and $z_2 = c + di$, where a, b, c, and d are real numbers. Then by (1.51),

$$z_1 + z_2 = (a + c) + (b + d)i.$$

By Definition 1.12.5 of the conjugate,

$$\overline{z_1 + z_2} = (a + c) - (b + d)i.$$

which we may rewrite as

$$\overline{z_1 + z_2} = (a - bi) + (c - di)$$
$$= \overline{z_1} + \overline{z_2},$$

and (1.56) is proved.

Equation (1.52) shows that

$$z_1 \cdot z_2 = (ac - bd) + (ad + bc)i,$$

and the conjugate is

$$\overline{z_1 \cdot z_2} = (ac - bd) - (ad + bc)i.$$

Since $\overline{z_1} = a - bi$ and $\overline{z_2} = c - di$,

$$\overline{z_1} \cdot \overline{z_2} = (a - bi)(c - di)$$
$$= (ac - bd) - (ad + bc)i.$$

Hence

$$\overline{z_1 \cdot z_2} = \overline{z_1} \cdot \overline{z_2},$$

which proves (1.57).

EXERCISES

Perform the operations indicated in Exercises 1 to 22, and write each answer in the form $a + bi$.

1. $(2 + 4i) + (7 + 2i)$

2. $(3 + 2i) + (2 + 5i)$

3. $(4 + 5i) + (2 - 6i)$

4. $(\frac{2}{3} + \frac{1}{2}i) + (\frac{1}{2} + \frac{1}{3}i)$

5. $(1 + 2i) - (2 + 3i)$

6. $(4 + i) - (3 + 4i)$

7. $(7 - 4i) - (4 + 3i)$

8. $(\frac{5}{3} + 2i) - (\frac{1}{4} - \frac{1}{2}i)$

9. $(3 + 12i) - (2 - 15i)$

10. $(5 + 3i) + (2 + 5i) - (9 - 4i)$

11. $(2 + i) + (3 - i) + 3i$

12. $(3 - 4i) + (3 + 4i) + (2 + i)$

13. $(3 + i)(2 + 3i)$

14. $(\frac{1}{2} + i)(\frac{1}{3} + i)$

15. $(4 - 2i)(5 + 2i)$

16. $(2 + 5i)(3 + 4i)$

17. $(6 - 5i)(6 + 5i)$

18. $(3 + 4i)(5 + 6i)$

19. $(\frac{3}{2} + 4i)i$

20. $(4 + i)(1 + i)(2i)$

21. $(1 + 2i)^2 i$

22. $(3 + 5i)^2(1 + i)$

Find the conjugate in Exercises 23 to 26.

23. $4 + 7i$

24. $3 - 2i$

25. $\frac{1}{2} + \frac{1}{3}i$

26. $\frac{1}{4} - \frac{2}{3}i$

Express Exercises 27 to 40 in the form $a + bi$.

27. $\dfrac{1 + i}{1 - i}$

28. $\dfrac{4 + 3i}{1 + 2i}$

29. $\dfrac{2 + 5i}{4 + 3i}$

30. $\dfrac{3 + 4i}{3 - 4i}$

31. $\dfrac{3 - 4i}{3 + 4i}$

32. $\dfrac{4 - 7i}{5 + 3i}$

33. $\dfrac{2 + i}{3i}$

34. $\dfrac{2}{3 + 2i}$

35. $\dfrac{1}{2i - 3}$

36. $\dfrac{3}{2 + 5i}$

37. $\dfrac{4i}{4 + 7i}$

38. $\dfrac{i}{1 + i}$

39. $\dfrac{\sqrt{2}i}{\sqrt{2} + i}$

40. $\dfrac{2 + \sqrt{3}i}{\sqrt{3} + 2i}$

41. Prove that if z_1 and z_2 are complex numbers, then

$$\overline{z_1 - z_2} = \overline{z_1} - \overline{z_2}.$$

42. Prove that if z_1, z_2, and z_3 are complex numbers, then

$$\overline{z_1 \cdot z_2 \cdot z_3} = \overline{z_1} \cdot \overline{z_2} \cdot \overline{z_3}.$$

KEY WORDS

natural numbers	*set*	*algebraic expression*
counting numbers	*subset*	*constant*
commutativity	*element*	*monomial*
associativity	*variable*	*polynomial*
distributive property	*open interval*	*degree*
multiplicative property	*closed interval*	*leading coefficient*
integers	*half-open interval*	*leading term*
additive identity	*absolute value*	*factoring*
additive inverse	*exponent*	*irreducible*
rational numbers	*power*	*synthetic division*
multiplicative inverse	*base*	*rational expression*
irrational numbers	*exponential*	*partial fraction*
decimal representation	*principal nth root*	*linear factor*
real numbers	*radical*	*repeated linear factor*
number line	*radicand*	*complex numbers*
coordinate of a point	*index*	*imaginary number*
graph of the number	*rationalizing*	*conjugate of a complex number*
one-to-one correspondence	*variable*	
inequalities	*domain*	

KEY FORMULAS

- $\dfrac{a}{b} \cdot \dfrac{c}{d} = \dfrac{ac}{bd}$

- $\dfrac{a}{c} + \dfrac{b}{c} = \dfrac{a+b}{c}$

- $\dfrac{a}{b} + \dfrac{c}{d} = \dfrac{ad+bc}{bd}$

- $\dfrac{a}{b} - \dfrac{c}{d} = \dfrac{ad-bc}{bd}$

- $\dfrac{a}{b} \div \dfrac{c}{d} = \dfrac{ad}{bc}$

- $x^m \cdot x^n = x^{m+n}$

- $\dfrac{x^m}{x^n} = \begin{cases} x^{m-n} & \text{if } m > n \\ \dfrac{1}{x^{n-m}} & \text{if } m < n, x \neq 0 \\ 1 & \text{if } m = n \end{cases}$

- $(x^m)^n = x^{mn}$

- $\left(\dfrac{x}{y}\right)^m = \dfrac{x^m}{y^m}$

- $x^0 = 1, \qquad x \neq 0$

- $x^{-m} = \dfrac{1}{x^m}, \qquad x \neq 0$

- $\left(\dfrac{a}{b}\right)^{-m} = \left(\dfrac{b}{a}\right)^m$

- $(ax + b)(cx + d) = acx^2 + (ad + bc)x + bd$
- $(x + b)(x + d) = x^2 + (b + d)x + bd$
- $(x + b)^2 = x^2 + 2bx + b^2$
- $(x - b)^2 = x^2 - 2bx + b^2$
- $(x + b)(x - b) = x^2 - b^2$
- $x^3 + a^3 = (x + a)(x^2 - ax + a^2)$
- $x^3 - a^3 = (x - a)(x^2 + ax + a^2)$

If $z_1 = a + bi$ and $z_2 = c + di$, where a, b, c, and d are real numbers, then

- $z_1 + z_2 = (a + c) + (b + d)i,$
- $z_1 - z_2 = (a - c) + (b - d)i,$
- $z_1 \cdot z_2 = (ac - bd) + (ad + bc)i.$

The conjugate of complex number $z = a + bi$ is

- $\bar{z} = a - bi.$

If $z_1 = a + bi$ and $z_2 = c + di$, then

- $\dfrac{z_1}{z_2} = \dfrac{a + bi}{c + di} = \dfrac{(a + bi)(c - di)}{(c + di)(c - di)}$

  $= \left(\dfrac{ac + bd}{c^2 + d^2}\right) + \left(\dfrac{bc - ad}{c^2 + d^2}\right)i.$

EXERCISES FOR REVIEW

In Exercises 1 to 10, classify each statement as true or false.

1. Every integer is a natural number.
2. Every natural number is a rational number.
3. For every integer a, there is an identity element under the operation of addition.
4. For every nonzero rational number, there is a multiplicative inverse.
5. $\sqrt{2}$ is an irrational number.
6. The irrational numbers are closed under multiplication.
7. The real numbers are closed under subtraction.
8. Every rational number can be expressed in the form of $a + bi$, where $a, b \in R$.
9. The set of real numbers is a subset of complex numbers.
10. There exists a one-to-one correspondence between the real numbers and the points on the real number line.

Let $A = \{-4, -2, 0, 1, \frac{2}{3}, \sqrt{2}, 5, \pi, 1/\sqrt{3}, 2\pi/3\}$. List the elements in Exercises 11 to 14.

11. Set of integers in A

12. Set of rational numbers in A

13. Set of irrational numbers in A

14. Set of real numbers in A

In Exercises 15 to 18, express each statement, using the appropriate symbol: $<$, $\leq$, $>$, or $\geq$.

15. x is positive.

16. $4x$ is less than 12.

17. $2x$ is greater than or equal to 13.

18. x is between 5 and 7, inclusive.

In Exercises 19 to 22, represent each set, using interval notation.

19. $\{x \mid 1 \leq x \leq 3\}$

20. $\{x \mid -1 < x \leq 4\}$

21. $\{x \mid -2 \leq x < 5\}$

22. $\{x \mid x \geq 1\}$

In Exercises 23 to 26, rewrite each expression without using absolute value.

23. $|3 - 7|$

24. $|3| - |7|$

25. $|-3| - |-7|$

26. $|x| < 2$

Simplify each expression in Exercises 27 to 30. All answers should be fractions.

27. $\dfrac{3^7 \cdot 2^{-4}}{2^2 \cdot 3^5}$

28. $(\frac{2}{3})^4 \cdot (\frac{4}{5})^{-2}$

29. $(\frac{32}{243})^{-2/5}$

30. $(216)^{2/3} \cdot (-128)^{-3/7}$

Simplify each expression in Exercises 31 to 36. All answers should have positive exponents.

31. $(3x^2 y^3 z)^2 (2xy^2 z^3)^2$

32. $\dfrac{(2a^3 b)^2}{3a^4 b^3}$

33. $(4x^{-3} y^{-2})^2 (3xy^{-3})^{-3}$

34. $(x^4 y^6 z^{-8})^{5/2}$

35. $\left(\dfrac{27x^3}{8y^3 z^6}\right)^{-4/3}$

36. $\left(\dfrac{125a^3 b^4}{27a^{-6} b}\right)^{1/3} (8a^3 b^6)^{2/3}$

Simplify each expression in Exercises 37 to 40.

37. $\sqrt{98}$

38. $\sqrt[4]{162}$

39. $\sqrt[3]{x^3 y^6}$

40. $\sqrt[5]{\dfrac{32x^5 y^{10}}{z^{15}}}$

Rationalize the denominator in each of the fractions in Exercises 41 and 42.

41. $\dfrac{1}{\sqrt{3} + \sqrt{2}}$

42. $\dfrac{1}{\sqrt{x + 2y} - \sqrt{x}}$

43. Let P and Q be two polynomials in x defined by

$$P(x) = x^2 + x + 1 \qquad Q(x) = 3x + 2.$$

Determine the following:

(a) $P(-1)$ **(b)** $P(2)$ **(c)** $Q(-2)$ **(d)** $Q(\frac{1}{2})$ **(e)** $P(Q(3))$ **(f)** $Q(P(2))$

Perform the indicated operations in Exercise 44 to 50.

44. $(2x^3 + 3x^2 - x + 5) + (4x^3 - x^2 + 3x + 1)$

45. $4xy^2(3x^2 + 2xy - y^2)$

46. $(2x + 3)(4x - 5)$

47. $(2x - 3)(4x^2 + 6x + 9)$

48. $(x + 2y + 3z)(x + 2y - 3z)$

49. $(2x + 3y)^2 - (2x - 3y)^2$

50. $(2x + y + 3z)(2x - y + 3z)$

Factor each expression in Exercises 51 to 56.

51. $x^2 + 5x - 36$

52. $2y^2 + 11y + 12$

53. $27x^3 - 64y^3$

54. $16x^4 - 81y^4$

55. $ab + a + b + 1$

56. $9x^2 - 24xy + 16y^2 - 25z^2$

Use synthetic division in Exercises 57 and 58 to find the quotient and the remainder when the first expression is divided by the second.

57. $x^3 - 2x^2 - 13x + 6$; $x + 3$

58. $3x^3 - 4x - 1$; $x - 2$

Reduce each expression in Exercises 59 to 62 to lowest terms.

59. $\dfrac{2xyz + 3xz^2}{4xy^2 + 12xyz + 9xz^2}$

60. $\dfrac{a + b}{a^3 + b^3}$

61. $\dfrac{x^2 + x - 2}{x^2 + 5x + 6} \cdot \dfrac{x^2 + 4x + 3}{x^2 + 3x - 4}$

62. $\dfrac{x^2 + xy}{x^2 + 2xy + y^2}$

Reduce each expression in Exercises 63 to 66 to a single fraction in lowest terms.

63. $\dfrac{x + y}{x + 2y} - \dfrac{y}{x + 3y}$

64. $\dfrac{y}{3x + 2y} + \dfrac{x - 2y}{3x + y}$

65. $\dfrac{\dfrac{1}{x} - \dfrac{1}{y}}{\dfrac{1}{x^2} - \dfrac{1}{y^2}}$

66. $\dfrac{1 - \dfrac{1}{x + 1}}{1 + \dfrac{1}{x + 1}}$

Reduce each expression in Exercises 67 to 70 to partial fractions.

67. $\dfrac{1}{x^2 + x}$

68. $\dfrac{x - 11}{x^2 + 2x - 15}$

69. $\dfrac{x - 2}{6x^2 - 7x + 2}$

70. $\dfrac{1}{x(x^2 - 2x + 1)}$

In Exercises 71 to 79, write each expression in the form of $a + bi$, where a and b are real numbers.

71. $(3 + 5i) + (1 - 2i)$

72. $(2 + 3i) - (1 - i)$

73. $(5 + 2i)(3 - 4i)$

74. $(\frac{1}{2} + 2i)(\frac{1}{5} - 2i)$

75. $\dfrac{i}{1 + i}$

76. $\dfrac{2 + i}{2 - i}$

77. $\dfrac{3 + i}{2 - 3i}$

78. $\dfrac{4 + 5i}{3 + 7i}$

79. $\dfrac{2 + i}{i}$

80. Prove that if z_1, z_2, and z_3 are complex numbers, then

$$\overline{z_1 + z_2 + z_3} = \bar{z}_1 + \bar{z}_2 + \bar{z}_3.$$

CHAPTER TEST

Simplify the expressions in Exercises 1 to 4. All answers should have positive exponents.

1. $(4x^3 y^5)(5x^{-4} y^3)$

2. $\dfrac{16x^2 y^2 z^{-4}}{40x^{-3} y^4 z}$

3. $\left(\dfrac{125x^3 y^4}{27x^{-6} y}\right)^{1/3}$

4. $\dfrac{(3a^2 b)^4}{4a^6 b^3}$

5. Rationalize the numerator:

$$\dfrac{\sqrt{x + h} - \sqrt{x}}{h}.$$

6. Given that P and Q are two polynomials in x defined by

$$P(x) = x^2 + 2x + 3 \qquad Q(x) = 2x + 5,$$

determine the following:
(a) $P(-1)$ (b) $Q(2)$ (c) $P(Q(3))$ (d) $Q(P(3))$
7. Expand the following and then simplify:
(a) $(2x + 3)(x^2 + 3x + 6)$ (b) $(x + 2y)(x - 2y)(x^2 + 4y^2)$
8. Factor the following:
(a) $4x^2 - 25y^2$ (b) $6x^2 - x - 12$
9. Use synthetic division to determine the quotient and the remainder by dividing

$$P(x) = 3x^3 + 2x^2 - 10x + 4$$

by $x - 2$.
10. Simplify the following expressions:
(a) $\dfrac{x^2 - 3x + 2}{x^2 + 5x + 6} \cdot \dfrac{x + 3}{x - 2}$ (b) $\dfrac{2}{x + 2} + \dfrac{3}{x - 2}$
11. Resolve into partial fractions:

$$\frac{1}{x(x + 2)}.$$

12. Perform the indicated operation in the following expressions, and write each answer in the form $a + bi$.
(a) $(4 + 3i)(3 - 2i)$ (b) $\dfrac{3 + 4i}{2 - 3i}$

EQUATIONS AND INEQUALITIES IN ONE VARIABLE

2

2.1

LINEAR EQUATIONS

Algebraic statements such as

$$3x + 4 = 7, \qquad x^2 = 6x - 8, \qquad \text{or} \qquad \sqrt{x} = 4$$

are called **equations** in the variable x. If the substitution of a real number for x in an equation results in a true statement, then that real number is a **solution,** or a **root,** of the equation. A number that is a solution of an equation is said to *satisfy* the given equation. The set of all real numbers that satisfy the given equation is called the **solution set.** For example, $x = 1$ is a solution of the equation

$$2x + 3 = 5$$

because $2(1) + 3 = 5$ is a true statement. The solution set is $\{1\}$. If we substitute any real number other than $x = 1$ in the equation $2x + 3 = 5$, we obtain a false statement. Similarly, $x = 2$ and $x = 4$ are solutions of the equation

$$x^2 = 6x - 8,$$

and the solution set is $\{2, 4\}$. Thus solving an equation is a matter of describing the solution set of the equation as explicitly as possible.

identity   The equation is called an **identity** if every real number is a solution of the equation. For example,

$$x^2 - 1 = (x - 1)(x + 1)$$

is an identity, since it is true for every real number x.

equivalent   Two equations are said to be **equivalent** if they have exactly the same solution set. The equations

$$x + 3 = 4 \qquad \text{and} \qquad 4x + 3 = 7$$

are equivalent, since both have the same solution set, namely $\{1\}$. The equations

$$x + 1 = 4 \qquad \text{and} \qquad x^2 + 2 = 11$$

are not equivalent. The first equation has $\{3\}$ as its solution set whereas the second equation has $\{-3, 3\}$ as its solution set.

To solve an equation, we generate a series of equivalent equations, each simpler than the one before, until we get an equation whose solution set is fairly obvious. The following proposition is an immediate consequence of the addition and multiplication properties of the real number system. It is used to generate a sequence of equivalent equations over the set of real numbers.

Proposition 2.1.1

If $P(x)$, $Q(x)$, and $R(x)$ are algebraic expressions, then for all values of x for which $P(x)$, $Q(x)$, and $R(x)$ are real numbers, the statement

$$P(x) = Q(x) \tag{2.1}$$

is equivalent to each of the following statements:

$$P(x) + R(x) = Q(x) + R(x) \tag{2.2}$$

$$P(x) - R(x) = Q(x) - R(x) \tag{2.3}$$

$$P(x) \cdot R(x) = Q(x) \cdot R(x) \qquad (R(x) \neq 0) \tag{2.4}$$

$$\frac{P(x)}{R(x)} = \frac{Q(x)}{R(x)} \qquad (R(x) \neq 0). \tag{2.5}$$

Proposition 2.1.1 allows us to add or subtract the same expression on both sides of a given equation without changing the solution set. Multiplying or dividing both sides of the equation by a nonzero expression also results in an equivalent equation. *The goal of these operations is simply to isolate the variable on one side of the equation and determine the solution set.*

Definition 2.1.1

An equation of the form

$$ax + b = 0,$$

where a and b are real numbers and $a \neq 0$, is called a *linear* equation.

Example 2.1

Solve the following equation for x:

$$3x - 4 = 11$$

Solution: We generate the following series of equivalent equations:

$$3x - 4 = 11$$

$$3x - 4 + 4 = 11 + 4 \qquad \text{[Equation (2.2)]}$$

$$3x = 15$$

$$\tfrac{1}{3}(3x) = \tfrac{1}{3}(15) \qquad \text{[Equation (2.4)]}$$

$$x = 5.$$

The equation $x = 5$ is equivalent to the original equation $3x - 4 = 11$. The solution set is $\{5\}$.

Let us now look at equations that can be reduced to linear equations.

Example 2.2

Solve the following equation for x:

$$\frac{2x}{3} - \frac{x}{4} = \frac{1}{2}.$$

Solution: First we determine the lowest common denominator of 2, 3, and 4. It is 12. Multiplying both sides of the equation by 12 and then using the distributive property of the real number system, we get

$$12\left(\frac{2x}{3} - \frac{x}{4}\right) = 12\left(\frac{1}{2}\right)$$

$$12\left(\frac{2x}{3}\right) - 12\left(\frac{x}{4}\right) = 6$$

$$8x - 3x = 6$$

$$5x = 6$$

$$x = \tfrac{6}{5}.$$

To safeguard against arithmetic errors, we should verify that $x = \tfrac{6}{5}$ is a solution of the given equation. Substituting $\tfrac{6}{5}$ for x in the left-hand side of the original equation, we obtain

$$\frac{2\left(\tfrac{6}{5}\right)}{3} - \frac{\tfrac{6}{5}}{4} = \frac{2}{3}\left(\frac{6}{5}\right) - \frac{1}{4}\left(\frac{6}{5}\right) = \frac{4}{5} - \frac{3}{10} = \frac{1}{2}.$$

Hence the solution set is $\{\tfrac{6}{5}\}$.

Example 2.3

Solve for x:

$$\frac{2}{x - 5} = \frac{1}{3x - 9}.$$

Solution: Observe that the equation is meaningless for $x = 3$ and $x = 5$ because these values of x involve division by zero, and division by zero is not defined in mathematics. We therefore exclude these values from the solution set. To clear fractions, we multiply both sides of the equation by $(x - 5)(3x - 9)$. This product is nonzero, since $x \neq 3$ and $x \neq 5$. Thus we have

$$(x - 5)(3x - 9)\left(\frac{2}{x - 5}\right) = (x - 5)(3x - 9)\left(\frac{1}{3x - 9}\right)$$

$$(3x - 9)(2) = x - 5$$

$$6x - 18 = x - 5.$$

Now we can generate a set of equivalent equations by using Proposition 2.1.1. First we add $(-x)$ to both sides of the equation. Thus

$$(-x) + 6x - 18 = (-x) + x - 5$$

$$5x - 18 = -5.$$

Next we add 18 to both sides of the equation, obtaining

$$5x - 18 + 18 = -5 + 18$$

$$5x = 13.$$

Finally, we multiply both sides of this equation by $\frac{1}{5}$, and we have

$$\tfrac{1}{5}(5x) = \tfrac{1}{5}(13)$$

$$x = \tfrac{13}{5}.$$

We now verify the solution by substituting $x = \frac{13}{5}$ into the given equation.

$$\frac{2}{x-5} = \frac{2}{\frac{13}{5} - 5} = \frac{2}{\frac{13}{5} - \frac{25}{5}} = \frac{2}{\frac{-12}{5}} = 2\left(-\frac{5}{12}\right) = -\frac{5}{6},$$

and

$$\frac{1}{3x - 9} = \frac{1}{3(\frac{13}{5}) - 9} = \frac{1}{\frac{39}{5} - \frac{45}{5}} = \frac{1}{\frac{-6}{5}} = -\frac{5}{6}.$$

Thus the solution set is $\{\frac{13}{5}\}$.

Sometimes, we may inadvertently multiply both sides of an equation by an algebraic expression which equals zero for some values of x. In that event, we may not obtain an equivalent equation. Consider, for instance, the following example.

Example 2.4 Solve for x:

$$\frac{2x}{x-3} + 3 = \frac{6}{x-3}.$$

Solution: We exclude $x = 3$ from the solution set because $x = 3$ involves division by zero. To eliminate fractions, we multiply both sides of the equation by $(x - 3)$ and obtain

$$(x - 3)\left(\frac{2x}{x-3} + 3\right) = (x - 3)\left(\frac{6}{x-3}\right)$$

$$2x + 3(x - 3) = 6$$

$$2x + 3x - 9 = 6$$

$$5x = 15$$

$$x = 3.$$

Since $x = 3$ has already been excluded from the solution set, we can conclude that the original equation has no solution. The solution set is the empty, or null, set.

Example 2.5   Solve for x:

$$(2x + 3)(3x + 1) = (x - 6)(6x - 1).$$

Solution: We generate the following sequence of equivalent equations:

$$(2x + 3)(3x + 1) = (x - 6)(6x - 1)$$
$$6x^2 + 11x + 3 = 6x^2 - 37x + 6$$
$$11x + 3 = -37x + 6$$
$$48x = 3$$
$$x = \tfrac{1}{16}.$$

We suggest that the reader substitute $x = \tfrac{1}{16}$ in both sides of the original equation to prove that $x = \tfrac{1}{16}$ is indeed the solution. Hence the solution set is $\{\tfrac{1}{16}\}$.

EXERCISES

In Exercises 1 to 8, determine which of the following pairs of equations are equivalent:

1. $x + 2 = 4$
  $3x + 4 = 10$

2. $4x + 3 = 15$
  $x = 3$

3. $x - 2 = 0$
  $x^2 - 4 = 0$

4. $2y - 3 = 1$
  $(0.5)y^2 = 2$

5. $3y - 4 = -1$
  $y^2 = 1$

6. $\dfrac{3x + 2}{4} = 1$
  $2x - 3 = 1$

7. $x = 3$
  $x^2 - 9 = 0$

8. $y = 5$
  $y^2 = 25$

In Exercises 9 to 55, solve for x.

9. $2x + 5 = 11$

10. $3x - 6 = 7$

11. $3x + 7 = 5x$

12. $3x + 5 = x + 2$

13. $5x - 3 = 4x + 11$

14. $2(x + 3) = 3(x - 2)$

15. $4(x - 3) = 5(x - 5)$

16. $6(2x + 5) = 5(3x - 1)$

17. $0.3(x - 2) = 0.4(2x - 1)$

18. $0.2(3x - 4) = 1.3$

19. $\dfrac{x}{3} + 1.5 = 2$

20. $0.8x + \tfrac{2}{3} = 4$

21. $\dfrac{x}{2} + \dfrac{x}{3} = 10$

22. $\dfrac{2x}{3} + \dfrac{x}{2} = 1$

23. $0.6x + 0.5x = 1$

24. $\dfrac{2x - 1}{3} + \dfrac{x + 1}{2} = 5$

25. $\dfrac{2x}{3} + \dfrac{3(x + 1)}{4} = 5$

26. $\dfrac{3x + 1}{4} + \dfrac{x}{5} = 2$

27. $\dfrac{x}{3} - 8 = 5 - \dfrac{3x}{4}$

28. $\dfrac{3x}{4} - 1 = 2 + \dfrac{x}{5}$

29. $\dfrac{3}{x} + \dfrac{4}{x} = 2$

30. $\dfrac{3}{4x} + \dfrac{1}{x} = \dfrac{7}{8}$

31. $\dfrac{2}{2x - 3} = \dfrac{3}{4x - 6}$

32. $\dfrac{1}{1 - x} = \dfrac{1}{x + 1}$

33. $\dfrac{1}{3x} = \dfrac{1}{x + 4}$

34. $\dfrac{1}{x + 3} = \dfrac{2}{x + 10}$

35. $\dfrac{1}{x - 1} + \dfrac{x}{x - 1} = \dfrac{3}{2}$

36. $\dfrac{8}{2x + 3} - \dfrac{3x}{2x + 3} = 2$

37. $\dfrac{3x}{2x + 1} - \dfrac{1}{2x + 1} = \dfrac{8}{2x + 1}$

38. $\dfrac{2}{x} + \dfrac{3}{x + 3} = \dfrac{4 - x}{x(x + 3)}$

39. $\dfrac{5}{2x} + \dfrac{2}{x - 1} = \dfrac{3x + 2}{2x(x - 1)}$

40. $\dfrac{3}{x + 1} - \dfrac{1}{x} = \dfrac{1}{x(x + 1)}$

41. $\dfrac{2}{x + 1} - \dfrac{3}{x - 1} = \dfrac{4x}{(x + 1)(x - 1)}$

42. $\dfrac{2x}{x - 1} = 5 + \dfrac{2}{x - 1}$

43. $\dfrac{3x}{x - 2} = 5 + \dfrac{6}{x - 2}$

44. $\dfrac{9x}{3x - 1} = 2 + \dfrac{3}{3x - 1}$

45. $\dfrac{1}{3x - 2} = \dfrac{2}{9x - 6}$

46. $\dfrac{2}{3x - 1} + 3 = 5$

47. $\dfrac{2x}{x - 1} + \dfrac{2}{x - 1} = 3$

48. $\dfrac{1}{x - 1} + \dfrac{3}{x - 1} = 5$

49. $(x - 1)(x - 2) = (x + 2)(x + 3)$

50. $3x(x + 2) = (3x - 1)(x - 2)$

51. $(2x + 3)(5x + 1) = (5x - 2)(2x - 9)$

52. $(x + 4)(2x - 1) = (x + 2)(2x + 1)$

53. $\dfrac{2x + 3}{6x - 1} = \dfrac{x - 6}{3x + 1}$

54. $\dfrac{2x - 1}{x + 2} = \dfrac{6(x + 2)}{3x + 2}$

55. $\dfrac{x}{2} - \dfrac{1}{x} = \dfrac{6x + 5}{12}$

Using a calculator, solve the equations in Exercises 56 to 58 for x.

56. $3.12(x + 2) + 4.13(2x + 3) = 35.70$

57. $2.06(3x + 2) + 3.15(2x + 1) = 6x + 23.47$

58. $2.59(2x + 1.37) - 4.03(x + 2.01) = 0.3355$

Using a calculator, solve the equations in Exercises 59 and 60 for x. Round your answers to four decimal places.

59. $\dfrac{1.5x + 2.3}{3.2} + \dfrac{0.7x - 6.5}{1.5} = 2$

60. $\dfrac{2.25x - 1.75}{0.2} + \dfrac{4.13x + 2.25}{0.3} = 1.50$

2.2

FORMULAS AND APPLIED PROBLEMS

Formulas or equations involving variables are used in all fields of study. The formula $A = \pi r^2$, for instance, describes the area A of a circle in terms of its radius r. The equation $s = 16t^2$ describes the law of gravity relating to falling bodies. If we know the time t (in seconds) after the release of a certain object, the formula $s = 16t^2$ gives us the distance s (in feet) the object has fallen from its starting point.

Sometimes it is necessary to express a particular variable in terms of other variables appearing in the equation. To accomplish this, we transform the original equation into an equivalent equation with the desired variable isolated on one side of the equation.

Let us consider some examples.

Example 2.6

If C represents the temperature on the Celsius scale and F the temperature on the Fahrenheit scale, then the relationship between the two is given by

$$C = \tfrac{5}{9}(F - 32).$$

Solve for F.

Solution: To isolate F on one side of the equation, we generate the following equivalent equations:

$$C = \tfrac{5}{9}(F - 32)$$
$$\tfrac{9}{5}C = \tfrac{9}{5} \cdot \tfrac{5}{9}(F - 32)$$
$$\tfrac{9}{5}C = F - 32$$
$$\tfrac{9}{5}C + 32 = F - 32 + 32$$
$$\tfrac{9}{5}C + 32 = F.$$

Thus

$$F = \tfrac{9}{5}C + 32.$$

Example 2.7

The formula

$$A = 2\pi r^2 + 2\pi rh$$

gives the area A of a cylinder in terms of its height h and radius r. Solve for h.

Solution: To isolate h on one side of the equation, we proceed as follows:

$$2\pi r^2 + 2\pi rh = A$$
$$2\pi r(r + h) = A$$
$$r + h = \frac{A}{2\pi r}$$
$$h = \frac{A}{2\pi r} - r.$$

Example 2.8

The formula

$$\frac{1}{R} = \frac{1}{r_1} + \frac{1}{r_2}$$

is used in the theory of electric circuits. Solve for R.

Solution: We generate the following equivalent equations:

$$\frac{1}{R} = \frac{1}{r_1} + \frac{1}{r_2},$$

$$\frac{1}{R} = \frac{r_2 + r_1}{r_1 r_2}.$$

Taking the reciprocal of both sides of the equation, we obtain

$$R = \frac{r_1 r_2}{r_2 + r_1}.$$

We now solve word problems that involve the use of mathematical tools. These problems occur frequently in the natural sciences, social sciences, and technology, as well as other fields, and are often called *applied problems*. Solving such problems requires an ability to translate given information to a mathematical statement, to use algebraic techniques to arrive at the solution, and then to interpret the answer as it relates to the original problem.

Because of their unlimited variety, there is no general rule for solving applied problems. However, we outline a general strategy for attacking them. The following suggestions may prove helpful.

1. Read the problem carefully, several times if necessary. It is important that you understand what information is contained in the problem and what questions are being asked.
2. Summarize the information you are given, and make a list of the questions that need to be answered.
3. Introduce a letter variable to represent the unknown quantity. Look for words such as *what, when, how far, how long, how much,* and *find,* in order to identify the unknown quantity.
4. Translate the given information to an equation that relates known and unknown quantities.
5. Solve the equation.
6. Check that the solution makes sense. For instance, if you are asked how many persons are required to complete a certain project and your answer is $3\frac{7}{8}$, you have probably made an error.

Remember that it takes a lot of practice to develop proficiency in solving word problems. Be patient and keep trying!

Let us now look at some examples.

Example 2.9

Susan has three test scores of 69, 78, and 71 in a mathematics course. What must her score be on the next test for her average to be 75?

Solution: Let x be Susan's score on her next test. The average of her four scores is their sum divided by 4. Thus the average of 69, 78, 71, and x is

$$\frac{69 + 78 + 71 + x}{4}.$$

Since this average is to be 75, we set up the equation

$$\frac{69 + 78 + 71 + x}{4} = 75$$

and solve for x. That is,

$$\frac{218 + x}{4} = 75$$

$$218 + x = 300$$

$$x = 82.$$

A score of 82 on the last test will give Susan an average of 75 for the four tests.

To check, we note that if the four test scores are 69, 78, 71, and 82, then the average of these scores is

$$\frac{69 + 78 + 71 + 82}{4} = 75.$$

Consider now the business-related problem of determining the interest on an investment of P dollars. If r is the annual rate of interest and t is the time in years, then the simple interest is given by the formula

$$\text{Simple interest} = Prt.$$

Suppose, for example, that $1500 is invested for 4 years at 6 percent annual rate of interest. Then the simple interest is $1500(0.06)(4) = \$360$.

Example 2.10

Jovita invested $7500, some at 5.5 percent and the remainder at 6 percent annual rate of interest. If she collected a total of $435 in interest at the end of the year, how much money was invested at each rate?

Solution: Let x be the money invested at 5.5 percent. Then the remaining $7500 - x$ was invested at 6 percent annual interest. Simple interest from the money invested at 5.5 percent is

$$x \cdot 0.055 \cdot 1,$$

and the interest from the 6 percent investment is

$$(7500 - x)(0.06)(1).$$

Since the total simple interest is $435, we have the equation

$$x(0.055) + (7500 - x)(0.06) = 435.$$

Solving for x, we obtain

$$0.055x + 7500(0.06) - 0.06x = 435$$

$$-0.005x + 450 = 435$$

$$-0.005x = -15$$

$$x = \frac{-15}{-0.005} = 3000.$$

Thus $3000 was invested at 5.5 percent and $4500 at 6 percent.

To check our answer, we see that the simple interest from $3000 invested at 5.5 percent is $(3000)(0.055) = \$165$. The interest from $4500 investment at 6 percent is $(4500)(0.06) = \$270$. Hence the total interest is $\$165 + \$270 = \$435$.

Motion problems in physics rely on the use of a distance formula that relates the average rate of speed of a moving object and the elapsed time to the distance traveled. If an object travels at an average rate of speed r for a time t, then the distance d traveled is given by the formula

$$d = rt.$$

Two related formulas are

$$r = \frac{d}{t} \quad \text{and} \quad t = \frac{d}{r}.$$

Example 2.11   A freight train takes 12 hours to travel the same distance that a passenger train travels in 8 hours. If the passenger train travels 20 miles per hour faster than the freight train, find the average rate of speed of each train.

Solution: Let r denote the average rate of speed of the freight train. Then $r + 20$ is the average rate of speed of the passenger train. The distance traveled by the freight train and that traveled by the passenger train are

$$12r \quad \text{and} \quad 8(r + 20),$$

respectively. Since both trains have traveled the same distance, we have the equation

$$12r = 8(r + 20).$$

Solving for r, we have

$$12r = 8r + 160$$

$$4r = 160$$

$$r = 40.$$

The freight train travels at an average rate of 40 miles per hour, and the passenger train at an average rate of 60 miles per hour.

 To check, we can see that if the freight train travels at an average rate of 40 miles per hour, then in 12 hours it travels a distance of $12 \times 40 = 480$ miles. Similarly, if the passenger train travels at an average rate of 60 miles per hour, then in 8 hours it also travels a distance of $8 \times 60 = 480$ miles.

Example 2.12

José and Ray live 12 miles apart. They start jogging at 7:00 A.M. They run toward each other on the same route at an average rate of 3.5 and 4.5 miles per hour, respectively. When will they pass each other? How far will each have run by that time?

Solution: Let t denote the time (in hours) elapsed at the moment of the meeting. Using the distance formula $d = rt$, we note that in t hours, José has covered a distance of $3.5t$ while Ray has run $4.5t$ miles. Together, they have covered a distance of 12 miles. We must solve the equation

$$3.5t + 4.5t = 12$$

for t.

$$3.5t + 4.5t = 12$$

$$8t = 12$$

$$t = \tfrac{12}{8} = 1.5 \text{ hours.}$$

Thus the two joggers pass each other at 8:30 A.M. Note that José has run a distance of $(3.5)(1.5) = 5.25$ miles while Ray has covered a distance of $(4.5)(1.5) = 6.75$ miles in 1.5 hours.

 To check, we see that José and Ray have traveled a total distance of $5.25 + 6.75 = 12$ miles.

Example 2.13

A freight train, traveling at an average rate of 40 miles per hour, leaves New York for Cleveland. An hour and a half later, an express train also leaves New York for Cleveland, on parallel tracks, traveling at an average rate of 65 miles per hour. How long after the express train leaves will it catch the freight train, and how far from New York will they be?

Solution: Let t represent the travel time of the express train. The freight train has already traveled for 1.5 hours when the express train leaves New York. Therefore the freight train will have traveled for $t + \frac{3}{2}$ hours when the express train catches up. Using the distance formula $d = rt$, we write an expression for the distance traveled by each train as shown in the following table:

	Average Rate r	Time t	Distance d
Freight train	40	$t + \frac{3}{2}$	$40(t + \frac{3}{2})$
Express train	65	t	$65t$

At the moment the express train overtakes the freight train, each will have traveled the same distance. This means that we must solve the equation

$$65t = 40(t + \tfrac{3}{2})$$

for t. Thus

$$65t = 40t + 60$$

$$25t = 60$$

$$t = \tfrac{60}{25} = 2\tfrac{2}{5} \text{ hours, or 2 hours 24 minutes.}$$

The express train will overtake the freight train 2 hours 24 minutes after it leaves New York. Both trains will be

$$(\tfrac{12}{5})(65) = 156 \text{ miles}$$

from New York at the moment the express train catches the freight train.

Work problems rely on the basic principle that if a project can be completed in time t, then $1/t$ of it can be completed in 1 unit of time. Thus, if a pipe can fill a swimming pool in 8 hours, then only $\frac{1}{8}$ of it can be filled in 1 hour. Similarly, if a house can be built in 40 days, then only $\frac{1}{40}$ of it can be completed in 1 day, and so on.

Example 2.14

Pam can finish a certain project in 10 days, and Leo can finish the same project in 15 days. How long will it take them to complete the project if they decide to work together?

Solution: In one day, Pam can finish one-tenth of the project. Similarly, Leo can complete one-fifteenth of the project in 1 day. Together, they can finish

$$\tfrac{1}{10} + \tfrac{1}{15}$$

of the project in 1 day.

If t represents the number of days it takes Pam and Leo working together to complete the project, then in 1 day they will complete $1/t$ of the task. Thus we have the equation

$$\frac{1}{10} + \frac{1}{15} = \frac{1}{t}.$$

Multiplying both sides of this equation by $30t$, we obtain

$$3t + 2t = 30$$
$$5t = 30$$
$$t = 6.$$

Hence Pam and Leo can complete the project together in 6 days.

Many applications of problems involve **mixing two substances** to obtain a prescribed mixture. Let us look at some which may be solved by using linear equations.

Example 2.15

How much water should be added to 3 liters of pure acid to obtain a solution which contains 15 percent acid?

Solution: Let x be the number of liters of water to be added to 3 liters of pure acid. Then the mixture contains $(x + 3)$ liters of solution which contains 15 percent acid. Since this acid content must equal 3 liters of pure acid, we have the equation

$$(x + 3)(0.15) = 3,$$

which must be solved for x.

$$0.15x + 0.45 = 3$$
$$0.15x = 2.55$$
$$x = \frac{2.55}{0.15} = 17.$$

Hence 17 liters of water should be added to 3 liters of pure acid.

To check the answer, we observe that if 17 liters of water is added to the pure acid, then the resulting mixture of 20 liters contains $(20)(0.15) = 3$ liters of pure acid.

Example 2.16

How many quarts of solution A containing 20 percent alcohol must be mixed with 50 quarts of solution B containing 65 percent alcohol to produce a solution containing 50 percent alcohol?

Solution: Let x be the number of quarts of the solution A. Then the mixture contains $(x + 50)$ quarts of the solution containing

$$(0.20)(x) + (0.65)(50) \qquad \text{quarts}$$

of pure alcohol. Since the alcohol content must be equal to $0.50(x + 50)$, we have

$$0.50(x + 50) = (0.20)x + (0.65)(50)$$

to be solved for x. So

$$0.50x + 25 = 0.20x + 32.5$$

$$0.30x = 7.5$$

$$x = \frac{7.5}{0.3} = 25.$$

Hence, 25 quarts of solution A must be added to solution B to obtain a mixture containing 50 percent alcohol. To check the answer, we note that 25 quarts of solution A contains $(25)(0.20) = 5.00$ quarts of alcohol. Similarly, 50 quarts of solution B contains $(50)(0.65) = 32.5$ quarts of alcohol. This means that 75 quarts of the mixture contains $5.00 + 32.5 = 37.5$ quarts of alcohol, which is 50 percent of 75 quarts.

EXERCISES

Solve each equation in Exercises 1 to 20 for the indicated variable.

1. $F = ma$ for a

2. $d = rt$ for r

3. $C = 2\pi r$ for r

4. $A = bh$ for b

5. $v = u + gt$ for u

6. $v = u + gt$ for t

7. $F = \dfrac{km_1 m_2}{d^2}$ for m_1

8. $V = \dfrac{\pi r^2 h}{3}$ for h

9. $A = \dfrac{h}{2}(b + c)$ for b

10. $F = \frac{9}{5}C + 32$ for C

11. $ax + by = ab$ for x

12. $2x - 3y = 5$ for y

13. $A = r(r + s)$ for s

14. $S = a + (n - 1)d$ for n

15. $d = \dfrac{s - a}{n - 1}$ for s

16. $d = \dfrac{s - a}{n - 1}$ for n

17. $s = ut + \frac{1}{2}gt^2$ for u

18. $s = ut + \frac{1}{2}gt^2$ for g

19. $\dfrac{1}{r} = \dfrac{1}{r_1} + \dfrac{1}{r_2}$ for r_1

20. $\dfrac{x}{a} + \dfrac{y}{b} = 1$ for x

21. The sum of two numbers is 165. One of the numbers is 39 more than the other. Find the numbers.

22. Find three consecutive numbers whose sum is 72.

23. A student in a certain course has three test scores of 85, 62, and 73. What must that student's score be on the next test so that the average is 76?

24. Anita has scores of 64, 71, 86, and 81 on four tests in her accounting course. What must her score on the next test be so that her average is 80?

25. Arnold invested $3500, some at 5.5 percent and the remainder at 6 percent annual interest. If he collected $202.50 at the end of 1 year, how much money was invested at each rate?

26. Ms. Talbot invested $8000, some at 6 percent and the remainder at 7 percent annual interest. If she collected $510 at the end of 1 year, how much money was invested at each rate?

27. A collection of coins consisting of nickels and dimes has a value of $4.20. If there are 49 coins in all, how many of each kind are in the collection?

28. A collection of coins consisting of dimes and quarters has a value of $15.55. If there are 19 more dimes than quarters, how many of each are in the collection?

29. The admission price at a theater performance was $3.50 for adults and $2 for children. If the total receipts for the performance were $850 for 275 paid admissions, how many adults and how many children attended the performance?

30. The admission price at a certain basketball game was $1.50 for adults and $0.75 for children. There were 400 fewer adult tickets sold than children's tickets. If the total receipts for the paid admissions were $3000, how many adults and how many children attended the game?

31. A boat travels 210 kilometers downstream in the same time it takes the boat to travel 140 kilometers upstream. The speed of the current is 5 kilometers per hour. Determine the speed of the boat in still water.

32. A passenger train travels 200 miles in the same time it takes a freight train to travel 140 miles. If the passenger train goes 15 miles per hour faster than the freight train, what is the rate of speed of each train?

33. Inez and Mike, who are 5 miles apart, start jogging at 6:00 P.M. If they run toward each other on the same route at uniform rates of 4 and 6 miles per hour, respectively, when will they meet? How far will each have run?

34. It took Bertha 20 minutes to drive her boat upstream to water-ski at her favorite spot. Returning downstream later in the day at the same speed took 15 minutes. If the speed of the current was 5 kilometers per hour, what was the speed of the boat?

35. Mr. Hawkes drove from one city to another at an average speed of 45 miles per hour. On his return trip, he averaged 40 miles per hour, and his elapsed time was 30 minutes longer. What was the total distance traveled?

36. Henry walked from his home to a movie theater at the rate of 4 miles per hour. He ran back home over the same route at the rate of 6 miles per hour. How far was the movie theater from his home if the total time going to and coming from the movies was 20 minutes?

37. A freight train leaves New York for Chicago, traveling at an average rate of 45 miles per hour. Two hours later, a passenger train leaves New York for Chicago, on parallel tracks, traveling at an average rate of 60 miles per hour. At what distance from New York does the passenger train pass the freight train?

38. Jane starts jogging from a certain point and runs at a uniform rate of 4 miles per hour. Ten minutes later, Paul starts from the same point and follows the same route running at a uniform rate of 6 miles per hour. How long will it take Paul to catch up with Jane? How far will each have run?

39. Pipe A can fill a tank in 2 hours, and pipe B can fill it in 3 hours. How long will it take to fill the tank if both pipes are opened at the same time?

40. Carrie can build a house in 30 days. She can build the same house in 20 days if she works with her son. How long will it take her son to build the same house if he decides to work alone, assuming that he has the know-how?

41. Allan can finish a certain project in 8 hours, Karen can complete it in 10 hours, and Bob can finish it in 20 hours. If Allan, Karen, and Bob decide to work together, how long will it take them to complete the project?

42. Pipe A can fill a swimming pool in 4 hours, and pipe B can fill it in 6 hours. Outlet pipe C can empty the pool in 8 hours. How long will it take to fill the pool if pipe C is left open inadvertently at the moment pipes A and B are opened?

43. How much water should be added to 8 liters of pure acid to obtain a solution which is 20 percent acid?

44. Six liters of a chemical solution contains a 40 percent concentration of an acid. How many liters of pure acid must be added to increase the concentration to 60 percent acid?

45. How many pounds of an alloy containing 20 percent silver must be melted with 30 pounds of an alloy containing 35 percent silver to obtain an alloy containing 30 percent silver?

46. How many liters of solution A containing 75 percent alcohol must be mixed with 15 liters of solution B containing 45 percent alcohol to prepare a solution containing 70 percent alcohol?

47. A chemical solution A contains 40 percent acid. A second solution B contains 20 percent acid. How much of each chemical solution should be mixed to obtain 80 liters of a solution containing 35 percent acid?

48. Antifreeze A contains 18 percent alcohol. Antifreeze B contains 12 percent alcohol. How many liters of each should be mixed to get 30 liters of a mixture that contains 16 percent alcohol?

2.3

QUADRATIC EQUATIONS

An equation of the form

$$ax^2 + bx + c = 0,$$

quadratic equation   where a, b, and c are real numbers and $a \neq 0$, is called a *quadratic equation.* To solve quadratic equations, it is necessary to introduce the following property of real number system.

Proposition 2.3.1 | If a and b are real numbers and $ab = 0$, then either $a = 0$ or $b = 0$.

Let us consider some illustrative examples.

Example 2.17 | Solve the equation

$$7x^2 - 19x + 10 = 0.$$

Solution: Observe that if we factor the left-hand side, we obtain

$$(7x - 5)(x - 2) = 0.$$

By Proposition 2.3.1, either

$$7x - 5 = 0 \quad \text{or} \quad x - 2 = 0.$$

Solving each of these linear equations separately, we find that

$$x = \tfrac{5}{7} \quad \text{or} \quad x = 2.$$

Hence, the solution set is $\{\tfrac{5}{7}, 2\}$.

Example 2.18 | Solve the equation

$$x^2 - k = 0 \qquad (k \geq 0).$$

Solution: When we factor the left-hand side, we observe that

$$(x - \sqrt{k})(x + \sqrt{k}) = 0.$$

By Proposition 2.3.1, either $x - \sqrt{k} = 0$ or $x + \sqrt{k} = 0$; and

$$x - \sqrt{k} = 0 \qquad \text{implies that} \qquad x = \sqrt{k}$$

$$x + \sqrt{k} = 0 \qquad \text{implies that} \qquad x = -\sqrt{k}.$$

The solution set of the quadratic equation $x^2 - k = 0$ is $\{-\sqrt{k}, \sqrt{k}\}$.

We have shown that if $x^2 = k$, then $x = \pm\sqrt{k}$. This means that when we take the square root of both sides of the equation

$$x^2 = k,$$

we take *both* the positive and the negative square roots of k, and not just the principal square root, as discussed in Chapter 1. If $k < 0$, then the roots are imaginary.

Example 2.19 | Solve

$$(x - 2)^2 = -9.$$

Solution: Taking the square root of both sides, we obtain

$$x - 2 = \sqrt{-9} \qquad \text{or} \qquad x - 2 = -\sqrt{-9}$$

That is,

$$x - 2 = 3i \qquad \text{or} \qquad x - 2 = -3i,$$

where $i = \sqrt{-1}$. Thus $x = 2 + 3i$ and $x = 2 - 3i$ are the solutions of the equation $(x - 2)^2 = -9$. The solution set is $\{2 + 3i, 2 - 3i\}$.

Example 2.20 | Solve the equation

$$x^2 - 16x + 64 = 0.$$

Solution: Factoring the left-hand side, we note that the original equation is equivalent to

$$(x - 8)(x - 8) = 0.$$

The solution of the linear equation $x - 8 = 0$ is given by $x = 8$. The solution set is $\{8\}$.

Since $x - 8$ appears as a factor twice in the above equation, the real number 8 is called **a double root, or a root of multiplicity 2** of the quadratic equation $x^2 - 16x + 64 = 0$.

In the above examples, we were able to determine the solution sets of the equations by factoring. Since this method is not always successful, we develop *completing the square* a general method for solving quadratic equations, called **completing the square.**

Observe that expressions such as $x^2 + 4x + 4$ and $4x^2 - 12x + 9$ are perfect squares, since

$$x^2 + 4x + 4 = (x + 2)^2$$

and

$$4x^2 - 12x + 9 = (2x - 3)^2.$$

Now suppose we are given $x^2 + 10x$. What should be added to $x^2 + 10x$ to get a perfect square? By adding 25, we obtain $x^2 + 10x + 25$, which equals $(x + 5)^2$. Thus by adding 25 to $x^2 + 10x$, we have completed the square. *In general, for an expression of the form $x^2 + kx$, we add the square of one-half the coefficient of x, that is, $(k/2)^2$, so that the resulting expression is a perfect square.*

Example 2.21

The equation

$$x^2 + 6x - 1 = 0$$

cannot be solved by factoring. We therefore write the given equation in the form

$$x^2 + 6x = 1.$$

To complete the square of the left-hand side, we add the square of one-half the coefficient of x, that is, $(6/2)^2 = 9$, to both sides of the equation. We obtain

$$x^2 + 6x + 9 = 1 + 9$$

$$x^2 + 6x + 9 = 10$$

$$(x + 3)^2 = 10.$$

Taking the square root of each side of the equation, we get

$$x + 3 = \sqrt{10} \quad \text{or} \quad x + 3 = -\sqrt{10}$$

$$x = -3 + \sqrt{10} \quad \text{or} \quad x = -3 - \sqrt{10}.$$

Thus the solution set is $\{-3 - \sqrt{10}, -3 + \sqrt{10}\}$.

We shall now apply the method of completing the square to the general quadratic equation

$$ax^2 + bx + c = 0, \quad a \neq 0. \tag{2.6}$$

Adding $-c$ to both sides of the equation, we obtain

$$ax^2 + bx = -c. \tag{2.7}$$

Dividing the equation (2.7) by a, the coefficient of x^2, we have

$$x^2 + \frac{b}{a}x = -\frac{c}{a}. \qquad (2.8)$$

We next add $[b/(2a)]^2$, the square of one-half the coefficient of x, to both sides of (2.8) to get

$$x^2 + \frac{b}{a}x + \left(\frac{b}{2a}\right)^2 = \left(\frac{b}{2a}\right)^2 - \frac{c}{a} = \frac{b^2}{4a^2} - \frac{c}{a}.$$

That is,

$$\left(x + \frac{b}{2a}\right)^2 = \frac{b^2 - 4ac}{4a^2}. \qquad (2.9)$$

Taking the square root of both sides, we have

$$x + \frac{b}{2a} = \frac{\pm\sqrt{b^2 - 4ac}}{2a}. \qquad (2.10)$$

quadratic formula

Finally, we solve for x and obtain the *quadratic formula*

$$x = \frac{-b \pm \sqrt{b^2 - 4ac}}{2a}. \qquad (2.11)$$

We have shown that the roots of the equation $ax^2 + bx + c = 0$ $(a \neq 0)$ are given by

$$\frac{-b \pm \sqrt{b^2 - 4ac}}{2a}$$

discriminant

where a, b, and c are real numbers. The quantity $b^2 - 4ac$ in the quadratic formula (2.11) is called the **discriminant** of the quadratic equation. It can be used to determine the nature of the roots of the quadratic equation as outlined below:

1. If $b^2 - 4ac = 0$, the quadratic equation has a double root.
2. If $b^2 - 4ac$ is positive, then the roots are real and unequal. Further, if $b^2 - 4ac$ is a perfect square, then the roots are rational and unequal.
3. If $b^2 - 4ac$ is negative, then the roots are complex.

Example 2.22

Use the discriminant to determine the nature of the roots of the following quadratic equations:

(a) $x^2 + 16x + 64 = 0$ (b) $3x^2 - 5x - 6 = 0$
(c) $x^2 - x - 12 = 0$ (d) $3x^2 - 5x + 4 = 0$

Solution:

(a) Here, $a = 1$, $b = 16$, and $c = 64$. So

$$b^2 - 4ac = (16)^2 - 4(1)(64) = 256 - 256 = 0.$$

Since the discriminant $b^2 - 4ac = 0$, the quadratic equation has a double root.

(b) Here, $a = 3$, $b = -5$, and $c = -6$, and so

$$b^2 - 4ac = (-5)^2 - 4(3)(-6) = 25 + 72 = 97.$$

Since the discriminant is positive, the equation has real and unequal roots.

(c) Observe that $a = 1$, $b = -1$, and $c = -12$. Here,

$$b^2 - 4ac = (-1)^2 - 4(1)(-12) = 1 + 48 = 49,$$

which is a perfect square. Hence, the roots are rational and unequal.

(d) We have $a = 3$, $b = -5$, and $c = 4$. Thus,

$$b^2 - 4ac = (-5)^2 - 4(3)(4) = 25 - 48 = -23,$$

which is negative. Hence, the roots are complex.

Example 2.23

Solve the equation

$$2x^2 - 3x - 4 = 0$$

by using the quadratic formula.

Solution: Here, $a = 2$, $b = -3$, and $c = -4$. By using the quadratic formula, we have

$$x = \frac{-(-3) \pm \sqrt{(-3)^2 - 4(2)(-4)}}{2(2)}$$

$$= \frac{3 \pm \sqrt{9 + 32}}{4}$$

$$= \frac{3 \pm \sqrt{41}}{4}.$$

The solution set is $\{(3 - \sqrt{41})/4, (3 + \sqrt{41})/4\}$.

Example 2.24

Solve the equation

$$x^2 - 8x + 41 = 0$$

by using the quadratic formula.

Solution: Substituting $a = 1$, $b = -8$, and $c = 41$ in the quadratic formula, we get

$$x = \frac{-(-8) \pm \sqrt{(-8)^2 - 4(1)(41)}}{2(1)}$$

$$= \frac{8 \pm \sqrt{64 - 164}}{2}$$

$$= \frac{8 \pm \sqrt{-100}}{2}$$

$$= \frac{8 \pm 10i}{2} = \frac{2(4 \pm 5i)}{2} = 4 \pm 5i.$$

Hence the solutions are $x = 4 - 5i$ and $x = 4 + 5i$. The solution set is $\{4 - 5i, 4 + 5i\}$.

Let us now consider a problem which leads to a quadratic equation.

Example 2.25   Solve the equation

$$\frac{6}{10 + x} + \frac{6}{10 - x} = \frac{5}{4}.$$

Solution: We must exclude $x = -10$ and $x = 10$ from the solution set because these values of x involve division by zero, which is not defined. To eliminate fractions, we multiply both sides of the equation by $4(10 + x)(10 - x)$, the least common multiple of 4, $10 - x$, and $10 + x$. We obtain the following sequence of equivalent equations:

$$4(10 + x)(10 - x)\left(\frac{6}{10 + x}\right) + 4(10 + x)(10 - x)\left(\frac{6}{10 - x}\right)$$

$$= 4(10 + x)(10 - x)\left(\frac{5}{4}\right)$$

$$24(10 - x) + 24(10 + x) = 5(10 + x)(10 - x)$$

$$240 - 24x + 240 + 24x = 5(100 - x^2)$$

$$480 = 500 - 5x^2$$

$$5x^2 = 20$$

$$x^2 = 4$$

$$x = \pm 2.$$

Thus $x = -2$ and $x = 2$ are the solutions of the original equation, and the solution set is $\{-2, 2\}$.

Next we consider some **applied problems** that lead to quadratic equations.

Example 2.26

A ball is thrown vertically into the air with an initial velocity of 128 feet per second. Its height h (in feet) after t seconds is given by

$$h = 128t - 16t^2.$$

When will the ball be at 192 feet above the ground?

Solution: We want to determine at what time t the ball will be at a height of 192 feet. This means that we must solve the equation

$$192 = 128t - 16t^2$$

for t. This equation may be equivalently expressed as

$$16t^2 - 128t + 192 = 0$$

$$t^2 - 8t + 12 = 0.$$

Factoring the left-hand side, we observe that

$$(t - 2)(t - 6) = 0.$$

Therefore, $t - 2 = 0$ or $t - 6 = 0$, which means that $t = 2$ or $t = 6$. Substituting these values of t in the equation $h = 128t - 16t^2$, we obtain

$$h = 128(2) - 16(2)^2 = 256 - 64 = 192$$

and

$$h = 128(6) - 16(6)^2 = 768 - 576 = 192.$$

We conclude that the ball will be 192 feet above the ground on its way up after 2 seconds and will again be at this height on its way down after 6 seconds in the air.

Example 2.27

A woman can row 18 miles downstream and back in 5 hours 36 minutes. If her rate of rowing in still water is 7 miles per hour, what is the speed of the current?

Solution: Let r be the speed of the current. Then $7 + r$ is the total speed of the boat downstream and $7 - r$ is its speed upstream. If t_1 and t_2 represent the time the boat needs to travel a distance of 18 miles upstream and 18 miles downstream, respectively, then

$$t_1 = \frac{18}{7 + r} \qquad \text{and} \qquad t_2 = \frac{18}{7 - r}.$$

Since $t_1 + t_2 = \frac{28}{5}$ hours, we have the equation

$$\frac{18}{7 + r} + \frac{18}{7 - r} = \frac{28}{5}.$$

Solving this equation for r, we find that $r = -2$ or $r = 2$. Since the speed of the current cannot be negative, we conclude that the speed of the current is 2 miles per hour.

Example 2.28

Sunil can complete a certain project in 5 more hours than his brother John. If they work together, they can finish the job in 6 hours. How long would it take each one to complete the project?

Solution: Let t be the time (in hours) it takes John to finish the job. Then $t + 5$ hours is the time it takes Sunil to finish it. Further, John can finish $1/t$ of the project in 1 hour, while Sunil can do $1/(t + 5)$ of the work in 1 hour. Working together, they can complete $\frac{1}{6}$ of the project in 1 hour. Therefore

$$\frac{1}{t} + \frac{1}{t + 5} = \frac{1}{6}.$$

Multiplying both sides of the equation by $6t(t + 5)$, the least common multiple of 6, t, and $t + 5$, we get

$$6(t + 5) + 6t = t(t + 5)$$
$$6t + 30 + 6t = t^2 + 5t$$
$$12t + 30 = t^2 + 5t,$$

or

$$t^2 - 7t - 30 = 0$$
$$(t - 10)(t + 3) = 0.$$

This means that $t = 10$ or $t = -3$. Since t cannot be negative, we conclude that John and Sunil can finish the project in 10 and 15 hours, respectively.

EXERCISES

Solve Exercises 1 to 12 by factoring.

1. $x^2 - 4x + 3 = 0$
2. $x^2 - 2x - 8 = 0$
3. $2x^2 + 9x + 4 = 0$
4. $6x^2 + 7x - 3 = 0$
5. $x^2 + 3x + 2 = 0$
6. $x^2 + 5x + 6 = 0$
7. $x^2 + 7x + 12 = 0$
8. $x^2 + 11x + 28 = 0$
9. $x^2 - 5x + 6 = 0$
10. $x^2 - x - 2 = 0$
11. $x^2 - 3x - 40 = 0$
12. $x^2 + x - 20 = 0$

Solve Exercises 13 to 20 by completing the square.

13. $x^2 - x - 6 = 0$
14. $x^2 - 2x - 8 = 0$
15. $2x^2 + x - 1 = 0$
16. $2x^2 - 5x - 3 = 0$
17. $2x^2 - 3x + 1 = 0$
18. $5x^2 - 12x + 4 = 0$
19. $6x^2 - 5x - 4 = 0$
20. $6x^2 + 19x + 10 = 0$

Determine the nature of the roots in Exercises 21 to 30.

21. $2x^2 - 4x + 5 = 0$ **22.** $3x^2 + 9x + 4 = 0$ **23.** $6x^2 + 13x + 5 = 0$

24. $4x^2 - 4x + 1 = 0$ **25.** $3x^2 - 20x + 25 = 0$ **26.** $4x^2 - 9x + 3 = 0$

27. $4x^2 - 3x - 2 = 0$ **28.** $2x^2 + x + 1 = 0$ **29.** $x^2 - 2\sqrt{3}x + 3 = 0$

30. $x^2 - 8x + 41 = 0$

Solve Exercises 31 to 46 by using the quadratic formula.

31. $x^2 + x - 12 = 0$ **32.** $x^2 - x - 6 = 0$ **33.** $x^2 + 5x + 3 = 0$

34. $x^2 - 7x - 8 = 0$ **35.** $2x^2 - 3x - 9 = 0$ **36.** $3x^2 - 5x + 1 = 0$

37. $3x^2 + 10x - 8 = 0$ **38.** $3x^2 + 23x + 14 = 0$ **39.** $5x^2 + 7x - 6 = 0$

40. $x^2 + x - 1 = 0$ **41.** $x^2 + x + 1 = 0$ **42.** $x^2 - 2x + 2 = 0$

43. $x^2 - 4x + 5 = 0$ **44.** $3x^2 + 7x - 1 = 0$ **45.** $16x^2 - 16x + 13 = 0$

46. $x^2 - 6x + 13 = 0$

Solve by any method the equations in Exercises 47 to 54.

47. $1 - \dfrac{2}{x} = \dfrac{35}{x^2}$ **48.** $\dfrac{19}{x} = 7 + \dfrac{10}{x^2}$ **49.** $x + \dfrac{1}{x} = \dfrac{5}{2}$

50. $x - \dfrac{1}{x} = \dfrac{8}{3}$ **51.** $\dfrac{1}{x} + \dfrac{6}{x + 4} = 1$ **52.** $\dfrac{1}{2x + 3} - \dfrac{3}{2x - 3} = 2$

53. $\dfrac{4x}{x - 3} - \dfrac{3x - 1}{x + 3} = 3$ **54.** $\dfrac{7}{x - 1} - \dfrac{6}{x^2 - 1} = 5$

55. If r_1 and r_2 are the roots of the equation

$$ax^2 + bx + c = 0,$$

then show that $r_1 + r_2 = -\dfrac{b}{a}$ and $r_1 r_2 = \dfrac{c}{a}$.

56. If one root of the equation

$$2x^2 + kx - 3 = 0$$

is -3, find the other root.

57. Find the value of k so that the equation

$$2x^2 + 4x + k = 0$$

has equal roots.

58. Find the value of k so that the equation

$$kx^2 - 2x - 5 = 0$$

has roots whose product equals $-\tfrac{5}{3}$.

Solve Exercises 59 to 62 for the indicated variable.

59. $s = 16t^2$ for t **60.** $V = \pi r^2 h$ for r

61. $s = ut + \tfrac{1}{2}gt^2$ for t **62.** $V = 2\pi r(r + h)$ for r

63. The sum of two numbers is 27, and their product is 180. Find the numbers.

64. Find two consecutive numbers whose product is 210.

65. A ball is thrown vertically into the air with an initial velocity of 96 feet per second. Its height h (in feet) after t seconds is given by

$$h = 96t - 16t^2.$$

When will the ball be 128 feet above the ground?

66. A projectile is shot vertically into the air with an initial velocity of 80 feet per second. Its height h (in feet) after t seconds is given by

$$h = 80t - 16t^2.$$

When will the projectile be at a height of 84 feet?

67. A rectangular piece of cardboard is 4 inches longer than its width. A 5-inch square is cut from each corner, and the sides are folded to form an open box. The box contains 700 cubic inches. Find the dimensions of the box.

68. A 2-inch square is cut from each corner of a rectangular piece of cardboard whose length exceeds its width by 3 inches. The sides are turned up, and an open box is formed. What are the dimensions of the box if the volume is 260 cubic inches?

69. Two freight trains A and B leave a certain station at the same time at right angles to each other. Train A travels 12 miles per hour faster than train B. Find the speed of each train if they are 120 miles apart after 2 hours.

70. Two airplanes are flying at different altitudes at right angles to each other. Airplane A is traveling 40 miles per hour faster than airplane B. Determine the speed of each plane if they are 600 miles apart after 3 hours.

71. A woman can row 30 miles upstream and back in a total of 8 hours. If her rate of rowing in still water is 8 miles per hour, what is the speed of the current?

72. Ellen drives 5 miles per hour faster than Peter. Both leave New York at the same time for Boston, a distance of about 225 miles. It takes Peter 0.5 hour longer than Ellen to make the trip. Determine Ellen's average speed.

73. Anita drove a distance of 180 miles. If she had traveled 10 miles per hour faster, she would have made the trip in $1\frac{1}{2}$ hours less time. What was her speed?

74. Washington can complete a job in 3 hours more time than his sister Jeanne. Together they can finish the project in 3 hours 36 minutes. How long would it take each to finish the project?

75. Bonnie can finish a job in 15 more days than Jeff. If they can finish it together in 18 days, how long would it take each to complete it?

76. Kenneth can paint his house in 6 days less time than his brother Henry. Working together, they can do the job in 4 days. How long would it take each to paint the house working alone?

2.4

EQUATIONS CONTAINING FRACTIONAL EXPONENTS

To find solution sets for equations involving radical or fractional exponents, we use the method of raising both sides of the equation to a positive integral power. As an illustration, consider the following example.

Example 2.29　Find the solution set of

$$\sqrt{x + 6} - \sqrt{x - 6} = 2.$$

Solution: The given equation is equivalent to

$$\sqrt{x + 6} = 2 + \sqrt{x - 6}$$

Squaring both sides, we have

$$x + 6 = 4 + 4\sqrt{x - 6} + (x - 6),$$

which, upon simplification, yields

$$2 = \sqrt{x - 6}.$$

To eliminate the radical sign, we again square both sides of the equation and obtain

$$4 = x - 6, \quad \text{or} \quad x = 10.$$

Substituting $x = 10$ in the original equation, we get

$$\sqrt{10 + 6} - \sqrt{10 - 6} = 2$$
$$\sqrt{16} - \sqrt{4} = 2$$
$$4 - 2 = 2,$$

which is a true statement. We therefore conclude that the solution set is $\{10\}$.

The method of solving equations that involve radical or fractional exponents is based on the following proposition, which we state without proof.

Proposition 2.4.1

If $P(x)$ and $Q(x)$ are algebraic expressions in x, then the solution set of the equation $P(x) = Q(x)$ is a subset of the solution set of the equation $[P(x)]^n = [Q(x)]^n$, for any natural number n.

What does this mean? The proposition says that every solution of the equation $P(x) = Q(x)$ is also a solution of the equation $[P(x)]^n = [Q(x)]^n$, but the converse may not be true. In other words, not every solution of the equation $[P(x)]^n = [Q(x)]^n$ is a solution of the equation $P(x) = Q(x)$. As a simple illustration, suppose that we start with the equation $x = 1$. Squaring both sides, we obtain $x^2 = 1$ which has $\{-1, 1\}$ as its solution set. Note that the solution set $\{1\}$ of the equation $x = 1$ is contained in the solution set $\{-1, 1\}$ of the equation $x^2 = 1$. However, the solution set $\{-1, 1\}$ includes -1, which is not a solution of the original equation $x = 1$. Any such solution of $[P(x)]^n = [Q(x)]^n$ that is

extraneous solution not a solution of the given equation $P(x) = Q(x)$ is called an **extraneous solution** of that equation.

Since the equation $[P(x)]^n = [Q(x)]^n$ may have additional solutions that are not a solution of the original equation $P(x) = Q(x)$, it is essential that every solution be substituted for the variable x in the original equation to check its validity.

Let us now consider some examples.

Example 2.30

Solve for x:

$$\sqrt[3]{x^2 + 2} = 3.$$

Solution: Raising both sides of the equation to third power, we obtain

$$(\sqrt[3]{x^2 + 2})^3 = 3^3$$

$$x^2 + 2 = 27$$

$$x^2 = 25$$

$$x = 5 \qquad \text{or} \qquad x = -5.$$

Substituting $x = 5$ in the original equation, we obtain $\sqrt[3]{27} = 3$, which is a true statement. A similar statement holds for $x = -5$. The solution set of the given equation is $\{-5, 5\}$.

Example 2.31

Solve for x:

$$x = \sqrt{2x + 1} + 1.$$

Solution: The given equation is equivalent to

$$x - 1 = \sqrt{2x + 1}.$$

Observe that one side of the equation contains a radical, and to eliminate it, we need to square both sides. We get

$$x^2 - 2x + 1 = 2x + 1$$

$$x^2 - 4x = 0$$

$$x(x - 4) = 0$$

$$x = 0 \qquad \text{or} \qquad x = 4.$$

Substituting $x = 0$ in the original equation, we get

$$0 = \sqrt{1} + 1$$

$$0 = 2,$$

which is false. Thus $x = 0$ is not a solution of the given equation. Substituting $x = 4$ in the given equation, we have

$$4 = \sqrt{8 + 1} + 1$$

$$4 = 3 + 1,$$

which is a true statement. Hence the solution set is $\{4\}$.

Example 2.32

Solve for x:

$$\sqrt{2x + 1} + \sqrt{x + 4} = 3.$$

Solution: First we rewrite the given equation in an equivalent form

$$\sqrt{2x + 1} = 3 - \sqrt{x + 4}.$$

To eliminate the radical signs, we square both sides and obtain

$$2x + 1 = 9 - 6\sqrt{x + 4} + (x + 4),$$

or

$$x - 12 = -6\sqrt{x + 4}.$$

One side of the equation still contains a radical. To eliminate it, we square both sides again, thus obtaining

$$x^2 - 24x + 144 = 36x + 144,$$

which, upon simplification, yields

$$x^2 - 60x = 0$$

$$x(x - 60) = 0$$

$$x = 0 \quad \text{or} \quad x = 60.$$

We leave it for the reader to verify that $x = 60$ is an extraneous solution of the given equation. The solution set is $\{0\}$.

Next we solve quadratic equations of the type

$$ax^{2n} + bx^n + c = 0.$$

If we set $x^n = z$, then the given equation reduces to a quadratic equation in z, that is,

$$az^2 + bz + c = 0.$$

Example 2.33

Solve

$$x^{2/3} + x^{1/3} - 2 = 0.$$

Solution: Let $z = x^{1/3}$. Then the given equation reduces to

$$z^2 + z - 2 = 0.$$

Factoring the left-hand side, we get

$$(z + 2)(z - 1) = 0$$

$$z = -2 \quad \text{or} \quad z = 1.$$

Since $z = x^{1/3}$, we have

$$x^{1/3} = -2 \qquad \text{or} \qquad x^{1/3} = 1.$$

Raising each of these equations to the third power, we get

$$x = -8 \qquad \text{or} \qquad x = 1.$$

When $x = -8$, the original equation yields

$$[(-8)^{1/3}]^2 + (-8)^{1/3} - 2 = 0$$
$$4 - 2 - 2 = 0,$$

which is true. Therefore $x = -8$ is a solution. It can be shown that $x = 1$ also satisfies the given equation. Hence the solution set is $\{-8, 1\}$.

Example 2.34

Solve for x:

$$x^6 - 9x^3 + 8 = 0.$$

Solution: Setting $z = x^3$, we reduce the given equation to the quadratic form

$$z^2 - 9z + 8 = 0.$$

That is,

$$(z - 8)(z - 1) = 0.$$
$$z = 8 \qquad \text{or} \qquad z = 1.$$

Since $z = x^3$, we have

$$x^3 = 8 \qquad \text{or} \qquad x^3 = 1.$$

Taking the cube root of these equations, we obtain

$$x = 2 \qquad \text{or} \qquad x = 1.$$

We leave it for the reader to verify that $x = 1$ and $x = 2$ are the solutions of the original equation. The solution set is $\{1, 2\}$.

EXERCISES

Find the solution set for each equation in Exercises 1 to 44.

1. $\sqrt{x} = 3$

2. $3\sqrt{x} = 5$

3. $\sqrt{x} + 2 = 6$

4. $\sqrt{x} - 3 = 1$

5. $\sqrt{x + 1} = 4$

6. $\sqrt{x + 4} = 3$

7. $\sqrt{2x + 1} = 5$

8. $\sqrt{3x + 4} = 4$

9. $\sqrt{2x + 1} = 2\sqrt{x}$

10. $2\sqrt{x+4} = 3\sqrt{x}$

11. $\sqrt{x+2} + 4 = x$

12. $\sqrt{3x+1} + 3 = x$

13. $\sqrt{6x+7} - 2 = x$

14. $\sqrt{5x-4} + 2 = x$

15. $\sqrt[3]{x-1} = -1$

16. $\sqrt[3]{x-1} = 1$

17. $\sqrt[3]{x^2 - 1} = 2$

18. $(5x+2)^{1/3} = (2x)^{1/3}$

19. $[x(x-1)]^{1/4} = 2^{1/4}$

20. $[x(x-6)]^{1/4} = 2$

21. $\sqrt{x+20} - \sqrt{x+4} = 2$

22. $\sqrt{x+6} - \sqrt{x-6} = 3$

23. $\sqrt{x-4} + \sqrt{x-7} = 3$

24. $\sqrt{5x+6} - \sqrt{x+3} = 3$

25. $\sqrt{x+4} + \sqrt{9-x} = 5$

26. $\sqrt{2x-3} = \sqrt{x+7} - 2$

27. $\sqrt{2x+1} + \sqrt{3x+2} = \sqrt{5x+3}$

28. $\sqrt{x+7} + \sqrt{x+2} = \sqrt{6x+13}$

29. $\sqrt{x+5} + \sqrt{x+12} = \sqrt{2x+41}$

30. $\sqrt{x+5} + \sqrt{x+21} = \sqrt{6x+40}$

31. $x^{2/3} - 7x^{1/3} + 12 = 0$

32. $x^{2/3} + x^{1/3} - 6 = 0$

33. $x^{2/3} - 2x^{1/3} - 8 = 0$

34. $x^{2/3} - 2x^{1/3} - 35 = 0$

35. $2x^{2/3} - x^{1/3} - 6 = 0$

36. $\dfrac{1}{x} + \dfrac{1}{\sqrt{x}} - 6 = 0$

37. $x^4 - 5x^2 + 6 = 0$

38. $2x^4 - 11x^2 + 12 = 0$

39. $4x^4 - 39x^2 + 27 = 0$

40. $7x^{-2} + 19x^{-1} = 6$

41. $x^{-4} - 5x^{-2} + 6 = 0$

42. $x^{-1/3} = x^{-2/3}$

43. $(2x+3)^{-1/3} = (4x+1)^{-1/3}$

44. $2x^{2/3} = x^{1/3}$

2.5

LINEAR INEQUALITIES

Mathematical statements such as

$$ax > b, \qquad ax \geq b, \qquad ax < b, \qquad ax \leq b,$$

where a and b are real numbers and $a \neq 0$, are called *linear inequalities*. A solution for an inequality is a real number that satisfies the inequality, and the solution set is the set of all such real numbers. Thus solving an inequality is a matter of describing the solution set of the inequality as explicitly as possible.

As in the case of linear equations, we solve an inequality by generating a series of equivalent inequalities (inequalities having the same solution set) until we arrive at one for which the solution set is obvious. To solve an inequality, we must know the following proposition. Its proof follows directly from the properties of the real number system.

Proposition 2.5.1

If $P(x)$, $Q(x)$, and $R(x)$ are algebraic expressions, then for all values of x for which $P(x)$, $Q(x)$, and $R(x)$ are real numbers, the statement

$$P(x) < Q(x) \tag{2.12}$$

is equivalent to each of the following statements:

$$P(x) + R(x) < Q(x) + R(x) \tag{2.13}$$

$$P(x) - R(x) < Q(x) - R(x) \tag{2.14}$$

$$P(x) \cdot R(x) < Q(x) \cdot R(x) \qquad (R(x) > 0) \tag{2.15}$$

$$P(x) \cdot R(x) > Q(x) \cdot R(x) \qquad (R(x) < 0). \tag{2.16}$$

The properties of the relation $<$ may be extended to the relation $\leq$. Let us look at some examples.

Example 2.35

Solve the following inequality for x, and represent the solution set graphically:

$$5x + 16 \leq 3x + 25.$$

Solution: Adding $(-3x)$ to both sides of the inequality, we obtain

$$(-3x) + 5x + 16 \leq (-3x) + 3x + 25$$
$$2x + 16 \leq 25.$$

Next we add (-16) to both sides:

$$2x + 16 + (-16) \leq 25 + (-16)$$
$$2x \leq 9.$$

Finally, we multiply both sides by $\frac{1}{2}$, and we have

$$x \leq \tfrac{9}{2}.$$

Thus the solution consists of all real numbers less than or equal to $\frac{9}{2}$. In interval notation, the solution set is $(-\infty, \frac{9}{2}]$. It is represented graphically in Figure 2.1. Note that the real number $\frac{9}{2}$ is included in the solution set, as shown by the solid circle at $\frac{9}{2}$.

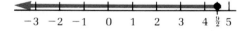

FIGURE 2.1 -3 -2 -1 0 1 2 3 4 $\tfrac{9}{2}$ 5

Example 2.36

Solve the following inequality for x, and represent the solution set graphically:

$$4x - 32 \leq 7x - 17.$$

Solution: Adding 32 to both sides, we obtain

$$4x \leq 7x + 15.$$

Now we add $(-7x)$ to both sides, and we get

$$-3x \leq 15.$$

Finally, we multiply both sides of the inequality by $-\frac{1}{3}$. Property (2.16) implies that multiplication by a negative number reverses the direction of the inequality.

Thus we obtain
$$x \geq -5.$$
In interval notation, the solution set is $[-5, \infty)$, as illustrated in Figure 2.2.

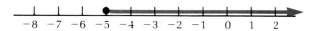

FIGURE 2.2 $-8 \ -7 \ -6 \ -5 \ -4 \ -3 \ -2 \ -1 \ \ 0 \ \ 1 \ \ 2$

Example 2.37   Solve the following inequality for x, and represent the solution set graphically:
$$-11 < 4x - 3 \leq 9.$$

Solution: Adding 3 to both inequalities, we have
$$-11 + 3 < 4x - 3 + 3 \leq 9 + 3$$
$$-8 < 4x \leq 12$$
Multiplying both inequalities by $\frac{1}{4}$, we have
$$-2 < x \leq 3$$
as the solution of the given inequality. In interval notation, the solution set is $(-2, 3]$, as illustrated in Figure 2.3.

FIGURE 2.3 $-3 \ -2 \ -1 \ \ 0 \ \ 1 \ \ 2 \ \ 3 \ \ 4 \ \ 5$

EXERCISES

Solve each of the inequalities in Exercises 1 to 20 for x, and represent the solution, using interval notation.

1. $x + 3 < 5$
2. $x - 2 < 4$
3. $x - 5 > -2$
4. $x + 1 > -3$
5. $2x - 1 < 9$
6. $3x - 2 \leq 4$
7. $4x - 5 \leq 11$
8. $5x - 4 \geq 6$
9. $-3x + 7 < -2$
10. $-4x + 3 \geq 7$
11. $\frac{1}{3}x + 2 \leq 5$
12. $\frac{1}{2}x + 1 \geq 4$
13. $3x - 8 \leq 5(2 - x)$
14. $13 - 5x < 1 + x$
15. $3(1 - 2x) \leq 2(x + 5)$
16. $3(5 - x) \leq 4(2 + x)$
17. $7(2 - 3x) \geq 3x + 8$
18. $\frac{1}{3}(x - 5) \geq \frac{3}{4}(2x + 1)$
19. $\frac{3x - 4}{2} < x + 4$
20. $\frac{8 - x}{2} < \frac{1}{4}x + 7$

Solve each of the inequalities in Exercises 21 to 26 for x, and represent the solution set graphically.

21. $-6 < 3x < 15$
22. $-7 < 2x + 3 \leq 5$
23. $1 \leq 5x - 4 \leq 11$
24. $2 \leq \frac{3x - 2}{2} < 7$
25. $1 \leq \frac{7 - 2x}{3} \leq 3$
26. $-3 \leq \frac{4 - 4x}{3} \leq -1$

2.6

QUADRATIC, POLYNOMIAL, AND RATIONAL INEQUALITIES

An inequality of the form

$$ax^2 + bx + c < 0,$$

where a, b, and c are real numbers and $a \neq 0$, is called a **quadratic inequality.** The relation $<$ can be replaced with $\leq$, $>$, or $\geq$. In this section, we consider the solution of the inequalities that uses a **graphical procedure.**

Example 2.38   Solve the following inequality for x:

$$x(x - 3) > 0.$$

Solution: First we find the values of x that make the left-hand side equal to zero. Clearly, these values are $x = 0$ and $x = 3$. The factor x is positive to the right of zero and negative to the left of zero, as shown in Figure 2.4. Similarly, the factor $x - 3$ is positive if

$$x - 3 > 0$$

$$x > 3$$

and negative if $x < 3$, as illustrated in Figure 2.5.

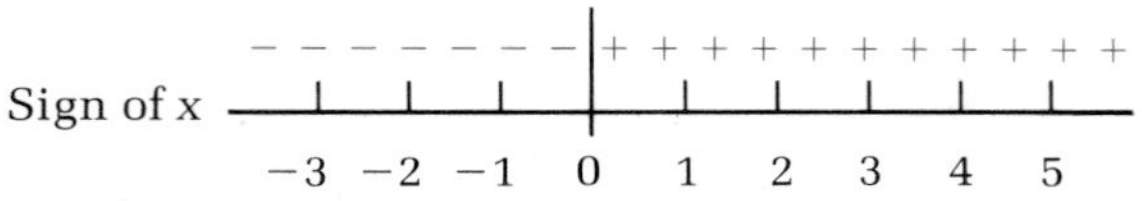

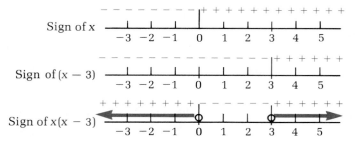

FIGURE 2.4

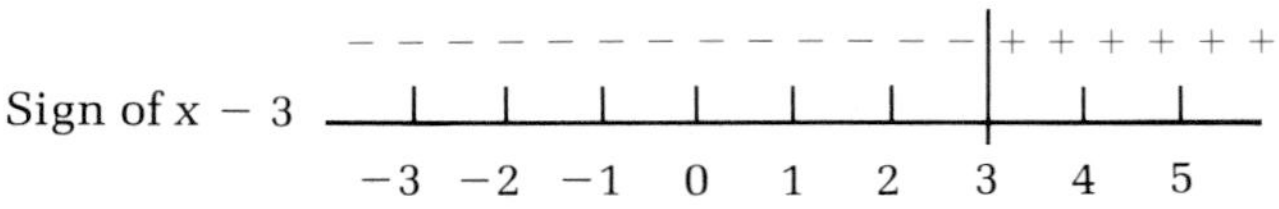

FIGURE 2.5

Now we consider the product $x(x - 3)$ in each of the three regions

$$(-\infty, 0), \qquad (0, 3), \qquad \text{and} \qquad (3, \infty).$$

Observe that in Figure 2.6 both factors are negative in the interval $(-\infty, 0)$; therefore, their product is positive in the interval $(-\infty, 0)$. In the interval $(0, 3)$,

FIGURE 2.6

x is positive while $x - 3$ is negative; therefore, the product $x(x - 3)$ is negative. In the interval $(3, \infty)$, both factors are positive, therefore, their product is positive. Thus, $x(x - 3)$ is positive in the intervals $(-\infty, 0)$ and $(3, \infty)$. Since both intervals constitute the solution set, we may write the solution as the **union*** of the two intervals

$$(-\infty, 0) \cup (3, \infty),$$

as displayed in Figure 2.6. Note that the endpoints 0 and 3 are *not* included in the solution set.

Example 2.39

Solve for x:

$$x^2 - 6x + 8 < 0.$$

Solution: First we factor the left-hand side of the inequality and obtain

$$(x - 2)(x - 4) < 0.$$

Figure 2.7 shows the signs of each factor along with the sign of the product. The factor $(x - 2)$ is positive for $x > 2$ and negative for $x < 2$. Similarly, $(x - 4)$ is positive for $x > 4$ and negative for $x < 4$. The polynomial $x^2 - 6x + 8$ is negative when the product of its factors is negative, that is, for the interval $(2, 4)$ as displayed in Figure 2.7.

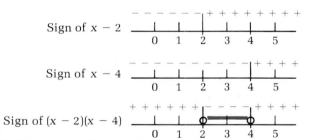

FIGURE 2.7

Example 2.40

Solve the following inequality for x, and represent the solution set, using interval notation:

$$x^2 + 4x - 5 \geq 0.$$

Solution: Factoring the left-hand side of the inequality, we obtain

$$(x + 5)(x - 1) \geq 0.$$

---

* The union of two sets A and B, written $A \cup B$, is defined as

$$A \cup B = \{x \mid x \in A \text{ or } x \in B\}.$$

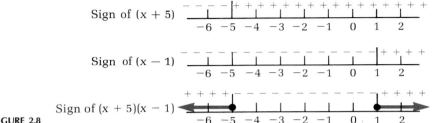

FIGURE 2.8

Figure 2.8 shows the signs of each factor along with the sign of the product. The polynomial $x^2 + 4x - 5$ is greater than or equal to 0 in the intervals $(-\infty, 5]$ and $[1, \infty)$. Since both intervals belong to the solution set, we write the solution as the union of two intervals:

$$(-\infty, -5] \cup [1, \infty),$$

as illustrated in Figure 2.8.

Polynomial Inequalities

We can extend the graphical method to determine the solution of inequalities with more than two factors.

Example 2.41

Solve the following inequality for x, and represent the solution set graphically:

$$x(x - 2)(x - 4) < 0.$$

Solution: Clearly, $x = 0$, $x = 2$, and $x = 4$ are the values that make the left-hand side equal to zero. The factor x is positive to the right of zero and negative to the left of zero. Similarly, the factor $x - 2$ is positive if

$$x - 2 > 0$$

$$x > 2$$

and negative if $x < 2$. The factor $x - 4$ is positive if

$$x - 4 > 0$$

$$x > 4$$

and negative if $x < 4$. The product $x(x - 2)(x - 4)$ is positive or negative depending on the signs of the factors x, $(x - 2)$, and $(x - 4)$. The signs of each factor along with the sign of the product are shown in Figure 2.9. The product of three factors is negative if and only if it has an odd number of negative

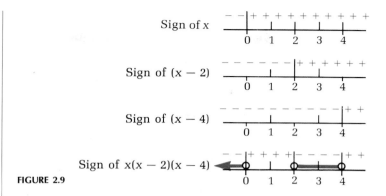

FIGURE 2.9

factors. From Figure 2.9 we observe that $x(x-2)(x-4)$ is negative if and only if $x < 0$ or $2 < x < 4$. In the interval notation, the solution set is

$$(-\infty, 0) \cup (2, 4).$$

Rational Inequalities

We can also use the graphical method to solve inequalities that are quotients of polynomials. Such inequalities are called **rational inequalities.**

Example 2.42　Solve the following inequality for x:

$$\frac{3x - 1}{x + 2} > 1.$$

Solution:　We subtract 1 from both sides of the inequality to obtain zero on one side and then combine the terms on the left into a single fraction. That is,

$$\frac{3x - 1}{x + 2} - 1 > 0$$

$$\frac{(3x - 1) - (x + 2)}{x + 2} > 0$$

$$\frac{2x - 3}{x + 2} > 0.$$

The quotient is positive or negative depending on the signs of $2x - 3$ and $x + 2$. Figure 2.10 shows the signs for $2x - 3$ and $x + 2$ along with the sign of the quotient $(2x - 3)/(x + 2)$. The endpoints -2 and $\frac{3}{2}$ are *not* included in the

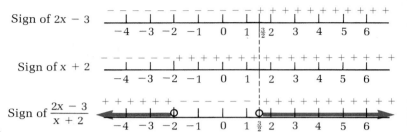

Sign of $2x - 3$

Sign of $x + 2$

Sign of $\dfrac{2x - 3}{x + 2}$

FIGURE 2.10

solution set. We observe that the rational inequality is positive when $x < -2$ or $x > \frac{3}{2}$. In interval notation, the solution set is

$$(-\infty, -2) \cup (\tfrac{3}{2}, \infty)$$

EXERCISES

Solve each inequality in Exercises 1 to 32 for x, and represent the solution set, using interval notation.

1. $x(x - 1) > 0$ 2. $x(2 - x) < 0$ 3. $x(4 - x) > 0$
4. $x(2x - 1) > 0$ 5. $x^2 - 3x + 2 > 0$ 6. $x^2 + 3x - 4 < 0$
7. $x^2 - 5x + 6 < 0$ 8. $x^2 - 5x - 14 \geq 0$ 9. $x^2 - x - 2 \geq 0$
10. $x^2 - 2x - 8 \leq 0$ 11. $4x(1 - x) \leq 1$ 12. $x^2 - 6x + 9 > 9$
13. $x(2x + 3) \leq 5$ 14. $x(x - 2) \leq 24$ 15. $x(x - 1) \leq 6$
16. $x(x + 1) < 2$ 17. $4x^2 > 9x$ 18. $x^2 + 3 \geq 4x$
19. $x^3 - x^2 - 2x < 0$ 20. $x^3 - 9x \leq 0$ 21. $x(x - 1)(x - 4) > 0$

22. $(x - 1)(x - 2)(x - 4) > 0$ 23. $\dfrac{x + 2}{x - 3} < 0$ 24. $\dfrac{2x - 1}{x + 3} > 0$

25. $\dfrac{2x + 1}{2 - x} > 0$ 26. $\dfrac{3 - x}{x + 1} < 0$ 27. $\dfrac{x + 1}{x + 2} > 3$

28. $\dfrac{x + 4}{2x - 1} < 3$ 29. $\dfrac{3}{5 - 3x} < 1$ 30. $\dfrac{13x + 3}{2x + 1} < 6$

31. $\dfrac{x + 1}{x - 2} > 3$ 32. $\dfrac{x + 3}{x - 4} < 1$

33. A ball thrown vertically upward attains a height h (in feet) given by

$$h = 96t - 16t^2$$

   after t seconds. During what period is the ball higher than 80 feet?

34. A projectile is fired vertically upward from ground level. After t seconds, its height h (in feet) is given by

$$h = 320t - 16t^2.$$

   During what time period is the projectile higher than 1200 feet?

2.7 EQUATIONS AND INEQUALITIES INVOLVING ABSOLUTE VALUE

Recall from Chapter 1 that

$$|x| = \begin{cases} x & \text{if } x \geq 0 \\ -x & \text{if } x < 0. \end{cases}$$

The concept of absolute value can be used to measure the distance between any two points, say, A and B, on the real number line.

Definition 2.7.1

If x_1 and x_2 are the coordinates for two points A and B on the real number line, then the distance between A and B is $|x_1 - x_2|$.

Thus if $x_1 = -2$ and $x_2 = 6$ are the coordinates of points A and B, respectively, then the distance between A and B, as illustrated in Figure 2.11, is given by

$$|x_1 - x_2| = |-2 - 6| = |-8| = 8.$$

FIGURE 2.11

Observe that if we had interchanged x_1 and x_2 so that $x_1 = 6$ and $x_2 = -2$, then the result would have been the same. That is,

$$|x_1 - x_2| = |6 - (-2)| = |8| = 8.$$

We now discuss methods of solving equations and inequalities involving absolute value.

Example 2.43

Solve for x:

$$|x - 5| = 4.$$

Solution: Using Definition 1.2.6 of absolute value, we have

$$x - 5 = 4 \qquad \text{or} \qquad x - 5 = -4$$
$$x = 9 \qquad \text{or} \qquad x = 1.$$

The solution set is $\{1, 9\}$. Note that $x = 1$ and $x = 9$ are both 4 units from $x = 5$ on the number line.

Example 2.44   Solve for x:

$$|3x + 1| = |x - 2|.$$

Solution: The equations in absolute value must either be equal to each other or be negatives of each other. Therefore, we have

$$3x + 1 = x - 2 \qquad \text{or} \qquad 3x + 1 = -(x - 2)$$

$$2x = -3 \qquad \text{or} \qquad 3x + 1 = -x + 2$$

$$x = -\tfrac{3}{2} \qquad \text{or} \qquad 4x = 1$$

$$x = \tfrac{1}{4}.$$

The solution set is $\{-\tfrac{3}{2}, \tfrac{1}{4}\}$.

Next we use Definition 1.2.6 to determine real numbers that satisfy an inequality such as $|x| < 3$. Note that if $x \geq 0$, than $|x| = x$ and $x < 3$. Thus the numbers $0 \leq x < 3$ satisfy the inequality. If $x < 0$, then $|x| = -x$ and $-x < 3$. Thus $-3 < x < 0$. Combining these two results, we obtain

$$-3 < x < 3.$$

Thus all real numbers greater than -3 and less than 3 satisfy the inequality $|x| < 3$. This can be interpreted geometrically as shown in Figure 2.12. Note that the numbers -3 and 3 are *not* included in the solution set.

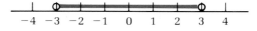

FIGURE 2.12 $-4 \quad -3 \quad -2 \quad -1 \quad 0 \quad 1 \quad 2 \quad 3 \quad 4$

The inequality $|x| \leq 3$, on the other hand, consists of all real numbers between -3 and 3 including the numbers -3 and 3, as shown in Figure 2.13.

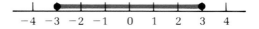

FIGURE 2.13 $-4 \quad -3 \quad -2 \quad -1 \quad 0 \quad 1 \quad 2 \quad 3 \quad 4$

This discussion leads us to the following proposition.

Proposition 2.7.1   Let a be any positive real number. Then the inequality $|x| \leq a$ holds if and only if $-a \leq x \leq a$.

This proposition implies that x is located within a units of the origin 0 (zero) on the number line. Let us now illustrate its use in the following examples.

Example 2.45

Find all real numbers x satisfying the inequality

$$|x - 4| \leq 3.$$

Solution: The inequality $|x - 4| \leq 3$ is satisfied if and only if

$$-3 \leq x - 4 \leq 3.$$

Adding 4 to each term, we get

$$1 \leq x \leq 7.$$

Thus all real numbers between 1 and 7, inclusive, satisfy the given inequality. In interval notation, the solution set is $[1, 7]$.

Example 2.46

Solve for x:

$$|4 - 5x| < 6.$$

Solution: This inequality is equivalent to

$$-6 < 4 - 5x < 6$$

$$-10 < -5x < 2.$$

Multiplying each term by $-\frac{1}{5}$ reverses the direction of each of the inequalities, and we have

$$2 > x > -\tfrac{2}{5},$$

or

$$-\tfrac{2}{5} < x < 2.$$

This implies that all real numbers between $-\frac{2}{5}$ and 2 satisfy the inequality. In interval notation, the solution set is $(-\frac{2}{5}, 2)$.

How do we interpret an inequality like $|x| \geq 4$? Note that if $x \geq 0$, then $|x| = x$ and $x \geq 4$. Thus the numbers for which $x \geq 4$ satisfy the inequality $|x| \geq 4$. If $x < 0$, then $|x| = -x$ and $-x \geq 4$. This means that $x \leq -4$ also satisfies the inequality. Combining these two results gives us

$$x \leq -4 \qquad \text{or} \qquad x \geq 4.$$

This can be interpreted geometrically as shown in Figure 2.14. Geometrically, x is *at least* 4 units away from the origin.

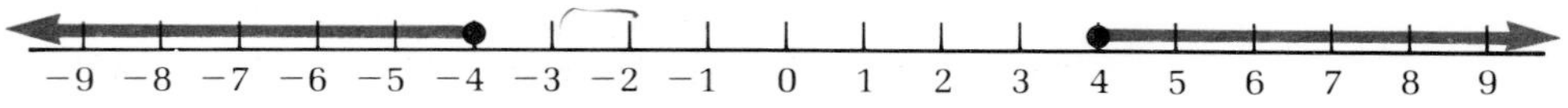

FIGURE 2.14

The inequality $|x| > 4$, on the other hand, consists of all real numbers less than -4 or greater than 4, as shown in Figure 2.15. The endpoints -4 and 4 are not included in the solution set. Intuitively, we can say that x is located *more than* 4 units away from the origin 0 on the number line.

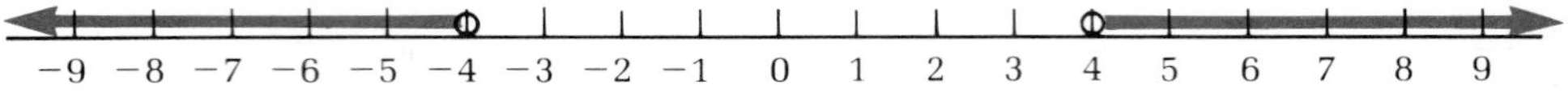

FIGURE 2.15
$$-9 \quad -8 \quad -7 \quad -6 \quad -5 \quad -4 \quad -3 \quad -2 \quad -1 \quad 0 \quad 1 \quad 2 \quad 3 \quad 4 \quad 5 \quad 6 \quad 7 \quad 8 \quad 9$$

This discussion leads us to the following proposition.

Proposition 2.7.2

> Let a be any positive real number. Then the inequality $|x| \geq a$ holds if and only if $x \leq -a$ or $x \geq a$.

Example 2.47

Solve the following inequality for all real numbers x, and represent the solution graphically:

$$|3x - 4| \geq 5.$$

Solution: According to Proposition 2.7.2, the inequality $|3x - 4| \geq 5$ holds if and only if

$$3x - 4 \leq -5 \quad \text{or} \quad 3x - 4 \geq 5.$$

Solving each of these inequalities for x, we have $x \leq -\frac{1}{3}$, or $x \geq 3$, respectively. This may be represented geometrically as shown in Figure 2.16. All real numbers less than or equal to $-\frac{1}{3}$ or greater than or equal to 3 satisfy the given inequality. In interval notation, the solution set is $(-\infty, -\frac{1}{3}] \cup [3, \infty)$.

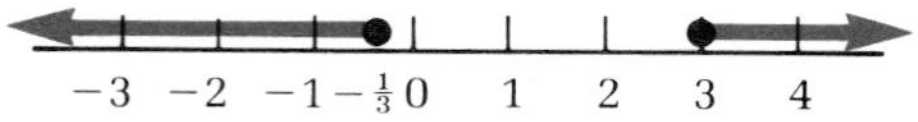

FIGURE 2.16
$$-3 \quad -2 \quad -1-\tfrac{1}{3}\,0 \quad 1 \quad 2 \quad 3 \quad 4$$

EXERCISES

Solve for x in Exercises 1 to 20.

1. $|x - 2| = 3$

2. $|2x - 3| = 4$

3. $|3x - 2| = 5$

4. $|3x - 2| = 6$

5. $|8 - 3x| = 4$

6. $|3x - 8| = 4$

7. $\left|\dfrac{x - 3}{4}\right| = 2$

8. $\left|\dfrac{2x - 1}{4}\right| = 3$

9. $\left|\dfrac{3x - 1}{2}\right| = 4$

10. $\left|\dfrac{4x - 3}{5}\right| = 1$

11. $|2x + 3| = |x|$

12. $|3x + 1| = |x - 1|$

13. $|3x - 2| = |2x + 1|$

14. $|4x - 5| = |3x - 4|$

15. $4|x - 2| = 3|3x - 1|$

16. $|2x - 1| = 2|3x - 5|$ **17.** $\left|\dfrac{2x + 5}{3}\right| = \left|\dfrac{x - 2}{2}\right|$ **18.** $\left|\dfrac{3x + 2}{4}\right| = \left|\dfrac{x}{5}\right|$

19. $\left|\dfrac{3x + 1}{2}\right| = 2|x|$ **20.** $\left|\dfrac{2x + 3}{3}\right| = |x - 1|$

Solve each inequality in Exercises 21 to 34 for x, and represent the solution set, using interval notation.

21. $|x| \le 1$ **22.** $|2x| \le 1$ **23.** $|3x| > 1$ **24.** $|x - \frac{1}{2}| \ge 1$

25. $|x - 2| \le 3$ **26.** $|2x - 3| \le 5$ **27.** $|x - 3| > 2$ **28.** $|3x - 6| > 12$

29. $|2x - 4| > 6$ **30.** $|3x - 4| < 8$ **31.** $\left|\dfrac{x + 2}{3}\right| \le 2$ **32.** $\left|\dfrac{x}{2} - 3\right| < 1$

33. $\left|\dfrac{x}{3} - 2\right| \le 2$ **34.** $\left|\dfrac{x}{4} + 1\right| \ge 3$

KEY WORDS

equation   *linear equation*   *discriminant*
solution   *quadratic equation*   *root of multiplicity 2*
root   *completing the square*   *extraneous solution*
identity   *double root*   *linear inequality*
equivalent equations   *quadratic formula*   *quadratic inequality*

KEY FORMULAS

- If a and b are real numbers and $ab = 0$, then either $a = 0$ or $b = 0$.

- An equation of the form $ax^2 + bx + c = 0$, where a, b, and c are real numbers and $a \ne 0$, has roots given by

$$x = \frac{-b \pm \sqrt{b^2 - 4ac}}{2a}$$

This formula is known as the **quadratic formula.**

- The quantity $b^2 - 4ac$ in the quadratic formula is called the **discriminant.**

EXERCISES FOR REVIEW

Solve each equation in Exercises 1 to 22 for x.

1. $5x - 4 = 3x - 6$

2. $2x - 3(1 - 2x) = 4$

3. $\dfrac{x - 1}{2} + \dfrac{x - 2}{3} = \dfrac{x + 2}{10} + \dfrac{5}{6}$

4. $\dfrac{3}{x - 6} = \dfrac{1}{2(x - 2)}$

5. $\dfrac{2x+2}{3x-5} = \dfrac{4x+1}{6x-7}$

6. $\dfrac{2x}{x-1} = 3 - \dfrac{2}{x-1}$

7. $x^2 - 9x + 20 = 0$

8. $4x^2 - 24x + 11 = 0$

9. $x^2 - 4x + 1 = 0$

10. $x^2 + x + 1 = 0$

11. $x^2 + 2ix - 2 = 0$

12. $\dfrac{1}{x} + \dfrac{1}{x+5} = \dfrac{1}{6}$

13. $\dfrac{1}{x+1} + \dfrac{3}{x-1} = 5$

14. $\dfrac{2}{x^2-9} + \dfrac{4x}{x+3} = 1$

15. $\sqrt{x+5} = x - 1$

16. $\sqrt{x+6} + \sqrt{x+1} = 5$

17. $\sqrt{2x+1} - \sqrt{x+4} = 3$

18. $\sqrt{x+2} = \sqrt[4]{5x+6}$

19. $x^4 + x^2 - 6 = 0$

20. $x^{1/2} - x^{1/4} - 2 = 0$

21. $\left| \dfrac{2x-1}{3} \right| = 2$

22. $|3x+4| = |x-7|$

Find the value of k in Exercises 23 to 26 so that the equations have the given solutions.

23. $2x^2 + kx - 3 = 0$ has one root equal to -3.

24. $3x^2 + kx + 3 = 0$ has equal roots.

25. $kx^2 - 11x - 6 = 0$ has roots whose sum is $\frac{11}{2}$.

26. $kx^2 - 2x - 7 = 0$ has roots whose product is $-\frac{7}{4}$.

Solve each inequality in Exercises 27 to 38 for x.

27. $2x - 3 \geq 4x + 7$

28. $\dfrac{5x-4}{2} < x + \dfrac{11}{4}$

29. $-5 < 2x + 3 < 13$

30. $-3 \leq \dfrac{3-2x}{5} \leq 1$

31. $x^2 \leq 9$

32. $x^2 + 3x - 10 < 0$

33. $x^2 - 5x + 4 \geq 0$

34. $\dfrac{1}{x+1} > \dfrac{2}{x}$

35. $\dfrac{x}{8x-3} > 0$

36. $(x-1)(x-2)(x-3) > 0$

37. $|2x-3| \leq 2$

38. $\left| \dfrac{x}{2} - \dfrac{1}{3} \right| > 1$

39. Ming invested $7500, some at 6 percent and the remainder at 9 percent annual interest. If Ming collected $630 at the end of 1 year, how much money was invested at each rate?

40. Two cars leave New Haven at 6:00 A.M. traveling in opposite directions. The car traveling north has an average speed which is 5 miles per hour faster than the car traveling south. After 3 hours, they are 285 miles apart. How fast is each car traveling?

41. Pipe A can fill a swimming pool in 2 hours, pipe B can fill it in 3 hours, and outlet pipe C can empty the pool in 6 hours. How long will it take to fill the pool if pipe C is left open inadvertently at the moment pipes A and B are opened?

42. Chemical solution A contains 10 percent acid. A second solution B contains 15 percent acid. How much of each solution should be mixed to obtain 10 liters of a solution containing 12 percent acid?

43. In a given right triangle, the sum of the lengths of the sides is 14 centimeters, and the hypotenuse is 10 centimeters. Find the length of the shorter side.

44. A uniform border is added to two adjacent sides of a rug that is 12 feet long and 9 feet wide. The border adds 72 square feet to the area of the rug. Find the width of the border.

45. A plane flies 500 miles with the wind and returns the same distance against the wind. The wind is blowing at 25 miles per hour, and the plane's total travel time is 5 hours 50 minutes. Find the speed of the plane in still air.

46. Pipe A can fill a tank in 2 more hours time than pipe B. Together they can fill the tank in 2 hours 24 minutes. How long would it take each pipe alone to fill the tank?

CHAPTER TEST

Solve the following equations for x:

1. $4(x - 3) = 5(x - 5)$

2. $2x - 3(1 - 2x) = 4$

3. $3x(x + 2) = (3x - 1)(x + 2)$

4. $\dfrac{2(x + 1)}{3x - 5} = \dfrac{4x + 1}{6x - 7}$

Solve for x:

5. $ax + by = ab$

6. $\dfrac{1}{x} + \dfrac{1}{y} = \dfrac{1}{z}$

Solve the following equations for x:

7. $3x^2 + 7x - 1 = 0$

8. $\sqrt{x + 6} - \sqrt{x - 6} = 2$

9. $2x^{2/3} - x^{1/3} = 6$

10. $(x^2 - 1)^{1/3} = 2$

Solve the following inequalities for x, and represent the solution set (a) graphically and (b) by using the interval notation.

11. $4x^2 - 9x > 0$

12. $\dfrac{x + 3}{x - 4} < 1$

Solve the following for x:

13. $4|x - 2| = 3|2x - 1|$

14. $|2x - 3| \le 5$

15. Diane can finish a certain project in 10 days, and Susan can finish the same project in 15 days. How long will it take them to complete the project if they decide to work together?

16. Judy walked from her home to a movie theater at the rate of 4 miles per hour. She ran back home over the same route at the rate of 6 miles per hour. How far was the movie theater from her home if the total time going to and coming from the movie theater was 20 minutes?

17. Chemical solution A contains 10 percent acid. A second solution B contains 15 percent acid. How much of each solution should be mixed to obtain 10 liters of a solution containing 12 percent acid?

18. Sohail drove a distance of 180 miles. If he had traveled 10 miles faster, he would have made the trip in 1 hour 30 minutes less time. What was his speed?

FUNCTIONS

3

INTRODUCTION

Every field of study establishes relationships between quantities that can be measured or observed. In the physical sciences, the relation between the pressure and the volume of a gas at a certain temperature is expressed by Boyle's law. Civil engineers develop formulas that predict the number of inches a beam will bend when subjected to certain weights or loads. In education, researchers study how an IQ measure relates to scholastic achievements in the elementary grades, high school, or college. Economists strive to determine the relationships between the demand and the price of a product, between consumption of a commodity and its supply in the market. In this chapter, we introduce the coordinate system in two dimensions as a useful means of depicting a relationship. Then we concentrate on the essential features of such quantitative relationships that are central to the study of mathematics.

3.1

COORDINATE SYSTEM IN TWO DIMENSIONS

In Section 1.1, we stated that a one-to-one correspondence exists between the real number system and the points on the real number line. We now examine a one-to-one correspondence between the points in a plane and the set of ordered pairs of real numbers. An ordered pair, written as (a, b), is a listing of two real numbers a and b in a certain order. The elements a and b are called the **first** and the **second** components, respectively, in the ordered pair. Two ordered pairs (a, b) and (c, d) are equal if and only if $a = c$ and $b = d$.

There are many uses for ordered pairs. Consider, for example, a table which records the heights and weights of first-year students. For each pair of real numbers, the first and the second numbers represent, the height (in inches) and weight (in pounds), respectively, of a student. Thus the entry (62, 108) means that some student who is 62 inches tall weighs 108 pounds. Note that the entry (108, 62) means that some student who is 108 inches tall weighs 62 pounds—an impossibility. As another illustration, consider an auditor who wants to describe the financial status of an airline by listing a pair of numbers in which the first and second numbers represent the total assets and liabilities, respectively, in millions of dollars. The pair (8, 0) means that the airline is in an excellent financial position, while the pair (0, 8) implies that the airline is on the verge of bankruptcy. Thus the pairs (8, 0) and (0, 8) convey very different information because of the order of the elements.

Remember that the symbols { } and ( ) are distinct notions: one denotes a set, and the other represents an ordered pair. Unfortunately, the notations for an open interval and for an ordered pair are the same, but you will be able to distinguish them by the context.

We choose an arbitrary point 0 in the plane and call it the origin. Through 0, we draw a pair of perpendicular lines, one horizontal, and the other vertical. The horizontal line is called the x axis, and the vertical line is called the y axis. *coordinate axis* The axes together are called the **coordinate axes,** and the plane in which the *coordinate plane* axes lie is called the **coordinate plane.** An arbitrary unit of length is selected,

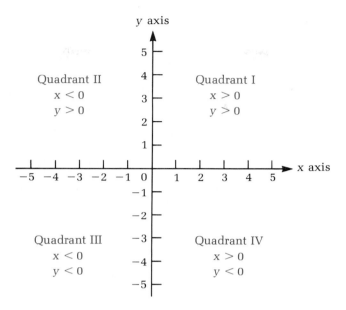

FIGURE 3.1

and a number scale is marked on the x axis, positive to the right and negative to the left of the origin 0. Similarly, a scale is marked along the vertical axis with positive numbers marked upward and negative numbers marked downward. Usually, the units measured along the x axis have the same length as the units along the y axis, as shown in Figure 3.1. However, this is not necessary in every instance; for convenience, different scales may be used on the axes. The coordinate axes divide the plane into four parts, called **quadrants,** numbered counterclockwise as indicated in Figure 3.1.

quadrant

We are now ready to set up a one-to-one correspondence between the set of ordered pairs of real numbers and the points of the plane. Let P be a point in the plane (see Figure 3.2). We draw two lines through P, one perpendicular to the x axis and the other perpendicular to the y axis. The intersection with

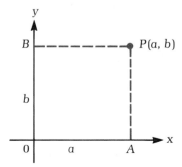

FIGURE 3.2

the horizontal axis is at point A, while the intersection with the vertical axis is at point B. If a is the real number associated with point A on the horizontal axis, then $a > 0$ if A is to the right of the origin and $a < 0$ if A is to the left of the origin. Similarly, if b is the real number associated with point B on the vertical axis, then B is b units above the x axis if $b > 0$ and below the x axis if $b < 0$. The ordered pair associated with point P in the plane is (a, b). This is usually written as $P(a, b)$. The number a is called the x *coordinate*, or **abscissa,** of P, and the number b is called the y coordinate, or **ordinate,** of P. Thus P is a point with coordinates (a, b).

abscissa
ordinate

Figure 3.3 shows the location of points $P(3, 4)$, $Q(-2, 3)$, $R(-4, -3)$, and $S(4, -5)$ in the plane. Since the coordinate axes divide the plane into four quadrants, we can determine the quadrant of a point from the signs of its coordinates (see Figure 3.1). A point with two positive coordinates lies in quadrant I; a point $P(c, d)$ with negative c and positive d lies in quadrant II; a point with both coordinates negative is located in quadrant III; and a point with positive c and negative d is located in quadrant IV. Thus if (c, d) is any ordered pair of real numbers, we start at the origin and go along the x axis to the right if c is positive and to the left if c is negative until we reach $|c|$ units from the origin 0. Then we go $|d|$ units along a line parallel to the y axis upward if d is positive and downward if d is negative. The point so located has coordinates (c, d). We noted earlier that to each point in the plane there corresponds a unique ordered pair of real numbers, and, conversely, corresponding to each ordered pair of real numbers there exists exactly one point in the plane.

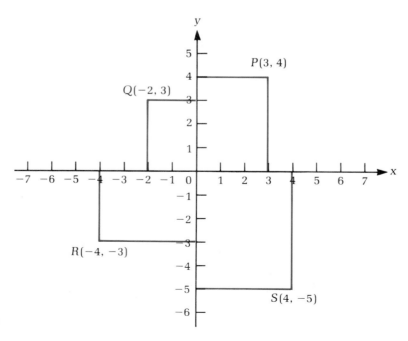

FIGURE 3.3

Let us now use the coordinate system to determine the distance between two points. Let x_1 and x_2 be the x coordinates of two points having the same y coordinate. The horizontal distance between them is, by Definition 2.7.1, $|x_1 - x_2|$. Thus the distance between $A(2, 3)$ and $B(7, 3)$ is $|2 - 7| = |-5| = 5$ units, and the distance between $C(-5, 5)$ and $D(6, 5)$ is $|-5 - 6| = |-11| = 11$ units (see Figure 3.4). Similarly, if y_1 and y_2 are the y coordinates of two points having the same x coordinate, then the distance between them is $|y_1 - y_2|$. Thus the vertical distance between $P(-2, 2)$ and $Q(-2, 6)$ is $|2 - 6| = |-4| = 4$ units, while the distance between $R(4, 4)$ and $S(4, -5)$ is $|4 - (-5)| = |9| = 9$ units. Recall that the distance is always nonnegative.

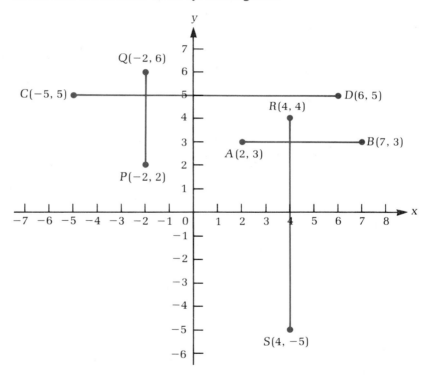

FIGURE 3.4

Let us now extend the concept of distance between points to two points $A(x_1, y_1)$ and $B(x_2, y_2)$ in a plane that have neither the same x coordinate nor the same y coordinate (see Figure 3.5).

Draw a line through A parallel to the x axis and another line through B parallel to the y axis. If N is the point of intersection, then the coordinates of N are (x_2, y_1). Triangle ANB is a right triangle having a right angle at N, and sides AN and NB are of length $|x_2 - x_1|$ and $|y_2 - y_1|$, respectively. If d_{AB} denotes the distance from A to B, then by the Pythagorean theorem, we have

$$d_{AB}^2 = d_{AN}^2 + d_{NB}^2,$$

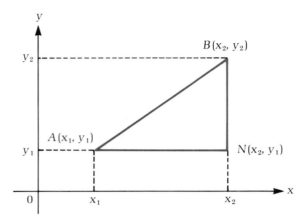

FIGURE 3.5

or

$$d_{AB} = \sqrt{d_{AN}^2 + d_{NB}^2}$$
$$= \sqrt{|x_2 - x_1|^2 + |y_2 - y_1|^2}$$
$$= \sqrt{(x_2 - x_1)^2 + (y_2 - y_1)^2}.$$

Hence we have established the following proposition.

Proposition 3.1.1

The distance from $A(x_1, y_1)$ to $B(x_2, y_2)$ is equal to

$$\sqrt{(x_2 - x_1)^2 + (y_2 - y_1)^2}.$$

It can be shown that this proposition holds for $x_1 = x_2$ or $y_1 = y_2$. We leave this as an exercise for the reader.

Example 3.1

Find the distance between $A(8, 10)$ and $B(20, 15)$.

Solution: Letting $x_1 = 8$, $y_1 = 10$, $x_2 = 20$, and $y_2 = 15$ in Proposition 3.1.1, we have

$$d_{AB} = \sqrt{(20 - 8)^2 + (15 - 10)^2}$$
$$= \sqrt{(12)^2 + (5)^2}$$
$$= \sqrt{144 + 25} = \sqrt{169} = 13.$$

Note that had we interchanged (x_1, y_1) and (x_2, y_2), d_{AB} would have been the same.

Example 3.2   Show that the points $A(-6, 3)$, $B(3, -5)$, and $C(-1, 5)$ are the vertices of a right triangle.

Solution: From the Pythagorean theorem in plane geometry, we know that a triangle is a right triangle if and only if the sum of the squares of two of its sides is equal to the square of the third side. Using the distance formula, we find

$$d_{AB} = \sqrt{[3 - (-6)]^2 + (-5 - 3)^2} = \sqrt{9^2 + (-8)^2}$$
$$= \sqrt{81 + 64} = \sqrt{145},$$

$$d_{BC} = \sqrt{(-1 - 3)^2 + [5 - (-5)]^2} = \sqrt{(-4)^2 + 10^2}$$
$$= \sqrt{16 + 100} = \sqrt{116},$$

and

$$d_{AC} = \sqrt{[-1 - (-6)]^2 + (5 - 3)^2} = \sqrt{(-5)^2 + 2^2}$$
$$= \sqrt{25 + 4} = \sqrt{29}.$$

Since

$$d_{AB}^2 = d_{BC}^2 + d_{AC}^2,$$

we conclude that the triangle is a right triangle.

Now we develop a formula for the coordinates of the **midpoint** of a line segment. Let $A(x_1, y_1)$ and $B(x_2, y_2)$ be two points in a plane, and let $M(x, y)$ be the midpoint of the line segment AB, as shown in Figure 3.6. The lines drawn through A, M, and B parallel to the y axis intersect the x axis at $L(x_1, 0)$, $P(x, 0)$, and $N(x_2, 0)$, respectively. From plane geometry, it follows that a line drawn through M (parallel to the y axis) bisects the line segment LN at point

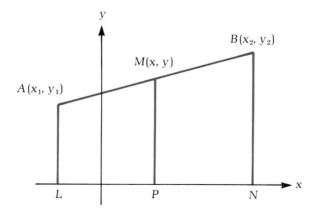

FIGURE 3.6

P. That is,

$$d_{LP} = d_{PN}.$$

Since $x_1 < x$, we have $x - x_1 > 0$ and hence

$$d_{LP} = x - x_1.$$

Similarly, $x < x_2$ implies that $x_2 - x > 0$ and hence

$$d_{PN} = x_2 - x.$$

Because P is the midpoint of the line segment LN and $d_{LP} = d_{PN}$, we have

$$x - x_1 = x_2 - x,$$

or

$$2x = x_1 + x_2$$

$$x = \frac{x_1 + x_2}{2}.$$

In a similar manner, it can be shown that the ordinate of point M is

$$y = \frac{y_1 + y_2}{2}.$$

Hence we have established the following proposition.

Proposition 3.1.2

The coordinates of the midpoint of the line segment from $A(x_1, y_1)$ to $B(x_2, y_2)$ are

$$\left(\frac{x_1 + x_2}{2}, \frac{y_1 + y_2}{2} \right).$$

Example 3.3

Find the coordinates of the midpoint of the line segment from $A(-5, 3)$ to $B(7, 1)$.

Solution: Letting $x_1 = -5$, $y_1 = 3$, $x_2 = 7$, and $y_2 = 1$ in Proposition 3.1.2, we have

$$x = \frac{-5 + 7}{2} = \frac{2}{2} = 1 \qquad \text{and} \qquad y = \frac{3 + 1}{2} = \frac{4}{2} = 2.$$

The coordinates of the midpoint of the line segment AB are $(1, 2)$.

EXERCISES

Graph the points in Exercises 1 to 8.

1. $(3, 5)$ 2. $(-2, 4)$ 3. $(-3, -4)$ 4. $(3, 0)$
5. $(0, 3)$ 6. $(12, 5)$ 7. $(5, 12)$ 8. $(-5, -12)$

Using Figure 3.7, determine the coordinates of the points in Exercises 9 to 14.

9. Point A 10. Point B 11. Point C
12. Point D 13. Point P 14. Point Q

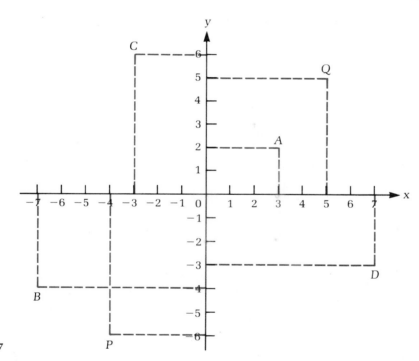

FIGURE 3.7

Determine the distance between each pair of points in Exercises 15 to 20 without using Proposition 3.1.1.

15. $(2, 4)$ and $(8, 4)$ 16. $(3, 5)$ and $(3, -4)$ 17. $(8, 6)$ and $(8, 12)$
18. $(8, 7)$ and $(8, -7)$ 19. $(12, 5)$ and $(-8, 5)$ 20. $(4, -5)$ and $(-10, -5)$

Determine the distance between each pair of points in Exercises 21 to 28.

21. $(0, 0)$ and $(3, 4)$ 22. $(6, 2)$ and $(10, 2)$ 23. $(3, 4)$ and $(9, 12)$
24. $(8, 5)$ and $(8, -10)$ 25. $(-2, 1)$ and $(10, 6)$ 26. $(-8, -9)$ and $(16, -2)$
27. $(5, 6)$ and $(5, 12)$ 28. $(4, -3)$ and $(20, 9)$

29. A shipment of household furniture is to be transported from town A to town B. It can be shipped directly by train at a cost of $1.25 per mile or by moving van via town C at a cost of $1 per mile. Assuming that towns A, B, and C are located as shown in Figure 3.8, determine which mode of transportation is cheaper.

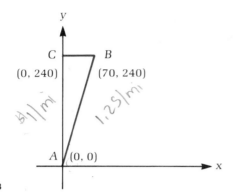

FIGURE 3.8

30. A new automobile is to be shipped to town C from a plant located in town A. It can be shipped directly by freight train at a cost of $1.30 per mile or by truck via town B at a cost of $1.25 per mile. Assuming that the coordinates of A, B, and C are (20, 11), (200, 11), and (200, 30), respectively, determine which mode of transportation is cheaper.

31. Show that the points (12, 8), $(-2, 6)$, and (6, 0) are the vertices of a right triangle. (*Hint:* Use the Pythagorean theorem.)

32. Show that the points (8, 5), $(1, -2)$, and $(-3, 2)$ are the vertices of a right triangle.

Determine the coordinates of the midpoint of the line segment joining each pair of points in Exercises 33 to 36.

33. $A(-8, 6)$ and $B(8, 8)$ **34.** $A(3, 9)$ and $B(9, 3)$

35. $A(-4, -6)$ and $B(-8, -10)$ **36.** $A(-5, -7)$ and $B(11, 13)$

37. Given $A(1, 9)$, find the coordinates of point B such that $M(3, 1)$ is the midpoint of the line segment AB.

38. Given $A(-5, 7)$, find the coordinates of point B such that $M(2, 6)$ is the midpoint of the line segment AB.

3.2

FUNCTIONS AND EQUATIONS

The equation $A = \pi r^2$ expresses the relationship between the area A of a circle and its radius r. This formula or rule, which associates with each positive real number r a unique value of A, is a **function.**

As another illustration, consider the equation $s = 16t^2$ describing the law of a falling body. If an object is released from rest, then 1 second after its release it has traveled 16 feet down from its starting point, 2 seconds after its release it has traveled 64 feet, 3 seconds after its release it has traveled 144 feet, and so on. Thus if we know the time t (in seconds) after the release of an object, the formula $s = 16t^2$ gives the distance s the object has fallen from its starting point. Observe that for *each* value of t, there is one and only one value of s corresponding to that value of t.

Let us now consider an equation $y = 4x + 3$. For each real number x, we obtain one and only one real number y corresponding to that value of x. For example, if $x = \frac{1}{2}$, then $y = 5$; if $x = 1$, then $y = 7$; if $x = \frac{5}{4}$, then $y = 8$; if $x = 2$, then $y = 11$; if $x = 5$, then $y = 23$; if $x = -3$, then $y = -9$; if $x = -7$, then $y = -25$; and so on. Thus the equation $y = 4x + 3$ is a rule which assigns to each value of x one and only one value of y. Essentially, the rule for this equation is to take 4 times the value of x and add 3 to it. From such practical origins comes the abstract mathematical idea of a function, which we define as follows.

Definition 3.2.1

> Let A and B be two nonempty sets. A *function* from A to B is a rule which assigns to each element x in A one and only one element y in B. This element y is called the *image* of x.

domain
range

Figure 3.9 illustrates the concept of a function. The set A is called the **domain** of the function, while the subset of B which contains images of *all* elements of A is called the **range** of the function. We may use any kind of objects to make up the domain and the range of the function. However, most of the functions we encounter in mathematics are real-valued functions of a real variable; that is, the domain and the range of the functions are sets of real numbers. The letter x, which represents an arbitrary number from the domain, is an

independent variable
dependent variable

independent variable. The corresponding letter y representing an element in the range is the **dependent variable,** since its value depends on the number assigned to x. When two variables are related in this manner, we say that y *is a function of* x. Thus the area of a circle is a function of its radius, the distance traveled by an object when released from rest is a function of time, and so on.

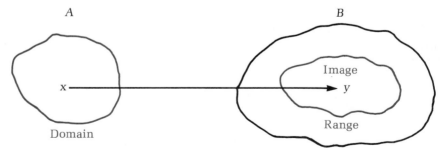

FIGURE 3.9

We denote functions by lowercase letters: $f, g, h, \ldots$. If f is a function, then the number which f assigns to an element x in the domain is denoted by $f(x)$, or simply by y. The symbol $f(x)$ is read "f of x" or "f evaluated at x" or "the value of f at the point x." To appreciate the usefulness of this mathematical symbolism, suppose that we want to describe the demand for a commodity in a symbolic form. If p denotes the price of a commodity and q represents the quantity of that commodity demanded by consumers, then we can express this quantitative relationship by $q = f(p)$. This simply means that the quantity in demand depends, in some way, on the price of the commodity. Likewise, we may write $A = f(r)$, where A denotes the area of a circle and r designates its radius. The symbolic expression for the distance traveled as a function of the time may be written $d = g(t)$.

To specify a function, we must give its domain D_f and the rule f which assigns to each element x from its domain one and only one value $f(x)$, thus determining the range R_f. Generally, the rule of correspondence is expressed in the form of a mathematical equation, as shown in the next few examples.

Example 3.4

Let f be a function defined by

$$f(x) = 3x + 2$$

with the set $\{1, 2, 3, 4\}$ as its domain. Then

$$f(1) = 3(1) + 2 = 5 \qquad f(3) = 3(3) + 2 = 11$$
$$f(2) = 3(2) + 2 = 8 \qquad f(4) = 3(4) + 2 = 14.$$

The range R_f is the set $\{5, 8, 11, 14\}$.

Example 3.5

Let f be a function defined by

$$f(x) = x^2 + 3$$

with the set $\{-2, -1, 0, 1, 2, 3\}$ as its domain. Then

$$f(-2) = (-2)^2 + 3 = 7 \qquad f(1) = 1^2 + 3 = 4$$
$$f(-1) = (-1)^2 + 3 = 4 \qquad f(2) = 2^2 + 3 = 7$$
$$f(0) = 0^2 + 3 = 3 \qquad f(3) = 3^2 + 3 = 12.$$

The range R_f is the set $\{3, 4, 7, 12\}$.

Example 3.6

Let f be a function defined by

$$f(x) = x^3 + 4$$

with the set $\{-3, -2, -1, 0, 1, 2, 3, 4\}$ as its domain. Then

$$f(-3) = (-3)^3 + 4 = -23 \qquad f(1) = 1^3 + 4 = 5$$
$$f(-2) = (-2)^3 + 4 = -4 \qquad f(2) = 2^3 + 4 = 12$$
$$f(-1) = (-1)^3 + 4 = 3 \qquad f(3) = 3^3 + 4 = 31$$
$$f(0) = 0^3 + 4 = 4 \qquad f(4) = 4^3 + 4 = 68.$$

The range R_f is the set $\{-23, -4, 3, 4, 5, 12, 31, 68\}$.

Example 3.7

Let f be a function defined by

$$f(x) = 3x + 1$$

whose domain is the set of real numbers. Then for some arbitrary choices from the domain, we have

$f(-4) = 3(-4) + 1 = -11$	$f(1) = 3(1) + 1 = 4$
$f(-3) = 3(-3) + 1 = -8$	$f(\sqrt{2}) = 3\sqrt{2} + 1$
$f(-1) = 3(-1) + 1 = -2$	$f(2) = 3(2) + 1 = 7$
$f(0) = 3(0) + 1 = 1$	$f(3) = 3(3) + 1 = 10$
$f(\frac{1}{3}) = 3(\frac{1}{3}) + 1 = 2$	$f(\pi) = 3\pi + 1$
$f(\frac{2}{3}) = 3(\frac{2}{3}) + 1 = 3$	$f(5) = 3(5) + 1 = 16$

and so on. Intuitively, the range is the set of all real numbers.

Although the domain is an essential part of the function, it is frequently not mentioned, especially when the rule of correspondence between sets is given by a mathematical equation. **The general convention is that unless the domain is stated explicitly, it is understood to consist of all real numbers for which the expression in x is defined.** The rule f that associates to each element x in the domain one and only one value $f(x)$ determines the range.

Example 3.8

Let f be a function defined by

$$f(x) = \sqrt{9 - x^2}.$$

Observe that this rule of correspondence will yield a real number only if $9 - x^2 \geq 0$. The solution of this inequality shows that the domain D_f is the set of real numbers between -3 and 3, inclusive. The rule f assigns to each x in the domain D_f one and only one real number $\sqrt{9 - x^2}$ in the range. Since $x^2 \geq 0$ for each real number x, it follows that $9 - x^2 \leq 9$ and, consequently, $\sqrt{9 - x^2} \leq 3$. Observe that the range R_f consists of nonnegative real numbers between 0 and 3, inclusive.

It may be instructive to think of a function as a machine that produces a unique output for a given input (see Figure 3.10). If f is a function, then we can interpret the domain of the function as the set of all those elements that can be entered into the machine. If we enter an element x in the machine, it produces $f(x)$ as its output. The range of the machine is the set of all possible output. As an example, consider the function

$$f(x) = 3x^2 + 2$$

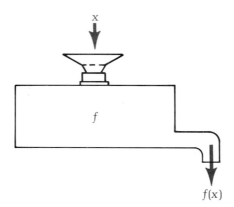

FIGURE 3.10 $f(x)$

with the set of real numbers as its domain. If a real number x is entered, the machine squares its input, multiplies it by 3, and then adds 2 to the number thus obtained. For some choices from the domain, we have

$$f(-4) = 3(-4)^2 + 2 = 50 \qquad f(1) = 3(1)^2 + 2 = 5$$
$$f(-3) = 3(-3)^2 + 2 = 29 \qquad f(\sqrt{2}) = 3(\sqrt{2})^2 + 2 = 8$$
$$f(-2) = 3(-2)^2 + 2 = 14 \qquad f(\sqrt{3}) = 3(\sqrt{3})^2 + 2 = 11$$
$$f(-1) = 3(-1)^2 + 2 = 5 \qquad f(2) = 3(2)^2 + 2 = 14$$
$$f(0) = 3(0)^2 + 2 = 2 \qquad f(\pi) = 3(\pi)^2 + 2.$$

Since x^2 is nonnegative for each real number x, it follows that $3x^2 \geq 0$ and, consequently, $3x^2 + 2 \geq 2$. Hence the range of the function is the set of all real numbers greater than or equal to 2.

Notice that in the notation $f(x)$, the independent variable x is a dummy variable. For example, if $f(x) = 3x^2 + 2$, then $f(a) = 3a^2 + 2$, $f(b) = 3b^2 + 2$, $f(c) = 3c^2 + 2$, $f(2d) = 3(2d)^2 + 2 = 12d^2 + 2$, $f(h + 1) = 3(h + 1)^2 + 2$, and so on. Thus, no matter what is put in the machine, that quantity is squared, the result is multiplied by 3, and 2 is added. It is sometimes helpful to think of this function as

$$f(\ \ ) = 3(\ \ )^2 + 2,$$

where any symbol representing a number in the domain of the function may be placed in both the parentheses.

Example 3.9 | Let f be a function defined by

$$f(x) = \sqrt{x} + 4.$$

Determine

(a) $f(\pi)$ **(b)** $f(9)$
(c) $f(a+b)$ **(d)** $f(a)+f(b)$

Solution:

(a) $f(\pi) = \sqrt{\pi} + 4$
(b) $f(9) = \sqrt{9} + 4 = 3 + 4 = 7$
(c) $f(a+b) = \sqrt{a+b} + 4$
(d) $f(a) + f(b) = \sqrt{a} + 4 + \sqrt{b} + 4 = \sqrt{a} + \sqrt{b} + 8.$

Keep in mind that in general

$$f(a+b) \neq f(a) + f(b).$$

Example 3.10

Let f be a function defined by the equation

$$f(x) = x^2 + 3x + 4.$$

If a and h are real numbers and $h \neq 0$, find

$$\frac{f(a+h) - f(a)}{h}.$$

Solution:

$$f(a+h) = (a+h)^2 + 3(a+h) + 4$$
$$= (a^2 + 2ah + h^2) + (3a + 3h) + 4$$
$$f(a) = a^2 + 3a + 4$$
$$f(a+h) - f(a) = (a^2 + 2ah + h^2) + (3a + 3h) + 4 - (a^2 + 3a + 4)$$
$$= 2ah + 3h + h^2$$

Thus

$$\frac{f(a+h) - f(a)}{h} = \frac{2ah + 3h + h^2}{h} = 2a + 3 + h \qquad (h \neq 0).$$

Functions whose domains are restricted occur frequently in business and economics. Consider, for instance, Example 3.11.

Example 3.11

The market demand function of a product is given by the equation

$$f(p) = 200 - 10p,$$

where $f(p)$ denotes the quantity demanded per year in thousands of units and p denotes the price per unit. Since p is an element of D_f and the corresponding element of the range is denoted by $f(p)$, we have for some choices of p

$$f(0) = 200 \qquad f(5.5) = 145$$
$$f(1) = 190 \qquad \cdots$$
$$f(1.5) = 185 \qquad f(10.2) = 98$$
$$f(2) = 180 \qquad f(12) = 80$$
$$f(3) = 170 \qquad f(19.5) = 5$$
$$f(4) = 160 \qquad f(20) = 0.$$

Since the demand is never negative, it follows that $f(p)$ must be nonnegative. Because the price per unit is charged in dollars and cents, the price p may assume the values in the set $\{0, 0.01, 0.02, \ldots, 19.98, 19.99, 20.00\}$. If $p > 20$, then $10p > 200$, and, consequently, $200 - 10p$ is a negative number. Thus the domain D_f of the function f consists of two-decimal-place numbers between 0 and 20, inclusive. It can be shown that the range R_f is the set $\{0, 0.1, 0.2, \ldots, 199.8, 199.9, 200\}$.

Observe that for each value of p in the set D_f in Example 3.11, there is one and only one value $f(p)$ in the set R_f; and, conversely, for each value $f(p)$ in the set R_f, there is one and only one element in the domain D_f. Thus the relationship between the elements of these two sets constitutes what is commonly known as a **one-to-one function.** Other examples of one-to-one functions are the relationships between husbands and wives in a monogamous society, employees of a company and their social security numbers, college students and their IDs, the 50 states of the United States and their governors, and the set of all real numbers and the points on the line. Before we get carried away, let us pause for a formal definition of a one-to-one function.

one-to-one function

Definition 3.2.2

Let A and B be two nonempty sets. A function from A to B is called a one-to-one function if for each $f(x)$ in the range R_f, there is one and only one value of x in the domain D_f.

In other words, whenever $f(a) = f(b)$ for two elements a and b in the domain of f, then $a = b$.

Example 3.12 | If $f(x) = 5x + 4$, where x is real, show that f is one-to-one.

Solution: Suppose that a and b are two real numbers in the domain of the function f. Then $5a + 4$ and $5b + 4$ are the corresponding images in the range. If $f(a) = f(b)$, then

$$5a + 4 = 5b + 4$$

$$5a = 5b$$

$$a = b.$$

Therefore f is one-to-one.

Let us now give an example of a function which is not one-to-one.

Example 3.13

If $f(x) = 3x^2 + 2$, where x is real, show that f is not one-to-one.

Solution: Let a and $(-a)$ be two nonzero real numbers in the domain of f. Then

$$f(a) = 3a^2 + 2$$

and

$$f(-a) = 3(-a)^2 + 2 = 3a^2 + 2.$$

Since **different** real numbers, a and $(-a)$, in D_f have the **same** image, namely, $3a^2 + 2$, we recognize that f is not a one-to-one function.

To give a concrete example of a function which is not one-to-one, consider the set of 30 students who have taken a final examination in a chemistry course. After the tests have been graded, the instructor assigns to each test a letter grade of A, B, C, D, or F. For each test there corresponds one and only one letter grade, so the relationship is a function. The domain is the set of 30 tests, and the range consists of only those letter grades that have actually been assigned. The function is not one-to-one because **more than one test** corresponds to the **same** letter grade in the range.

As an illustration of a real-valued function which is not one-to-one, consider a travel agency in Boston which advertises an all-expense-paid 5-day cruise to Nova Scotia for special groups. The agency makes reservations for 50 couples on the boat. The charge is $600 per couple, with an additional charge of $20 per couple for each subsequent cancellation. If x is the number of cancellations received prior to the departure of the boat, then $(50 - x)$ couples plan to be on the boat, and each pays $600 plus $20x$. Thus the total receipts of the travel agency are given by the function

$$f(x) = \$(50 - x)(600 + 20x),$$

which on simplification yields

$$f(x) = \$20(1500 + 20x - x^2).$$

The domain D_f is the set

$$\{0, 1, 2, \ldots, 49, 50\}.$$

Let us now compute $f(x)$ for some values of $x \in D_f$. For example,

$$f(0) = \$30,000 \qquad f(15) = \$31,500$$
$$f(5) = \$31,500 \qquad f(20) = \$30,000$$
$$f(10) = \$32,000 \qquad f(25) = \$27,500$$

and so on. Again, we can verify that for each value x in the domain, there is a unique value $f(x)$, and the definition of the function is satisfied. However, two values of x in the domain, for example, $x = 0$ and $x = 20$, correspond to the same value $f(x) = \$30,000$. Similarly, $x = 5$ and $x = 15$ in the domain correspond to the same value $f(x) = \$31,500$ in the range. Thus we have some elements in the range of the function which correspond to more than one element in the domain. This shows that the function under consideration is not one-to-one.

So far, we have considered a function as a rule of correspondence that pairs with each element of its domain one and only one element of its range. This shows that **a function is simply a set of ordered pairs such that no two ordered pairs have the same first element.** The following is therefore an **alternate** definition.

Definition 3.2.3

> Let A and B be two nonempty sets. A function from A to B is a set of ordered pairs (x, y), where $x \in A$ and $y \in B$, such that every element of A appears as a first element of exactly one ordered pair.

Example 3.14

The set of ordered pairs

$$\{(1, 5), (2, 7), (3, 9), (4, 11), (5, 13), (6, 15)\}$$

is a function having the set $A = \{1, 2, 3, 4, 5, 6\}$ as its domain and the set $B = \{5, 7, 9, 11, 13, 15\}$ as its range. Figure 3.11 describes this function. Another way to describe this function is

$$\{(x, f(x)) \mid f(x) = 2x + 3, x \in A\}$$

where $A = \{1, 2, 3, 4, 5, 6\}$. Note that this function is one-to-one.

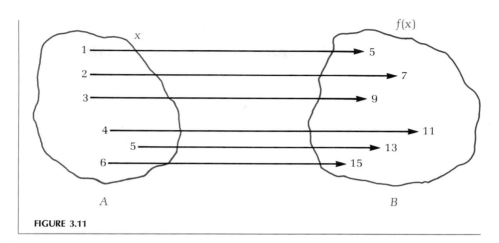

FIGURE 3.11

Example 3.15

The set of ordered pairs

$$\{(-3, 19), (-1, 3), (1, 3), (3, 19)\}$$

is a function with the set $A = \{-3, -1, 1, 3\}$ as its domain and the set $B = \{3, 19\}$ as its range. The rule of correspondence that pairs elements of the domain with the elements of the range is pictured in Figure 3.12. In a mathematical equation, we may express this function as

$$\{(x, f(x)) \mid f(x) = 2x^2 + 1, x \in A\}.$$

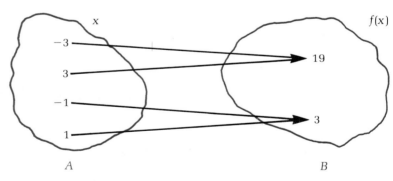

FIGURE 3.12

Note that $f(-3) = f(3) = 19$, but $3 \neq -3$; hence f is not a one-to-one function.

Example 3.16

The set of ordered pairs

$$\{(2, 4), (3, 5), (4, 6), (3, 7)\}$$

is *not* a function because an element 3 in the domain is paired with two elements 5 and 7 in the range, as illustrated in Figure 3.13. This violates Definition 3.2.3.

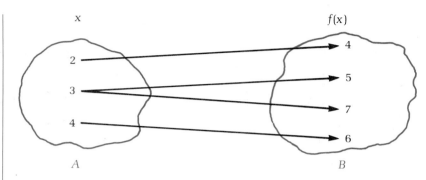

FIGURE 3.13

Notice that in order for a set or ordered pairs to be a function from set *A* to set *B*, each element of *A* must appear as a first element in one and only one of the ordered pairs.

EXERCISES

Decide whether the situations described in Exercises 1 to 8 constitute functions.
1. The correspondence between the state colleges in California and their presidents
2. The correspondence between each book in a library and the number of pages in the book
3. The correspondence between the states and their representatives in the House of Representatives
4. The correspondence between a driver and her or his driving license
5. The correspondence between the products sold in a department store and their prices
6. The correspondence between a human being and body weight
7. The correspondence between the set of ordered pairs of real numbers and the points in the plane
8. The correspondence between the residents of Philadelphia and their telephone numbers
9. State which of the functions in Exercises 1 to 8 are one-to-one.

Determine whether the rules of correspondence from set *A* to set *B* in Exercises 10 to 13 are functions.

10.   *A*   *B*   **11.**   *A*   *B*

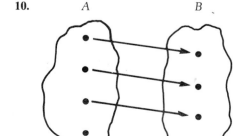   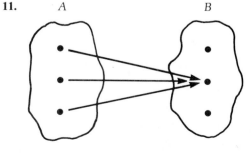

12. A   B   **13.** A   B

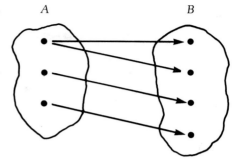

Determine whether the sets of ordered pairs in Exercises 14 to 21 are functions.

14. $\{(1, 1), (2, 2), (3, 3), (4, 4), (5, 5), (6, 6)\}$ **15.** $\{(-1, 1), (1, 1), (-2, 4), (2, 4), (3, 9)\}$

16. $\{(1, -1), (1, 1), (4, -2), (4, 2), (9, 3)\}$ **17.** $\{(1, 1), (2, 1), (3, 1), (4, 1), (5, 1), (6, 1)\}$

18. $\{(1, 1), (2, 8), (3, 27), (4, 64)\}$ **19.** $\{(1, 1), (8, 2), (27, 3), (64, 4)\}$

20. $\{(2, 3), (3, 4), (4, 5), (5, 6)\}$ **21.** $\{(4, 5), (5, 6), (5, 7), (5, 8), (5, 9)\}$

22. State which of the functions in Exercises 14 to 21 are one-to-one.

23. Given that

$$f(x) = 2x^2 + 3,$$

determine the following:

(a) $f(0)$ **(b)** $f(-1)$ **(c)** $f(1)$ **(d)** $f(2)$ **(e)** $f(3)$ **(f)** $f(4)$

24. Given that

$$f(x) = 3x^2 + 2x + 4,$$

determine the following:

(a) $f(-1)$ **(b)** $f(1)$ **(c)** $f(2)$ **(d)** $f(\frac{1}{2})$ **(e)** $f(-\frac{1}{2})$ **(f)** $f(\sqrt{2})$

25. Given that

$$g(x) = \frac{2x + 1}{3x - 1}$$

determine the following:

(a) $g(2)$ **(b)** $g(3)$ **(c)** $g(-3)$ **(d)** $g(a)$ **(e)** $g(a + h)$ **(f)** $g(a + h) - g(a)$

26. Given that

$$g(x) = 3x + 5,$$

determine the following:

(a) $g(a)$ **(b)** $g(a + h)$ **(c)** $g(a) + g(h)$ **(d)** $g\left(\dfrac{1}{a + h}\right)$ **(e)** $g\left(\dfrac{1}{a}\right)$ **(f)** $g\left(\dfrac{1}{a}\right) + g\left(\dfrac{1}{h}\right)$

27. Given that

$$f(x) = x^2 + x + 2,$$

determine the following:

(a) $f(-a)$ **(b)** $f(\sqrt{a})$ **(c)** $f(a)$ **(d)** $f(a + h)$ **(e)** $f(a) + f(h)$ **(f)** $\dfrac{f(a + h) - f(a)}{h}$

28. Given that

$$f(x) = x^3$$

determine and simplify the following:

(a) $f(a + h)$ (b) $f(a + h) - f(a)$ (c) $f(a) + f(h)$ (d) $\dfrac{f(a + h) - f(a)}{h}$

29. The research department of ABC Company determines that the demand for their product is given by

$$d(p) = 8000 - 40p, \qquad 100 \leq p \leq 200,$$

where p is the price in dollars per unit. Determine the demand $d(p)$ for the following values of p:

(a) $p = 100$ (b) $p = 120$ (c) $p = 150$ (d) $p = 170$ (e) $p = 190$ (f) $p = 200$

30. The consumption function for the heating oil in an apartment complex is found to be

$$f(t) = 25,000 - 500t,$$

where $f(t)$ is the number of gallons remaining in the tank t days after the delivery of oil. Determine the oil level in the tank after

(a) 10 days   (b) 15 days   (c) 18 days   (d) 25 days   (e) 45 days   (f) 50 days.

Assuming that the domain is the set $\{-2, -1, 0, 1, 2\}$, determine the range for the functions in Exercises 31 to 36.

31. $f(x) = 2x + 1$ **32.** $f(x) = x^2$ **33.** $f(x) = |x|$

34. $f(x) = x^3$ **35.** $f(x) = \dfrac{1}{x + 3}$ **36.** $f(x) = \sqrt{4 - x^2}$

In Exercises 37 to 42, determine the largest subset of real numbers that can serve as the domain of the function.

37. $f(x) = 3x$ **38.** $f(x) = x^2 + 1$ **39.** $f(x) = |x| + 2$

40. $f(x) = \dfrac{1}{x - 2}$ **41.** $f(x) = \dfrac{1}{x^2 - 9}$ **42.** $f(x) = \sqrt{x}$

43. The market demand function of a product is given by

$$d(p) = 10,000 - 400p,$$

where $d(p)$ denotes the quantity demanded and p denotes the price per unit. Determine the domain and the range of d.

44. A car-leasing agency owns a fleet of 120 cars. Experience has shown that all the cars are rented at a rate of $30 per day per car but for each $1.00 increase in rent, 3 fewer cars are rented. Express the revenue function $R(x)$ as a function of x, where x is the additional cost of renting the car above $30.

45. A resort hotel will provide dinner, dancing, and drinks for a local organization. The charge is $40 per couple if 100 couples attend; then it decreases by $0.20 for each additional couple above the minimum of 100. Express the total revenue function $R(x)$ as a function of x, the number of additional couples above the minimum.

46. A travel agency advertises an all-expense-paid tour to Washington, D.C., for college students. The agency charters one train car with a seating capacity of 50. The fare is $100 per student plus an additional charge of $5 for each empty seat. Express the total revenue function $R(x)$ as a function of x, the number of empty seats on the train.

47. A restaurant owner plans to build a new restaurant having a seating capacity between 90 and 125. Experience has shown that the profit averages $12 per week per chair if the seating capacity is 90; it decreases by $0.10 per chair for each additional seat in excess of 90. Express the profit function $P(x)$ as a function of x, the number of additional seats above the minimum of 90.

48. A farmer has 3600 feet of barbed wire to enclose a rectangular pasture. One side of the plot is along the bank of the river and needs no fence. Find the area of the pasture $A(x)$, where x is the width of the pasture (in feet).

49. A farmer has 3000 feet of fence to enclose a pasture. Her plan includes fencing the entire area and then subdividing it by running a fence across the width as shown in Figure 3.14. Find the area of the pasture $A(x)$, where x is the width of the pasture (in feet).

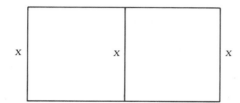

FIGURE 3.14

50. An open box is to be made from a rectangular sheet of tin by cutting out equal squares from each corner and folding up the sides. If the sheet is 21 inches long and 16 inches wide, express the volume $V(x)$ of the box as a function of x, where x is the length of the square to be cut from each corner of the sheet, as shown in Figure 3.15.

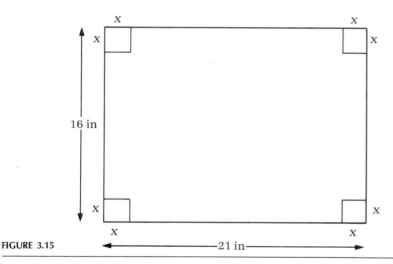

FIGURE 3.15

3.3

GRAPH OF A FUNCTION

In this section, we consider certain elementary functions and their graphs. Assuming that a function is given by a mathematical equation, we can learn the simultaneous behavior of x and y by plotting a set of points (x, y), where x represents an arbitrary value selected from the domain of the function and

y is the corresponding value determined by the rule of correspondence. *We follow the convention that if the domain consists of only a few points, the points plotted in the plane should represent the entire graph. In the case where the domain consists of real numbers in an interval, the graph will consist of a solid curve drawn through the plotted points. The graph will be drawn as a dotted curve if the domain of the function consists of a finite number of points or of rational numbers within a specified interval.*

Example 3.17

Consider the function

$$f(x) = 2x + 3$$

having as a domain the set of integers between -2 and 2, inclusive. Below is a set of ordered pairs given in a table, the first row containing elements of the domain and the second row containing the corresponding elements of the range.

x:	-2	-1	0	1	2
$f(x)$:	-1	1	3	5	7

Plotting the ordered pairs $(-2, -1)$, $(-1, 1)$, $(0, 3)$, $(1, 5)$, and $(2, 7)$, we have the graph in Figure 3.16. Had the domain instead consisted of all real numbers, then the graph would consist of the solid line drawn through the plotted points, as shown in Figure 3.17.

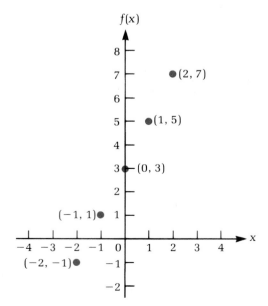

FIGURE 3.16

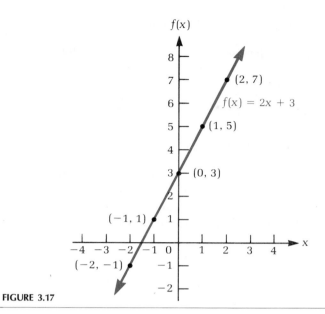

$f(x)$

$(2, 7)$

$f(x) = 2x + 3$

$(1, 5)$

$(0, 3)$

$(-1, 1)$

$(-2, -1)$

x

FIGURE 3.17

Example 3.18

The demand for a certain product is given by

$$f(x) = 100 - 20x,$$

where $f(x)$ represents the demand of the product at a unit price of x dollars. Graph $f(x)$.

Solution: Since the price per unit is charged in dollars and cents, the variable x may assume the values 0, 0.01, 0.02, 0.03, . . . , 4.98, 4.99, 5.00. If $x > 5$, then $20x > 100$, and, consequently, $100 - 20x$ is a negative number. Thus the domain of the function f consists of only two-decimal-place numbers between 0 and 5, inclusive. The values of $f(x)$ for some values of x are given in the following table.

x	$f(x) = 100 - 20x$
0	100
1	80
2	60
3	40
4	20
5	0

The related set of ordered pairs (0, 100), (1, 80), (2, 60), (3, 40), (4, 20), and (5, 0) determines a set of points that are plotted in Figure 3.18. The dotted line represents the corresponding graph of $f(x)$.

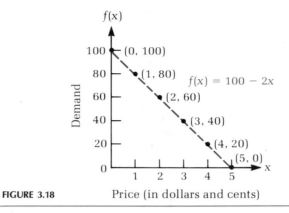

FIGURE 3.18   Price (in dollars and cents)

Example 3.19 | Plot the graph of $f(x) = |x|$.

Solution: The domain D_f of the function f is the set of all real numbers, while R_f is the set of nonnegative real numbers. (Why?) The values of $f(x)$ for some arbitrary values of x are given in the table below.

x	$f(x) = |x|$
-3	3
-2	2
-1	1
0	0
1	1
2	2
3	3

Plotting the ordered pairs $(-3, 3)$, $(-2, 2)$, ..., $(3, 3)$, we have the graph in Figure 3.19.

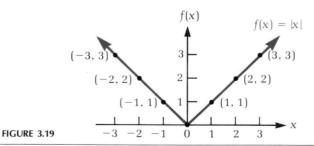

FIGURE 3.19

Example 3.20

Plot the graph of $f(x) = \sqrt{x}$.

Solution: The domain D_f of the function f is the set of nonnegative real numbers, since this rule of correspondence does not yield a real number for $x < 0$. The values of $f(x)$ for some values of x are given in the table below:

x:	0	1	2	3	4	5
$f(x)$:	0	1	$\sqrt{2}$	$\sqrt{3}$	2	$\sqrt{5}$

When we plot the ordered pairs $(0, 0), (1, 1), \ldots, (5, \sqrt{5})$, we get the graph shown in Figure 3.20. Since the graph lies above the x axis, we conclude that the range of the function f is the set of nonnegative real numbers.

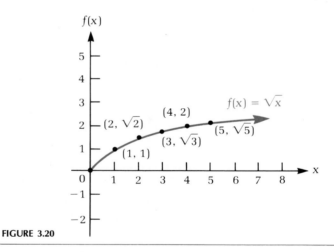

FIGURE 3.20

Example 3.21

Plot the graph of $f(x) = \sqrt{9 - x^2}$.

Solution: Recall from Example 3.8 that the domain of the function f is the set of real numbers between -3 and 3, inclusive, while the range R_f is the set of real numbers between 0 and 3, inclusive. Thus the graph of f must be in the first two quadrants. The following table lists some points $(x, f(x))$ on the graph shown in Figure 3.21.

x:	-3	-2	-1	0	1	2	3
$f(x)$:	0	$\sqrt{5}$	$\sqrt{8}$	3	$\sqrt{8}$	$\sqrt{5}$	0

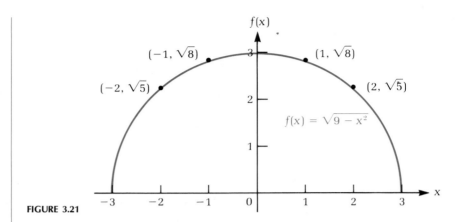

FIGURE 3.21

Observe that the graph of the function f is an upper semicircle.

greatest-integer function   So far, we have considered functions which are relatively easy to graph and have somewhat familiar forms. Consider next the **greatest-integer function**

$$f(x) = [\![x]\!],$$

where $[\![x]\!]$ denotes the greatest integer not exceeding x. This means that

$$[\![\tfrac{1}{3}]\!] = 0, \qquad [\![\tfrac{4}{3}]\!] = 1, \qquad [\![2.4]\!] = 2, \qquad [\![3]\!] = 3, \qquad [\![-\tfrac{1}{3}]\!] = -1$$

$$[\![-\tfrac{5}{2}]\!] = -3, \qquad [\![-3.4]\!] = -4, \qquad [\![-4.5]\!] = -5, \qquad [\![-6.7]\!] = -7,$$

and so on. The graph of this function for $-4 \le x \le 4$ is shown in Figure 3.22.

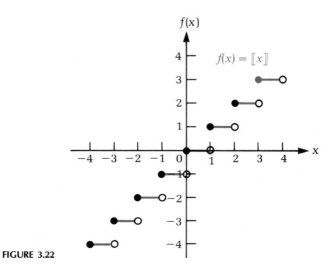

FIGURE 3.22

Note that the open circle placed at the right end of each line segment means that the right endpoint of that segment is not a part of the graph. The graph consists of short intervals of horizontal lines. Note that the function has a constant value in each interval and that each such interval is closed on the left and open on the right.

As a final illustration, consider a graph where $f(x)$ remains constant for some values of x.

Example 3.22

The management of a parking lot charges its customers daily according to the formula

$$f(x) = \begin{cases} \$1.50 + \$0.50 \, (x-2) & (2 \le x \le 8) \\ \$4.50 & (8 < x \le 24), \end{cases}$$

where x is the number of hours or part thereof that the car is parked and $f(x)$ is the amount assessed for parking the car for x hours. There is a minimum charge of \$1.50 and a maximum charge of \$4.50.

(a) Determine the cost for parking up to 2, 3, 4, . . . , 24 hours.

(b) Graph $f(x)$.

(c) Find from the graph the costs for parking for $3\frac{1}{2}$ and $4\frac{3}{4}$ hours.

Solution:

(a) Apparently, the management charges \$1.50 for the first 2-hour period and \$0.50 for each additional hour or fraction thereof. The parking schedule is given in the table.

Time	Parking Cost $f(x)$
First 2 hours	\$1.50
Third hour	2.00
Fourth hour	2.50
Fifth hour	3.00
Sixth hour	3.50
Seventh hour	4.00
Eighth through twenty-fourth hour	4.50

(b) The graph of $f(x)$ is given in Figure 3.23.

(c) The parking for $3\frac{1}{2}$ hours costs \$2.50 while the cost for 4 hours 45 minutes parking is \$3.

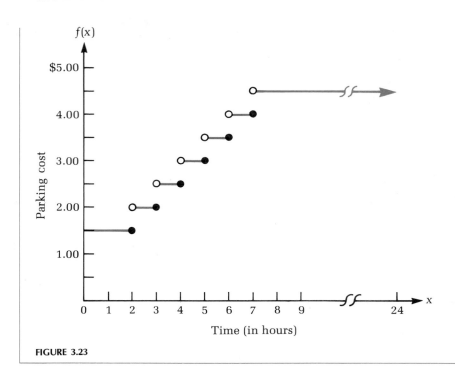

FIGURE 3.23

EXERCISES

Graph the function in Exercises 1 to 12 by plotting points.

1. $f(x) = -x$
2. $f(x) = 4x + 5$
3. $f(x) = \sqrt{x - 1}$
4. $f(x) = \sqrt{4 - x}$
5. $f(x) = x^2$
6. $f(x) = 6$
7. $f(x) = \sqrt{16 - x^2}$
8. $f(x) = \sqrt{4 - x^2}$
9. $f(x) = x^3$
10. $f(x) = |2 - x|$
11. $f(x) = \dfrac{1}{x}, \; x \neq 0$
12. $f(x) = \dfrac{|x|}{x}, \; x \neq 0$

13. The market demand function for a product is given by

$$d(p) = 480 - 40p,$$

where $d(p)$ denotes the demand of the product at price p (in dollars) per unit.
(a) Determine the demand for $p = 1, 2, 3, 4, 5, 6, 8, 10,$ and 12.
(b) Graph $d(p)$ for the same price range.

14. The market supply function for beer in New York is given by

$$s(p) = 200 + 40p,$$

where $s(p)$ denotes the quantity of beer supplied in thousands of gallons per year at price p (in dollars) per gallon.

(a) Determine the supply for beer for $p = 1, 2, 3, 4, 5, 6, 7, 8$.

(b) Graph $s(p)$ for the same price range.

15. The demand for a certain product is given by

$$g(p) = \frac{100}{p}, \qquad p > 0,$$

where $g(p)$ is the demand in units at price p.

(a) Determine the demand for $p = 2, 5, 10, 20$, and 50.

(b) Graph $g(p)$ for the same price range.

16. In an experiment on motivation, J. S. Brown trained a group of rats to run down a narrow passage in a cage to receive food in a box. He then connected the rats, using a harness, to an overhead wire that was attached to a spring scale. A rat was placed at distance d (in centimeters) from the food box, and the pull g (in grams) of the rat toward the food was measured. Brown found that the relationship between these variables could be approximated by the function

$$g(d) = -\frac{d}{5} + 70, \qquad 30 \le d \le 175.$$

(a) Determine the pull g for $d = 30, 50, 100, 125, 150$, and 175.

(b) Graph $g(d)$ for the same values of d.

17. The manager of a chain of bookstores has determined that the supply $s(p)$ of certain paperbacks is given approximately by

$$s(p) = 30 - \frac{60}{p - 1},$$

where $s(p)$ denotes the supply at the price p (in dollars) per carton of books.

(a) Determine the supply $s(p)$ for $p = 3, 4, 5, 6, 7, 11, 16, 21$, and 31.

(b) Graph $s(p)$ by plotting the corresponding points obtained in (a) and joining them by a dotted curve.

18. The market demand function for unskilled labor in the United States is approximated by

$$f(x) = \frac{400}{x^2},$$

where $f(x)$ represents thousands of labor hours per month and x denotes the wage rate per hour in dollars.

(a) Determine the demand for $x = 1, 2, 4, 5, 10$, and 20.

(b) Graph $f(x)$ by plotting the corresponding points obtained in (a) and joining them by a dotted curve.

19. A travel agency advertises an all-expense-paid 5-day cruise to Nova Scotia for certain special groups. The agency makes reservations for 50 couples on the boat. The charge is $600 per couple with an additional charge of $20 for each subsequent cancellation. The total receipts of the travel agency are given by

$$f(x) = 20(1500 + 20x - x^2), \qquad x = 0, 1, 2, \ldots, 50,$$

where x is the number of cancellations received prior to the departure of the boat.

(a) Determine $f(x)$ for $x = 0, 5, 10, 15, 20, 25, 30, 35, 40, 45$, and 50.

(b) Graph $f(x)$ by plotting the corresponding points obtained in (a) and joining them by a dotted curve.

20. A charter flight charges a fare of $200 per person plus $5 for each unsold seat on the plane. The plane has a seating capacity of 100 passengers.

(a) Determine the revenue function $R(x)$ as a function of x, the number of empty seats on the plane.

(b)   Determine $R(x)$ for $x = 0, 10, 20, 30, 40, 50, 60, 70,$ and 80.

(c)   Graph $R(x)$ by plotting the corresponding points obtained in (b) and joining them by a dotted curve.

21.   A direct-dial call from New Haven to Miami during a weekday from 8.00 A.M. to 5.00 P.M. costs $0.64 for the first minute and $0.44 for each additional minute or fraction thereof. What will it cost Mike if he talks for

(a)   10 minutes?   **(b)**   30 minutes?   **(c)**   45 minutes?

22.   An attendant in a parking lot charges $3 for the first hour and $0.75 for each additional half-hour or fraction thereof. What will it cost Mrs. Nowlan if she parks her car for

(a)   0.5 hour?   **(b)**   2 hours?   **(c)**   3 hours?

(d)   3 hours 45 minutes?   **(e)**   4 hours 15 minutes?   **(f)**   4.5 hours?

23.   A cab charges $0.50 for the first quarter mile of transportation and $0.20 for each additional quarter mile or fraction thereof. What will a cab cost a passenger who needs to travel

(a)   0.5 mile?   **(b)**   1 mile?   **(c)**   1.5 miles?

(d)   2 miles?   **(e)**   2.5 miles?   **(f)**   3 miles?

3.4

LINEAR FUNCTIONS

Definition 3.4.1

Any function of the form $(x, f(x))$, where $f(x) = mx + b$, is called a *linear* function.

The word *linear* means that the graph of this function is a straight line. Conversely, the equation of any nonvertical straight line can be expressed in the form $f(x) = mx + b$.

The coefficient of x determines the *slope* of the line. This concept is important because it measures the rate at which changes are taking place—the rate at which a car is going uphill, the rate at which enrollment is dropping in liberal arts colleges, the rate at which the economy is recovering from a recession, and so on. Consider for example, the sales for three companies A, B, and C, and compare the rate at which their volumes of sales have increased or decreased over the period from 1981 to 1984. The following table shows the sales volumes of these companies in 1981 and 1984.

Company	Sales in 1981	Sales in 1984
A	$145,000	$163,000
B	$135,000	$141,000
C	$140,000	$122,000

Assuming that the sales volume for each of these companies follows a straight-line trend, we plot two points corresponding to each company and then connect them by means of a straight line, as displayed in Figure 3.24. Observe that two points are sufficient to graph a straight line.

The sales for company A have increased from $145,000 in 1981 to $163,000 in 1984. Thus the average increase in sales is $18,000/3 = $6000. The sales for

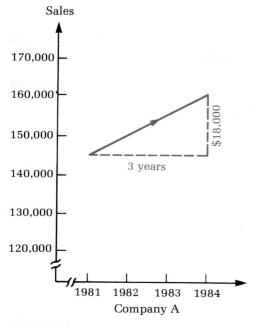

FIGURE 3.24(a)

FIGURE 3.24(b)

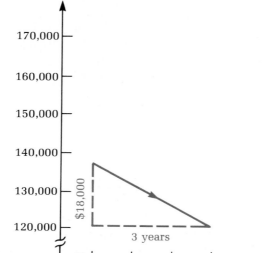

FIGURE 3.24(c)

company B have increased by $6000 over this 3-year period; thus, the average annual increase in sales is $2000. This accounts for the fact that the slope of the linear function for company B [see Figure 3.24(b)] is not as steep as that of the linear function for company A [see Figure 3.24(a)]. Nevertheless, the upward trend of companies A and B reflects their growth. The sales for company C, on the other hand, have decreased from $140,000 in 1981 to $122,000 in 1984. This is reflected by the line segment for company C, which sinks downward, thus showing a drop in sales during the 3-year period [see Figure 3.24(c)].

Formally, we define the slope of a line as follows.

Definition 3.4.2

Let $\mathscr{L}$ be a nonvertical line joining two distinct points A and B with coordinates (x_1, y_1) and (x_2, y_2), respectively. Then the *slope m* of the line $\mathscr{L}$ is defined by

$$m = \frac{y_2 - y_1}{x_2 - x_1}, \qquad x_1 \neq x_2. \tag{3.1}$$

Note that had we used (x_2, y_2) and (x_1, y_1) as coordinates of A and B, respectively, the result would have been the same. That is,

$$\frac{y_1 - y_2}{x_1 - x_2} = \frac{y_2 - y_1}{x_2 - x_1}.$$

If a line rises sharply to the right as illustrated in Figure 3.24(a), the difference in the y coordinates of two points is large in comparison to the difference in corresponding x coordinates. This situation yields a large value for the slope m. When a line sinks downward from left to right [see Figure 3.24(c)], the differences in the x and y coordinates of any two points on this line have opposite signs; consequently, the slope of such a line is negative. If a line is parallel to the x axis (see Figure 3.25), then all points on this horizontal line have the same y coordinates, that is, $y_1 = y_2$. This means that the difference in y coordinates of all points on this line segment is zero; consequently, the slope m is zero. A vertical line, on the other hand, does not have any slope. This is clear from the fact that the x coordinates of any two points

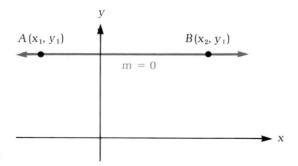

FIGURE 3.25

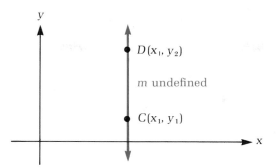

FIGURE 3.26

on a vertical line are the same (see Figure 3.26), that is, $x_1 = x_2$. Hence the denominator in $(y_2 - y_1)/(x_2 - x_1)$ is zero, and the slope m of this vertical line is not defined. We wish to emphasize that a horizontal line has slope zero, while a vertical line has undefined slope.

Example 3.23

Find the slope of the line through A and B for the following pairs of points:

(a) $A(3, 5)$, $B(5, 11)$

(b) $A(8, -10)$, $B(3, -2)$

(c) $A(6, 8)$, $B(-15, 8)$

Solution:

(a) Let $x_1 = 3$, $y_1 = 5$; $x_2 = 5$, $y_2 = 11$. Then

$$m = \frac{y_2 - y_1}{x_2 - x_1} = \frac{11 - 5}{5 - 3} = \frac{6}{2} = 3.$$

Note that had we used $(5, 11)$ as (x_1, y_1) and $(3, 5)$ as (x_2, y_2), the resulting slope m would have been the same.

(b) Let $x_1 = 8$, $y_1 = -10$; $x_2 = 3$, $y_2 = -2$. Then

$$m = \frac{y_2 - y_1}{x_2 - x_1} = \frac{-2 - (-10)}{3 - 8} = -\frac{8}{5}.$$

(c) Let $x_1 = 6$, $y_1 = 8$; $x_2 = -15$, $y_2 = 8$. Then

$$m = \frac{y_2 - y_1}{x_2 - x_1} = \frac{8 - 8}{-15 - 6} = -\frac{0}{21} = 0.$$

The slope of a line does not depend on what pair of points on the line is used to calculate it. Let A, B, and C be any three points on a nonvertical straight line (see Figure 3.27). For convenience, we assume the slope to be positive; an analogous argument for negative slope also holds.

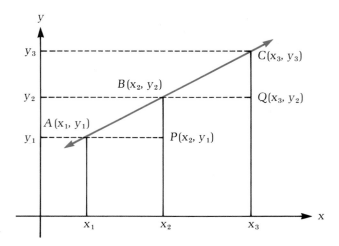

FIGURE 3.27

Observe that the right triangles APB and BQC are similar. (Why?) Thus the lengths of the corresponding sides have the same ratio. Hence

$$\frac{d_{PB}}{d_{AP}} = \frac{d_{QC}}{d_{BQ}}.$$

Now

$$d_{PB} = y_2 - y_1, \qquad d_{AP} = x_2 - x_1$$
$$d_{QC} = y_3 - y_2, \qquad d_{BQ} = x_3 - x_2.$$

Thus

$$\frac{y_2 - y_1}{x_2 - x_1} = \frac{y_3 - y_2}{x_3 - x_2}.$$

But the left- and right-hand sides of the equation are the slopes of AB and BC, respectively. Therefore the slopes of AB and BC are equal. Further, the slope of a line is independent of the pair of points chosen.

Equation of a Straight Line

Finding the slope of a line is just one phase of determining the equation of the line. Now we see how to determine that equation. We begin with lines that are parallel to one of the axes.

Proposition 3.4.1

The equation of a straight line parallel to the x axis and $|b|$ units from it is $y = b$. If b is positive, the line is above the x axis; if b is negative, the line is below the x axis.

This means that the equation $y = 2$ represents a line parallel to the x axis and 2 units above it. Similarly, the equation $y = -2$ is a line also parallel to the x axis but 2 units below it (see Figure 3.28).

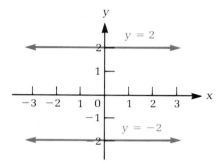

FIGURE 3.28

Lines parallel to the y axis are treated in a similar manner. The equation $x = 2$ represents a line parallel to the y axis and 2 units to the right of it. Similarly, the equation $x = -2$ is a line parallel to the y axis but 2 units to the left of it (see Figure 3.29).

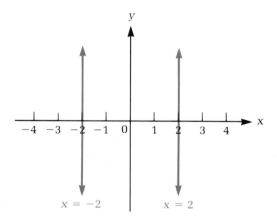

FIGURE 3.29

In general, we have the following proposition.

Proposition 3.4.2

The equation of a straight line parallel to the y axis and $|a|$ units from it is $x = a$. If a is positive, the line is to the right of the y axis; if a is negative, the line is to the left of the y axis.

Next we determine the equation of a nonvertical line. This line meets the y axis at some point where the x coordinate is zero. The y coordinate of this point is called the **y intercept** of the line and is generally denoted by b. Figure 3.30 illustrates that the coordinates of the y intercept of AB, namely point C, are $(0, b)$. If $D(x, y)$ is any arbitrary point on this line, then the slope m is obtained by applying Definition 3.4.1 to points $C(0, b)$ and $D(x, y)$.

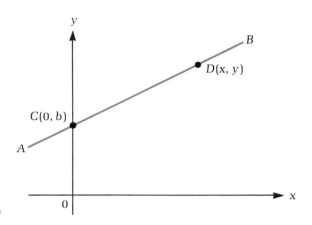

FIGURE 3.30

Thus

$$m = \frac{y - b}{x - 0};$$

that is,

$$y - b = mx$$

or

$$y = mx + b.$$

Conversely, suppose that we are given any point (x_1, y_1) satisfying the equation $y = mx + b$. Observe that $(0, b)$ also satisfies the equation and that the slope of the line connecting $(0, b)$ and (x_1, y_1) is

$$\frac{y_1 - b}{x_1 - 0} = \frac{(mx_1 + b) - b}{x_1 - 0} = \frac{mx_1}{x_1} = m \qquad (x_1 \neq 0).$$

Because any other point (x_2, y_2) satisfying $y = mx + b$ will also lie on the line of slope m through $(0, b)$, we conclude that *all* points satisfying $y = mx + b$ lie on the straight line with slope m and y intercept b. Thus we have established the following proposition.

Proposition 3.4.3

Let $\mathscr{L}$ be the line passing through $(0, b)$ and having slope m. Then a point (x, y) lies on $\mathscr{L}$ if and only if it satisfies the equation

$$y = mx + b. \qquad (3.2)$$

Note that any nonvertical line has a slope m and a y intercept b. Thus the equation of the line passing through $(0, 3)$ with slope 4 (that is, $m = 4$ and $b = 3$) is given by

$$y = 4x + 3.$$

slope intercept Since m and b define this equation, (3.2) is called the **slope-intercept** form of the equation of a straight line. Observe that if the nonvertical line passes through the origin, then $b = 0$, and Equation (3.2) becomes $y = mx$.

Example 3.24

Determine the slope and the y intercept of

$$3y = 2x + 6.$$

Solution: By rewriting this equation in the slope-intercept form, (3.2) gives us

$$y = \frac{2}{3}x + 2.$$

Comparing this equation with (3.2), we see that the slope is $\frac{2}{3}$ and the y intercept is $(0, 2)$. In other words, $m = \frac{2}{3}$ and $b = 2$.

Proposition 3.4.4

Two distinct (nonvertical) lines are parallel if and only if their slopes are equal.

Example 3.25

Determine whether the following lines are parallel:
(a) $4x + 5y - 8 = 0$ and $3x - 2y + 12 = 0$
(b) $2x - 3y + 4 = 0$ and $6x - 9y + 25 = 0$

Solution:

(a) Rewriting $4x + 5y - 8 = 0$ and $3x - 2y + 12 = 0$ in the slope-intercept form (3.2), we have

$$y = -\frac{4}{5}x + \frac{8}{5} \quad \text{and} \quad y = \frac{3}{2}x + 6,$$

respectively. These lines have slopes $-\frac{4}{5}$ and $\frac{3}{2}$, respectively. Because these slopes are not equal, the lines are **not** parallel.

(b) Expressing $2x - 3y + 4 = 0$ in the form (3.2), we get

$$y = \frac{2}{3}x + \frac{4}{3}$$

which implies that slope of this line is $\frac{2}{3}$. Similarly, solving $6x - 9y + 25 = 0$ for y, we have

$$y = \frac{6}{9}x + \frac{25}{9},$$

and this implies that the slope of this line is $\frac{6}{9} = \frac{2}{3}$. Because the slopes of these lines are equal, we conclude that the lines are parallel.

Next, we state (without proof) the following proposition.

Proposition 3.4.5

Two distinct (nonvertical) lines are perpendicular to each other if and only if the product of their slopes equals (-1).

Example 3.26

Determine whether the following lines are perpendicular:

(a) $4x + 3y - 6 = 0$ and $3x - 4y + 12 = 0$
(b) $3x + 2y + 8 = 0$ and $5x - 3y + 10 = 0$

Solution:

(a) Solving $4x + 3y - 6 = 0$ for y, we get

$$y = -\frac{4}{3}x + 2.$$

Thus the line has slope $m_1 = -\frac{4}{3}$. Similarly, the line $3x - 4y + 12 = 0$ can be written as

$$y = \frac{3}{4}x + 3$$

so that this line has slope $m_2 = \frac{3}{4}$. Because $m_1 m_2 = (-\frac{4}{3})(\frac{3}{4}) = -1$, the lines are perpendicular to each other.

(b) The lines $3x + 2y + 8 = 0$ and $5x - 3y + 10 = 0$, expressed in the slope-intercept form (3.2), are

$$y = -\frac{3}{2}x - 4 \qquad \text{and} \qquad y = \frac{5}{3}x + \frac{10}{3}.$$

The slopes of the lines are $m_1 = -\frac{3}{2}$ and $m_2 = \frac{5}{3}$, respectively. Since $m_1 m_2 \neq -1$, we conclude that these lines are not perpendicular to each other.

Suppose that the slope m of a straight line and a point (x_1, y_1) on the line are known. If (x, y) is any other point on the line, then using (3.1), we have

$$\frac{y - y_1}{x - x_1} = m \qquad (x \neq x_1),$$

and

$$y - y_1 = m(x - x_1).$$

Thus we have established the following proposition.

Proposition 3.4.6

The equation of a straight line passing through (x_1, y_1) and having slope m is

$$y - y_1 = m(x - x_1). \tag{3.3}$$

point slope

Since equation (3.3) uses a given point (x_1, y_1) and the slope m of the line, it is called the **point-slope** form of the equation of the line.

Example 3.27

Find the equation of the line which passes through the point $(3, 5)$ and has slope $m = 2$.

Solution: Note that

$$x_1 = 3, \qquad y_1 = 5, \qquad \text{and} \qquad m = 2.$$

We use (3.3) to get the required equation of the line,

$$y - 5 = 2(x - 3).$$

This simplifies to

$$y = 2x - 1.$$

Example 3.28

Find the equation of the line passing through $A(5, 8)$ and $B(3, 2)$.

Solution: The slope of the line joining two points $A(5, 8)$ and $B(3, 2)$ is

$$m = \frac{2 - 8}{3 - 5} = \frac{-6}{-2} = 3.$$

Choosing one of the points as (x_1, y_1), say, $(5, 8)$, we use (3.3) and obtain

$$y - 8 = 3(x - 5),$$

which simplifies to

$$y = 3x - 7.$$

Note that had we used (3, 2) instead as (x_1, y_1), the resulting equation would have been the same.

We have considered equations of straight lines in all possible positions. We have shown that a line either is parallel to the x axis, is parallel to the y axis, or meets the axes at fixed points. In the first case, its equation is of the form $y = b$ (Proposition 3.4.1); in the second case, its equation is of the form $x = a$ (Proposition 3.4.2); and in the third case, it is represented by $y = mx + b$, where m and b are fixed real numbers (Proposition 3.4.3). All these equations can be rewritten in the general form

$$Ax + By + C = 0$$

where A, B, and C are any real numbers and A and B cannot both be zero. Note that the first equation is in y alone, the second is in x alone, and the third contains both x and y.

Next we reverse the procedure and start out with an equation $Ax + By + C = 0$ and show that every equation of this form represents a straight line.

Proposition 3.4.7

Every equation of the form

$$Ax + By + C = 0 \qquad (A \neq 0 \quad \text{or} \quad B \neq 0), \tag{3.4}$$

where A, B, and C have fixed values, represents a straight line.
    To prove this assertion, we consider two possible cases.

Case A

Suppose $B \neq 0$. Solving (3.4) for y, we have

$$y = -\frac{A}{B}x - \frac{C}{B},$$

which is an equation of a straight line in slope-intercept form with $m = -A/B$ and $b = -C/B$. If $A = 0$, then

$$By + C = 0 \qquad \text{or} \qquad y = -\frac{C}{B} \qquad (B \neq 0),$$

and this represents an equation of the straight line parallel to the x axis.

Case B

Suppose $B = 0$ and $A \neq 0$. Then

$$Ax + C = 0 \qquad \text{or} \qquad x = -\frac{C}{A},$$

which represents an equation of the straight line parallel to the y axis. Hence $Ax + By + C = 0$ represents a straight line as long as $A \neq 0$ or $B \neq 0$.

EXERCISES

Find the slope of the line passing through each pair of points A and B in Exercises 1 to 8.
1. $A(3, 4)$ and $B(4, 3)$
2. $A(-2, 2)$ and $B(5, 5)$
3. $A(-5, 2)$ and $B(5, 6)$
4. $A(6, 2)$ and $B(6, 6)$
5. $A(4, 5)$ and $B(8, 13)$
6. $A(3, 2)$ and $B(6, 8)$
7. $A(8, 4)$ and $B(10, 4)$
8. $A(9, 1)$ and $B(6, 5)$

Find the equation of each of the lines in Exercises 9 to 12.
9. Parallel to the x axis and 5 units above it
10. Parallel to the x axis and 3 units below it
11. Parallel to the y axis and 2 units to the right of it
12. Parallel to the y axis and 1 unit to the left of it

Find the slope of the lines in Exercises 13 to 18.
13. $x = 5$
14. $y = 2x + 7$
15. $x - 2y = 4$
16. $y = 3$
17. $y = 3x + 7$
18. $4x - 5y = 0$

Find the slope and the y intercept of the lines in Exercises 19 to 24.
19. $2x - 3y + 6 = 0$
20. $3x + 4y + 12 = 0$
21. $5x + 6y = 30$
22. $2x - 3y = 0$
23. $4x + y = 0$
24. $x + 2y + 4 = 0$

Determine whether the pairs of lines in Exercises 25 to 32 are parallel, perpendicular, or neither.
25. $x + 2y - 4 = 0$ and $3x + 6y + 13 = 0$
26. $2x - 3y - 7 = 0$ and $4x - 5y + 8 = 0$
27. $2x + 3y - 6 = 0$ and $3x - 2y + 12 = 0$
28. $3x + y + 6 = 0$ and $6x - 2y + 3 = 0$
29. $4x - 5y + 20 = 0$ and $5x + 4y - 30 = 0$
30. $3x - 4y + 12 = 0$ and $4x + 3y - 24 = 0$
31. $5x + 6y - 30 = 0$ and $6x + 5y + 42 = 0$
32. $2x - 5y + 25 = 0$ and $4x + 3y - 26 = 0$

In Exercises 33 to 36, find the slope-intercept form of the equation of the line which passes through the given point with the given slope.
33. $A(0, 3)$ and $m = 4$
34. $A(0, -3)$ and $m = -\frac{3}{2}$
35. $A(3, 4)$ and $m = 0$
36. $A(1, 2)$ and $m = -\frac{1}{3}$

In Exercises 37 to 40, find the slope-intercept form of the equation which has the given slope and given y intercept.
37. $m = -4$ and y intercept 5
38. $m = \frac{3}{4}$ and y intercept 3
39. $m = 2$ and y intercept -4
40. $m = -5$ and y intercept -3

In Exercises 41 to 44, find the slope-intercept form of the equation which passes through the given point.
41. $P(4, 5)$ and $m = 1$
42. $P(5, -2)$ and $m = \frac{2}{3}$
43. $P(4, 5)$ and parallel to the line $2x - 3y - 3 = 0$
44. $P(2, 3)$ and perpendicular to the line $x - 2y + 4 = 0$

Find the equation of the line which passes through each pair of points A and B in Exercises 45 to 48.
45. $A(5, 0)$ and $B(4, -1)$
46. $A(0, 3)$ and $B(4, 5)$
47. $A(4, -4)$ and $B(6, 5)$
48. $A(-3, 6)$ and $B(-5, 10)$
49. Find the equation of the line passing through each pair of points in Exercises 1 to 8.

50. If a line has nonzero x and y intercepts a and b, respectively, show that its equation can be expressed in the form

$$\frac{x}{a} + \frac{y}{b} = 1. \tag{3.5}$$

In Exercises 51 to 56, use Equation (3.5) in Exercise 50 to find the equation of the line whose intercepts on the x axis and y axis are, respectively,

51. 3 and 4   **52.** 4 and 5   **53.** -1 and 2

54. -2 and -3 **55.** $\frac{1}{2}$ and $\frac{1}{3}$ **56.** $\frac{2}{3}$ and $-\frac{3}{4}$

57. Find the equation of the line which passes through the point $A(2, 3)$ and has equal nonzero x and y intercepts.

APPLICATIONS

Linear functions play an important role in the process called cost analysis used by business firms to determine the cost of production. A **cost function** can often be expressed as a linear equation of the form

$$C(x) = mx + b.$$

Here b is the fixed cost and generally includes rent, salaries, interest, insurance, advertising, and equipment costs; m refers to the costs that can be charged directly to the product the firm produces. These production-related costs, referred to as *variable* costs, include the cost of raw material, fuel, supplies, packaging, freight, and direct labor. Consider, for instance, that the cost of making x executive chairs in a furniture company is approximated by the linear equation

$$C(x) = 100x + 6000.$$

Note that $6000 is the overhead cost of the plant and $100 represents the cost of raw material, labor, and so on, used in the production of each chair. Thus the total cost of producing 30 chairs is

$$C(30) = (100)(30) + 6000 = \$9000$$

whereas the cost of producing 50 chairs is

$$C(50) = (100)(50) + 6000 = \$11,000.$$

As another illustration, consider a car rental company at an international airport that leases automobiles and charges $29.95 a day plus $0.15 per mile. Clearly, $29.95 is a fixed charge, and $0.15 per mile is a variable cost related directly to the number of miles the car is driven. If x represents the distance traveled in a day, then the cost function may be expressed as

$$C(x) = 0.15x + 29.95.$$

Thus if a customer drives 100 miles in a day, he is charged

$$C(100) = (0.15)(100) + 29.95 = \$44.95;$$

but if he travels 200 miles in a day, then he must pay

$$C(200) = (0.15)(200) + 29.95 = \$59.95.$$

Depreciation of apartment buildings, office equipment, furniture, machines, and other business assets is another illustration of the application of a linear function. The loss in value of such items is generally computed by using the **linear depreciation method** approved by the Internal Revenue Service. As the name suggests, the method assumes that business property with a life span of n years loses $1/n$ of its original cost every year. Thus if C is the original cost of an asset depreciated linearly over n years, then the total value after x years is

linear depreciation method

$$y = C - \left(\frac{C}{n}\right) \cdot x$$

$$= C\left(1 - \frac{x}{n}\right) \qquad (0 \le x \le n). \qquad \textbf{(3.6)}$$

Example 3.29

A commercial building worth $100,000 in 1983 is assumed to depreciate linearly over its 20-year life span. Determine the value in

(a) 1990

(b) 1997.

Solution: Here,

$$C = \$100,000, \qquad n = 20 \text{ years}.$$

The value of the property in 1990 is obtained by letting $x = 7$ in (3.6). Thus

$$y = 100,000\left(1 - \frac{7}{20}\right) = 65,000.$$

Similarly, the value of the building in 1997 is obtained by substituting $x = 14$ in (3.6). This yields

$$y = 100,000\left(1 - \frac{14}{20}\right) = 30,000.$$

Note that linear depreciation does not necessarily reflect the actual rate at which business and other assets depreciate in value. Some assets, such as automobiles, lose money more rapidly in the first few years than in the latter part of their life. For this reason, some nonlinear methods are also used for depreciating property.

Another application of linear functions involves determining **trend lines,** which are useful in forecasting sales, inventory, and population growth.

Example 3.30

The sales of a certain department store follow an approximate straight-line trend. The sales were $45,000 in 1978 and $75,000 in 1984. Determine the equation of the straight line representing this sales pattern. Assuming that this trend continues, estimate the sales in 1987.

Solution: Let x represent time in years, with $x = 0$ representing 1978. Then $x = 6$ represents the year 1984. Since sales are linearly related to time, we need to determine the equation of the straight line passing through $(0, 45,000)$ and $(6, 75,000)$. The slope of this line is

$$m = \frac{75,000 - 45,000}{6 - 0} = \frac{30,000}{6} = 5000.$$

Thus the equation of the straight line is

$$y = 5000x + 45,000,$$

where y denotes the sales in year x.

To find the estimated sales in 1987, we substitute $x = 9$ in the above equation. Then

$$y = (5000)(9) + 45,000$$
$$= 90,000.$$

EXERCISES

1. A community college is offering a course in principles of real estate. The operating cost for this course is $900 plus $20 for each student enrolled.
   (a) Determine the cost function $C(x)$, where x is the number of students enrolled.
   (b) What is the operating cost if 25 students are registered in the course?

2. A local newspaper is conducting a survey to study the voting habits of the residents in a certain community. The operating cost of the survey is $1500 plus $2.50 for each family polled.
   (a) Determine the cost function $C(x)$, where x is the number of families polled in the community.
   (b) What is the total cost if 2400 families are polled?
   (c) How many families can be polled if $9000 is available in the budget?

3. A car rental agency in a town charges $32.95 per day plus $0.20 per mile to rent an automobile.
    (a)   Determine the cost function $C(x)$, where x is the number of miles a customer drives in a day.
    (b)   What is the total charge if the automobile is used for 45 miles in a day?

4. A new machine was installed in a manufacturing plant in 1982 at an expense of $150,000. Assuming that the machine depreciates linearly over 25 years, determine the equation of the straight line which expresses the value of the machine after x years. What will be its value in 1995?

5. An apartment building worth $400,000 built in 1980 is being depreciated linearly over 40 years.
    (a)   Determine the equation of the straight line which expresses the value of the building after x years.
    (b)   What would be the value of the building after 12 years?

6. A personal computer purchased on December 1, 1983, for $4500 is being depreciated linearly over 3 years.
    (a)   Determine the equation of the straight line which expresses the value of the computer after x months.
    (b)   What would be the value of this computer on June 1, 1985?

7. New equipment purchased for $75,000 on January 1, 1984 has a scrap value of $3000 at the end of 1989.
    (a)   Assuming that the equipment depreciates linearly, determine the equation of the straight line which expresses the value of the machine at the end of x months.
    (b)   What would be the value of the machine after 26 months?

8. An on-line registration system set up in a community college for $200,000 in 1983 is assumed to have a scrap value of $20,000 at the end of 10 years.
    (a)   Determine the equation of the straight line which relates the value of this equipment in dollars at the end of x years.
    (b)   What would be the value of this system in 1990?

9. The population of a small town is approximately linearly related to time. The population was 4000 in 1975 and 5000 in 1983.
    (a)   Determine the linear function representing the increase in the population after x years.
    (b)   What would be the estimated population of this town in 1988?

10. An investment of $10,000 in a certain bond fund yields an annual income of $1000, but an investment of $50,000 in the same fund yields an annual income of $6000.
    (a)   Assuming that income y is a linear function of fund investment x, determine the linear function that expresses the annual income y in terms of the investment x.
    (b)   What is the annual income if $70,000 is invested in the fund?

11. An apartment complex has a storage tank which holds its heating oil. The tank was filled with 25,000 gallons of oil on December 1, 1983. Ten days later, the oil level in the tank dropped to 20,000 gallons.
    (a)   Assuming that the consumption of heating oil is related linearly to the number of days, determine the equation of the straight line which expresses the consumption of oil y after x days.
    (b)   What would be the oil level in the tank after 23 days?

12. A publishing company must set a price for a new book. The past sales experience of the company on similar books reveals the following information:

Number of Copies Sold	Price of Book
15,000	$20
20,000	$18

    (a)   Assuming that the relationship is linear, determine the equation of the line that expresses the price of book y in terms of the copies sold x.
    (b)   What price should be set for the book if the management expects to sell 35,000 copies?

3.6

QUADRATIC FUNCTIONS

Definition 3.6.1

Any function of the form

$$y = ax^2 + bx + c, \qquad a \neq 0,$$

where a, b, and c are real numbers, is called a *quadratic function*.

The restriction $a \neq 0$ is necessary because if $a = 0$, then we have

$$y = bx + c$$

which is a linear function. Further, we may note in passing that we have used the notation $f(x)$ to represent a function of x. We may also write y to represent a function, since $y = f(x)$.

The simplest types of quadratic functions are $y = x^2$, where $a = 1$, $b = 0$, $c = 0$, and $y = -x^2$, where $a = -1$ and $b = c = 0$.

Example 3.31

Sketch the graphs of the following equations:

(a) $y = x^2$ **(b)** $y = -x^2$

Solution: Since we cannot list all possible pairs of real numbers satisfying each of these equations, we use a partial list of points to determine the shape of each curve. The values of y for some values of x for $y = x^2$ and $y = -x^2$ are given in Tables 3.1 and 3.2, respectively.

(a) $y = x^2$

TABLE 3.1

x	-4	-3	-2	-1	0	1	2	3	4
y	16	9	4	1	0	1	4	9	16

(b) $y = -x^2$

TABLE 3.2

x	-4	-3	-2	-1	0	1	2	3	4
y	-16	-9	-4	-1	0	-1	-4	-9	-16

Plotting some of the related ordered pairs and joining them by a smooth curve, we have the graphs of $y = x^2$ and $y = -x^2$ in Figures 3.31 and 3.32, respectively.

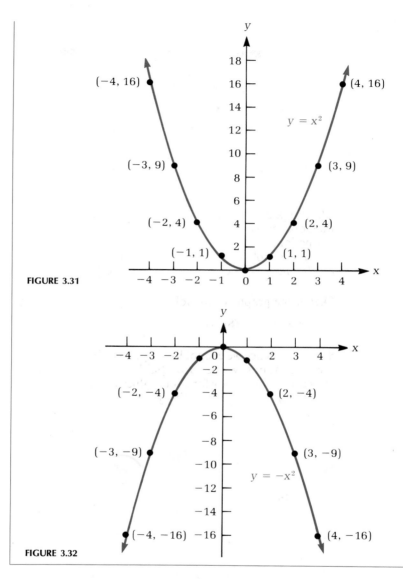

FIGURE 3.31

FIGURE 3.32

Example 3.32

Plot the graphs of the following equations:

(a) $y = x^2 + 3$ **(b)** $y = x^2 - 2$

Solution: The graph of $y = x^2$ was sketched in Figure 3.31. To graph $y = x^2 + 3$, we simply increase the y coordinate of each point on the graph of $y = x^2$ by 3, as shown in Figure 3.33. To graph $y = x^2 - 2$, we decrease the y-coordinate of each point on the graph of $y = x^2$ by 2 as shown in Figure 3.34.

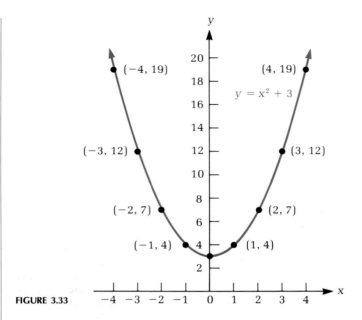

FIGURE 3.33

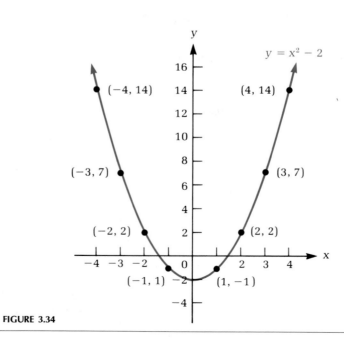

FIGURE 3.34

To obtain the graph of

$$y = ax^2,$$

we multiply the y coordinate of each point on the graph of $y = x^2$ by the real number a.

To graph the function

$$y = ax^2 + c,$$

we shift the graph of $y = ax^2$ up or down by $|c|$ units, depending on whether $c > 0$ or $c < 0$.

parabola   The graphs shown in Figures 3.31, 3.32, 3.33, and 3.34 are called **parabolas.** Generally, the graph of every quadratic function

$$y = ax^2 + bx + c,$$

where a, b, and c are real numbers and $a \neq 0$, is a parabola that opens **upward** if $a > 0$ and **downward** if $a < 0$. Every parabola is *symmetric* about a line called *axis of symmetry*   the **axis of symmetry** for the parabola. This axis separates the parabola into two equal and symmetric parts so that each part is the mirror image of the other. For example, if (x_1, y_1) is a point on the graph of $y = x^2$, then so is $(-x_1, y_1)$, as shown in Figure 3.31. Similar remarks apply to the graph of $y = -x^2$ in Figure 3.32.

Example 3.33   Plot the graph of $y = x^2 + 2x + 3$.

Solution: To graph this function, we first rewrite this equation in the form

$$y = (x - h)^2 + k$$

by *completing the square*. First we write

$$y = (x^2 + 2x \quad) + 3.$$

Adding the square of one-half the coefficient of x to the expression inside the parentheses and subtracting this quantity from the expression outside the parentheses, we obtain

$$y = (x^2 + 2x + 1) - 1 + 3$$
$$= (x + 1)^2 + 2.$$

We are now able to determine an important characteristic of the parabola's graph. Since $(x + 1)^2 \geq 0$ for all real values of x, we observe that $y = 2$ is the lowest value of y, and it is obtained when $x = -1$. This lowest point $(-1, 2)$ *vertex*   on the graph is called the **vertex** of the parabola, and $x = -1$ is called the *axis of symmetry*   **axis of symmetry** for the parabola.

The values of y for some values of x are given in Table 3.3.

TABLE 3.3

x	-4	-3	-2	-1	0	1	2	3
y	11	6	3	2	3	6	11	18

Plotting the ordered pairs $(-4, 11)$, $(-3, 6)$, $(-2, 3)$, $(-1, 2)$, $(0, 3)$, $(1, 6)$, $(2, 11)$, and $(3, 18)$, we obtain the graph in Figure 3.35.

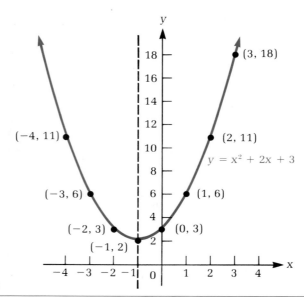

FIGURE 3.35

Example 3.34

Plot the graph of $y = 2x^2 + 12x + 3$.

Solution: First we rewrite this equation in the form

$$y = a(x - h)^2 + k$$

by completing the square. To accomplish this, we write

$$y = 2(x^2 + 6x \qquad) + 3.$$

To convert the expression $x^2 + 6x$ inside the parentheses, we add, as before, the square of one-half the coefficient of x, that is, $[\frac{1}{2}(6)]^2 = 9$; then we subtract

$(2)(9) = 18$ from the expression outside the parentheses. Thus

$$y = 2(x^2 + 6x + 9) + 3 - 18$$
$$= 2(x + 3)^2 - 15.$$

Observe that $(x + 3)^2 \geq 0$ for all real values of x. Consequently, $y = -15$ is the lowest possible value of y, and it is obtained when $x = -3$. Thus the **vertex** of the parabola is $(-3, -15)$, and $x = -3$ is the **axis of symmetry** for the parabola.

The values of y for some values of x are given in Table 3.4.

TABLE 3.4

x	-6	-5	-4	-3	-2	-1	0
y	3	-7	-13	-15	-13	-7	3

Plotting the ordered pairs $(-6, 3)$, $(-5, -7)$, $(-4, -13)$, $(-3, -15)$, $(-2, -13)$, $(-1, -7)$, and $(0, 3)$, we obtain the graph displayed in Figure 3.36.

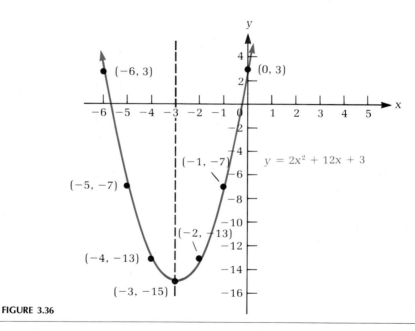

FIGURE 3.36

Example 3.35   Plot the graph of $y = -2x^2 + 8x - 5$.

Solution: To express this equation in the form

$$y = a(x - h)^2 + k,$$

we first rewrite the given equation in the form

$$y = -2(x^2 - 4x \quad) - 5.$$

Then we add to the expression inside the parentheses the square of one-half the coefficient of x, that is, $[\frac{1}{2}(-4)]^2 = 4$, and subtract $(-2)(4) = -8$ from the expression outside the parentheses. We obtain

$$y = -2(x^2 - 4x + 4) - 5 - (-8)$$
$$= -2(x - 2)^2 + 3.$$

Note that $-2(x - 2)^2 \leq 0$ for all real values of x. Hence $y = 3$ is the highest possible value of y, and it is obtained when $x = 2$. This means that the vertex $(2, 3)$ is the highest point of the parabola, and $x = 2$ is the axis of symmetry for the parabola.

The values of y for some arbitrary choices of x are given in Table 3.5.

TABLE 3.5

x	-1	0	1	2	3	4	5
y	-15	-5	1	3	1	-5	-15

Plotting the ordered pairs $(-1, -15)$, $(0, -5)$, $(1, 1)$, $(2, 3)$, $(3, 1)$, $(4, -5)$, and $(5, -15)$, we obtain the graph shown in Figure 3.37.

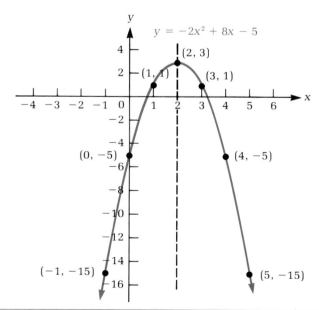

FIGURE 3.37

We now prove the following result in general.

Proposition 3.6.1

Any quadratic function

$$y = ax^2 + bx + c, \tag{3.7}$$

where a, b, and c are real numbers and $a \neq 0$, can be expressed in the form

$$y = a(x - h)^2 + k, \tag{3.8}$$

where

$$h = -\frac{b}{2a} \quad \text{and} \quad k = \frac{4ac - b^2}{4a}.$$

Proof: Let us first rewrite

$$y = ax^2 + bx + c$$

as

$$y = a\left(x^2 + \frac{b}{a}x\right) + c.$$

Adding $[b/(2a)]^2$, the square of one-half the coefficient of x, to the expression inside the parentheses and subtracting a $[b/(2a)]^2 = b^2/(4a)$ from the expression outside the parentheses, we obtain

$$y = a\left(x^2 + \frac{b}{a}x + \frac{b^2}{4a^2}\right) + c - \frac{b^2}{4a}$$

$$= a\left(x + \frac{b}{2a}\right)^2 + \frac{4ac - b^2}{4a},$$

which is of the form

$$y = a(x - h)^2 + k,$$

where

$$h = -\frac{b}{2a} \quad \text{and} \quad k = \frac{4ac - b^2}{4a}.$$

Observe that in the quadratic function

$$y = a(x - h)^2 + k, \qquad a \neq 0,$$

the expression $(x - h)^2 \geq 0$ for all real values of x.

Note that

1. the point (h, k) is the vertex of the parabola;
2. the line $x = h$ is the axis symmetry;
3. if $a > 0$, then the parabola opens upward and (h, k) is the lowest point on the graph; and
4. if $a < 0$, then the parabola opens downward and (h, k) is the highest point on the graph.

Example 3.36

Find the vertex and the axis of symmetry for the parabola

$$y = -3x^2 + 6x + 5.$$

Solution: Comparing this function with the general quadratic form (3.7), we observe that

$$a = -3, \qquad b = 6, \qquad \text{and} \qquad c = 5.$$

Thus

$$h = -\frac{b}{2a} = 1 \qquad \text{and} \qquad k = \frac{4ac - b^2}{4a} = \frac{-60 - 36}{-12} = 8.$$

Hence the vertex of the parabola is (1, 8). The line $x = 1$, which passes through the vertex, is the axis of symmetry for the parabola.

Applications

Since the vertex is the highest or lowest point on the graph of a parabola, it determines the maximum or minimum value of the quadratic function. We will see that the location of the vertex of the parabola is an important component in solving some interesting applied problems.

Example 3.37

The total profit made by selling x units of a single product is given by

$$P(x) = 160x - x^2 - 300.$$

Determine the number of units that maximize the profit.

Solution: Comparing the function $P(x)$ with the general quadratic function (3.7), we see that

$$a = -1, \qquad b = 160, \qquad \text{and} \qquad c = -300.$$

The vertex of the parabola is given by (h, k), where

$$h = -\frac{b}{2a} = 80 \qquad \text{and} \qquad k = \frac{4ac - b^2}{4a} = 6100.$$

Since $a < 0$, the parabola opens downward, and the vertex (80, 6100) is the highest point on the graph. The maximum profit is therefore \$6100 and is obtained by selling 80 units.

Example 3.38

A product's cost function is given by

$$g(x) = 4x^2 - 1800x + 250{,}000,$$

where x is the number of products the industry produces in 1 month. Determine the number of products the industry must produce to minimize cost.

Solution: Comparing $g(x)$ with the quadratic function (3.7), we see that

$$a = 4, \qquad b = -1800, \qquad \text{and} \qquad c = 250{,}000.$$

The vertex of the parabola is (h, k) where

$$h = -\frac{b}{2a} = 225 \qquad \text{and} \qquad k = \frac{4ac - b^2}{4a} = 47{,}500.$$

Since $a > 0$, the parabola opens upward, and the vertex $(225, 47{,}500)$ is the lowest point on the graph. The industry must therefore produce 225 units to minimize the cost.

Example 3.39

A farmer has 3000 feet of barbed wire to enclose a rectangular pasture. He plans to fence the entire area and then to subdivide it by running a fence across the middle, as shown in Figure 3.38. Determine the dimensions of the pasture that would enclose the maximum area.

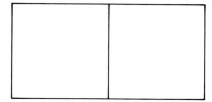

FIGURE 3.38

Solution: Let x be the width of the pasture. Then the length of the field to be fenced is $(3000 - 3x)/2$. The area to be maximized is

$$A(x) = x\left(\frac{3000 - 3x}{2}\right)$$

$$= 1500x - \tfrac{3}{2}x^2.$$

Comparing $A(x)$ with the general quadratic function (3.7), we observe that

$$a = -\tfrac{3}{2}, \qquad b = 1500, \qquad \text{and} \qquad c = 0.$$

The vertex of the parabola is (h, k), where

$$h = -\frac{b}{2a} = 500 \qquad \text{and} \qquad k = \frac{4ac - b^2}{4a} = 375{,}000.$$

The parabola opens downward, and the point $(500, 375{,}000)$ is the highest point on the graph of the parabola. (Why?) The maximum area of 375,000 square feet is fenced when the width $x = 500$ feet and the length of the field is $375{,}000/500 = 750$ feet.

Example 3.40

A travel agency has chartered a flight to London for high school students. The charge is $200 per student, with an additional charge of $2.50 for each subsequent cancellation. The plane has a seating capacity of 100 passengers. Determine the number of cancellations that would maximize the revenue of the travel agency.

Solution: Let x be the number of cancellations received before the departure of the flight; then $100 - x$ passengers are on the flight, and each pays $200 plus $2.50x$. Thus the total receipts for the travel agency are

$$R(x) = (100 - x)(200 + 2.50x)$$
$$= 20{,}000 + 50x - 2.50x^2 \qquad (0 \le x \le 100).$$

Comparing $R(x)$ with the general quadratic function (3.7), we note that

$$a = -2.50, \qquad b = 50, \qquad \text{and} \qquad c = 20{,}000.$$

The vertex of the parabola is (h, k), where

$$h = -\frac{b}{2a} = 10 \qquad \text{and} \qquad k = \frac{4ac - b^2}{4a} = 20{,}250.$$

Since $a < 0$, the parabola opens downward, and the vertex $(10, 20{,}250)$ is the highest point on the graph. We conclude that the travel agency receives a maximum of $20,250 when 10 cancellations are received prior to the departure of the plane.

EXERCISES

Graph the parabolas in Exercises 1 to 14.

1. $y = 2x^2$
2. $y = 3x^2$
3. $y = \frac{3}{2}x^2$
4. $y = \frac{1}{2}x^2$
5. $y = -x^2 + 2$
6. $y = -x^2 - 4$
7. $y = x^2 - 4x + 3$
8. $y = x^2 - 3x - 4$
9. $y = 2x^2 - 5x + 8$
10. $y = 2x^2 - 4x + 5$
11. $y = 4 - 2x - x^2$
12. $y = -3x^2 - 4x + 1$
13. $y = 8x - x^2$
14. $y = 100 - \frac{1}{2}x^2$

Determine the vertex and the axis of symmetry for the parabolas in Exercises 15 to 22.

15. $y = x^2 + 2x + 3$
16. $y = x^2 - 12x + 24$
17. $y = 2x^2 + 8x$
18. $y = 2x^2 + 10x + 3$
19. $y = -2x^2 + 4x + 7$
20. $y = 4x^2 - 9$
21. $y = 3x^2 + 6x + 15$
22. $y = 80x - 4x^2$
23. Find the output x that maximizes the profit function

$$P(x) = 200x - \frac{1}{2}x^2.$$

24. The number of air conditioners that the ABC department store expects to sell next summer is given by the function

$$s(x) = 15 + 20x - x^2, \qquad 0 \le x \le 15,$$

where x is the number of hot days in a given summer. Determine the number of hot days that will maximize the sales of air conditioners next summer.

25. Because of raw material shortages and higher labor costs, it has become increasingly expensive to drill for oil in relatively inaccessible locations. In fact, the profit (in millions of dollars) is determined by

$$P(x) = 80x - 10x^2,$$

where x is the number of millions of barrels of oil produced. Determine the level of production x that will yield maximum profit.

26. An automobile dealer has determined that a commercial advertisement on the local television channel has an appreciable impact on the sales of new-model cars. It is estimated that the number of cars sold $s(x)$ and the expenditure for advertisement x (in thousands of dollars) are related by the function

$$s(x) = 50 + 36x - 3x^2.$$

Determine the amount that must be spent on television commercials to maximize car sales.

27. The cost function in a certain industry is given by

$$c(x) = 0.5x^2 - 10x + 200,$$

where x is the number of items the industry produces in 1 week. Determine the value of x for which $c(x)$ is minimum.

28. The cost function of a product in an industry is given by

$$c(x) = 0.01x^2 - 50x + 100,000,$$

where x is the number of items the industry produces in 1 month. Determine the number of units that must be produced to minimize cost.

29. A projectile is shot vertically in the air with an initial velocity of 32 feet per second. Its height after t seconds is given by

$$h = 32t - 16t^2.$$

Find the maximum height the projectile will attain.

30. The height of an object thrown vertically in the air from the top of a cliff 160 feet high at time $t = 0$ is given by

$$h = 160 + 48t - 16t^2,$$

where time t is in seconds. Find the maximum height the object will attain.

31. The number of bacteria (in thousands) present in a certain culture at time t is given approximately by the function

$$f(t) = 80t - 8t^2 + 20,$$

where t is recorded in hours. Determine the value of t for which the number of bacteria is maximum.

32. A psychologist has prepared a list of nonsense words for use in a memory experiment. The number of words $N(t)$ that a person is able to memorize after t hours of practice is given by

$$N(t) = 75t - 15t^2, \qquad 0 \le t \le 4.$$

Determine the number of hours that a person should practice so as to maximize $N(t)$.

33. The distance d (in miles) between two ships at sea is a function of time t (in hours). If

$$d(t) = 256 - 70t + 5t^2,$$

find the time at which the distance between the ships is minimum. Determine the minimum distance between the ships.

34. The concentration of sugar in the bloodstream of a patient is approximated by the function

$$Q(t) = 5.2 + 0.1t - 0.1t^2,$$

where $Q(t)$ is the quantity of sugar (in suitable units) in the bloodstream t hours after a sugar metabolism test is performed. After how many hours will the concentration reach its maximum level? What is the maximum amount of sugar in the bloodstream?

35. A farmer has 4800 feet of fence to enclose a rectangular pasture. One side of the plot is to be along the bank of a river and needs no fence. Determine the dimensions of this plot so as to enclose the maximum area.

36. Suppose that the farmer in Problem 35 wishes to enclose the plot away from the river and needs to fence all four sides of the plot. Determine the dimensions of the plot that would enclose the maximum area.

37. A photographer has a thin piece of wood 40 inches long. How should she cut the wood to make a picture frame that would enclose the maximum area?

38. A farmer has 6400 feet of barbed wire to enclose a rectangular pasture. Her plan includes fencing the entire area and then subdividing it into three equal plots with the fence running across the width as shown in Figure 3.39. Determine the dimensions of the pasture which would enclose the maximum area.

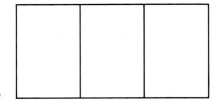

FIGURE 3.39

39. A resort hotel will provide dinner, dancing, and drinks for a local organization. The charge is $40 per couple if 100 couples attend; it then decreases by $0.20 for each additional couple above the minimum of 100. Determine the number of couples that will generate maximum revenue for the hotel.

40. A car-leasing agency needs to determine the rate it should charge for renting a car. The agency owns a fleet of 120 cars. Experience has shown that all the cars are rented at a rate of $30 per day per car. For each $1 increase in rent, 3 fewer cars are rented. How much should the agency charge to maximize the total revenue?

41. A high school has arranged a dance for the senior class. The charge is $10 per couple if 50 couples attend the dance; it then decreases by $0.10 for each additional couple above the minimum of 50 couples. Determine the number of couples that would generate the maximum revenue. What is the maximum revenue?

42. An alumni group makes plans for a charter flight to Key West, Florida. The plane fare is $100 per person if at least 40 persons go and will decrease by $0.50 for every additional person above the minimum of 40. Determine the number of passengers that will maximize the revenue for the alumni group. What is the maximum revenue?

43. A farmer estimates that if he begins picking his peaches today, he will have a crop of 500 pounds and can make a profit of $0.30 a pound. He can wait to pick them up for as long as 1 week. For each day he waits, he will have 25 extra pounds of peaches, but his profit will decrease by $0.01 per pound per day. When should he pick his crop so as to maximize the profit? Determine also his maximum earnings.

44. A restaurant owner plans to build a new restaurant with a seating capacity between 100 and 140. The owner has estimated that her profit will average $16 per week per seat if the seating capacity is 100, then will decrease by $0.10 per chair for each additional seat in excess of 100. Determine the number of seats that must be planned so as to maximize the profit. What is the maximum profit?

COMPOSITE AND INVERSE FUNCTIONS

composite function
When f and g are two real-valued functions, we can obtain a new function called the **composite function** of g on f as long as the range of the function f is contained in the domain of the function g.

Definition 3.7.1

> The *composite function* of g on f, denoted by $g \circ f$, is defined by
>
> $$(g \circ f)(x) = g(f(x))$$
>
> for all x in D_f for which $f(x)$ is in D_g.

The definition implies that we first determine $f(x)$ under the rule of correspondence f and then apply rule g to the element thus obtained. Let A and B be the domains of the functions f and g, respectively. Since x is in the set A, $f(x) \in B$, which is the domain of the function g. Thus the domain of the composite function $g \circ f$ consists of only those elements x for which $f(x)$ is in the domain of g (see Figure 3.40).

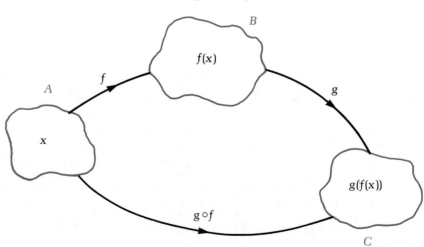

FIGURE 3.40

Consider Figure 3.41. If an element x is entered into machine f, the output from this machine becomes the input for machine g. In turn, machine g yields $g(f(x))$ as the final output corresponding to every x fed into machine f.

x

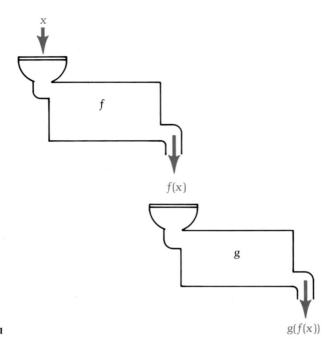

f

f(x)

g

FIGURE 3.41   g(f(x))

Example 3.41   Let f be a function defined by

$$f(x) = x + 2$$

with the set $A = \{1, 2, 3, 4, 5\}$ as its domain. Determine $g(f(x))$ for each $x \in A$ if $g(x) = 2x + 5$ and the domain of g is the range of f.

Solution: First we compute $f(x)$ for each $x \in A$. Observe that

$$f(1) = 3, \qquad f(2) = 4, \qquad f(3) = 5, \qquad f(4) = 6, \qquad \text{and} \qquad f(5) = 7.$$

Therefore the range of the function f is the set $B = \{3, 4, 5, 6, 7\}$ as illustrated in Figure 3.42. Now g is a function from set B to set C. We compute $g(x)$ for each element in B. Since $g(x) = 2x + 5$, we have

$$g(3) = 11, \qquad g(4) = 13, \qquad g(5) = 15, \qquad g(6) = 17, \qquad \text{and} \qquad g(7) = 19.$$

Thus the range of the function $g \circ f$ is the set $C = \{11, 13, 15, 17, 19\}$.

Alternatively, we can determine a formula for the composite function $g \circ f$ and then compute $g(f(x))$ for each $x \in A$. Recall that since $g(x) = 2x + 5$, we have

$$\begin{aligned}
(g \circ f)(x) = g(f(x)) &= 2[f(x)] + 5 \\
&= 2(x + 2) + 5 \\
&= 2x + 9.
\end{aligned}$$

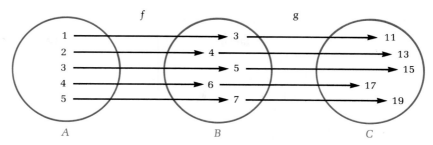

FIGURE 3.42

Now

$$g(f(1)) = 2(1) + 9 = 11, \qquad g(f(2)) = 2(2) + 9 = 13,$$
$$g(f(3)) = 2(3) + 9 = 15, \qquad g(f(4)) = 2(4) + 9 = 17,$$

and

$$g(f(5)) = 2(5) + 9 = 19.$$

Thus the range of the composite function $h = g(f(x))$ is precisely the same that we established in Figure 3.42.

As a practical application of the composite function, assume that the demand of a product is given by $d(p)$, where p is the price at which the product is sold. However, this selling price p depends, in turn, on the cost of producing that product. Thus the demand is a function of price, and price is a function of cost. In other words,

$$(d \circ p)c = d(p(c)).$$

Example 3.42

The demand for a product is given by

$$d(p) = 100 - p^2 \qquad (0 < p \le 10),$$

where p is the selling price of the product. If p is given by the linear function

$$p = 3c + 5,$$

then the demand in terms of cost c is given by

$$(d \circ p)(c) = d(p(c))$$
$$= 100 - (3c + 5)^2$$
$$= 100 - (9c^2 + 30c + 25)$$
$$= 3(25 - 10c - 3c^2).$$

Example 3.43

Let $f(x) = x^2 + 1$ and $g(x) = 3x + 4$. Determine the following:

(a) $g(f(x))$

(b) $f(g(x))$

Solution:

(a) $g(f(x)) = 3[f(x)] + 4$

$\qquad\qquad = 3(x^2 + 1) + 4$

$\qquad\qquad = 3x^2 + 7$

(b) $f(g(x)) = [g(x)]^2 + 1$

$\qquad\qquad = (3x + 4)^2 + 1$

$\qquad\qquad = 9x^2 + 24x + 17.$

Example 3.43 demonstrates the fact that in general

$$g(f(x)) \neq f(g(x)).$$

However, if the function f is one-to-one, we can define a function g that has a special relationship to the original function f (see Figure 3.43). Let A be the domain of the function f. For each $a \in A$, if f carries $a \in A$ to $f(a) \in B$, then g is defined to be the function which brings $f(a) \in B$ back to $a \in A$. This implies that $g(f(a)) = a$ for each $a \in A$.

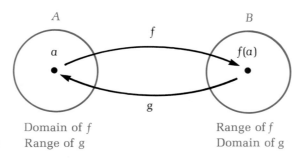

FIGURE 3.43

Domain of f Range of f
Range of g Domain of g

Definition 3.7.2

Let f be a one-to-one function having as its domain the set A and as its range the set B. Then a function g having as its domain the set B and as its range the set A is called the *inverse* of f if

$$g(f(x)) = x \qquad \text{for } x \in A.$$

The *inverse function g* of the function f is denoted by f^{-1}. Thus

$$f^{-1}(f(x)) = x \qquad \text{for } x \in A.$$

We can also show that

$$f(f^{-1}(x)) = x \qquad \text{for } x \in B.$$

Example 3.44

The functions $f(x) = 3x - 2$ and $g(x) = (x + 2)/3$ are inverses of each other because

$$g(f(x)) = \frac{f(x) + 2}{3} = \frac{(3x - 2) + 2}{3} = x$$

and

$$f(g(x)) = 3[g(x)] - 2 = 3\left(\frac{x + 2}{3}\right) - 2 = x.$$

Thus the inverse function of f is

$$f^{-1}(x) = \frac{x + 2}{3}.$$

Similarly,

$$g^{-1}(x) = 3x - 2.$$

Example 3.45

Show that

$$f(x) = \sqrt{x - 2} \qquad (x \geq 2),$$

and

$$g(x) = x^2 + 2 \qquad (x \geq 0),$$

are inverses of each other.

Solution: We need to check whether

$$g(f(x)) = f(g(x)) = x.$$

Observe that

$$\begin{aligned} g(f(x)) &= [f(x)]^2 + 2 \\ &= (\sqrt{x - 2})^2 + 2 \\ &= (x - 2) + 2 \\ &= x \end{aligned}$$

and

$$f(g(x)) = \sqrt{g(x) - 2}$$
$$= \sqrt{(x^2 + 2) - 2}$$
$$= x.$$

This proves that f and g are inverses of each other.

Example 3.46

Let $f(x) = 2x + 3$. Does $f^{-1}(x)$ exist? If so, find f^{-1}.

Solution: To show that f^{-1} exists, it is necessary to demonstrate that f is one-to-one. Suppose that a and b are two elements in the domain of f and $f(a) = f(b)$. Then

$$2a + 3 = 2b + 3,$$

which implies that $a = b$. Thus f is a one-to-one function.
   If f^{-1} is to be the inverse function of f, then

$$f(f^{-1}(x)) = x$$

which implies that

$$2[f^{-1}(x)] + 3 = x.$$

Solving for $f^{-1}(x)$, we have

$$f^{-1}(x) = \frac{x - 3}{2}.$$

   Let us now compare the graphs of $f(x) = 2x + 3$ and its inverse function $f^{-1}(x) = (x - 3)/2$ as shown in Figure 3.44. The ordered pairs

$$(-3, -3), \qquad (0, 3), \qquad (2, 7), \qquad \text{and} \qquad (3, 9)$$

are on the graph of the function f, while the ordered pairs

$$(-3, -3), \qquad (3, 0), \qquad (7, 2), \qquad \text{and} \qquad (9, 3)$$

are on the graph of f^{-1}. Note that each ordered pair (a, b) of the function f has been replaced by an ordered pair (b, a) in the inverse function f^{-1}. Geometrically, the graphs of $f(x) = 2x + 3$ and $f^{-1}(x) = (x - 3)/2$ are symmetric about the line $y = x$. In other words, the graph of $f^{-1}(x)$ is the reflection of the graph of $f(x)$ in the line $y = x$. **Thus, if f is a one-to-one function, then its inverse f^{-1} is obtained by simply interchanging the roles of x and y in the**

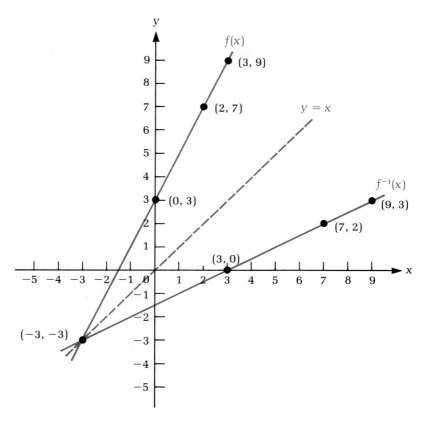

FIGURE 3.44

original equation and then solving for y. More specifically, we outline the following steps:

> *Step 1:* Write $y = f(x)$.
>
> *Step 2:* Interchange x and y in the equation $y = f(x)$.
>
> *Step 3:* Solve for y.
>
> *Step 4:* Replace y by $f^{-1}(x)$.

Example 3.47 For the following functions, determine f^{-1}:

(a) $f(x) = 3x - 4$ **(b)** $f(x) = x^3$ **(c)** $f(x) = \dfrac{x + 3}{x + 2}$

Solution:

(a)
$$f(x) = 3x - 4$$

Step 1:
$$y = 3x - 4.$$

Step 2: Interchange x and y in the equation $y = 3x - 4$. We have
$$x = 3y - 4.$$

Step 3: Solving for y, we get
$$y = \frac{x + 4}{3}.$$

Step 4: Therefore,
$$f^{-1}(x) = \frac{x + 4}{3}.$$

(b)
$$f(x) = x^3$$

Step 1:
$$y = x^3$$

Step 2: Interchanging variables x and y, we have
$$x = y^3.$$

Step 3: Solving for y, we get
$$y = x^{1/3}.$$

Step 4: Hence,
$$f^{-1}(x) = x^{1/3}.$$

(c)
$$f(x) = \frac{x + 3}{x + 2}, \qquad x \neq -2.$$

Step 1:
$$y = \frac{x + 3}{x + 2}.$$

Step 2: Interchanging variables x and y in the above equation, we have
$$x = \frac{y + 3}{y + 2}, \qquad y \neq -2.$$

Step 3: Now we solve for y:
$$x(y + 2) = y + 3$$
$$xy + 2x = y + 3$$
$$xy - y = 3 - 2x,$$

or

$$y(x - 1) = 3 - 2x.$$

Thus,

$$y = \frac{3 - 2x}{x - 1}$$

Step 4: Hence,

$$f^{-1}(x) = \frac{3 - 2x}{x - 1}, \qquad x \neq 1.$$

We must warn you that only one-to-one functions have inverses.

Example 3.48

Show that $f(x) = x^2$, where x is any real number, does not have an inverse.

Solution: To show that f^{-1} exists, we must demonstrate that f is one-to-one. By Definition 3.2.2, we must show that for each $f(x)$ in the range R_f, there is one and only one value in the domain D_f.

Let b and $(-b)$ be two nonzero real numbers in the domain D_f. Then

$$f(b) = b^2 \qquad \text{and} \qquad f(-b) = (-b)^2 = b^2.$$

Since different real numbers, b and $(-b)$, in D_f have the same image, namely, b^2, we conclude that f is *not* one-to-one. Hence f^{-1} does not exist.

Example 3.49

Let $f(x) = x^2$, $x \geq 0$. Determine f^{-1} and then sketch the graphs of f and f^{-1} in the same coordinate system.

Solution: To show that f^{-1} exists, we must show that f is one-to-one. Let a and b be two elements in the domain of f. Then

$$f(a) = a^2 \qquad \text{and} \qquad f(b) = b^2.$$

If $f(a) = f(b)$, then

$$a^2 = b^2$$

$$a^2 - b^2 = 0$$

$$(a + b)(a - b) = 0$$

$$a = -b \qquad \text{or} \qquad a = b.$$

Since $(-b)$ is not in the domain of f, we conclude that $a = b$. Therefore, f is one-to-one. Hence f^{-1} exists. To find f^{-1}, we proceed as follows:

Step 1:
$$y = x^2, \qquad x \geq 0$$

Step 2:   Interchanging variables x and y, we have

$$x = y^2.$$

Step 3:   Solving for y, we get

$$y = \sqrt{x}$$

Step 4:   Hence,

$$f^{-1}(x) = \sqrt{x}.$$

The graphs of $f(x) = x^2$, $x \geq 0$, and $f^{-1}(x) = \sqrt{x}$ are shown in Figure 3.45. Observe that for any real numbers a and b, if the ordered pair (a, b) is on the graph of $f(x) = x^2$, then the ordered pair (b, a) is on the graph of $f^{-1}(x) = \sqrt{x}$. We can, therefore, obtain the graph of f^{-1} simply by reflecting all the points on the graph of f through the line $y = x$. Notice that the graphs of $f(x)$ and $f^{-1}(x)$ in Figure 3.45 are symmetric with respect to the line $y = x$.

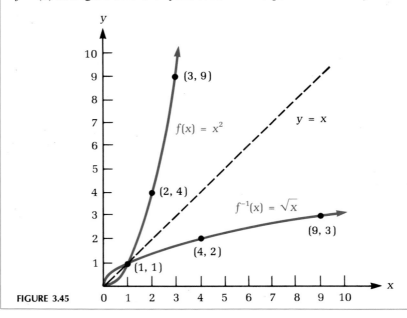

FIGURE 3.45

EXERCISES

1. Given that $f(x) = 3x + 2$ and $g(x) = 2x + 3$, determine the value of $g(f(x))$ for $x = 1, 2, 3, 4, 5,$ and 6.
2. Given that $f(x) = 4x + 5$ and $g(x) = 5x + 2$, determine the value of $f(g(x))$ for $x = 1, 2, 3, 4, 5,$ and 6.
3. Given that $f(x) = 3x + 4$ and $g(x) = 4x + 3$, determine the value of $g(f(x))$ and $f(g(x))$ for $x = 1, 2, 3, 4, 5,$ and 6.

4. Determine $d(p(c))$ in the following:
   (a) $d(p) = 100 - 4p^2$, $p(c) = 5c$ (b) $d(p) = 1000 - p^3$, $p(c) = c + 4$

Determine $g(f(x))$ and $f(g(x))$ in Exercises 5 to 20.

5. $f(x) = 3x + 5$, $g(x) = x - 5$

6. $f(x) = 2x + 1$, $g(x) = 2x - 3$

7. $f(x) = 3x + 1$, $g(x) = \dfrac{x + 4}{3}$

8. $f(x) = x$, $g(x) = \dfrac{1}{x}$

9. $f(x) = x^2 + 1$, $g(x) = 3x$

10. $f(x) = x^2$, $g(x) = 1 - x$

11. $f(x) = 2x^2 + 5$, $g(x) = 3x - 1$

12. $f(x) = x + \dfrac{1}{x}$, $g(x) = 6x - 5$

13. $f(x) = x^2 + 2x + 3$, $g(x) = 1 - x^3$

14. $f(x) = x^2 - 3$, $g(x) = \sqrt{x + 3}$

15. $f(x) = x^2 + 2$, $g(x) = \sqrt{x - 1}$

16. $f(x) = x^2$, $g(x) = \dfrac{x}{x + 1}$

17. $f(x) = \sqrt{x}$, $g(x) = x^2$

18. $f(x) = x^3$, $g(x) = x^{1/3}$

19. $f(x) = x^3 + 2x^2 + x + 4$, $g(x) = \dfrac{1}{x - 2}$

20. $f(x) = \dfrac{x^2}{x + 1}$, $g(x) = x - 1$

In Exercises 21 to 24, show that the functions f and g are inverses of each other.

21. $f(x) = 2x + 3$, $g(x) = \dfrac{x - 3}{2}$

22. $f(x) = x^2 + 4$ $(x \geq 0)$, $g(x) = \sqrt{x - 4}$ $(x \geq 4)$

23. $f(x) = x^3 - 1$, $g(x) = (x + 1)^{1/3}$

24. $f(x) = \dfrac{2x}{x + 1}$ $(x \neq -1)$, $g(x) = \dfrac{x}{2 - x}$ $(x \neq 2)$

In Exercises 25 to 30, decide whether each function f is one-to-one; if so, determine its inverse f^{-1}.

25. $f(x) = 4x - 3$

26. $f(x) = 3x + 3$

27. $f(x) = \sqrt{x - 5}$

28. $f(x) = \sqrt{x} - 5$

29. $f(x) = x^3 + 1$

30. $f(x) = 2 - x^3$

In Exercises 31 to 44, determine the inverse function f^{-1}.

31. $f(x) = 2x$

32. $f(x) = 2x - 1$

33. $f(x) = 3 - 4x$

34. $f(x) = 4x^2$, $x \geq 0$

35. $f(x) = \sqrt{x - 3}$, $x \geq 3$

36. $f(x) = \sqrt{x^2 - 4}$, $x \geq 2$

37. $f(x) = \sqrt{4 - x^2}$, $0 \leq x \leq 2$

38. $f(x) = x^3 + 1$

39. $f(x) = \dfrac{1}{x - 2}$, $x \neq 2$

40. $f(x) = \dfrac{1}{2x - 3}$, $x \neq \dfrac{3}{2}$

41. $f(x) = \dfrac{x + 2}{x}$, $x \neq 0$

42. $f(x) = \dfrac{x + 2}{x - 2}$, $x \neq 2$

43. $f(x) = \dfrac{x - 1}{x + 1}$, $x \neq -1$

44. $f(x) = \dfrac{1}{\sqrt{x}}$, $x > 0$

In Exercises 45 to 50, sketch the graphs of f and f^{-1} on the same coordinate system.

45. $f(x) = 3x$

46. $f(x) = 2x - 1$

47. $f(x) = \sqrt{x - 1}$, $x \geq 1$

48. $f(x) = \sqrt{x} - 1$

49. $f(x) = \dfrac{1}{x}$, $x \neq 0$

50. $f(x) = -\dfrac{1}{x}$, $x \neq 0$

51. Show that $f(x) = |x|$ does not have an inverse function.

52. Show that $f(x) = 2x^2 + 3$, where x is real, does not have an inverse function.

3.8

VARIATION

In almost every field, scientists seek to establish relationships between two variables when an increase in one variable produces an increase or decrease in the other variable.

Definition 3.8.1

Let x and y be two variables. Then y is said to *vary directly* with x, or y is *directly proportional* to x, if $y = kx$ for some nonzero real number k. This real number k is called the *constant of variation*, or *constant of proportionality*.

Example 3.50

The formula

$$C = 2\pi r$$

for the circumference of the circle expresses the fact that the circumference C varies directly with the radius r of the circle. The constant of proportionality is 2π.

Example 3.51

Express y in terms of x if y is directly proportional to x and $y = 4$ when $x = 5$.

Solution: Since y is directly proportional to x, we know from Definition 3.8.1 that $y = kx$. The problem is to determine the constant k. If we substitute 5 for x and 4 for y in the equation $y = kx$, we find that

$$4 = 5k \qquad \text{or} \qquad k = \tfrac{4}{5}.$$

Hence the variables x and y satisfy the equation

$$y = \tfrac{4}{5}x.$$

Example 3.52

Hooke's law states that the extension of an elastic spring beyond its natural length is directly proportional to the force applied. If a weight of 9 pounds stretches a spring 1.5 inches, how much weight is required to stretch the spring 2.5 inches beyond its natural length?

Solution: Let e denote the extension of an elastic spring when it is pulled by a weight of w pounds. Then

$$e = kw,$$

where k is the constant of proportionality. We know that $e = 1.5$ inches when $w = 9$ pounds. Hence

$$1.5 = 9k \qquad \text{or} \qquad k = \tfrac{1}{6}.$$

Thus the relationship between e and w is

$$e = \tfrac{1}{6}w.$$

If $e = 2.5$ inches, then

$$2.5 = \frac{w}{6} \qquad \text{or} \qquad w = 6(2.5) = 15 \text{ pounds.}$$

Sometimes one variable varies directly with a power of the other variable.

Definition 3.8.2

Let x and y be two variables, and let n be a positive real number. Then y varies directly with the nth power of x if $y = kx^n$ for some nonzero real number k.

Example 3.53

The formula

$$A = \pi r^2$$

for the area of a circle expresses the fact that the area A varies directly with the square of the radius r. The constant of proportionality is π. Similarly, the formula

$$V = \frac{4}{3}\pi r^3$$

for the volume of a sphere asserts that the volume V varies directly with the cube of the radius r of the sphere. In this case, the constant of proportionality is $\tfrac{4}{3}\pi$.

Example 3.54

If an object is released from rest in a vacuum near the surface of the earth, then the distance it falls varies directly with the square of the elapsed time t. If $s = 400$ feet when $t = 5$ seconds, express s in terms of t. How long will it take an object to fall if it is dropped from a height of 1600 feet?

Solution: Since the distance s is directly proportional to the square of the time t, we have

$$s = kt^2,$$

where k is some constant. Further, $s = 400$ when $t = 5$. Thus

$$400 = k(5)^2 \qquad \text{or} \qquad k = 16.$$

The relationship between s and t is therefore

$$s = 16t^2.$$

If $s = 1600$ feet, then

$$t^2 = \frac{1600}{16} = 100 \qquad \text{or} \qquad t = 10 \text{ seconds.}$$

It will take 10 seconds for an object to fall a distance of 1600 feet.

In a number of practical situations, an increase in one variable leads to a decrease in the other. The demand for a certain product depends largely on its price. Generally speaking, as the price increases, the quantity demanded by customers decreases. On the other hand, if a department store reduces the price of a certain item, the demand for that product increases. If a gas at a certain temperature is compressed, the pressure of the gas increases as its volume decreases. These examples and many others can be described in terms of **inverse variation** which we now consider.

Definition 3.8.3

Let x and y be two variables. Then y is said to vary *inversely* with x if

$$y = \frac{k}{x}$$

for some nonzero constant k.

Example 3.55

Express y in terms of x if y varies inversely with x and $y = 5$ when $x = 4$.

Solution: Since y varies inversely with x, we know from Definition 3.8.3 that

$$y = \frac{k}{x},$$

where k is the constant to be determined. If we substitute 4 for x and 5 for y, we find that

$$5 = \frac{k}{4} \qquad \text{or} \qquad k = 20.$$

Hence the variables x and y satisfy the equation

$$y = \frac{20}{x}.$$

Example 3.56

The number of days required to finish a certain project varies inversely with the number of workers on the project. If six workers are needed to finish the job in 16 days, how many workers must be employed to complete the job in 12 days?

Solution: Let x be the number of days required to complete the project and y be the number of workers on the project. Since y is inversely proportional to x, we know that

$$y = \frac{k}{x}.$$

Further, $y = 6$ when $x = 16$. Hence

$$6 = \frac{k}{16} \qquad \text{or} \qquad k = 96.$$

Thus the relationship between x and y is given by

$$y = \frac{96}{x}.$$

If $x = 12$, then

$$y = \frac{96}{12} = 8.$$

If the job is to be completed in 12 days, eight workers are needed.

Just as y can vary directly with a power of x, it can also vary inversely with a power of x.

Definition 3.8.4

Let x and y be two variables. Then y varies inversely with the nth power of x if for some nonzero real number k,

$$y = \frac{k}{x^n}.$$

Example 3.5.7

The intensity of illumination (measured in candles) produced by a source of light varies inversely with the square of the distance from the source. If the light has an intensity of 16 candles at a distance of 5 feet from its source, what will be the intensity of the light 10 feet from the source?

Solution: If y is the intensity of the light placed at a distance x from its source, then

$$y = \frac{k}{x^2}.$$

Since $y = 16$ when $x = 5$, we have

$$16 = \frac{k}{25} \qquad \text{or} \qquad k = (16)(25) = 400.$$

Thus the relationship between x and y is

$$y = \frac{400}{x^2}.$$

If $x = 10$, then

$$y = \frac{400}{100} = 4.$$

The intensity of light is 4 candles placed 10 feet from the source.

There are several situations in which one variable varies directly with the product of two or more independent variables.

Definition 3.8.5

> Let u, v, w, and y be the variables. Then y is said to vary jointly with u and v if $y = kuv$ for some nonzero constant k. Similarly, y varies jointly with u, v, and w if $y = kuvw$ for some nonzero constant k.

Sometimes one variable varies jointly with various combinations of powers of several independent variables. For example, if y varies jointly with the square of u and the cube of v, then $y = ku^2v^3$ for some constant k. Finally, combinations of direct and inverse variations often occur in the same problem. For instance, if y varies directly with u and inversely with v, then

$$y = \frac{ku}{v}$$

for some nonzero constant k. Again, if y varies jointly with u and the square of v and inversely with the cube of w, then

$$y = \frac{kuv^2}{w^3}$$

for some nonzero constant k.

Let us look at some examples.

Example 3.58

The volume of a right circular cone varies jointly with the square of the radius of the base and the height of the cone. If the radius of the base is 7 feet and the height of the cone is 15 feet, the volume is 245π cubic feet. Determine the height of the cone if its volume is 42π cubic feet and its base has a radius of 3 feet.

Solution: Let h and r denote the height of the cone and the radius of the base, respectively, and let V be the volume of the cone. Then

$$V = kr^2h,$$

where k is a constant to be determined. Because $V = 245\pi$ cubic feet, where $r = 7$ feet and $h = 15$ feet, we have

$$245\pi = k(7^2)(15)$$

$$k = \frac{245\pi}{(49)(15)}$$

$$k = \frac{\pi}{3}.$$

Thus the relationship among V, r, and h is

$$V = \frac{\pi r^2 h}{3}.$$

If $V = 42\pi$ and $r = 3$, then

$$42\pi = \frac{\pi 3^2 h}{3}$$

or

$$h = 14,$$

and therefore the height is 14 feet.

Example 3.59

The resistance of an electric wire varies directly with its length and inversely with the square of its diameter. If a wire 2000 feet long with diameter 0.2 inch has a resistance of 32 ohms, determine the resistance in a wire made of the same material with diameter 0.4 inch and 5000 feet long.

Solution: Let R, t, and d denote the resistance (in ohms), the length of the wire (in inches), and the diameter (in inches), respectively. Then

$$R = \frac{kt}{d^2},$$

where k is the constant of proportionality. Since $R = 32$ ohms when $t = 2000$ feet, or 24,000 inches, and $d = 0.2$ inch, then

$$32 = \frac{k(24,000)}{(0.2)^2}$$

or

$$k = \frac{32(0.04)}{24,000} = \frac{1}{18,750}.$$

Therefore the relationship among R, t, and d is

$$R = \frac{t}{18{,}750d^2}.$$

It $t = 5000$ feet, or 60,000 inches, and $d = 0.4$ inch, then

$$R = \frac{60{,}000}{18{,}750(0.4)^2}$$

$$= 20 \text{ ohms.}$$

The resistance is 20 ohms.

EXERCISES

In Exercises 1 to 10, determine the constant of proportionality from the specified conditions.

1. y varies directly with x and $y = 8$ when $x = 4$.
2. v varies directly with u and $v = 10$ when $u = 3$.
3. y varies directly with x^2 and $y = 75$ when $x = 5$.
4. s is directly proportional to t^2 and $s = 144$ when $t = 3$.
5. v is inversely proportional to u and $v = 8$ when $u = 5$.
6. r is inversely proportional to t^2 and $r = 10$ when $t = 2$.
7. v varies directly with t and inversely with p and $v = 100$ when $t = 200$ and $p = 50$.
8. y varies directly with x^2 and inversely with $\sqrt{z}$. If $x = 5$ and $z = 9$, then $y = 6$.
9. v varies jointly with h and r^2. If $h = 4$ and $r = 6$, then $v = 144\pi$.
10. F varies jointly with m and M and inversely with d^2. If $m = 100$, $M = 50$, and $d = 25$, then $F = 240$.
11. Suppose that y varies directly with x^2. If $y = 45$ when $x = 3$, find the value of y when $x = 4$.
12. Suppose that r varies inversely with t. If $r = 6$ when $t = 3$, find the value of r when $t = 5$.
13. Suppose that r varies directly with m and inversely with d^2. If $r = 40$ when $m = 500$ and $d = 2$, find r when $m = 3000$ and $d = 4$.
14. Suppose that y varies jointly with u^2 and v and inversely with w. If $y = 144$ when $u = 4$, $v = 6$, and $w = 8$, find y when $u = 5$, $v = 8$, and $w = 15$.
15. The velocity of a body falling from rest is directly proportional to the time it falls. If a body attains a speed of 80 feet per second after 2.5 seconds, how fast will it be falling 1.5 seconds later?
16. For an object falling freely from rest, the distance s (in feet) that the object falls varies directly with the square of the elapsed time t (in seconds). If $s = 144$ feet when $t = 3$ seconds, find the formula for s in terms of t. How long will it take for an object to fall 576 feet?
17. The volume of the sphere varies directly with the cube of its radius. If a sphere having a radius of 3 inches has a volume of 36π cubic inches, what will the volume of a sphere be if its radius is 6 inches?
18. Boyle's law states that the volume V of a gas varies inversely with its pressure, provided that the temperature remains constant. If the volume is 200 cubic inches when the pressure is 50 pounds per square inch, what will the volume be when the pressure is increased to 80 pounds per square inch?

19. The electric resistance of a wire of given length and material is inversely proportional to its cross-sectional area. The resistance in a circuit composed of the wire with a cross section of 82 square millimeters is 210 ohms. What would be the resistance if the wire had a cross section of 70 square millimeters?

20. The volume V of a gas varies directly with its temperature t and inversely with its pressure p. If the volume is 50 cubic feet when the temperature is 300° (absolute) and the pressure is 40 pounds per square inch, what will the volume be if the temperature is raised to 315° (absolute) while the pressure is decreased to 30 pounds per square inch?

21. The intensity of illumination received from the source of light varies inversely with the square of the distance from the source. How far from a 200-candle light would a screen have to be placed to receive the same illumination as a screen placed 30 feet from a 50-candle light source?

22. The electric resistance of a wire varies directly with the length of the wire and inversely with the square of its diameter. If the resistance of a wire 40 feet long and 0.02 inches in diameter is 1.8 ohms, what would the resistance be of a wire 50 feet long and 0.03 inches in diameter?

23. The kinetic energy E of a moving body varies jointly with its mass M and the square of its velocity. If a body weighing 20 pounds and moving with a velocity of 40 feet per second has a kinetic energy of 600 foot-pounds, find the kinetic energy of a 30-pound object moving with a velocity of 20 feet per second.

24. Kepler's third law states that the time in which a planet revolves about the sun varies directly with the $\frac{3}{2}$ power of the maximum radius of its orbit. Using 93 million miles as the maximum radius of the earth's orbit and 142 million miles as the maximum radius of the orbit of Mars, in how many days will Mars make one revolution about the sun?

KEY WORDS

ordered pair	*one-to-one function*	*parabola*
coordinate axis	*greatest integer function*	*vertex*
coordinate plane	*linear function*	*axis of symmetry*
quadrant	*slope*	*completing the square*
abscissa	*equation of a line*	*composite function*
ordinate	*slope-intercept*	*inverse function*
function	*parallel*	*vary directly*
image	*perpendicular*	*directly proportional*
independent variable	*point-slope*	*constant of proportionality*
dependent variable	*x and y intercepts*	*vary inversely*
domain	*linear depreciation method*	*vary jointly*
range	*quadratic function*	*constant of variation*

KEY FORMULAS

• The distance from $A(x_1, y_1)$ to $B(x_2, y_2)$ is given by

$$\sqrt{(x_2 - x_1)^2 + (y_2 - y_1)^2}$$

• The coordinates of the midpoint of the line seg-

ment from $A(x_1, y_1)$ to $B(x_2, y_2)$ are

$$\left(\frac{x_1 + x_2}{2}, \frac{y_1 + y_2}{2} \right)$$

- Let $\mathscr{L}$ be a nonvertical line joining two distinct points A and B with coordinates (x_1, y_1) and (x_2, y_2), respectively. Then the slope m of the line $\mathscr{L}$ is given by

$$m = \frac{y_2 - y_1}{x_2 - x_1}$$

- The equation of a straight line parallel to the x axis and $|b|$ units from it is $y = b$. If $b > 0$, the line is above the x axis; if $b < 0$, the line is below the x axis.

- The equation of a straight line parallel to the y axis and $|a|$ units from it is $x = a$. If $a > 0$, the line is to the right of y axis; if $a < 0$, the line is to the left of y axis.

- Let $\mathscr{L}$ be the line passing through $(0, b)$ and having slope m. Then a point (x, y) lies on $\mathscr{L}$ if and only if it satisfies

$$y = mx + b \qquad \text{(slope-intercept form)}.$$

- Two distinct (nonvertical) lines are parallel if and only if their slopes are equal.
- Two distinct (nonvertical) lines are perpendicular if and only if the product of their slopes equals (-1).

The equation of a straight line passing through (x_1, y_1) and having slope m is

$$y - y_1 = m(x - x_1) \qquad \text{(point-slope form)}.$$

- Every equation of the form $Ax + By + C = 0$ ($A \neq 0$ or $B \neq 0$), where A, B, and C have fixed values, represents a straight line.

- If a line has nonzero x and y intercepts a and b, respectively, then its equation can be expressed in the form

$$\frac{x}{a} + \frac{y}{b} = 1$$

- Any quadratic function $y = ax^2 + bx + c$, where a, b, and c are real numbers and $a \neq 0$, can be expressed in the form $y = a(x - h)^2 + k$, where

$$h = -\frac{b}{2a} \quad \text{and} \quad k = \frac{4ac - b^2}{4a}.$$

The point (h, k) is the vertex of the parabola, and the line $x = h$ is the axis of symmetry.

- If f is a one-to-one function, then its inverse f^{-1} is obtained by simply interchanging the roles of x and y in the original equation and then solving for y.

EXERCISES FOR REVIEW

1. Show that the points $A(9, 6)$, $B(-1, 2)$, and $C(1, -3)$ are the vertices of a right triangle.
2. Given that $f(x) = 2x^2 + 3x - 1$, determine the following:
   (a) $f(2)$ (b) $f(3.5)$ (c) $f(-4)$ (d) $f(a + h)$ (e) $f(\sqrt{a})$ (f) $f(1/a)$
3. Given that $f(x) = 3x^2 + 5x$, determine the following and simplify the result:

$$\frac{f(a + h) - f(a)}{h}$$

4. Determine whether the following sets of ordered pairs are functions:
   (a) $\{(1, 2), (2, 3), (2, 4)\}$ (b) $\{(1, 1), (2, 1), (3, 1)\}$

In Exercises 5 to 10, determine the largest subset of real numbers that can serve as the domain of the following functions:

5. $f(x) = 2x$ 6. $f(x) = \sqrt{2x - 1}$ 7. $f(x) = \sqrt{x^2 - 1}$
8. $f(x) = |x| + 2$ 9. $f(x) = 1/x$ 10. $f(x) = \sqrt{1 - x^2}$

Sketch the graphs of the functions in Exercises 11 to 16.

11. $f(x) = 3x + 2$ **12.** $f(x) = x^2 - 2$ **13.** $f(x) = 2 - x^2$

14. $f(x) = |x| + 1$ **15.** $f(x) = |x + 1|$ **16.** $f(x) = \sqrt{x + 1}$

Sketch the graphs of the quadratic functions in Exercises 17 to 20.

17. $y = x^2 + 2x + 2$ **18.** $y = 6x - x^2$

19. $y = 2x^2 + 12x + 3$ **20.** $y = -2x^2 + 8x - 5$

In Exercises 21 and 22, find the slope and y intercept:

21. $y = 3x + 5$ **22.** $2x - 3y + 6 = 0$

Determine whether the pairs of lines in Exercises 23 to 26 are parallel, perpendicular, or neither.

23. $3x - 3y + 6 = 0$ and $3x + 2y - 12 = 0$

24. $5x + 6y - 30 = 0$ and $10x + 12y - 15 = 0$

25. $x - 3y + 4 = 0$ and $3x + 9y + 2 = 0$

26. $4x - 5y + 20 = 0$ and $5x + 4y - 16 = 0$

In Exercises 27 to 31, find the equation of the line which passes through

27. $P(2, -7)$ with slope $m = 4$ **28.** $P(3, 5)$ and $Q(6, 8)$ **29.** $P(2, 3)$ and $Q(2, 8)$

30. $P(-1, 4)$ and parallel to the line $2x + 3y - 6 = 0$

31. $P(5, 3)$ and perpendicular to the line $x - 2y + 8 = 0$

32. Determine the vertex and the axis of symmetry for the parabola

$$y = -2x^2 + 4x + 7.$$

Determine $g(f(x))$ and $f(g(x))$ in Exercises 33 to 36.

33. $f(x) = 2x^2 + x + 5, \quad g(x) = 3x + 1$ **34.** $f(x) = x + 1, \quad g(x) = x^2$

35. $f(x) = \sqrt{x}, \quad g(x) = x^2$ **36.** $f(x) = x^2 + 2, \quad g(x) = \sqrt{x - 1}$

Determine the inverse function f^{-1}, if it exists, in Exercises 37 to 40.

37. $f(x) = |x|$ **38.** $f(x) = 2x^2, \, x \geq 0$

39. $f(x) = x^3 - 1$ **40.** $f(x) = \sqrt{x - 1}, \, x \geq 1$

41. A rental agency in a town leases cars and charges \$32.95 a day plus \$0.25 per mile.
  (a) Determine the cost function $C(x)$, where x is the number of miles a customer drives in a day.
  (b) What is the total charge if the car is used for 50 miles in a day?

42. The normal weight in pounds of a person is linearly related to height in inches. The normal weight for a 60-inch-tall person is 110 pounds and that for a 70-inch-tall person is 160 pounds. Determine the linear function representing weight of a person in terms of height.

43. A projectile is shot vertically in the air with an initial velocity of 400 feet per second. The height of the projectile after t seconds is

$$h = 400t - 16t^2.$$

Find the maximum height h the projectile will attain.

44. A grocery store has a price list, which the store expects its new checkers to memorize. The number of items N whose prices a checker is able to memorize after t hours of practice is

$$N = 60t - 12t^2.$$

Determine the number of hours a checker should practice to maximize N. What is the maximum value of N?

45. A high school is planning to build a rectangular playground to enclose the maximum possible area. Find the dimensions of the playground if the school has 1920 feet of fencing, assuming that one side borders another building and needs no fence.

46. Suppose that r varies directly with m^2 and inversely with d^3. If $r = 4$ when $m = 3$ and $d = 2$, find r when $m = 2$ and $d = 3$.

47. The number of days required to finish a certain project varies inversely with the number of workers on the project. If five workers are needed to complete the job in 12 days, how many must be employed to finish the job in 10 days?

48. Ohm's law states that the electric current in a circuit varies directly with the voltage and inversely with the resistance. Given that the current in a 20-ohm resistor is 0.5 ampere when a voltage of 6 volts is applied, what is the current in the resistor when the voltage applied is 12 volts?

49. The force of gravitation of two oppositely charged bodies is inversely proportional to the square of the distance between them. If two bodies are 10 centimeters apart, the force of attraction is 40 dynes. How far apart should the bodies be moved to make the force of gravitation 16 dynes?

CHAPTER TEST

1. Find the distance between the points $A(-2, 1)$ and $B(10, 6)$.

2. Determine whether or not the following sets of ordered pairs are functions:
    (a) $\{(1, 1), (1, 2), (1, 3)\}$ (b) $\{(1, 1), (2, 1), (3, 1)\}$

3. Determine the largest subset of real numbers that can serve as the domain of each of the following functions:
    (a) $f(x) = \sqrt{x}$ (b) $f(x) = 3x$ (c) $f(x) = \dfrac{1}{x}$

4. Given that $f(x) = x^2 + 3x$, determine the following and simplify:

$$\frac{f(2 + h) - f(2)}{h}$$

In Exercises 5 to 8, find the equation of the line which passes through

5. the point $P(3, 4)$ and has slope $m = 2$.

6. the points $P(7, 6)$ and $Q(9, 8)$

7. the points $P(7, 6)$ and $Q(9, 6)$

8. the point $P(3, 5)$ and is perpendicular to the line $5x + 6y - 30 = 0$

9. Given that $f(x) = 2x^2 + 5$ and $g(x) = 3x$, determine each of the following:
    (a) $g[f(x)]$ (b) $f[g(x)]$

10. Determine the inverse function f^{-1}, given that $f(x) = 2x^3$.

11. Determine the vertex and axis of symmetry for the parabola

$$y = 3x^2 + 5x + 12.$$

12. The cost function in a certain industry is given by

$$c(x) = 0.5x^2 - 10x + 200$$

where x is the number of items the industry produces in 1 week. Determine the value of x for which $c(x)$ is minimum.

13. Determine the constant of proportionality if s is directly proportional to t^2 and $s = 144$ when $t = 3$.

14. Find two positive numbers that have a sum of 78 and are such that their product is as large as possible.

15. Graph
   (a) $f(x) = x^2 + 2x + 2$ **(b)** $f(x) = |x| + 1$

POLYNOMIAL AND
RATIONAL FUNCTIONS

4

INTRODUCTION

The concept of polynomial was introduced in Chapter 1. At that time, we discussed basic operations such as adding, subtracting, multiplying, and factoring polynomials. Division of polynomials and synthetic division were also introduced. In this chapter, we consider polynomials in depth and discuss the solutions of polynomial equations. Specifically, we approximate real solutions of polynomials, if they exist. Then we explore techniques for graphing polynomial and rational functions.

4.1

POLYNOMIAL FUNCTIONS

Definition 4.1.1

Any function of the form

$$P(x) = a_n x^n + a_{n-1} x^{n-1} + \cdots + a_2 x^2 + a_1 x + a_0 \qquad (a_n \neq 0),$$

where $a_0, a_1, \ldots, a_n$ are real numbers and n is a nonnegative integer, is called a *polynomial function of degree n.*

Some examples of polynomial functions are

$$P(x) = x^2 + 2x + 5 \qquad \text{of degree 2,}$$

$$P(x) = 2x^3 - 3x^2 + 1 \qquad \text{of degree 3,}$$

$$P(x) = \tfrac{1}{2}x^4 + 3x^3 - \tfrac{1}{3}x^2 + x + 1 \qquad \text{of degree 4,}$$

constant polynomial   and so on. If $P(x) = c$, then we have a polynomial of degree zero, and we call it a **constant polynomial function.**

Definition 4.1.2

The *zeros of a polynomial* function $P(x)$ are the values of x for which $P(x) = 0$.

This definition suggests that zeros of $P(x)$ are identical with the solutions of $P(x) = 0$. The zeros of $P(x) = x^2 - 5x + 6$, for example, are obtained by solving the equation $x^2 - 5x + 6 = 0$ for x. Thus the zeros of $P(x)$ are 2 and 3.

Example 4.1

Determine the zeros of
$$P(x) = x(x^2 - x - 2).$$

Solution: Since zeros of $P(x)$ are the solutions of $P(x) = 0$, we must solve for x the equation

$$x(x^2 - x - 2) = 0.$$

When we factor the left-hand side, we observe that

$$x(x + 1)(x - 2) = 0.$$

Hence the zeros of $P(x)$ are 0, -1, and 2.

Since it may be time-consuming to determine zeros of a polynomial function of a larger degree, we may rely on synthetic division to examine whether a certain value of x is a zero of the polynomial. Recall from Section 1.9 that when we divide a polynomial $P(x)$ of degree n by a polynomial of the form $x - k$, where k is a real number, we obtain a polynomial $Q(x)$ of degree $n - 1$ for a quotient and r as a remainder. For example, we may divide $P(x) = 3x^3 + 2x^2 - 10x + 4$ by $x - 2$, using synthetic division as follows:

$$
\begin{array}{r|rrrr}
2 & 3 & 2 & -10 & 4 \\
  & \downarrow\ 6 & & 16 & 12 \\
\hline
  & 3 & 8 & 6 & 16
\end{array}
$$

That is,

$$\frac{3x^3 + 2x^2 - 10x + 4}{x - 2} = 3x^2 + 8x + 6 + \frac{16}{x - 2}$$

or

$$3x^3 + 2x^2 - 10x + 4 = (x - 2) \cdot (3x^2 + 8x + 6) + 16$$

$$P(x) \qquad\quad = (x - k) \cdot \qquad Q(x) \qquad + r.$$

Notice that when $P(x) = 3x^3 + 2x^2 - 10x + 4$ is divided by $x - 2$, the remainder $r = 16$. If we substitute $x = 2$ directly in $P(x)$, we have

$$
\begin{aligned}
P(2) &= 3(2^3) + 2(2^2) - 10(2) + 4 \\
&= 3(8) + 2(4) - 10(2) + 4 \\
&= 24 + 8 - 20 + 4 \\
&= 16.
\end{aligned}
$$

This agrees with the result obtained by synthetic division and suggests the following proposition.

Proposition 4.1.1

If a polynomial $P(x)$ is divided by $x - k$, then the remainder $r = P(k)$.

Proof: Division of $P(x)$ by $x - k$ yields a polynomial $Q(x)$ and a remainder r such that

$$P(x) = (x - k)Q(x) + r.$$

Since this statement is true for all values of x, it must be true for $x = k$. Thus

$$
\begin{aligned}
P(k) &= (k - k)Q(k) + r \\
&= (0)Q(k) + r \\
&= r.
\end{aligned}
$$

remainder theorem

This proposition, known as the **remainder theorem,** asserts that when $P(x)$ is divided by $x - k$, the remainder r is equal to the value of $P(x)$ at $x = k$.

Example 4.2

Determine the remainder when

$$P(x) = x^3 - x^2 - 3x + 9$$

is divided by $x + 2$.

Solution: We may rewrite $x + 2$ as

$$x + 2 = x - (-2).$$

Using synthetic division, we obtain

$$
\begin{array}{r|rrrr}
-2 & 1 & -1 & -3 & 9 \\
   & \downarrow & -2 & 6 & -6 \\
\hline
   & 1 & -3 & 3 & 3
\end{array}
$$

Hence $P(-2) = 3$ is the remainder.
   If we use remainder theorem and substitute $x = -2$ in $P(x)$, then

$$
\begin{aligned}
P(-2) &= (-2)^3 - (-2)^2 - 3(-2) + 9 \\
&= -8 - 4 + 6 + 9 \\
&= 3
\end{aligned}
$$

as before.

Example 4.3

Given that

$$P(x) = x^3 - 2x^2 + x - 5$$

find $P(3)$.

Solution: Using synthetic division, we observe that

$$
\begin{array}{r|rrrr}
3 & 1 & -2 & 1 & -5 \\
  & \downarrow & 3 & 3 & 12 \\
\hline
  & 1 & 1 & 4 & 7
\end{array}
$$

Hence $P(3) = 7$. Notice that if we substitute $x = 3$ in $P(x)$, we also have

$$
\begin{aligned}
P(3) &= 3^3 - 2(3^2) + 3 - 5 \\
&= 27 - 18 + 3 - 5 \\
&= 7.
\end{aligned}
$$

Sometimes it is more expedient to find the remainder by using synthetic division than by direct substitution.

Example 4.4

Given that

$$P(x) = x^4 - 2x^3 - 3x^2 + x + 6,$$

find $P(5)$.

Solution: Using synthetic division, we have

$$
\begin{array}{r|rrrrr}
5 & 1 & -2 & -3 & 1 & 6 \\
  & \downarrow & 5 & 15 & 60 & 305 \\
\hline
  & 1 & 3 & 12 & 61 & 311
\end{array}.
$$

Hence $P(5) = 311$.

Notice that the method of determining $P(5)$ by substituting $x = 5$ in $P(x)$ would be cumbersome.

Let us now consider dividing a polynomial having complex numbers as coefficients by a polynomial of the form $x - k$, where k is a complex number. Remember that synthetic division is carried out as usual except that i^2, wherever it occurs, is replaced by the real number -1.

Example 4.5

Let

$$P(x) = 2x^3 + 3x^2 - x + 6.$$

Find $P(1 - i)$.

Solution: Using synthetic division as usual, we obtain

$$
\begin{array}{r|rrrr}
1 - i & 2 & 3 & -1 & 6 \\
      & \downarrow & 2 - 2i & 3 - 7i & -5 - 9i \\
\hline
      & 2 & 5 - 2i & 2 - 7i & 1 - 9i
\end{array}.
$$

Thus $P(1 - i) = 1 - 9i$. We suggest that you compare this procedure to that of substituting $x = 1 - i$ in the polynomial!

factor theorem

The following proposition, called the **factor theorem,** is an immediate consequence of the remainder theorem.

Proposition 4.1.2

Factor Theorem: The polynomial $P(x)$ with complex coefficients has a factor $x - k$ if and only if $P(k) = 0$.

Proof: From the remainder theorem, we have

$$P(x) = (x - k) \cdot Q(x) + P(k).$$

If $P(k) = 0$, then

$$P(x) = (x - k) \cdot Q(x).$$

Hence $x - k$ is a factor of $P(x)$. Conversely, if $x - k$ is a factor of $P(x)$, then we have

$$P(x) = (x - k) \cdot Q(x)$$

which implies that the remainder $P(k) = 0$.

Example 4.6

Show that $x + 1$ is a factor of

$$P(x) = x^3 + 3x^2 - 4x - 6.$$

Solution: If $P(-1) = 0$, then by Proposition 4.1.2, $x + 1$ is a factor of $P(x)$. Using synthetic division, we observe that

$$
\begin{array}{r|rrrr}
-1 & 1 & 3 & -4 & -6 \\
   & \downarrow & -1 & -2 & 6 \\
\hline
   & 1 & 2 & -6 & 0
\end{array}
$$

Since the remainder is 0, we conclude that $x + 1$ is a factor of $P(x)$.

The converse of the factor theorem, which we now state and prove, provides a useful tool for obtaining zeros of a polynomial $P(x)$.

Proposition 4.1.3

If $x - k$ is a factor of the polynomial $P(x)$, then $x = k$ is a zero of $P(x)$.

Proof: Since $x - k$ is a factor of $P(x)$, we have

$$P(x) = (x - k) \cdot Q(x),$$

and the remainder r obtained by dividing $P(x)$ by $x - k$ is 0. By the remainder theorem, we have $P(k) = 0$, which means that $x = k$ is a zero of the polynomial $P(x)$.

Example 4.7

Given that $x = 2$ is a zero of the polynomial

$$P(x) = x^3 - 3x^2 - 4x + 12,$$

find all zeros of $P(x)$.

Solution: The zeros of a polynomial $P(x)$ are the solutions of $P(x) = 0$. If $P(x)$ is divided by $x - 2$, we notice that

$$
\begin{array}{r|rrrr}
2 & 1 & -3 & -4 & 12 \\
  & \downarrow & 2 & -2 & -12 \\
\hline
  & 1 & -1 & -6 & 0
\end{array}
$$

Thus the quotient is

$$Q(x) = x^2 - x - 6.$$

Observe that $Q(x)$ can be factored into $(x - 3)(x + 2)$. The zeros of $Q(x)$ are -2 and 3, and so the zeros of $P(x)$ are 2, -2, and 3.

EXERCISES

In Exercises 1 to 12, find the remainder $P(k)$ when $P(x)$ is divided by a polynomial of the form $x - k$.

1. $P(x) = x^2 + 2x + 5$; $x - 1$

2. $P(x) = 2x^2 - 3x + 1$; $x - 2$

3. $P(x) = x^3 + 2x^2 - 3x - 4$; $x - 2$

4. $P(x) = x^3 - 2x^2 + x + 1$; $x + 2$

5. $P(x) = 2x^3 + 3x^2 - 4x + 5$; $x + 1$

6. $P(x) = 3x^3 - 4x^2 + x + 3$; $x + 3$

7. $P(x) = x^3 + 27$; $x + 3$

8. $P(x) = x^3 - 64$; $x - 4$

9. $P(x) = x^3 + 2ix^2 + 2x + (1 + i)$; $x - 2i$

10. $P(x) = x^3 + 3x^2 - ix + 1$; $x + i$

11. $P(x) = 2x^2 - 3x + 4$; $x - (1 + i)$

12. $P(x) = 2x^3 + x^2 - 3x + 5$; $x - (1 + 2i)$

Find the value of the polynomial $P(x)$ for the given value of x in Exercises 13 to 22.

13. $P(x) = 2x^2 - 3x + 1$; 2

14. $P(x) = 3x^2 - x + 4$; -3

15. $P(x) = x^3 + x^2 + x + 1$; -1

16. $P(x) = 2x^3 - 6x^2 + x + 1$; 3

17. $P(x) = x^3 - 2x^2 + 3x + 4$; i

18. $P(x) = x^3 - x^2 + 2x + 3$; $2i$

19. $P(x) = x^4 + ix^3 - x^2 + ix + (2 - 3i)$; $-i$

20. $P(x) = x^4 - 2x^3 + x^2 + 3$; $-2i$

21. $P(x) = x^2 + 4x - 6$; $1 - i$

22. $P(x) = x^2 + x + 1$; $2 + i$

In Exercises 23 to 32, determine whether the polynomial of the form $x - k$ is a factor of the polynomial $P(x)$. Use the factor theorem.

23. $P(x) = x^3 + 2x^2 - 12x - 9$; $x - 3$

24. $P(x) = x^3 - 4x^2 + 2x + 3$; $x + 1$

25. $P(x) = 5x^3 + 12x^2 - 36x - 16$; $x - 2$

26. $P(x) = x^3 - 4x^2 - 18x + 19$; $x + 3$

27. $P(x) = 4x^3 - 4x^2 - 11x + 6$; $x - \frac{1}{2}$

28. $P(x) = 2x^4 + 5x^3 + 3x^2$; $x + \frac{3}{2}$

29. $P(x) = x^3 - x^2 + 4x - 4$; $x - 2i$

30. $P(x) = x^3 - x^2 + 4x - 4$; $x + 2i$

31. $P(x) = 2x^3 - 5x^2 + 6x - 2$; $x - (1 + i)$

32. $P(x) = x^3 - 6x^2 + 13x - 10$; $x - (2 + i)$

In Exercises 33 to 38, determine the value of k such that the second polynomial is a factor of the first.

33. $P(x) = 2x^3 + 3x^2 - x + k$; $x - 1$

34. $P(x) = x^3 - 6x^2 + 5x + k$; $x - 3$

35. $P(x) = x^3 - 4x^2 + 2kx + 4$; $x - 2$

36. $P(x) = 2x^4 + 5x^3 - 2x^2 + kx + 3$; $x + 3$

37. $P(x) = kx^3 - 2x^2 + kx + 6$; $x - 2$

38. $P(x) = kx^4 + 3x^3 - 2x^2 + 3kx - 20$; $x + 4$

39. Given that $x = -2$ is a zero of the polynomial

$$P(x) = x^3 + 4x^2 + x - 6,$$

find all zeros of $P(x)$.

40. Given that $x = 1$ is a zero of the polynomial

$$P(x) = x^3 - 2x^2 - x + 2,$$

find all zeros of $P(x)$.

41. Given that $P(x) = 1.2x^2 + 0.03x + 0.078$, find $P(2.1)$ by
   (a) synthetic division and   **(b)** direct substitution.

42. Given that

$$P(x) = 2.3x^3 + 1.21x^2 - 0.912x + 0.7854,$$

find $P(1.5)$ by   **(a)** synthetic division and   **(b)** direct substitution.

43. Given that

$$P(x) = 3.4x^3 + 2.43x^2 - 1.534x + 1.9854,$$

find $P(1.7)$ by   **(a)** synthetic division and   **(b)** direct substitution.

44. Given that

$$P(x) = 0.2x^4 + 1.19x^3 + 2.435x^2 + 0.986x - 2.34587,$$

find $P(3.2)$ by   **(a)** synthetic division and   **(b)** direct substitution.

4.2

COMPLEX ZEROS OF POLYNOMIAL FUNCTIONS

In this section, we consider the number of zeros to be expected from a polynomial of degree n. A first-degree polynomial

$$P(x) = ax + b \qquad (a \neq 0),$$

has exactly one zero, namely, $-b/a$. A second-degree polynomial

$$P(x) = ax^2 + bx + c \qquad (a \neq 0),$$

has at least one and at most two zeros. These observations lead to the following generalization, which we state without proof.

Propostion 4.2.1   Every polynomial $P(x)$ of degree $n \geq 1$ has at least one zero.

fundamental theorem   This proposition, known as the **fundamental theorem of algebra,** was first
of algebra   proved by Gauss in 1799. The proposition asserts that a zero of a polynomial

exists but gives no indication of how to find it. We use this proposition to prove the following statement.

Proposition 4.2.2

Every polynomial $P(x)$ of degree $n \geq 1$ can be factored into n linear factors.

Proof: Suppose that $P(x)$ is a polynomial of degree n. By Proposition 4.2.1, we know that $P(x)$ has at least one zero. Let us label this zero as k_1. By the factor theorem, $x - k_1$ is a factor of $P(x)$, and we write

$$P(x) = (x - k_1)Q_1(x),$$

where $Q_1(x)$ is the quotient obtained by dividing $P(x)$ by $x - k_1$ and is of degree $n - 1$. Now $Q_1(x)$ has a zero, say, k_2, and we write

$$Q_1(x) = (x - k_2)Q_2(x).$$

This means that

$$P(x) = (x - k_1)(x - k_2)Q_2(x),$$

where $Q_2(x)$ is a polynomial of degree $n - 2$. Repeating this factoring process n times, we obtain

$$P(x) = (x - k_1)(x - k_2) \cdots (x - k_n)Q_n(x),$$

where $Q_n(x)$ is a polynomial of degree 0 and consists of the leading constant coefficient $a_n \neq 0$.

We now use Proposition 4.2.2 to prove the following.

Proposition 4.2.3

Every polynomial $P(x)$ of degree $n \geq 1$ has at most n zeros.

Proof: We use an indirect proof here. Suppose a polynomial of degree $n \geq 1$ has $n + 1$ distinct zeros, which we label as $k_1, k_2, \ldots, k_n$, and k. By Proposition 4.2.2, we may use n zeros $k_1, k_2, \ldots, k_n$ and write

$$P(x) = a_n(x - k_1)(x - k_2) \cdots (x - k_n) \qquad (a_n \neq 0).$$

Observe that since k is also a zero of $P(x)$, we have $P(k) = 0$. Thus

$$a_n(k - k_1)(k - k_2) \cdots (k - k_n) = 0 \qquad (a_n \neq 0).$$

The fact that $k_1, k_2, \ldots, k_n$ and k are $n + 1$ distinct zeros of $P(x)$ makes each factor on the left-hand side different from 0. Since the product of nonzero numbers cannot be zero, we have a contradiction. Hence our assumption that a polynomial $P(x)$ of degree n has $n + 1$ distinct zeros is false.

Example 4.8

The polynomial $P(x) = x^2 - x - 2$ is of degree 2 and therefore has at most two zeros. We see that

$$x^2 - x - 2 = 0$$

$$(x - 2)(x + 1) = 0.$$

Hence the zeros of $P(x)$ are -1 and 2.

Given the zeros of a polynomial and the value of $P(c)$ where c is not a root, we can find the polynomial.

Example 4.9

Find the polynomial $P(x)$ of degree 3 with zeros 1, -2, and 3, given that $P(2) = -20$.

Solution: By the factor theorem, $P(x)$ has factors $x - 1$, $x + 2$, and $x - 3$. Hence $P(x)$ is of the form

$$P(x) = a(x - 1)(x + 2)(x - 3)$$

for some nonzero number a. Since $P(2) = -20$, we see that

$$a(2 - 1)(2 + 2)(2 - 3) = -20$$

$$-4a = -20$$

$$a = 5.$$

Thus

$$P(x) = 5(x - 1)(x + 2)(x - 3)$$
$$= 5(x^3 - 2x^2 - 5x + 6).$$

Sometimes two or more zeros of a polynomial are equal. If each zero is counted as many times as its appears, we have a total of exactly n zeros.

Definition 4.2.1

If a factor $x - k$ appears m times in a factorization of the polynomial $P(x)$, then k is said to be a *zero of multiplicity m.*

Example 4.10

The polynomial $P(x) = x^3 - 2x^2 + x$ is of degree 3 and therefore has three zeros. We write

$$P(x) = x(x^2 - 2x + 1) = x(x - 1)(x - 1).$$

The two distinct zeros are 0 and 1. The real number 1 is a zero of multiplicity 2 because the factor $x - 1$ appears twice in the factorization of $P(x)$.

Example 4.11

Find a polynomial $P(x)$ of degree 4 that has three distinct zeros 2, 3, and -4, given that 3 is a zero of multiplicity 2.

Solution: By the factor theorem, $P(x)$ has factors $x - 2$, $x - 3$, and $x + 4$. Since 3 is a zero of multiplicity 2, the factor $x - 3$ must appear twice in the factorization of $P(x)$. This means that $P(x)$ must be of the form

$$P(x) = a(x - 2)(x - 3)^2(x + 4),$$

where a is any nonzero number. Setting $a = 1$, we have

$$P(x) = (x - 2)(x - 3)^2(x + 4)$$
$$= x^4 - 4x^3 - 11x^2 + 66x - 72.$$

If a zero of multiplicity m is counted m times, then Proposition 4.2.2 assures us that a polynomial $P(x)$ of degree $n \geq 1$ has at least n zeros. Proposition 4.2.3 guarantees that $P(x)$ has at most n zeros. Hence we have the following result.

Proposition 4.2.4

If a zero of multiplicity m is counted m times, then a polynomial $P(x)$ of degree $n \geq 1$ has exactly n zeros.

Example 4.12

The polynomial $P(x) = x^2 - 4x + 13$ is of degree 2. Using the quadratic formula, we see that the zeros of the polynomial are

$$x = \frac{-(-4) \pm \sqrt{16 - 52}}{2}$$

$$= \frac{4 \pm 6i}{2}$$

$$= 2 \pm 3i.$$

Thus we have two distinct zeros, $2 - 3i$ and $2 + 3i$.

Notice that the two zeros in Example 4.12 are alike except for the signs of the imaginary parts. Recall from Chapter 1 that complex numbers that differ only in the signs of imaginary parts are called **conjugate complex numbers.** For instance, if $z = a + bi$, then $\bar{z} = a - bi$ is the complex conjugate of $z = a + bi$. Each real number a is its own conjugate, since $a + 0i = a - 0i = a$. For any

conjugate complex
numbers

polynomial $P(x)$ with real coefficients, it can be shown that conjugate complex zeros always occur in pairs. We prove this result in the following proposition.

Proposition 4.2.5

If $P(x)$ is a polynomial of degree n with real coefficients and if the complex number z is a zero of $P(x)$, then its conjugate $\bar{z}$ is also a zero of $P(x)$.

Proof: Let

$$P(x) = a_nx^n + a_{n-1}x^{n-1} + \cdots + a_1x + a_0 \qquad (a_n \neq 0),$$

where $a_0, a_1, \ldots, a_n$ are real numbers. If z is a zero of $P(x)$, then $P(z) = 0$, and we have

$$a_nz^n + a_{n-1}z^{n-1} + \cdots + a_1z + a_0 = 0.$$

Taking conjugates of both sides, we have

$$\overline{a_nz^n + a_{n-1}z^{n-1} + \cdots + a_1z + a_0} = \overline{0}.$$

Since the conjugate of the sum (product) is the sum (product) of their conjugates, by Proposition 1.12.1 we obtain

$$\overline{a_nz^n} + \overline{a_{n-1}z^{n-1}} + \cdots + \overline{a_1z} + \overline{a_0} = \overline{0}$$

and

$$\overline{a_n}\,\overline{z^n} + \overline{a_{n-1}}\,\overline{z^{n-1}} + \cdots + \overline{a_1}\,\overline{z} + \overline{a_0} = \overline{0}.$$

Each real number is its own conjugate. This means that

$$\overline{a_n} = a_n, \overline{a_{n-1}} = a_{n-1}, \ldots, \overline{a_1} = a_1, \overline{a_0} = a_0, \qquad \text{and} \qquad \overline{0} = 0.$$

Thus

$$a_n\overline{z^n} + a_{n-1}\overline{z^{n-1}} + \cdots + a_1\overline{z} + a_0 = 0.$$

Since

$$\overline{z^n} = (\bar{z})^n, \overline{z^{n-1}} = (\bar{z})^{n-1}, \ldots, \overline{z} = z$$

by Proposition 1.12.1, we have

$$a_n(\bar{z})^n + a_{n-1}(\bar{z})^{n-1} + \cdots + a_1(\bar{z}) + a_0 = 0,$$

or

$$P(\bar{z}) = 0.$$

Hence, $\bar{z}$ is a zero of $P(x)$.

Note the requirement that the polynomial $P(x)$ have real coefficients. For instance, observe that $P(x) = x - (2 + 3i)$ has $2 + 3i$ as a zero, but its conjugate complex $2 - 3i$ is not a zero of $P(x)$. Further, Proposition 4.2.5 is important in helping us predict the number of real zeros of a polynomial with real coefficients. A polynomial $P(x)$ of odd degree ($n \geq 1$) has at least one real zero because conjugate complex zeros occur in pairs. On the other hand, a polynomial of even degree ($n \geq 2$) need not have real zeros but may have up to n real zeros.

Example 4.13

Find the polynomial $P(x)$ of lowest degree that has 2 and $3 + i$ as its zeros.

Solution: By Proposition 4.2.5, $3 - i$ is another zero of $P(x)$. Thus the polynomial $P(x)$ has three distinct zeros, 2, $3 + i$, and $3 - i$. By the factor theorem, $P(x)$ has factors $x - 2$, $x - (3 + i)$, and $x - (3 - i)$. Hence $P(x)$ is of the form

$$P(x) = a(x - 2)(x - 3 - i)(x - 3 + i),$$

where a is a nonzero real number. Setting $a = 1$ and performing the multiplication, we have

$$P(x) = x^3 - 8x^2 + 22x - 20.$$

Example 4.14

Given that $2 + 3i$ is a zero of $P(x) = x^4 - 8x^3 + 32x^2 - 64x + 39$, find all zeros of $P(x)$.

Solution: Since $2 + 3i$ is a zero of $P(x)$, by Proposition 4.2.5, $2 - 3i$ is also a zero of $P(x)$. Using synthetic division, we have

$$
\begin{array}{r|ccccc}
2 + 3i & 1 & -8 & 32 & -64 & 39 \\
 & \downarrow & 2 + 3i & -21 - 12i & 58 + 9i & -39 \\
\hline
 & 1 & -6 + 3i & 11 - 12i & -6 + 9i & 0
\end{array},
$$

so that

$$P(x) = (x - 2 - 3i)Q(x)$$

where

$$Q(x) = x^3 + (-6 + 3i)x^2 + (11 - 12i)x + (-6 + 9i).$$

Since $2 - 3i$ is another zero of $P(x)$, it must also be a zero of $Q(x)$. Using synthetic division again, we obtain

$$
\begin{array}{r|cccc}
2 - 3i & 1 & -6 + 3i & 11 - 12i & -6 + 9i \\
 & \downarrow & 2 - 3i & -8 + 12i & 6 - 9i \\
\hline
 & 1 & -4 & 3 & 0
\end{array}.
$$

This means that

$$
\begin{aligned}
Q(x) &= (x - 2 + 3i)(x^2 - 4x + 3) \\
&= (x - 2 + 3i)(x - 1)(x - 3).
\end{aligned}
$$

Hence

$$P(x) = (x - 2 - 3i)(x - 2 + 3i)(x - 1)(x - 3)$$

and the zeros of $P(x)$ are $2 + 3i$, $2 - 3i$, 1, and 3.

EXERCISES

Find the zeros of the polynomials in Exercises 1 to 6.

1. $P(x) = x^2 - 5x + 6$ **2.** $P(x) = 7x^2 - 19x + 10$ **3.** $P(x) = x^2 + 6x - 1$

4. $P(x) = 2x^2 - 3x - 4$ **5.** $P(x) = x(x^2 + 1)$ **6.** $P(x) = x^3 - 4x^2 + 13x$

Find the zeros of the polynomials in Exercises 7 to 12 and state the multiplicity of each.

7. $P(x) = x^2(x^2 - 2x + 1)$ **8.** $P(x) = x^3(x^2 - x - 2)$ **9.** $P(x) = (x - 1)^2(x + 2)$

10. $P(x) = (x^2 - 6x + 8)^2$ **11.** $P(x) = (x - 1)^3(2x + 1)$ **12.** $P(x) = (2x - 3)(3x + 2)^3$

Find the polynomial with the indicated zeros and satisfying the given conditions in Exercises 13 to 18.

13. $-1, -3, 2; P(0) = -6$ **14.** $1, -1, 2; P(3) = 40$ **15.** $-2, i, -i; P(1) = 6$

16. $2 + i, 2 - i, 3; P(0) = -15$ **17.** $-1, -2, 3, 2; P(1) = 12$ **18.** $-1, 7, 2i, -2i; P(1) = -60$

19. Find a polynomial $P(x)$ of degree 4 such that 1 is a zero of multiplicity 3 and 2 is a zero of multiplicity 1.

20. Find a polynomial $P(x)$ of degree 4 such that -3 is a zero of multiplicity 1, 2 is a zero of multiplicity 2, and 1 is a zero of multiplicity 1.

21. Find a polynomial $P(x)$ of degree 4 that has three distinct zeros $2 + i$, $2 - i$, and 3, given that 3 is a zero of multiplicity 2.

22. Find a polynomial $P(x)$ of degree 4 that has three distinct zeros $1 + 2i$, $1 - 2i$, and 4, given that 4 is a zero of multiplicity 2.

In Exercises 23 to 28, find the real polynomial of lowest degree with the given numbers as some of its zeros.

23. $-2, i$ **24.** $3, 2i$ **25.** $2, 1 + i$

26. $4, 3 + 2i$ **27.** $-1, 7, 2i$ **28.** $1, 2, 2 + 3i$

One zero is given for each polynomial function in Exercises 29 to 36. Find other zeros of $P(x)$.

29. $P(x) = x^3 + 2x^2 - 5x - 6; 2$ **30.** $P(x) = x^3 + 2x^2 - x - 2; -1$

31. $P(x) = x^3 + 2x^2 + x + 2; i$ **32.** $P(x) = x^3 - 4x^2 + x - 4; i$

33. $P(x) = x^3 - 5x^2 + 7x + 13; 3 + 2i$ **34.** $P(x) = x^4 - x^3 + 2x^2 - 4x - 8; 2i$

35. $P(x) = x^4 - 2x^3 - 7x^2 + 8x + 12; 3$ **36.** $P(x) = x^4 - 6x^3 - 3x^2 - 24x - 28; 7$

37. Given that 1.1 is a zero of $P(x) = x^3 - 3.6x^2 + 4.31x + 1.716$, find all zeros of $P(x)$.

38. Given that 2.3 is a zero of $P(x) = x^3 - 7.3x^2 + 16.94x + 12.512$, find all zeros of $P(x)$.

4.3

RATIONAL ZEROS OF POLYNOMIAL FUNCTIONS

In this section, we consider polynomials with integral coefficients. Such polynomials have zeros, of course, but these may or may not be rational. For example, $x^2 - 1 = 0$ has rational zeros, but $x^2 - 2 = 0$ does not. We will determine whether a polynomial with integral coefficients has rational zeros and, if so, how to find them.

Proposition 4.3.1

Let

$$P(x) = a_n x^n + a_{n-1} x^{n-1} + \cdots + a_2 x^2 + a_1 x + a_0 \qquad (a_n \neq 0),$$

be a polynomial with integral coefficients. If a rational number p/q (reduced to lowest terms) is a zero of $P(x)$, then p is a factor of a_0 and q is a factor of a_n.

Proof: Since p/q is a zero of $P(x)$, it must satisfy the equation $P(x) = 0$. This means that

$$a_n \left(\frac{p}{q} \right)^n + a_{n-1} \left(\frac{p}{q} \right)^{n-1} + \cdots + a_2 \left(\frac{p}{q} \right)^2 + a_1 \left(\frac{p}{q} \right) + a_0 = 0. \qquad (4.1)$$

We multiply both sides of the equation by q^n to obtain

$$a_n p^n + a_{n-1} p^{n-1} q + \cdots + a_2 p^2 q^{n-2} + a_1 p q^{n-1} + a_0 q^n = 0 \qquad (4.2)$$

which we rewrite as

$$a_n p^n + a_{n-1} p^{n-1} q + \cdots + a_2 p^2 q^{n-2} + a_1 p q^{n-1} = -a_0 q^n \qquad (4.3)$$

or

$$p(a_n p^{n-1} + a_{n-1} p^{n-2} q + \cdots + a_1 q^{n-1}) = -a_0 q^n. \qquad (4.4)$$

Since the left-hand side of (4.4) has p as a factor, p must also be a factor of $-a_0 q^n$. Further, since p and q have no common factor other than 1, we have p as a factor of a_0. Similarly, we can rewrite (4.2) in the form

$$q(a_{n-1} p^{n-1} + a_{n-2} p^{n-2} q + \cdots + a_1 p q^{n-2} + a_0 q^{n-1}) = -a_n p^n. \qquad (4.5)$$

This equation shows that q is a factor of $-a_n p^n$, but q and p have no common factor other than 1; so we conclude that q must be a factor of a_n.

Notice what the proposition says. It does not determine whether $P(x)$ has any rational zeros. The proposition merely says that if there are rational zeros, then they must be of a specific form.

Example 4.15

Find the rational zeros (if any) of the polynomial

$$P(x) = 2x^3 + 3x^2 + 2x + 3.$$

Solution: If the polynomial has a rational zero of the form p/q, then p must be a factor of 3, and q must be a factor of 2. Possible choices of p are

$$1, \qquad -1, \qquad 3, \qquad \text{and} \qquad -3,$$

and possible choices of q are

$$1, \qquad -1, \qquad 2, \qquad \text{and} \qquad -2.$$

Therefore the set of rational zeros of $P(x)$ must be a subset of the set

$$\left\{1, -1, 3, -3, \frac{1}{2}, -\frac{1}{2}, \frac{3}{2}, -\frac{3}{2}\right\}.$$

Further, since all coefficients of $P(x)$ are positive, no positive values of x, when substituted in $P(x)$, will make $P(x) = 0$. This means that no positive rational can be a zero, and the only possibilities that remain are -1, -3, $-\frac{1}{2}$, and $-\frac{3}{2}$. To determine which of these (if any) are zero, we must calculate $P(-1)$, $P(-3)$, and so on. We find that $P(-\frac{3}{2}) = 0$, since by synthetic division, we get

$$
\begin{array}{r|rrrr}
-\frac{3}{2} & 2 & 3 & 2 & 3 \\
 & \downarrow & -3 & 0 & -3 \\
\hline
 & 2 & 0 & 2 & 0
\end{array}
$$

Thus

$$P(x) = \left(x + \frac{3}{2}\right)(2x^2 + 2)$$

$$= 2\left(x + \frac{3}{2}\right)(x^2 + 1).$$

The other two zeros of $P(x)$ are zeros of the quadratic $x^2 + 1$, namely, i and $-i$. Hence the only rational zero of $P(x)$ is $-\frac{3}{2}$.

Example 4.16

Find the rational zeros (if any) of the polynomial

$$P(x) = x^4 + x^3 - 7x^2 - x + 6.$$

Solution: If p/q is a rational zero of $P(x)$, then p must be a factor of 6, and q must be a factor of 1. The possible choices of p are

$$1, \quad -1, \quad 2, \quad -2, \quad 3, \quad -3, \quad 6, \quad \text{and} \quad -6.$$

and the possible choices of q are

$$1 \quad \text{and} \quad -1.$$

Thus the set of rational zeros of $P(x)$ must be a subset of the set

$$\{1, -1, 2, -2, 3, -3, 6, -6\}.$$

Obviously, not all the eight members of this set can be zeros of $P(x)$, since the polynomial is of degree 4 and can therefore have *at most* four zeros. By synthetic division, we find that $P(1) = 0$, since

$$
\begin{array}{r|rrrrr}
1 & 1 & 1 & -7 & -1 & 6 \\
 & \downarrow & 1 & 2 & -5 & -6 \\
\hline
 & 1 & 2 & -5 & -6 & 0
\end{array}
$$

and we have

$$P(x) = (x - 1)Q(x),$$

where

$$Q(x) = x^3 + 2x^2 - 5x - 6.$$

Notice that any rational zero of $Q(x)$ must again be a subset of the set

$$\{1, -1, 2, -2, 3, -3, 6, -6\}.$$

We see that $Q(-1) = 0$ because

$$
\begin{array}{r|rrrr}
-1 & 1 & 2 & -5 & -6 \\
   & \downarrow & -1 & -1 & 6 \\
\hline
   & 1 & 1 & -6 & 0
\end{array}.
$$

Therefore,

$$Q(x) = (x + 1)(x^2 + x - 6) = (x + 1)(x - 2)(x + 3).$$

Thus

$$P(x) = (x - 1)(x + 1)(x - 2)(x + 3),$$

indicating that the four rational zeros of $P(x)$ are $1, -1, 2,$ and -3.

Although Proposition 4.3.1 pertains specifically to polynomials with integral coefficients, it may also be used to investigate the rational zeros of a polynomial whose coefficients are rational. First we multiply both sides of the polynomial by the least common multiple of the denominators to ensure that all fractional coefficients are converted to integers. Then we use Proposition 4.3.1.

Example 4.17 Find all rational zeros of

$$P(x) = \frac{2}{3} x^3 - \frac{1}{2} x^2 + \frac{2}{3} x - \frac{1}{2}.$$

Solution: The least common multiple of the denominators is 6. Multiplying both sides of the polynomial by 6, we get

$$6P(x) = 4x^3 - 3x^2 + 4x - 3.$$

Now we can study the rational zeros of $P(x)$ by investigating the rational zeros of $6P(x)$. The rational zeros of this polynomial are members of the set

$$\{1, -1, 3, -3, \tfrac{1}{2}, -\tfrac{1}{2}, \tfrac{3}{2}, -\tfrac{3}{2}, \tfrac{1}{4}, -\tfrac{1}{4}, \tfrac{3}{4}, -\tfrac{3}{4}\}.$$

To determine which of these, if any, are zeros of $P(x)$, we calculate $P(1), P(-1),$ $\ldots, P(-\tfrac{3}{4})$ by using synthetic division. We find that $\tfrac{3}{4}$ is the only rational zero of $6P(x)$, and we have

$$6P(x) = (x - \tfrac{3}{4})(4x^2 + 4) = 4(x - \tfrac{3}{4})(x^2 + 1).$$

The other two zeros of the quadratic $x^2 + 1$ are i and $-i$. Hence the only rational zero of $P(x)$ is $\tfrac{3}{4}$.

Example 4.18   Show that the polynomial $P(x) = x^3 + 3x - 5$ has no rational zeros.

Solution: Since the leading coefficient of $P(x)$ is 1, the only possible choices for p/q is a subset of the set

$$\{1, -1, 5, -5\}.$$

We find that when we substitute these numbers in $P(x)$, none are zeros of $P(x)$. Hence $P(x)$ has no rational zeros.

EXERCISES

Find the rational zeros, if they exist, in Exercises 1 to 24.

1. $P(x) = x^3 + 1$
2. $P(x) = x^3 - 1$
3. $P(x) = 8x^3 + 27$
4. $P(x) = 8x^3 - 27$
5. $P(x) = x^3 - 3x - 2$
6. $P(x) = x^3 + x^2 - x - 10$
7. $P(x) = x^3 - 8x^2 + 17x - 6$
8. $P(x) = x^3 + 2x^2 - x - 2$
9. $P(x) = x^3 + 3x^2 - x - 3$
10. $P(x) = 2x^3 - 7x^2 - 27x - 18$
11. $P(x) = x^3 - x^2 - 4x - 3$
12. $P(x) = x^3 + 9x^2 - 14x - 24$
13. $P(x) = 2x^3 + x^2 - 18x - 20$
14. $P(x) = x^4 - 9x^2 + 2x + 6$
15. $P(x) = 3x^3 - 5x^2 - 14x - 4$
16. $P(x) = 2x^3 + x^2 - 3$
17. $P(x) = x^4 - 2x^3 - 7x^2 + 8x + 12$
18. $P(x) = x^4 + 2x^3 + 2x^2 - 4x - 8$
19. $P(x) = x^3 + 2x^2 + x + 2$
20. $P(x) = x^4 - 6x^3 - 3x^2 - 24x - 28$
21. $P(x) = 48x^4 - 52x^3 + 13x - 3$
22. $P(x) = 12x^4 + 4x^3 - 3x^2 - x$
23. $P(x) = \tfrac{2}{3}x^3 + \tfrac{17}{6}x^2 - \tfrac{55}{6}x + 5$
24. $P(x) = 4x^4 - 4x^3 - 25x^2 + x + 6$
25. Show that the polynomial $P(x) = x^4 + x^2 + 2x + 6$ has no rational zeros.
26. Show that the polynomial $P(x) = 2x^4 - 3x^3 - 8x^2 - 5x - 3$ has no rational zeros.

4.4

REAL ZEROS OF POLYNOMIAL FUNCTIONS

We have seen that Proposition 4.3.1 provides a tool for finding all possible rational zeros. However, it gives no information about the irrational zeros. We state a proposition that will help us determine an interval in which the real zeros will occur, if they exist. The proof is beyond our current scope.

Proposition 4.4.1

> Let $P(x)$ be a polynomial with real coefficients, and let the coefficient of the highest power of x be positive. If $k_1 \geq 0$ and if the terms in the third row of synthetic division of $P(x)$ by $x - k_1$ are all nonnegative, then k_1 is an upper bound of the real zeros of $P(x)$. If $k_2 \leq 0$ and if the terms in the third row of synthetic division of $P(x)$ by $x - k_2$ alternate in sign, then k_2 is a lower bound of the real zeros of $P(x)$.

Example 4.19

Find an upper bound and a lower bound of the real zeros of

$$P(x) = 3x^3 + 7x^2 + x - 4.$$

Solution: If p/q is a rational zero, then p must divide -4, and q must divide 3. The possibilities for p are

$$1, \quad -1, \quad 2, \quad -2, \quad 4, \quad \text{and} \quad -4,$$

and the possibilities for q are

$$1, \quad -1, \quad 3, \quad \text{and} \quad -3.$$

Hence the possible rational zeros of $P(x)$ are members of the set

$$\{1, -1, 2, -2, 4, -4, \tfrac{1}{3}, -\tfrac{1}{3}, \tfrac{2}{3}, -\tfrac{2}{3}, \tfrac{4}{3}, -\tfrac{4}{3}\}.$$

By Proposition 4.4.1, we find that 1 is an upper bound of all the real zeros of $P(x)$. This follows from the fact that if we divide $P(x)$ by $x - 1$, we have

$$
\begin{array}{r|rrrr}
1 & 3 & 7 & 1 & -4 \\
& \downarrow & 3 & 10 & 11 \\
\hline
& 3 & 10 & 11 & 7
\end{array},
$$

and all entries in the third row are nonnegative. In searching for real zeros, we will therefore not test any zero greater than 1. Hence we rule out the values $\tfrac{4}{3}$, 2, and 4. Similarly, if we divide $P(x)$ by $x + 3$, we find that

$$
\begin{array}{r|rrrr}
-3 & 3 & 7 & 1 & 4 \\
& \downarrow & -9 & 6 & -21 \\
\hline
& 3 & -2 & 7 & -25
\end{array}
$$

and entries in the third row alternate in sign. By Proposition 4.4.1, it follows that -3 is a lower bound of all the real zeros of $P(x)$. We therefore eliminate -4 as a possible real zero.

Descartes' rule of signs

Another proposition that we state, but do not prove, is **Descartes' rule of signs.** This proposition helps us determine the number of positive and negative real zeros we can expect from a polynomial $P(x)$. We simply count the number of *variations in sign* occurring in $P(x)$. Let us first see what is meant by variations in sign.

variations in sign

A variation in sign is counted when the coefficients of two successive terms in a polynomial differ in sign. For example, the polynomial $P(x) = 6x^4 - 23x^3 + 19x^2 + 5x - 3$ has three variations in sign, as shown below:

$$+\,6x^4 - 23x^3 + 19x^2 + 5x - 3.$$

The polynomial $P(x) = x^4 + x^3 + x^2 + x + 1$ does not have any variation in sign.

Proposition 4.4.2

Descartes' Rule of Signs: Let $P(x)$ be a polynomial with real coefficients. Then the following apply.

(a) The number of positive real zeros is either equal to the number of variations in sign occurring in the coefficients of $P(x)$ or less than the number of variations by a positive even integer.

(b) The number of negative real zeros is either equal to the number of variations in sign occurring in the coefficients of $P(-x)$ or less than the number of variations by a positive even integer.

Example 4.20

The polynomial $P(x) = x^4 - 2x^3 - x^2 + 2x - 3$ has three variations in sign shown:

$$P(x) = +x^4 - 2x^3 - x^2 + 2x - 3.$$

Thus by Proposition 4.4.2, $P(x)$ has either three or one positive real zero. Now observe that

$$P(-x) = x^4 + 2x^3 - x^2 - 2x - 3$$

has only one variation in sign. Therefore we expect one negative real zero for $P(x)$.

Example 4.21

Find the real zeros of

$$P(x) = 6x^4 - 23x^3 + 19x^2 + 5x - 3.$$

Solution: Since the degree of the polynomial is 4, we know that $P(x)$ has at most four zeros. There are three variations in sign for $P(x)$. Hence $P(x)$ has either three or one positive real zero. Since $P(-x)$ has only one variation in sign, we expect $P(x)$ to have one negative real zero. Thus we might have three positive and one negative real zeros or one positive and one negative real zeros and two complex zeros for $P(x)$.

We leave it for the reader to verify that the possible rational zeros of $P(x)$ are members of the set

$$\{1, -1, 3, -3, \tfrac{1}{2}, -\tfrac{1}{2}, \tfrac{3}{2}, -\tfrac{3}{2}, \tfrac{1}{3}, -\tfrac{1}{3}, \tfrac{1}{6}, -\tfrac{1}{6}\}$$

By using synthetic division, we see that -1 is a lower bound for the real zeros of $P(x)$:

$$
\begin{array}{r|rrrrr}
-1 & 6 & -23 & 19 & 5 & -3 \\
 & \downarrow & -6 & 29 & -48 & 43 \\
\hline
 & 6 & -29 & 48 & -43 & 40
\end{array}
$$

Hence we rule out the value -3 as a possible zero. Further, we find that $\frac{1}{3}$ is a rational zero of $P(x)$, since

$$
\begin{array}{r|rrrrr}
\frac{1}{3} & 6 & -23 & 19 & 5 & -3 \\
 & \downarrow & 2 & -7 & 4 & 3 \\
\hline
 & 6 & -21 & 12 & 9 & 0
\end{array}
$$

This means that

$$P(x) = (x - \tfrac{1}{3})(6x^3 - 21x^2 + 12x + 9).$$

The quotient $Q(x) = 6x^3 - 21x^2 + 12x + 9$ has $\frac{3}{2}$ as a rational zero:

$$
\begin{array}{r|rrrr}
\frac{3}{2} & 6 & -21 & 12 & 9 \\
 & \downarrow & 9 & -18 & -9 \\
\hline
 & 6 & -12 & -6 & 0
\end{array}
$$

Hence

$$Q(x) = (x - \tfrac{3}{2})(6x^2 - 12x - 6).$$

The quadratic equation

$$6x^2 - 12x - 6 = 0$$

or

$$x^2 - 2x - 1 = 0$$

has solutions

$$x = \frac{2 \pm \sqrt{8}}{2} = 1 \pm \sqrt{2}.$$

Thus the real zeros of $P(x)$ are $\dfrac{1}{3}, \dfrac{3}{2}, 1 + \sqrt{2}$, and $1 - \sqrt{2}$.

location theorem   When irrational zeros occur, they are usually approximated by using graphing or numerical methods. The following proposition, known as the **location theorem,** which we state without proof, gives a method for narrowing our search for the real zeros.

Propostion 4.4.3

If $P(x)$ is a real polynomial and a and b are real numbers such that $P(a)$ and $P(b)$ are opposite in sign, then there is at least one real zero c in the open interval (a, b) such that $P(c) = 0$.

Example 4.22

Approximate the real zeros of

$$P(x) = x^3 - 3x^2 - 9x + 23$$

to one decimal place.

Solution: There are two variations in sign for $P(x)$. Therefore the polynomial has either two or no positive real zeros. Further, $P(-x)$ has only one variation in sign. Consequently, we expect $P(x)$ to have one negative real zero. This means that $P(x)$ has two positive and one negative real zeros. Observe that there are no rational zeros, since the only possible rational zeros, 23 and -23, do not work. To approximate irrational zeros, we make a table of function values by using either substitution or synthetic division.

x	-3	-2	-1	0	1	2	3	4	5
$P(x)$	-4	21	28	23	12	1	-4	3	28

Observe that $P(-3)$ and $P(-2)$ have opposite signs. Thus the negative real zero lies between -3 and -2. We can get a better approximation for the negative real zero of $P(x)$ by observing that

$$P(-2.9) = -0.519 \quad \text{and} \quad P(-2.8) = 2.728.$$

Since $P(-2.9) < 0$ and $P(-2.8) > 0$, we know that the negative real zero lies between -2.9 and -2.8. Since $P(-2.9)$ is closer to zero than $P(-2.8)$, we are probably safe in saying that to one decimal place of accuracy, $x = -2.9$ is a negative real zero of $P(x)$. Similarly, we can see that there are two changes in sign of $P(x)$ for positive values of x, that is,

$$P(2) = 1 \quad \text{and} \quad P(3) = -4; \qquad P(3) = -4 \quad \text{and} \quad P(4) = 3.$$

Hence there are two positive real zeros to be found. It can be shown that to one decimal place of accuracy, 2.2 and 3.8 are positive real zeros of $P(x)$.

EXERCISES

Find an upper bound and a lower bound for the real zeros of each polynomial in Exercises 1 to 10.

1. $P(x) = x^3 - 3x + 1$
2. $P(x) = x^3 - 8x + 2$
3. $P(x) = x^3 - x^2 - x - 4$
4. $P(x) = x^3 + 2x^2 - 7x - 1$

5. $P(x) = 8x^3 + 12x^2 - 26x - 15$

6. $P(x) = 8x^3 - 12x^2 - 66x + 35$

7. $P(x) = 6x^3 - 7x^2 + 11x + 35$

8. $P(x) = 2x^4 - 5x^3 + 5x - 2$

9. $P(x) = 2x^4 + 3x^3 + 70x^2 + 108x - 72$

10. $P(x) = 2x^5 + x^4 - 2x - 1$

Find the number of positive and negative real zeros in Exercises 11 to 18.

11. $P(x) = x^3 + 2x^2 - x - 2$

12. $P(x) = x^3 - x^2 - 3x - 3$

13. $P(x) = x^3 - 3x^2 - 2x - 6$

14. $P(x) = x^3 + 3x^2 - 2x - 6$

15. $P(x) = 2x^4 - 5x^3 + 5x - 2$

16. $P(x) = x^4 - 2x^3 - 7x^2 + 10x + 10$

17. $P(x) = 4x^5 - 16x^4 + 17x^3 - 19x^2 + 13x - 3$

18. $P(x) = 24x^5 - 22x^4 + 25x^3 - 20x^2 + x + 2$

Show that the polynomial has a real zero between the numbers specified in Exercises 19 to 26.

19. $P(x) = x^3 - x^2 - x - 4$; 2 and 3

20. $P(x) = 8x^3 + 12x^2 - 26x - 15$; 1 and 2

21. $P(x) = x^3 - 4x + 1$; 1 and 2

22. $P(x) = 2x^3 + x^2 + 5x - 3$; 0 and 1

23. $P(x) = 8x^3 - 12x^2 + 2x + 1$; 1 and 2

24. $P(x) = 8x^4 + 2x^3 + 15x^2 + 4x - 2$; 0 and 1

25. $P(x) = x^3 - x^2 - x - 4$; 2.2 and 2.3

26. $P(x) = x^3 - 3x^2 - 9x + 23$; -2.89 and -2.88

Find the real zeros of each polynomial in Exercises 27 to **34**.

27. $P(x) = 2x^3 - 9x^2 + 2x + 20$

28. $P(x) = x^3 - 4x^2 + 3x + 2$

29. $P(x) = 2x^3 - 9x^2 + 6x - 1$

30. $P(x) = x^3 + 3x^2 - 2x - 6$

31. $P(x) = x^4 - 9x^2 + 2x + 6$

32. $P(x) = 3x^4 + 14x^3 + 14x^2 - 8x - 8$

33. $P(x) = 9x^4 + 15x^3 - 20x^2 - 20x + 16$

34. $P(x) = 8x^4 + 6x^3 - 15x^2 - 12x - 2$

Approximate the positive real zeros in Exercises 35 to 38 to one decimal place.

35. $P(x) = x^3 - 3x - 1$

36. $P(x) = x^3 - 3x + 1$

37. $P(x) = x^3 - 4x + 1$

38. $P(x) = x^3 + 2x^2 - 1$

Approximate the real zeros in Exercises 39 and 40 to one decimal place.

39. $P(x) = x^3 - 3x + 1$

40. $P(x) = x^4 - 5x^2 + 6$

41. Show that the polynomial $P(x) = x^4 + x^2 + 1$ has no real zeros.

42. Show that the polynomial $P(x) = x^4 - 6x^2 + 10$ has no real zeros.

4.5

GRAPHING POLYNOMIAL FUNCTIONS

We graphed polynomial functions of degree 1 and 2 in Chapter 3. Now we consider graphs of polynomial functions of degree 3 and 4. First we consider a special case.

Definition 4.5.1

A function of the form

$$f(x) = ax^n \qquad (a \neq 0)$$

is called a *power function*.

For $n = 1$, the graph of $f(x)$ is a straight line passing through the origin. For $n = 2$, the graph of $f(x) = ax^2$ is a parabola that opens upward if $a > 0$ and downward if $a < 0$. Let us now examine the graph of $f(x) = ax^3$ for $a = 1$ and $a = -1$.

Example 4.23

Sketch the graph of

(a) $f(x) = x^3$ **(b)** $g(x) = -x^3$.

Solution: Setting $f(x)$ and $g(x)$ each equal to zero, we see that $x = 0$ is the only possible rational zero of these functions. Further, both $f(x)$ and $g(x)$ have as their domain the set of real numbers. Since we are unable to consider all possible ordered pairs satisfying the above equations, we use only a partial list of points to determine the shape of these curves. The values of $f(x)$ and $g(x)$ for some values of x are given here:

x	-2	$-\frac{3}{2}$	-1	$-\frac{1}{2}$	$-\frac{1}{4}$	0	$\frac{1}{4}$	$\frac{1}{2}$	1	$\frac{3}{2}$	2
$f(x)$	-8	$-\frac{27}{8}$	-1	$-\frac{1}{8}$	$-\frac{1}{64}$	0	$\frac{1}{64}$	$\frac{1}{8}$	1	$\frac{27}{8}$	8
$g(x)$	8	$\frac{27}{8}$	1	$\frac{1}{8}$	$\frac{1}{64}$	0	$-\frac{1}{64}$	$-\frac{1}{8}$	-1	$-\frac{27}{8}$	-8

Plotting the related ordered pairs and joining them by a smooth curve, we have the graphs of $f(x)$ and $g(x)$ in Figures 4.1 and 4.2, respectively.

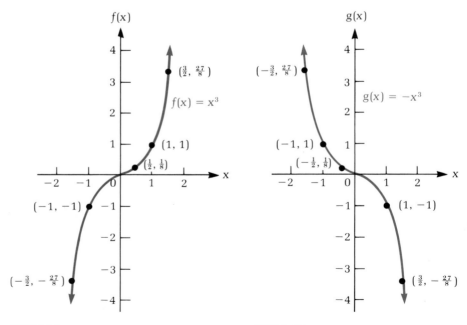

FIGURE 4.1   **FIGURE 4.2**

Geometrically, the graph of $f(x) = ax^3$ is *symmetric with respect to the origin* in the sense that if (c, d) is on the graph, then so is $(-c, -d)$. This is true in general for $f(x) = ax^n$, n odd and $n \geq 1$. To determine whether the graph is similar to that in Figure 4.1 or 4.2, we examine the coefficient of a; if $a > 0$, then the graph is similar to that in Figure 4.1; but if $a < 0$, then the graph is similar to that in Figure 4.2. We suggest that you sketch the graphs of $f(x) = x^5$ and $g(x) = -x^5$ to be convinced that the graphs of these curves are similar to those of $f(x) = x^3$ and $g(x) = -x^3$, respectively.

How sharply the graph of $f(x) = ax^3$ rises or falls depends on the numerical value of the coefficient a. For example, $f(x) = 4x^3$, then

$$f(-3) = -108, \qquad f(-2) = -32, \qquad f(-1) = -4, \qquad f(-\tfrac{1}{2}) = -\tfrac{1}{2},$$

$$f(\tfrac{1}{2}) = \tfrac{1}{2}, \qquad f(1) = 4, \qquad f(2) = 32, \qquad f(3) = 108,$$

and so on. If $f(x) = \tfrac{1}{2}x^3$, then

$$f(-2) = -4, \qquad f(-\tfrac{3}{2}) = -\tfrac{27}{16},$$

$$f(-1) = -\tfrac{1}{2}, \qquad f(-\tfrac{1}{2}) = -\tfrac{1}{16}, \qquad f(\tfrac{1}{2}) = \tfrac{1}{16},$$

$$f(1) = \tfrac{1}{2}, \qquad f(\tfrac{3}{2}) = \tfrac{27}{16}, \qquad f(2) = 4,$$

and so on. Plotting a few of the above ordered pairs, we have the graphs of $f(x) = 4x^3$ and $f(x) = \tfrac{1}{2}x^3$ in Figures 4.3 and 4.4, respectively. Observe that the greater the numerical value of coefficient a, the more sharply the graph rises.

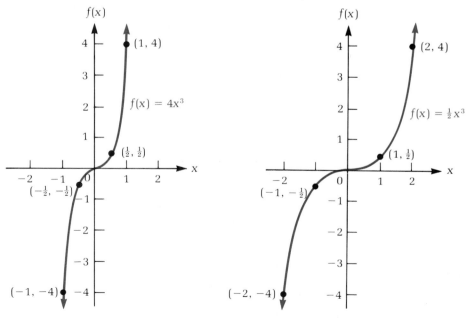

FIGURE 4.3   FIGURE 4.4

The graph of $f(x) = ax^n$, for n even and $n \geq 2$, is symmetric with respect to the y axis in the sense that if (c, d) is on the graph, then so is $(-c, d)$. For example, if $f(x) = x^4$, then

$$f(-2) = 16, \qquad f(-\tfrac{3}{2}) = \tfrac{81}{16}, \qquad f(-1) = 1, \qquad f(0) = 0,$$

$$f(1) = 1, \qquad f(\tfrac{3}{2}) = \tfrac{81}{16}, \qquad f(2) = 16,$$

and so on. Plotting some of the ordered pairs, we have the graph in Figure 4.5.

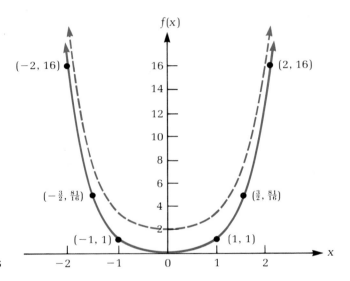

FIGURE 4.5

The graph of $f(x) = ax^n + c, c \neq 0$, is obtained by simply raising or lowering by $|c|$ units the graph of $f(x) = ax^n$, depending on whether $c > 0$ or $c < 0$. For example, the graph of

$$f(x) = x^4 + 2$$

is obtained by shifting vertically by 2 units the graph of $f(x) = x^4$ (see the dashed curve in Figure 4.5).

The graph of a general polynomial function

$$P(x) = a_n x^n + a_{n-1} x^{n-1} + \cdots + a_2 x^2 + a_1 x + a_0 \qquad (a_n \neq 0)$$

requires careful analysis. For $n = 1$, the polynomial

$$P(x) = a_1 x + a_0 \qquad (a_1 \neq 0)$$

is of first degree. These polynomials are linear functions, which we discussed in Section 3.4. For $n = 2$, the polynomial

$$P(x) = a_2 x^2 + a_1 x + a_0 \qquad (a_2 \neq 0)$$

is of second degree. Such polynomials are quadratic functions. We have shown in Section 3.5 that the graph of a quadratic function is a parabola that opens upward if $a_2 > 0$ and downward if $a_2 < 0$.

As the degree of the polynomial function increases, sketching its graph becomes increasingly difficult. We will certainly use the information regarding possible zeros of the polynomial function as an aid in sketching its graph. However, we restrict ourselves to functions whose graphs are smooth curves drawn on the basis of a relatively few but sufficient number of points.

Example 4.24

Sketch the graph of $y = P(x) = x^3 - 3x^2 - 9x + 20$.

Solution: By Descartes' rule of signs, we expect $P(x)$ to have two or no positive real zeros and one negative real zero. The possible rational zeros are

$$1, \quad -1, \quad 2, \quad -2, \quad 4, \quad -4, \quad 5, \quad -5, \quad 10, \quad -10, \quad 20, \quad -20.$$

Using synthetic division, we see that 5 is an upper bound and -3 is a lower bound of the real zeros of $P(x)$. Hence we can rule out values that are less than -3 as well as those that are greater than 5. Furthermore, 4 is a rational zero, since by synthetic division we have

$$
\begin{array}{r|rrrr}
4 & 1 & -3 & -9 & 20 \\
  & \downarrow & 4 & 4 & -20 \\
\hline
  & 1 & 1 & -5 & 0
\end{array}.
$$

Thus

$$P(x) = (x - 4)(x^2 + x - 5).$$

The quadratic equation $x^2 + x - 5 = 0$ has solutions given by

$$x = \frac{-1 \pm \sqrt{21}}{2}.$$

Thus the zeros of $P(x)$ are 4, $(-1 - \sqrt{21})/2$, and $(-1 + \sqrt{21})/2$. Next we need several ordered pairs that will enable us to sketch the graph. We use synthetic division to evaluate, say, $P(-2)$:

$$
\begin{array}{r|rrrr}
-2 & 1 & -3 & -9 & 20 \\
   & \downarrow & -2 & 10 & -2 \\
\hline
   & 1 & -5 & 1 & 18
\end{array}.
$$

Thus $P(-2) = 18$, and the point $(-2, 18)$ lies on the graph. Again, if we are to evaluate, say, $P(-1)$, we use synthetic division:

$$
\begin{array}{r|rrrr}
-1 & 1 & -3 & -9 & 20 \\
   & \downarrow & -1 & 4 & 5 \\
\hline
   & 1 & -4 & -5 & 25
\end{array}.
$$

Hence the point $(-1, 25)$ is also on the graph of $P(x)$. Since we need several such points, we use a shortened form of synthetic division. Notice in the following array that each row after the first is the bottom row in the respective synthetic division involved.

x	$1 - 3 - \;\;\;9 + 20$	Ordered Pairs
-3	$1 - 6 + \;\;\;9 - \;\;7$	$(-3, -7)$
-2	$1 - 5 + \;\;\;1 + 18$	$(-2, 18)$
-1	$1 - 4 - \;\;\;5 + 25$	$(-1, 25)$
0	$1 - 3 - \;\;\;9 + 20$	$(0, 20)$
1	$1 - 2 - 11 + \;\;9$	$(1, 9)$
2	$1 - 1 - 11 - \;\;2$	$(2, -2)$
3	$1 + 0 - \;\;\;9 - \;\;7$	$(3, -7)$
4	$1 + 1 - \;\;\;5 + \;\;0$	$(4, 0)$
5	$1 + 2 + \;\;\;1 + 25$	$(5, 25)$

Plotting the ordered pairs $(-3, -7)$, $(-2, 18)$, $(-1, 25)$, . . . , $(5, 25)$ and joining them by a smooth curve, we have the graph of the polynomial $y = P(x)$ in Figure 4.6.

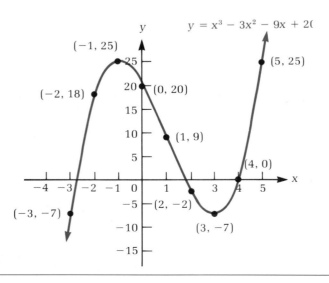

FIGURE 4.6

If the polynomial function can be expressed as a product of first-degree polynomials, we can easily determine zeros of the polynomial.

Example 4.25

Sketch the graph of

$$y = P(x) = x^3 + 2x^2 - 16x - 32.$$

Solution: Using synthetic division, we find that $P(-2) = 0$, since

$$
\begin{array}{r|rrrr}
-2 & 1 & 2 & -16 & -32 \\
   & \downarrow & -2 & 0 & 32 \\
\hline
   & 1 & 0 & -16 & 0
\end{array}
$$

and we have

$$
\begin{aligned}
P(x) &= (x + 2)(x^2 - 16) \\
     &= (x + 2)(x + 4)(x - 4).
\end{aligned}
$$

The rational zeros of $P(x)$ are therefore -2, -4, and 4. Next we need several points to sketch the graph. We use again the shortened form of synthetic division:

x	$1 + 2 - 16 - 32$	Ordered Pairs
-5	$1 - 3 - \ \ 1 - 27$	$(-5, -27)$
-4	$1 - 2 - \ \ 8 + \ \ 0$	$(-4, 0)$
-3	$1 - 1 - 13 + \ \ 7$	$(-3, 7)$
-2	$1 + 0 - 16 + \ \ 0$	$(-2, 0)$
-1	$1 + 1 - 17 - 15$	$(-1, -15)$
0	$1 + 2 - 16 - 32$	$(0, -32)$
1	$1 + 3 - 13 - 45$	$(1, -45)$
2	$1 + 4 - \ \ 8 - 48$	$(2, -48)$
3	$1 + 5 - \ \ 1 - 35$	$(3, -35)$
4	$1 + 6 + \ \ 8 + \ \ 0$	$(4, 0)$
5	$1 + 7 + 19 + 63$	$(5, 63)$

Plotting the ordered pairs $(-5, -27)$, $(-4, 0)$, ..., $(4, 0)$ and then joining them by a smooth curve, we have the graph of $P(x)$ in Figure 4.7.

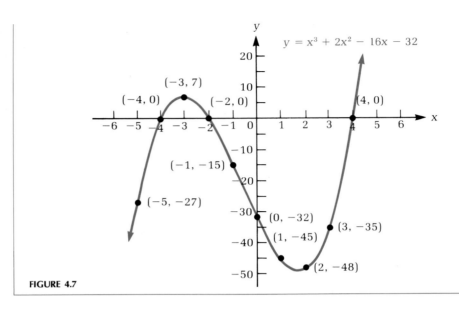

FIGURE 4.7

Example 4.26

Sketch the graph of

$$y = P(x) = x^4 - x^3 - 2x^2 + 3x - 3.$$

Solution: By Descartes' rule of signs, we expect $P(x)$ to have one or three positive real zeros and one negative real zero. The possible rational zeros are 1, -1, 3, and -3. Using synthetic division, we find that none of the rational values are zeros of the polynomial. However, we see that -3 is a lower bound and 3 is an upper bound of the real zeros of $P(x)$. This suggests that we evaluate $P(x)$ at each integer from -3 to 3 to get some ordered pairs that we may use to sketch the graph. As before, we use a shortened form of synthetic divsion:

x	1 - 1 - 2 + 3 - 3	Ordered Pairs
-3	1 - 4 + 10 - 27 + 78	$(-3, 78)$
-2	1 - 3 + 4 - 5 + 7	$(-2, 7)$
-1	1 - 2 + 0 + 3 - 6	$(-1, -6)$
0	1 - 1 - 2 + 3 - 3	$(0, -3)$
1	1 + 0 - 2 + 1 - 2	$(1, -2)$
2	1 + 1 + 0 + 3 + 3	$(2, 3)$
3	1 + 2 + 4 + 15 + 42	$(3, 42)$

The location theorem guarantees that the polynomial has a negative real zero between -2 and -1 and a positive real zero between 1 and 2. Plotting the ordered pairs $(-2, 7), (-1, -6), (0, -3), (1, -2),$ and $(2, 3)$ and joining them by a smooth curve, we have the graph in Figure 4.8.

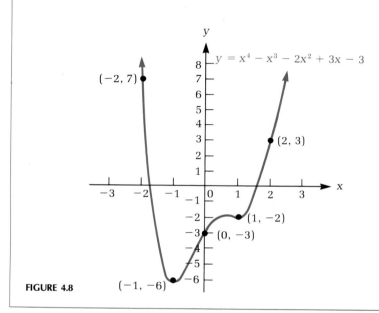

FIGURE 4.8

EXERCISES

Sketch the graph of the polynomials in Exercises 1 to 18.

1. $P(x) = x^3 - 1$
2. $P(x) = x^3 + 2$
3. $P(x) = -x^3 + 1$
4. $P(x) = -x^3 - 3$
5. $P(x) = (x - 1)^3$
6. $P(x) = (x + 2)^3$
7. $P(x) = x^3 - 6x^2 + 9x + 5$
8. $P(x) = x^3 - 3x^2$
9. $P(x) = x^3 - 3x^2 - 9x + 10$
10. $P(x) = 2x^3 - 9x^2 + 12x - 3$
11. $P(x) = 2x^3 + 3x^2 - 12x$
12. $P(x) = x^3 + x^2 - 4x - 4$
13. $P(x) = x^3 - 3x^2 - x + 3$
14. $P(x) = x^3 - 6x^2 + x + 14$
15. $P(x) = x^4 - 8x^2$
16. $P(x) = x^4 - 4x^2$
17. $P(x) = x^4 + 8x^3 + 18x^2 - 5$
18. $P(x) = 2x^4 - 3x^3 + 4x^2 + 5x - 1$

4.6

GRAPHING RATIONAL FUNCTIONS

Definition 4.6.1

Let $P(x)$ and $Q(x)$ be two polynomials. Then a function of the form

$$y = f(x) = \frac{P(x)}{Q(x)}$$

is called a *rational function*. The domain of $f(x)$ is the set of all real numbers except those for which $Q(x) = 0$.

Some examples of rational functions are

$$f(x) = \frac{2x + 1}{x - 1} \qquad (x \neq 1),$$

$$f(x) = \frac{3x^2 - 4x + 5}{2x - 3} \qquad \left(x \neq \frac{3}{2}\right),$$

$$f(x) = \frac{1}{x + 2} \qquad (x \neq -2).$$

Remember that we must exclude from the domain of the function all those real numbers for which the denominator is zero.

Example 4.27

Sketch the graph of

$$f(x) = \frac{1}{x + 1}$$

Solution: The function $f(x)$ has as its domain the set of real numbers except $x = -1$. At $x = -1$, the denominator is zero, and the function is not defined. However, we can examine the behavior of $f(x)$ for values of x close to -1. For $x > -1$, $x + 1 > 0$, and so $f(x) > 0$. Further,

$$f(99) = 0.01 \qquad\qquad f(0) = 1$$
$$f(49) = 0.02 \qquad\qquad f(-0.5) = 2$$
$$f(9) = 0.10 \qquad\qquad f(-0.9) = 10$$
$$f(3) = 0.25 \qquad\qquad f(-0.99) = 100$$
$$f(1) = 0.50 \qquad\qquad f(-0.999) = 1000.$$

This means that as x gets closer and closer to -1 from the right, $x + 1$ gets closer and closer to zero, and $f(x)$ gets larger and larger. Now let us examine what happens if x approaches -1 from the left. For $x < -1$, $x + 1 < 0$, and hence $f(x) < 0$. Also

$$f(-101) = -0.01 \qquad f(-2) = -1$$
$$f(-51) = -0.02 \qquad f(-1.5) = -2$$
$$f(-26) = -0.04 \qquad f(-1.1) = -10$$
$$f(-5) = -0.25 \qquad f(-1.01) = -100$$
$$f(-3) = -0.50 \qquad f(-1.001) = -1000.$$

vertical asymptote

As x approaches -1 from the left, $x + 1$ approaches zero, and $f(x)$ becomes a large negative number. Because $|f(x)|$ becomes large as x approaches -1, we call the vertical line $x = -1$ a **vertical asymptote** for the graph.

horizontal asymptote

Notice that as x increases and moves to the right away from -1, $f(x)$ decreases and approaches zero. Similarly, as x moves farther away from -1 in the negative direction, $f(x)$ again approaches zero. Since $f(x)$ gets closer and closer to zero as $|x|$ increases, we call the x axis a **horizontal asymptote** for the graph. The graph of $f(x)$ is displayed in Figure 4.9.

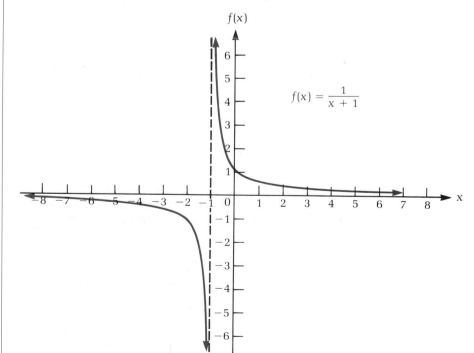

FIGURE 4.9

Example 4.28   Sketch the graph of

$$f(x) = \frac{1}{x^2}.$$

Solution: Observe that the function is defined for all real numbers except $x = 0$. We will see that $x = 0$ is a vertical asymptote for the graph of $f(x)$. To examine the behavior of $f(x)$ for values of $x < 0$, note that

$$f(-4) = 0.0625 \qquad f(-0.2) = 25$$
$$f(-2) = 0.25 \qquad f(-0.1) = 100$$
$$f(-1) = 1 \qquad f(-0.01) = 10,000$$
$$f(-0.5) = 4 \qquad f(-0.001) = 1,000,000.$$

Now we observe how $f(x)$ behaves for values of $x > 0$. Notice that

$$f(0.001) = 1,000,000 \qquad f(0.5) = 4$$
$$f(0.01) = 10,000 \qquad f(1) = 1$$
$$f(0.1) = 100 \qquad f(2) = 0.25$$
$$f(0.2) = 25 \qquad f(4) = 0.0625.$$

Thus $f(x)$ approaches zero as $|x|$ increases. This means that the x axis is a horizontal asymptote for the graph. Further, since all function values are positive, the graph of $f(x)$ is above the x axis. The graph is displayed in Figure 4.10.

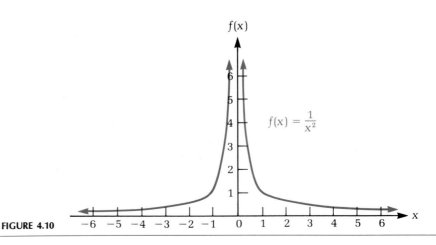

FIGURE 4.10

Let us now pause and investigate, in general, the problem of finding asymptotes. Consider the rational function

$$y = \frac{P(x)}{Q(x)} \quad [Q(x) \neq 0].$$

If the function has **vertical asymptotes,** they may be found by setting $Q(x) = 0$ and then solving this equation for x. The values of x for which $Q(x) = 0$ are those of the vertical asymptotes for the graph of the rational function.

Example 4.29

(a) The graph of the function

$$y = \frac{2x + 3}{x - 4}$$

has $x = 4$ as a vertical asymptote, since the denominator $x - 4$ is zero at $x = 4$.

(b) The graph of the function

$$y = \frac{x}{(x - 3)(x + 2)}$$

has $x = -2$ and $x = 3$ as vertical asymptotes, since the denominator is zero at these values.

Horizontal asymptotes for the rational function

$$y = \frac{P(x)}{Q(x)} \quad [Q(x) \neq 0]$$

occur when the degree of $P(x)$ is less than or equal to the degree of $Q(x)$. First we consider a function for which the degree of $P(x)$ is less than that of $Q(x)$.

Example 4.30

Determine the horizontal asymptotes, if they exist, for

$$y = \frac{2x + 1}{x^2 - 1}.$$

Solution: Dividing the numerator and the denominator by x^2, the highest power of x in the fraction, we obtain

$$y = \frac{\dfrac{2}{x} + \dfrac{1}{x^2}}{1 - \dfrac{1}{x^2}}.$$

As $|x|$ becomes larger and larger, each expression with x or x^2 in its denominator becomes smaller and smaller. Consequently, for large values of $|x|$, the numerator

$$\frac{2}{x} + \frac{1}{x^2}$$

approaches zero, while the denominator

$$1 - \frac{1}{x^2}$$

approaches 1. Hence y approaches $0/1 = 0$, and $y = 0$ (x axis) is the horizontal asymptote.

Next we consider a rational function for which the degree of the numerator and the denominator is the same.

Example 4.31

Determine the horizontal asymptotes, if any, for the rational function

$$y = \frac{2x^2 + 3x + 1}{x^2 - x + 6}.$$

Solution: As before, we divide the numerator and the denominator by x^2, the highest power of x in the fraction; we obtain

$$y = \frac{2 + \dfrac{3}{x} + \dfrac{1}{x^2}}{1 - \dfrac{1}{x} + \dfrac{6}{x^2}}.$$

For large values of $|x|$, the numerator approaches 2, and the denominator approaches 1. The function values approach $2/1 = 2$. Hence $y = 2$ is a horizontal asymptote.

In Examples 4.30 and 4.31, we determined horizontal asymptotes for the rational function

$$y = \frac{P(x)}{Q(x)} \qquad [Q(x) \neq 0],$$

where the degree of $P(x)$ is less than or equal to the degree of the polynomial $Q(x)$. We have shown that the horizontal asymptote is the x axis when the degree of $P(x)$ is less than the degree of $Q(x)$. If $P(x)$ and $Q(x)$ have the same degree, then the horizontal asymptote is $y = a/b$, where a and b are the leading coefficients of $P(x)$ and $Q(x)$, respectively. Now we consider the case in which the degree of $P(x)$ is 1 more than the degree of $Q(x)$.

Example 4.32

Determine the asymptotes, if any, for the rational function

$$y = \frac{x^2 - 5x + 5}{x - 2}.$$

Solution: First we notice that $x = 2$ is a vertical asymptote, since the denominator is zero at this value. There are no horizontal asymptotes, because as x gets large, so does y. However, if we use long division, we may rewrite the above equation in the form

$$y = x - 3 - \frac{1}{x - 2}.$$

oblique asymptote

Now as $|x|$ gets larger and larger, $1/(x - 2)$ gets smaller and smaller. Consequently, for large values of $|x|$, $1/(x - 2)$ approaches zero, and the y value approaches $x - 3$. In other words, the graph of the rational function approaches the graph of $y = x - 3$. The line $y = x - 3$ is called an **oblique asymptote.**

Asymptotes are often helpful in graphing rational functions. Zeros of the function are another useful aid in graphing. Remember that zeros of a function occur at the points where the graph crosses the x axis.

Example 4.33

Sketch the graph of

$$y = \frac{x - 1}{x + 1}.$$

Solution: Observe that $x = -1$ is a vertical asymptote. To determine horizontal asymptotes, if any, we divide the numerator and the denominator by x, the highest power of x in the fraction, and obtain

$$y = \frac{1 - \dfrac{1}{x}}{1 + \dfrac{1}{x}}.$$

For large values of $|x|$, $1/x$ approaches zero. Thus both the numerator and the denominator approach 1. Hence $y = 1$ is the horizontal asymptote.

Let us now determine some specific points for the graph. Plotting some of the ordered pairs from the table below gives us the graph in Figure 4.11.

x	-4	-3	-2	$-\frac{3}{2}$	$-\frac{1}{2}$	0	$\frac{1}{2}$	1	2	3
y	$\frac{5}{3}$	2	3	5	-3	-1	$-\frac{1}{3}$	0	$\frac{1}{3}$	$\frac{1}{2}$

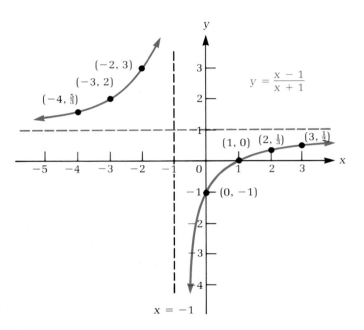

FIGURE 4.11

Example 4.34

Sketch the graph of

$$y = \frac{x^2}{x^2 - 1}$$

Solution: There are two vertical asymptotes, $x = -1$ and $x = 1$. To determine the horizontal asymptotes, if any, we divide both the numerator and the denominator by x^2, thus obtaining

$$y = \frac{1}{1 - \dfrac{1}{x^2}}$$

As $|x|$ gets large, $1/x^2$ approaches zero. Thus $y = 1$ is a horizontal asymptote. Further, the graph is symmetric with respect to the y axis in the sense that if (c, d) is on the graph, then so is $(-c, d)$. The table below shows some values which may aid in sketching the graph.

x	-3	-2	$-\frac{3}{2}$	$-\frac{5}{4}$	$-\frac{1}{2}$	0	$\frac{1}{2}$	$\frac{5}{4}$	$\frac{3}{2}$	2	3
y	$\frac{9}{8}$	$\frac{4}{3}$	$\frac{9}{5}$	$\frac{25}{9}$	$-\frac{1}{3}$	0	$-\frac{1}{3}$	$\frac{25}{9}$	$\frac{9}{5}$	$\frac{4}{3}$	$\frac{9}{8}$

Plotting some of the ordered pairs, we have the graph in Figure 4.12.

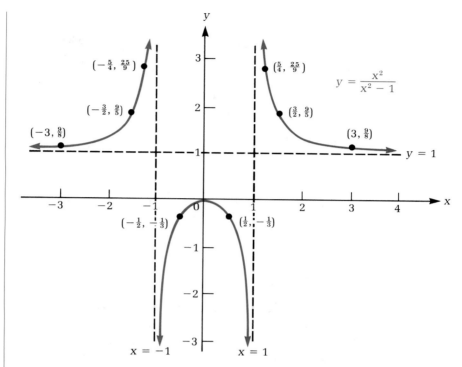

FIGURE 4.12

Example 4.35

Sketch the graph of

$$y = \frac{x^2 - 5x + 5}{x - 2}.$$

Solution: This is the same function we discussed in Example 4.32. We observed that $x = 2$ is a vertical asymptote, that there are no horizontal asymptotes, and that $y = x - 3$ is an oblique asymptote. This table shows some specific points for the graph:

x	0	1	3	4
y	$-\frac{5}{2}$	-1	-1	$\frac{1}{2}$

The graph is displayed in Figure 4.13.

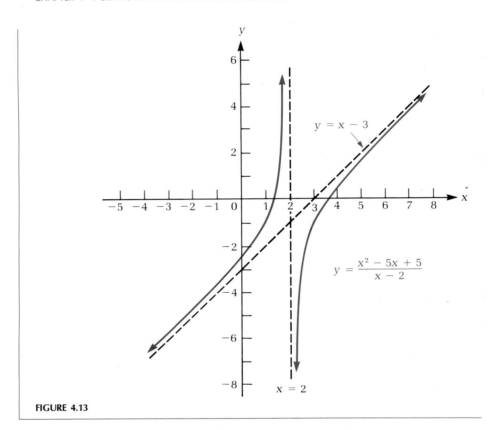

FIGURE 4.13

EXERCISES

Determine the vertical asymptotes of the graph in Exercises 1 to 6.

1. $y = \dfrac{1}{x - 2}$

2. $y = \dfrac{1}{x + 3}$

3. $y = \dfrac{1}{x^2 - 4}$

4. $y = \dfrac{1}{x^2 - 5x + 6}$

5. $y = \dfrac{x}{x^2 - x - 6}$

6. $y = \dfrac{x^2}{x^2 - x - 2}$

Determine horizontal, vertical, and oblique asymptotes (if any) of the graphs of each function in Exercises 7 to 14.

7. $y = \dfrac{2x}{x - 1}$

8. $y = \dfrac{1}{x^2 - 4}$

9. $y = \dfrac{x}{(x + 1)(x + 2)}$

10. $y = \dfrac{2x^2}{x^2 - 6x + 8}$

11. $y = \dfrac{4x^2 + 3}{2x^2 - x - 3}$

12. $y = \dfrac{x^2 + 1}{x}$

13. $y = \dfrac{x^2 - 4}{x - 1}$

14. $y = \dfrac{x^2 + 1}{x - 2}$

Sketch the graph of each function in Exercises 15 to 30.

15. $y = \dfrac{1}{x - 1}$ **16.** $y = \dfrac{1}{x + 2}$ **17.** $y = \dfrac{1}{(x + 1)^2}$

18. $y = \dfrac{1}{(x - 2)^2}$ **19.** $y = \dfrac{x - 1}{x + 4}$ **20.** $y = \dfrac{x - 1}{x + 2}$

21. $y = \dfrac{3x}{x^2 - x - 2}$ **22.** $y = \dfrac{3x}{x^2 + 5x + 4}$ **23.** $y = \dfrac{x - 1}{x^2 - 2x - 3}$

24. $y = \dfrac{x}{x^2 - x - 6}$ **25.** $y = \dfrac{1}{x^2 - 4}$ **26.** $y = \dfrac{5x}{4 - x^2}$

27. $y = \dfrac{2x + 1}{x}$ **28.** $y = \dfrac{x^2 - 4}{x - 1}$ **29.** $y = \dfrac{x^2 + 1}{x - 2}$

30. $y = \dfrac{x^2 - 5}{x + 2}$

KEY WORDS

polynomial function of degree n	*fundamental theorem of algebra*	*power function*
constant polynomial	*upper bound of real zeros*	*symmetric with respect to the origin*
zeros of a polynomial	*lower bound of real zeros*	*symmetric with respect to the y axis*
remainder theorem	*variations in sign*	*vertical asymptote*
factor theorem	*Descartes' rule of signs*	*horizontal asymptote*
zeros of multiplicity m	*location theorem*	*oblique asymptote*

KEY IDEAS

* The zeros of a polynomial function $P(x)$ are the values of x for which $P(x) = 0$.

* The *remainder theorem* asserts that when $P(x)$ is divided by $x - k$, the remainder r is equal to the value of $P(x)$ at $x = k$.

* The *factor theorem* states that the polynomial $P(x)$ with complex coefficients has a factor $x - k$ if and only if $P(k) = 0$.

* If $x - k$ is a factor of the polynomial $P(x)$, then $x = k$ is a zero of $P(x)$.

* Every polynomial $P(x)$ of degree $n \geq 1$ has at least one zero.

* Every polynomial $P(x)$ of degree $n \geq 1$ can be factored into n linear factors.

* Every polynomial $P(x)$ of degree $n \geq 1$ has at most n zeros.

* If a factor $x - k$ appears m times in a factorization of the polynomial $P(x)$, then k is a *zero of multiplicity m*.

* If a zero of multiplicity m is counted m times, then a polynomial $P(x)$ of degree $n \geq 1$ has exactly n zeros.

* If $P(x)$ is a polynomial of degree n with real coefficients and if the complex number z is a zero of $P(x)$, then its conjugate $\bar{z}$ is also a zero of $P(x)$.

* Let $P(x) = a_n x^n + a_{n-1} x^{n-1} + \cdots + a_2 x^2 + a_1 x + a_0$, $a_n \neq 0$, be a polynomial with integral coefficients. If a rational number p/q is a zero of $P(x)$, then p is a factor of a_0 and q is a factor of a_n.

* Let $P(x)$ be a polynomial with real coefficients, and let the coefficient of the highest power of x be positive. If $k_1 \geq 0$ and if the terms in the third row

of synthetic division of $P(x)$ by $x - k_1$ are all non-negative, then k_1 is an upper bound of the real zeros of $P(x)$. If $k_2 \leq 0$ and if the terms in the third row of synthetic division of $P(x)$ by $x - k_2$ alternate in sign, then k_2 is a lower bound of the real zeros of $P(x)$.

- *Descartes' rule of signs.* Let $P(x)$ be a polynomial with real coefficients. Then the number of positive real zeros is either equal to the number of variations in sign occurring in the coefficients of $P(x)$

or less than the number of variations by a positive even integer, and the number of negative real zeros is either equal to the number of variations in sign occurring in the coefficients of $P(-x)$ or less than the number of variations by a positive even integer.

- *Location theorem.* If $P(x)$ is a real polynomial and a and b are real numbers such that $P(a)$ and $P(b)$ are opposite in sign, then there is at least one real zero c in the open interval (a, b) such that $P(c) = 0$.

EXERCISES FOR REVIEW

Determine the zeros of the polynomials in Exercises 1 to 6.

1. $P(x) = x^2 - 9x + 14$ **2.** $P(x) = x^2 + 4x - 21$ **3.** $P(x) = 4x^2 + 12x - 7$

4. $P(x) = 10x^2 - 17x + 3$ **5.** $P(x) = x^3 + x$ **6.** $P(x) = x^3 + x^2 - 2x$

In Exercises 7 to 10, find the remainder $P(k)$ when $P(x)$ is divided by the polynomial of the form $x - k$.

7. $P(x) = x^3 + 3x^2 + 3x + 5; x + 1$ **8.** $P(x) = x^3 - 7x^2 - 13x + 3; x + 2$

9. $P(x) = x^3 - 4x^2 - 18x + 19; x + 3$ **10.** $P(x) = x^3 - 2ix^2 + ix + 5; x - i$

Find the value of the polynomial $P(x)$ for the given value of x in Exercises 11 to 14.

11. $P(x) = 3x^2 - x + 4; -3$ **12.** $P(x) = 2x^3 - 6x^2 + x + 1; 3$

13. $P(x) = x^3 - 2x^2 + 3x + 4; i$ **14.** $P(x) = x^2 + 4x - 6; 1 - i$

In Exercises 15 to 18, determine whether the polynomial of the form $x - k$ is a factor of the polynomial $P(x)$. Use the factor theorem.

15. $P(x) = x^3 + 3x^2 - 4x - 6; x + 1$ **16.** $P(x) = x^3 - 4x^2 - 18x + 19; x + 3$

17. $P(x) = 2x^3 - 5x^2 + 6x - 2; x - (1 + i)$ **18.** $P(x) = x^3 - 6x^2 + 13x - 10; x - (2 + i)$

In Exercises 19 and 20, determine the value of k such that the second polynomial is a factor of the first.

19. $x^3 - 4x^2 + 2kx + 4; x - 2$ **20.** $kx^4 + 3x^3 - 2x^2 + 3kx - 20; x + 4$

In Exercises 21 and 22, determine the polynomial $P(x)$ of lowest degree with given numbers as some of its zeros.

21. $2, 3 + i$ **22.** $1, 2, 2 + 3i$

23. Find the polynomial $P(x)$ of degree 4 which has three distinct zeros, 1, -2, and 3, given that 3 is a zero of multiplicity 2.

24. Find a polynomial of degree 4 that has three distinct zeros, $1 + 2i$, $1 - 2i$, and 4, given that 4 is a zero of multiplicity 2.

In Exercises 25 to 28, one zero is given for each polynomial function. Find other zeros of $P(x)$.

25. $P(x) = x^3 + 2x^2 - x - 2; -1$ **26.** $P(x) = x^3 - 5x^2 + 7x + 13; 3 + 2i$

27. $P(x) = x^4 - x^3 + 2x^2 - 4x - 8; 2i$ **28.** $P(x) = x^4 - 6x^3 - 3x^2 - 24x - 28; 7$

In Exercises 29 and 30, find the rational zeros (if they exist) of the polynomial.

29. $P(x) = x^3 - 4x^2 + x + 6$ **30.** $P(x) = 2x^3 + 3x^2 + 2x + 3$

In Exercises 31 to 34, determine the real zeros of the polynomial.

31. $P(x) = x^3 + 3x^2 - 2x - 6$

32. $P(x) = 2x^3 - 9x^2 + 2x + 20$

33. $P(x) = x^4 - 9x^2 + 2x + 6$

34. $P(x) = 6x^4 - 23x^3 + 19x^2 + 5x - 3$

In Exercises 35 and 36, approximate the real zeros of polynomial to one decimal place.

35. $P(x) = x^3 - 3x^2 + 3$

36. $P(x) = x^4 - 5x^2 + 6$

Graph the functions in Exercises 37 to 40.

37. $y = x^3 - 3x^2 - 9x + 10$

38. $y = x^3 - 9x^2 + 24x$

39. $y = x + \dfrac{1}{x}$

40. $y = \dfrac{x}{x + 1}$

CHAPTER TEST

1. Determine the zeros of the polynomials:
  (a) $P(x) = x^3 + x$ **(b)** $P(x) = x(4x^2 + 12x - 7)$

2. Using synthetic division, find the remainder $P(k)$ when
$$P(x) = x^3 - 2x^2 + x + 1$$
is divided by $x + 2$.

3. Determine whether $x - (2 + i)$ is a factor of the polynomial
$$P(x) = x^3 - 6x^2 + 13x - 10.$$

4. Find a polynomial $P(x)$ of degree 4 such that 1 is a zero of multiplicity 3 and 2 is a zero of multiplicity 1.

5. A polynomial $P(x) = x^3 - 3x^2 - 4x + 12$ has 2 as one of its zeros; find other zeros of $P(x)$.

6. Find the real polynomial of lowest degree that has -2 and i as its zeros.

7. Find an upper bound and a lower bound for the real zeros of the polynomial
$$P(x) = x^3 - x^2 - x - 4.$$

8. Find the number of positive and negative real zeros of the polynomial
$$P(x) = x^3 - x^2 - 3x - 3.$$

9. Determine the horizontal, vertical, and oblique asymptotes (if any) of the graph of
$$y = \frac{2x^2}{x^2 - 6x + 8}.$$

10. Graph the following functions:
  (a) $y = (x - 1)^3$ **(b)** $y = \dfrac{1}{x - 1}$

EXPONENTIAL AND LOGARITHMIC FUNCTIONS

5

INTRODUCTION

Exponential functions and the closely related logarithmic functions have widespread applications. Exponential functions are particularly useful in describing various processes of growth and decay in biology, chemistry, economics, psychology, and life sciences. Logarithmic functions, which are predominantly used to simplify numerical calculations, represent a remarkable time-saving advance in scientific knowledge. In this chapter, we define these functions and then study some of their applications.

5.1

EXPONENTIAL FUNCTIONS

In Chapter 1, we studied expressions of the type x^p, where the base x is an independent variable and the exponent p is a real number. Now we turn our attention to situations in which the base is constant and the exponent is an independent variable. Consider, for example, an equation of the form $y = b^x$, where b is any positive number not equal to 1 and the variable x appears as an exponent. This equation is referred to as an *exponential function*, since after the base b is selected, the value of y depends only on x.

Definition 5.1.1

A function $f(x) = b^x$, where $b > 0$, $b \neq 1$, and x is any real number, is called an *exponential function*.

We wish to emphasize that the base b must be a positive real number; if it were negative, b^x would not be a real number for x equal to $\frac{1}{2}$, $\frac{1}{4}$, or any other fraction with an even denominator. In the trivial case of $b = 1$, the function b^x equals 1 for all values of x. The domain of $f(x) = b^x$ is the set of all real numbers, and the range is the set of all positive real numbers if $b \neq 1$ and $b > 0$.

The following examples illustrate several properties of exponential functions.

Example 5.1

Graph the function $f(x) = 2^x$.

Solution: The table below shows the values of $f(x)$ corresponding to some values of x.

x	-4	-3	-2	-1	0	1	2	3
$f(x)$	$\frac{1}{16}$	$\frac{1}{8}$	$\frac{1}{4}$	$\frac{1}{2}$	1	2	4	8

Plotting the ordered pairs $(x, f(x))$ from the table and connecting these points in the plane, we obtain the graph of $f(x) = 2^x$ in Figure 5.1.

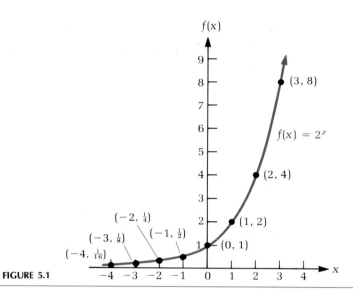

FIGURE 5.1

Example 5.2

Graph the function $g(x) = 4^x$.

Solution: Rather than plotting the ordered pairs $(x, g(x))$ and then connecting them with a smooth curve, we observe that

$$g(x) = 4^x = (2^2)^x = 2^{2x}.$$

Comparing the function $g(x) = 2^{2x}$ with the function $f(x)$ in Example 5.1, we note that in $g(x)$, the value of x is only half that required in $f(x)$ to obtain the same value of y. For example,

$$g(-2) = f(-4) = \tfrac{1}{16}$$
$$g(-1) = f(-2) = \tfrac{1}{8}$$
$$g(1) = f(2) = 4,$$

and so on. Hence, for a given y value, we move each point on the graph of $f(x) = 2^x$ half as far from the y axis to obtain the corresponding point on $g(x)$, as shown in Figure 5.2.

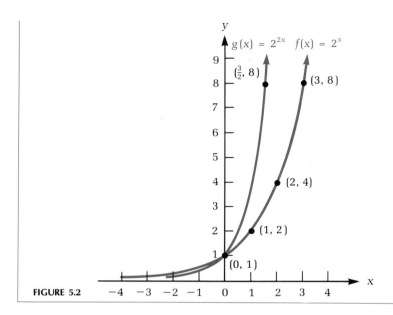

FIGURE 5.2

Example 5.3   Graph the function $f(x) = 3^x$.

Solution: Plotting the ordered pairs $(x, f(x))$ given in the table below and connecting the points with a smooth curve, we obtain the graph of the function in Figure 5.3.

x	-3	-2	-1	0	1	2
$f(x)$	$\frac{1}{27}$	$\frac{1}{9}$	$\frac{1}{3}$	1	3	9

An irrational number e, approximately equal to 2.71828, plays an important role in mathematics. Specifically, the function e^x is used in applications that require an exponential-type function to describe a physical phenomenon. As a result, e^x is called the **natural** exponential function. Table 1 in the Appendix lists the values of e^x and e^{-x} from $x = 0$ to $x = 9.9$ in steps of 0.1. **Also scientific hand-held calculators calculate e^x at the push of a button.**

Example 5.4   Assuming that $e = 2.71828$, sketch the graph of $f(x) = e^x$.

Solution: The following table shows the approximate values of $f(x)$ corresponding to some values of x.

x	-4	-3	-2	-1	0	1	1.5	2	2.5	3
$f(x)$	0.02	0.05	0.14	0.37	1.00	2.72	4.48	7.39	12.18	20.09

Plotting the ordered pairs $(x, f(x))$, we obtain the graph of e^x in Figure 5.4.

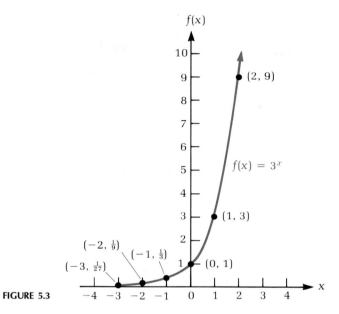

FIGURE 5.3

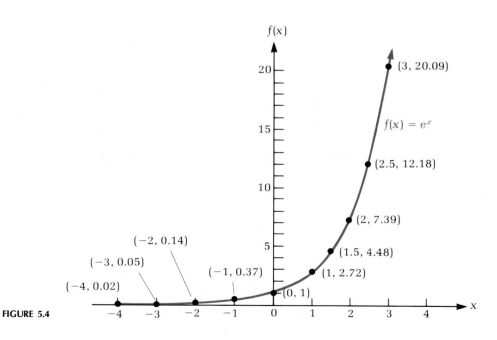

FIGURE 5.4

The preceding examples suggest the following observations concerning the graph of b^x when $b > 1$.

1. Since $b^0 = 1$, the graph of b^x passes through the point $(0, 1)$.
2. Here $b^x > 0$ for all values of x; hence the graph lies above the x axis.
3. As x increases, the value of b^x increases; hence the graph rises to the right.
4. As x decreases along the negative x axis, the value of b^x decreases and gets closer and closer to zero.
5. For $b \neq 1$, $b^x = b^y$ if and only if $x = y$; thus $f(x) = b^x$ is a one-to-one function.

Next we examine the graph of b^x for $0 < b < 1$. Here the appearance is different.

Example 5.5

Graph the function $g(x) = (\frac{1}{2})^x$.

Solution: Observe that $g(x) = (\frac{1}{2})^x = 2^{-x}$.

Comparing the function $g(x) = (\frac{1}{2})^x$ with the function $f(x) = 2^x$ in Example 5.1, we note that $g(x) = f(-x)$. For example,

$$g(3) = f(-3) = \tfrac{1}{8} \qquad g(2) = f(-2) = \tfrac{1}{4} \qquad g(1) = f(-1) = \tfrac{1}{2}$$
$$g(-1) = f(1) = 2 \qquad g(-2) = f(2) = 4$$

and so on. This means that the graph of $g(x) = (\frac{1}{2})^x$ is a reflection of $f(x) = 2^x$ in the y axis. Using Figure 5.1, we can, therefore, sketch the graph of $g(x) = (\frac{1}{2})^x$ as shown in Figure 5.5.

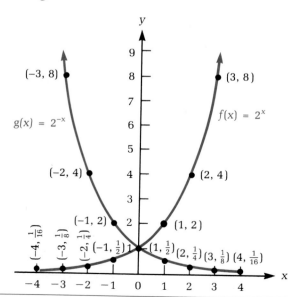

FIGURE 5.5

In summary, the general properties of exponential functions $f(x) = b^x, b > 0$, $b \neq 1$, are the following.

1. For $x = 0$, $b^x = 1$.

2. For all values of x, $b^x > 0$.

3. As x increases, the value of b^x increases for $b > 1$ and decreases for $0 < b < 1$.

4. As x decreases, the value of b^x decreases for $b > 1$ and increases for $0 < b < 1$.

5. $f(x) = b^x$ is a one-to-one function.

Let us now look at some other exponential functions.

Example 5.6

Graph the function

$$f(x) = 2^{|x|}.$$

Solution: The following table shows values of $f(x)$ corresponding to some values of x.

x	-3	-2	-1	0	1	2	3
$f(x)$	8	4	2	1	2	4	8

Plotting the ordered pairs $(x, f(x))$, we obtain the graph of the function in Figure 5.6.

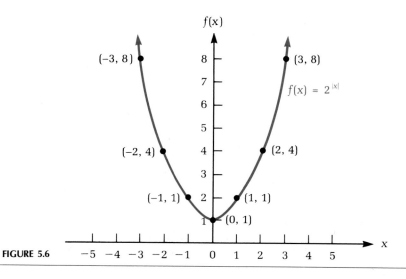

FIGURE 5.6

Example 5.7

Graph the function $f(x) = 2^{-x^2}$.

Solution: The following table shows values of $f(x)$ corresponding to some values of x.

x	-3	-2	-1	0	1	2	3
$f(x)$	$\frac{1}{512}$	$\frac{1}{16}$	$\frac{1}{2}$	1	$\frac{1}{2}$	$\frac{1}{16}$	$\frac{1}{512}$

Observe that for all values of x, $f(x) \le 1$. Thus, we may choose a different scale for axis of x and axis of y. Plotting the ordered pairs $(x, f(x))$, we obtain the graph of $f(x)$ in Figure 5.7.

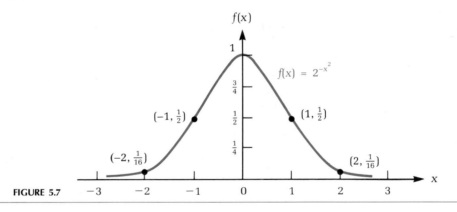

FIGURE 5.7

EXERCISES

Sketch the graph of each function in Exercises 1 to 22.

1. $y = \left(\frac{3}{2}\right)^x$
2. $y = 3^{x+1}$
3. $y = 2^{x-1}$
4. $y = 3^{x-1}$
5. $y = 2^{x+1}$
6. $y = 3^{1-x}$
7. $y = 2^x + 1$
8. $y = 2^{2x} + 2$
9. $y = \left(\frac{1}{3}\right)^x$
10. $y = \left(\frac{1}{3}\right)^{x+1}$
11. $y = 2(3^x)$
12. $y = 3(2^x)$
13. $y = 3\left(\frac{1}{2}\right)^x$
14. $y = 2\left(\frac{1}{3}\right)^x$
15. $y = 3^x - 1$
16. $y = 2^x - 2^{-x}$
17. $y = \left(\frac{1}{2}\right)^x + \left(\frac{1}{2}\right)^{-x}$
18. $y = \dfrac{2^x + 2^{-x}}{2}$
19. $y = 2^{|x-1|}$
20. $y = 2^{|x|+1}$
21. $y = 3^{-x^2}$
22. $y = 3^{1-x^2}$

23. Sketch the graph of $y = x^2$ and $y = 2^x$ on the same coordinate axes.
24. Given that

$$f(n) = \left(1 + \frac{1}{n}\right)^n$$

use a scientific calculator to determine each of the following values:

(a) $f(1)$ (b) $f(10)$ (c) $f(100)$ (d) $f(1,000)$

(e) $f(10,000)$ (f) $f(100,000)$ (g) $f(1,000,000)$ (h) $f(10,000,000)$.

Show that as n gets larger and larger, $f(n)$ gets closer and closer to $e = 2.71828 \ldots$.

Using a scientific hand-held calculator, sketch the graph of the following functions in Exercises 25 to 30.

25. $y = e^{2x}$ 26. $y = 2e^x$ 27. $y = 2e^{-x}$

28. $y = 2e^{-x+1}$ 29. $y = \dfrac{e^x + e^{-x}}{2}$ 30. $y = \dfrac{e^x - e^{-x}}{2}$

5.2

EXPONENTIAL EQUATIONS

Definition 5.2.1

> An equation that uses a variable as an exponent is referred to as an *exponential equation*.

Exponential equations are solved by using the fundamental property that if $b^x = b^y$, then $x = y$. The following examples illustrate a simple method for solving these equations.

Example 5.8

Solve $2^x = 16$ for x.

Solution: In order to have the same bases on both sides of the equation, we express 16 as a power of base 2. Thus we have

$$2^x = 2^4,$$

from which it follows that $x = 4$.

Example 5.9

Solve $5^{-2x} = 625$ for x.

Solution: Since $625 = 5^4$, we may rewrite the given equation as

$$5^{-2x} = 5^4.$$

Thus $-2x = 4$, or $x = -2$.

Example 5.10

Solve $4^x = 32$ for x.

Solution: We must first rewrite both sides of the equation so that they have the same base. Since 32 is not an integral power of 4, we consider the common base 2 and write $4 = 2^2$ and 32 as 2^5. Thus

$$(2^2)^x = 2^5$$

$$2^{2x} = 2^5$$

which implies that $2x = 5$, or $x = \frac{5}{2}$.

Example 5.11 | Solve for x:

$$\left(\tfrac{3}{4}\right)^{-x} = \tfrac{243}{1024}.$$

Solution: The equation

$$\left(\frac{3}{4}\right)^{-x} = \frac{243}{1024}$$

may also be written as

$$\left(\frac{3}{4}\right)^{-x} = \left(\frac{3}{4}\right)^{5}$$

from which it follows that $x = -5$.

EXERCISES

Solve the equations in Exercises 1 to 32 for x.

1. $2^x = 4$
2. $2^x = 1$
3. $2^x = \tfrac{1}{8}$
4. $\left(\tfrac{1}{2}\right)^x = 16$

5. $3^x = 9$
6. $3^x = \tfrac{1}{27}$
7. $\left(\tfrac{1}{3}\right)^x = 81$
8. $\left(\tfrac{2}{3}\right)^x = \tfrac{4}{9}$

9. $\left(\tfrac{2}{3}\right)^x = \tfrac{16}{81}$
10. $\left(\tfrac{3}{5}\right)^x = \tfrac{27}{125}$
11. $4^x = 64$
12. $4^{-2x} = \tfrac{1}{256}$

13. $4^x = 2$
14. $16^x = 128$
15. $5^x = \tfrac{1}{25}$
16. $5^{2x} = 125$

17. $2^{2-x} = 16$
18. $5^{3-x} = 25$
19. $5^{2x+1} = 125$
20. $3^{2x} = 243$

21. $27^x = 9$
22. $10^x = 0.0001$
23. $5^x = 0.008$
24. $7^x = \tfrac{1}{343}$

25. $\left(\tfrac{2}{3}\right)^{-x} = \tfrac{8}{27}$
26. $\left(\tfrac{3}{4}\right)^{-2x} = \tfrac{256}{81}$
27. $\left(\tfrac{5}{6}\right)^{-x} = \tfrac{125}{216}$
28. $\left(\tfrac{4}{5}\right)^{-x} = \tfrac{125}{64}$

29. $\left(\tfrac{6}{7}\right)^{2x} = \tfrac{216}{343}$
30. $\left(\tfrac{6}{7}\right)^{-x} = \tfrac{49}{36}$
31. $\left(\tfrac{3}{5}\right)^{-x} = \tfrac{3125}{243}$
32. $\left(\tfrac{8}{9}\right)^{-2x} = \tfrac{729}{512}$

5.3

COMPOUND INTEREST

Suppose that Ms. Charles opens a 2-year savings account with $600 and the bank pays her simple interest at the annual rate of 6 percent. If she withdraws the interest earned each year, she will have an annual income of

$$\$600(0.06) = \$36,$$

and the balance in the account remains at $600. If she does not withdraw the interest, then the interest for the second year is computed on $636 in the account at the beginning of the second year. The interest for the second year would then be

$$\$636(0.06) = \$38.16$$

compounding

instead of $36. The practice of reinvesting the interest to earn additional interest is called **compounding.** Thus the theory of **compound interest** assumes that the principal and the interest earned continue to be reinvested each year.

To develop a basic formula for compound interest, consider an investment of P dollars at the rate of interest i per year. Since the interest during the first year is Pi, the balance at the end of 1 year will be

$$P + Pi = P(1 + i).$$

This balance considered as principal at the beginning of the second year earns interest

$$P(1 + i)i$$

during the second year. The balance at the end of second year is then

$$P(1 + i) + P(1 + i)i = P(1 + i)(1 + i) = P(1 + i)^2.$$

Now $P(1 + i)^2$ is considered as principal at the beginning of the third year. This principal earns interest

$$P(1 + i)^2 i$$

during the third year, which brings the total on deposit at the end of the third year to

$$P(1 + i)^2 + P(1 + i)^2 i = P(1 + i)^2(1 + i) = P(1 + i)^3.$$

Thus the general formula for the continued growth of P dollars can be shown to be

$$A(t) = P(1 + i)^t, \tag{5.1}$$

accumulated value

where $A(t)$ is the **accumulated value** of the principal P at the end of t years. The values of $(1 + i)^t$ may be computed by using a hand-held calculator.

Example 5.12

If $600 is deposited at 6 percent, compounded annually, what will be the total on deposit at the end of 5 years? How much of that will be the interest?

Solution: Here

$$P = \$600, \qquad i = 0.06, \qquad \text{and} \qquad t = 5.$$

Using (5.1), we obtain

$$A(5) = \$600(1 + 0.06)^5$$
$$= \$600(1.06)^5$$
$$= \$600(1.338226)$$
$$= \$802.94.$$

The accumulated value will be $802.94, and the interest will be

$$\$802.94 - \$600 = \$202.94.$$

We have shown that an investment of P dollars will accumulate to $P(1 + i)^t$ at the end of t years at an annual interest rate i. Sometimes it may be important to determine how much money a person must invest so as to have a specified sum at the end of t years. In that case, we solve (5.1) for P and obtain

$$P = A(t)(1 + i)^{-t}. \qquad (5.2)$$

present value   This value of P is called the **present value** of the amount $A(t)$. The values of $(1 + i)^{-t}$ may be obtained by using a hand-held calculator.

Example 5.13   Mr. Spark wishes to have $4000 available in a bank account for his daughter's first-year college expenses. What sum of money must he invest now at 6.5 percent, compounded annually, if his daughter is to start college 4 years hence?

Solution: Using (5.2), we have

$$P = \$4000(1.065)^{-4}$$
$$= \$4000(0.777323)$$
$$= \$3109.29.$$

Mr. Spark must invest $3109.29.

In recent years, it has become an established practice among commercial banks, savings and loan institutions, credit unions, and other approved financial companies to pay interest more frequently than once a year. Generally, the rate of interest is quoted on an annual basis but is, in practice, compounded quarterly, monthly, or even daily. Conceivably, this compounding frequency could progress to every hour, every minute, every second, and so on. The frequency with which interest is compounded and paid or reinvested to earn additional interest is *conversion period*   called the **conversion period.**
effective rate of interest   The term **effective rate of interest** is the rate that, compounded annually, would produce the same dollar amount as the quoted or published rate compounded m times a year. An interest rate of 6 percent per year compounded monthly does not mean that 6 percent is credited to an account after each month but rather that an interest $\frac{1}{12}$ of 6 percent, or 0.5 percent, is credited *nominal rate of interest*   monthly to the existing balance. The 6 percent is called the **nominal rate of interest.** The effective rate of interest is 6.168 percent, as evidenced by the fact that each $1.00 of principal compounded monthly at 6 percent yields

$$(1 + 0.005)^{12} = 1.061678,$$

so that annual interest earned is

$$1.061678 - 1 = 0.061678,$$

or approximately 6.168 percent.

As another illustration, suppose that a savings bank advertisement states that $1000 deposited with the bank for a period of $2\frac{1}{2}$ or more years at 12.00 percent interest, compounded daily, results in an actual yield of 12.747 percent. The **effective rate of interest** is thus 12.747 percent as compared to the **nominal rate of interest** of 12.00 percent.

Let i_m denote a nominal rate of interest, where interest is compounded m times a year. Clearly, i_m/m is interest paid at the end of each of the m conversion periods. By an argument similar to the one used in developing (5.1), it can be shown that the accumulated value of P dollars at the end of t years is given by

$$A = P\left(1 + \frac{i_m}{m}\right)^{mt} \tag{5.3}$$

What happens to expression (5.3) if m gets larger and larger? The following example provides an intuitive insight.

Example 5.14   Find the accumulated value after 10 years if $2000 is invested at 6 percent compounded **(a)** annually, **(b)** semiannually, **(c)** quarterly, **(d)** monthly, and **(e)** daily.

Solution:

(a) Here $P = \$2000$, $i = 0.06$, and $t = 10$. Using (5.1), we obtain

$$A = \$2000(1.06)^{10}$$
$$= \$2000(1.790848)$$
$$= \$3581.70.$$

(b) Here $P = \$2000$, $i_m = 0.06$, $m = 2$; $i_m/m = 0.06/2 = 0.03$, and $t = 10$. Using (5.3), we have

$$A = \$2000(1.03)^{20}$$
$$= \$2000(1.806111)$$
$$= \$3612.22.$$

(c) Here $P = \$2000$, $i_m = 0.06$, $m = 4$, $i_m/m = 0.06/4 = 0.015$, and $t = 10$. Hence using (5.3), we obtain

$$A = \$2000(1.015)^{40}$$
$$= \$2000(1.814018)$$
$$= \$3628.04.$$

(d) Here $P = \$2000$, $i_m = 0.06$, $m = 12$, $i_m/m = 0.06/12 = 0.005$, and $t = 10$. Hence

$$A = \$2000(1.005)^{120}$$
$$= \$2000(1.819396)$$
$$= \$3638.79.$$

(e) Again $P = \$2000$, $i_m = 0.06$, $m = 365$,

$$\frac{i_m}{m} = \frac{0.06}{365} = 0.000164384,$$

and $t = 10$. We are assuming that there are 365 days in a year; thus there are $10 \times 365 = 3650$ conversion periods. Hence

$$A = \$2000(1.000164384)^{3650}.$$

Using an electronic calculator, we observe that

$$(1.000164384)^{3650} = 1.822032.$$

Thus

$$A = \$2000(1.822032)$$
$$= \$3644.06$$

The information obtained in this example is summarized in Table 5.1. Note that the more often interest is compounded, the greater is the accumulated value of the principal invested.

TABLE 5.1 $P = \$2000$, $i_m = 0.06$, $t = 10$ years			
Compounded	Value of m	Accumulated Value of P	Successive Differences
Annually	1	$3581.70	
Semiannually	2	3612.22	$30.52
Quarterly	4	3628.04	15.82
Monthly	12	3638.79	10.75
Daily	365	3644.06	5.27

When interest is compounded continuously, then the amount of money accumulated A after t years is

$$A = Pe^{i_m t}, \tag{5.4}$$

where $e = 2.71828\ldots$. The actual development of e requires some exposure to calculus and is therefore beyond our present scope. Thus, if $P = \$2000$, $i_m = 0.06$, and $t = 10$ years and interest is compounded continuously, then

$$A = \$2000e^{(0.06)10}$$
$$= \$2000e^{0.60}$$
$$= \$2000(1.8221)$$
$$= \$3644.20,$$

which is slightly more than the amount obtained by daily compounding.

Example 5.15

Determine how much money should be deposited in the bank if the money is to accumulate to $2500 in 5 years at 6 percent interest compounded (a) monthly, (b) daily, and (c) continuously.

Solution: Observe that Equation (5.2), converted to nominal rates of interest, becomes

$$P = A(t)\left(1 + \frac{i_m}{m}\right)^{-mt}.$$

(a) Here $A = \$2500$, $m = 12$, $i_m = 0.06$, $i_m/m = 0.005$, and $t = 5$. Using the above equation, we obtain

$$P = \$2500(1.005)^{-60}$$
$$= \$2500(0.741372)$$
$$= \$1853.43.$$

(b) Assuming there are 365 days in a year, we have $5 \times 365 = 1825$ conversion periods in 5 years. Here $A = \$2500$, $i_m = 0.06$, $m = 365$, and

$$\frac{i_m}{m} = \frac{0.06}{365} = 0.000164384.$$

Hence

$$P = \$2500(1.000164384)^{-1825}.$$

Using an electronic calculator, we observe that

$$(1.000164384)^{-1825} = 0.740836.$$

Thus

$$P = \$2500(0.740836)$$
$$= \$1852.09.$$

(c) From (5.4), we have

$$A = Pe^{i_m \cdot t}$$

or

$$P = Ae^{-i_m \cdot t}.$$

Here $A = \$2500$, $i_m = 0.06$, and $t = 5$ years. Hence

$$P = \$2500e^{-(0.06) \cdot 5}$$
$$= \$2500e^{-0.30}$$
$$= \$2500(0.74082)$$
$$= \$1852.05.$$

We have shown that if interest i is compounded annually, then the accumulated value of the principal P at the end of 1 year is

$$A = P(1 + i).$$

If interest i_m/m is compounded m times a year, then the accumulated value of the principal P is

$$A = P\left(1 + \frac{i_m}{m}\right)^m.$$

Therefore, the interest rate i compounded annually that will yield the same accumulated value A for a principal P invested at an interest of i_m/m compounded m times a year is given by

$$1 + i = \left(1 + \frac{i_m}{m}\right)^m.$$

It follows that the effective annual rate of interest is given by

$$i = \left(1 + \frac{i_m}{m}\right)^m - 1. \tag{5.5}$$

If interest i is compounded continuously, then the effective rate of interest is given by

$$i = e^{i_m} - 1. \tag{5.6}$$

Example 5.16

Determine the effective rate of interest if the nominal rate of 9 percent is compounded **(a)** semiannually, **(b)** quarterly, **(c)** monthly, and **(d)** continuously.

Solution:

(a) We have $i_m = 0.09$, $m = 2$; $i_m/m = 0.09/2 = 0.045$. Using (5.5), we obtain

$$i = (1 + 0.045)^2 - 1$$
$$= (1.045)^2 - 1$$
$$= 1.092025 - 1$$
$$= 0.092025, \text{ or } 9.20 \text{ percent approximately.}$$

(b) Here $i_m = 0.09$, $m = 4$; $i_m/m = 0.09/4 = 0.0225$. Using (5.5), we have

$$i = (1 + 0.0225)^4 - 1$$
$$= (1.0225)^4 - 1$$
$$= 1.093083 - 1$$
$$= 0.093083, \text{ or } 9.31 \text{ percent approximately.}$$

(c) Here $i_m = 0.09$, $m = 12$; $i_m/m = 0.09/12 = 0.0075$. Using (5.5), we have

$$i = (1 + 0.0075)^{12} - 1$$
$$= (1.0075)^{12} - 1$$
$$= 1.093807 - 1$$
$$= 0.093807, \text{ or } 9.381 \text{ percent approximately.}$$

(d) Here $i_m = 0.09$ and interest is compounded continuously. We use (5.6) and obtain

$$i = e^{0.09} - 1$$
$$= 1.094174 - 1$$
$$= 0.094174, \text{ or } 9.417 \text{ percent approximately.}$$

EXERCISES

Find the accumulated value A for the deposits in Exercises 1 to 16.
1. $750 at 5.50 percent compounded annually for 4 years.
2. $1200 at 5.75 percent compounded annually for 5 years.
3. $1000 at 6 percent compounded semiannually for 6 years.
4. $1500 at 7.50 percent compounded semiannually for 4 years.
5. $1500 at 7.50 percent compounded quarterly for 5 years.
6. $2000 at 8 percent compounded quarterly for 3 years.
7. $1800 at 6 percent compounded monthly for 6 years.
8. $2500 at 7.50 percent compounded monthly for 3 years.
9. $3000 at 9 percent compounded monthly for 4 years.
10. $3500 at 6 percent compounded daily for 5 years.
11. $1800 at 9 percent compounded daily for 4 years.
12. $1650 at 7.50 percent compounded daily for 3 years.
13. $4500 at 7.54 percent compounded continuously for 5 years.
14. $3000 at 13.42 percent compounded continuously for 3 years.
15. $3200 at 9.90 percent compounded continuously for $2\frac{1}{2}$ years.
16. $2000 at 11.08 percent compounded continuously for 2 years.

In Exercises 17 to 20, determine the amount of interest earned for each deposit.
17. $3000 at 7.536 percent compounded continuously for 3 years.
18. $10,000 at 13.20 percent compounded monthly for 5 years.
19. $6000 at 6 percent compounded daily for 4 years.
20. $4500 at 11.56 percent compounded continuously for $2\frac{1}{2}$ years.

In Exercises 21 to 30, determine the present value for the accumulated amounts.

21. $1800 at 8 percent compounded quarterly for 3 years.

22. $2500 at 6 percent compounded annually for 5 years.

23. $4000 at 7 percent compounded quarterly for 4 years.

24. $5000 at 5.50 percent compounded quarterly for 5 years.

25. $2500 at 6 percent compounded daily for 5 years.

26. $3600 at 9 percent compounded monthly for 5 years.

27. $4200 at 15 percent compounded monthly for $2\frac{1}{2}$ years.

28. $2400 at 6.55 percent compounded continuously for 6 years.

29. $3200 at 9.325 percent compounded continuously for 4 years.

30. $10,000 at 11.32 percent compounded continuously for 5 years.

In Exercises 31 to 36, determine the effective rate of interest.

31. 15 percent compounded monthly.   **32.** 10 percent compounded quarterly.

33. 18 percent compounded monthly.   **34.** 12 percent compounded continuously.

35. 19.68 percent compounded monthly.   **36.** 21 percent compounded continuously.

37. Ms. Hawkes wishes to have $4500 available for a European trip she is planning to take 3 years from now. What sum of money must she place on deposit now if interest is 10 percent compounded continuously?

38. Susan wishes to have $25,000 available for her daughter who will enter college 4 years from now. What sum of money she must place on deposit now if interest is 11.56 percent compounded continuously?

39. Mike wishes to have $12,000 available for a new car he is planning to buy $2\frac{1}{2}$ years from now. What sum of money must he invest now if interest is 12 percent compounded **(a)** quarterly, **(b)** monthly, **(c)** daily, and **(d)** continuously?

40. Henry wishes to have $2500 for a video recording unit he is planning to buy 2 years from now. What sum of money must he deposit now if interest is 17 percent compounded continuously?

APPLICATIONS OF EXPONENTIAL FUNCTIONS

Many practical situations involving growth and decay can be described by an exponential function of the form

$$f(t) = Ae^{kt},$$

where A reflects the amount of substance present at time $t = 0$, k is a nonzero constant, and $f(t)$ denotes the amount present at time t. At time $t = 0$,

$$f(0) = Ae^{k \cdot 0} = Ae^0 = A.$$

Thus,

$$f(t) = f(0)e^{kt}. \tag{5.7}$$

When $k > 0$, the function $f(t)$ increases with time, and we speak of **exponential growth.** Few simple examples of exponential growth are exhibited by the growth of bacteria in a culture, growth of fish population in a lake, the growth of fruit fly population in a laboratory container, the growth of population in a city, and the like.

Example 5.17

The number of bacteria $N(t)$ present in a certain culture at time t is given by

$$N(t) = 10{,}000e^{0.3t},$$

where t is measured in hours. Find the number of bacteria when **(a)** $t = 0$, **(b)** $t = 2$, **(c)** $t = 4$, **(d)** $t = 6$, and **(e)** $t = 8$. Sketch the graph of $N(t)$.

Solution:

(a) The number of bacteria at time $t = 0$ is given by

$$
\begin{aligned}
N(0) &= 10{,}000e^{(0.30)0} \\
&= 10{,}000e^{0} \\
&= 10{,}000(1) \\
&= 10{,}000.
\end{aligned}
$$

(b) The number of bacteria at time $t = 2$ is

$$
\begin{aligned}
N(2) &= 10{,}000e^{(0.3)2} \\
&= 10{,}000e^{0.6} \\
&= 10{,}000(1.8221) = 18{,}221.
\end{aligned}
$$

(c) The number of bacteria at $t = 4$ hours is

$$
\begin{aligned}
N(4) &= 10{,}000e^{(0.3)4} \\
&= 10{,}000e^{1.2} \\
&= 10{,}000(3.3201) = 33{,}201.
\end{aligned}
$$

(d) The number of bacteria at time $t = 6$ hours is

$$
\begin{aligned}
N(6) &= 10{,}000e^{(0.3)6} \\
&= 10{,}000e^{1.8} \\
&= 10{,}000(6.0496) = 60{,}496.
\end{aligned}
$$

(e) The number of bacteria at time $t = 8$ hours is

$$
\begin{aligned}
N(8) &= 10{,}000e^{(0.3)8} \\
&= 10{,}000e^{2.4} \\
&= 10{,}000(11.0232) = 110{,}231.
\end{aligned}
$$

Plotting $(t, N(t))$ for $t = 0, 2, 4, 6,$ and 8 and connecting the points with a smooth curve, we obtain the graph in Figure 5.8.

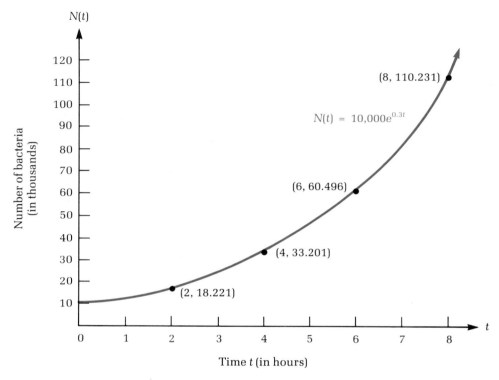

FIGURE 5.8

Exponential Decay

When $k < 0$, the function $f(t)$ in (5.7) decreases with time t, and we have a negative exponential growth, or **exponential decay.** An example of exponential decay is given by the disintegration of radioactive debris such as strontium 90 that falls on plants and grass, thereby contaminating the food supply of animals. This is one of the serious problems associated with the aboveground nuclear explosions. Further, it is widely known that atmospheric pressure, measured in torr, decreases with increasing altitude according to a certain mathematical model. In marketing, the sales of a particular product will decline if advertising and other promotion of that product are stopped.

Example 5.18

The quantity $Q(t)$ of strontium (measured in pounds) present at time t is given by

$$Q(t) = 10e^{-0.024t},$$

where t is measured in years. Determine the quantity of strontium remaining after $t = 0, 10, 20, 30, 40, 50, 60, 70, 80, 90,$ and 100 years. Sketch the graph of $Q(t)$.

Solution: Using a scientific calculator or Appendix Table 1, we have

$$Q(0) = 10e^0 = 10$$
$$Q(10) = 10e^{-0.024(10)} = 10e^{-0.24} = 10(0.7866) = 7.866$$
$$Q(20) = 10e^{-0.024(20)} = 10e^{-0.48} = 10(0.6188) = 6.188$$
$$Q(30) = 10e^{-0.024(30)} = 10e^{-0.72} = 10(0.4868) = 4.868$$
$$Q(40) = 10e^{-0.024(40)} = 10e^{-0.96} = 10(0.3829) = 3.829$$
$$Q(50) = 10e^{-0.024(50)} = 10e^{-1.20} = 10(0.3012) = 3.012$$
$$Q(60) = 10e^{-0.024(60)} = 10e^{-1.44} = 10(0.2369) = 2.369$$
$$Q(70) = 10e^{-0.024(70)} = 10e^{-1.68} = 10(0.1864) = 1.864$$
$$Q(80) = 10e^{-0.024(80)} = 10e^{-1.92} = 10(0.1466) = 1.466$$
$$Q(90) = 10e^{-0.024(90)} = 10e^{-2.16} = 10(0.1153) = 1.153$$
$$Q(100) = 10e^{-0.024(100)} = 10e^{-2.40} = 10(0.0907) = 0.907$$

Plotting $(t, Q(t))$ for given values of t, we obtain the graph in Figure 5.9.

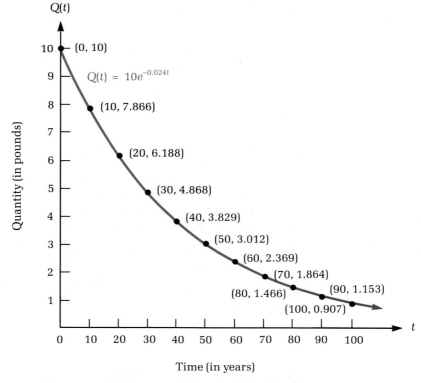

FIGURE 5.9

Social Diffusion Curve

Sociologists recognize a phenomenon called **social diffusion.** This phenomenon deals with the spreading of a piece of information such as a technological innovation, result of an election, or some other cultural fad among a given population. Given a population of fixed size N, if $f(t)$ is the number of people who have received the information by time t, then $N - f(t)$ is the number of those who have not yet heard the information. If the information is broadcast frequently by radio and television, then it seems reasonable to assume that the rate of diffusion of this information is directly proportional to the number $N - f(t)$ who have yet to hear the information. Sociologists have determined that a good mathematical model for $f(t)$ is given by

$$f(t) = N(1 - e^{-kt}), \qquad\qquad (5.8)$$

where k is a constant. If we plot $f(t)$ against time t for a fixed value of N and k, then we obtain a curve of the type shown in Figure 5.10.

At time $t = 0$, $e^{-kt} = 1$, which implies that $f(0) = 0$. In other words, at time $t = 0$, no one in the population has heard any information. As t increases, e^{-kt} becomes small and $f(t)$ will approach N. This means that when sufficient time has elapsed, nearly everyone in the population will get the information.

Learning Curve

The study of learning has been a basic concern of psychologists for more than a century, and there is a vast literature of books and articles developing different theories or presenting the results of learning involving human and animal subjects. In most of these studies, psychologists agree that in many learning

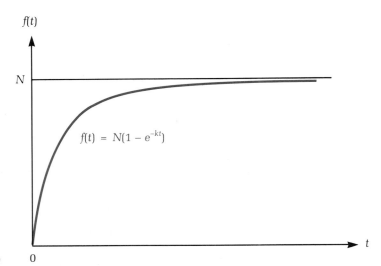

FIGURE 5.10   0

situations, such as learning a foreign language or a new subject or a sport, one progresses faster in the beginning, then slows down, and finally levels off. Suppose, for example, that a psychologist has prepared a list of N nonsense words for use in a memory experiment. By giving this list to a subject, one can determine an empirical relationship between the number of nonsense words memorized accurately and the time spent to reach that level of performance. Psychologists have determined that a mathematical model for this experiment is approximated by an exponential equation of the form

$$f(t) = N(1 - e^{-kt}),$$

where $f(t)$ is the number of words memorized accurately after time t and k is some appropriate positive constant. If we plot $f(t)$ against time t for some fixed values of N and k, then we obtain a curve similar to the one shown in Figure 5.11. This curve is called the **learning curve.**

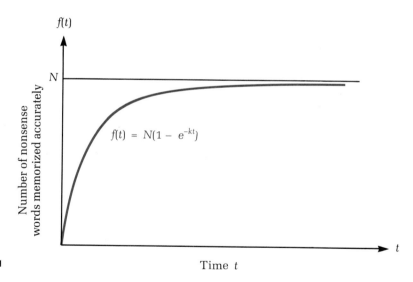

FIGURE 5.11

Let us look at an example.

Example 5.19

A psychologist has prepared a list of nonsense words for use in a memory experiment. The number of words $N(t)$ that a person is able to memorize after t hours of practice is given by

$$N(t) = 100(1 - e^{-0.2t}).$$

Determine $N(t)$ for $t = 1, 2, 3, 4, 5, 6, 8, 10,$ and 16. Sketch also the graph of $N(t)$.

Solution: The values of $N(t)$ for different values of t are given in Table 5.2.

TABLE 5.2

Time t (in hours)	$N(t)$
1	$100(1 - e^{-0.2}) = 100(1 - 0.81873) = 18.127 \approx 18$
2	$100(1 - e^{-0.4}) = 100(1 - 0.67032) = 32.968 \approx 33$
3	$100(1 - e^{-0.6}) = 100(1 - 0.54881) = 45.119 \approx 45$
4	$100(1 - e^{-0.8}) = 100(1 - 0.44933) = 55.067 \approx 55$
5	$100(1 - e^{-1.0}) = 100(1 - 0.36788) = 63.212 \approx 63$
6	$100(1 - e^{-1.2}) = 100(1 - 0.30119) = 69.881 \approx 70$
8	$100(1 - e^{-1.6}) = 100(1 - 0.20190) = 79.810 \approx 80$
10	$100(1 - e^{-2.0}) = 100(1 - 0.13534) = 86.466 \approx 86$
16	$100(1 - e^{-3.2}) = 100(1 - 0.04076) = 95.924 \approx 96$

Plotting the ordered pairs $(t, N(t))$ and connecting these points with a smooth curve, we obtain the graph in Figure 5.12. Note that as time t increases, $N(t)$ approaches the maximum level.

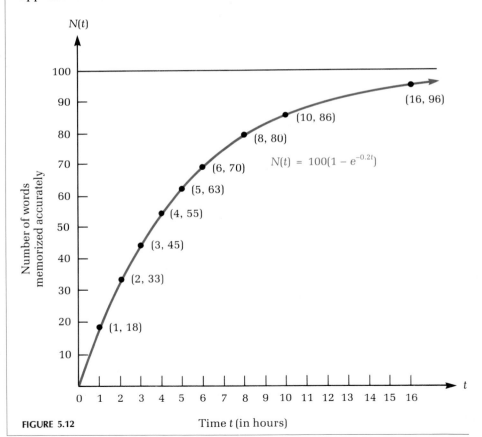

FIGURE 5.12

Let us now look at an example from business.

Example 5.20

The annual profit for a publisher from the sales of a certain book is given by

$$P(t) = \$5000 + \$10,000e^{-0.4(t-1)},$$

where t is the number of years the text is on the market. Determine the annual profit at the end of 1, 2, 3, 4, 5, and 10 years. Graph $P(t)$.

Solution: Table 5.3 shows the values of $P(t)$ for different values of t. Plotting the ordered pairs $(t, P(t))$, we obtain the graph shown in Figure 5.13. This shows that profit tends to level off with time and gradually approaches a level of \$5000.

TABLE 5.3

Times (in Years)	Profit (in Dollars)
1	$5000 + 10,000e^0 = 15,000$
2	$5000 + 10,000(e^{-0.4}) = 5000 + 10,000(0.67032) = 11,703$
3	$5000 + 10,000(e^{-0.8}) = 5000 + 10,000(0.44933) = 9493$
4	$5000 + 10,000(e^{-1.2}) = 5000 + 10,000(0.30119) = 8012$
5	$5000 + 10,000(e^{-1.6}) = 5000 + 10,000(0.20190) = 7019$
6	$5000 + 10,000(e^{-2.0}) = 5000 + 10,000(0.13534) = 6353$
10	$5000 + 10,000(e^{-3.6}) = 5000 + 10,000(0.02732) = 5273$

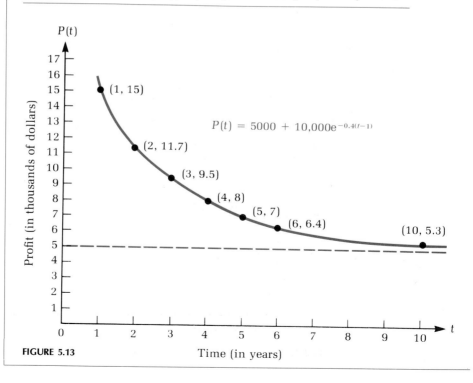

FIGURE 5.13

Time (in years)

EXERCISES

1. The population $P(t)$ of a certain city is given by

$$P(t) = 15{,}000e^{0.04t},$$

   where t represents time measured in years. Find the population of the city when **(a)** $t = 5$, **(b)** $t = 10$, and **(c)** $t = 15$.

2. The number of bacteria $N(t)$ present in a certain culture at time t is

$$N(t) = 500e^{0.3t},$$

   where t is in hours. Find the number of bacteria present at time **(a)** $t = 0$, **(b)** $t = 2$, and **(c)** $t = 5$.

3. The quantity $Q(t)$ of a radioactive substance (measured in grams) present at time t is

$$Q(t) = 500e^{-0.05t},$$

   where t is recorded in days. Find the quantity of the radioactive substance when **(a)** $t = 6$, **(b)** $t = 12$, and **(c)** $t = 18$.

4. The quantity of radium (a substance subject to radioactive decay) $Q(t)$, measured in milligrams, at time t is given by

$$Q(t) = 1000e^{-0.04t},$$

   where t is measured in centuries. Determine the quantity present when **(a)** $t = 5$, **(b)** $t = 10$, **(c)** $t = 15$, and **(d)** $t = 20$.

5. The concentration of a drug in the bloodstream of a patient is approximated by

$$Q(t) = 5e^{-0.3t},$$

   where $Q(t)$ is the quantity of drug, measured in milligrams, in the bloodstream t hours after it is injected. Determine $Q(t)$ for $t = 1, 2, 3, 4, 5, 6, 8$, and 10 hours.

6. The quantity $Q(t)$ of strontium 90 (measured in pounds) present at time t is

$$Q(t) = 100e^{-0.0244t},$$

   where t is measured in years. Determine the quantity of strontium 90 remaining after $t = 0, 28, 56, 84, 112, 140, 168$, and 196 years. Sketch the graph of $Q(t)$.

7. The annual raise in salaries for a university's faculty is determined by the formula

$$S(t) = \$20{,}000e^{-1.2t}$$

   after t years of service. Determine the raise in salaries after $t = 1, 2, 3, 4$, and 5 years. Sketch the graph of $S(t)$.

8. The monthly sales of new model of automobile in Vermont are approximated by

$$S(t) = 4000e^{-0.3t},$$

   where t is the number of months this model has been in production. Determine $S(t)$ for $t = 1, 2, \ldots, 12$. Sketch the graph of $S(t)$.

9. A student's record of learning a foreign language is given by the exponential equation

$$N(t) = 150(1 - e^{-0.32t}),$$

where $N(t)$ is the number of new words the student learns after t hours of practice. Determine $N(t)$ for $t = 1$, 2, 3, 4, 5, 6, 8, 10, 12, 14, and 16 hours. Sketch the graph of $N(t)$.

10. Miguel's record of learning to type on a word-processing unit is given by

$$N(t) = 75(1 - e^{-0.09t}),$$

where $N(t)$ is the number of words per minute he is able to type after t weeks of instruction. Determine $N(t)$ for $t = 5$, 10, 15, 20, 25, 30, 35, and 40 weeks. Sketch the graph of $N(t)$.

11. A manufacturer's annual profit from the sales of a certain product is given by

$$P(t) = \$20{,}000(1 + e^{-0.3t}),$$

where t is the number of years the product is on the market. Determine $P(t)$ for $t = 1, 2, 3, 4, 5, 6, 8$, and 10 years.

12. The sales of a new product of the ABC Company in Chicago are approximated by

$$S(t) = 2000(5 + 4e^{-0.5t}),$$

where t is the number of months the product is advertised. Determine $S(t)$ for $t = 1, 2, 3, 4, 5, 6, 7, 8, 10$, and 12. Sketch the graph of $S(t)$.

5.5 LOGARITHMIC FUNCTIONS

Thus far, we have discussed exponential functions represented by the equation $y = b^x$, where $b > 0$, $b \neq 1$, and x is any real number. Although the exponential equation $y = b^x$ is not usually easy to solve for x when y and b are known, one would expect the equation to have a solution. Further, the graphs of $y = b^x$ for $b > 1$ in Figure 5.1 and for $0 < b < 1$ in Figure 5.5 represent the one-to-one correspondence between the set of all real numbers x and the set of all positive real values of y. This means that the exponential function *logarithmic function* $y = b^x$ has an inverse function. We call this inverse the **logarithmic function;** that is, if $y = b^x$, then we say that $x = \log_b y$. The number x is called the logarithm of y to the base b.

Definition 5.5.1

For $b > 0$, $b \neq 1$, and $y > 0$,

$$x = \log_b y \quad \text{if and only if} \quad y = b^x.$$

In words, the logarithm of a positive number to a given base is the exponent to which the base must be raised to yield that number. If we write $3^4 = 81$, then the logarithm of 81 to the base 3 is 4; that is, $\log_3 81 = 4$. If we write $10^3 = 1000$, then $\log 1000 = 3$. For further illustrations, let us consider the following examples.

Example 5.21

Determine $\log_2 64$.

Solution: Let $x = \log_2 64$. Using the definition of logarithmic function, we obtain the exponential equation

$$2^x = 64.$$

Since $64 = 2^6$, we see that $x = 6$, and so

$$\log_2 64 = 6.$$

Example 5.22

Determine $\log_3 \frac{1}{27}$.

Solution: Let $x = \log_3 \frac{1}{27}$. Then

$$3^x = \tfrac{1}{27}.$$

Expressing $\frac{1}{27}$ as a power of the base 3, we have

$$(\tfrac{1}{3})^3 = 3^{-3}.$$

Thus

$$3^x = 3^{-3},$$

which implies that $x = -3$ and $\log_3 \frac{1}{27} = -3$.

Example 5.23

Find $\log_8 16$.

Solution: If $x = \log_8 16$, then $8^x = 16$. Since $8 = 2^3$ and $16 = 2^4$, we have

$$2^{3x} = 2^4,$$

which implies that $3x = 4$ or $x = \tfrac{4}{3}$; therefore $\log_8 16 = \tfrac{4}{3}$.

We consider it worthwhile to reinforce the fact that the functions $y = \log_b x$ and $y = b^x$ are inverses of each other. If we want to sketch the graph of $y = \log_b x$ for $b > 1$, we need only sketch the graph of its inverse function $y = b^x$ for $b > 1$ and then graph its mirror image in the line $y = x$ (see Figure 5.14).

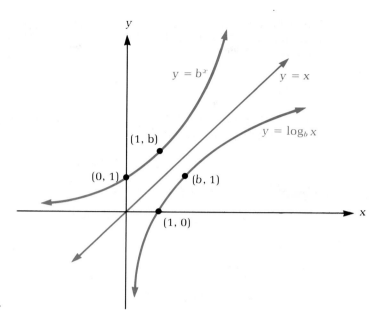

FIGURE 5.14

From the graph, we observe the following.

1. $\log_b x$ is defined only for $x > 0$.
2. $\log_b x < 0$ for $0 < x < 1$.
3. $\log_b x = 0$ for $x = 1$.
4. $\log_b x > 0$ for $x > 1$.
5. $\log_b x = 1$ for $x = b$.

EXERCISES

Express the equation in Exercises 1 to 8 in logarithmic form.

1. $2^4 = 16$	**2.** $3^4 = 81$	**3.** $4^5 = 1024$	**4.** $6^3 = 216$
5. $(36)^{1/2} = 6$	**6.** $(32)^{3/5} = 8$	**7.** $(216)^{2/3} = 36$	**8.** $\left(\frac{2}{3}\right)^{-3} = \frac{27}{8}$

Express the quantities in Exercises 9 to 16 in exponential form.

9. $\log_3 27 = 3$	**10.** $\log_8 4 = \frac{2}{3}$	**11.** $\log_2 64 = 6$	**12.** $\log_{32} 2 = \frac{1}{5}$
13. $\log_4 256 = 4$	**14.** $\log_6 1 = 0$	**15.** $\log_6 36 = 2$	**16.** $\log_{10} 0.001 = -3$

Solve the equations in Exercises 17 to 26.

17. $\log_3 x = 4$	**18.** $\log_2 x = 7$	**19.** $\log_8 512 = x$	**20.** $\log_x 25 = -2$
21. $\log_5 125 = x$	**22.** $\log_x 16 = 4$	**23.** $\log_5 x = -2$	**24.** $\log_9 3 = x$
25. $\log_6 216 = x$	**26.** $\log_x 343 = 3$		

5.6

FUNDAMENTAL PROPERTIES OF LOGARITHMIC FUNCTIONS

There is a definitive relationship between the properties of the exponential and logarithmic functions. The following logarithmic properties are immediate consequences of the exponent properties discussed in Section 1.3. From now on, we assume that all bases are well defined—that is, positive and not equal to 1.

Proposition 5.6.1

If m and n are positive integers and b is any base, then

(a) $\log_b mn = \log_b m + \log_b n$

(b) $\log_b \dfrac{m}{n} = \log_b m - \log_b n.$

Proof: Let $\log_b m = p$ and $\log_b n = q$. Using Definition 5.5.1, we have

$$m = b^p \quad \text{and} \quad n = b^q.$$

(a) Observe that

$$mn = b^p b^q = b^{p+q}$$

and by Definition 5.5.1 it follows that

$$\log_b mn = p + q = \log_b m + \log_b n.$$

(b) Also

$$\frac{m}{n} = \frac{b^p}{b^q} = b^{p-q},$$

and using Definition 5.5.1, we obtain

$$\log_b \frac{m}{n} = p - q = \log_b m - \log_b n.$$

For an application of this proposition, consider Example 5.24.

Example 5.24

Given that $\log_b 2 = 0.69$, $\log_b 3 = 1.10$, and $\log_b 5 = 1.61$, find the following:

(a) $\log_b 6$

(b) $\log_b 15$

(c) $\log_b \frac{3}{2}$

(d) $\log_b \frac{10}{3}$

Solution:

(a) $\log_b 6 = \log_b (2 \times 3) = \log_b 2 + \log_b 3$
$$= 0.69 + 1.10 = 1.79$$

(b) $\log_b 15 = \log_b (3 \times 5) = \log_b 3 + \log_b 5$
$$= 1.10 + 1.61 = 2.71$$

(c) $\log_b \frac{3}{2} = \log_b 3 - \log_b 2$
$$= 1.10 - 0.69 = 0.41$$

(d) $\log_b \dfrac{10}{3} = \log_b \dfrac{5 \times 2}{3}$

$$= \log_b 5 + \log_b 2 - \log_b 3$$
$$= 1.61 + 0.69 - 1.10 = 1.20$$

Proposition 5.6.2

Let m be any positive real number, and let r be any real number; then

$$\log_b m^r = r \log_b m.$$

Proof: Let $\log_b m = p$. Then $m = b^p$ and $m^r = b^{pr}$. Hence

$$\log_b m^r = pr = rp = r \log_b m.$$

Example 5.25

Given that $\log_b 2 = 0.69$ and $\log_b 3 = 1.10$, determine the following:

(a) $\log_b \sqrt{2}$

(b) $\log_b 72$

Solution:

(a) $\log_b \sqrt{2} = \log_b 2^{1/2}$
$$= \tfrac{1}{2} \log_b 2$$
$$= \tfrac{1}{2}(0.69)$$
$$= 0.345$$

(b) $\log_b 72 = \log_b (2^3 \times 3^2) = \log_b 2^3 + \log_b 3^2$ (Proposition 5.6.1)
$$= 3 \log_b 2 + 2 \log_b 3 \qquad\qquad\quad \text{(Proposition 5.6.2)}$$
$$= 3(0.69) + 2(1.10)$$
$$= 4.27$$

Proposition 5.6.3

Let a and b be two bases, and let m be any positive number; then

$$\log_a m = \frac{\log_b m}{\log_b a}.$$

Proof: Let $\log_a m = p$, which implies that $m = a^p$. In a like manner, if $\log_b a = n$, then $a = b^n$. Thus

$$m = a^p = (b^n)^p = b^{np}.$$

This statement implies that

$$\log_b m = np = (\log_b a)(\log_a m)$$

which proves our assertion that

$$\log_a m = \frac{\log_b m}{\log_b a}.$$

Example 5.26

Given that $\log_b 2 = 0.69$, $\log_b 3 = 1.10$, and $\log_b 5 = 1.61$, determine the following:

(a) $\log_3 5$ **(b)** $\log_2 3$

Solution:

(a) $\log_3 5 = \dfrac{\log_b 5}{\log_b 3} = \dfrac{1.61}{1.10} = 1.46$

(b) $\log_2 3 = \dfrac{\log_b 3}{\log_b 2} = \dfrac{1.10}{0.69} = 1.59$

Example 5.27

Express

$$\log_b \sqrt[5]{\frac{xy^2}{z^3}}$$

as the sum or difference of logarithms of x, y, and z.

Solution:

$$\log_b \sqrt[5]{\frac{xy^2}{z^3}} = \log_b \left(\frac{xy^2}{z^3}\right)^{1/5} = \frac{1}{5} \log_b \frac{xy^2}{z^3} \qquad \text{(Proposition 5.6.2)}$$

$$= \tfrac{1}{5}(\log_b xy^2 - \log_b z^3) \qquad \text{(Proposition 5.6.1)}$$

$$= \tfrac{1}{5}(\log_b x + \log_b y^2 - \log_b z^3) \qquad \text{(Proposition 5.6.1)}$$

$$= \tfrac{1}{5}(\log_b x + 2\log_b y - 3\log_b z) \qquad \text{(Proposition 5.6.2)}$$

Example 5.28	Express $\frac{1}{3}(\log_b x^2 - \log_b y)$ as a single logarithm.

Solution:

$$\frac{1}{3}(\log_b x^2 - \log_b y) = \frac{1}{3}\log_b \frac{x^2}{y} \qquad \text{(Proposition 5.6.1)}$$

$$= \log_b \left(\frac{x^2}{y}\right)^{1/3} \qquad \text{(Proposition 5.6.2)}$$

EXERCISES

Given that $\log_b 2 = 0.69$, $\log_b 3 = 1.10$, $\log_b 5 = 1.61$, and $\log_b 7 = 1.95$, determine the expressions in Exercises 1 to 12.

1. $\log_b \frac{2}{7}$
2. $\log_b 21$
3. $\log_b 35$
4. $\log_b 25$
5. $\log 128$
6. $\log_b 343$
7. $\log_b 105$
8. $\log_b 98$
9. $\log_b 42$
10. $\log_b 315$
11. $\log_b 245$
12. $\log_b 147$

Using properties of logarithmic functions and a scientific calculator, determine the expressions in Exercises 13 to 24.

13. $\log_{10} \frac{21}{2}$
14. $\log_{10} \frac{14}{3}$
15. $\log_{10} \sqrt{2}$
16. $\log_{10} \sqrt{3}$
17. $\log_{10} \sqrt{6}$
18. $\log_{10} \sqrt{21}$
19. $\log_{10} (21)^{1/3}$
20. $\log_{10} (12)^{1/3}$
21. $\log_3 7$
22. $\log_2 3$
23. $\log_2 7$
24. $\log_3 2$

Express the terms in Exercises 25 to 34 as the sum or difference of logarithms of x, y, and z.

25. $\log_b xy^2$
26. $\log_b (xy)^2$
27. $\log_b (xyz)^3$
28. $\log_b (x^2yz)^2$
29. $\log_b \frac{x^2}{y}$
30. $\log_b \frac{x^2y}{z}$
31. $\log_b \sqrt{xy}$
32. $\log_b \sqrt[3]{xy^2}$
33. $\log_b \left(\frac{x^2y}{z}\right)^{1/3}$
34. $\log_b \sqrt[4]{\frac{xy^2}{z^3}}$

Express the terms in Exercises 35 to 44 as a single logarithm.

35. $\log_b x + \log_b y$
36. $\log_b x + 2\log_b y$
37. $2\log_b x - \log_b y$
38. $\log_b 2x + 3\log_b y$
39. $3\log_b x + 4\log_b y - \log_b z$
40. $2\log_b x + \log_b y - 3\log_b z$
41. $\frac{1}{2}\log_b x + \frac{1}{3}\log_b y$
42. $\frac{1}{3}\log_b x - \frac{2}{3}\log_b y$
43. $\frac{1}{4}\log_b x + \log_b y - 3\log_b z$
44. $\frac{1}{2}\log_b x + \frac{1}{3}\log_b y - \frac{1}{4}\log_b z$
45. Prove that $(\log_a b)(\log_b a) = 1$.
46. Prove that if m, n, and p are positive numbers and b is any base, then

$$\log_b mnp = \log_b m + \log_b n + \log_b p.$$

5.7

COMMON AND NATURAL LOGARITHMS

Common Logarithms

Logarithms to the **base 10** are called **common logarithms.** In the past, these logarithms were used extensively for simplifying numerical work in mathematics and other sciences. Recent developments in digital computers and hand-held calculators have diminished significantly the importance of logarithms as a computing device. Nevertheless, a basic understanding of the relationship between logarithmic and exponential functions remains important.

Briefly, the common logarithm of a number is the exponent to which the base 10 must be raised to yield that number. That is,

$$\text{If } 10^x = y, \quad \text{then } \log_{10} y = x.$$

For the sake of brevity, we abbreviate $\log_{10} y$ to $\log y$. Since $10^1 = 1$, $10^2 = 100$, $10^3 = 1000$, and $10^4 = 10,000$, it follows that the common logarithm of 10 is 1, the common logarithm of 100 is 2, the common logarithm of 1000 is 3, and so on. Thus the following statements are equivalent:

$$10^1 = 10 \quad \text{and} \quad \log 10 = 1$$
$$10^2 = 100 \quad \text{and} \quad \log 100 = 2$$
$$10^3 = 1000 \quad \text{and} \quad \log 1000 = 3.$$

Further,

$$10^{-1} = 0.1 \quad \text{so} \quad \log 0.1 = -1$$
$$10^{-2} = 0.01 \quad \text{so} \quad \log 0.01 = -2$$
$$10^{-3} = 0.001 \quad \text{so} \quad \log 0.001 = -3.$$

Thus, the logarithms of the integral powers of 10 are precisely those integers. Evidently, if

$$10^x = 83,$$

then x must be a number between 1 and 2, since $10^1 = 10$ and $10^2 = 100$. That is, x is the integer 1 plus a decimal fraction. In fact, every logarithm may be expressed as a sum of an integer and a decimal fraction.

Definition 5.7.1

> The integral part of a common logarithm is called the *characteristic*. The decimal part of a common logarithm is called the *mantissa*.

Any positive number x can be expressed in scientific notation as

$$x = n \times 10^c,$$

where c is an integer (which may be positive, zero, or negative) and n satisfies the inequality $1 \leq n < 10$. For example,

$$23 = 2.3 \times 10 \qquad\qquad 0.71 = 7.1 \times 10^{-1}$$

$$475 = 4.75 \times 10^2 \qquad 0.0420 = 4.2 \times 10^{-2}$$

$$5430 = 5.43 \times 10^3 \qquad 0.0083 = 8.3 \times 10^{-3}$$

and so on.

Since $x = n \times 10^c$, it follows that

$$
\begin{aligned}
\log x &= \log (n \times 10^c) \\
&= \log n + \log 10^c \qquad \text{(Proposition 5.6.1)} \\
&= \log n + c \log 10 \qquad \text{(Proposition 5.6.2)}
\end{aligned}
$$

where c is the characteristic and $\log n$ is the mantissa. Further, because

$$1 \leq n < 10$$

and $\log n$ increases as n increases, we have

$$\log 1 \leq \log n < \log 10$$

$$0 \leq \log n < 1.$$

Hence $\log n$ is a number between 0 and 1. Thus any positive real number x can be written in the **standard** form

$$\log x = (\log n) + c, \tag{5.9}$$

where c is an integer and $\log n$ is a positive decimal fraction.

The mantissa, $\log n$, is obtained from Appendix Table 2, which lists four-decimal-place approximations for the logarithm of the numbers from 1.00 to 9.99 in steps of 0.01. Consider, for instance, $\log 2.35$. Referring to Appendix Table 2, we locate the row for 2.3 and then move across the row until we are under the column headed by 5 and find the entry 0.3711. We may use this table to determine that

$$\log 3.59 = 0.5551,$$

$$\log 5.47 = 0.7380,$$

$$\log 7.82 = 0.8932,$$

and so on.

Alternatively, most of the scientific calculators are equipped with a log key, and we can determine log x at the push of a button.

Example 5.29

Find log 3480 by **(a)** using tables and **(b)** using a scientific calculator.

Solution:

(a) Since

$$3480 = 3.48 \times 10^3,$$

it follows that $c = 3$ and $n = 3.48$. Using Appendix Table 2, we find that log 3.48 = 0.5416. Thus

$$\log 3480 = 0.5416 + 3 = 3.5416.$$

(b) We simply go through the following steps on a scientific calculator:

Enter	Press	Display
3480	log	3.5415792.

Thus,

$$\log 3480 = 3.5416 \qquad \text{(rounded to four decimal places).}$$

Example 5.30

Find log 0.0534 by **(a)** using tables and **(b)** using a calculator.

Solution:

(a) Since

$$0.0534 = 5.34 \times 10^{-2},$$

we have $c = -2$ and $n = 5.34$. Thus

$$\begin{aligned} \log 0.0534 &= (\log 5.34) + (-2) \\ &= 0.7275 + (-2) \\ &= -1.2725 \end{aligned}$$

(b) Using a scientific calculator, we proceed as follows:

Enter	Press	Display
0.0534	log	−1.2724587.

Thus,

$$\log 0.0534 = -1.2725 \qquad \text{(rounded to four decimal places).}$$

We now know how to find $\log x$ for a given positive number x by using tables as well as a calculator. But if $\log x$ is given, how do we use tables to determine x? First, we write $\log x$ in the standard form

$$\log x = (\log n) + c,$$

where c is the characteristic and $\log n$ is the mantissa. Since this statement is equivalent to

$$x = n \times 10^c,$$

we determine x by finding n and then multiplying it by 10^c. Note that the value of $\log n$ is always a number between 0 and 1 and is listed in the body of Appendix Table 2.

Example 5.31

Find x if $\log x = 3.8669$.

Solution: Since $\log x = 0.8669 + 3$, it follows that the characteristic $c = 3$ and $\log n = 0.8669$. Referring to the body of Appendix Table 2, we find that mantissa 0.8669 corresponds to the number $n = 7.36$. Thus

$$x = n \times 10^c = 7.36 \times 10^3 = 7360.$$

Example 5.32

Find x if $\log x = -2.1232$.

Solution: First we express $\log x$ in the standard form

$$(\log n) + c,$$

where $\log n$ is a positive decimal fraction. Since Appendix Table 2 gives only positive values, we obtain the mantissa by a simple algebraic step. We add and subtract 3, thus obtaining

$$\log x = \underbrace{-2.1232 + 3} + (-3) = 0.8768 + (-3),$$

which implies that $c = -3$ and $\log n = 0.8768$. Referring to Appendix Table 2, we see that the mantissa 0.8768 corresponds to $n = 7.53$. Thus

$$x = n \times 10^c = 7.53 \times 10^{-3} = 0.00753.$$

How do we use a scientific calculator to determine x if $\log x$ is given? Let us look at some examples.

Example 5.33

Find x if $\log x = 3.8669$.

Solution: Remember that $\log x = 3.8669$, when written in exponential form, is

$$x = 10^{3.8669}.$$

To determine the value of x, we proceed as follows:

Enter	Press	Display
10	y^x	10.
3.8669	=	7360.37599.

If a calculator is also equipped with a 10^x key, then we may obtain the answer in one step as follows:

Enter	Press	Display
3.8669	10^x	7360.37599.

This, when rounded to four decimal places, shows that $x = 7360.376$, and this is approximately the same answer we obtained in Example 5.31.

Example 5.34

Find x if $\log x = -2.1232$.

Solution: The expression $\log x = -2.1232$, written in exponential form, is equivalent to $x = 10^{-2.1232}$. To determine the value of x, we proceed as follows:

Enter	Press	Display
2.1232	$\pm$	-2.1232
	10^x	0.0075301.

This, when rounded to five decimal places, yields

$$x = 0.00753,$$

and this agrees with the answer we obtained in Example 5.32.

We use logarithms to solve certain equations in which the variable appears in an exponential or logarithmic expression.

Example 5.35 | Solve the equation $2^x = 12$.

Solution: Taking the common logarithms of both sides, we get

$$\log 2^x = \log 12$$

$$x \log 2 = \log 12 \qquad \text{(Proposition 5.6.2)}$$

$$x = \frac{\log 12}{\log 2}.$$

Using a calculator or Appendix Table 2, we have

$$\log 12 = 1.0792$$

$$\log 2 = 0.3010.$$

Thus

$$x = \frac{1.0792}{0.3010} = 3.5854.$$

Example 5.36 | Solve the equation $3^{2x-1} = 5^{x+1}$.

Solution: Taking the common logarithms of both sides, we have

$$\log 3^{2x-1} = \log 5^{x+1}$$

$$(2x - 1) \log 3 = (x + 1) \log 5 \qquad \text{(Proposition 5.6.2)}$$

$$2x \log 3 - \log 3 = x \log 5 + \log 5$$

$$2x \log 3 - x \log 5 = \log 5 + \log 3$$

$$x(2 \log 3 - \log 5) = \log 5 + \log 3.$$

Solving for x, we get

$$x = \frac{\log 5 + \log 3}{2 \log 3 - \log 5}.$$

Using a calculator or Appendix Table 2, we see that

$$\log 3 = 0.4771, \qquad \log 5 = 0.6990.$$

Thus

$$x = \frac{0.6990 + 0.4771}{2(0.4771) - 0.6990} = \frac{1.1761}{0.2552} = 4.6085.$$

Example 5.37

Solve the equation
$$\log (3x - 1) = 2 + \log (x - 2).$$

Solution: This equation is equivalent to
$$\log (3x - 1) - \log (x - 2) = 2$$
$$\log \frac{3x - 1}{x - 2} = 2. \qquad \text{(Proposition 5.6.1)}$$

Using the definition of logarithm, we obtain
$$\frac{3x - 1}{x - 2} = 10^2 = 100$$
$$3x - 1 = 100(x - 2).$$

Solving for x, we get
$$x = \tfrac{199}{97}.$$

Example 5.38

Solve the equation
$$\log (x + 2) = 1 - \log (x - 1).$$

Solution: This equation is equivalent to
$$\log (x + 2) + \log (x - 1) = 1$$
$$\log [(x + 2)(x - 1)] = 1 \qquad \text{(Proposition 5.6.1)}$$
$$(x + 2)(x - 1) = 10^1 = 10$$
$$x^2 + x - 2 = 10$$
$$x^2 + x - 12 = 0$$
$$(x + 4)(x - 3) = 0$$
$$x = -4 \quad \text{or} \quad x = 3.$$

Substituting $x = -4$ in the original equation, we find that $\log (x + 2) = \log (-2)$ and $\log (x - 1) = \log (-5)$. Since negative numbers do not have logarithms, we conclude that $x = -4$ is not a solution. The solution set is $\{3\}$.

Natural Logarithms

The natural logarithm of a number is the exponent to which **base e** must be raised to yield that number. The notation $\ln x$ is to be interpreted as the natural logarithm of x, while $\log x$ will continue to be interpreted as logarithm of x to base 10. Appendix Table 3 lists the values of x from 1.00 to 9.99 in steps of 0.01. Consider, for example, $\ln 6.43$. Referring to Appendix Table 3, we locate the row for 6.4 and then move across until we are under the column headed by 0.03, and we find the entry 1.8610. We may use this table to determine that

$$\ln 2.68 = 0.9858, \qquad \ln 7.67 = 2.0373, \qquad \ln 8.84 = 2.1793,$$

and so on. To determine ln 10, we use the fact

$$\log_e 10 = \frac{\log_{10} 10}{\log_{10} e}. \qquad \text{(Proposition 5.6.3)}$$

Assuming that $e = 2.71828$, it can be shown that $\log e = 0.4343$. Thus

$$\ln 10 = \frac{1}{0.4343} = 2.3026.$$

Observe that

$$\ln 100 = \ln 10^2 = 2 \ln 10 = 2(2.3026) = 4.6052,$$
$$\ln 1000 = \ln 10^3 = 3 \ln 10 = 3(2.3026) = 6.9078,$$
$$\ln 10,000 = \ln 10^4 = 4 \ln 10 = 4(2.3026) = 9.2104,$$

and so on. Alternatively, we can also determine ln x by pressing the ln key on a scientific calculator.

Example 5.39   Find ln 4730 by using **(a)** tables and **(b)** a calculator.

Solution:

(a)
$$\begin{aligned} \ln 4730 &= \ln (4.73 \times 10^3) \\ &= \ln 4.73 + \ln 10^3 &&\text{(Proposition 5.6.1)} \\ &= \ln 4.73 + 3 \ln 10 &&\text{(Proposition 5.6.2)} \end{aligned}$$

From Appendix Table 3, we find

$$\ln 4.73 = 1.5539.$$

Thus,

$$\begin{aligned} \ln 4730 &= 1.5539 + 3(2.3026) \\ &= 8.4617. \end{aligned}$$

(b) To use a scientific calculator, we proceed as follows:

Enter	Press	Display
4730	ln	8.4616805.

Rounding to four decimal places, we have

$$\ln 4730 = 8.4617,$$

as before.

Example 5.40

Find ln 0.689 by using **(a)** tables and **(b)** a calculator.

Solution:

(a) $\ln 0.689 = \ln (6.89 \times 10^{-1})$

$\qquad\qquad = \ln 6.89 + \ln 10^{-1}$ (Proposition 5.6.1)

$\qquad\qquad = \ln 6.89 - \ln 10$ (Proposition 5.6.2)

$\qquad\qquad = 1.9301 - 2.3026$

$\qquad\qquad = -0.3725$

(b) Using a scientific calculator involves the following steps:

Enter	Press	Display
0.689	ln	$-0.372514008.$

This, when rounded to four decimal places, shows that

$$\ln 0.689 = -0.3725,$$

as before.

Given a positive number x, we now know how to find ln x. How do we use a calculator to determine x if ln x is given?

Example 5.41

Find x if $\ln x = 1.2698$.

Solution: Remember that $\ln x = 1.2698$, when written in exponential form, is

$$x = e^{1.2698}.$$

To determine the value of x, we proceed as follows:

Enter	Press	Display
1.2698	e^x	3.5601405.

This, when rounded to four decimal places, is 3.5601. Hence,

$$x = 3.5601.$$

Example 5.42

Find x if $\ln x = -2.0731$.

Solution: The expression $\ln x = -2.0731$, written in exponential form, is equivalent to

$$x = e^{-2.0731}.$$

Now we proceed as follows:

Enter	Press	Display
2.0731	$\pm$	-2.0731
	e^x	0.1257952.

This, when rounded to four decimal places, shows that $x = 0.1258$.

We now use natural logarithms to solve certain equations in which the variable appears in an exponential form with base e.

Example 5.43

Given that $e^{3x} = 25$, determine x.

Solution: Taking natural logarithms, we have

$$\ln e^{3x} = \ln 25.$$

Using a scientific calculator, we see that $\ln 25 = 3.2189$. Thus,

$$\ln e^{3x} = 3.2189.$$

Observe that

$$\ln e^{3x} = 3x \ln e \qquad \text{(Proposition 5.6.2)}$$
$$= 3x. \qquad \text{(ln } e = 1\text{)}$$

That is,

$$3x = 3.2189$$

$$x = \frac{3.2189}{3} = 1.0730.$$

Example 5.44

Given that $e^{1-2x} = 12$, determine x.

Solution: Again, we take natural logarithms of both sides. We have

$$\ln e^{1-2x} = \ln 12 = 2.4849.$$

Also observe that

$$\ln e^{1-2x} = (1 - 2x) \ln e = 1 - 2x. \qquad \text{(ln } e = 1\text{)}$$

Thus, we get

$$1 - 2x = 2.4849.$$

Solving for x, we get

$$x = -0.74245.$$

EXERCISES

Using Appendix Table 2 or a scientific calculator, find the common logarithms of the numbers in Exercises 1 to 10. Round your answers to four decimal places.

1. 234   **2.** 2340   **3.** 23.4   **4.** 2.34

5. 0.234   **6.** 0.0234   **7.** 0.00583   **8.** 0.000871

9. 0.0000678   **10.** 0.0000945

Using Appendix Table 3 or a scientific calculator, find the natural logarithms of the numbers in Exercises 11 to 22. Round your answers to four decimal places.

11. 2650   **12.** 265   **13.** 26.5   **14.** 0.0265

15. 0.00265   **16.** 0.265   **17.** 372   **18.** 6350

19. 0.0943   **20.** 0.00625   **21.** 0.00034   **22.** 0.000034

Using a scientific calculator, find the value of x in Exercises 23 to 42. Round your answers to four decimal places.

23. $\log x = 0.4048$ **24.** $\log x = 1.6972$ **25.** $\log x = 3.1614$ **26.** $\log x = -1.0783$

27. $\log x = -2.342$ **28.** $\log x = -1.6180$ **29.** $\log x = 1.4393$ **30.** $\log x = 3.9284$

31. $\log x = -2.3261$ **32.** $\log x = -2.3958$ **33.** $\ln x = 2.3026$ **34.** $\ln x = 1.2698$

35. $\ln x = 1.4564$ **36.** $\ln x = 1.7029$ **37.** $\ln x = 2.2279$ **38.** $\ln x = 4.2679$

39. $\ln x = -1.4410$ **40.** $\ln x = -5.9915$ **41.** $\ln x = -3.6613$ **42.** $\ln x = -0.3719$

Using a calculator, solve the exponential equations in Exercises 43 to 70 for x. Round your answers to four decimal places.

43. $10^x = 37$ **44.** $10^x = 78$ **45.** $10^x = 87$ **46.** $10^x = 346$

47. $10^x = 34.6$ **48.** $10^x = 0.00346$ **49.** $10^x = \frac{126}{457}$ **50.** $10^{-x} = \frac{123}{231}$

51. $2^x = 15$ **52.** $2^x = 67$ **53.** $3^x = 4$ **54.** $4^x = 25$

55. $5^x = 4$ **56.** $3^{2x-1} = 26$ **57.** $2^{2x+1} = 5^{1-3x}$ **58.** $3^{2x-1} = 4^{x+2}$

59. $4^{x+2} = 8^{2x-3}$ **60.** $4^{2x+3} = 5^{x-2}$ **61.** $6^{x+3} = 9^x$ **62.** $2^{3x-1} = 7^x$

63. $e^x = 10$ **64.** $e^x = 24.3$ **65.** $2e^x = 130$ **66.** $e^{3x} = 64$

67. $e^{2x} = 121$ **68.** $2e^{2x} = 50$ **69.** $e^{3x+1} = 67.8$ **70.** $e^{1-4x} = 13.24$

Solve the logarithmic equations in Exercises 71 to 80 for x.

71. $\log(2x + 3) = 1$ **72.** $\log(3x + 4) = 2$

73. $\log(x^2 - 3x + 2) - \log(x - 1) = 0$ **74.** $\log(x + 2) - 2\log x = 0$

75. $\log(3x - 1) = 1 + \log(x - 2)$ **76.** $\log(2x + 1) = 1 + \log(x - 2)$

77. $\log x - \log(x - 5) = 2$ **78.** $\log x + \log(x + 3) = 1$

79. $\log x + \log(x - 9) = 1$ **80.** $\log(2x - 1) - \log(x - 3) = 2$

5.8

FURTHER EXPONENTIAL MODELS

Problems involving compound interest frequently employ logarithmic equations. For example, if a principal amount of P is invested at an annual rate of interest i, compounded continuously, then the accumulated value of P after t years is given by

$$A = Pe^{it}.$$

Let us look at an example.

Example 5.45

How long will it take for a principal amount of money to double if it is invested at 5.50 percent annual rate of interest, compounded continuously?

Solution: Let P be the principal amount of money invested. Then we must determine t such that

$$Pe^{0.055t} = 2P$$

$$e^{0.055t} = 2.$$

Taking natural logarithms, we get

$$0.055t = \ln 2$$

$$t = \frac{\ln 2}{0.055} = \frac{0.6931}{0.055} = 12.60 \text{ years.}$$

The accumulated principal and interest will be slightly more than doubled if the principal is invested for 13 years.

Another type of problem that incorporates natural logarithms involves exponential growth and decay. In many practical situations, the rate of growth or decay of a substance is proportional to the amount of substance present at any time t. We observed earlier that the amount of substance $A(t)$ present at a given time t is given by

$$A(t) = A(0)(e^{kt}),$$

where $A(0)$ is the amount of substance present at time $t = 0$ and k is a constant. **When $k > 0$, we have exponential growth; and when $k < 0$, we speak of exponential decay.**

Example 5.46

Radioactive strontium 90 is used in nuclear reactors and decays according to the formula

$$Q(t) = Q(0)(e^{-0.0244t}),$$

where $Q(t)$ is the amount present at time t and t is measured in years. Determine the half-life of strontium 90; that is, find t at which $Q(t) = \frac{1}{2}Q(0)$.

Solution: The problem reduces to finding t so that

$$Q(0)(e^{-0.0244t}) = \frac{1}{2} Q(0)$$

$$e^{-0.0244t} = \frac{1}{2}.$$

Taking natural logarithms, we have

$$-0.0244t = \ln \frac{1}{2} = \ln 0.5$$

$$-0.0244t = -0.6931$$

$$t = \frac{0.6931}{0.0244} = 28.4057$$

This means that strontium loses one-half of its initial quantity in a little more than 28 years.

Example 5.47

The number of bacteria in a culture was 1000 initially and 8000 two hours later. Express the number of bacteria present at time t as an exponential function. Assuming ideal conditions for growth, how many bacteria are there after 3 hours?

Solution: Here $A(0) = 1000$ and $A(2) = 8000$. Thus

$$8000 = 1000e^{2k}$$

$$e^{2k} = 8.$$

Taking natural logarithms, we obtain

$$2k = \ln 8$$

$$k = \frac{1}{2} \ln 8 = \frac{1}{2} (2.0794) = 1.0397.$$

Thus the exponential equation that describes the growth of bacteria is

$$A(t) = 1000e^{1.0397t}.$$

Consequently, the number of bacteria after 3 hours is

$$A(3) = 1000e^{(1.0397)(3)} = 1000e^{3.1191}.$$

Using an electronic calculator, we obtain

$$e^{3.1191} = 22.626.$$

Thus

$$A(3) = 1000(22.626) = 22{,}626.$$

Example 5.48

A student's record of learning a foreign language is given by the exponential equation

$$N(t) = 150(1 - e^{-0.32t}),$$

where $N(t)$ is the number of new words the student learns after t hours of practice. Estimate the number of hours this student will take to learn 90 words.

Solution: The problem is to find t so that

$$150(1 - e^{-0.32t}) = 90$$

$$1 - e^{-0.32t} = \tfrac{90}{150} = 0.60$$

$$e^{-0.32t} = 1 - 0.60$$

$$e^{-0.32t} = 0.40$$

Taking natural logarithms, we have

$$-0.32t = -0.9163$$

$$t = \frac{0.9163}{0.32} = 2.8634 \text{ hours}$$

This means that the student will learn 90 words in approximately 2.86 hours, or 2 hours 52 minutes.

Example 5.49

After a jury convicted a member of Congress for accepting bribes for awarding certain defense contracts, the radio and television stations began to broadcast the news frequently. Within an hour, 20 percent of the citizens heard the news. Using the exponential model (5.8)

$$f(t) = N(1 - e^{-kt}),$$

how long will it take for the news to reach 95 percent of the citizens?

Solution: First, we must find the value of the constant k in the above model. Using the information that 20 percent of the citizens heard the news within 1 hour, we have

$$N(1 - e^{k \cdot 1}) = 0.20N$$

$$1 - e^{k} = 0.20$$

$$e^{k} = 0.80.$$

Taking natural logarithms, we have

$$k = \ln 0.80 = -0.2231.$$

The model for spreading the news in this case is

$$f(t) = N(1 - e^{-0.2231t}).$$

Now we want to determine t so that

$$N(1 - e^{-0.2231t}) = 0.95N$$

which means that

$$1 - e^{-0.2231t} = 0.95$$

$$e^{-0.2231t} = 0.05.$$

Taking natural logarithms, we get

$$-0.2231t = -2.9957$$

$$t = \frac{2.9957}{0.2231} = 13.4276.$$

Thus, the news will reach 95 percent of the citizens in approximately 13 hours 30 minutes.

Logistic Growth Curve

The simple exponential model (5.7)

$$f(t) = f(0)(e^{kt})$$

that we used in Section 5.4 for describing the growth of bacteria in a culture or the growth of fish population in a lake or the growth of population in a city is grossly inadequate for practical purposes. The model assumes that we have sufficient resources to sustain any level of population for any length of time so that there is no overcrowding or any kind of interference between individuals in the population. These assumptions are not very realistic. In fact, every species of organism has some restricted environment, with a finite amount of space and a limited supply of resources. Further, the environment has a "maximal capacity," an upper limit on the number of individuals that can exist on the available resources. As the size of the population approaches its maximal capacity, its rate of growth slows down and finally tapers to zero. The **logistic growth model** takes into account some of the effects of the environment on a population.

The equation expressing the logistic growth model is

$$f(t) = \frac{N}{1 + ce^{-Nkt}}, \tag{5.10}$$

where N is the maximal capacity of the environment and c and k are some positive constants. If we plot $f(t)$ against time t for fixed values of N, c, and k, then we obtain a curve of the type shown in Figure 5.15.

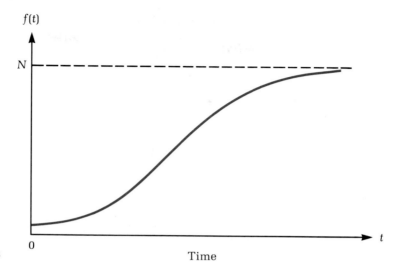

$f(t)$

N

0

t

Time

FIGURE 5.15

Example 5.50

An ecologist is observing the growth of fish in a lake. The number of fish in the lake was 100 initially and 400 six months later. Assuming that the lake has the maximal capacity of 2000 fish, determine the logistic growth model for the number of fish $f(t)$ at time t. Estimate the number of fish after **(a)** $t = 12$ months and **(b)** $t = 24$ months.

Solution: Here $N = 2000$. Therefore, we have

$$f(t) = \frac{2000}{1 + ce^{-2000kt}}. \tag{5.11}$$

Using the fact that at time $t = 0$ there are 100 fish, we see that

$$100 = \frac{2000}{1 + c},$$

which means that

$$c + 1 = 20$$

$$c = 19.$$

Further, we know that at $t = 6$, there are 400 fish. Using the information that $f(6) = 400$ and $c = 19$, we have

$$400 = \frac{2000}{1 + 19e^{-12,000k}}$$

$$1 + 19e^{-12,000k} = 5$$

$$19e^{-12,000k} = 4$$

$$e^{-12,000k} = \tfrac{4}{19} = 0.2105.$$

Taking natural logarithms, we get

$$-12{,}000k = -1.5583$$

$$k = 0.00013.$$

Hence, the logistic growth model is

$$f(t) = \frac{2000}{1 + 19e^{-0.26t}}. \tag{5.12}$$

Setting $t = 12$ in Equation (5.12), we have

$$f(12) = \frac{2000}{1 + 19e^{-3.12}}$$

$$= \frac{2000}{1 + 19(0.044157)}$$

$$= \frac{2000}{1.838983}$$

$$= 1087.55,$$

which, when rounded, yields $f(12) = 1088$. Similarly, substituting $t = 24$ in Equation (5.12), we get

$$f(24) = \frac{2000}{1 + 19e^{-6.24}}$$

$$= \frac{2000}{1 + 19(0.0019498)}$$

$$= \frac{2000}{1 + 0.037046}$$

$$= \frac{2000}{1.037046}$$

$$= 1928.55,$$

which, when rounded, yields $f(24) = 1929$ fish.

A Simple Epidemic Model

We now present a simple mathematical model of the spread of infectious diseases. This model, which is a slight variation of the logistic growth model, will serve as an introduction to the rapidly growing field of mathematical epidemiology where mathematicians and biologists are working together to gain a better understanding of the spread and control of epidemics.

Consider an individual who may be infected by some contagious disease. The organism may enter the body through an agent spread by coughing or sneezing, by drinking contaminated water, or through intimate personal contact with another infected person. The affected individual may continue to transmit the disease to others without either being aware that the individual is infected. As the disease spreads through the community (with population N), the number of unaffected persons continues to decrease, and everyone in the population will eventually contract the disease.

We assume that the rate of contact is directly proportional to the number $f(t)$ who are currently infected and the number $N - f(t)$ of those who have not yet been infected but may become so in due course. Based on this assumption that is basic to almost all mathematical models of epidemics, the equation that expresses the simple epidemic model is

$$f(t) = \frac{N}{1 + ce^{-Nkt}},$$

where N is the size of the population and c and k are positive constants to be determined from the characteristics of the epidemic. If we plot $f(t)$ against time t for fixed values of N, c, and k, then we obtain a graph similar to the one shown in Figure 5.15.

At time $t = 0$, $e^{-Nkt} = 1$. For convenience we assume that the epidemic starts at time $t = 0$ with a single infected person which means that $f(0) = 1$. Using this information, we get

$$1 = \frac{N}{1 + c},$$

which implies that $c = N - 1$. Thus, the epidemic model is modified so that we have

$$f(t) = \frac{N}{1 + (N - 1)(e^{-Nkt})} \tag{5.13}$$

As t increases, e^{-Nkt} becomes small and $f(t)$ approaches N. Thus the model predicts that eventually everyone in the population will get infected. We must warn you that the model (5.13) is based on the assumption that once an individual exhibits some symptoms of a contagious disease, public health officials make no attempt to control the disease. This is highly unrealistic because widespread emergency measures will be introduced to end the epidemic when only a small number of the potentially susceptible members of the community have been affected. This is reinforced by the fact that a mere possibility that some strain of influenza virus might attack the United States in 1976 led the government to plan for the vaccination of the entire nation of more than 200 million people.

Despite some limitations, an epidemic model has several applications. Sociologists recognize this model and use it to describe the spread of a rumor. Scientists use it to describe the spread of a technological innovation. In economics the model is used to spread the knowledge about a new product in the market. Remember that this model assumes that the information is spread by the members of the population themselves, which is in sharp contrast to the **social diffusion model** where the information is broadcast by radio, television stations, and other means of communications.

Example 5.51

A number of people in a small community of 2000 are coming down with a certain type of influenza. Public health officials report 53 new cases after the end of 2 weeks. Assuming that this trend continues, what will be the number of residents infected after 4 weeks?

Solution: Here $N = 2000$, $t = 2$, and $f(2) = 53$. Substituting this information in (5.13), we have

$$53 = \frac{2000}{1 + 1999e^{-4000k}}$$

$$1 + 1999e^{-4000k} = \frac{2000}{53}$$

$$1 + 1999e^{-4000k} = 37.73585$$

$$1999e^{-4000k} = 36.73585$$

$$e^{-4000k} = 0.018377.$$

Taking natural logarithms, we get

$$-4000k = -3.99666,$$

or

$$k = 0.001$$

approximately. Thus, the epidemic model for this situation is

$$f(t) = \frac{2000}{1 + 1999e^{-2t}}.$$

Setting $t = 4$ in this equation, we get

$$f(4) = \frac{2000}{1 + 1999e^{-8}} = \frac{2000}{1.670589} = 1197.$$

Hence, 1197 individuals are infected after 4 weeks.

▦ EXERCISES

1. How long will it take a principal amount to double if it is invested at 6 percent annual rate of interest compounded continuously?

2. How long will it take a principal amount to double if it is invested at 9.3865 percent annual rate of interest compounded continuously?

3. The quantity of radium $Q(t)$, measured in milligrams, at time t is given by

$$Q(t) = 1000e^{-0.1393t},$$

where t is measured in centuries. Determine the half-life of radium; that is, find t at which $Q(t) = 500$ milligrams.

4. The number of bacteria $N(t)$ present in a certain culture at time t is

$$N(t) = N(0)(e^{kt}),$$

where $N(0)$ is the amount present at time $t = 0$ and t is measured in hours. If the number of bacteria doubles in 10 hours, determine the growth constant k.

5. The number of fruit flies $N(t)$ grows according to

$$N(t) = N(0)(e^{kt}),$$

where $N(0)$ is the number of fruit flies at time t and t is measured in days. If the number of fruit flies doubles in 15 days, determine the growth constant k.

6. Radioactive strontium 90 is used in nuclear reactors and decays according to the formula

$$Q(t) = Q(0)(e^{-kt}),$$

where $Q(t)$ is the amount present at time t and t is measured in centuries. If the strontium 90 loses half of its quantity in 28 years, what is the decay constant?

7. Radioactive carbon 14 has a half-life of 5700 years. Determine its decay constant.

8. An Egyptian mummy is found to have lost 65 percent of its original carbon 14. Assuming that its decay constant is 0.00012, determine approximately when this person died.

9. A skull uncovered in an archaeological site has lost 90 percent of the original amount of carbon 14. Assuming that the decay constant is 0.00012, estimate the age of the skull.

10. Atmospheric pressure P, measured in torr, decreases with increasing altitude h, in meters, according to the equation

$$P(h) = 760e^{-kh},$$

where k is a positive constant. If the pressure at 10,000 meters above sea level is 225 torr, what will be the pressure at 15,000 meters above sea level?

11. The number of bacteria in a culture was 2000 initially and 8000 after 1 hour. Assuming ideal conditions for growth, how many bacteria will there be after **(a)** 2 hours and **(b)** 4 hours?

12. The population of a certain city was 475,000 in 1970 and 510,000 in 1980. Find the value of the growth constant k, and then estimate the population of the city in 1985.

13. A student's record of learning to type is given by the exponential equation

$$N(t) = 80(1 - e^{-0.08t}),$$

where $N(t)$ is the number of words the student can type per minute after t weeks of instruction. Estimate the number of weeks this student will take to learn to type 40 words per minute.

14. After a jury convicted a senator for driving under the influence of liquor, the radio and television stations began to broadcast the news frequently. Within 1 hour, 25 percent of the citizens heard the news. Using the exponential model

$$f(t) = N(1 - e^{-kt}),$$

determine how long will it take for the news to reach 90 percent of the citizens.

15. The number of residents $N(t)$ suffering from influenza t days after its outbreak is approximated by

$$N(t) = \frac{500}{1 + 499e^{-0.3t}}.$$

Determine the number of residents who have contracted influenza after **(a)** 5 days, **(b)** 10 days, and **(c)** 20 days.

16. A biologist is studying the growth of a colony of fruit flies. Let the number of fruit flies $N(t)$ after t days be approximated by the logistic growth equation

$$N(t) = \frac{1060}{1 + 80e^{-0.2t}}.$$

Determine the number of fruit flies after **(a)** $t = 5$ days, **(b)** $t = 10$ days, and **(c)** $t = 20$ days.

17. A rumor tends to spread through a community of population 800 at a rate directly proportional to the number $N(t)$ of residents who heard the rumor and the number, $800 - N(t)$, of residents who have not heard the rumor after t days. Assuming that the logistic growth model (5.13) holds and 268 residents heard the rumor after 5 days, estimate the number of residents who heard the rumor after 10 days.

18. An ecologist is observing the growth of fish in a lake. The number of fish in the lake was 100 initially and 200 two months later. Assuming that the lake has the maximal capacity of 1500 fish, determine the logistic growth model for the number of fish $N(t)$ at time t. Estimate the number of fish after **(a)** $t = 4$ months and **(b)** $t = 6$ months.

KEY WORDS

exponential function
exponential equation
compound interest
compounding
accumulated value
present value
conversion period

effective rate of interest
nominal rate of interest
exponential growth
exponential decay
logarithmic function
logarithmic equation
common logarithm

natural logarithm
half-life
logistic growth curve
simple epidemic model
social diffusion model

KEY FORMULAS

● Given that i_m/m is interest paid at the end of each of the m conversion periods in 1 year, the accumulated value of P dollars at the end of t years is

$$A = P\left(1 + \frac{i_m}{m}\right)^{mt}.$$

● If interest is compounded continuously, then the accumulated value of P dollars at the end of t years is

$$A = Pe^{it},$$

where i is the annual rate of interest.

- If interest is compounded at a rate of i_m/m for each of m conversion periods in 1 year, then the effective rate of interest is

$$i = \left(1 + \frac{i_m}{m}\right)^m - 1.$$

- If interest is compounded continuously, then the effective rate of interest is

$$i = e^{im} - 1.$$

LAWS OF LOGARITHMS

- $\log_b mn = \log_b m + \log_b n$
- $\log_b \dfrac{m}{n} = \log_b m - \log_b n$
- $\log_b m^r = r \log_b m$

- $\log_a m = \dfrac{\log_b m}{\log_b a}$
- $(\log_a b)(\log_b a) = 1$

EXERCISES FOR REVIEW

Solve the equations in Exercises 1 to 8 for x.

1. $3^{2x} = 81$
2. $27^x = 81$
3. $5^x = 0.008$
4. $10^x = 0.0001$
5. $\log_3 x = 4$
6. $\log_x 16 = 2$
7. $\log_2 32 = x$
8. $\log_8 4 = x$

Given that $\log_b 2 = 0.3010$, $\log_b 3 = 0.4771$, and $\log_b 5 = 0.6990$, find the expressions in Exercises 9 to 14.

9. $\log_b 6$
10. $\log_b 30$
11. $\log_b \dfrac{15}{2}$
12. $\log_b \sqrt{3}$
13. $\log_5 3$
14. $\log_3 5$

Using a scientific calculator, find the common logarithms of the numbers in Exercises 15 to 18. Round your answers to four decimal places.

15. 5140
16. 24.3
17. 0.0243
18. 0.00243

Using a scientific calculator, find the natural logartihms of the numbers in Exercises 19 to 22. Round your answers to four decimal places.

19. 567
20. 31.8
21. 0.234
22. 0.00234

Using a scientific calculator, find the value of x in Exercises 23 to 28. Round your answers to four decimal places.

23. $\log x = 1.8476$
24. $\log x = 0.5527$
25. $\log x = -1.2692$
26. $\ln x = 1.6658$
27. $\ln x = 3.8199$
28. $\ln x = -0.5674$

Using a scientific calculator, solve the exponential equations in Exercises 29 to 38. Round your answers to four decimal places.

29. $10^x = 27$
30. $10^x = 3.7$
31. $10^{-x} = \dfrac{341}{563}$
32. $2^x = 14$
33. $3^{2x} = 14.89$
34. $2^{2x-1} = 17.68$
35. $5^{3x+2} = 23.54$
36. $e^{2x} = 8.30$
37. $e^{-x} = 0.548$
38. $3^{x+2} = 5^{2x-1}$

Solve the logarithmic equations in Exercises 39 and 40 for x.

39. $\log(x + 2) - 2\log x = 0$ **40.** $\log(2x + 1) = 1 + \log(x - 2)$.

41. The number of bacteria $N(t)$ in a certain culture is given by

$$N(t) = 20{,}000e^{-0.01t},$$

where t is recorded in hours. Find the number of bacteria present at time **(a)** $t = 0$, **(b)** $t = 10$, and **(c)** $t = 20$.

42. A certain industrial machine depreciates so that its value after t years is approximated by

$$Q(t) = Q(0)(e^{-0.05t}).$$

If the machine is worth $6065.31 after 10 years, what was its original value?

43. A company's gross income is given by

$$P(t) = \$10{,}000e^{0.5(t-1)},$$

where t is recorded in years. Determine the gross income of the company for $t = 1, 2, 3, 4, 5$, and 6 years. Sketch the graph of $P(t)$.

44. The annual profit for a publisher from the sales of a certain book is given by

$$P(t) = \$5000 + \$10{,}000e^{-0.4(t-1)},$$

where t is the number of years the text is on the market. Determine the annual profit at the end of 1, 2, 3, 4, 5, 6, and 10 years. Sketch the graph of $P(t)$.

45. Anita's record of learning to type is given by

$$N(t) = 80(1 - e^{-0.08t}),$$

where $N(t)$ is the number of words per minute she is able to type after t weeks of instruction. Determine $N(t)$ for $t = 5, 10, 15, 20, 25$, and 30 weeks. Sketch the graph of $N(t)$.

46. The market research division of a frozen food company estimates that the monthly demand for chickens (in thousands) is given by

$$D(t) = 8 - 7e^{-0.8(t-1)},$$

where t is recorded in months. Calculate the estimated sales after $t = 1, 2, 3, 4, 5, 6, 8, 10$, and 12 months. Sketch the graph of $D(t)$.

47. Find the accumulated amount and the compound interest if $3500 is invested for 3 years at 12 percent annual interest, compounded **(a)** quarterly, **(b)** monthly, and **(c)** continuously.

48. Mike wishes to have $30,000 available for his son who will enter college in 8 years. What sum of money must he invest now if the current rate of interest is 12.35 percent, compounded continuously?

49. How long will it take for the principal to double itself if it is invested at 9.50 percent rate of interest, compounded continuously?

50. The population of a certain country was 50 million in 1975 and 75 million in 1980. Assuming that the population grows exponentially, what will the population of this country be in 1990?

51. A radioactive substance decays exponentially. If 500 grams of the substance was present intially and 300 grams was present after 25 years, what is its decay constant? How many grams of the substance will remain after 100 years?

52. Find the half-life of a radioactive substance if it loses 10 percent of its initial quantity in 5 years.

53. A student's record of learning a foreign language is given by the exponential equation

$$N(t) = 120(1 - e^{-0.4t}),$$

where $N(t)$ is the number of new words the student learns after t hours of practice. Estimate the number of hours this student will need to learn 75 words.

54. After a jury convicted a mayor of a certain town for "fixing up" and then awarding some city contracts to his son, the radio and television stations began to broadcast the news frequently. Within an hour, 25 percent of the town residents heard the news. Using the exponential model (5.8), determine how long it will take for the news to reach 90 percent of the town residents.

55. A number of people in a small community of 1000 are coming down with a certain type of infectious disease. Public health officials report 40 cases after the end of 3 weeks. Using the epidemic model (5.13), determine the number of residents infected after 6 weeks.

56. A rumor tends to spread through a community of 600 people at a rate directly proportional to the number $N(t)$ of residents who heard the rumor and the number, $600 - N(t)$, of residents who have not heard the rumor after t days. Assuming that the logistic growth model (5.13) holds and 78 residents heard the rumor after 1 day, estimate the number of residents who heard the rumor after 2 days.

Sketch the graph of the exponential functions in Exercises 57 to 60.

57. $y = 2^x$ 58. $y = 3^x + 1$ 59. $y = 2^x + 2^{-x}$ 60. $y = 2^{|x-1|}$

CHAPTER TEST

1. Solve the following for x.
   (a) $2^{-x} = 32$ (b) $3^x = 243$ (c) $4^x = 128$
   (d) $\log_5 x = 3$ (e) $\log_x 128 = 7$ (f) $\log_3 81 = x$

2. Given that $\log_b 2 = 0.69$, $\log_b 3 = 1.10$, and $\log_b 5 = 1.61$, find (a) $\log_b 15$, (b) $\log_b 50$, and (c) $\log_3 5$.

3. Solve the exponential equation $3^x = 15$.

4. Solve for x:
$$\log(x + 2) = 2 \log x.$$

5. Find the accumulated value if $5000 is deposited in a bank for 4 years at 12 percent compounded (a) quarterly, (b) monthly, and (c) continuously.

6. The quantity $Q(t)$ of a radioactive substance (measured in grams) present at time t is given by
$$Q(t) = Q(0)(e^{-kt}).$$
If this radioactive material loses half of its quantity in 28 years, what is the decay constant?

7. How long will it take for a principal amount of money to double if it is invested at 10 percent annual interest compounded continuously

8. The population of a certain city was 475,000 in 1970 and 510,000 in 1980. Find the value of the growth constant k, and then estimate the population of the city in 1990.

Sketch the graph of the following functions:

9. $y = 2^x + 2^{-x}$

10. $y = 3^{x-1}$.

TRIGONOMETRIC FUNCTIONS

6

INTRODUCTION

In this chapter we define the trigonometric functions, study their properties, and explore their applications in several fields. The historic basis of the word *trigonometry* is triangle measurement, a subject that is essential to the study of navigation, surveying, and astronomy. Even though the study of triangles dates at least to the early Egyptian civilization, the first known textbook bearing the title *Trigonometry* was published around 1600 by a professor of mathematics in Germany. As the study of the subject has progressed, its applications have expanded. Today, a knowledge of trigonometric functions is essential in the study of sound, mechanics, electronics, and vibration.

trigonometry

Classical trigonometry is concerned with functions of angles defined as ratios of the sides of a right triangle. Therefore we define angles and their measure at the onset of our work in trigonometry. Later we consider trigonometric functions of a general angle and of real numbers. Trigonometric applications are included throughout the chapter.

6.1

ANGLES AND THEIR MEASURE

ray

From geometry we know that an angle is defined in terms of rays. A *ray* is a half line including its endpoint, as shown in Figure 6.1. If the ray *m* rotates about its endpoint *O* from its original position *OA* to a position *OC*, then it sweeps out, or generates, the angle *AOC*, as shown in Figure 6.2. Thus an *angle* consists of two rays radiating from the same endpoint, called the **vertex.**

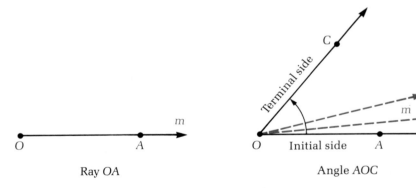

Ray *OA*   Angle *AOC*

FIGURE 6.1   **FIGURE 6.2**

initial side
terminal side

The rays *OA* and *OC* are called the sides of the angle *AOC*, where *OA* is the **initial side** and *OC* is the **terminal side.**

If the rotation of the generating ray is in the *counterclockwise* direction, the angle is considered *positive;* if the rotation is in the *clockwise* direction, the angle is considered *negative.* A curved arrow indicates the direction of rotation (see Figure 6.3).

Angles are usually labeled by naming a point on the initial side followed by the name of the vertex and then the name of a point on the terminal side.

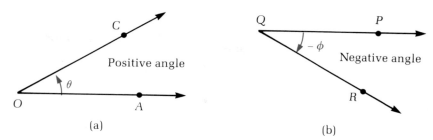

FIGURE 6.3   (a)   (b)

Thus the positive angle in Figure 6.3 would be named $\angle AOC$, and the negative angle would be named $\angle PQR$. Angles may also be labeled by the vertex alone ($\angle O$ or $\angle Q$) when there is no confusion about which angle is being referred to. Another common method for labeling angles is by a Greek letter written within the angle next to the direction curve, such as $\angle \theta$ (angle theta) or $\angle \phi$ (angle phi), as shown in Figure 6.3.

The amount of rotation required to move a ray from its initial position to its terminal position is defined as the *measure* of the angle formed. This measure is usually given in degrees. A complete rotation of a ray in either direction back to its starting position generates a circle, as shown in Figure 6.4. In this case, the initial side and the terminal side are the same.

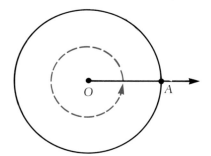

FIGURE 6.4

A **degree** is defined as $\frac{1}{360}$ of a complete revolution of a ray about its endpoint; thus there are 360° in a circle. A degree is divided into 60 smaller units called *minutes*, and each minute is divided into 60 smaller units called *seconds:*

$$1° = 60 \text{ minutes } (60')$$

and

$$1' = 60 \text{ seconds } (60'').$$

Therefore, 360° is equivalent to 359°59′60″.

In trigonometry, we do not place a limit on the magnitude of an angle. The ray may continue to rotate, thus sweeping out two revolutions and generating angles from 360° to 720°, and so on. The measure of angle *AOC*, therefore, depends not only on the position of the sides, but also on the direction and extent of rotation of the initial ray. However, any two measures of angle AOC will always differ by some multiple of 360°.

Example 6.1

Angle AOC in Figure 6.5 may have any one of these values:

$$\alpha = 25°, \qquad \beta = 385°, \qquad \gamma = -335°.$$

Note that

$$\beta - \alpha = 385° - 25° = 360°$$

$$\alpha - \gamma = 25° - (-335°) = 360°$$

$$\gamma - \beta = -335° - 385° = -720° = 2(-360°).$$

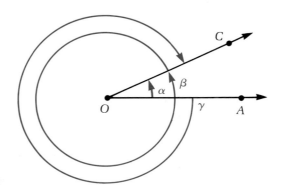

FIGURE 6.5

If an angle is placed on a coordinate system so that the initial side lies on the positive x axis with the vertex at the origin, then the angle is said to be in **standard position.** Figure 6.6 illustrates four different angles in standard position.

standard position

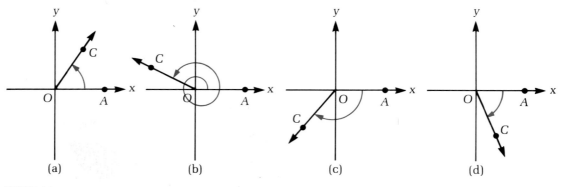

FIGURE 6.6

quadrant　　　The *quadrants* of the coordinate system are labeled with Roman numerals, I, II, III, IV, going counterclockwise as shown in Figure 6.7. An angle is said to be in a particular quadrant if its terminal side lies in that quadrant. An angle *quadrantal angle* whose terminal side lies along either of the coordinate axes is called a **quadrantal angle.** For example, $\angle AOB$ in Figure 6.7 is in quadrant I, $\angle AOC$ is in quadrant III, $\angle AOD$ is in quadrant III, and $\angle AOE$ is a quadrantal angle.

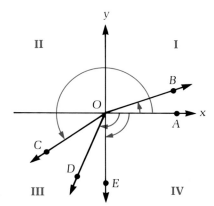

FIGURE 6.7

Angles that have the same initial side and the same terminal side are called **coterminal angles.** For instance, angles $150°$ and $-210°$ in Figure 6.8 are coterminal angles. Angles α, β, and γ in Figure 6.5 are also coterminal angles.

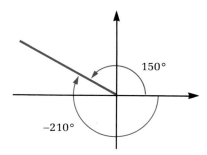

FIGURE 6.8

radian　　　The degree is not the only unit for measuring the size of an angle. Another unit of angular measure, called the **radian,** is more useful in advanced mathematics, especially in problems involving the methods of calculus.

Definition 6.1.1 | One *radian* is the measure of a central angle that subtends on the circumference of a circle an *arc* of length equal to the length of the circle's radius.

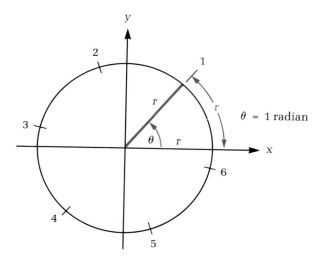

FIGURE 6.9

Since the circumference of a circle is equal to $2\pi r$, there are $2\pi r/r$, or 2π, arcs of 1 radian on a circle, that is, approximately 6.28 arcs of 1 radian (see Figure 6.9). Note that a radian is an angle of fixed magnitude. In circles of different radii, as shown in Figure 6.10, the radian is independent of the size of the circle. There are always 2π radians in a circle, just as there are always 360° in a circle. Note also that like the length of an arc, a radian is a real number.

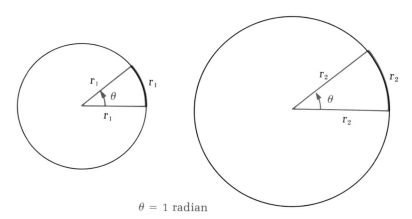

FIGURE 6.10 $\theta = 1$ radian

Since angles may be measured in either degrees or radians, we need to be able to convert from one form of measure to the other. There are 360° in a circle, and there are 2π radians in a circle. This provides the basis for the relationship between degrees and radians:

$$2\pi \text{ radians} = 360°$$

$$\pi \text{ radians} = 180° \tag{6.1}$$

$$1 \text{ radian} = \frac{180°}{\pi} \tag{6.2}$$

$$1 \text{ radian} = 57.2958°, \text{ or } 57°17.75' \text{ approximately.} \tag{6.2a}$$

On the other hand,

$$1° = \frac{\pi}{180} \text{ radian} \tag{6.3}$$

$$1° = 0.017453 \text{ radian approximately.} \tag{6.3a}$$

When no other unit of angular measure is indicated, we assume that the angle is expressed in radian measure.

Example 6.2

Convert the following angles from degree to radian measure: **(a)** 30°, **(b)** 135°, **(c)** 270°.

Solution: Since $1° = \pi/180$ radian, by Equation (6.3),

(a) $30° = 30 \cdot \dfrac{\pi}{180} = \dfrac{\pi}{6}$

(b) $135° = 135 \cdot \dfrac{\pi}{180} = \dfrac{3\pi}{4}$

(c) $270° = 270 \cdot \dfrac{\pi}{180} = \dfrac{3\pi}{2}.$

From Equation (6.3a), since $1° = 0.017453$ radian,

(a) $30° = 30(0.017453) = 0.52359$ radian

(b) $135° = 135(0.017453) = 2.356155$ radians

(c) $270° = 270(0.017453) = 4.71231$ radians.

Example 6.3

Convert the following angles from radian measure to degree measure: **(a)** $\pi/4$, **(b)** $5\pi/6$, **(c)** $4\pi/3$.

Solution: Using (6.2), we find

(a) $\dfrac{\pi}{4} = \dfrac{\pi}{4} \cdot \dfrac{180}{\pi} = 45°$

(b) $\dfrac{5\pi}{6} = \dfrac{5\pi}{6} \cdot \dfrac{180}{\pi} = 150°$

(c) $\dfrac{4\pi}{3} = \dfrac{4\pi}{3} \cdot \dfrac{180}{\pi} = 240°.$

Example 6.4

Convert an angle of 0.6 radian to degrees, expressed to the nearest minute.

Solution: Since 1 radian $= 180/\pi$ degrees $= 57.2958°$ (approximately), multiply 57.2958 by 0.6 to obtain 34.3775°. Since there are 60′ in a degree, multiply 0.3775 by 60 to obtain 22.65′. Thus 0.6 radian is equal to 34°23′ to the nearest minute.

One advantage of radian measure is the ease with which we can locate an integral multiple of a special angle by counting locations around the circle. There are some locations, such as π, that are integral multiples of several angles. For $\pi/6$ (30°), there are 12 locations that represent the integral multiples of the angle (see Figure 6.11). For $\pi/4$ (45°), there are 8 locations that represent integral multiples of the angle (see Figure 6.12). For $\pi/3$ (60°), there are 6 locations that represent integral multiples of the angle (see Figure 6.13).

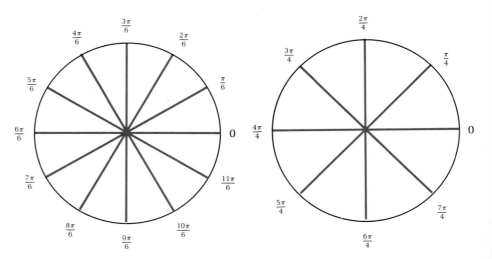

FIGURE 6.11   FIGURE 6.12

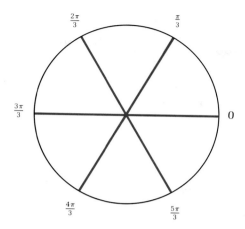

FIGURE 6.13

Arc Length

We may now establish a relationship between a central angle and the subtended arc. In Figure 6.14, the central angle θ subtends an arc of length s on the circle of radius r. We know from plane geometry that in a circle, arcs are proportional to their central angles. If angle AOB is 1 radian, then the length of arc AB is r:

$$\frac{s}{\theta} = \frac{r}{1} \qquad (\theta \text{ in radians})$$

Therefore, the subtended arc is proportional to the radius:

$$s = r\theta \qquad (\theta \text{ in radians}) \tag{6.4}$$

or

$$\theta\,(\text{radians}) = \frac{s\,(\text{arc length})}{r\,(\text{radius})} \tag{6.5}$$

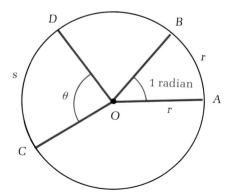

FIGURE 6.14

The radian measure of an angle is the ratio of the length of the subtended arc to the radius of the circle. It is frequently a matter of personal preference whether an angle is measured in degrees or radians. Note, however, that formulas (6.4) and (6.5) are valid only when θ is measured in radians.

Example 6.5

Find the length of the arc on a circle having a radius of 7 centimeters (cm) which subtends a central angle of 45°, as shown in Figure 6.15.

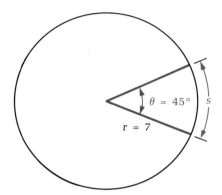

FIGURE 6.15

Solution: To use formula (6.4), first convert 45° to radian measure:

$$45° = 45 \cdot \frac{\pi}{180} = \frac{\pi}{4}.$$

Rewriting formula (6.4) gives

$$s = \theta r = \frac{\pi}{4} \cdot 7 \text{ cm} = \frac{7\pi}{4} \text{ cm}$$

$$= 5.5 \text{ cm approximately.}$$

Example 6.6

A pendulum swings through an arc of 30° describing an arc 3 meters (m) long. How long is the pendulum, as shown in Figure 6.16?

Solution: Using formula (6.4) and converting 30° to radian measure, we have

$$s = \theta r$$

$$r = \frac{s}{\theta} = \frac{3 \text{ m}}{\pi/6} = \frac{18 \text{ m}}{\pi}$$

$$= 5.73 \text{ m approximately.}$$

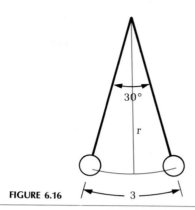

FIGURE 6.16

Example 6.7

What central angle will subtend an arc of 10 centimeters (cm) on a circle of radius 1 meter (m)? Express the answer in degrees and minutes.

Solution: Using formula (6.5), we get

$$\theta = \frac{s}{r} = \frac{10 \text{ cm}}{1 \text{ m}} = \frac{10 \text{ cm}}{100 \text{ cm}} = 0.1 \text{ radian.}$$

Using formula (6.2) yields

$$(0.1)\left(\frac{180°}{\pi}\right) = 5.73° = 5°44'.$$

EXERCISES

If $A + B = 360°$, find B for the given A in Exercises 1 to 15.

1. 45°	**2.** −60°	**3.** 400°	**4.** 215°
5. −305°	**6.** 127°	**7.** 152°10′	**8.** 236°40′
9. 562°15′	**10.** −65°45′	**11.** −125°42′	**12.** 325°20′35″
13. 179°41′4″	**14.** −110°32′12″	**15.** 314°15′20″	

In Exercises 16 to 27, find the least positive angle that is coterminal with the given angle.

16. 530°	**17.** 640°	**18.** 450°	**19.** 890°
20. 720°	**21.** 1180°	**22.** $\dfrac{20\pi}{3}$	**23.** $\dfrac{16\pi}{3}$
24. $\dfrac{-4\pi}{3}$	**25.** $\dfrac{-20\pi}{3}$	**26.** $\dfrac{17\pi}{6}$	**27.** $\dfrac{15\pi}{4}$

In Exercises 28 to 33, determine the quadrant in which the terminal side of the angle lies if the angle is in standard position.

28. $\dfrac{26\pi}{3}$

29. $\dfrac{27\pi}{5}$

30. 2 radians

31. 3 radians

32. $-630°$

33. $-1245°$

Convert the angles in Exercises 34 to 45 from degree to radian measure. Leave your answer in terms of π.

34. $45°$ **35.** $245°$ **36.** $4°$ **37.** $100°$

38. $60°$ **39.** $-225°$ **40.** $-30°$ **41.** $-135°$

42. $300°$ **43.** $18°$ **44.** $36°$ **45.** $330°$

Convert the expressions in Exercises 46 to 54 from radian to degree measure, expressed to the nearest 10′.

46. $\dfrac{5\pi}{9}$

47. $\dfrac{5\pi}{3}$

48. 1.9

49. $-\dfrac{5\pi}{4}$

50. $-\dfrac{\pi}{12}$

51. 0.19

52. $\dfrac{11\pi}{6}$

53. $\dfrac{3\pi}{4}$

54. 0.58

55. A central angle θ subtends an arc 8 centimeters long on a circle having a radius of 5 centimeters. Find the radian measure of θ.

56. A central angle θ subtends an arc 5 centimeters long on a circle having a radius of 4 centimeters. Approximate the measure of θ in (a) radians and (b) degrees, to the nearest 10′.

57. Approximate the length of an arc subtended by a central angle of $60°$ on a circle whose radius is 9 centimeters.

58. Approximate the length of an arc subtended by a central angle of $70°$ on a circle whose radius is 10 centimeters.

59. Find the radius of the circle in which a central angle of $20°$ subtends an arc having a length of 5 meters.

60. Find the radius of the circle in which a central angle of $80°$ subtends an arc having a length of 20 centimeters.

61. A heavy weight is pulled by a cord wound around a drum, as shown in Figure 6.17. If the radius of the drum is 3 meters, how far will the weight be moved when the drum is rotated through an angle of $90°$?

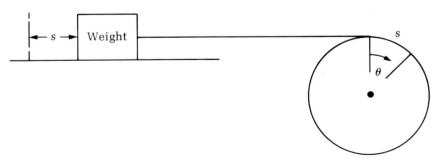

FIGURE 6.17

62. A wheel on a car has a 15-inch radius. Through what angle (in radians) does the wheel turn while the car travels 1 mile?

63. A wheel on a car has a 13-inch radius. Through what angle (in radians) does the wheel turn while the car travels 1 mile?

64. In a 24-hour period, the earth rotates 360°. Through how many radians does it rotate in **(a)** 8 hours, **(b)** 1 hour, **(c)** 1 minute, **(d)** 1 second?

65. The length of a pulley belt is 12 meters. If the pulley is 0.5 meter in diameter, find how many revolutions the pulley makes as the belt makes 1 complete revolution.

66. A boat moved in a circular course around an anchored buoy, as shown in Figure 6.18. While the boat traveled a distance of 2.5 kilometers, the angle between the lines of sight from the boat to the buoy was 110°. How far was the boat from the buoy?

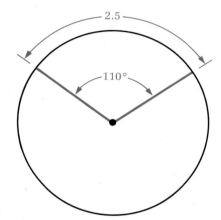

FIGURE 6.18

67. A bucket of water is drawn from a well by pulling a rope over a pulley. If the bucket is raised 10 feet while the pulley is turned through 7 revolutions, find the radius of the pulley.

68. A pulley with a 3-inch radius is used to draw water from a well. How many feet is the bucket raised if the pulley revolves 10 times?

69. Assume that the earth is a perfect sphere with a radius of 3960 miles. How many miles does a 50° longitude angle cut off at the equator? (See Figure 6.19.)

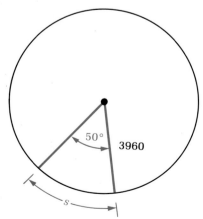

FIGURE 6.19

70. Assume that the earth is a perfect sphere with a radius of 3960 miles. Find the distance from the equator of a point located at 41°30'N on the surface of the earth. (See Figure 6.20.)

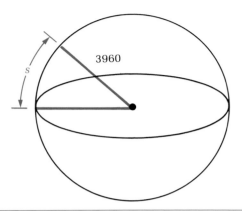

FIGURE 6.20

6.2

RIGHT TRIANGLES AND TRIGONOMETRIC RATIOS

A right triangle is formed when two sides are perpendicular to each other and is denoted by the symbol shown in Figure 6.21(*a*). The angles and sides of a right triangle are usually labeled as shown in Figure 6.21(*b*), where capital letters identify the angles and corresponding lowercase letters identify the sides opposite the angles. It is also customary to label the right angle $\angle C$ and the side opposite the right angle, *c*. The sides adjacent to the right angle are called the *legs* of the right triangle, and the side opposite the right angle is called the *hypotenuse*. Opposite $\angle A$ is side *a*, and opposite $\angle B$ is side *b*.

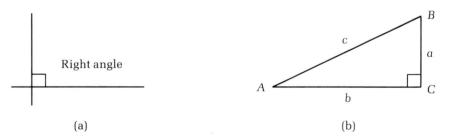

FIGURE 6.21   (a)   (b)

We know from geometry that the sum of the angles in any triangle is 180°. Since the right angle measures 90°, the two acute angles must add to the remaining 90°. Therefore, if the measure of angle *A*, or $m\angle A$, is α, then the measure of angle *B*, or $m\angle B$, is $90° - \alpha$. Thus the shape of a right triangle is determined when one acute angle is given (see Figure 6.22). Observe that the largest angle in a right triangle is the right angle and that the longest side is the hypotenuse.

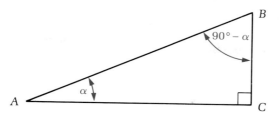

FIGURE 6.22

A special kind of right triangle is one in which the legs are the same length $(b = a)$ and the acute angles are equal, in measure, namely 45°. This is called an **isosceles right triangle.** If we know the length of the sides, then a unique triangle is determined, and we can calculate the length of the hypotenuse by the Pythagorean theorem, $c^2 = a^2 + b^2$ (see Figure 6.23):

$$c^2 = a^2 + a^2 \qquad (b = a)$$
$$= 2a^2$$
$$c = \sqrt{2}a.$$

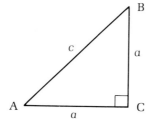

FIGURE 6.23

| **Example 6.8** | Find the length of the hypotenuse of an isosceles right triangle whose legs are 10 centimeters each.

Solution: Using the Pythagorean theorem, we find

$$c^2 = a^2 + b^2$$
$$= 10^2 + 10^2$$
$$= 100 + 100 = 200$$
$$c = 10\sqrt{2}.$$

Now let us explore the relationships between two right triangles ABC and $A'B'C'$ in which $m\angle A = m\angle A'$. Since all right angles are equal in measure, $m\angle C = m\angle C'$, and we are forced to conclude that the $m\angle B = m\angle B'$. In a right triangle if α denotes the angle at A, then $90° - \alpha$ is the measure of the angle at B. When corresponding angles in two triangles are equal in measure, the triangles are similar (see Figure 6.24). Similar triangles have the same shape

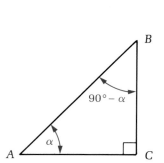

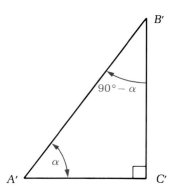

FIGURE 6.24

but not necessarily the same size. However, their corresponding sides are in proportion. In other words,

$$\frac{a}{a'} = \frac{b}{b'} = \frac{c}{c'}.$$

Example 6.9 | Given right triangles ABC and $A'B'C'$ shown in Figure 6.25, find the length of side b'.

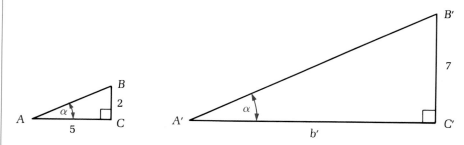

FIGURE 6.25

Solution: Since $m \angle A = m \angle A'$, triangle ABC is similar to triangle $A'B'C'$ and the corresponding sides are in proportion. Thus,

$$\frac{a}{a'} = \frac{b}{b'}$$

$$\frac{2}{7} = \frac{5}{b'}$$

$$b' = \frac{35}{2} = 17.5.$$

The following three proportions also hold true for any similar triangles:

$$\frac{a}{b} = \frac{a'}{b'} \quad \text{or} \quad \frac{a}{c} = \frac{a'}{c'} \quad \text{or} \quad \frac{b}{c} = \frac{b'}{c'}.$$

Notice that the solution to Example 6.9 can be reached by using the proportion

$$\frac{a}{b} = \frac{a'}{b'}$$

$$\frac{2}{5} = \frac{7}{b'}$$

$$b' = \frac{35}{2} = 17.5.$$

Example 6.10

At a certain time of the day, a meter stick casts a shadow of 0.62 meter while a tree nearby casts a shadow 18 meters long. How high is the tree?

Solution: Make a sketch of the information, as shown in Figure 6.26, and solve the proportion of the corresponding sides:

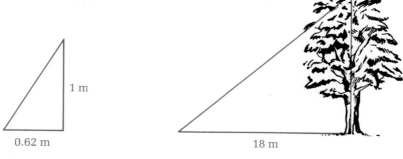

1 m

0.62 m

18 m

FIGURE 6.26

$$\frac{1}{0.62} = \frac{x}{18}$$

$$x = 29 \text{ meters.}$$

Since the shape of a right triangle is determined by one of the acute angles, say $\angle A$, the various ratios of sides a, b, and c depend on the size of $\angle A$. The trigonometric functions are defined in terms of these ratios. Note that there are three basic relationships—sine, cosine, and tangent—and their respective reciprocals—cosecant, secant, and cotangent.

Definition 6.2.1

The triangle definitions of the trigonometric functions for any right triangle ABC, where C is the right angle (shown in Figure 6.27), are as follows:

$$\text{sine } A = \sin A = \frac{\text{length of side opposite } A}{\text{length of the hypotenuse}} = \frac{a}{c}$$

$$\text{cosine } A = \cos A = \frac{\text{length of side adjacent to } A}{\text{length of the hypotenuse}} = \frac{b}{c}$$

$$\text{tangent } A = \tan A = \frac{\text{length of side opposite } A}{\text{length of side adjacent to } A} = \frac{a}{b}$$

$$\text{cosecant } A = \csc A = \frac{\text{length of hypotenuse}}{\text{length of side opposite } A} = \frac{c}{a}$$

$$\text{secant } A = \sec A = \frac{\text{length of hypotenuse}}{\text{length of side adjacent to } A} = \frac{c}{b}$$

$$\text{cotangent } A = \cot A = \frac{\text{length of side adjacent to } A}{\text{length of side opposite } A} = \frac{b}{a}$$

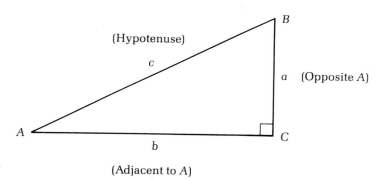

FIGURE 6.27   (Adjacent to A)

Comment

Since c, the hypotenuse, is the longest side in the triangle, the sine and cosine of either acute angle must be less than 1 and the cosecant and secant of either acute angle must be greater than 1, or

$$c > a \qquad\qquad c > b$$

$$\frac{a}{c} < 1 \quad \text{and} \quad \frac{b}{c} < 1$$

$$\sin A < 1 \qquad\qquad \cos A < 1.$$

Similarly,

$$\frac{c}{a} > 1 \qquad \text{and} \qquad \frac{c}{b} > 1$$

$$\csc A > 1 \qquad\qquad \sec A > 1.$$

Also, for a hypotenuse of length c, as angle A increases in size ($0° < A < 90°$), side a increases in size and side b decreases in size (see Figure 6.28). Therefore, when an acute angle A increases in size, $\sin A$, $\tan A$ and $\sec A$ increase in value, whereas $\cos A$, $\cot A$, and $\csc A$ decrease in value. The same statement holds true for angle B.

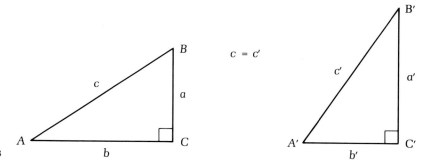

FIGURE 6.28

Example 6.11 State the six trigonometric functions of angle A defined by the triangle in Figure 6.29.

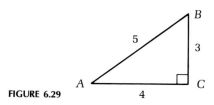

FIGURE 6.29

Solution:

$$\sin A = \tfrac{3}{5} \qquad \tan A = \tfrac{3}{4} \qquad \sec A = \tfrac{5}{4}$$

$$\cos A = \tfrac{4}{5} \qquad \cot A = \tfrac{4}{3} \qquad \csc A = \tfrac{5}{3}.$$

Example 6.12 Given the right triangle ABC with $\cos A = \tfrac{15}{17}$, find the length of the third side. Then find the other five trigonometric ratios of the angle.

Solution: Make a sketch of the triangle as in Figure 6.30, and find the length of side a:

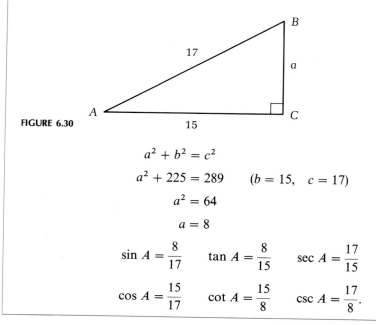

FIGURE 6.30

$$a^2 + b^2 = c^2$$

$$a^2 + 225 = 289 \qquad (b = 15, \quad c = 17)$$

$$a^2 = 64$$

$$a = 8$$

$$\sin A = \frac{8}{17} \qquad \tan A = \frac{8}{15} \qquad \sec A = \frac{17}{15}$$

$$\cos A = \frac{15}{17} \qquad \cot A = \frac{15}{8} \qquad \csc A = \frac{17}{8}.$$

If we apply the statements of Definition 6.2.1 to the angle at B instead of the angle at A in Figure 6.31 and then compare the ratios with those at A, we find some very interesting results:

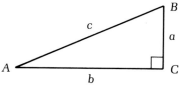

FIGURE 6.31

$$\sin B = \frac{b}{c} = \cos A \qquad \csc B = \frac{c}{b} = \sec A$$

$$\cos B = \frac{a}{c} = \sin A \qquad \sec B = \frac{c}{a} = \csc A$$

$$\tan B = \frac{b}{a} = \cot A \qquad \cot B = \frac{a}{b} = \tan A.$$

Since A and B are the acute angles of a right triangle and their sum is $90°$, angles A and B are **complementary angles.** Since $\sin A = \cos B$ and $\cos A = \sin B$, the sine and cosine functions are called **cofunctions.** In a similar manner, the tangent and cotangent are cofunctions, as are the secant and cosecant. This leads us to the following proposition.

Proposition 6.2.1

If A is an acute angle of a right triangle, then any trigonometric function of A is equal to the corresponding cofunction of the complement of A.

Example 6.13

Given a right triangle ABC and $\sin A = \frac{5}{13}$, find the six trigonometric functions of angles A and B.

Solution: Make a sketch of the triangle, as in Figure 6.32. Find the length of the third side:

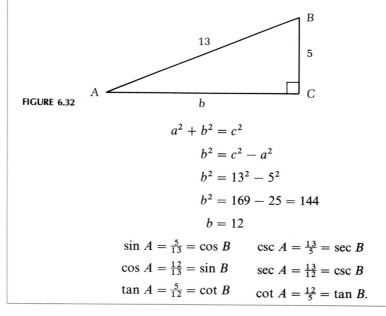

FIGURE 6.32

$$a^2 + b^2 = c^2$$

$$b^2 = c^2 - a^2$$

$$b^2 = 13^2 - 5^2$$

$$b^2 = 169 - 25 = 144$$

$$b = 12$$

$\sin A = \frac{5}{13} = \cos B$ $\csc A = \frac{13}{5} = \sec B$

$\cos A = \frac{12}{13} = \sin B$ $\sec A = \frac{13}{12} = \csc B$

$\tan A = \frac{5}{12} = \cot B$ $\cot A = \frac{12}{5} = \tan B.$

Special Angles

In this section we study the special angles: $30°$, $45°$, and $60°$. In Sections 6.5 and 6.6 we study the quadrantal angles—$0°$, $90°$, $180°$, and $270°$—as well as integral multiples of the special angles, such as $120°$, $225°$, and $330°$.

45° Angle

We already know that the hypotenuse of an isosceles right triangle is equal to $\sqrt{2}$ times the length of its side. If we let the side $a = 1$, then we have the triangle shown in Figure 6.33. Using the trigonometric ratios, we get

$$\sin 45° = \frac{1}{\sqrt{2}} = \frac{1}{2}\sqrt{2} \qquad \csc 45° = \frac{\sqrt{2}}{1} = \sqrt{2}$$

$$\cos 45° = \frac{1}{\sqrt{2}} = \frac{1}{2} \cdot \sqrt{2} \qquad \sec 45° = \frac{\sqrt{2}}{1} = \sqrt{2}$$

$$\tan 45° = \frac{1}{1} = 1 \qquad \cot 45° = \frac{1}{1} = 1.$$

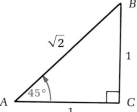

FIGURE 6.33

60° Angle and 30° Angle

Draw an equilateral triangle ABC whose sides are 2 units in length. Drop a perpendicular line from vertex C to side AB at point D. Line CD bisects side AB. Thus AD and DB are 1 unit in length. Line CD also bisects angle ACB, forming two 30° angles [see Figure 6.34(a)]. Since two congruent right triangles have been formed, one triangle is shown in Figure 6.34(b). Using the Pythagorean theorem, we calculate the length of CD, or b:

$$b^2 = 2^2 - 1^2 = 4 - 1 = 3$$
$$b = \sqrt{3}.$$

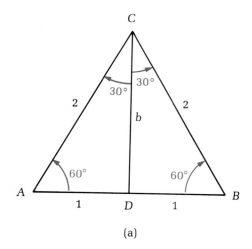

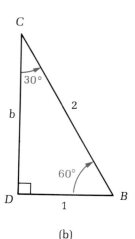

FIGURE 6.34   (a)   (b)

Using the trigonometric ratios, we find

$$\sin 60° = \frac{\sqrt{3}}{2} \qquad \tan 60° = \frac{\sqrt{3}}{1} \qquad \sec 60° = \frac{2}{1}$$

$$\cos 60° = \frac{1}{2} \qquad \cot 60° = \frac{\sqrt{3}}{3} \qquad \csc 60° = \frac{2\sqrt{3}}{3}$$

and

$$\sin 30° = \frac{1}{2} \qquad \tan 30° = \frac{\sqrt{3}}{3} \qquad \sec 30° = \frac{2\sqrt{3}}{3}$$

$$\cos 30° = \frac{\sqrt{3}}{2} \qquad \cot 30° = \frac{\sqrt{3}}{1} \qquad \csc 30° = \frac{2}{1}.$$

Take note of the cofunction relationships that are apparent here: $\sin 60° = \cos 30°$, $\tan 60° = \cot 30°$, and $\sec 60° = \csc 30°$.

Example 6.14

Show that

$$\cos 30° = \sqrt{\frac{1 + \cos 60°}{2}}$$

is true.

Solution: We know $\cos 30° = \sqrt{3}/2$, and

$$\sqrt{\frac{1 + \cos 60°}{2}} = \sqrt{\frac{1 + \frac{1}{2}}{2}} = \sqrt{\frac{3}{4}} = \frac{\sqrt{3}}{2}.$$

Therefore,

$$\cos 30° = \sqrt{\frac{1 + \cos 60°}{2}}.$$

Example 6.15

Show that $\sin^2 60° + \cos^2 60° = 1$ is true. [Note that with trigonometric functions, the notation $\sin^2 \theta$ means $(\sin \theta)^2$. It does *not* mean $\sin \theta^2$.]

Solution:

$$\sin^2 60° + \cos^2 60° = \left(\frac{\sqrt{3}}{2}\right)^2 + \left(\frac{1}{2}\right)^2$$

$$= \frac{3}{4} + \frac{1}{4} = 1$$

Example 6.16   Find the numerical value for the expression $\sin 30° + \sin 60°$.

Solution:

$$\sin 30° = \frac{1}{2} \qquad \sin 60° = \frac{\sqrt{3}}{2}$$

$$\sin 30° + \sin 60° = \frac{1}{2} + \frac{\sqrt{3}}{2} = \frac{1 + \sqrt{3}}{2}.$$

Warning: Note that $\sin 30° + \sin 60° \neq \sin(30° + 60°)$ because $\sin 30° + \sin 60° = (1 + \sqrt{3})/2$, but $\sin(30° + 60°) = \sin 90° = 1$.

EXERCISES

Given the measure of one acute angle of a right triangle, find the measure of the other acute angle in Exercises 1 to 8.

1.   38°   **2.**   55°   **3.**   22°   **4.**   60°

5.   49°   **6.**   68°   **7.**   81.5°   **8.**   19.5°

Find the six trigonometric functions of angles A and B for the triangles in Exercises 9 to 14.

9.

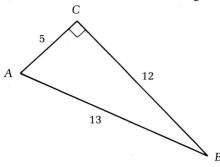

10.

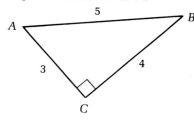

11.

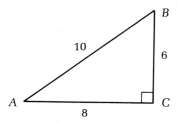

12.

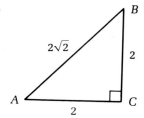

13.

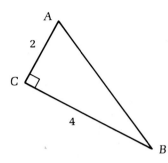

14.

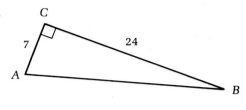

Find the six trigonometric functions of angle θ in Exercises 15 to 18.

15.

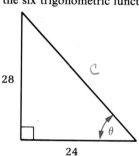

16.

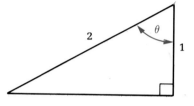

17.

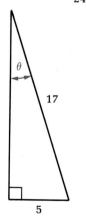

18.

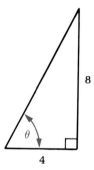

Given a right triangle ABC and one of its trigonometric ratios, find the other five trigonometric ratios in Exercises 19 to 30.

19. $\sin A = \dfrac{3}{5}$

20. $\sin B = \dfrac{\sqrt{3}}{2}$

21. $\cos B = \dfrac{2}{5}$

22. $\cos A = \dfrac{1}{2}$

23. $\tan A = \dfrac{3}{2}$

24. $\tan B = \dfrac{\sqrt{3}}{2}$

25. $\sec B = \dfrac{5}{3}$

26. $\sec A = \dfrac{13}{5}$

27. $\csc B = 5$

28. $\csc A = \dfrac{4}{1}$

29. $\cot A = \dfrac{12}{5}$

30. $\cot B = 1$

31. Complete the table with the exact value of the function.

θ	$\sin \theta$	$\cos \theta$	$\tan \theta$	$\csc \theta$	$\sec \theta$	$\cot \theta$
30°						
45°						
60°						

Show that the statements in Exercises 32 to 37 are true. Note that the notation $\sin^2 \theta$ means $(\sin \theta)^2$.

32. $\sin^2 45° + \cos^2 45° = 1$ **33.** $\sin 30° = \sqrt{\dfrac{1 - \cos 60°}{2}}$ **34.** $1 + \tan^2 30° = \sec^2 30°$

35. $1 + \cot^2 60° = \csc^2 60°$ **36.** $\tan 60° = \dfrac{2 \tan 30°}{1 - \tan^2 30°}$ **37.** $\cos 60° = 1 - 2 \sin^2 30°$

Find the numerical values for each of the expressions in Exercises 38 to 60. Give the answers in the simplest radical form when radicals are involved.

38. $(\tan 60°)(\cot 60°)$

39. $\tan 45° + \cot 45°$

40. $(\sin 45°)(\cos 45°)$

41. $\sin^2 30° + \cos^2 30°$

42. $2(\sin 30°)(\cos 30°)$

43. $(\cos 60°)(\tan 60°)$

44. $\sin 45° + \cos 45°$

45. $\cos 30° + \cos 60°$

46. $2(\sin 60°)(\cos 60°)$

47. $\cos^2 30° - \sin^2 30°$

48. $\cos^2 60° - \sin^2 60°$

49. $\dfrac{\tan 30° + \tan 60°}{1 - (\tan 30°)(\tan 60°)}$

50. $\dfrac{1 - \cos 60°}{\sin 60°}$

51. $(\sin 45°)(\cot 45°) - \cos 45°$

52. $(\sin 30°)(\cos 60°) + (\cos 30°)(\sin 60°)$

53. $(\cos 30°)(\sin 60°) - (\sin 30°)(\cos 60°)$

54. $\dfrac{2 \tan 30°}{1 - (\tan 30°)^2}$

55. $(\cos 30°)(\sec 30°)$

56. $\dfrac{\sin 60°}{1 + \cos 60°}$

57. $1 - 2 \sin^2 30°$

58. $1 - 2 \sin^2 45°$

59. $2 \cos^2 45° - 1$

60. $2 \cos^2 30° - 1$

6.3

TABLES AND CALCULATORS

In the last section we learned the trigonometric functions for certain special angles. The general method for finding values of the trigonometric functions of any angle is based on calculus and so is beyond the scope of this discussion. Fortunately these computations have been done for us by mathematicians and can be found in Appendix Table 4 for angles measured in both degrees and radians. In this section we first discuss the procedures, using the tables and then using a scientific calculator.

If you now turn to Appendix Table 4, you will find the values of the trigonometric functions of angles from 0° to 90° (0 to 1.5708 radians). These values have been rounded to four decimal places and are given at intervals of 10′. The table includes only the more commonly used functions of sin θ, cos θ, tan θ, and cot θ. Reprinted in Table 6.1 is a portion of Appendix Table 4 for use in the

TABLE 6.1

Angle	Radians	Sin	Tan	Cot	Cos		
9°00′	.1571	.1564	.1584	6.314	.9877	1.4137	**81°00′**
10′	600	.1593	.1614	6.197	.9872	108	**50′**
20′	629	.1622	.1644	6.084	.9868	079	**40′**
30′	.1658	.1650	.1673	5.976	.9863	1.4050	**30′**
40′	687	.1679	.1703	5.871	.9858	1.4021	**20′**
50′	716	.1708	.1733	5.769	.9853	1.3992	**10′**
10°00′	.1745	.1736	.1763	5.671	.9848	1.3963	**80°00′**
10′	774	.1765	.1793	5.576	.9843	934	**50′**
20′	804	.1794	.1823	5.485	.9838	904	**40′**
30′	.1833	.1822	.1853	5.396	.9833	1.3875	**30′**
40′	862	.1851	.1883	5.309	.9827	846	**20′**
50′	891	.1880	.1914	5.226	.9822	817	**10′**
11°00′	.1920	.1908	.1944	5.145	.9816	1.3788	**79°00′**
10′	949	.1937	.1974	5.066	.9811	759	**50′**
40′	560	.2532	.2617	3.821	.9674	148	**20′**
50′	589	.2560	.2648	3.776	.9667	119	**10′**
15°00′	.2618	.2588	.2679	3.732	.9659	1.3090	**75°00′**
10′	647	.2616	.2711	3.689	.9652	061	**50′**
20′	676	.2644	.2742	3.647	.9644	032	**40′**
30′	.2705	.2672	.2773	3.606	.9636	1.3003	**30′**
40′	734	.2700	.2805	3.566	.9628	1.2974	**20′**
50′	763	.2728	.2836	3.526	.9621	945	**10′**
16°00′	.2793	.2756	.2867	3.487	.9613	1.2915	
10′	822	.2784	.2899	3.450	.9605	886	**50′**
20′	851	.2812	.2931	3.412	.9596	857	**40′**
30′	.2880	.2840	.2962	3.376	.9588	1.2828	**30′**
40′	909	.2868	.2994	3.340	.9580	799	**20′**
50′	938	.2896	.3026	3.305	.9572	770	**10′**
17°00′	.2967	.2924	.3057	3.271	.9563	1.2741	**73°00′**
10′	996	.2952	.3089	3.237	.9555	712	**50′**
20′	.3025	.2979	.3121	3.204	.9546	683	**40′**
30′	.3054	.3007	.3153	3.172	.9537	1.2654	**30′**
40′	083	.3035	.3185	3.140	.9528	625	**20′**
50′	113	.3062	.3217	3.108	.9520	595	**10′**
18°00′	.3142	.3090	.3249	3.078	.9511	1.2566	**72°00′**
		Cos	Cot	Tan	Sin	Radians	Angle

examples that follow. Notice that as the size of θ increases, $\sin \theta$ and $\tan \theta$ also increase whereas $\cot \theta$ and $\cos \theta$ decrease.

To find the value of a certain function of an angle from $0°$ to $45°$, inclusive, look for the angle in the far left-hand column of the table and find the value of the function opposite that entry under the appropriate heading. To find the value of a function from $45°$ to $90°$, inclusive, look for the angle in the far right-hand column (reading upward) and use the heading printed at the bottom of the page. To find an angle whose trigonometric function is known, we reverse the process.

| Example 6.17 | Use Table 6.1 to find **(a)** $\cos 16°20'$, **(b)** $\tan 74°40'$, and **(c)** $\sin 1.3817$. |

Solution:

(a) Locate $16°20'$ in the left column under the cosine heading and read 0.9596.
(b) Locate $74°40'$ in the right-hand column, and follow across the tangent column at the bottom of the table to read the value 3.647.
(c) Since there is no degree symbol given, the angle is in radian measure. Using the right-hand column in Table 6.1 with the word *radians* on the bottom, we locate 1.3817 and read across to the column with sine at the bottom to find the value 0.9822.

| Example 6.18 | Find, to the nearest $10'$, the acute angle θ such that **(a)** $\cos \theta = 0.9822$ and **(b)** $\cot \theta = 0.3153$. |

Solution:

(a) Locate 0.9822 in the cosine column, and read the left-hand column in degrees to find $10°50'$.
(b) Locate 0.3153 in the cotangent column at the bottom of the table. Read the right-hand column of degrees to find $72°30'$.

To find the trigonometric functions of angles other than to the nearest $10'$, we recommend the use of a scientific calculator. Since calculators provide values for decimal parts of a degree, it will be necessary to change any angle given in degrees, minutes, and/or seconds by using the fact that

$$1' = \left(\frac{1}{60}\right)° \quad \text{and} \quad 1'' = \left(\frac{1}{60}\right)' = \left(\frac{1}{3600}\right)°.$$

We briefly discuss some keystroke instructions for an algebraic logic calculator and illustrate how to call up the trigonometric values.

Example 6.19

Use a scientific calculator to find **(a)** cos 45°, **(b)** tan 39°14′, **(c)** sin 68°17′38″, **(d)** cot (π/6).

Solution: Be sure that the calculator is in the proper mode, and follow these instructions.

(a) Select degree mode:

Enter	Press	Display
45	cos	0.70710678.

(b) Select degree mode:

Enter	Press	Display
14	÷	14
60	=	0.23333333
	+	0.23333333
39	=	39.233333
	tan	0.8165493.

(c) Select degree mode:

$$68°17′38″ = 68° + [17(\tfrac{1}{60})]° + [38(\tfrac{1}{3600})]°$$

Enter	Press	Display
38	÷	38
3600	=	0.01055556
	+	0.01055556
17	÷	17
60	=	0.29388889
	+	0.29388889
68	=	68.293889
	sin	0.92909313.

(d) Select radian mode:

Enter	Press	Display
π	÷	3.1415927
6	=	0.52359878
	cot	1.7320508.

Example 6.20   Use a scientific calculator to find angle θ in both degrees and radians: **(a)** $\sin \theta = 0.6275$ and **(b)** $\tan \theta = -3.72$.

Solution:

(a) Since $\sin \theta = 0.6275$,

Enter	Press	Display
0.6275	INV SIN	38.865917 (degrees)
0.6275	DRG	0.6275
	INV SIN	0.67833822 (radian)

(b) Since $\tan \theta = -3.72$,

Enter	Press	Display
3.72	$+/-$	-3.72
	INV TAN	-74.953608 (degrees)
3.72	DRG	-3.72
	INV TAN	-1.3081872 (radians)

EXERCISES

Use Appendix Table 4 to find the values in Exercises 1 to 10.

1. $\sin 13°20'$ **2.** $\cos 80°40'$ **3.** $\cos 56°30'$ **4.** $\sin 2°10'$

5. $\tan 0°10'$ **6.** $\sin 54°40'$ **7.** $\cot 70°10'$ **8.** $\tan 10°$

9. $\cos 71°50'$ **10.** $\cos 10°$

11–20.   Use a hand-held calculator to find the values of Exercises 1 to 10, respectively.

Find, to the nearest 10', the acute angle θ in Exercises 21 to 30 (use Appendix Table 4).

21. $\cos \theta = 0.4147$ **22.** $\sin \theta = 0.2447$ **23.** $\tan \theta = 0.5095$ **24.** $\cot \theta = 4.638$

25. $\cot \theta = 0.3217$ **26.** $\cos \theta = 0.9605$ **27.** $\sin \theta = 0.0553$ **28.** $\tan \theta = 0.2004$

29. $\cos \theta = 0.6905$ **30.** $\sin \theta = 0.7528$

31–40.   Use a hand-held calculator to find, to the nearest hundredth of a degree, the acute angle θ in Exercises 21 to 30.

Use Appendix Table 4 to find the trigonometric functions of angles given in radian measure for Exercises 41 to 46.

41. $\sin 1.0443$ **42.** $\cot 0.7010$ **43.** $\tan 0.3956$

44. $\cos 0.7679$ **45.** $\cot 0.4014$ **46.** $\sin 1.1025$

47–52.   Use a hand-held calculator to find the trigonometric values of the angles given in Exercises 41 to 46.

Find the values of the functions in Exercises 53 to 61.

53. sin 13°22'

54. cos 80°45'

55. cos 56°37'

56. sin 92°16'

57. tan 0°08'

58. cot 70°11'

59. tan 10°06'5"

60. cos 71°54'47"

61. sin 162°14'22"

Find the acute angle θ in radians in Exercises 62 to 70.

62. $\cos \theta = 0.4137$

63. $\sin \theta = 0.2457$

64. $\tan \theta = 0.5075$

65. $\cot \theta = 4.688$

66. $\cot \theta = 0.3235$

67. $\cos \theta = 0.9602$

68. $\sin \theta = 0.0558$

69. $\tan \theta = 0.2014$

70. $\cos \theta = 0.6915$

6.4   RIGHT TRIANGLES AND THEIR APPLICATIONS

We defined the trigonometric functions as ratios of the lengths of the legs and hypotenuse of a right triangle in Section 6.2. In this section we begin with a review of right-triangle relationships and then study applications in geometry, surveying, and navigation.

right triangle   Besides its right angle, a **right triangle** has five parts, namely, its two acute angles and its three sides. If two of these parts are known and at least one of them is a side, it is possible to determine the other three parts. Determining the unknown parts constitutes finding a solution of the triangle.

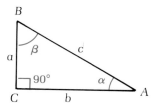

FIGURE 6.35

For the solution of the right triangle ABC (see Figure 6.35), we have the following formulas, which you should verify:

$$\alpha + \beta = 90° \qquad (6.6)$$

$$a^2 + b^2 = c^2 \qquad (6.10)$$

$$\sin \alpha = \frac{a}{c} = \cos \beta \qquad (6.7)$$

$$\csc \alpha = \frac{c}{a} = \sec \beta \qquad (6.11)$$

$$\cos \alpha = \frac{b}{c} = \sin \beta \qquad (6.8)$$

$$\sec \alpha = \frac{c}{b} = \csc \beta \qquad (6.12)$$

$$\tan \alpha = \frac{a}{b} = \cot \beta \qquad (6.9)$$

$$\cot \alpha = \frac{b}{a} = \tan \beta \qquad (6.13)$$

Example 6.21

Solve the right triangle ABC, given that $c = 21.4$ and $\alpha = 32°12'$.

Solution: Using formulas (6.2) and (6.3) and a calculator, we find

$$\sin \alpha = \frac{a}{c} \qquad \text{therefore} \qquad a = c \sin \alpha$$
$$= 21.4 \sin 32.2° = 11.4$$

$$\cos \alpha = \frac{b}{c} \qquad \text{therefore} \qquad b = c \cos \alpha$$
$$= 21.4 \cos 32.2° = 18.1$$

$$\alpha + \beta = 90° \qquad \text{therefore} \qquad \beta = 90° - \alpha$$
$$= 90° - 32°12' = 57°48'.$$

Example 6.22

A radio antenna is to be stretched from the corner of a house to the top of a pole 62 meters away. The wire will make an angle of 15° with the horizontal. How long must the wire be if 8 percent is to be added for sag?

Solution: Make a representative drawing of the information given, as in Figure 6.36. Then

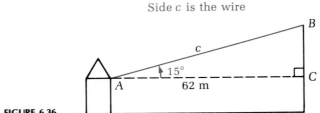

Side c is the wire

FIGURE 6.36

$$\cos 15° = \frac{62}{c},$$

$$c = \frac{62}{\cos 15°} = \frac{62}{0.9659} = 64.2,$$

$$8\% \text{ of } 64.2 = 0.08 \times 64.2 = 5.1,$$

$$\text{Length of wire} = 64.2 + 5.1 = 69.3 \text{ meters}.$$

Example 6.23

A uniform grade of 2.8 percent on a highway means that for every 100 meters along a horizontal path, the road rises 2.8 meters. The distance between two points on such a hill is 1400 meters, as shown in Figure 6.37. Find the difference in elevation between these two points.

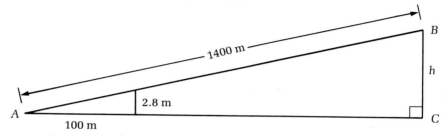

FIGURE 6.37

Solution: Using formula (6.9) and a calculator, we get

$$\tan A = \frac{2.8}{100} = 0.028,$$

$$A = 1.6038628°.$$

Using formula (6.7) and a calculator, we find

$$\sin A = \frac{h}{1400},$$

$$h = 1400 \sin 1.6038628° = 39 \text{ meters}.$$

When an object is observed above the horizontal plane, the angle between the line of sight and the horizontal is called the **angle of elevation.** In Figure 6.38, the observer is at point O, the horizontal is represented by the line OH, and the object being sighted is at point E. Angle θ is the angle of elevation. When the object being

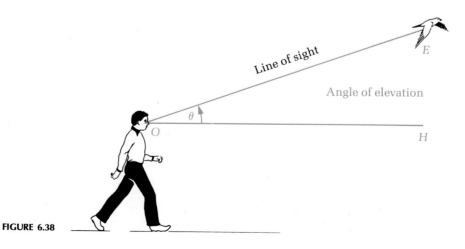

FIGURE 6.38

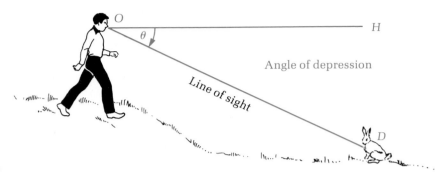

FIGURE 6.39

observed is below the horizontal plane of the observer, the angle is called the **angle of depression.** In Figure 6.39 the object sighted is at point *D*.

Example 6.24

From a point 1150 meters from the foot of a cliff, the angle of elevation of the top is 20.4°. How high is the cliff?

Solution: Make a sketch as in Figure 6.40. Then

$$\tan 20.4° = \frac{h}{1150},$$
$$h = (1150)(0.37189667) = 428 \text{ meters.}$$

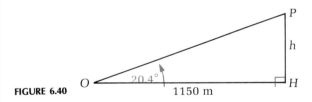

FIGURE 6.40

Example 6.25

Find the height of a hot-air balloon directly above town *A* if the angle of depression of town *B* is 15°20′ and town *B* is 6.5 kilometers from *A*.

Solution: Make a sketch as in Figure 6.41. From geometry, we know that the

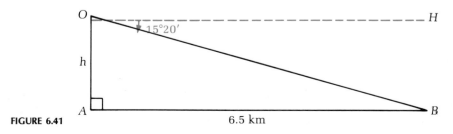

FIGURE 6.41

measure of angle $B = 15°20'$. So

$$\tan 15°20' = \frac{h}{6.5},$$

$$h = (6.5)(0.2742) = 1.78 \text{ kilometers.}$$

In surveying, the direction, or **bearing,** of a line is the acute angle made with the north-south line. The north or south designation is stated first, followed by the acute angle, then the east or west designation. See Figure 6.42 for illustrations of a bearing in the four quadrants.

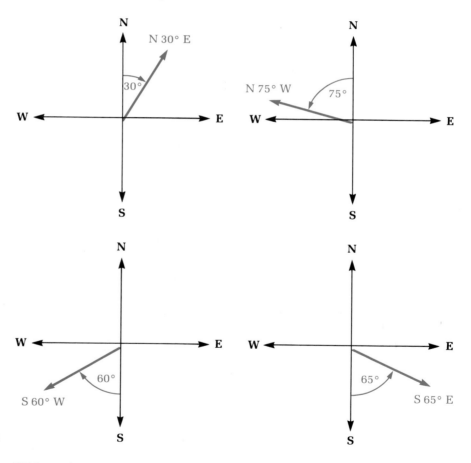

FIGURE 6.42

Example 6.26   Whitney Avenue in Hamden runs north-south, and 75 feet north of the corner where it intersects Skiff Street is a large gas station. If a restaurant on Skiff Street catches on fire and sparks are thrown 120 feet, determine whether the station is in danger from the sparks. The bearing of the restaurant from the gas station is S40°W.

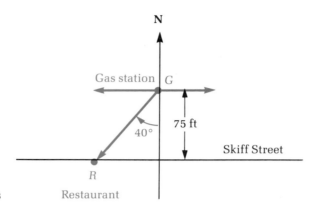

FIGURE 6.43

Solution: Make a sketch of the information, as shown in Figure 6.43. Using formula (6.8) and letting d = distance from G to R, we find

$$\cos 40° = \frac{75}{d},$$

$$d = \frac{75}{\cos 40°} = \frac{75}{0.7660} = 98 \text{ feet.}$$

Therefore, the gas station is in danger from the sparks.

In ship and plane navigation, the reference line for measuring directions is the north line, drawn as a vertical line with an arrow at the top. The *course C_n* of

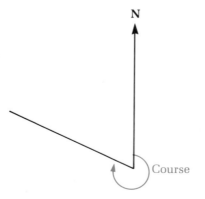

FIGURE 6.44

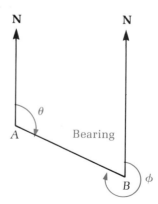

FIGURE 6.45

a ship is the angle measured clockwise from north (see Figure 6.44). *The bearing of point B from point A is the angle measured clockwise from the north to line AB*. This is shown as angle θ in Figure 6.45. The bearing of A from B is angle ϕ in Figure 6.45.

Example 6.27

A ship sails 24 nautical miles on a course 134°. It then sails on a course 224° for 45 nautical miles. What course must the ship sail if it is to return to its starting point?

Solution: First make a sketch representing the given condition, as in Figure 6.46.

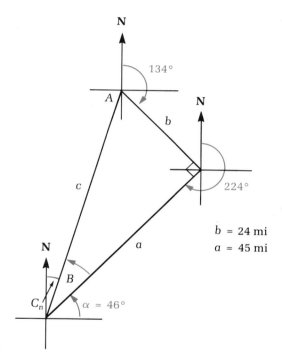

FIGURE 6.46

From geometry we know we have a right triangle and angle $\alpha = 46°$. To find angle B, we use formula (6.9):

$$\tan B = \frac{b}{a} = \frac{24}{45} = 0.5333,$$

$$B = 28° \text{ to nearest degree.}$$

To find C_n, we note that

$$C_n = 90° - (46° + B)$$
$$= 90° - (46° + 28°)$$
$$= 16°.$$

To find the distance, we use formula (6.10):

$$a^2 + b^2 = c^2$$
$$45^2 + 24^2 = c^2$$
$$2601 = c^2$$
$$c = 51 \text{ nautical miles.}$$

Example 6.28

The height of a bridge can be estimated by making two sightings, as shown in Figure 6.47. If points A and B are 400 feet apart, how tall is the bridge?

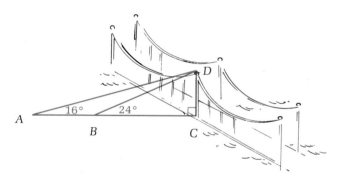

FIGURE 6.47

Solution: Using triangle ACD in Figure 6.48, we find

$$\tan 16° = \frac{h}{400 + x},$$
$$h = (400 + x)(\tan 16°).$$

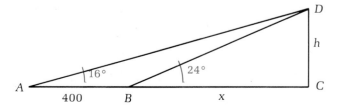

FIGURE 6.48

From the triangle BCD in Figure 6.48,

$$\cot 24° = \frac{x}{h},$$

$$x = h \cot 24°.$$

Substituting yields

$$h = (400 + h \cot 24°)(\tan 16°)$$
$$= 400 \tan 16° + h \cot 24° \tan 16°.$$

Solving for h, we get

$$h = \frac{400 \tan 16°}{1 - \cot 24° \tan 16°}.$$

Using a calculator gives

$$h = 322 \text{ feet (approximately).}$$

EXERCISES

Solve the right triangles in Exercises 1 to 30, finding angle measures to the nearest 10'. Here A, B, and C are the angles with the right angle at C; a, b, and c are the sides opposite these angles, respectively. Use a calculator for Exercises 17 to 30.

1. $c = 16.4$, $A = 24°50'$
2. $A = 78°40'$, $b = 1340$
3. $a = 24.6$, $b = 40.4$
4. $B = 56°30'$, $c = 0.0447$
5. $a = 25.3$, $A = 38°40'$
6. $a = 16$, $c = 20$
7. $B = 12°40'$, $b = 98.1$
8. $b = 100$, $c = 450$
9. $a = 2$, $A = 32°$
10. $b = 3$, $A = 29°$
11. $a = 12.5$, $b = 9.1$
12. $a = 16.3$, $B = 38°30'$
13. $b = 0.64$, $A = 62°40'$
14. $b = 709$, $B = 18°10'$
15. $c = 6.29$, $A = 10°20'$
16. $c = 6.18$, $B = 30°10'$
17. $a = 127$, $c = 164$
18. $A = 44°15'$, $a = 843.2$
19. $a = 27.46$, $c = 38.82$
20. $a = 81.48$, $b = 72.19$
21. $a = 14.92$, $A = 65°30'$
22. $b = 1.927$, $A = 40°$
23. $B = 47°26'$, $c = 4217$
24. $B = 6°12'$, $c = 1720$
25. $a = 14.72$, $b = 13.14$
26. $b = 1.241$, $c = 3.265$
27. $A = 13°30'$, $c = 627$
28. $a = 101$, $b = 120$
29. $b = 0.0353$, $c = 0.0589$
30. $A = 11°36'$, $b = 0.4214$

31. A ladder 12 meters long reaches from the ground to a window 11 meters high. How far from the wall is the foot of the ladder, and what angle does it make with the horizontal?

32. A radio antenna is stretched from a point on a house 10.9 meters high to the top of a pole 17.1 meters high. The horizontal distance between the points is 21.2 meters. Find the angle that the antenna makes with the horizontal and the length of the antenna if 0.2 meter is added to allow for contraction.

33. In each step of a certain stairway, the tread (horizontal part) is 28 centimeters, and the riser (vertical part) is 20 centimeters. What angle does the banister make with the horizontal?

34. The radius of a circle is 20 centimeters. Find the length of a side of an inscribed regular pentagon

35. A boat is anchored near a bridge. From a point on its deck 5 meters above the water, the angle of elevation of the top of the bridge is 39°, while the angle of depression of the bridge's image reflected in the water is 54°. Find the height of the bridge above the water and the distance of the boat from the bridge.

36. From a lighthouse 25.1 meters above the level of the water, the angle of depression of a boat is 23°40′. How far is the boat from a point at water level directly under the point of observation?

37. From a mountain 1780 meters high, the angle of depression of a point on the nearer shore of a river is 48°40′, and the angle of depression of a point directly across on the opposite side is 22°20′. What is the width of the river between these two points?

38. From a point on the ground 100 meters from the foot of a flagstaff, the angle of elevation of the top is 30°. How high is the flagstaff?

39. If the angle of elevation of the sun is 30°, how long a shadow will be cast on the ground by a child 1 meter tall?

40. From the top of a building 28 meters high, the angle of depression of an automobile is 29°10′. How far is the automobile from the foot of the building measured horizontally?

41. The diagonal of a rectangle is 7.25 centimeters, and the shorter side is 3.2 centimeters. Find **(a)** the angle the diagonal makes with the longer side and **(b)** the length of the longer side.

42. In landing an airplane, a pilot wants to clear a 60-foot wall by 50 feet and land 750 feet beyond the wall. What should the angle of descent measure?

43. From a certain point P, the angle of elevation of the top of a cliff is 46°. At a point 50 meters farther away, the angle of elevation is 27°. Find the height of the cliff.

44. A painting 4 meters high is hung on a wall so that the bottom edge is 3.4 meters above the floor. If the viewer's eyes are 152 centimeters above the floor and the angle of elevation of the top of the painting is 44°, how far is the viewer from the wall?

45. An airplane with an air speed of 450 miles per hour has an angle of descent of 21°. How fast is the plane approaching the ground?

46. A ship sails 18.5 nautical miles on course 110°. It then sails 45 nautical miles on course 200°. Find **(a)** how far the ship is from its starting point and **(b)** what course it must sail to return to its starting point.

47. A map shows that point A is 5.6 kilometers due south of point C. Point B is 10.7 kilometers due east of point C. Find the bearing of A from B.

48. From the top of a tower 40 meters high, the angles of depression to two points in an easterly direction are 18° and 28°. Find the distance between the two points.

49. A radar station tracking a rocket indicates the angle of elevation to be 25° and the line-of-sight distance (referred to as the *slant range*) to be 50 kilometers. Find the altitude and the horizontal range of the rocket at that moment.

50. Mrs. Smith wishes to build a footbridge across a shallow river on her property. From a point A at the edge of the river bank, she locates a point B on the other side of the river with a bearing of N52°E. Then she walks east along the edge of the river to a point C which is 84 meters from A, which she finds is due south of B. How wide is the river?

51. A tree 40 feet high casts a shadow 48 feet long. Find the angle of elevation of the sun.

52. A ladder 16 feet long rests against a house, making an angle of 23° with the house. How far is the bottom of the ladder from the base of the house?

53. The length of the path going diagonally across the New Haven green is about 195 meters. The path makes an angle of 28° with the longer side of the rectangular green. How much fencing is required to enclose the green?

54. A swimmer starts out at Lighthouse Point and swims due west toward Savin Rock. Fifteen minutes after starting, the swimmer is observed by someone at City Point with a transit. At that moment the bearing of the swimmer is S20°E (see Figure 6.49). Ten minutes later the swimmer is due south of the observer, and the distance between them is 3000 yards. Find the rate of the swimmer.

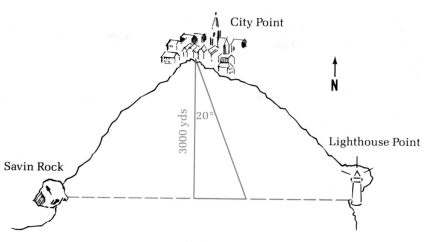

City Point

N

3000 yds   20°

Lighthouse Point

Savin Rock

FIGURE 6.49

Long Island Sound

6.5

TRIGONOMETRIC FUNCTIONS OF A GENERAL ANGLE

In this section we define the six basic trigonometric functions with respect to any angle θ placed in standard position on a coordinate system. Recall Definition 6.2.1 in which the trigonometric functions were defined in terms of ratios of the sides of a right triangle. This definition will be extended to a triangle in reference to coordinate axes.

Given a point P in any quadrant of a rectangular coordinate plane, a unique right triangle is determined by dropping a perpendicular line from P to the x axis and connecting P to the origin (see Figure 6.50). Therefore, given an angle θ in standard position, we may choose any convenient point on the

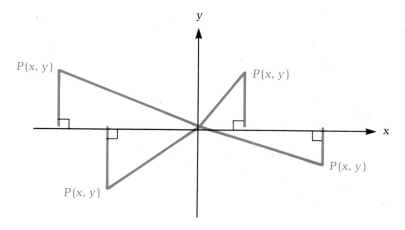

FIGURE 6.50

terminal side of θ and determine a right triangle. Let the coordinates of the point be (x, y) and the distance from P to the origin be r. *The coordinates (x, y) represent the lengths of the sides of the right triangle, and the radius r corresponds to the hypotenuse.* The radius r is always positive, but x and y may be positive, negative, or zero depending on the location of P in the plane (see Figure 6.51).

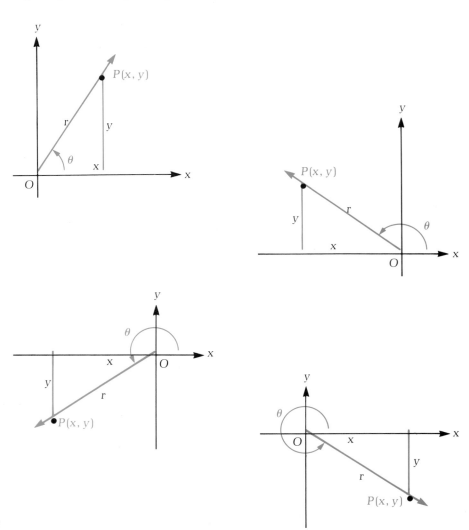

FIGURE 6.51

If P lies on an axis, a triangle is not determined, but we may still identify values for x, y, and r. If P lies on the x axis, the y coordinate is zero and $r = |x|$. If P lies on the y axis, the x coordinate is zero and $r = |y|$ (see Figure 6.52 on p. 357).

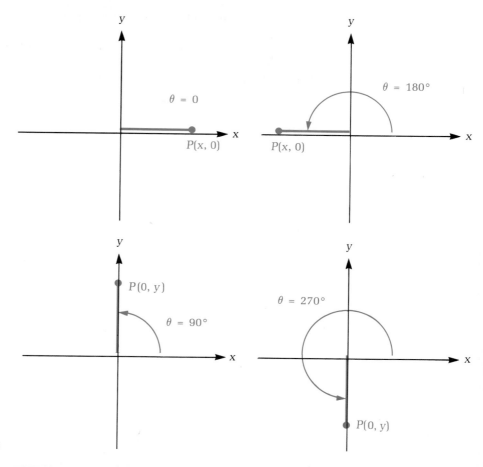

FIGURE 6.52

Definition 6.5.1

For any angle θ in standard position on the rectangular coordinate plane and any point $P(x, y)$ on the terminal side of θ, the *trigonometric functions* are defined by the ratios in Table 6.2.

TABLE 6.2   Definitions of Trigonometric Functions

Function	Abbreviation	Ratio (in Letters)	Ratio (in Words)
Sine of θ	$\sin \theta$	y/r	y coordinate/radius
Cosine of θ	$\cos \theta$	x/r	x coordinate/radius
Tangent of θ	$\tan \theta$	y/x	y coordinate/x coordinate
Cosecant of θ	$\csc \theta$	r/y	radius/y coordinate
Secant of θ	$\sec \theta$	r/x	radius/x coordinate
Cotangent of θ	$\cot \theta$	x/y	x coordinate/y coordinate

These definitions are fundamental to the study of trigonometry and should be memorized. By the Pythagorean theorem, the equation

$$x^2 + y^2 = r^2$$

expresses the relationship among the variables x, y, and r.

Example 6.29   Find the trigonometric functions of the angle AOC if the coordinates of point P are $(3, 4)$.

Solution: First sketch angle AOC in standard position with point P on the terminal side, as shown in Figure 6.53. By the Pythagorean theorem,

$$x^2 + y^2 = r^2$$
$$3^2 + 4^2 = r^2 \qquad (x = 3, \quad y = 4)$$
$$9 + 16 = r^2$$
$$25 = r^2$$
$$5 = r.$$

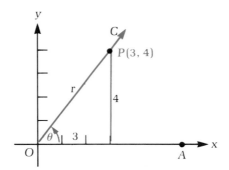

FIGURE 6.53

Therefore,

$$\sin \theta = \tfrac{4}{5} \qquad \tan \theta = \tfrac{4}{3} \qquad \sec \theta = \tfrac{5}{3}$$
$$\cos \theta = \tfrac{3}{5} \qquad \cot \theta = \tfrac{3}{4} \qquad \csc \theta = \tfrac{5}{4}.$$

Example 6.30   Find the trigonometric functions of a positive angle A if the point $(-12, 5)$ is on the terminal side.

Solution: First make a sketch of angle A, as shown in Figure 6.54. Then compute

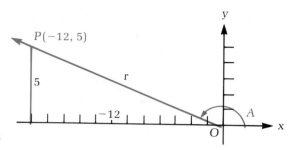

FIGURE 6.54

the value of r:

$$x^2 + y^2 = r^2$$

$$(-12)^2 + 5^2 = r^2 \qquad (x = -12, \quad y = 5)$$

$$144 + 25 = r^2 \qquad 169 = r^2 \qquad 13 = r.$$

Therefore $\quad \sin A = \frac{5}{13} \qquad \tan A = -\frac{5}{12} \qquad \csc A = \frac{13}{5}$

$\cos A = -\frac{12}{13} \qquad \cot A = -\frac{12}{5} \qquad \sec A = -\frac{13}{12}$

Example 6.31

If θ is a positive angle in the third quadrant and $\tan \theta = \frac{4}{5}$, find the other five functions of the angle.

Solution: Since θ is in the third quadrant, x and y must both be negative:

$$\tan \theta = \frac{y}{x} = \frac{-4}{-5} = \frac{4}{5}.$$

The point $P(-5, -4)$ lies on the terminal side of θ, as shown in Figure 6.55:

$$x^2 + y^2 = r^2$$

$$(-5)^2 + (-4)^2 = r^2 \qquad (x = -5, \quad y = -4)$$

$$25 + 16 = r^2 \qquad 41 = r^2 \qquad \sqrt{41} = r.$$

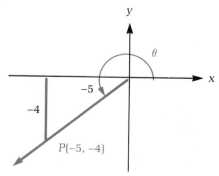

FIGURE 6.55

Therefore,

$$\sin \theta = \frac{-4}{\sqrt{41}} = \frac{-4\sqrt{41}}{41} \qquad \csc \theta = -\frac{\sqrt{41}}{4}$$

$$\cos \theta = \frac{-5}{\sqrt{41}} = \frac{-5\sqrt{41}}{41} \qquad \sec \theta = -\frac{\sqrt{41}}{5}$$

$$\tan \theta = \frac{-4}{-5} = \frac{4}{5} \qquad \cot \theta = \frac{-5}{-4} = \frac{5}{4}.$$

Example 6.32

Find the trigonometric functions of a positive angle A if the point $(0, -6)$ is on the terminal side.

Solution: Make a sketch of angle A, as shown in Figure 6.56. Use Definition 6.3.1 with $x = 0$, $y = -6$, and $r = |y| = 6$:

$$\sin A = \frac{-6}{6} = -1 \qquad \csc A = \frac{6}{-6} = -1$$

$$\cos A = \frac{0}{6} = 0 \qquad \sec A = \frac{6}{0} \quad \text{(undefined)}$$

$$\tan A = \frac{-6}{0} \quad \text{(undefined)} \qquad \cot A = \frac{0}{-6} = 0.$$

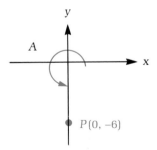

FIGURE 6.56

The tangent and secant functions of this angle are undefined because division by zero has no meaning.

EXERCISES

Find the value of r for each of the positions of P, in Exercises 1 to 15; then find the values of the six trigonometric functions of an angle in standard position with P on the terminal side. Sketch the angle on a coordinate system.

1. $(3, -4)$	**2.** $(1, \sqrt{3})$	**3.** $(5, 0)$	**4.** $(-8, 6)$
5. $(-3, -3)$	**6.** $(-2, 4)$	**7.** $(1, 1)$	**8.** $(-4, 0)$
9. $(5, -8)$	**10.** $(0, 2)$	**11.** $(12, 32)$	**12.** $(-3, -1)$
13. $(5, -12)$	**14.** $(-8, 15)$	**15.** $(-2, 2\sqrt{3})$	

16. In the table below, fill in each blank with the proper sign ($+$ or $-$) for the trigonometric function of an angle in standard position with terminal side in the indicated quadrant.

Quadrant	sin A	cos A	tan A	csc A	sec A	cot A
I						
II						
III						
IV						

17. The following table illustrates the signs of two of the trigonometric functions for each of eleven angles (a to k) as shown. State the quadrant of the angle and the signs of the other trigonometric functions.

	a	b	c	d	e	f	g	h	i	j	k
Quadrant											
sin θ	+			+						+	
cos θ			−		+	+		+			
tan θ	+	+	+		−			+			
csc θ		−					−		+		+
sec θ				−		+				−	
cot θ					+				+		−

Find the other five functions of the angle which satisfy the given conditions in Exercises 18 to 31.

18. $\tan \theta = \frac{2}{3}$, θ is in quadrant I

19. $\sin \theta = \frac{3}{5}$, θ is in quadrant II

20. $\cos \theta = -\frac{1}{2}$, θ is in quadrant III

21. $\sin B = \frac{1}{4}$, B is in quadrant I

22. $\cos B = -\frac{3}{4}$, B is in quadrant II

23. $\sec A = 2$, A is in quadrant IV

24. $\csc B = -\sqrt{3}$, B is in quadrant III

25. $\tan \theta = 2$, θ is in quadrant III

26. $\cos \theta = \frac{8}{17}$, θ is in quadrant IV

27. $\tan \theta = \frac{8}{15}$, $\cos \theta$ is negative

28. $\tan B = -\frac{1}{3}$, B is in quadrant IV

29. $\sin A = -\frac{4}{5}$, $\tan A$ is positive

30. $\cos A = \frac{4}{5}$, $\sin A$ is positive

31. $\cot B = -1$, $\sin B$ is positive

6.6　RELATED ANGLES AND THE UNIT CIRCLE

First recall that the values of the trigonometric functions of an angle depend on only the position of the terminal side of the angle. Given the coordinates of a point P on the terminal side of the angle, the x and y coordinates are known and the radius can be determined. Suppose that the coordinates of P are not known, but the size of the angle is known. We limit our discussion at this time to angles that are integral multiples of the special angles. For angles in the second, third, and fourth quadrants, we define an acute angle α related to the given angle θ.

Definition 6.6.1

The *related angle* α associated with any given angle θ is the positive acute angle in the right triangle with its vertex at the origin. The x axis always forms one side of the related angle (see Figure 6.57).

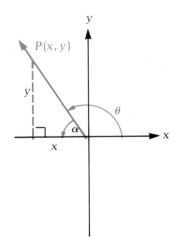

$\alpha = 180° - \theta$

$x < 0$

$y > 0$

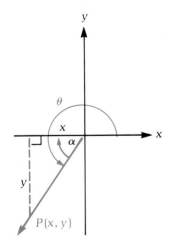

$\alpha = \theta - 180°$

$x < 0$

$y < 0$

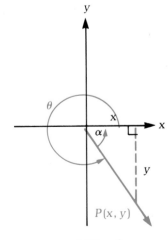

$\alpha = 360° - \theta$

$x > 0$

$y < 0$

FIGURE 6.57

Angles in the second quadrant are related to $180° - \theta$, angles in the third quadrant are related to $\theta - 180°$, and angles in the fourth quadrant are related to $360° - \theta$. After the value of α is determined, it is only necessary to determine the proper signs (positive or negative) associated with the x and y coordinates in the given quadrant (see Figure 6.57).

Example 6.33

Find the six basic trigonometric functions of $150°$.

Solution: Since $150°$ is in the second quadrant, the related angle $\alpha = 180° - 150° = 30°$. Recall the special $30°$ angle relationship, and apply the appropriate signs for the x and y coordinates in the second quadrant, as shown in Figure 6.58.

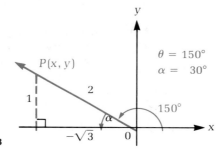

FIGURE 6.58

Then

$$\sin 150° = \frac{1}{2} \qquad\qquad \csc 150° = 2$$

$$\cos 150° = \frac{-\sqrt{3}}{2} \qquad\qquad \sec 150° = \frac{-2}{\sqrt{3}} \quad \text{or} \quad \frac{-2\sqrt{3}}{3}$$

$$\tan 150° = -\frac{1}{\sqrt{3}} \quad \text{or} \quad \frac{-\sqrt{3}}{3} \qquad \cot 150° = -\sqrt{3}.$$

Example 6.34

Find the six basic trigonometric ratios for $225°$.

Solution: The related angle in the third quadrant is $\theta - 180°$; therefore the related angle is $225° - 180° = 45°$. Recall the special $45°$ angle relationship, and apply the appropriate signs for x and y coordinates in the third quadrant (see Figure 6.59).

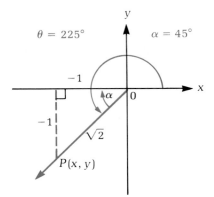

FIGURE 6.59

Then

$$\sin 225° = \frac{-1}{\sqrt{2}} = \frac{-\sqrt{2}}{2} \qquad \csc 225° = -\sqrt{2}$$

$$\cos 225° = \frac{-1}{\sqrt{2}} = \frac{-\sqrt{2}}{2} \qquad \sec 225° = -\sqrt{2}$$

$$\tan 225° = \frac{-1}{-1} = 1 \qquad \cot 225° = \frac{-1}{-1} = 1.$$

Example 6.35

Find the six basic trigonometric ratios for 300°.

Solution: The related angle in the fourth quadrant is found by taking $360° - \theta$. Therefore, $\alpha = 360° - 300° = 60°$. Recall the special 60° angle relationships, and apply the appropriate signs for the x and y coordinates in the fourth quadrant, as shown in Figure 6.60.

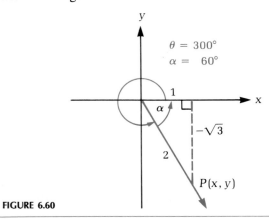

FIGURE 6.60

Then

$$\sin 300° = \frac{-\sqrt{3}}{2} \qquad \csc 300° = \frac{-2}{\sqrt{3}} = \frac{-2\sqrt{3}}{3}$$

$$\cos 300° = \frac{1}{2} \qquad \sec 300° = 2$$

$$\tan 300° = -\sqrt{3} \qquad \cot 300° = \frac{-1}{\sqrt{3}} = \frac{-\sqrt{3}}{3}.$$

Example 6.36   Find the six basic trigonometric ratios for $-120°$.

Solution: The angle $-120°$ is equivalent to the positive angle $\theta = 360° - 120° = 240°$. Therefore, the related angle $\alpha = 240° - 180° = 60°$. Sketch the angle as shown in Figure 6.61.

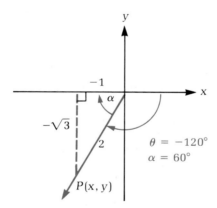

FIGURE 6.61

Then

$$\sin(-120°) = \frac{-\sqrt{3}}{2} \qquad \csc(-120°) = \frac{-2}{\sqrt{3}} = \frac{-2\sqrt{3}}{3}$$

$$\cos(-120°) = \frac{-1}{2} \qquad \sec(-120°) = -2$$

$$\tan(-120°) = \sqrt{3} \qquad \cot(-120°) = \frac{1}{\sqrt{3}} = \frac{\sqrt{3}}{3}.$$

Unit Circle

We have observed that the values of the trigonometric functions are uniquely determined by the coordinates of any point P on the terminal side of the angle. Referring to Figure 6.62, we see that for two distinct points P and P' on the same terminal side of the angle, two right triangles are formed, POA and $P'OA'$. Geometrically these two right triangles are similar because they have congruent acute angles; therefore, the corresponding sides of these two triangles are in proportion.

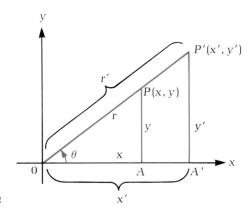

$$\frac{y}{r} = \frac{y'}{r'} \qquad \frac{r}{y} = \frac{r'}{y'}$$

$$\frac{x}{r} = \frac{x'}{r'} \qquad \frac{r}{x} = \frac{r'}{x'}$$

$$\frac{y}{x} = \frac{y'}{x'} \qquad \frac{x}{y} = \frac{x'}{y'}$$

FIGURE 6.62

Since the trigonometric functions are independent of the *radius*, we can simplify our computations if we choose point P such that the radius is 1 unit *unit circle* in length. The set of all such points P generates the *unit circle*, as shown in

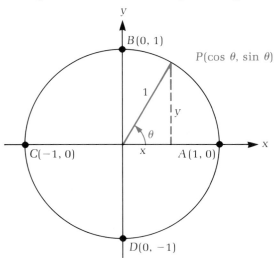

FIGURE 6.63

Figure 6.63. Thus the following simple relationships evolve:

$$\sin \theta = \frac{y}{r} = \frac{y}{1} = y \qquad \cos \theta = \frac{x}{r} = \frac{x}{1} = x.$$

Therefore the trigonometric coordinates of the point P on the unit circle are $(\cos \theta, \sin \theta)$. The trigonometric functions are referred to as **circular functions** because of their relationship with the unit circle. Example 6.37 shows that for the quadrantal angles, the values of their trigonometric functions may be immediately determined.

Example 6.37

The angle $0°$ has coordinates $(1, 0)$; hence

$$\sin 0° = 0 \qquad \csc 0° = \frac{1}{0} \quad \text{undefined}$$

$$\cos 0° = 1 \qquad \sec 0° = \frac{1}{1} = 1$$

$$\tan 0° = \frac{0}{1} = 0 \qquad \cot 0° = \frac{1}{0} \quad \text{undefined.}$$

The functional values of the other quadrantal angles can also be obtained by using the coordinates of the point on the terminal side of the angle on the unit circle:

Angle θ	Coordinates	$\sin \theta$	$\cos \theta$	$\tan \theta$
$90°$	$(0, 1)$	1	0	undefined
$180°$	$(-1, 0)$	0	-1	0
$270°$	$(0, -1)$	-1	0	undefined

From this information we observe that the sine and cosine functions vary from a maximum of 1 to a minimum of -1. This variation is summarized by quadrants:

Quadrant	θ Varies from	$\sin \theta$ Varies from	$\cos \theta$ Varies from
I	$0°$ to $90°$	0 to 1	1 to 0
II	$90°$ to $180°$	1 to 0	0 to -1
III	$180°$ to $270°$	0 to -1	-1 to 0
IV	$270°$ to $360°$	-1 to 0	0 to 1

For the unit circle, the tangent is defined as the ratio of the y coordinate to the x coordinate; this represents the ratio of $\sin \theta$ to $\cos \theta$. Therefore, the tangent can have for its value any positive or negative number or zero. Notice what happens to $\tan \theta$ as angle θ approaches 90°. $\sin \theta$ gets closer to the value 1, while $\cos \theta$ gets closer to the value 0; thus the ratio of $\sin \theta / \cos \theta$ increases without limit, and $\tan \theta$ is undefined at $\theta = 90°$.

Special Angles on the Unit Circle

Shown in Figures 6.64 to 6.66 are the special angles, 30°, 45°, and 60°, as they arise in each quadrant of the unit circle. The coordinates of any point P on the

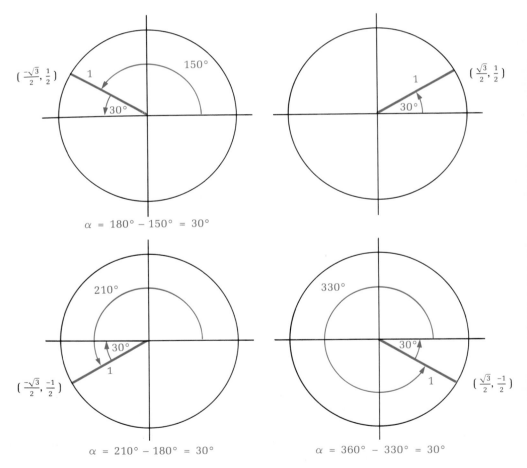

FIGURE 6.64

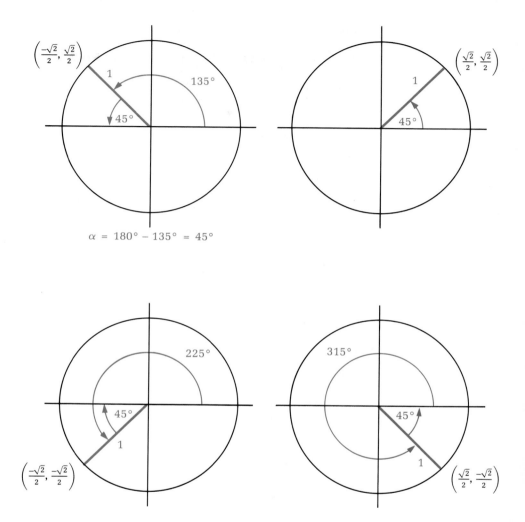

FIGURE 6.65

unit circle are ($\cos \theta$, $\sin \theta$). Since $\cos \theta$ corresponds to the x coordinate and $\sin \theta$ corresponds to the y coordinate, the coordinates of P take the appropriate signs ($+$ or $-$) of the quadrant. For the 30° reference angle, the coordinates are ($\sqrt{3}/2$, $\frac{1}{2}$) (see Figure 6.64). For the 45° reference angle, the coordinates are ($\sqrt{2}/2$, $\sqrt{2}/2$) (see Figure 6.65). For the 60° reference angle, the coordinates are ($\frac{1}{2}$, $\sqrt{3}/2$) (see Figure 6.66).

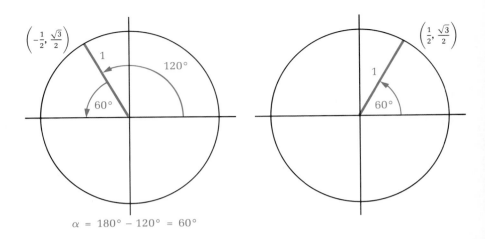

$$\alpha = 180° - 120° = 60°$$

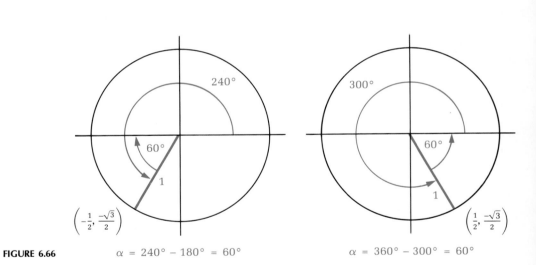

FIGURE 6.66 $\alpha = 240° - 180° = 60°$ $\alpha = 360° - 300° = 60°$

EXERCISES

Show that the statements in Exercises 1 to 30 are true.

1. $\sin^2 135° + \cos^2 135° = 1$

2. $\sin^2 225° + \cos^2 225° = 1$

3. $1 + \tan^2 120° = \sec^2 120°$

4. $1 + \cot^2 150° = \csc^2 150°$

5. $2(\sin 60°)(\cos 60°) = \sin 120°$

6. $2(\sin 90°)(\cos 90°) = \sin 180°$

7. $\cos 300° = 2 \cos^2 150° - 1$

8. $\cos^2 210° + \sin^2 210° = 1$

9. $\cos^2 135° - \sin^2 135° = \cos 270°$

10. $\cos^2 180° - \sin^2 180° = \cos 360°$

11. $\cos 60° = \sqrt{\dfrac{1 + \cos 120°}{2}}$

12. $\cos 45° = \sqrt{\dfrac{1 + \cos 90°}{2}}$

13. $\tan 135° = \dfrac{1 - \cos 270°}{\sin 270°}$

14. $\cot 120° = \dfrac{\sin 240°}{1 - \cos 240°}$

15. $1 + \cot^2 330° = \csc^2 330°$

16. $1 + \tan^2 150° = \sec^2 150°$

17. $\sin 45° = \sqrt{\dfrac{1 - \cos 90°}{2}}$

18. $\sin 60° = \sqrt{\dfrac{1 - \cos 120°}{2}}$

19. $\sin 420° = 2(\sin 210°)(\cos 210°)$

20. $\sin 270° = 2(\sin 135°)(\cos 135°)$

21. $\cos(-120°) = \cos 120°$

22. $\cos(-135°) = \cos 135°$

23. $\tan(-150°) = -\tan 150°$

24. $\tan(-210°) = -\tan 210°$

25. $\sin^2 150° + \cos^2 150° = 1$

26. $\sin^2 300° + \cos^2 300° = 1$

27. $\tan 120° = \dfrac{\sin 120°}{\cos 120°}$

28. $\cot 315° = \dfrac{\cos 315°}{\sin 315°}$

29. $1 + \cot^2 300° = \csc^2 300°$

30. $1 + \cot^2 210° = \csc^2 210°$

For each of the angles in Exercises 31 to 42, find the related angle and the values of the six basic trigonometric functions of the angle.

31. $120°$

32. $-30°$

33. $300°$

34. $210°$

35. $240°$

36. $570°$

37. $480°$

38. $660°$

39. $-120°$

40. $135°$

41. $-150°$

42. $-180°$

6.7

GRAPHS OF BASIC TRIGONOMETRIC FUNCTIONS

In this section, we consider the graphs of the equations $y = \sin \theta$, $y = \cos \theta$, and $y = \tan \theta$. In each equation, θ is the angle measured in degrees or radians, and y is the value of the designated trigonometric function.

We can find many ordered pairs (θ, y) that satisfy these equations by selecting pairs of values from Appendix Table 4. However, to be independent of tables, we use the special angles $0°$, $30°$, $60°$, and $90°$ and their integral multiples. By doing so, we may compute the numerical values of the acute angles, using the same values with the appropriate sign for the integral multiple values. These values are tabulated to the third decimal place in Table 6.3. Since $\pi = 3.14$ approximately, a proper scale for graphing would assign an ordinate value of 1 to about one-third of the distance from 0 to π.

TABLE 6.3

θ, degrees	θ, radians	$y = \sin\theta$	$y = \cos\theta$	$y = \tan\theta$
0	0	0	1	0
30	$\dfrac{\pi}{6}$	0.5	0.866	0.577
60	$\dfrac{\pi}{3}$	0.866	0.5	1.732
90	$\dfrac{\pi}{2}$	1	0	undefined
120	$\dfrac{2\pi}{3}$	0.866	-0.5	-1.732
150	$\dfrac{5\pi}{6}$	0.5	-0.866	-0.577
180	π	0	-1	0
210	$\dfrac{7\pi}{6}$	-0.5	-0.866	0.577
240	$\dfrac{4\pi}{3}$	-0.866	-0.5	1.732
270	$\dfrac{3\pi}{2}$	-1	0	undefined
300	$\dfrac{5\pi}{3}$	-0.866	0.5	-1.732
330	$\dfrac{11\pi}{6}$	-0.5	0.866	-0.577
360	2π	0	1	0

Sine Function

Plotting the points for $y = \sin\theta$ and connecting the points with a smooth curve, we have the graph shown in Figure 6.67. Let us consider the periodicity of this function.

Definition 6.7.1

A function f is *periodic* if there is some real number k such that $f(t + k) = f(t)$ for all t. The least such positive real number k (if it exists) is the *period* of f.

For any real number θ on the unit circle, the number $\theta \pm 2\pi$ has the same y coordinate. Actually, $\sin\theta = \sin(\theta + 2k\pi)$, where k is an integer. The basic sine

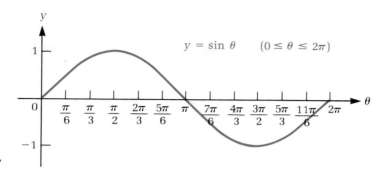

FIGURE 6.67

curve shown in Figure 6.67 repeats itself to the right and left, as shown in Figure 6.68.

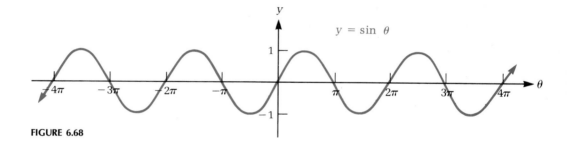

FIGURE 6.68

Properties of Sine Function

1. The domain is the set of all real numbers. The range is the set of real numbers between $+1$ and -1, inclusive.
2. The function is periodic, with a period of 2π, and continuous everywhere.
3. The sine function is an odd function and symmetric with respect to the origin. Thus $\sin(-\theta) = -\sin\theta$.
4. If we add or subtract π from any value θ, the resulting sine value is the additive inverse of the original sine value. That is, $\sin(\theta \pm \pi) = -\sin\theta$.

Cosine Function

Plotting the points for $y = \cos\theta$ given in Table 6.3 and connecting the points with a smooth curve, we obtain the graph shown in Figure 6.69.

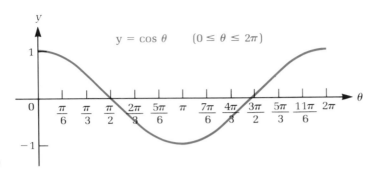

FIGURE 6.69

For any real number θ on the unit circle, the number $\theta \pm 2\pi$ has the same x coordinate. Therefore $\cos \theta = \cos (\theta + 2k\pi)$, where k is any integer, as shown in Figure 6.70.

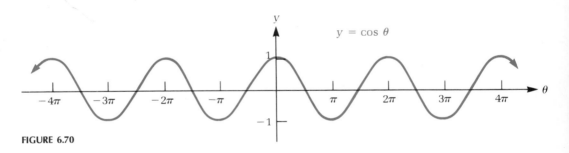

FIGURE 6.70

Properties of Cosine Function

1. The domain is the set of all real numbers. The range is the set of real numbers between $+1$ and -1, inclusive.
2. The function is periodic, with a period of 2π, and continuous everywhere.
3. The cosine function is an even function and is symmetric with respect to the y axis. Thus $\cos (-\theta) = \cos \theta$.
4. If we add or subtract π from any value θ, the resulting cosine value is the additive inverse of the original cosine value. In general, $\cos (\theta \pm \pi) = -\cos \theta$.
5. The graph of $y = \cos \theta$ can be obtained by shifting the graph of $y = \sin \theta$ to the left $\pi/2$ units. Thus $\cos \theta = \sin (\theta + \pi/2)$ (see Figure 6.71).

sinusoidal   Any curve that has the same shape as the sine function is called **sinusoidal.** Thus the graph of the cosine function is sinusoidal, since it is merely the sine

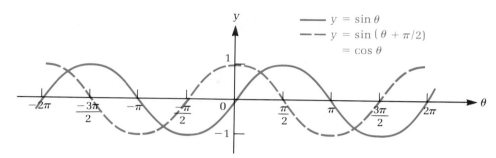

FIGURE 6.71

function shifted 90° (or $\pi/2$ radians). One period of a sinuosidal function is
cycle   called a **cycle.**
        We now graph a few variations and sums of the sine and cosine functions.
Thus far we have used Greek letters such as θ, α, β, and γ to represent angles.
However, any variable is feasible, particularly since the trigonometric functions
can be considered as functions of real numbers. In the next set of examples, we
use the variable x. In other examples, we may use t, for time. This is appropriate,
for instance, when the motion of a particle is a sinusoidal curve and the position
of the particle is a function of time.

Example 6.38   Graph the equation $y = 2 + \sin x$.

Solution: The graph of this equation is a shifting of the graph of $y = \sin x$ upward two units (see Figure 6.72).

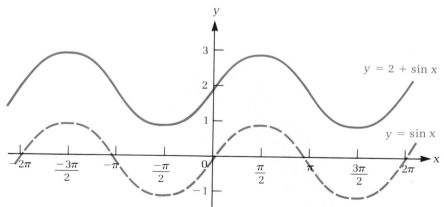

FIGURE 6.72

Example 6.39   Compare the graphs of $y = \sin x$, $y = \sin (x + \pi/3)$, and $y = \sin (x - \pi/3)$.

Solution: Each function has a period of 2π. The effect of adding or subtracting $\pi/3$ to the variable x is to shift the graph of $y = \sin x$ to the left or to the right, respectively. The value of $\sin (x + \pi/3)$ is zero when $x + \pi/3$ is zero, that is, for $x = -\pi/3$; thus there is a shift to the left. The value of $\sin (x - \pi/3)$ is zero when $x - \pi/3$ is zero, that is, for $x = +\pi/3$; thus there is a shift to the right of $\pi/3$ units (see Figure 6.73).

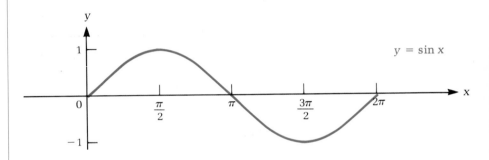

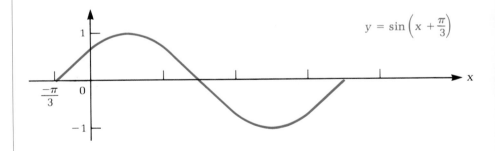

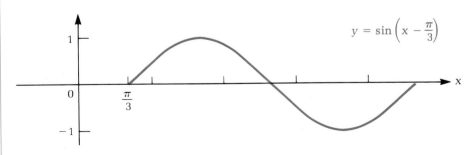

FIGURE 6.73

Example 6.40

Graph the function $y = -1 + \cos(x - \pi/4)$ for one cycle.

Solution: This function represents a downward shift of the basic cosine function by 1 unit and a horizontal shift to the right by $\pi/4$ unit, as shown in Figure 6.74.

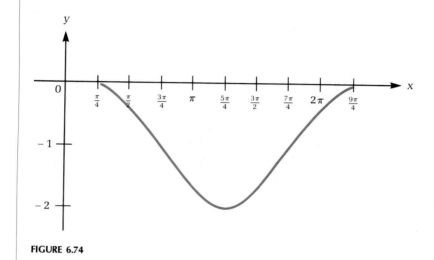

FIGURE 6.74

Tangent Function

Plotting the points for $y = \tan\theta$ given in Table 6.3, we obtain the graph shown in Figure 6.75.

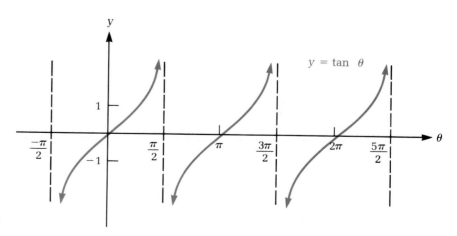

FIGURE 6.75

Since the tangent of a number is the quotient of its sine and cosine values, the tangent function is undefined when the cosine value is zero. But the cosine function is zero for all odd multiples of $\pi/2$; thus as θ approaches $\pi/2$, the cosine value becomes very small and, accordingly, the tangent value becomes very large. We indicate this with a vertical asymptote to the curve wherever the cosine is zero.

Properties of Tangent Function

1. The domain is the set of real numbers excluding odd multiples of $\pi/2$; that is, $\pi/2 + k\pi$, where k is any integer.
2. The range is the set of all real numbers.
3. The function is periodic with a period of π.
4. The function is symmetric with respect to the origin. Thus $\tan(-\theta) = -\tan\theta$.
5. If we add or subtract π from any value θ, the resulting tangent value is equal to the given tangent value. That is, $\tan(\theta \pm \pi) = \tan\theta$.
6. The tangent function is increasing on the interval $\pi/2 + k\pi < \theta < \pi/2 + (k + 1)\pi$.
7. For the equation $y = \tan k\theta$, the period is π/k.

Example 6.41   Graph $y = -\tan 3x$ for three periods.

Solution: Since $k = 3$, the period is $\pi/3$, and a complete cycle can be graphed between $-\pi/6$ and $\pi/6$. The negative sign produces a reflection across the x axis. Note that a table of values is not needed (see Figure 6.76).

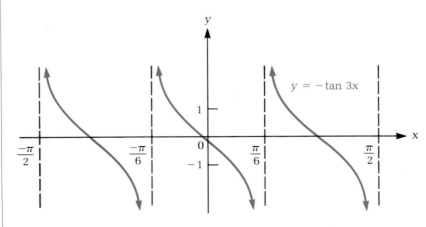

FIGURE 6.76

Reciprocal Functions

The graphs of these functions can easily be sketched by using the basic trigono-metric function graphs. Recall that the cosecant, secant, and cotangent functions are the reciprocals of the sine, cosine, and tangent functions, respectively. A function and its reciprocal are positive together and negative together. They also have the same period and pattern of symmetry. Whenever a function is increasing, its reciprocal is decreasing. Whenever the value of a function is zero, its reciprocal function is undefined. (Why?)

Since the tangent function has a period of π, the cotangent function has a period of π. The tangent function is an increasing function, and the cotangent function is a decreasing function. The tangent function has a value of zero at integral multiples of π; therefore, the cotangent function is undefined at these points (see Figure 6.77).

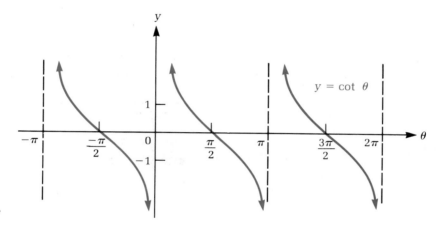

FIGURE 6.77

Example 6.42

Graph the cosecant function and describe its properties.

Solution: We first make a sketch of the sine curve. Wherever the sine function passes through zero, draw a vertical asymptote line. Since the reciprocal of 1 is 1, mark those points where the sine function is $+1$ or -1 as points of the cosecant function. Where the sine function approaches zero from the positive side, the cosecant function increases without limit, approaching the asymptote line. Where the sine function approaches zero from the negative side, the cosecant function decreases without limit, again approaching the asymptote line (see Figure 6.78).

The cosecant function is periodic with a period of 2π; therefore, $\csc (x + 2\pi) = \csc x$. Its domain is the set of all real numbers except for multiples of π. The range consists of all real numbers greater than or equal to 1 or less than or equal to -1. The graph is symmetric with respect to the origin; thus $\csc (-x) = -\csc x$.

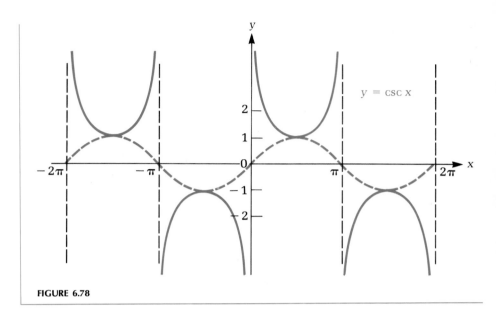

FIGURE 6.78

EXERCISES

In Exercises 1 to 24, sketch the curves for at least one period.

1. $y = \sin x + 3$

2. $y = \sin x - \frac{1}{2}$

3. $y = \cos x + \frac{3}{2}$

4. $y = \cos x - 2$

5. $y = \tan\left(x - \frac{\pi}{2}\right)$

6. $y = \tan\left(x + \frac{\pi}{4}\right)$

7. $y = \sin(-x)$

8. $y = \cos(-x)$

9. $y = \cot x$

10. $y = \sin\left(\frac{3\pi}{2} - x\right)$

11. $y = \sin\left(x + \frac{\pi}{4}\right) + 1$

12. $y = \cos\left(x - \frac{\pi}{2}\right) - 1$

13. $y = \sin\left(x + \frac{\pi}{3}\right)$

14. $y = \sin\left(x - \frac{\pi}{2}\right)$

15. $y = \cos\left(x - \frac{\pi}{2}\right)$

16. $y = \sin(x + \pi)$

17. $y = \cos\left(x + \frac{3\pi}{4}\right)$

18. $y = \cos\left(x - \frac{4\pi}{3}\right)$

19. $y = |\cos x|$

20. $y = |(\sin x) + 1|$

21. $y = \tan\left(x + \frac{\pi}{2}\right)$

22. $y = \tan\left(x - \frac{\pi}{3}\right)$

23. $y = \cot(x - \pi)$

24. $y = \cot\left(x + \frac{\pi}{4}\right)$

25. Graph $y = \sec x$ by using a sketch of $y = \cos x$ for values of x from -2π to 2π. Describe the properties of $y = \sec x$, including domain, range, period, symmetry, and asymptotes.

In Exercises 26 to 29, sketch the curve for one period.

26. $y = (\sec x) + 1$ **27.** $y = (\csc x) - 1$ **28.** $y = \csc\left(x - \dfrac{\pi}{6}\right)$ **29.** $y = \sec\left(x + \dfrac{\pi}{6}\right)$

6.8

GRAPHS OF SINUSOIDAL FUNCTIONS

In the previous section we graphed the basic trigonometric functions and vertical and horizontal shifts of the graphs. In this section we graph functions of the type $y = A \sin(Bx + C)$, called **sinusoidal functions.** We shall see that the effect of A is to cause a vertical expansion or contraction and that B affects the horizontal expansion and contraction of the basic function.

Definition 6.8.1

For any equation of the form $y = A \sin \theta$, the absolute value of the coefficient A determines the *amplitude* of the function. In other words, $|A|$ represents the maximum y coordinate value. The zeros of the function are the same as those of $y = \sin \theta$.

Example 6.43

Graph the equation $y = 2 \sin x$.

Solution: The coefficient 2 produces a vertical stretching of the graph of $y = \sin x$ (see Figure 6.79).

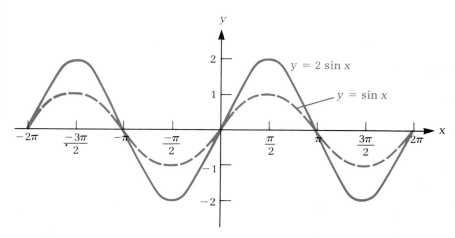

FIGURE 6.79

Example 6.44

Graph the equation $y = -\frac{1}{2}\sin x$.

Solution: The effect of the negative sign is to cause a reflection of the graph of $y = \sin x$ with respect to the x axis. The effect of the coefficient $\frac{1}{2}$ is to produce a vertical shrinking of the graph (see Figure 6.80).

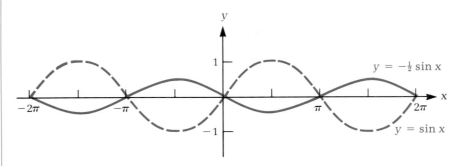

FIGURE 6.80

Example 6.45

Graph the following equations:

(a) $y = \sin 2x$

(b) $y = \sin (x/2)$.

Solution:

(a) The graph of this function is a horizontal shrinking of the graph of $y = \sin x$. The effect of the coefficient 2 is to halve the period, as shown in the graph in Figure 6.81.

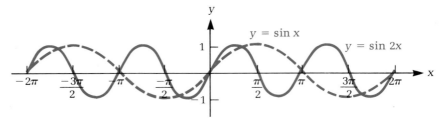

FIGURE 6.81

(b) The graph of this function is a horizontal stretching of the graph of $y = \sin x$. For each value of x, we are plotting the $\sin (x/2)$; thus the period of the function is doubled, as shown in Figure 6.82.

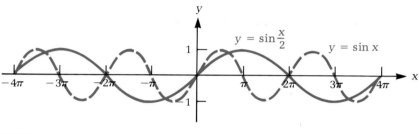

FIGURE 6.82

Example 6.46

Graph the function $y = 3 \cos 2\theta$ for one complete period.

Solution: The amplitude of this function is 3. We observe that when $\theta = \pi$, $y = 3 \cos 2\pi$. Therefore a full period is completed as θ varies from 0 to π. Thus by changing the coefficient of θ, we have altered the period of the function, as shown in Figure 6.83.

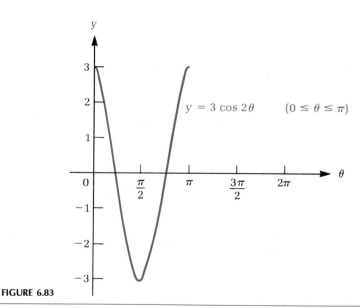

FIGURE 6.83

Definition 6.8.2

In equations of the form $y = A \sin B\theta$ or $y = A \cos B\theta$, where A and B are constants, the period of the function is given by

$$P = \frac{2\pi}{|B|}.$$

Since $|B|\theta$ varies from 0 to 2π, angle θ varies from 0 to $2\pi/|B|$.

So far, we have explored the effect of multiplying a sinusoidal function, $y = \sin x$ or $y = \cos x$, by a constant A, which produced a change in the amplitude, and the effect of multiplying the variable x of the function by a constant B, which produced a change in the period. Let us now examine the effect of adding a constant to the variable—specifically, in functions of the type $y = A \sin(Bx + C)$, that is, $y = A \sin B(x + C/B)$.

Example 6.47

Graph one cycle of the function $y = \frac{1}{2} \sin (2x + \pi/4)$.

Solution: The amplitude is $\frac{1}{2}$, and the period is $2\pi/2 = \pi$. Since the sine function is equal to zero at 0, π, and 2π, the graph of this function crosses the x axis when the sum $2x + \pi/4$ is equal to 0, π, or 2π. That happens when x equals $-\pi/8$, $3\pi/8$, and $7\pi/8$. Thus the graph is shifted $\pi/8$ unit to the left (see Figure 6.84). The value $x = -\pi/8$ is called the *phase shift* of this sine function.

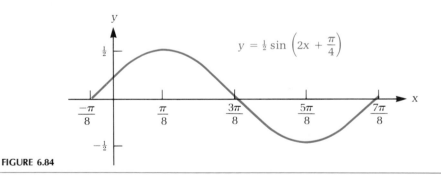

FIGURE 6.84

To summarize, a function of the form $y = A \sin (Bx + C)$ has an amplitude of $|A|$, a period of $2\pi/|B|$, and zero values when $x = -C/B$. The value of x for which the argument of the sine function is zero is the phase shift. A similar observation holds true for $y = A \cos (Bx + C)$.

Example 6.48

If the displacement s of a moving body is related to the time t by the equation $s = A \sin (Bt + C)$, then the motion is defined as *simple harmonic motion*. Many physical phenomena, from pendulum motion to the vibration of atoms, are *simple harmonic motion*

characterized by *simple harmonic motion.* Graph one cycle of the function $s = 3 \sin(2t + \pi/2)$.

Solution: See Figure 6.85. Then

$$|A| = 3, \qquad \frac{2\pi}{|B|} = \frac{2\pi}{|2|} = \pi, \qquad x = -\frac{C}{B} = -\frac{\pi}{4}.$$

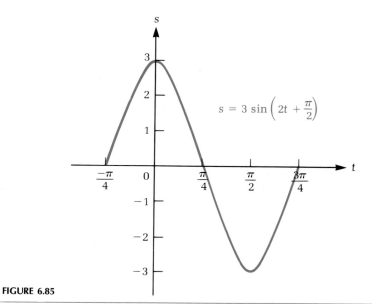

FIGURE 6.85

Let us now turn our attention to the graph of a function that is the sum of two sinusoidal functions. One method of graphing such a function is to construct a table of values and then plot the points. However, a simpler technique is a method called the **addition of ordinates.** In this method, we graph the two functions separately but on the same coordinate axes. We then add the ordinates (second coordinates, or y values) graphically.

addition of ordinates

Example 6.49

Use the method of addition of ordinates to graph the function

$$y = \cos x + \sin 2x.$$

Solution: We graph $y = \cos x$ and $y = \sin 2x$ separately, but we use the same axes. Then we graphically add a few ordinates to obtain points representing the sum of the two functions (see Figure 6.86). Pay particular attention to those

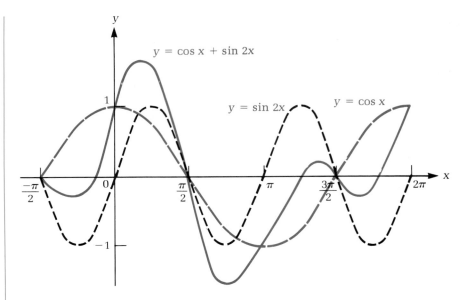

FIGURE 6.86

points where the individual functions are at zero or an extreme value. A compass or a pair of dividers is most helpful.

EXERCISES

In Exercises 1 to 24, sketch the graph of the function for one cycle. State the amplitude and period of the function.

1. $y = 3 \sin x$
2. $y = \frac{1}{4} \sin x$
3. $y = -4 \sin x$

4. $y = -2 \sin x$
5. $y = \cos \dfrac{x}{4}$
6. $y = \sin \dfrac{x}{3}$

7. $y = -2 \cos x$
8. $y = -3 \cos x$
9. $y = \sin \dfrac{3x}{2}$

10. $y = \cos \dfrac{5x}{3}$
11. $y = 4 \cos \dfrac{\pi x}{2}$
12. $y = 2 \cos \dfrac{5\pi x}{2}$

13. $y = \frac{1}{3} \sin 2\beta$
14. $y = \frac{1}{2} \cos 4x$
15. $y = \dfrac{1}{2} \cos \left(\dfrac{x}{2} - \dfrac{\pi}{3} \right)$

16. $y = 2 \sin (3x + \pi)$
17. $y = -3 \sin \left(x + \dfrac{\pi}{6} \right)$
18. $y = 3 \cos \left(2x + \dfrac{\pi}{2} \right)$

19. $y = 2 \sin x + \sin 2x$
20. $y = 2 \cos x + \cos 2x$
21. $y = \sin \theta + \cos \theta$

22. $y = 2 \sin \theta + \cos 2\theta$
23. $y = 3 \cos x - \sin 2x$
24. $y = 3 \sin x - \cos 2x$

KEY WORDS

trigonometry   *sine*   *arc*
angle   *cosine*   *right triangle*
ray   *tangent*   *angle of elevation*
initial side   *cosecant*   *angle of depression*
terminal side   *secant*   *course*
vertex   *cotangent*   *bearing*
degree   *reciprocal function*   *periodic*
minute   *complementary angles*   *period*
second   *cofunction*   *sinusoidal*
standard position   *related angle*   *cycle*
quadrant   *radius*   *amplitude*
quadrantal angle   *unit circle*   *phase shift*
abscissa   *circular function*   *simple harmonic motion*
ordinate   *radian*   *addition of ordinates*

KEY FORMULAS

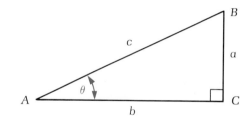

$$\sin \theta = \frac{a}{c} \qquad \tan \theta = \frac{a}{b} \qquad \sec \theta = \frac{c}{b}$$

$$\cos \theta = \frac{b}{c} \qquad \cot \theta = \frac{b}{a} \qquad \csc \theta = \frac{c}{a}$$

45°, 45°, 90° triangle   30°, 60°, 90° triangle

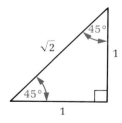   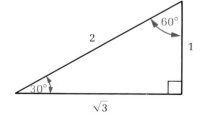

$$1° = 60' = 3600''$$

$$1° = \frac{\pi}{180} \text{ radian} \qquad 1 \text{ radian} = \frac{180}{\pi} \text{ degrees}$$

$$\text{Arc length } s = r \cdot \theta \qquad \theta \text{ is in radians}$$

EXERCISES FOR REVIEW

1. With the aid of a protractor, construct an angle of $-135°$ in standard position on a coordinate system.
2. What is the related angle of $-135°$?
3. Determine the coordinates of a point P on the terminal side of the angle $-135°$ drawn on the unit circle.
4. Find the values of all six trigonometric functions of the angle $-135°$.
5–8. Repeat Exercises 1 to 4 for the angle $330°$.
9. If $A + B = 360°$ and $A = 114°36'17''$, find B.
10. If $A + B = 360°$ and $A = -46°22'$, find B.
11. Let $P(5, -7)$ be a point on the terminal side of an angle in standard position. Find the value of r and the six trigonometric functions.
12. Let $P(-2, -4)$ be a point on the terminal side of an angle in standard position. Find the value of r and the six trigonometric functions.
13. If $\sin \theta = 3/4$ and θ is in quadrant II, find the other five functions of the angle.
14. If $\tan \theta = \sqrt{3}$ and θ is in quadrant III, find the other five functions of the angle.
15. Complete the table with the exact values of the functions.

θ	0°	90°	150°	180°	240°	270°	315°
$\sin \theta$							
$\cos \theta$							
$\tan \theta$							

16. Show that the following is true:

$$\sin^2 30° + \cos^2 30° = 1.$$

17. Show that the following is true:

$$\sin 90° = 2 \sin 45° \cos 45°.$$

18. Find a numerical value for $2 \sin 60° \cos 60°$.
19. Find a numerical value for

$$\cos 30° \cos 60° - \sin 30° \sin 60°.$$

20. Show that the following is true:

$$1 + \cot^2 210° = \csc^2 210°.$$

21. For the following angles, find the related angle and the values of the six basic trigonometric functions: **(a)** $225°$ **(b)** $-135°$ **(c)** $600°$.
22. Convert $90°$ to radian measure, using π.
23. Convert $135°$ to radian measure, using π.
24. Convert $300°$ to radian measure, using π.
25. Convert $5\pi/12$ to degree measure.
26. Convert $-\pi/4$ to degree measure.
27. Convert 0.42 radian to degree measure, to the nearest degree.
28. Find the trigonometric value of the angle given in radian measure: $\sin 0.9483$.

29. Find the length of the arc on a circle having a radius of 8 centimeters that subtends a central angle of 60°.

30. A pendulum swings through an arc of 80°, describing an arc which is 50 centimeters long. How long is the pendulum?

Solve the right triangles, ABC, in Exercises 31 to 35. The right angle is at C.

31. $A = 49°30'$, $c = 425$ **32.** $B = 21°10'$, $a = 300$ **33.** $a = 6000$, $b = 8000$

34. $b = 0.043$, $c = 0.07$ **35.** $a = 0.707$, $A = 65°20'$

36. What is the angle of elevation of the sun if an object 40 meters high casts a shadow 60 meters long?

37. From a point 1150 meters from the foot of a cliff, the angle of elevation of the top is 20°. How high is the cliff?

38. How high is a kite when the length of the string is 600 feet and the angle of elevation is 30°?

39. Two buoys are directly in line with a cliff that stands 282 feet above the water. Their angles of depression from the top of the cliff are 42°30' and 37°20'. How far apart are the buoys, and what are the horizontal distances from the cliff?

For each of the functions in Exercises 40 to 51, determine the period of the function and graph at least one cycle.

40. $y = -2 \sin 2\theta$ **41.** $y = -3 \sin 3x$ **42.** $y = \dfrac{1}{2} \cos \dfrac{\theta}{2}$ **43.** $y = \dfrac{1}{3} \cos \dfrac{x}{3}$

44. $y = 3 \sin \dfrac{\theta}{3}$ **45.** $y = 4 \tan 2x$ **46.** $y = \cot 3\theta$ **47.** $y = 2 \sin \left(x + \dfrac{\pi}{4} \right)$

48. $y = -\cos \left(x - \dfrac{\pi}{4} \right)$ **49.** $y = \sin x + \cos x$ **50.** $y = \csc 2x$ **51.** $y = 3 \sec 2x$

CHAPTER TEST

1. Determine the coordinates of a point P on the terminal side of an angle 300° drawn on the unit circle.

2. Find the values of the six trigonometric functions of the angle $\theta = -30°$.

3. Let $P(-2, 5)$ be a point on the terminal side of an angle in standard position. Find the value of r and the six trigonometric functions.

4. If $\cos \theta = \frac{2}{3}$ and θ is in quadrant IV, find the other five trigonometric functions of the angle.

5. Find a numerical value for

$$2 \sin 60° \cos 60°.$$

6. Find the values of the six basic trigonometric functions of the angle 630°.

7. Convert the following angles from degree to radian measure, using π: **(a)** 175°, **(b)** 340°.

8. Convert the following angles from radian to degree measure: **(a)** $5\pi/18$, **(b)** $-\pi/12$.

9. Find the angle subtended by an arc 12 centimeters on a circle of radius 7.4 centimeters.

10. Determine the period and amplitude of the function, and graph one cycle: **(a)** $y = 2 \cos 4\theta$, **(b)** $y = -\sin(x + \pi/2)$.

11. Solve the right triangle ABC where $A = 41.6°$ and $c = 160$.

12. Find the angle of elevation of the sun if a tree 15 meters high casts a shadow 25 meters long.

TRIGONOMETRIC EQUATIONS, IDENTITIES, AND INVERSES

7

INTRODUCTION

In previous chapters we learned to solve algebraic, exponential, and logarithmic equations. In this chapter we study two types of trigonometric equations. Those of the first type are called *identity equations*, and as with algebraic identities, these statements are true for all permissible replacements of the variable, that is, for any angle for which the function is defined. Examples of identity equations are $\sec^2 \theta = \tan^2 \theta + 1$ and

$$\tan (x - y) = \frac{\tan x - \tan y}{1 + \tan x \tan y}.$$

The second type of trigonometric equation is called a *conditional equation* because it is true only for specific replacements of the variable. Examples of conditional equations are $4 \cos^2 x = 1$ and $\sin 2x + \sin x = 0$.

Identities are often needed to aid in the solution of conditional equations. Thus in Section 7.1 we learn to prove identity equations, and in Section 7.2 we solve conditional equations. As we study more complex trigonometric functions later in this chapter, we also solve conditional equations using those functions.

7.1

RELATIONS AND IDENTITIES OF TRIGONOMETRY

From the definition of the six basic trigonometric functions, it is possible to identify several relations among these functions. Each function involves a combination of x, y, and r, and we can combine the functions in several ways. The set of basic relations is called the *fundamental trigonometric identities*. There are three general groups of trigonometric relations: the *reciprocal relations*, the *quotient relations*, and the *Pythagorean relations*.

Reciprocal Relations

We have already observed that it is possible to determine $\csc \theta$, $\sec \theta$, and $\cot \theta$ by taking the reciprocals of $\sin \theta$, $\cos \theta$, and $\tan \theta$, respectively. Since $\sin \theta = y/r$ and $\csc \theta = r/y$, then

$$\sin \theta = \frac{1}{\csc \theta} \quad \text{or} \quad \csc \theta = \frac{1}{\sin \theta}. \tag{7.1}$$

In a like manner,

$$\cos \theta = \frac{1}{\sec \theta} \quad \text{or} \quad \sec \theta = \frac{1}{\cos \theta}, \tag{7.2}$$

and

$$\tan \theta = \frac{1}{\cot \theta} \quad \text{or} \quad \cot \theta = \frac{1}{\tan \theta}. \tag{7.3}$$

From these three statements it follows that

$$\sin \theta \cdot \csc \theta = 1 \tag{7.1a}$$

$$\cos \theta \cdot \sec \theta = 1 \tag{7.2a}$$

$$\tan \theta \cdot \cot \theta = 1 \tag{7.3a}$$

Quotient Relations

Starting again with the basic definitions of $\sin \theta = y/r$ and $\cos \theta = x/r$ and then taking their quotient, we find

$$\frac{\sin \theta}{\cos \theta} = \frac{y/r}{x/r} = \frac{y}{x} = \tan \theta. \tag{7.4}$$

Similarly,

$$\frac{\cos \theta}{\sin \theta} = \frac{x/r}{y/r} = \frac{x}{y} = \cot \theta. \tag{7.5}$$

Pythagorean Relations

The Pythagorean theorem states that the sum of the squares of the legs of a right triangle is equal to the square of the hypotenuse. In our work in trigonometry, this translates to

$$x^2 + y^2 = r^2.$$

We take this fundamental relationship and divide through by r^2 (a number which is always positive) to obtain

$$\frac{x^2}{r^2} + \frac{y^2}{r^2} = \frac{r^2}{r^2},$$

or

$$\left(\frac{x}{r}\right)^2 + \left(\frac{y}{r}\right)^2 = 1,$$

which trigonometrically is equivalent to

$$\cos^2 \theta + \sin^2 \theta = 1. \tag{7.6}$$

This statement is true for all values of θ and is known as the *fundamental trigonometric identity*.

If we divide this fundamental identity by $\cos^2 \theta$, the result is

$$1 + \tan^2 \theta = \sec^2 \theta. \tag{7.7}$$

However, $\cos \theta = 0$ for all odd multiples of $\pi/2$, and the tangent function is undefined at those points. Thus this identity is restricted at $\theta \neq (2k + 1)\pi/2$, where k is an integer.

If we divide the fundamental identity by $\sin^2 \theta$, the result is

$$\cot^2 \theta + 1 = \csc^2 \theta. \tag{7.8}$$

However, $\sin \theta = 0$ for all integral multiples of π, and the cotangent function is undefined at those points. Therefore this identity is restricted to $\theta \neq k\pi$, where k is an integer.

Note that we may use the addition and multiplication principles of equality to rewrite statements (7.1) to (7.8) in alternate forms. For instance,

$$\cos^2 \theta + \sin^2 \theta = 1$$

is equivalent to each of the following statements:

(a) $\sin^2 \theta + \cos^2 \theta = 1$
(b) $\cos^2 \theta = 1 - \sin^2 \theta$
(c) $\sin^2 \theta = 1 - \cos^2 \theta$
(d) $\cos \theta = \pm\sqrt{1 - \sin^2 \theta}$
(e) $\sin \theta = \pm\sqrt{1 - \cos^2 \theta}$

Example 7.1

Verify the Pythagorean relations for $\theta = 45°$.

Solution:

(a) $$\sin^2 \theta + \cos^2 \theta = 1$$
$$(\sin 45°)^2 + (\cos 45°)^2 = 1$$
$$\left(\frac{\sqrt{2}}{2}\right)^2 + \left(\frac{\sqrt{2}}{2}\right)^2 = 1$$
$$\frac{2}{4} + \frac{2}{4} = 1$$

(b) $\quad 1 + \tan^2\theta = \sec^2\theta$

$$1 + (\tan 45°)^2 = (\sec 45°)^2$$

$$1 + 1^2 = \left(\frac{2}{\sqrt{2}}\right)^2$$

$$2 = \frac{4}{2}$$

(c) $\quad \cot^2\theta + 1 = \csc^2\theta$

$$(\cot 45°)^2 + 1 = (\csc 45°)^2$$

$$1^2 + 1 = \left(\frac{2}{\sqrt{2}}\right)^2$$

$$2 = \frac{4}{2}$$

Example 7.2

Assuming that x is an angle in the first quadrant and $\tan x = 3$, determine the value of

$$(\sec^2 x - 1)(\cot x).$$

Solution:

$$(\sec^2 x - 1)(\cot x) = (\tan^2 x + 1 - 1)(\cot x) \qquad \text{(substituting for } \sec^2 x\text{)}$$

$$= (\tan^2 x)\left(\frac{1}{\tan x}\right) \qquad \text{(substituting for } \cot x\text{)}$$

$$= \tan x \qquad \text{(algebraic simplification)}$$

$$= 3. \qquad \text{(substituting for } \tan x\text{)}$$

Many complex trigonometric expressions can be simplified by using the fundamental trigonometric relations. This procedure is illustrated in Example 7.3.

Example 7.3

Use formulas (7.1) to (7.8) to prove the following statements by transforming the more complicated right side until it can be simplified to the basic function on the left side:

(a) $\quad \cos x = \dfrac{\sin x}{\tan x}$ $\qquad$ **(b)** $\quad \csc x = \dfrac{\cot x}{\cos x}$

(c) $\quad \sin x = \dfrac{\tan x}{\sqrt{1 + \tan^2 x}}, \sec x > 0$

Solution: In each case, we transform the right-hand side to the form on the left.

(a) $\cos x = \dfrac{\sin x}{\tan x}$

$\qquad = \dfrac{\sin x}{\dfrac{\sin x}{\cos x}}$ $\big[$using (7.4)$\big]$

$\qquad = \sin x \cdot \dfrac{\cos x}{\sin x}$ (inverting and multiplying)

$\qquad = \cos x$ (simplifying)

(b) $\csc x = \dfrac{\cot x}{\cos x}$

$\qquad = \dfrac{\dfrac{\cos x}{\sin x}}{\cos x}$ $\big[$using (7.5)$\big]$

$\qquad = \dfrac{\cos x}{\sin x} \cdot \dfrac{1}{\cos x}$ (inverting and multiplying)

$\qquad = \dfrac{1}{\sin x}$ (simplifying)

$\qquad = \csc x$ $\big[$using (7.1)$\big]$

(c) $\sin x = \dfrac{\tan x}{\sqrt{1 + \tan^2 x}}$

$\qquad = \dfrac{\tan x}{\sqrt{\sec^2 x}}$ $\big[$using (7.7)$\big]$

$\qquad = \dfrac{\tan x}{\sec x}$ (taking the square root)

$\qquad = \dfrac{\dfrac{\sin x}{\cos x}}{\dfrac{1}{\cos x}}$ $\big[$using (7.4) and (7.2)$\big]$

$\qquad = \dfrac{\sin x}{\cos x} \cdot \cos x$ (inverting and multiplying)

$\qquad = \sin x$ (simplifying)

Example 7.4 Write the following expressions as a single fraction in terms of sine and/or cosine functions, and simplify the result.

(a) $(\sin A)(\cot^2 A)(\sec^2 A)$ **(b)** $(\sec^2 A - 1)(\cos^2 A)$

Solution:

(a) $(\sin A)(\cot^2 A)(\sec^2 A) = \sin A \cdot \dfrac{\cos^2 A}{\sin^2 A} \cdot \dfrac{1}{\cos^2 A}$

$$= \frac{1}{\sin A}$$

(b) $(\sec^2 A - 1)(\cos^2 A) = \left(\dfrac{1}{\cos^2 A} - 1\right)(\cos^2 A)$

$$= \frac{\cos^2 A}{\cos^2 A} - \cos^2 A$$

$$= 1 - \cos^2 A$$

$$= \sin^2 A.$$

We are now prepared to begin proving identities. Any identity can be proved by three different procedures: (1) the left-hand member may be reduced to the right-hand member, (2) the right-hand member may be reduced to the left-hand member, or (3) each side may be separately reduced to the same form as long as the steps are reversible. The series of steps used in the simplification process is by no means unique. Therefore, no matter which side you start on, there may be a choice of next steps. Naturally, some choices result in a shorter proof than others. Practice alone helps. However, here are some suggestions:

1. Memorize the eight fundamental relations, (7.1) to (7.8), so that you recognize one when part of it appears in the equation.

2. It is usually better to start with the side of the equation that appears to be more complicated.

3. Look for opportunities to use the algebraic manipulations. Be cautious; just as in algebra, division by zero is undefined.

4. If both sides of the equation appear to be equally complicated express each function in terms of sines or cosines and simplify.

5. Keep in mind where you are heading and what steps might get you there.

Example 7.5 Prove the identity

$$\frac{\cos A}{1 + \sin A} + \frac{\cos A}{1 - \sin A} = 2 \sec A.$$

Solution: Here we start with the more complicated left-hand side:

$$\frac{(\cos A)(1 - \sin A) + (\cos A)(1 + \sin A)}{(1 + \sin A)(1 - \sin A)} \qquad \text{(combining terms using a common denominator)}$$

$$= \frac{\cos A - \sin A \cos A + \cos A + \sin A \cos A}{1 - \sin^2 A} \qquad \text{(multiplying)}$$

$$= \frac{2 \cos A}{1 - \sin^2 A} \qquad \text{(adding)}$$

$$= \frac{2 \cos A}{\cos^2 A} \qquad \text{[using (7.6)]}$$

$$= \frac{2}{\cos A} \qquad \text{(simplifying)}$$

$$= 2 \sec A. \qquad \text{[using (7.2)]}$$

In Example 7.6, we have an identity in which there is no clear indication of which side is more complicated. Since the right-hand side is expressed in sines and cosines, we start with the left-hand side and express the given functions in terms of sines and cosines.

Example 7.6 | Prove the identity

$$\frac{1}{\csc \theta + \cot \theta} = \frac{1 - \cos \theta}{\sin \theta}.$$

Solution:

$$\frac{1}{\csc \theta + \cot \theta} = \frac{1}{\dfrac{1}{\sin \theta} + \dfrac{\cos \theta}{\sin \theta}} \qquad \text{[using (7.1) and (7.5)]}$$

$$= \frac{1}{\dfrac{1 + \cos \theta}{\sin \theta}} = \frac{\sin \theta}{1 + \cos \theta}$$

$$= \frac{\sin \theta}{1 + \cos \theta} \cdot \frac{1 - \cos \theta}{1 - \cos \theta} = \frac{(\sin \theta)(1 - \cos \theta)}{1 - \cos^2 \theta}$$

$$= \frac{(\sin \theta)(1 - \cos \theta)}{\sin^2 \theta} \qquad \text{[using (7.6)]}$$

$$= \frac{1 - \cos \theta}{\sin \theta}. \qquad \text{(canceling)}$$

Remember that the steps for proving the identity in Example 7.6 are by no means unique. Here is an alternate solution.

Example 7.7 Prove the identity

$$\frac{1}{\csc\theta + \cot\theta} = \frac{1 - \cos\theta}{\sin\theta}.$$

Solution: Here we start with the left-hand side and multiply by the conjugate of the denominator:

$$\frac{1}{\csc\theta + \cot\theta} = \frac{1}{\csc\theta + \cot\theta} \cdot \frac{\csc\theta - \cot\theta}{\csc\theta - \cot\theta}$$

$$= \frac{\csc\theta - \cot\theta}{\csc^2\theta - \cot^2\theta}$$

$$= \frac{\csc\theta - \cot\theta}{\cot^2\theta + 1 - \cot^2\theta} \qquad \left[\text{using (7.8)}\right]$$

$$= \frac{\csc\theta - \cot\theta}{1}$$

$$= \frac{1}{\sin\theta} - \frac{\cos\theta}{\sin\theta} \qquad \left[\text{using (7.1) and (7.5)}\right]$$

$$= \frac{1 - \cos\theta}{\sin\theta} \qquad \text{(combining terms)}$$

Example 7.8 Prove that $\csc^4 A - \cot^4 A = \csc^2 A + \cot^2 A$.

Solution: We start with the left-hand side and factor the difference of squares:

$$\csc^4\theta - \cot^4\theta = (\csc^2 A + \cot^2 A)(\csc^2 A - \cot^2 A) \qquad \text{(factoring)}$$
$$= (\csc^2 A + \cot^2 A)(\cot^2 A + 1 - \cot^2 A) \qquad \left[\text{using (7.8)}\right]$$
$$= (\csc^2 A + \cot^2 A)(1) \qquad \text{(algebra)}$$
$$= \csc^2 A + \cot^2 A$$

Cofunction Identities

In Chapter 6, in our study of right-triangle trigonometry, we pointed out that cofunctions of complementary angles are equal. To review, if θ is an acute angle, then θ and $90° - \theta$ are complementary angles. Two trigonometric functions that have equal values for complementary angles are called **cofunctions.** The cosine function is the cofunction of the sine function. Similarly, the tangent and cotangent are cofunctions, as are the secant and cosecant. This relationship can be generalized by stating

$$f(\theta) = \text{cof}\,(90° - \theta), \tag{7.9}$$

where f is the trigonometric function and cof is the corresponding cofunction. Specifically,

$$\sin \theta = \cos (90° - \theta) \qquad \cos \theta = \sin (90° - \theta)$$
$$\tan \theta = \cot (90° - \theta) \qquad \cot \theta = \tan (90° - \theta)$$
$$\sec \theta = \csc (90° - \theta) \qquad \csc \theta = \sec (90° - \theta).$$

Other identities involving trigonometric functions of $\theta + 90°$ or $\theta - 90°$ can be readily developed by referring to the unit circle in Figure 7.1.

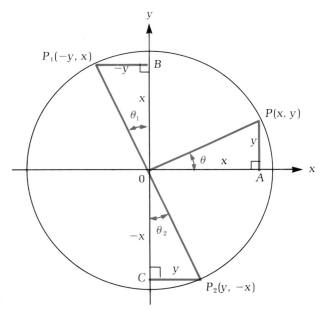

FIGURE 7.1

Let P be determined by an angle θ; then $\cos \theta = x$ and $\sin \theta = y$. Let P_1 be determined by an angle $\theta + 90°$. Since angles POP_1 and AOB are right angles, the measure of angle θ_1 equals the measure of angle θ. With $\theta = \theta_1$ and equal hypotenuse values of 1, right triangles PAO and P_1BO are congruent, and their corresponding sides are equal in measure. Specifically, $d_{PA} = d_{P_1B}$ and $d_{OA} = d_{OB}$. Since the unit circle is superimposed on the coordinate system, we may denote P as (x, y) and P_1 as $(-y, x)$. Recalling that the first coordinate tells us the cosine of the angle, whereas the second tells us the sine, we find

$$\sin (\theta + 90°) = \cos \theta \qquad (7.10)$$

and

$$\cos (\theta + 90°) = -\sin \theta. \qquad (7.11)$$

Since the tangent is the ratio of the sine to the cosine, we have

$$\tan (\theta + 90°) = -\cot \theta. \qquad (7.12)$$

Let the point P_2 be determined by an angle $\theta - 90°$. By a similar argument, we can show that right triangles PAO and P_2CO are congruent, and the following identities hold:

$$\sin (\theta - 90°) = -\cos \theta \qquad (7.13)$$

$$\cos (\theta - 90°) = \sin \theta \qquad (7.14)$$

$$\tan (\theta - 90°) = -\cot \theta \qquad (7.15)$$

Negative-Angle Identities

We can arrive at statements for negative-angle identities geometrically. We have defined a positive angle θ as one which rotates in a counterclockwise direction and a negative angle θ as one which rotates in a clockwise direction. Since θ can be an acute angle or an obtuse angle, we consider both possibilities (see Figure 7.2). Note that the point $P(x, y)$ on the terminal side of θ has a reflection point $P'(x, -y)$ across the x axis; P and P' have the same x coordinates but opposite y coordinates. Observe that $\sin \theta = y/r$ and

$$\sin (-\theta) = -\frac{y}{r} = -\sin \theta. \qquad (7.16)$$

Similarly, $\cos \theta = x/r$ and

$$\cos (-\theta) = \frac{x}{r} = \cos \theta. \qquad (7.17)$$

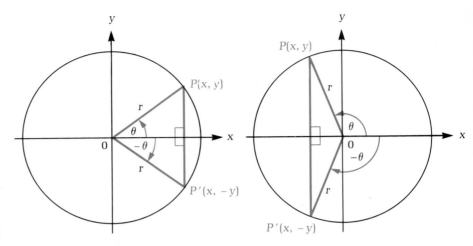

FIGURE 7.2

And $\tan \theta = y/x$ and

$$\tan(-\theta) = -\frac{y}{x} = -\tan \theta. \tag{7.18}$$

Example 7.9   Use negative-angle identities to rewrite the following in terms of functions of positive angles:

(a) $\dfrac{\sin(-\theta)}{\tan(-\theta)}$

(b) $\cos\left(\dfrac{-\pi}{8}\right) + \sin\left(\dfrac{-\pi}{8}\right)$

Solution: Using (7.16) to (7.18), we get

(a) $\sin(-\theta) = -\sin \theta$ and $\tan(-\theta) = -\tan \theta,$

therefore,

$$\frac{\sin(-\theta)}{\tan(-\theta)} = \frac{-\sin \theta}{-\tan \theta} = \frac{\sin \theta}{\tan \theta}.$$

(b) $\cos\left(\dfrac{-\pi}{8}\right) = \cos\dfrac{\pi}{8}$ and $\sin\left(\dfrac{-\pi}{8}\right) = -\sin\dfrac{\pi}{8}$

therefore,

$$\cos\left(\frac{-\pi}{8}\right) + \sin\left(\frac{-\pi}{8}\right) = \cos\frac{\pi}{8} - \sin\frac{\pi}{8}.$$

Contradiction

A statement that is false for all values of the variable is called a **contradiction.** In the algebra of real numbers, sentences such as $x = x + 1$ and $y^2 = -5$ are examples of contradictions. It is impossible for a real number to be equal to its successor, and a squared number is always positive in the real number system. You can recognize a statement that is a contradiction because it violates an algebraic property or a function definition. Example 7.10 illustrates some trigonometric statements that are contradictions.

Example 7.10

Explain why the following statements are contradictions: **(a)** $\sin\theta = \frac{3}{2}$, **(b)** $\cos^2\theta = -1$, **(c)** $\csc x = 0.5$, and **(d)** $\sec x = 0$.

Solution:

(a) Recall from Section 6.7 that $-1 \le \sin\theta \le 1$. Therefore $\sin\theta = \frac{3}{2}$ is false for all values of θ.

(b) As in the algebraic properties of real numbers, a squared number cannot be negative. Therefore the statement $\cos^2\theta = -1$ is a contradiction.

(c) The cosecant function is the reciprocal of the sine function, and its range is the set of all real numbers greater than or equal to 1 or less than or equal to -1. Since the value 0.5 does not fall in this range, the statement $\csc x = 0.5$ is a contradiction.

(d) The secant function is the reciprocal of the cosine function and has the same range as the cosecant function. Therefore, the secant function can never be equal to zero, and the statement $\sec x = 0$ is a contradiction.

The following chart may be helpful in recalling the range of a trigonometric function.

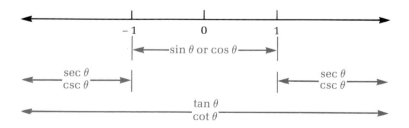

EXERCISES

Use formulas (7.1) to (7.8) to obtain the relations in Exercises 1 to 8.

1. $\tan x \cot x = 1$

2. $(\cos x)(\tan x + \sec x) = \sin x + 1$

3. $\tan x = \dfrac{\sec x}{\csc x}$

4. $(1 + \tan x)(1 - \tan x) = 2 - \sec^2 x$

5. $\cos x = \dfrac{\sin x}{\tan x}$

6. $(\tan x)(\cot x + \csc x) = 1 + \sec x$

7. $\cos x = \sin x \cot x$

8. $(1 + \cos x)(1 - \cos x) = \sin^2 x$

Assuming that x is an angle in the first quadrant, find the value of the expressions in Exercises 9 to 11 algebraically.

9. $(\sin^2 x - \cos^2 x)(\csc x)$, given that $\sin x = \frac{1}{2}$

10. $(\tan^2 x + \cot^2 x)(\sec^2 x + \csc^2 x)$, given that $\tan x = 2$

11. $(1 - \csc^2 x)(\tan x)$, given that $\sin x = \frac{1}{3}$

Prove the identities in Exercises 12 to 42.

12. $\cos \theta = \sin \theta \cot \theta$

13. $(1 + \cos \theta)(1 - \cos \theta) = \sin^2 \theta$

14. $(\sin \theta + \cos \theta)^2 = 1 + 2 \sin \theta \cos \theta$

15. $\dfrac{\sec^2 x}{\sin^2 x} = \csc^2 x + \sec^2 x$

16. $\dfrac{\sin x}{1 + \cos x} + \dfrac{1 + \cos x}{\sin x} = 2 \csc x$

17. $(\csc^2 x)(1 - \cos^2 x) = 1$

18. $\dfrac{\tan x + \cot x}{\tan x - \cot x} = \dfrac{\tan^2 x + 1}{\tan^2 x - 1}$

19. $\sin^3 x + \cos^3 x = (\sin x + \cos x)(1 - \sin x \cos x)$

20. $\dfrac{1}{1 - \sin B} + \dfrac{1}{1 + \sin B} = 2 \sec^2 B$

21. $\dfrac{\cot B}{\sec B - \tan B} - \dfrac{\cos B}{\sec B + \tan B} = \sin B + \csc B$

22. $\dfrac{(2 \cos x)(1 - \cos^2 x)}{2 \sin x \cos x} = \sin x$

23. $\dfrac{\cos \theta}{1 + \sin \theta} = \sec \theta - \tan \theta$

24. $\dfrac{\csc x + \cot x}{\sin x + \tan x} = \csc x \cot x$

25. $\sec^4 x = \sec^2 x + \tan^2 x \sec^2 x$

26. $\sin^3 x = \sin x - \sin x \cos^2 x$

27. $\cos^4 x = \cos^2 x - \cos^2 x \sin^2 x$

28. $\cos^2 x - \sin^2 x = 2 \cos^2 x - 1$

29. $\dfrac{\sin x}{1 - \cos x} = \dfrac{1 + \cos x}{\sin x}$

30. $\dfrac{1 + \sin x}{\cos x} = \dfrac{\cos x}{1 - \sin x}$

31. $\dfrac{1}{\sec x - \tan x} = \tan x + \sec x$

32. $\dfrac{1}{\csc x + \cot x} = \csc x - \cot x$

33. $\dfrac{1 + \cos x}{\sin x} + \dfrac{\sin x}{\cos x} = \dfrac{\cos x + 1}{\sin x \cos x}$

34. $\dfrac{1 - \sin x}{\cos x} = \dfrac{\cos x}{1 + \sin x}$

35. $\dfrac{1 + \tan x}{1 + \cot x} = \dfrac{\sec x}{\csc x}$

36. $\dfrac{\sin x + \cos x}{\sec x + \csc x} = \dfrac{\sin x}{\sec x}$

37. $\dfrac{1 + \tan x}{1 - \tan x} + \dfrac{1 + \cot x}{1 - \cot x} = 0$

38. $\dfrac{1}{\sin x \cos x} - \dfrac{\cos x}{\sin x} = \dfrac{\sin x \cos x}{1 - \sin^2 x}$

39. $\dfrac{1 - \cos x}{\sin x} = \dfrac{\sin x}{1 + \cos x}$

40. $\dfrac{\cot x - 1}{1 - \tan x} = \dfrac{\csc x}{\sec x}$

41. $\dfrac{\sin x - \cos x}{\sec x - \csc x} = \dfrac{\cos x}{\csc x}$

42. $\dfrac{\cos^2 x + \cot x}{\cos^2 x - \cot x} = \dfrac{\cos^2 x \tan x + 1}{\cos^2 x \tan x - 1}$

Use formulas (7.10) to (7.15) to find an identity for the expressions in Exercises 43 to 48.

43. $\sec(\theta + 90°)$ **44.** $\csc(\theta + 90°)$ **45.** $\cot(\theta + 90°)$

46. $\sec(\theta - 90°)$ **47.** $\csc(\theta - 90°)$ **48.** $\cot(\theta - 90°)$

In Exercises 49 to 54, rewrite each expression in terms of functions of positive angles.

49. $\cot(-12°) + \sec(-12°)$ **50.** $\sin(-17°)\cot(-17°)$ **51.** $\cos\left(\dfrac{-\pi}{12}\right)\tan\left(\dfrac{-\pi}{12}\right)$

52. $\tan\left(\dfrac{-4\pi}{5}\right) - \cot\left(\dfrac{-4\pi}{5}\right)$ **53.** $\csc^2(-2) + \sec^2(-2)$ **54.** $\cos^2(-3) - \sin^2(-3)$

Find the exact value of each expression in Exercises 55 to 60.

55. $\cos\left(\dfrac{-2\pi}{3}\right) + \tan\left(\dfrac{-\pi}{6}\right)$ **56.** $\sin\left(\dfrac{-\pi}{6}\right) - \cos\left(\dfrac{-5\pi}{4}\right)$ **57.** $\sin(-150°) + \cos(-210°)$

58. $\cos(-135°) - \sec(-240°)$ **59.** $\tan\left(\dfrac{-\pi}{4}\right) - \cot\left(\dfrac{-3\pi}{4}\right)$ **60.** $\sec\left(\dfrac{-\pi}{3}\right) - \csc\left(\dfrac{-5\pi}{6}\right)$

Explain why the statements in Exercises 61 to 66 are contradictions.

61. $\sin \theta = -2$ **62.** $\cos \theta = 5$ **63.** $\cos^2 \theta = -1$

64. $\sec \theta = \sqrt{0.09}$ **65.** $\csc^2 \theta = \frac{1}{4}$ **66.** $\tan \theta = \sqrt{-16}$

7.2

TRIGONOMETRIC EQUATIONS

trigonometric equation   A *trigonometric equation* is not unlike an algebraic equation. The solution set of a trigonometric equation consists of all the angles that satisfy the equation. Solving an algebraic equation usually involves the manipulation of the equation to isolate a given variable. However, the nature of trigonometric equations makes isolation of the variable difficult or impossible. A trigonometric equation usually contains more than one trigonometric function, but by using identities, a trigonometric equation can be transformed equivalently into one containing a single function. At this point of simplification, the values of the variable can be obtained. Much of the solution involves the same kind of algebraic manipulation that was used to prove identities. Guidelines for the solution of trigonometric equations are given here:

1. Transpose the entire expression to one side of the equation.

2. Use the fundamental relations [(7.1) to (7.8)] to express the statement in terms of a single function, if possible.

3. Use algebraic procedures such as factoring to write the expression as a product.

4. *Warning: Canceling a factor on both sides of an equation can lead to the loss of a part of the solution.*

5. *Warning: Squaring both sides of an equation can introduce extraneous roots. When squaring is used in solving the problem, the answers must be checked in the original equation.*

6. *Warning: Multiplying both sides of an equation by a factor in order to clear fractions can introduce an extraneous solution. Check answers in the original equation.*

Example 7.11

Solve the following equation for positive values of $x < 360°$:

$$\sin^2 x = \sin x \cos x.$$

Solution: If we divide by $\sin x$, we simplify the equation to $\sin x = \cos x$. However, we have lost part of the solution, as is shown below. As an alternate method of solution, transpose all terms to one side of the equation and factor:

$$(\sin x)(\sin x - \cos x) = 0,$$

$$\sin x = 0 \qquad \text{or} \qquad \sin x - \cos x = 0.$$

Now $\sin x = 0$ when x equals $0°$ and $180°$ (these solutions would have been lost by the first technique), and $\sin x = \cos x$ when x equals $45°$ and $225°$. Therefore, the solution set is $\{0°, 45°, 180°, 225°\}$.

Example 7.12

Solve the equation $\cos x = -\sin x$ for positive values of x less than $360°$.

Solution: Dividing both sides of the equation by $\cos x$ and using (7.4), we get

$$\cos x = -\sin x$$

$$1 = \frac{-\sin x}{\cos x}$$

$$1 = -\tan x$$

$$\tan x = -1.$$

Thus,

$$x = 135° \text{ or } 315°.$$

Example 7.13

Solve the trigonometric equation $\cot^2 x - 5 \cot x + 4 = 0$ over the interval $0° < x < 180°$. Round answers to the nearest degree.

Solution: Factor the left side and set each factor equal to zero:

$$\cot^2 x - 5 \cot x + 4 = (\cot x - 1)(\cot x - 4) = 0$$

$$\cot x - 1 = 0 \quad \text{or} \quad \cot x - 4 = 0$$

$$\cot x = 1 \quad \text{or} \quad \cot x = 4.$$

Using a scientific calculator or Appendix Table 4 we see that $x = 45°$ or $x = 14°$. Therefore the solution set within the given interval is $\{14°, 45°\}$.

Example 7.14

Solve the following equation for $0 \le x < 2\pi$:

$$\sin x + \cos x = 1.$$

Solution: Square both sides of the equation and simplify:

$$\sin x + \cos x = 1$$

$$(\sin x + \cos x)^2 = 1$$

$$\sin^2 x + 2 \sin x \cos x + \cos^2 x = 1$$

$$(\sin^2 x + \cos^2 x) + 2 \sin x \cos x = 1$$

$$1 + 2 \sin x \cos x = 1$$

$$2 \sin x \cos x = 0$$

$$\sin x \cos x = 0.$$

Therefore,

$$\sin x = 0 \quad \text{or} \quad \cos x = 0.$$

And

$$\sin x = 0 \quad \text{when } x = 0 \quad \text{and} \quad x = \pi,$$

$$\cos x = 0 \quad \text{when } x = \frac{\pi}{2} \quad \text{and} \quad x = \frac{3\pi}{2}.$$

Since squaring does not yield equivalent equations, we have to check the solutions in the original equation. We find that $x = \pi$ and $x = 3\pi/2$ do not check. Therefore the solution set is $\{0, \pi/2\}$.

If several functions of an angle are involved in an equation, a recommended trigonometric technique is to express all functions in terms of a single function by using algebraic or geometric methods. Avoid the use of radicals whenever possible. After solving, check that the trigonometric functions of the original equation are defined at the solution values.

Example 7.15

Solve the equation $(\sec^2 x)(1 + \cos x \tan x) = 2$.

Solution.   We express all functions in terms of sines and cosine.

$$\left(\frac{1}{\cos^2 x}\right)\left[1 + (\cos x)\left(\frac{\sin x}{\cos x}\right)\right] = 2$$

$$1 + \sin x = 2\cos^2 x \qquad \text{(multiplying by } \cos^2 x)$$

$$1 + \sin x = 2(1 - \sin^2 x) \qquad \text{[substituting (7.6)]}$$

$$2\sin^2 x + \sin x - 1 = 0 \qquad \text{(transposing)}$$

$$(2\sin x - 1)(\sin x + 1) = 0 \qquad \text{(factoring)}$$

$$2\sin x - 1 = 0 \qquad \text{or} \qquad \sin x + 1 = 0$$

$$\sin x = \tfrac{1}{2} \qquad \text{or} \qquad \sin x = -1$$

$$x = 30°, 150° \qquad \text{or} \qquad x = 270°.$$

The original equation contains the secant, cosine, and tangent functions. We must reject the solution $x = 270°$ because $\tan x$ is not defined at $x = 270°$ and secant x is not defined at $x = 270°$. We introduced this extraneous solution when we multiplied both sides of the equation by $\cos^2 x$, which is equal to zero when $x = 270°$. Therefore, the solution set is $\{30°, 150°\}$.

Example 7.16

Solve the following equation for $0° < x < 180°$:

$$\tan x + 3\sec^2 x = 5.$$

Solution:

$$\tan x + 3(\tan^2 x + 1) = 5 \qquad \text{[substituting (7.7)]}$$

$$3\tan^2 x + \tan x - 2 = 0 \qquad \text{(multiplying and transposing)}$$

$$(3\tan x - 2)(\tan x + 1) = 0 \qquad \text{(factoring)}$$

$$\tan x = \tfrac{2}{3} \qquad \text{or} \qquad \tan x = -1.$$

Letting $\tan x = 0.6667$, we find to the nearest 10′ that $x = 33°40'$. Letting $\tan x = -1$, we see that $x = 135°$. Thus, the solution set is $\{33°40', 135°\}$.

Example 7.17

Solve for all values of x:

$$4 \tan x \sec x - \tan x = 0.$$

Solution: Factor and set each factor equal to zero:

$$4 \tan x \sec x - \tan x = 0$$

$$\tan x(4 \sec x - 1) = 0$$

$$\tan x = 0 \quad \text{or} \quad 4 \sec x - 1 = 0.$$

Now $\tan x = 0$ when $x = n\pi$, where n is an integer. And $\sec x = \frac{1}{4}$ is a contradiction. There is no solution from this factor. Therefore, the solution set is $\{n\pi\}$, where n is an integer.

Example 7.18

Use a calculator to solve

$$\sin^2 x + 2\cos^2 x = 1.4.$$

Solution: Use (7.6) to express the equation in terms of only the sine function:

$$\sin^2 x + 2\cos^2 x = 1.4$$

$$\sin^2 x + 2(1 - \sin^2 x) = 1.4$$

$$\sin^2 x + 2 - 2\sin^2 x = 1.4$$

$$\sin^2 x = 0.6$$

$$\sin x = \pm 0.77459667.$$

To the nearest minute, the solution set is $\{50°46', 129°14', 230°46', 309°14'\}$.

EXERCISES

Solve for angles in the interval $[0, 360°)$ in Exercises 1 to 60.

1. $\sin^2 x = \frac{1}{4}$
2. $\tan^2 x = 1$
3. $(\sin \theta + 1)(2 \cos \theta - 1) = 0$
4. $(\tan^2 x - 3)(\csc x - 2) = 0$
5. $\sqrt{2} \tan \theta \sin \theta - \tan \theta = 0$
6. $1 - \sin x = \sqrt{3} \cos x$
7. $\csc x - 1 = \cot x$
8. $2 \tan^2 \theta - \sec^2 \theta = 0$
9. $2 \sin^2 x + \sin x - 1 = 0$
10. $(2 \sin^2 x - 1)(\sin x - 3) = 0$
11. $\sqrt{3} \tan x - 1 = \sec x$
12. $4 \csc^2 x - 7 \cot^2 x = 3$
13. $2 \cos^2 x + 11 \cos x + 5 = 0$
14. $8 \tan \theta + 2 \sec^2 \theta + 3 = 0$

15. $5 \cos x - 5 = -\sec x$
16. $4 \cos^2 x = 1$
17. $2 \cos x = 1$
18. $2 \sin^2 x = 1$
19. $8 \cos^2 x - 2 \cos x = 1$
20. $8 \cos^2 x + 2 \cos x = 1$
21. $2 \cos^2 x + \cos x = 0$
22. $2 \tan x + 3 = 0$
23. $\cot^2 x - 3 = 0$
24. $\cos^2 x + 2 \cos x = 3$
25. $4 \sin^3 x - \sin x = 0$
26. $\csc^2 x - 4 = 0$
27. $2 \cos^2 x + 3 \cos x = -1$
28. $2 \sin^2 x - \sin x = 3$
29. $2 \cos^2 x - \sqrt{3} \cos x = 0$
30. $2 \sin^2 x - 5 \sin x + 2 = 0$
31. $2 \sin x \tan x + \tan x - 2 \sin x - 1 = 0$
32. $\sin x + \cos x = 1$
33. $\tan^2 x \cos x - \cos x = 0$
34. $2 \cos^2 x \tan x - \tan x = 0$
35. $\tan^2 x + \sec x - 1 = 0$
36. $\tan x \sin x - \tan x = 0$
37. $2 \sec x \tan x + 2 \sec x + \tan x + 1 = 0$
38. $\sec^2 x = 4 \tan^2 x$
39. $\sec^2 x + 3 \tan x - 11 = 0$
40. $\sqrt{3} \cos x - \sin x = 1$
41. $\sqrt{2} \cos x - \sqrt{2} \sin x = 2$
42. $2 \cos x + 2 \sin x = \sqrt{2}$
43. $\tan^2 x + 4 = 2 \sec^2 x + \tan x$
44. $\sec^2 x - 2 \tan^2 x = 0$
45. $2 \csc x \cos x - 4 \cos x - \csc x + 2 = 0$
46. $\sin^2 x - 7 \sin x = 0$
47. $2 \cos^2 x = 1 + \cos x$
48. $2 \sin^3 x - \sin x = 0$
49. $4 \cos^3 x - 3 \cos x = 0$
50. $2 \cos x \tan x - 2 \cos x + \tan x = 1$
51. $9 \cos^2 x - 21 \cos x - 8 = 0$
52. $2\sqrt{3} \sin^2 x = \cos x$
53. $\sin^2 \theta - 2 \cos \theta + \tfrac{1}{4} = 0$
54. $3 \sec^2 \theta = 2 \csc \theta$
55. $\tan^2 \theta + \cot^2 \theta = 2$
56. $3(\sec^2 \theta + \tan^2 \theta) = 5$
57. $\tan \theta + \cot \theta = 2 \csc \theta$
58. $3 \tan \theta + \cot \theta = 5 \csc \theta$
59. $\csc x - \cot^2 x + 1 = 0$
60. $2 \sin x + \csc x - 3 = 0$

7.3

FUNCTIONS OF TWO ANGLES

If A, B, and C are real numbers, we know that $A(B + C) = AB + AC$ by the distributive property of multiplication over addition. For the trigonometric function $\cos(A - B)$, we might expect that the same type of property holds true and that $\cos(A - B) = \cos A - \cos B$. However, a simple example shows that this is false. Let $A = 0°$ and $B = 0°$; then $\cos A = 1$ and $\cos B = 1$, and $\cos A - \cos B = 0$. But $\cos(A - B) = \cos(0° - 0°) = \cos 0° = 1$. In this section, we develop formulas for the trigonometric function of the sum and difference of two angles.

To develop the correct formula for $\cos(A - B)$, let us draw A and B in standard position on the unit circle such that $A > B$. Let $P(x_1, y_1)$ and $Q(x_2, y_2)$ be the coordinates of corresponding points on the unit circle as shown in Figure 7.3, and let d be the distance between P and Q. By the distance formula,

$$d^2 = (x_2 - x_1)^2 + (y_2 - y_1)^2.$$

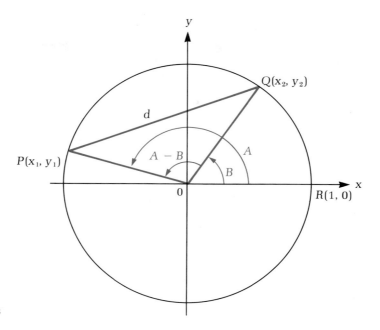

FIGURE 7.3

Since this is a unit circle,

$$x_1 = \cos A \qquad \text{and} \qquad y_1 = \sin A$$

$$x_2 = \cos B \qquad \text{and} \qquad y_2 = \sin B.$$

Substituting these values in the distance formula, we find

$$d^2 = (\cos B - \cos A)^2 + (\sin B - \sin A)^2$$
$$= \cos^2 B - 2 \cos B \cos A + \cos^2 A + \sin^2 B - 2 \sin B \sin A + \sin^2 A$$
$$= (\cos^2 B + \sin^2 B) + (\cos^2 A + \sin^2 A) - 2 \cos B \cos A - 2 \sin B \sin A.$$

Using the fundamental trigonometric identities to simplify, we find

$$d^2 = 1 + 1 - 2 \cos B \cos A - 2 \sin B \sin A,$$

or

$$d^2 = 2 - 2(\cos A \cos B + \sin A \sin B).$$

If we now rotate the entire figure in a clockwise direction through angle B, then Q would lie on $R(1, 0)$, and P would move to a new position $N(x_3, y_3)$. However, the distance would remain constant, and this represents the same distance between R and N, as shown in Figure 7.4. Since ON is on the terminal side of the angle $A - B$, the coordinates of N are

$$x_3 = \cos (A - B), \qquad y_3 = \sin (A - B).$$

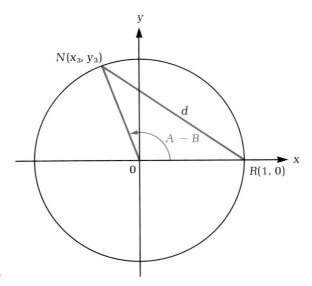

FIGURE 7.4

Using the distance formula again yields

$$
\begin{aligned}
d^2 &= (x_3 - 1)^2 + (y_3 - 0)^2 \\
&= x_3{}^2 + -2x_3 + 1 + y_3{}^2 \\
&= \cos^2 (A - B) - 2 \cos (A - B) + 1 + \sin^2 (A - B) \\
&= 1 + \left[\cos^2 (A - B) + \sin^2 (A - B)\right] - 2 \cos (A - B) \\
&= 2 - 2 \cos (A - B).
\end{aligned}
$$

Equating the two expressions for d^2, we get

$$2 - 2(\cos A \cos B + \sin A \sin B) = 2 - 2 \cos (A - B),$$

or

$$\cos (A - B) = \cos A \cos B + \sin A \sin B. \tag{7.19}$$

The value of $\cos (A + B)$ can easily be derived from formula (7.19):

$$
\begin{aligned}
\cos (A + B) &= \cos \left[A - (-B)\right] \\
&= \cos A \cos (-B) + \sin A \sin (-B) \\
&= \cos A \cos B + (\sin A)(-\sin B) \\
&= \cos A \cos B - \sin A \sin B
\end{aligned}
$$

$$\cos (A + B) = \cos A \cos B - \sin A \sin B. \tag{7.20}$$

addition formulas   Together, (7.19) and (7.20) make up the **addition formulas** for the cosine
function.

Example 7.19   Find the value of $\cos 75°$ by using (7.20) and special angle measures.

Solution:

$$\cos 75° = \cos (45° + 30°)$$
$$= \cos 45° \cos 30° - \sin 45° \sin 30°$$
$$= \frac{\sqrt{2}}{2} \cdot \frac{\sqrt{3}}{2} - \frac{\sqrt{2}}{2} \cdot \frac{1}{2}$$
$$= \frac{\sqrt{6}}{4} - \frac{\sqrt{2}}{4}$$
$$= \frac{\sqrt{6} - \sqrt{2}}{4}.$$

Example 7.20   If $\cos A = \frac{4}{5}$ and A is in quadrant I and if $\sin B = \frac{15}{17}$ and B is in quadrant II, find
$\cos (A + B)$.

Solution: Make a sketch of each angle on coordinate axes, and find the third
side of each reference triangle by using the Pythagorean theorem. Be sure to

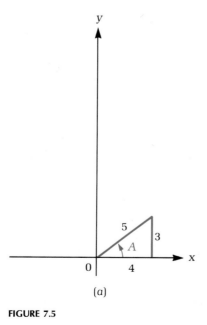

(a)

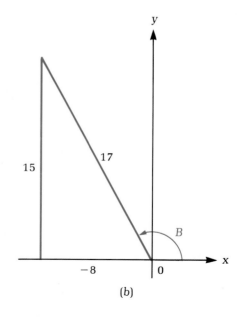

(b)

FIGURE 7.5

apply the proper signs that are associated with the given quadrants (see Figure 7.5):

$$\cos (A + B) = \cos A \cos B - \sin A \sin B$$

$$= \frac{4}{5} \cdot \frac{-8}{17} - \frac{3}{5} \cdot \frac{15}{17}$$

$$= \frac{-32}{85} - \frac{45}{85}$$

$$= \frac{-77}{85}.$$

Example 7.21

Reduce to a single term:

$$\cos \frac{4x}{5} \cos \frac{x}{5} - \sin \frac{4x}{5} \sin \frac{x}{5}.$$

Solution: Using (7.20) with $A = 4x/5$ and $B = x/5$, we find that the expression is equal to $\cos (4x/5 + x/5)$, or

$$\cos \frac{5x}{5} = \cos x.$$

Example 7.22

Verify the identity

$$\cos (A + B) \cos (A - B) = \cos^2 A - \sin^2 B.$$

Solution: Starting with the left-hand side, using (7.19) and (7.20), we get

$\cos (A + B) \cos (A - B)$
$= (\cos A \cos B - \sin A \sin B)$
$\quad \cdot (\cos A \cos B + \sin A \sin B)$
$= \cos^2 A \cos^2 B + \cos A \cos B \sin A \sin B$
$\quad - \cos A \cos B \sin A \sin B - \sin^2 A \sin^2 B$ (multiplying)
$= \cos^2 A \cos^2 B - \sin^2 A \sin^2 B$ (simplifying)
$= \cos^2 A \cos^2 B - (1 - \cos^2 A)(\sin^2 B)$ [using (7.5)]
$= \cos^2 A \cos^2 B - \sin^2 B + \cos^2 A \sin^2 B$ (multiplying)
$= (\cos^2 A)(\cos^2 B + \sin^2 B) - \sin^2 B$ (combining and factoring)
$= \cos^2 A - \sin^2 B.$

Just as the cosine function has addition formulas, so do the other basic trigonometric functions of sine and tangent. To derive the sine addition formulas, we need to first prove that $\cos (90° - \theta) = \sin \theta$:

$$\cos (90° - \theta) = \cos 90° \cos \theta + \sin 90° \sin \theta \qquad [\text{using (7.19)}]$$
$$= 0 \cdot \cos \theta + 1 \cdot \sin \theta$$
$$= \sin \theta.$$

If we replace θ by $90° - \theta$, we get

$$\cos [90° - (90° - \theta)] = \sin (90° - \theta),$$

or

$$\cos \theta = \sin (90° - \theta).$$

Using this result, we proceed by substituting $A + B$ for θ:

$$\sin \theta = \cos (90° - \theta)$$
$$\sin (A + B) = \cos [90° - (A + B)]$$
$$= \cos [(90° - A) - B]$$
$$= \cos (90° - A) \cos B + \sin (90° - A) \sin B$$
$$= \sin A \cos B + \cos A \sin B.$$

Thus,

$$\sin (A + B) = \sin A \cos B + \cos A \sin B. \qquad \textbf{(7.21)}$$

To find a formula for $\sin (A - B)$, we simply replace B by $-B$ in (7.21) to obtain

$$\sin [A + (-B)] = \sin A \cos (-B) + \cos A \sin (-B).$$

Or

$$\sin (A - B) = \sin A \cos B - \cos A \sin B. \qquad \textbf{(7.22)}$$

Formulas (7.21) and (7.22) are the addition formulas for the sine function.

To derive the formula for the tangent of the sum of two function angles, we use the ratio of the sine and cosine functions:

$$\tan (A + B) = \frac{\sin (A + B)}{\cos (A + B)}$$

$$= \frac{\sin A \cos B + \cos A \sin B}{\cos A \cos B - \sin A \sin B}.$$

We divide the numerator and denominator by $\cos A \cos B$:

$$\tan (A + B) = \frac{\dfrac{\sin A \cos B}{\cos A \cos B} + \dfrac{\cos A \sin B}{\cos A \cos B}}{\dfrac{\cos A \cos B}{\cos A \cos B} - \dfrac{\sin A \sin B}{\cos A \cos B}}.$$

Using (7.4) and simplifying, we find

$$\tan (A + B) = \frac{\tan A + \tan B}{1 - \tan A \tan B}.$$ **(7.23)**

Replacing B with $-B$ in (7.23), we obtain the formula for the tangent of the difference of two function angles:

$$\tan (A - B) = \frac{\tan A - \tan B}{1 + \tan A \tan B}.$$ **(7.24)**

Example 7.23   Use the exact values of the trigonometric functions at 30° and 45° to find the value of the sine and tangent at 15°.

Solution:

$$\begin{aligned}
\sin 15° &= \sin (45° - 30°) \\
&= \sin 45° \cos 30° - \cos 45° \sin 30° \\
&= \frac{\sqrt{2}}{2} \frac{\sqrt{3}}{2} - \frac{\sqrt{2}}{2} \cdot \frac{1}{2} \\
&= \frac{\sqrt{6}}{4} - \frac{\sqrt{2}}{4} \\
&= \frac{\sqrt{6} - \sqrt{2}}{4},
\end{aligned}$$

and

$$\begin{aligned}
\tan 15° &= \tan (45° - 30°) \\
&= \frac{\tan 45° - \tan 30°}{1 + \tan 45° \tan 30°} \\
&= \frac{1 - \sqrt{3}/3}{1 + \sqrt{3}/3} \\
&= \frac{3 - \sqrt{3}}{3 + \sqrt{3}} \cdot \frac{3 - \sqrt{3}}{3 - \sqrt{3}} \\
&= \frac{9 - 6\sqrt{3} + 3}{6} \\
&= 2 - \sqrt{3}.
\end{aligned}$$

The cofunction identity formulas, (7.10) to (7.15), were derived by using the unit circle. These same formulas can be derived as special cases of the addition formulas, (7.19) to (7.24).

Example 7.24

Use formulas (7.20) and (7.21) to derive formulas for **(a)** $\sin(\theta + 90°)$, **(b)** $\cos(\theta + 90°)$, and **(c)** $\tan(\theta + 90°)$. Use only the sine and cosine addition formulas.

Solution:

(a) $\sin(\theta + 90°) = \sin\theta\cos 90° + \cos\theta\sin 90°$
$$= (\sin\theta)(0) + (\cos\theta)(1)$$
$$= \cos\theta$$

(b) $\cos(\theta + 90°) = \cos\theta\cos 90° - \sin\theta\sin 90°$
$$= (\cos\theta)(0) - (\sin\theta)(1)$$
$$= -\sin\theta$$

(c) $\tan(\theta + 90°) = \dfrac{\sin(\theta + 90°)}{\cos(\theta + 90°)}$

$$= \dfrac{\cos\theta}{-\sin\theta}$$

$$= -\cot\theta.$$

Example 7.25

Use formulas (7.19) and (7.22) to derive formulas for **(a)** $\sin(\theta - 90°)$, **(b)** $\cos(\theta - 90°)$, and **(c)** $\tan(\theta - 90°)$. Use only the sine and cosine addition formulas.

Solution:

(a) $\sin(\theta - 90°) = \sin\theta\cos 90° - \cos\theta\sin 90°$
$$= (\sin\theta)(0) - (\cos\theta)(1)$$
$$= -\cos\theta$$

(b) $\cos(\theta - 90°) = \cos\theta\cos 90° + \sin\theta\sin 90°$
$$= (\cos\theta)(0) + (\sin\theta)(1)$$
$$= \sin\theta$$

(c) $\tan(\theta - 90°) = \dfrac{\sin(\theta - 90°)}{\cos(\theta - 90°)}$

$$= \dfrac{-\cos\theta}{\sin\theta}$$

$$= -\cot\theta.$$

Example 7.26

Construct a unit circle with an acute angle θ in quadrant I. Draw the reflection of angle θ about the axes in the other three quadrants. Determine the relationship between θ and its related angles in the other quadrants. Express $\sin \theta$, $\cos \theta$, and $\tan \theta$ in terms of the related angles.

Solution: Let $\sin \theta = y$ and $\cos \theta = x$. Label the corresponding lines in the other quadrants with the appropriate signs. Identify the related angles $\pi - \theta$, $\theta - \pi$, and $2\pi - \theta$, as shown in Figure 7.6. Read the appropriate relationships for the sine, cosine, and tangent functions:

$$\sin \theta = \sin (\pi - \theta) = -\sin (\theta - \pi) = -\sin (2\pi - \theta)$$

$$\cos \theta = -\cos (\pi - \theta) = -\cos (\theta - \pi) = \cos (2\pi - \theta)$$

$$\tan \theta = -\tan (\pi - \theta) = \tan (\theta - \pi) = -\tan (2\pi - \theta).$$

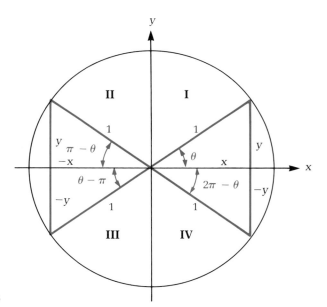

FIGURE 7.6

Another set of useful identities may be derived from a combination of the identities at hand. Let us add formulas (7.19) and (7.20), that is, add the left-hand and right-hand members:

$$\cos (A - B) = \cos A \cos B + \sin A \sin B$$
$$+ \cos (A + B) = \cos A \cos B - \sin A \sin B$$

or

$$\cos (A - B) + \cos (A + B) = 2 \cos A \cos B. \qquad (7.25)$$

product formulas The resulting equation is one of the **product formulas.** The product on the right can be written as the sum on the left. The other product formulas are derived now.

Subtracting formulas (7.19) and (7.20), we get

$$\cos (A - B) = \cos A \cos B + \sin A \sin B$$
$$- \cos (A + B) = \cos A \cos B - \sin A \sin B$$

or

$$\cos (A - B) - \cos (A + B) = 2 \sin A \sin B. \tag{7.26}$$

Now adding formulas (7.21) and (7.22), we get

$$\sin (A + B) = \sin A \cos B + \cos A \sin B$$
$$+ \sin (A - B) = \sin A \cos B - \cos A \sin B$$

or

$$\sin (A + B) + \sin (A - B) = 2 \sin A \cos B. \tag{7.27}$$

And subtracting formulas (7.21) and (7.22), we get

$$\sin (A + B) = \sin A \cos B + \cos A \sin B$$
$$- \sin (A - B) = \sin A \cos B + \cos A \sin B$$

or

$$\sin (A + B) - \sin (A - B) = 2 \cos A \sin B. \tag{7.28}$$

Example 7.27

Prove the identity

$$2 \sin 3x \cos 7x + \sin 4x = \sin 10x.$$

Solution: Use product formula (7.27), and let $A = 3x$ and $B = 7x$. Then

$$2 \sin A \cos B = \sin (A + B) + \sin (A - B)$$
$$2 \sin 3x \cos 7x = \sin 10x + \sin (-4x)$$
$$2 \sin 3x \cos 7x = \sin 10x - \sin 4x$$
$$2 \sin 3x \cos 7x + \sin 4x = \sin 10x.$$

EXERCISES

Use the exact values of the trigonometric functions at 30°, 45°, and 60° and their integral multiples to find the value of the sine, cosine, and tangent of the angles in Exercises 1 to 6.

1. 75°   **2.** 90°   **3.** 105°

4. 165°   **5.** 285°   **6.** 345°

Use formulas (7.19) to (7.24) to prove identities in Exercises 7 to 24.

7. $\cos (90° + \theta) = -\sin \theta$

8. $\cos (180° + \theta) = -\cos \theta$

9. $\cos (180° - \theta) = -\cos \theta$

10. $\cos (270° + \theta) = \sin \theta$

11. $\cos (270° - \theta) = -\sin \theta$

12. $\cos (360° - \theta) = \cos \theta$

13. $\sin (180° + A) = -\sin A$

14. $\sin (180° - A) = \sin A$

15. $\sin (90° + A) = \cos A$

16. $\sin (90° - A) = \cos A$

17. $\sin (270° - A) = -\cos A$

18. $\sin (270° + A) = -\cos A$

19. $\tan (180° - A) = -\tan A$

20. $\tan (180° + A) = \tan A$

21. $\tan (90° + A) = -\cot A$

22. $\tan (270° + A) = -\cot A$

23. $\tan (270° - A) = \cot A$

24. $\tan (360° - A) = -\tan A$

Given that $\sin A = \frac{3}{5}$ and $\sin B = \frac{4}{5}$ with both angles in the first quadrant, evaluate the functions in Exercises 25 to 30.

25. $\sin (A + B)$

26. $\sin (A - B)$

27. $\cos (A - B)$

28. $\cos (A + B)$

29. $\tan (A + B)$

30. $\tan (A - B)$

Use the given information to find $\cos (A + B)$ and $\cos (A - B)$ in Exercises 31 to 33.

31. $\sin A = \frac{5}{13}$, $\cos B = -\frac{4}{5}$, A and B are in quadrant II

32. $\cos A = -\frac{12}{13}$, A is in quadrant II, $\cot B = \frac{5}{12}$, B is in quadrant I

33. $\tan A = -\frac{15}{8}$, A is in quadrant IV, $\sec B = \frac{5}{4}$, B is in quadrant I

Use the given information to find $\sin (A + B)$ and $\tan (A - B)$ in Exercises 34 to 36.

34. $\sin A = \frac{5}{13}$, $\cos B = -\frac{4}{5}$, A and B are in quadrant II

35. $\cos A = -\frac{12}{13}$, A is in quadrant III, $\cot B = \frac{5}{12}$, B is in quadrant I

36. $\tan A = -\frac{15}{8}$, A is in quadrant IV, $\sec B = \frac{5}{4}$, B is in quadrant I

Express Exercises 37 to 40 in terms of functions of only θ.

37. $\sin (\theta + 60°)$

38. $\sin (\theta - 45°)$

39. $\tan \left(\theta + \dfrac{\pi}{6} \right)$

40. $\tan \left(\theta + \dfrac{\pi}{4} \right)$

Simplify Exercises 41 to 48.

41. $\sin 42° \cos 16° + \cos 42° \sin 16°$

42. $\cos 42° \cos 16° - \sin 42° \sin 16°$

43. $\dfrac{\tan 20° + \tan 15°}{1 - \tan 20° \tan 15°}$

44. $\cos (A + B) \cos B + \sin (A + B) \sin B$

45. $\sin (A - B) \cos B + \cos (A - B) \sin B$

46. $\sin A \cos (-B) + \cos A \sin (-B)$

47. $\cos A \cos (-B) + \sin A \sin (-B)$

48. $\cos (x + y) \cos (x - y) - \sin (x + y) \sin (x - y)$

Write the sum in Exercises 49 to 52 as a product of functions.

49. $\cos 3\theta + \cos \theta$

50. $\sin 11x + \sin 5x$

51. $6\left(\sin \dfrac{x}{2} - \sin \dfrac{x}{3} \right)$

52. $4(\cos 7x + \cos 5x)$

Reduce each expression in Exercises 53 and 54 to a single term.

53. $\cos 6A \cos 2A + \sin 6A \sin 2A$

54. $\cos (x - y) \cos y - \sin (x - y) \sin y$

55. Derive an identity for $\cot (A + B)$ in terms of $\cot A$ and $\cot B$.

56. Derive an identity for $\cot (A - B)$ in terms of $\cot A$ and $\cot B$.

57. Find an identity for $\sin 2\theta$. (*Hint:* $2\theta = \theta + \theta$.)

58. Find an identity for $\cos 2\theta$. (*Hint:* $2\theta = \theta + \theta$.)

Verify the identities in Exercises 59 to 68.

59. $\sin \left(\theta - \dfrac{\pi}{6} \right) + \cos \left(\theta - \dfrac{\pi}{3} \right) = \sqrt{3} \sin \theta$

60. $\tan \left(\theta + \dfrac{\pi}{4} \right) - \tan \left(\theta - \dfrac{3\pi}{4} \right) = 0$

61. $\cos (45° + x) - \cos (45° - x) = -\sqrt{2} \sin x$

62. $\cos \left(\theta - \dfrac{\pi}{4} \right) = \dfrac{\cos \theta + \sin \theta}{\sqrt{2}}$

63. $\tan (x + 45°) = \dfrac{\cos x + \sin x}{\cos x - \sin x}$

64. $\tan \left(x + \dfrac{\pi}{4} \right) = \dfrac{1 + \tan x}{1 - \tan x}$

65. $\cot (x + y) = \dfrac{1 - \tan x \tan y}{\tan x + \tan y}$

66. $\cot (x - y) = \dfrac{1 + \tan x \tan y}{\tan x - \tan y}$

67. $\tan 2x = \dfrac{2 \tan x}{1 - \tan^2 x}$

68. $\sin^2 x - \sin^2 y = \sin (x + y) \sin (x - y)$

Use a calculator to show that the statements in Exercises 69 to 78 are true.

69. $\sin (21° + 85°) = \sin 21° \cos 85° + \cos 21° \sin 85°$

70. $\cos (54° + 102°) = \cos 54° \cos 102° - \sin 54° \sin 102°$

71. $\cos (143° - 15°) = \cos 143° \cos 15° + \sin 143° \sin 15°$

72. $\sin (128° - 30°) = \sin 128° \cos 30° - \cos 128° \sin 30°$

73. $\tan (0.4 + 0.5) = \dfrac{\tan 0.4 + \tan 0.5}{1 - \tan 0.4 \tan 0.5}$

74. $\tan (2.8 - 1.3) = \dfrac{\tan 2.8 - \tan 1.3}{1 + \tan 2.8 \tan 1.3}$

75. $2 \sin 15° \cos 20° = \sin (15° + 20°) + \sin (15° - 20°)$

76. $2 \cos 200° \cos 75° = \cos (200° - 75°) + \cos (200° + 75°)$

77. $2 \sin 84° \sin 12° = \cos (84° - 12°) - \cos (84° + 12°)$

78. $2 \sin 1.5 \sin 2.2 = \cos (1.5 - 2.2) - \cos (1.5 + 2.2)$

7.4

MULTIPLE-ANGLE AND HALF-ANGLE FORMULAS

If we let $B = A$ in the identities for $\sin (A + B)$, $\cos (A + B)$, and $\tan (A + B)$, we have identities for $\sin 2A$, $\cos 2A$, and $\tan 2A$, respectively:

$$\sin (A + A) = \sin A \cos A + \cos A \sin A$$

so

$$\sin 2A = 2 \sin A \cos A. \qquad \textbf{(7.29)}$$

In a like manner, $\cos(A + A) = \cos A \cos A - \sin A \sin A$, so

$$\cos 2A = \cos^2 A - \sin^2 A, \tag{7.30}$$

or

$$\cos 2A = (1 - \sin^2 A) - \sin^2 A. \qquad [\text{using (7.6)}]$$

Thus,

$$\cos 2A = 1 - 2\sin^2 A. \tag{7.31}$$

Substituting $1 - \cos^2 A$ for $\sin^2 A$ in (7.30), we get

$$\cos 2A = \cos^2 A - (1 - \cos^2 A), \qquad [\text{using (7.6)}]$$

or

$$\cos 2A = 2\cos^2 A - 1. \tag{7.32}$$

Similarly,

$$\tan(A + A) = \frac{\tan A + \tan A}{1 - \tan A \tan A},$$

so

$$\tan 2A = \frac{2\tan A}{1 - \tan^2 A}. \tag{7.33}$$

Example 7.28   Given that $\sin A = \frac{8}{17}$ and A is in quadrant II, find **(a)** $\sin 2A$, **(b)** $\cos 2A$, and **(c)** $\tan 2A$.

Solution: Make a sketch representing angle A in standard position, and determine the third side of the related triangle. See Figure 7.7. Then find the value of

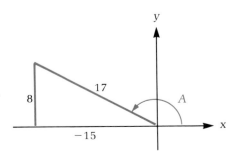

FIGURE 7.7

cos A and tan A. Finally, use formulas (7.29), (7.30), and (7.33), respectively:

$$\sin A = \frac{8}{17}, \qquad \cos A = \frac{-15}{17}, \qquad \tan A = \frac{-8}{15}.$$

(a) $\sin 2A = 2 \sin A \cos A$ [using (7.29)]

$$= 2\left(\frac{8}{17}\right)\left(\frac{-15}{17}\right) = \frac{-240}{289}$$

$$= -0.83 \text{ approximately}$$

(b) $\cos 2A = \cos^2 A - \sin^2 A$ [using (7.30)]

$$= \left(\frac{-15}{17}\right)^2 - \left(\frac{8}{17}\right)^2$$

$$= \frac{225}{289} - \frac{64}{289} = \frac{161}{289}$$

$$= 0.56 \text{ approximately}$$

(c) $\tan 2A = \dfrac{2 \tan A}{1 - \tan^2 A}$ [using (7.33)]

$$= \frac{2(\frac{-8}{15})}{1 - (\frac{-8}{15})^2}$$

$$= \frac{\frac{-16}{15}}{1 - \frac{64}{225}} = \frac{\frac{-16}{15}}{\frac{161}{225}}$$

$$= \frac{-240}{161}$$

$$= -1.49 \text{ approximately.}$$

Example 7.29 Find the solutions of the equation $\cos 2x + \cos x = 0$ from $0°$ to $360°$.

Solution:

$$\cos 2x + \cos x = 0$$

$$(2 \cos^2 x - 1) + \cos x = 0 \qquad \text{[using (7.32)]}$$

$$2 \cos^2 x + \cos x - 1 = 0 \qquad \text{(rearranging terms)}$$

$$(2 \cos x - 1)(\cos x + 1) = 0 \qquad \text{(factoring)}$$

$$\cos x = \tfrac{1}{2} \quad \text{or} \quad \cos x = -1. \qquad \text{(setting each factor equal to zero)}$$

The solution set is $\{60°, 180°, 300°\}$.

Example 7.30　　　Graph the function $y = \sin A \cos A$ for one cycle. Determine where the maximum and minimum values of the function occur.

Solution:　This function could be graphed by selecting values for A and then multiplying the values of $\sin A$ and $\cos A$ to determine the value of y. However, a simpler procedure results by utilizing the double-angle formula (7.29):

$$\sin 2A = 2 \sin A \cos A.$$

Therefore,

$$\tfrac{1}{2} \sin 2A = \sin A \cos A.$$

It is much easier to graph the equivalent function

$$y = \tfrac{1}{2} \sin 2A.$$

The amplitude is $\tfrac{1}{2}$, and the period is $2\pi/2$, of π. The maximum value of the sine function occurs when $2A = \pi/2$, or $A = \pi/4$. The minimum value occurs when $2A = 3\pi/2$, or $A = 3\pi/4$ (see Figure 7.8).

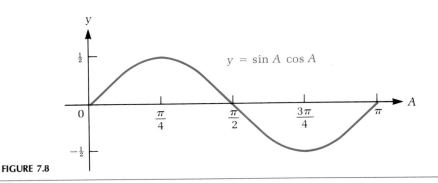

FIGURE 7.8

　　The formulas for functions of twice an angle can be used to develop formulas for half an angle. Using the identity for $\cos 2A$, (7.31), we solve for $\sin A$:

$$\cos 2A = 1 - 2 \sin^2 A$$

$$2 \sin^2 A = 1 - \cos 2A$$

$$\sin^2 A = \frac{1 - \cos 2A}{2}$$

$$\sin A = \pm \sqrt{\frac{1 - \cos 2A}{2}}.$$

Substituting $\theta/2$ for A, we have the half-angle formula for the sine function:

$$\sin \frac{\theta}{2} = \pm \sqrt{\frac{1 - \cos \theta}{2}}, \qquad (7.34)$$

where the sign is determined by the quadrant of $\theta/2$.

Similarly, we find the half-angle formula for the cosine function. Using the identity for $\cos 2A$, (7.32), we solve for $\cos A$:

$$\cos 2A = 2 \cos^2 A - 1$$

$$2 \cos^2 A = 1 + \cos 2A$$

$$\cos^2 A = \frac{1 + \cos A}{2}$$

$$\cos A = \pm \sqrt{\frac{1 + \cos 2A}{2}}.$$

Letting $A = \theta/2$, we have

$$\cos \frac{\theta}{2} = \pm \sqrt{\frac{1 + \cos \theta}{2}}, \qquad (7.35)$$

where the sign is determined by the quadrant of $\theta/2$.

A calculator with a square root key would be helpful in using formulas (7.34) and (7.35).

Half-angle formulas for the tangent function can be determined by taking the ratio of $\sin (\theta/2)$ to $\cos (\theta/2)$:

$$\tan \frac{\theta}{2} = \frac{\pm \sqrt{\dfrac{1 - \cos \theta}{2}}}{\pm \sqrt{\dfrac{1 + \cos \theta}{2}}} = \pm \sqrt{\frac{1 - \cos \theta}{1 + \cos \theta}}$$

$$= \sqrt{\frac{1 - \cos \theta}{1 + \cos \theta}} \cdot \sqrt{\frac{1 + \cos \theta}{1 + \cos \theta}}$$

$$= \sqrt{\frac{1 - \cos^2 \theta}{(1 + \cos \theta)^2}} = \sqrt{\frac{\sin^2 \theta}{(1 + \cos \theta)^2}}.$$

Or

$$\tan \frac{\theta}{2} = \frac{\sin \theta}{1 + \cos \theta}. \qquad (7.36)$$

An alternate form of the tangent half-angle formula is

$$\tan\frac{\theta}{2} = \frac{\sin\theta}{1+\cos\theta} = \frac{\sin\theta}{1+\cos\theta}\cdot\frac{1-\cos\theta}{1-\cos\theta}$$

$$= \frac{(\sin\theta)(1-\cos\theta)}{1-\cos^2\theta} = \frac{(\sin\theta)(1-\cos\theta)}{\sin^2\theta}.$$

Or

$$\tan\frac{\theta}{2} = \frac{1-\cos\theta}{\sin\theta}. \tag{7.37}$$

Example 7.31 Find the value of sin 15° without using Appendix Table 4 or a calculator.

Solution:

$$\sin 15° = \sin\frac{30°}{2} = \sqrt{\frac{1-\cos 30°}{2}}$$ [using (7.34) with the positive square root, since 15° is in quadrant I]

$$= \sqrt{\frac{1-\sqrt{3}/2}{2}}$$

$$= \frac{1}{2}\sqrt{2-\sqrt{3}}$$

$$\approx 0.2588 \text{ to four decimal places.}$$

Example 7.32 Find the value of cos (−22.5°) without using Appendix Table 4 or a calculator.

Solution:

$$\cos(-22.5°) = \cos\left(\frac{-45°}{2}\right)$$

$$= \sqrt{\frac{1+\cos(-45°)}{2}}$$ [using (7.35) with the positive square root, since −22.5° is in quadrant IV]

$$= \sqrt{\frac{1+\sqrt{2}/2}{2}}$$

$$= \frac{1}{2}\sqrt{2+\sqrt{2}}$$

$$\approx 0.9239 \text{ to four decimal places.}$$

Example 7.33

Find the value of $\tan 105°$ without using Appendix Table A or a calculator.

Solution:

$$\tan 105° = \tan \frac{210°}{2} = \frac{1 - \cos 210°}{\sin 210°}. \qquad [\text{using (7.37)}]$$

Since the angle $210°$ is in quadrant III, both the sine and cosine values are negative. The related angle is $30°$.

$$\tan 105° = \frac{1 - (-\sqrt{3}/2)}{\frac{-1}{2}}$$

$$= \frac{1 + \sqrt{3}/2}{\frac{-1}{2}}$$

$$= -(2 + \sqrt{3})$$

$$\approx -3.7321 \text{ to four decimal places.}$$

Example 7.34

Given that $\cos A = -\frac{8}{17}$ and A is in quadrant II, find **(a)** $\sin (A/2)$, **(b)** $\cos (A/2)$, and **(c)** $\tan (A/2)$.

Solution: Make a sketch representing angle A in standard position (see Figure 7.9), and determine the third side of the related triangle. Use (7.34) to (7.36) to find the desired half-angles:

$$x^2 + y^2 = r^2$$

$$y^2 = r^2 - x^2$$

$$y^2 = 289 - 64 = 225 \qquad r = 17, \quad x = -8$$

$$y = 15.$$

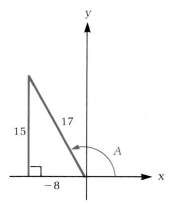

FIGURE 7.9

(a) $\sin \dfrac{A}{2} = \sqrt{\dfrac{1 - \cos A}{2}}$ [using (7.34) with the positive square root, since A being in quadrant II implies that $A/2$ is in Quadrant I]

$= \sqrt{\dfrac{1 - (-\frac{8}{17})}{2}}$

$= \sqrt{\dfrac{25}{34}}$

≈ 0.8575 to four decimal places.

(b) $\cos \dfrac{A}{2} = \sqrt{\dfrac{1 + \cos A}{2}}$ [using (7.35) with the positive square root, since $A/2$ is in quadrant I]

$= \sqrt{\dfrac{1 + (-\frac{8}{17})}{2}}$

$= \sqrt{\dfrac{9}{34}}$

≈ 0.5145 to four decimal places.

(c) $\tan \dfrac{A}{2} = \dfrac{\sin A}{1 + \cos A}$ [using (7.36)]

$= \dfrac{\frac{15}{17}}{1 + (-\frac{8}{17})}$

$= \dfrac{15}{9}$

≈ 1.6667 to four decimal places.

Note that $\tan (A/2)$ could have been found by simply taking the ratio of $\sin (A/2)$ to $\cos (A/2)$. This would omit the step of finding the third side of the triangle and would naturally yield the same result.

Example 7.35   Solve for all values of x in $[0, 2\pi]$:

$$\cos^2 \dfrac{x}{2} = \cos^2 x - 1.$$

Solution: Using formula (7.35) and squaring both sides of the formula, we get

$$\cos^2 \dfrac{x}{2} = \dfrac{1 + \cos x}{2}.$$

Therefore,

$$\frac{1 + \cos x}{2} = \cos^2 x - 1$$

and

$$2 \cos^2 x - \cos x - 3 = 0$$

$$(2 \cos x - 3)(\cos x + 1) = 0$$

$$2 \cos x - 3 = 0 \quad \text{or} \quad \cos x + 1 = 0$$

$$\cos x = \tfrac{3}{2} \quad \text{or} \quad \cos x = -1$$

Since the cosine function is not greater than 1, the equation $\cos x = \tfrac{3}{2}$ has no solution. The second equation $\cos x = -1$ yields $x = \pi$ as the only solution to the equation.

Other multiple-angle formulas can be developed with the help of (7.29) to (7.33). We now develop the formulas for $\sin 3x$, $\cos 3x$, and $\tan 3x$:

$$\begin{aligned}
\sin 3x &= \sin (2x + x) \\
&= \sin 2x \cos x + \cos 2x \sin x && [\text{using (7.21)}] \\
&= (2 \sin x \cos x)(\cos x) \\
&\quad + (1 - 2 \sin^2 x)(\sin x) && [\text{using (7.29) and (7.31)}] \\
&= 2 \sin x \cos^2 x + \sin x \\
&\quad - 2 \sin^3 x && (\text{multiplying}) \\
&= 2 \sin x (1 - \sin^2 x) + \sin x \\
&\quad - 2 \sin^3 x && [\text{using (7.6)}] \\
&= 2 \sin x - 2 \sin^3 x + \sin x \\
&\quad - 2 \sin^3 x && (\text{multiplying}) \\
&= 3 \sin x - 4 \sin^3 x && (\text{combining like terms})
\end{aligned}$$

$$\sin 3x = 3 \sin x - 4 \sin^3 x, \tag{7.38}$$

$$\begin{aligned}
\cos 3x &= \cos (2x + x) \\
&= \cos 2x \cos x - \sin 2x \sin x && [\text{using (7.20)}] \\
&= (1 - 2 \sin^2 x) \cos x \\
&\quad - (2 \sin x \cos x) \sin x && [\text{using (7.31) and (7.29)}] \\
&= \cos x - 2 \sin^2 x \cos x \\
&\quad - 2 \sin^2 x \cos x && (\text{distributing}) \\
&= \cos x - 4 \sin^2 x \cos x && (\text{combining})
\end{aligned}$$

$$\cos 3x = \cos x - 4 \sin^2 x \cos x, \tag{7.39}$$

$$\tan 3x = \tan (2x + x)$$

$$= \frac{\tan 2x + \tan x}{1 - \tan 2x \tan x} \qquad [\text{using } (7.23)]$$

$$= \frac{\dfrac{2 \tan x}{1 - \tan^2 x} + \tan x}{1 - \dfrac{2 \tan x}{1 - \tan^2 x} \cdot \tan x} \qquad [\text{using } (7.33)]$$

$$= \frac{2 \tan x + \tan x - \tan^3 x}{1 - \tan^2 x - 2 \tan^2 x} \qquad (\text{clearing fractions})$$

$$= \frac{3 \tan x - \tan^3 x}{1 - 3 \tan^2 x} \qquad (\text{combining})$$

$$\tan 3x = \frac{3 \tan x - \tan^3 x}{1 - 3 \tan^2 x}. \qquad\qquad \textbf{(7.40)}$$

Example 7.36   Prove the identity $\cos 3x = 4 \cos^2 x - 3 \cos x$.

Solution: Use formula (7.39):

$$\begin{aligned}
\cos 3x &= \cos x - 4 \sin^2 x \cos x \\
&= \cos x - 4(1 - \cos^2 x)(\cos x) \qquad [\text{using } (7.6)] \\
&= \cos x - 4 \cos x + 4 \cos^3 x \qquad (\text{distributing}) \\
&= 4 \cos^3 x - 3 \cos x. \qquad\qquad (\text{combining})
\end{aligned}$$

EXERCISES

Using exact values of trigonometric functions at 30°, 45°, and 60° and their integral multiples, find the exact values of the expressions in Exercises 1 to 12. (Do not use a calculator or Appendix Table 4.)

1. $\cos 15°$
2. $\tan 15°$
3. $\sin 22.5°$
4. $\tan 22.5°$
5. $\sin 75°$
6. $\cos 75°$
7. $\tan 75°$
8. $\cos 67.5°$
9. $\cos 112.5°$
10. $\sin 105°$
11. $\tan 157.5°$
12. $\sin (-112.5°)$

Find $\sin (A/2)$, $\cos (A/2)$, and $\tan (A/2)$, given the statements in Exercises 13 to 17. (Do not use a calculator or Appendix Table 4.)

13. $\csc A = \frac{25}{7}$, A in quadrant I
14. $\cot A = \frac{15}{8}$, A in quadrant III
15. $\cos A = -\frac{5}{13}$, A in quadrant II
16. $\cos A = \frac{12}{37}$, A in quadrant IV
17. $\csc A = \frac{5}{3}$, A in quadrant I

Prove the identities in Exercises 18 to 38.

18. $\cos^2 x = \sin^2 x + \cos 2x$
19. $\cos 4x = 1 - 8 \sin^2 x \cos^2 x$
20. $\cos^4 x - \sin^4 x = \cos 2x$
21. $\dfrac{1 + \tan^2 x}{1 - \tan^2 x} = \sec 2x$
22. $\tan 2x = (\tan x)(1 + \sec 2x)$
23. $1 - \sin 2x = (\sin x - \cos x)^2$

24. $\dfrac{\cos 3x}{\sin x} + \dfrac{\sin 3x}{\cos x} = 2 \cot 2x$ **25.** $2 \csc 2x = \sec x \csc x$ **26.** $2 \cot 2x = \cot x - \tan x$

27. $\sec 2x = \dfrac{\sec^2 x}{2 - \sec^2 x}$ **28.** $\sin 10A = 2 \sin 5A \cos 5A$ **29.** $\cos^2 3x - \sin^2 3x = \cos 6x$

30. $(\sin \theta + \cos \theta)^2 = 1 + \sin 2\theta$ **31.** $\cos 4\theta = 8 \cos^4 \theta - 8 \cos^2 \theta + 1$ **32.** $\cos^4 x - \sin^4 x = \cos 2x$

33. $\tan \theta + \cot \theta = 2 \csc 2\theta$ **34.** $2 \sin^2 2x + \cos 4x = 1$ **35.** $4 \sin \dfrac{\theta}{2} \cos \dfrac{\theta}{2} = 2 \sin \theta$

36. $\dfrac{\sin^2 2A}{\sin^2 A} = 4 - 4 \sin^2 A$ **37.** $\cot 2\theta = \dfrac{\cot^2 \theta - 1}{2 \cot \theta}$ **38.** $\csc 2x = \tfrac{1}{2}\sec x \csc x$

In Exercises 39 to 51, find all solutions of the equations in the interval $[0, 2\pi]$. Express the solutions in both degrees and radians.

39. $\sin 2x + \sin x = 0$ **40.** $\cos x + \cos 2x = 0$ **41.** $\cos A - \sin 2A = 0$

42. $\cos 2\theta - \tan \theta = 1$ **43.** $\tan 2x = \tan x$ **44.** $\tan 2A - 2 \cos A = 0$

45. $\sec 2x \csc 2x = 2 \csc 2x$ **46.** $\sin \dfrac{\theta}{2} + \cos \theta = 1$ **47.** $4 \sin^2 \dfrac{\theta}{2} + \cos^2 \theta = 2$

48. $\cos \dfrac{x}{2} = \cos x + 1$ **49.** $\sin^2 \dfrac{x}{2} = \sin^2 x$ **50.** $\sin 3x + \sin x = 0$

51. $\cos 3A + \sin 2A - \cos A = 0$

In Exercises 52 to 56, sketch a graph of the function by utilizing the formulas of this section to obtain an equivalent function of simpler form. Determine the maximum and minimum values of the function and where they occur.

52. $y = \sin 2x \cos 2x$ **53.** $y = \cos^2 x - \sin^2 x$ **54.** $y = \dfrac{1 - \cos x}{2}$ **55.** $y = \dfrac{\sin x}{1 + \cos x}$

Use a calculator to show that the statements in Exercises 56 to 65 are true.

56. $\sin 2(65°) = 2 \sin 65° \cos 65°$ **57.** $\sin 2(1.02) = 2 \sin 1.02 \cos 1.02$

58. $\cos^2 0.45 = \tfrac{1}{2}[1 + \cos 2(0.45)]$ **59.** $\cos^2 115° = \tfrac{1}{2}[1 + \cos 2(115°)]$

60. $\sin^2 1.7 = \tfrac{1}{2}[1 - \cos 2(1.7)]$ **61.** $\sin^2 15° = \tfrac{1}{2}[1 - \cos 2(15°)]$

62. $\cos 2(0.26) = \cos^2 0.26 - \sin^2 0.26$ **63.** $\tan \dfrac{1}{2}(3.0) = \dfrac{\sin 3.0}{1 + \cos 3.0}$

64. $\tan \dfrac{1}{2}(3.0) = \dfrac{1 - \cos 3.0}{\sin 3.0}$ **65.** $\sin 3(80°) = 3 \sin 80° - 4 \sin^3 80°$

7.5

inverse trigonometric
function

INVERSE TRIGONOMETRIC FUNCTIONS

Before we explore the concept of *inverse trigonometric functions*, let us review the basic concept of a function. In Section 3.2, we stated that a function is a relation that assigns one and only one image to each element of the domain. When the function also has the property that each image in the range corresponds to one and only one element of the domain, then the function is one-to-one, and we can define a new function that is the inverse of the original function.

Clearly, the sine function is not a one-to-one function. The domain is the set of real numbers, and the range is the interval $\{-1 \le y \le 1\}$. For every y in this

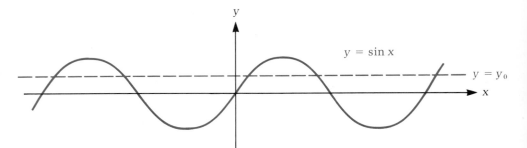

FIGURE 7.10

interval, the equation $y = \sin x$ has infinitely many solutions. This fact is apparent from the graph of the function. A horizontal line drawn through any range value y_0 intersects the curve repeatedly, as shown in Figure 7.10.

Nevertheless, there are times when we need to consider the angle x in the relation $y = \sin x$. In other words, we are looking specifically for the number x whose sine is y. This entails using the inverse of the sine function, which does not exist, strictly speaking. We must find a way to get around this difficulty. We do so by restricting the domain of the function and describing a new function $y = \text{Sin } x$ (read "cap-sine of x"). The capital letter S distinguishes this new function from the sine function. The range of this new function is the same as that of the original function, namely, $-1 \leq y \leq 1$. From the graph of this function, as shown in Figure 7.11, it is clear that the function is one-to-one and thus has an inverse function.

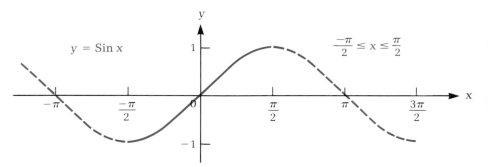

FIGURE 7.11

The concept of the inverse sine function was explored in the eighteenth century by the mathematicians Bernoulli and Euler. They introduced the notation $x = \text{Arcsin } y$ to denote an angle whose sine is y. Then in the early nineteenth century, the mathematician John Herschel introduced another notation which is in common use today, namely, $x = \text{Sin}^{-1} y$. Both statements mean that x is an angle whose sine is y. Note that f^{-1} is *not* exponential notation; $f^{-1} \neq 1/f$.

We obtain the graph of the inverse function by reflecting the original function $y = \sin x$ through the line $y = x$. The domain and range of the original function become the range and domain, respectively, of the inverse function (see Figure 7.12). Algebraically, we interchange x and y to obtain $x = \text{Sin } y$ and then express explicitly: $y = \text{Sin}^{-1} x$. Two important identities result from these new functions.

$$\sin (\text{Sin}^{-1} x) = x \qquad (-1 \le x \le 1)$$

$$\text{Sin}^{-1} (\sin y) = y \qquad \left(\frac{-\pi}{2} \le y \le \frac{\pi}{2}\right)$$

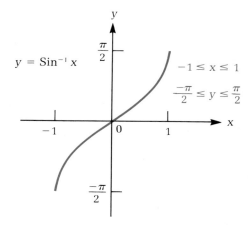

$y = \text{Sin}^{-1} x$

$-1 \le x \le 1$

$\dfrac{-\pi}{2} \le y \le \dfrac{\pi}{2}$

FIGURE 7.12

Example 7.37

Find Arcsin $\frac{1}{2}$.

Solution: Since $\sin 30° = \frac{1}{2}$,

$$\text{Arcsin } \frac{1}{2} = 30° \quad \text{or} \quad \frac{\pi}{6}.$$

Example 7.38

Find $\text{Sin}^{-1} (-0.2419)$.

Solution: Using a calculator or Appendix Table 4, we find that $\sin 14° = 0.2419$; therefore,

$$\text{Sin}^{-1} (-0.2419) = -14° \quad \text{or} \quad -0.2443 \text{ radian.}$$

The other inverse trigonometric functions are defined in a corresponding way. By restricting the domain of the cosine function, as shown in Figure 7.13, we have a new function $y = \text{Cos } x$, which is a one-to-one function.

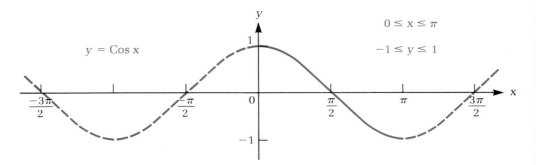

$$0 \leq x \leq \pi$$

$$-1 \leq y \leq 1$$

FIGURE 7.13

The inverse function $x = \text{Cos } y$, or explictly $y = \text{Cos}^{-1} x$, denotes an angle y whose cosine is x. The domain and range of the original function become the range and domain, respectively, of the inverse function, as shown in Figure 7.14.

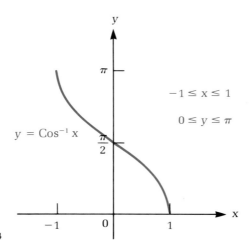

$$-1 \leq x \leq 1$$

$$0 \leq y \leq \pi$$

$$y = \text{Cos}^{-1} x$$

FIGURE 7.14

Example 7.39   Find Arccos $(\frac{-1}{2})$.

Solution: Let $\theta = \text{Arccos} \left(-\frac{1}{2}\right)$. Then

$$\cos \theta = -\frac{1}{2}$$

Hence,

$$\theta = 120° \quad \text{or} \quad \frac{2\pi}{3}.$$

Example 7.40

Find sin (Arccos $\frac{3}{5}$).

Solution: Let $\theta = $ Arccos $\frac{3}{5}$; then cos $\theta = \frac{3}{5}$. Since Arccos is defined from 0 to π, the angle θ is in the first quadrant. Drawing the corresponding right triangle as shown in Figure 7.15, we find $y = 4$. Therefore, sin (Arccos $\frac{3}{5}$) $= \frac{4}{5}$.

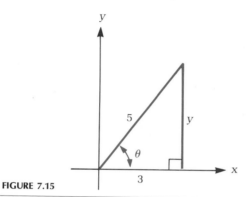

FIGURE 7.15

The tangent function has a period of π and completes a full cycle between $-\pi/2$ and $\pi/2$. By restricting the domain to these values, we have a one-to-one function $y = $ Tan x with asymptotes at $\pm \pi/2$ and a range of all real numbers, as shown in Figure 7.16.

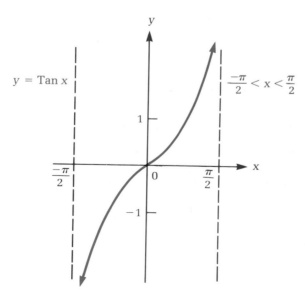

FIGURE 7.16

The inverse tangent function $x = \text{Tan } y$, or explicitly $y = \text{Tan}^{-1} x$, has a domain of all real numbers and a range of $-\pi/2 < y < \pi/2$, with asymptotes at $y = \pm\pi/2$. Its graph is shown in Figure 7.17.

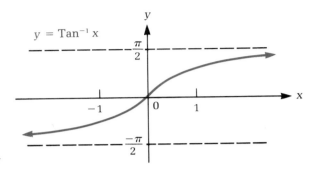

FIGURE 7.17

Example 7.41

Find $\text{Tan}^{-1} \sqrt{3}$.

Solution: Let $\theta = \text{Tan}^{-1} \sqrt{3}$; then

$$\tan \theta = \sqrt{3}.$$

Hence,

$$\theta = 60° \quad \text{or} \quad \frac{\pi}{3}.$$

The functions Arccot x, Arcsec x, and Arccsc x can be expressed in terms of the corresponding reciprocal function:

$$\text{Arccsc } x = \text{Arcsin } \frac{1}{x},$$

$$\text{Arcsec } x = \text{Arccos } \frac{1}{x},$$

$$\text{Arccot } x = \text{Arctan } \frac{1}{x}.$$

As a numerical example of each,

$$\text{Arccsc } 2 = \text{Arcsin } \frac{1}{2} = 30° \quad \text{or} \quad \frac{\pi}{6},$$

$$\text{Arcsec } \sqrt{2} = \text{Arccos } \frac{\sqrt{2}}{2} = 45° \quad \text{or} \quad \frac{\pi}{4},$$

$$\text{Arccot } \frac{\sqrt{3}}{3} = \text{Arctan } \sqrt{3} = 60° \quad \text{or} \quad \frac{\pi}{3}.$$

Example 7.42

Find the solutions of $2 \tan^2 x + 9 \tan x + 3 = 0$ in the interval $(-\pi/2, \pi/2)$.

Solution:

$$2 \tan^2 x + 9 \tan x + 3 = 0.$$

This equation may be considered as a quadratic equation in $\tan x$ ($2y^2 + 9y + 3 = 0$). Since the equation is not factorable, we use the quadratic formula:

$$\tan x = \frac{-9 \pm \sqrt{81 - 24}}{4}$$

$$= \frac{-9 \pm \sqrt{57}}{4}$$

$$= \frac{-9 \pm 7.5498}{4} \text{ approximately.}$$

Therefore

$$x = \text{Arctan}\,(-0.3626)$$

or

$$x = \text{Arctan}\,(-4.1375).$$

Using a calculator or Appendix Table 4, we find to the nearest degree

$$x = -20° \quad \text{and} \quad x = -76°.$$

Table 7.1 lists the restricted domains of the trigonometric functions of angles measured in radians or real numbers.

TABLE 7.1

Function	Domain	Range
$y = \text{Sin } x$	$-\dfrac{\pi}{2} \le x \le \dfrac{\pi}{2}$	$-1 \le y \le 1$
$y = \text{Cos } x$	$0 \le x \le \pi$	$-1 \le y \le 1$
$y = \text{Tan } x$	$-\dfrac{\pi}{2} < x < \dfrac{\pi}{2}$	all real numbers

In Table 7.2 are listed the definitions of the inverse trigonometric functions. Mathematicians do not all agree on the range of the inverse trigonometric functions. Those that are listed are the most common. These range values are also *principal values* referred to as the **principal values** of the function.

TABLE 7.2

Function	Domain	Range
$y = \text{Sin}^{-1} x$	$-1 \le x \le 1$	$-\dfrac{\pi}{2} \le y \le \dfrac{\pi}{2}$
$y = \text{Cos}^{-1} x$	$-1 \le x \le 1$	$0 \le y \le \pi$
$y = \text{Tan}^{-1} x$	all real numbers	$-\dfrac{\pi}{2} < y < \dfrac{\pi}{2}$

EXERCISES

Find the values of the expressions in Exercises 1 to 14 without using Appendix 4 or a calculator.

1. Arcsin 1
2. Arccos (-1)
3. Arcsin (-1)
4. Arctan 0

5. Arccos $(\sqrt{3}/2)$
6. Arctan 1
7. Arcsin $(-\frac{1}{2})$
8. Arccos $(-\sqrt{3}/2)$

9. $\text{Tan}^{-1} (-1)$
10. $\text{Cos}^{-1} 0$
11. $\text{Sin}^{-1} 1$
12. $\text{Sin}^{-1} 0$

13. $\text{Sin}^{-1} (-1)$
14. $\text{Sin}^{-1} (-\sqrt{3}/2)$

Find the values of the expressions in Exercises 15 to 24 by using Appendix Table 4 or a calculator.

15. Arctan 115
16. Arcsin 0.5592
17. Arctan 0.6128
18. Arccos 0.4536

19. Arcsin (-0.8572)
20. Arccos (-0.5275)
21. Arctan (-0.9657)
22. Arctan (-2.375)

23. $\text{Cos}^{-1} 0.3283$
24. $\text{Sin}^{-1} 0.6820$

Write the values of each expression in Exercises 25 to 46 without using Appendix Table 4 or a calculator.

25. $\sin [\text{Arcsin} (-1)]$
26. $\cos (\text{Arccos} \frac{1}{2})$
27. $\cos [\text{Arcsin} (-\frac{1}{2})]$

28. $\tan [\text{Arcsin} (-\frac{1}{2})]$
29. $\tan [\text{Arcsin} (-\sqrt{3}/2)]$
30. $\cot [\text{Arccos} (-\frac{1}{2})]$

31. $\tan [\text{Arccos} (-\frac{1}{2})]$
32. $\sec [\text{Arcsin} (-\frac{1}{2})]$
33. $\sec [\text{Arcsin} (-\sqrt{2}/2)]$

34. $\csc (\text{Arctan} 1)$
35. $\sin [\text{Arctan} (-1)]$
36. $\sin [\text{Arccos} (-\sqrt{3}/2)]$

37. $\text{Sin}^{-1} (\sin 45°)$
38. $\text{Cos}^{-1} (\cos 120°)$
39. $\text{Cos}^{-1} (\cos 210°)$

40. $\text{Tan}^{-1} (\sin 90°)$
41. $\text{Sin}^{-1} (\tan \pi)$
42. $\text{Tan}^{-1} [\sin (7\pi/2)]$

43. $\text{Cos}^{-1} (\cos 112°)$
44. $\tan (\text{Cos}^{-1} \frac{4}{5})$
45. $\sin (\text{Cos}^{-1} \frac{12}{13})$

46. $\tan (\text{Tan}^{-1} x)$

Find the values of the expressions in Exercises 47 to 50 by using Appendix Table 4 or a calculator.

47. $\sin (\text{Arccos} 0.2840)$
48. $\cos (\text{Arcsin} 0.9605)$
49. $\cos (\text{Arcsin} 0.8090)$
50. $\cos (\text{Arctan} 2.356)$

In Exercises 51 to 55, find the solutions of the equations in the given interval by using inverse trigonometric functions.

51. $3 \sin^2 \theta + 3 \sin \theta - 1 = 0$, $[-\pi/2, \pi/2]$
52. $3 \sin^2 \theta + 8 \sin \theta + 2 = 0$, $[-\pi/2, \pi/2]$
53. $35 \cos^4 \theta - 22 \cos^2 \theta + 3 = 0$, $[0, \pi]$
54. $10 \sin 2\theta - 12 \cos \theta + 15 \sin \theta - 9 = 0$, $(-\pi/2, \pi/2)$
55. $6 \tan^4 \theta - 31 \tan^2 \theta + 5 = 0$, $(-\pi/2, \pi/2)$

KEY WORDS

reciprocal relation
quotient relation
Pythagorean relation
trigonometric identity
complementary angles
cofunction identity
trigonometric equation

addition formulas
product formulas
multiple-angle formulas
half-angle formulas
inverse trigonometric function
Arcsin x
Sin^{-1} x

Cos^{-1} x
Arccos x
Tan^{-1}x
Arctan x
principal values

KEY FORMULAS

• *Basic Identities*

$$\csc \theta = \frac{1}{\sin \theta} \qquad \cot \theta = \frac{1}{\tan \theta}$$

$$\sec \theta = \frac{1}{\cos \theta} \qquad \sin^2 \theta + \cos^2 \theta = 1$$

$$\cot \theta = \frac{\cos \theta}{\sin \theta} \qquad 1 + \tan^2 \theta = \sec^2 \theta$$

$$\tan \theta = \frac{\sin \theta}{\cos \theta} \qquad 1 + \cot^2 \theta = \csc^2 \theta$$

• *Addition Identities*

$$\sin (A + B) = \sin A \cos B + \cos A \sin B$$

$$\sin (A - B) = \sin A \cos B - \cos A \sin B$$

$$\cos (A + B) = \cos A \cos B - \sin A \sin B$$

$$\cos (A - B) = \cos A \cos B + \sin A \sin B$$

$$\tan (A + B) = \frac{\tan A + \tan B}{1 - \tan A \tan B}$$

$$\tan (A - B) = \frac{\tan A - \tan B}{1 + \tan A \tan B}$$

• *Double-Angle Identities*

$$\sin 2A = 2 \sin A \cos A$$

$$\cos 2A = \cos^2 A - \sin^2 A = 2 \cos^2 A - 1$$

$$= 1 - 2 \sin^2 A$$

$$\tan 2A = \frac{2 \tan A}{1 - \tan^2 A}$$

• *Half-Angle Identities*

$$\sin \frac{A}{2} = \pm \sqrt{\frac{1 - \cos A}{2}}$$

$$\cos \frac{A}{2} = \pm \sqrt{\frac{1 + \cos A}{2}}$$

$$\tan \frac{A}{2} = \sqrt{\frac{1 - \cos A}{1 + \cos A}} = \frac{1 - \cos A}{\sin A} = \frac{\sin A}{1 + \cos A}$$

EXERCISES FOR REVIEW

1. Verify the Pythagorean relation for $\theta = 30°$.
2. Verify the Pythagorean relation for $\theta = 60°$.
3. Express $\cos x$ in terms of $\sin x$.
4. Express $\tan x$ in terms of $\sin x$.

Use formulas (7.1) to (7.8) to obtain the relations in Exercises 5 to 10.

5. $\csc x = \pm\sqrt{\cot^2 x + 1}$

6. $\cos x = \dfrac{\cot x}{\pm\sqrt{1 + \cot^2 x}}$

7. $\sin x = \dfrac{\tan x}{\pm\sqrt{1 + \tan^2 x}}$

8. $(\sec x + 1)(\sec x - 1) = \tan^2 x$

9. $(1 + \sin x)(1 - \sin x) = \cos^2 x$

10. $\csc x - \cos x \cot x = \sin x$

11. Assume that θ is an angle in the second quadrant. Find the value of $\cos \theta \csc^2 \theta$ when $\sin \theta = \frac{2}{3}$.

12. Assume that y is an angle in the fourth quadrant. Find the value of

$$\frac{\sin^2 y}{(\tan^2 y + 1)(\cos y)}$$

when $\cos y = \frac{3}{5}$.

Prove the identities in Exercises 13 to 34.

13. $\sin^2 \theta - \cos^2 \theta = 1 - 2 \cos^2 \theta$

14. $\tan x = \dfrac{\pm\sqrt{1 - \cos^2 x}}{\cos x}$

15. $2 \sin x + \tan x = \dfrac{2 + \sec x}{\csc x}$

16. $(1 + \cos x)(\csc x - \cot x) = \sin x$

17. $(\sin x + \cos x)(\sec x - \csc x) = \tan x - \cot x$

18. $\dfrac{\tan^2 x}{\sin^2 x} - \dfrac{\sin^2 x}{\cos^2 x} = 1$

19. $(1 - \sin x)(\sec x + \tan x) = \cos x$

20. $\dfrac{1}{\tan x + \cot x} = \sin x \cos x$

21. $\dfrac{1}{1 + \sin x} + \dfrac{1}{1 - \sin x} = 2 \sec^2 x$

22. $4 \sin^2 x \cos^2 x = 1 - \cos^2 2x$

23. $\cos 4x = 1 - 8 \sin^2 x \cos^2 x$

24. $\sec 2x = \dfrac{\sec^2 x}{2 - \sec^2 x}$

25. $2 \csc 2x = \sec x \csc x$

26. $\cos^4 x - \sin^4 x = \cos 2x$

27. $\tan 2x = (\tan x)(1 + \sec 2x)$

28. $2 \cot 2x = \cot x - \tan x$

29. $\sec^2 \dfrac{x}{2} = \dfrac{2 \tan x}{\tan x + \sin x}$

30. $\csc^2 \dfrac{x}{2} - 1 = \dfrac{\sec x + 1}{\sec x - 1}$

31. $\tan \dfrac{x}{2} = \csc x - \cot x$

32. $\tan x \sin 2x = 2 \sin^2 x$

33. $\dfrac{1 - \cos 2x}{\sin 2x} = \tan x$

34. $\sin^2 x = \dfrac{1 - \cos 2x}{2}$

Solve the equations in Exercises 35 to 40 for positive angles less than $360°$.

35. $\sin^2 \theta = \frac{1}{4}$

36. $\cos^2 x = \frac{1}{4}$

37. $2 \sin^4 x - 9 \sin^2 x + 4 = 0$

38. $\cos^2 x + \sin x = 1$

39. $2 \cos^2 \theta + \cos \theta - 3 = 0$

40. $\sqrt{3} \tan x \sec x - 2 = \sec x - 2\sqrt{3} \tan x$

Use the exact values of the trigonometric functions at $30°$, $45°$, and $60°$ and their integral multiples to find the values in Exercises 41 to 46.

41. $\sin 255°$

42. $\cos 22.5°$

43. $\tan(-165°)$

44. $\tan 255°$

45. $\sin 67.5°$

46. $\cos 105°$

Given that $\sin A = \frac{2}{3}$, A is in quadrant II, $\cos B = \frac{3}{5}$, and B is in quadrant IV, find the value of the expressions in Exercises 47 to 52.

47. $\sin (A - B)$ **48.** $\sin (A + B)$ **49.** $\cos (A + B)$

50. $\cos (A - B)$ **51.** $\tan (A + B)$ **52.** $\tan (A - B)$

Reduce the expressions in Exercises 53 and 54 to a single term.

53. $(1 + \cot^2 x) \tan^2 x$ **54.** $\sin x \, (\csc x - \sin x)$

55. Find **(a)** $\sin (A/2)$, **(b)** $\cos (A/2)$, and **(c)** $\tan (A/2)$, given that $\tan A = \frac{5}{12}$ and A is in quadrant III.

56. Show that $\text{Arctan} \frac{2}{3} + \text{Arctan} \frac{1}{5} = \pi/4$.

57. Show that $\text{Arctan} \, 2 + \text{Arctan} \, 3 = \pi - \text{Arctan} \, 1$.

58. Show that $\text{Arccos} \frac{1}{2} + 2 \, \text{Arcsin} \frac{1}{2} = 2\pi/3$.

Find the values of the expressions in Exercises 59 to 66 without using Appendix Table 4 or a calculator.

59. $\text{Arctan} \, (\sqrt{3}/3)$ **60.** $\text{Arctan} \, 0$ **61.** $\text{Cos}^{-1} \, (-\sqrt{2}/2)$ **62.** $\text{Tan}^{-1} \, (-\sqrt{3})$

63. $\text{Cos}^{-1} \, (-\frac{1}{2})$ **64.** $\sin (\text{Sin}^{-1} x)$ **65.** $\sin (\text{Cos}^{-1} x)$ **66.** $\text{Sin}^{-1} \, (\tan 0)$

Find the values of the expressions in Exercises 67 to 70 by using Appendix Table 4 or a calculator.

67. $\text{Sin}^{-1} \, (0.1564)$ **68.** $\text{Arccos} \, 0.7234$ **69.** $\text{Arctan} \, 2.300$ **70.** $\text{Cos}^{-1} \, (0.3090)$

71. Find the solution of the equation in the given interval by using inverse trigonometric functions:

$$2 \cos^2 \theta - \cos \theta - 1 = 0 \qquad [0, \pi].$$

CHAPTER TEST

1. Given that angle x is in the third quadrant and $\sin x = -\frac{3}{5}$, find the value of the expression

$$\frac{\tan x + 1}{1 - \cos x}.$$

2. Prove the identity

$$\frac{1 + \cos x}{\sin x} + \frac{\sin x}{1 + \cos x} = 2 \csc x.$$

3. Prove the identity

$$4 \sin \frac{x}{2} \cos \frac{x}{2} = 2 \sin x.$$

4. Solve the equation for positive angles less than 360°:
   (a) $4 \sin^2 \theta = 3$ **(b)** $3 \tan^2 x = 1$ **(c)** $2 \cos^2 t + 3 \cos t + 1 = 0$

5. Find the exact value of $\tan 285°$ without the use of tables or a calculator.

6. If angle A is in quadrant II, $\csc A = \frac{5}{3}$, angle B is in quadrant IV, and $\cos B = \frac{8}{17}$, find **(a)** $\sin (B - A)$ and **(b)** $\tan (A + B)$.

7. If angle A is in quadrant III and $\tan A = \frac{5}{12}$, find **(a)** $\sin (A/2)$, **(b)** $\tan (A/2)$, **(c)** $\cos 2A$.

8. Find the value of $\cos (\pi + \text{Sin}^{-1} \frac{1}{3})$.

TRIANGLES, VECTORS, AND COMPLEX NUMBERS

8

INTRODUCTION

In this chapter we learn to solve triangles that are not necessarily right triangles by using two well-known laws, then we use these laws to solve a variety of applications. In Section 8.4 we derive several formulas for determining the area of a triangular region. The chapter continues with a study of vectors and polar coordinates, culminating with the trigonometric form of complex numbers.

8.1

LAW OF SINES

A triangle that does not contain a right angle is called an *oblique triangle*. It has either three acute angles or one obtuse angle and two acute angles. In Section 6.4 we learned that a right triangle is uniquely determined when two of its five parts (other than the right angle) are given and at least one of those is the length of a side. To solve oblique triangles, we must be given three parts of the triangle, at least one of which is a side. Triangles may be classified according to the given information as follows (accepted abbreviations follow in parentheses).

1. Type 1: Two angles and one side (AAS)
2. Type 2: Two sides and the angle opposite one of them (SSA)
3. Type 3: Two sides and the included angle, or side-angle-side (SAS)
4. Type 4: Three sides (SSS)

We now state and prove a law that will enable us to solve triangles of types 1 and 2.

Law of Sines In any triangle, the sides are proportional to the sines of the opposite angles:

$$\frac{a}{\sin A} = \frac{b}{\sin B} = \frac{c}{\sin C}. \tag{8.1}$$

Proof: Place triangle ABC in standard position on a coordinate system. Figure 8.1(a) shows A as an acute (less than 90°) angle, and Figure 8.1(b) shows A

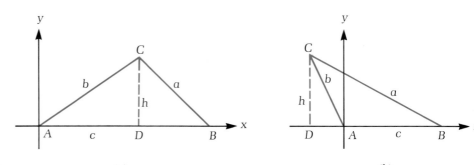

FIGURE 8.1   (a)   (b)

as an obtuse (greater than 90°) angle. The coordinates of C in either figure are $(b \cos A, b \sin A)$.

Since

$$\sin A = \frac{h}{b} \quad \text{or} \quad h = b \sin A$$

and

$$\sin B = \frac{h}{a} \quad \text{or} \quad h = a \sin B,$$

we know

$$b \sin A = a \sin B,$$

or

$$\frac{a}{\sin A} = \frac{b}{\sin B}.$$

By placing angle B in standard position, we can find the same ratio for c and $\sin C$, resulting in the following string of equalities:

$$\frac{a}{\sin A} = \frac{b}{\sin B} = \frac{c}{\sin C}.$$

Example 8.1
(Type 1: AAS)

Solve for the unknown parts of the triangle, given that $B = 30°$, $C = 45°$, and $b = 6$.

Solution:

$$A = 180° - (30° + 45°) = 105°.$$

And

$$\frac{a}{\sin A} = \frac{b}{\sin B},$$

therefore,

$$a = \frac{b \sin A}{\sin B} = \frac{6(0.9659)}{0.5000} = 11.5908.$$

Similarly,

$$\frac{c}{\sin C} = \frac{b}{\sin B},$$

therefore,

$$c = \frac{b \sin C}{\sin B} = \frac{6(0.7071)}{0.5000} = 8.4852.$$

There is a problem, however, in using the law of sines to solve type 2 problems. Since the sine of an angle is positive in both the first and second quadrants, when solving for an angle we do not know in which quadrant the angle belongs. Therefore, when two sides and the angle opposite one of them are given (SSA), this is referred to as the ambiguous case. Actually, with this given information, there are four possible cases to consider. For all four cases, we assume that we are given a, b, and A. From the data, we attempt to draw the triangle. First we construct the angle A and measure off the length b forming the segment AC. With C as a center we draw a radius a. The vertex B can be placed where this arc intersects the base line AD of the angle A.

Case 1 Here $a < b$ and a is equal to the perpendicular from C to AD. The triangle formed is a right triangle, $B = 90°$, and $\sin B = 1$ (see Figure 8.2). By the law of sines, $\sin B = (b \sin A)/a$. Therefore this case occurs whenever $\dfrac{b \sin A}{a} = 1$.

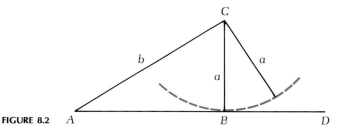

FIGURE 8.2

Example 8.2
(Type 2: SSA)

Given that $a = 4$, $b = 8$, and $A = 30°$, solve the triangle ABC.

Solution:

$$\sin B = \frac{b \sin A}{a} = \frac{8(0.5)}{4} = 1 \qquad \text{(case 1)}$$

$$B = 90° \qquad \text{(triangle } ABC \text{ is a unique right triangle)}$$

$$C = 60°$$

$$c = b \cos A$$

$$= \frac{8\sqrt{3}}{2}$$

$$= 6.9$$

Case 2 Here $a < b$ and a intersects AD in two points B and B', each of which is a vertex of a triangle that satisfies the conditions. Because the triangle is not uniquely determined, this case is referred to as the ambiguous case in the strict sense of the term. For a complete solution, it is necessary to solve both triangles.

In this case, a is longer for a given b and a given A than in the preceding case, so

$$\frac{b \sin A}{a} < 1.$$

This is expected, since B is not a right angle, and $\sin B$ must be less than 1 (see Figure 8.3).

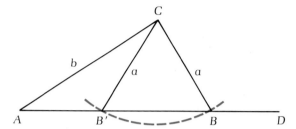

FIGURE 8.3

Example 8.3 (Type 2: SSA)

Given that $a = 21$, $b = 28$, and $A = 34°$, solve the triangle ABC.

Solution:

$$\sin B = \frac{b \sin A}{a} = \frac{28(0.5592)}{21} = 0.7456 < 1.$$

Therefore, this is case 2, so

$$B = 48°10' \qquad \text{and} \qquad B' = 180° - 48°10' = 131°50'.$$

We consider both possible triangle solutions.

Triangle ABC solution:

$$B = 48°10'$$

$$C = 180° - (34° + 48°10') = 97°50'$$

$$c = \frac{a \sin C}{\sin A} = \frac{21(0.9907)}{0.5592} = 37.2.$$

Triangle $AB'C'$ solution:

$$B' = 131°50'$$

$$C' = 180° - (34° + 131°50') = 14°10'$$

$$c = \frac{a \sin C}{\sin A} = \frac{21(0.2447)}{0.5592} = 9.2.$$

Case 3   Here $a < b$ and a is shorter than the perpendicular line from C to AD. In this case, no triangle exists satisfying the given conditions and

$$\frac{b \sin A}{a} > 1$$

(see Figure 8.4).

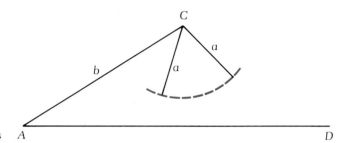

FIGURE 8.4

Example 8.4
(Type 2: SSA)

Given that $a = 5.6$, $b = 7.3$, and $A = 64°20'$, solve triangle ABC.

Solution:

$$\sin B = \frac{b \sin A}{a} = \frac{7.3(0.9013)}{5.6} = 1.17 > 1.$$

Therefore this is case 3. Since the sine of an angle cannot be greater than 1, there is no solution.

Case 4   Here $a \geq b$. In this case, the triangle always exists and is uniquely determined (see Figure 8.5).

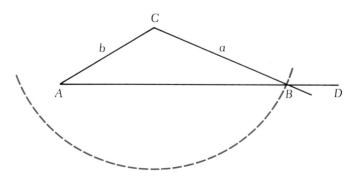

FIGURE 8.5

Example 8.5
(Type 2: SSA)

Given that $a = 25$, $b = 10$, and $A = 38°$, solve the triangle ABC.

Solution: Since $25 > 10$, this is case 4. So

$$\sin B = \frac{b \sin A}{a} = \frac{10(0.6157)}{25} = 0.2463$$

$$B = 14°21'$$

$$C = 180° - (38° + 14°20') = 127°40'$$

$$c = \frac{a \sin C}{\sin A} = \frac{25(0.7916)}{0.6157} = 32.14.$$

The law of sines cannot be directly used to solve triangles of type 3 (SAS) or type 4 (SSS). However, another set of relations known as the law of cosines is useful and is studied in Section 8.2.

EXERCISES

Solve the triangles in Exercises 1 to 10.

1. Given $a = 10$, $B = 45°$, $C = 60°$.
2. Given $b = 5$, $B = 42°$, $C = 28°$.
3. Given $a = 4$, $b = 6$, $B = 60°$.
4. Given $a = 8$, $A = 34°$, $b = 12$.
5. Given $b = 9$, $B = 24°$, $c = 9$.
6. Given $a = 65.6$, $A = 60°$, $c = 73.0$.
7. Given $b = 68$, $C = 72°$, $c = 42$.
8. Given $a = 12$, $b = 14$, $B = 38°$.
9. Given $b = 7073$, $B = 59°30'$, $c = 7836$.
10. Given $a = 22$, $c = 35$, $C = 50°10'$.

Use a hand-held calculator for Exercises 11–14.

11. Given $b = 27.36$, $B = 24°18'$, $C = 32°24'$, solve the triangle.
12. Given $a = 6425$, $B = 73°29'$, $C = 36°52'$, solve the triangle.
13. Given $c = 14.327$, $B = 68°22.7'$, $C = 29°26.3'$, solve the triangle.
14. Given $a = 4327.4$, $A = 36°16.8'$, $B = 42°24.6'$, solve the triangle.

8.2

LAW OF COSINES

Law of Cosines In any triangle, the square of any side is equal to the sum of the squares of the other two sides minus twice the product of those sides times the cosine of the included angle.

$$a^2 = b^2 + c^2 - 2bc \cos A,$$
$$b^2 = a^2 + c^2 - 2ac \cos B, \qquad (8.2)$$
$$c^2 = a^2 + b^2 - 2ab \cos C.$$

Proof: Let any triangle ABC be placed on a rectangular coordinate system as shown in Figure 8.6. Since angle A is placed in standard position, the coordinates of the vertex angle C are $(b \cos A, b \sin A)$, and the coordinates of the vertex angle B are $(c, 0)$.

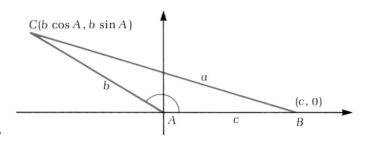

FIGURE 8.6

Using the formula for the squares of the distance between two points, we obtain

$$
\begin{aligned}
a^2 &= (b \cos A - c)^2 + (b \sin A - 0)^2 \\
&= b^2 \cos^2 A - 2bc \cos A + c^2 + b^2 \sin^2 A \\
&= b^2(\sin^2 A + \cos^2 A) + c^2 - 2bc \cos A \\
&= b^2 + c^2 - 2bc \cos A.
\end{aligned}
$$

Since the naming of the sides of the triangle is entirely arbitrary, the last two equations of the law of cosines are readily established.

The law of cosines may be used to find the third side of a triangle if two sides and the included angle are known (type 3, SAS). It may also be used to find an angle if three sides of a triangle are known (type 4, SSS).

Example 8.6   Given that $b = 4$, $c = 3$, and $A = 60°$, solve the triangle.

Solution:

$$
\begin{aligned}
a^2 &= b^2 + c^2 - 2bc \cos A \\
&= 16 + 9 - (2)(4)(3)(\tfrac{1}{2}) \\
&= 25 - 12 = 13
\end{aligned}
$$

$$
a = \sqrt{13} = 3.61
$$

$$
\frac{a}{\sin A} = \frac{b}{\sin B}
$$

$$
\frac{3.61}{\sin 60°} = \frac{4}{\sin B}
$$

$$\sin B = \frac{4(0.8660)}{3.61} = 0.9596$$

$$B = 73°40'$$

$$C = 180° - (A + B)$$
$$= 180° - (60° + 73°40')$$
$$= 46°20'.$$

Example 8.7

Given that $a = 6$, $b = 5$, and $c = 4$, find angle B.

Solution:

$$b^2 = a^2 + c^2 - 2ac \cos B$$

$$\cos B = \frac{-b^2 + a^2 + c^2}{2ac}$$

$$= \frac{-25 + 36 + 16}{(2)(6)(4)} = 0.5625$$

$$B = 55°50'.$$

EXERCISES

1. Given $a = 5$, $c = 3$, $B = 45°$, find b.
2. Given $a = 4$, $b = 2$, $c = 3$, find A.
3. Given $b = 12$, $c = 8$, $A = 44°$, find a.
4. Given $b = 10$, $c = 15$, $A = 112°$, find a.
5. Given $a = 3$, $b = 5$, $c = 6$, find C.
6. Given $a = 24$, $c = 32$, $B = 115°$, solve the triangle.
7. Given $a = 15$, $b = 12$, $C = 60°$, solve the triangle.
8. Given $b = 68$, $c = 14$, $A = 24°30'$, solve the triangle.
9. Show that the length of the diagonal d in the parallelogram in Figure 8.7 can be found by

$$d^2 = a^2 + b^2 + 2ab \cos \theta.$$

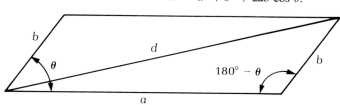

FIGURE 8.7

10. Show that in any triangle ABC,

$$c^2 = (a - b)^2 + 4 ab \sin^2 \frac{C}{2}.$$

11. Show that for triangle ABC,

$$\frac{a^2 + b^2 + c^2}{2abc} = \frac{\cos A}{a} + \frac{\cos B}{b} + \frac{\cos C}{c}.$$

12. Show that

$$2 \sin^2 \frac{A}{2} = \frac{(a + b - c)(a - b + c)}{2bc}.$$

13. Show that

$$2 \cos^2 \frac{A}{2} = \frac{(a + b + c)(b + c - a)}{2bc}.$$

14. Use the results of Exercises 12 and 13 to show that

$$\tan \frac{A}{2} = \frac{1}{s - a} \sqrt{\frac{(s - a)(s - b)(s - c)}{s}},$$

where $s = \frac{1}{2}(a + b + c)$.

8.3

APPLICATIONS

Any word problem that is depicted as a triangle and requires finding a side or an angle usually can be solved by using trigonometry. If the triangle is a right triangle, then we use the definitions of Section 6.4. If the triangle is an oblique triangle, then we use either the law of sines or the law of cosines or both to solve the problem. Deciding which law to use depends on the given information. If at least one angle and the side opposite to it are given, then use the law of sines. If two sides and the included angle are given, or if three sides are given, then use the law of cosines.

Example 8.8   Two friends live on opposite sides of a lake at points A and B, as shown in Figure 8.8. To buy communication equipment with the proper range, they must determine the approximate distance between their homes. They selected point C

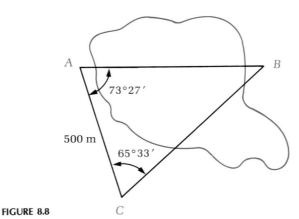

FIGURE 8.8

on the same side of the lake as A and 500 meters from A. By measurement, angle $CAB = 73°27'$ and angle $ACB = 65°33'$. Find AB, the distance between their homes.

Solution: Since two angles and a side are given, we use the law of sines. However, first we have to determine the measure of angle ABC:

$$m \angle ABC = 180° - (73°27' + 65°33') = 41°$$

$$\frac{c}{\sin C} = \frac{b}{\sin B}$$

$$c = \frac{b \sin C}{\sin B} = \frac{500 \sin 65°33'}{\sin 41°} = 694 \text{ meters.}$$

Example 8.9

When the elevation of the sun is such that a vertical rod 10 meters long casts a shadow 16 meters long, how long a shadow will be cast by a flagpole 24 meters long which is tipped 10° from the vertical away from the sun?

Solution: First make a sketch of the information, as shown in Figure 8.9. Then determine the angles of the triangle. Since only one side of the triangle is known, we use the law of sines:

$$\theta = \text{arc tan} \frac{10}{16} = 32°,$$

$$\phi = 180° - (32° + 80°) = 68°.$$

Using the law of sines, we see that

$$\frac{24}{\sin \theta} = \frac{AC}{\sin \phi},$$

$$AC = \frac{24 \sin 68°}{\sin 32°} = 42 \text{ meters.}$$

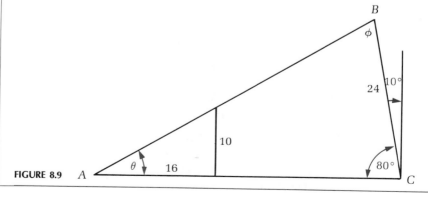

FIGURE 8.9

Example 8.10

A surveyor needs to determine the length of a tunnel to be cut through a hill from point A to point B. He locates a point C from which both A and B are visible and makes the following measurements: $AC = 348$ meters, $BC = 264$ meters, and angle $ACB = 102°24'$. Find the length of the tunnel.

Solution: Make a sketch of the information, as shown in Figure 8.10. Since two sides and the included angle are given, use the law of cosines:

$$c^2 = a^2 + b^2 - 2ab \cos C$$
$$= 348^2 + 264^2 - 2(348)(264)(\cos 102°24')$$

$$c = 480 \text{ meters.}$$

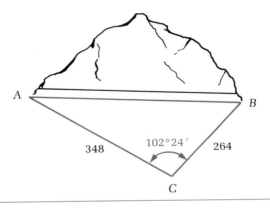

FIGURE 8.10

Example 8.11

Find the greatest angle of a triangle whose sides are 51.4, 42.6, and 45.7.

Solution: The greatest angle is always opposite the longest side. We use the law of cosines and let $\angle C$ be the angle opposite side 51.4:

$$c^2 = a^2 + b^2 - 2ab \cos C$$

$$\cos C = \frac{a^2 + b^2 - c^2}{2ab}$$

$$= \frac{(42.6)^2 + (45.7)^2 - (51.4)^2}{2(42.6)(45.7)}.$$

Using a calculator, we find

$$C = 71.1°.$$

Example 8.12

A vertical radio tower is built on a hillside that slants 11° from the horizontal. From a point 160 meters uphill from the tower the angle of elevation to the top of the tower is 14°. How high is the tower?

Solution: Make a sketch of the information, as shown in Figure 8.11. By parallel lines, we know that $\angle ACD$ is $11°$ and $\angle BAC$ is $79°$. Thus $\angle ACB = 14° + 11° = 25°$ and $\angle ABC = 180° - (79° + 25°) = 76°$. Using the law of sines, we find

$$\frac{c}{\sin C} = \frac{b}{\sin B}$$

$$c = \frac{b \sin C}{\sin B}$$

$$= \frac{160 \sin 25°}{\sin 76°}$$

$$= 70 \text{ meters.}$$

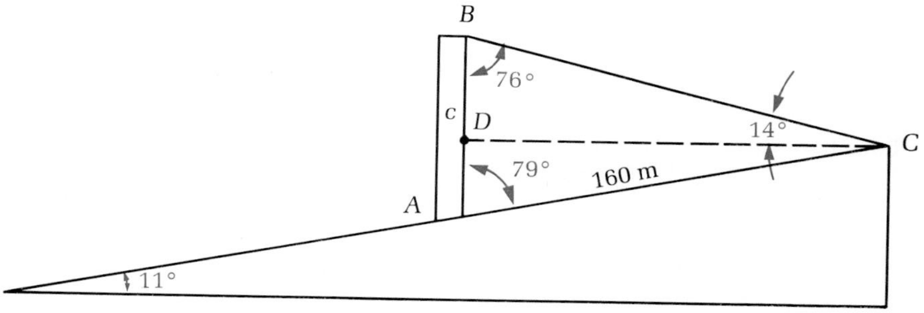

FIGURE 8.11

EXERCISES

1. An observer is 200 feet from one end of a pond and 300 feet from the other. She observes that the angle subtended by the pond is 120°. Find the length of the pond.

2. Sound travels at the rate of about 1100 feet per second. From a hill near a lake, the angle subtended by the line joining two boats on the lake is 15°. A pistol shot fired at the top of the hill is heard 1 second after firing in the nearer boat and $1\frac{3}{5}$ seconds after firing in the farther boat. How far apart are the boats?

3. Two sides of a parallelogram are 68 and 83 centimeters, and one of the diagonals is 42 centimeters. Find the angles of the parallelogram.

4. From a point in the same horizontal plane with the base of a building, the angles of elevation of the top and bottom of an antenna tower on top of the building are 64°40′ and 59°50′, respectively. If the building is 112 meters high, how tall is the antenna tower?

5. There is a house 220 meters from the bottom of a hill. From a point 100 meters from the base of the hill, a woman measures the angle of elevation of the house to be 12°20′. What angle does the hill make with the horizontal?

6. Two observers 3600 meters apart on a level plane saw, at the same time, an airplane flying between them and recorded the angles of elevation to be 36°30′ and 45°20′. How high was the airplane?

7. A ship sails 22 nautical miles on a bearing (heading) of 138° and then sails for 32 nautical miles at a heading of 215°. What course should it sail to return its starting point, and how many nautical miles must it go on this course?

8. Ridge Road and Hartford Turnpike meet at a 45° angle. If a lot extends 60 meters along Hartford Turnpike and 45 meters along Ridge Road, find the length of the back of the lot.

9. Whalley Avenue and Goffe Street meet at an angle of 40°. If a lot extends 30 meters along Goffe Street and 25 meters along Whalley Avenue, find the length of the back of the lot.

10. A ship sails 17 nautical miles on a heading of 327° and then sails for 24 nautical miles on a heading of 43°. What course should it sail to return to its starting point? How far is the ship from its starting point?

11. A distress signal, fired by a ship in Long Island Sound, is sighted from two points, A and B, on the Connecticut shoreline. Points A and B are 14 kilometers apart, and the line of sight from A to the ship makes an angle of 28°30′ with line AB, whereas the line of sight from B to the ship makes an angle of 40°10′ with line AB. How far is the ship from each point?

12. Find the measure of the smallest angle of a triangle whose sides measure 5.2, 6.1, and 7.3.

13. Find the measure of the smallest angle of a triangle whose sides measure 12.7, 15.2, and 20.4.

14. Find the measure of the largest angle of a triangle whose sides measure 6.4, 8.8, and 10.3.

15. Find the measure of the largest angle of a triangle whose sides measure 2.6, 4.1, and 5.2.

16. Two ships leave New Haven Harbor at the same time on courses that form an angle of 36°24′. After 1 hour, one ship has traveled 8 nautical miles, and the other has traveled 14 nautical miles. How far apart are the ships at this time?

17. A sailboat leaves Branford Harbor at noon and travels S64°E at 10 kilometers per hour. When will the boat be 8 kilometers from a Coast Guard station located 15 kilometers due east of the Branford pier?

18. Two boats leave Groton Long Point at the same time. The first travels at a rate of 18 nautical miles per hour on a bearing of S72°E while the second boat travels at 10 nautical miles per hour on a bearing of S5°W. How far apart are the boats after 3 hours?

19. A surveyor at point A sights two points, B and C, on the opposite side of a lake. If $AB = 331.7$ meters, $AC = 285.6$ meters, and angle $BAC = 120°40′$, find the distance BC.

20. The adjacent sides of a parallelogram are 20.62 and 24.73 centimeters. If one diagonal is 32.04 centimeters, is it the shorter or longer diagonal?

21. Three towns A, B, and C, are located such that $AB = 32$ kilometers, $BC = 18$ kilometers, and $AC = 20$ kilometers. The road from A to C is to be extended so that a new road from B will meet it at right angles at point D. How long are BD and CD?

22. Train A leaves Newtown traveling east at 85 miles per hour. Train B leaves Newtown 1 hour later on a track that makes an angle of 118° with the first track. If train B travels at 50 miles per hour, how far apart are the trains 3 hours after train A departed?

23. From a window 52 meters above the ground, the angle of elevation to the top of another building is 31°48′. From a point on the sidewalk directly below the window, the angle of elevation to the top of the same building is 54°36′. Find the height of the building.

24. A radio tower 42 meters high is on top of a hill. A cable 54 meters long stretches from the top of the tower to a point down the hill which is 24 meters from the base of the tower. Assume straight line distance. Find the angle that the hill makes with the horizontal.

25. A ship is en route from Orient Point to New London, a distance of 10 miles. After traveling 5 miles, the pilot discovers that she is 12° off course. At this point, how far is the ship from New London?

8.4

AREA OF A TRIANGULAR REGION

The formula for the *area of a triangular region* is one-half times the base times the height. For a right triangle, the base and the height are the sides of the triangle, and computing the area depends on determining these lengths. If an angle and the hypotenuse are given, the sides can be determined by using the sine and cosine relationships, as shown in the example below.

Example 8.13

Find the area of the right triangle ABC in which the right angle is at C, $c = 21$, and $A = 32°20'$.

Solution: Make a sketch of the triangle, as shown in Figure 8.12. Use the sine and cosine relationships to solve for a and b:

$$a = c \sin A = 21(0.5348) = 11.23,$$

$$b = c \cos A = 21(0.8450) = 17.75,$$

$$\text{Area} = \tfrac{1}{2} \cdot a \cdot b = \tfrac{1}{2}(17.75)(11.23) = 99.67.$$

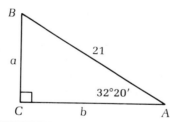

FIGURE 8.12

A general formula for the area of any triangular region can easily be developed. Place any triangle ABC on a coordinate axis with angle A in standard position, as shown in Figure 8.13 for (a) an acute angle A and (b) an obtuse angle A. Draw the altitude h of the triangle from C to the line containing AB. Since a right triangle is formed, $h = b \sin A$ (the y coordinate of C). The base of the right triangle formed is equal to $b \cos A$; thus the coordinates of point C are $(b \cos A, b \sin A)$. Thus

$$\text{Area of triangle } ABC = \frac{1}{2} \cdot \text{base} \cdot \text{height} = \frac{1}{2} \cdot c \cdot b \sin A.$$

We have shown that the area of a triangular region is one-half the product of the lengths of any two sides and the sine of the included angle:

$$\text{Area} = \frac{1}{2} bc \sin A = \frac{1}{2} ac \sin B = \frac{1}{2} ab \sin C. \tag{8.3}$$

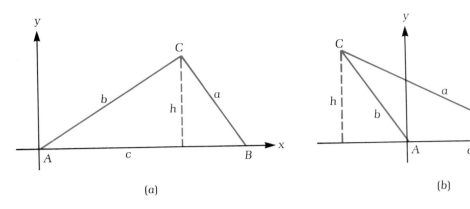

FIGURE 8.13

Example 8.14

Find the area of triangle ABC, given that $a = 9$ centimeters, $b = 10.5$ centimeters, and $C = 54°10'$.

Solution: Using formula (8.3), we get

$$\text{Area} = \frac{1}{2} ab \sin C$$

$$= \frac{1}{2} (9)(10.5)(\sin 54°10')$$

$$= \frac{1}{2} (9)(10.5)(0.8107)$$

$$= 38.3 \text{ square centimeters approximately.}$$

By using the law of sines, another formula can be obtained for the area of a triangular region in terms of the sines of all three angles and the length of one side. Referring to (8.3), we can see that $c = (b \sin C)/\sin B$. Substitute this expression for c in area $= \frac{1}{2}bc \sin A$:

$$\text{Area} = \frac{1}{2} \frac{b^2 \sin A \sin C}{\sin B}. \qquad (8.4)$$

Since $\sin B = \sin[180° - (A + C)] = \sin(A + C)$, a variation of this formula expresses the area in terms of two angles and their common side:

$$\text{Area} = \frac{1}{2} \frac{b^2 \sin A \sin C}{\sin(A + C)}. \qquad (8.4a)$$

Similar formulations can be derived in terms of two other angles and their common side.

Example 8.15

Find the area of triangle ABC, given that $a = 10$, $B = 40°$, and $C = 120°$.

Solution:

$$\text{Area} = \frac{1}{2}a^2 \frac{\sin B \sin C}{\sin(B + C)}$$

$$= \frac{1}{2}(10)^2 \frac{\sin 40° \sin 120°}{\sin 160°}$$

$$= \frac{1}{2}(10)^2 \frac{(0.6428)(0.8660)}{0.3420}$$

$$= 81.38 \text{ approximately.}$$

Still another formula can be developed that expresses the area in terms of the sides. We start with the basic area formula and square it:

$$(\text{Area})^2 = \frac{1}{4}b^2c^2 \sin^2 A.$$

Next we substitute for $\sin^2 A$:

$$(\text{Area})^2 = \frac{1}{4}b^2c^2(1 - \cos^2 A).$$

Then we substitute for $\cos^2 A$ from the law of cosines:

$$(\text{Area})^2 = \frac{b^2c^2}{4}\left[1 - \left(\frac{b^2 + c^2 - a^2}{2bc}\right)^2\right]$$

$$= \frac{b^2c^2}{4}\left[\frac{4b^2c^2 - (b^2 + c^2 - a^2)^2}{4b^2c^2}\right]$$

$$= \frac{(2bc + b^2 + c^2 - a^2)(2bc - b^2 - c^2 + a^2)}{16}$$

$$= \frac{[(b + c)^2 - a^2][a^2 - (b - c)^2]}{16}$$

$$= \frac{(b + c + a)(b + c - a)(a + b - c)(a - b + c)}{16}.$$

$$\text{Area} = \frac{1}{4}\sqrt{(a + b + c)(a + b - c)(b + c - a)(a + c - b)}. \tag{8.5}$$

A simpler form of this formula can be written by using the semiperimeter $s = \frac{1}{2}(a + b + c)$

$$2s = (a + b + c) \qquad 2(s - a) = (b + c - a)$$
$$2(s - c) = (a + b - c) \qquad 2(s - b) = (a + c - b)$$

and substituting these quantities in the formula above:

$$\text{Area} = \sqrt{s(s - a)(s - b)(s - c)} \qquad s = \tfrac{1}{2}(a + b + c). \qquad (8.6)$$

Example 8.16 Find the area of the triangular region ABC, given that $a = 67$ meters, $b = 80$ meters, and $c = 53$ meters.

Solution:

$$s = \frac{67 + 80 + 53}{2} = 100$$

$$s - a = 100 - 67 = 33$$
$$s - b = 100 - 80 = 20$$
$$s - c = 100 - 53 = 47$$
$$\text{Area} = \sqrt{100(33)(20)(47)}$$
$$= \sqrt{3,102,000}$$
$$= 1761 \text{ square meters approximately.}$$

EXERCISES

Find the area of the triangular regions, given the values in Exercises 1 to 14.

1. $a = 7, b = 6, C = 30°$.
2. $a = 6, b = 9, c = 5$.
3. $A = 30°, B = 45°, c = 20$.
4. $a = 869, b = 734, C = 56°30'$.
5. $b = 63, c = 41, B = 28°20'$.
6. $a = 743, A = 81°20', B = 62°50'$.
7. $a = 630, b = 308, c = 411$.
8. $a = 12, c = 8, B = 60°$.
9. $B = 40°, C = 120°, a = 10$.
10. $a = 13, b = 7, c = 8$.
11. $b = 21, c = 38, A = 112°$.
12. $a = 450, c = 375, A = 41°50'$.
13. $c = 425, A = 73°10', B = 55°30'$.
14. $a = 18, b = 3.4, c = 16.8$.
15. A triangular lot has frontages of 78 and 94 meters and a rear line 69 meters long. If the 78-meter frontage is increased to 109 meters, how much more will it cost to fence the new lot than the old one if the cost of fencing is $4 per meter?
16. A triangular mirror used for solar heating has sides 28, 28, and 3.6 centimeters. Find the area of the mirror.

17. A house is to be built on a quadrilateral lot, as shown in Figure 8.14. Find the area of the lot by dividing the quadrilateral into two triangles.

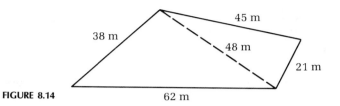

FIGURE 8.14

18. A machinist needs a triangular steel plate with dimensions of 14 by 18 by 24 inches. Find the area of the plate.

8.5

VECTORS AND POLAR COORDINATES

There are many applications of mathematics in physics for which quantities are described by specifying both a magnitude and a direction. These are known as *vector quantities*, or simply *vectors*, and are represented graphically as directed line segments, or arrows. Some examples of vector applications are displacement, velocity, and force. *Displacement* means that an object is moved a specified distance in a particular direction. For instance, a car travels 50 kilometers northwest; a person takes 10 steps due east. *Velocity* is the rate at which an object is moving in a particular direction. For instance, a car is traveling 90 kilometers per hour in a southerly direction; the wind is gusting at 50 miles per hour from the northeast. *Force* means that an object is pushed or pulled in a particular direction. A bike is pushed up a hill; a wagon is pulled along a sidewalk. In this section, we shall see how the trigonometric ratios are used in physical applications represented by vectors.

Definition 8.5.1

A *vector* $\overrightarrow{AB}$ is a directed line segment from an *initial point A* to a *terminal point B*. The length of line segment AB is proportional to the *magnitude ρ* of the vector and is denoted by $|\overrightarrow{AB}|$. The *direction* of the vector is described by an angle, θ (see Figure 8.15).

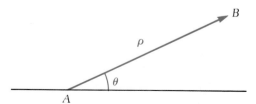

FIGURE 8.15

As an example of displacement of an object, or the path of a point, consider a person walking 5 kilometers up a mountain path to the top. The **displacement**

vector $\overrightarrow{AB}$ represents the movements of the person from point A to point B (see Figure 8.16).

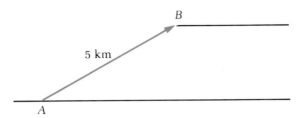

FIGURE 8.16

Now suppose that a jet plane is ascending at a constant rate of 200 miles per hour, and its line of flight makes an angle of 30° with the horizontal. This information can be represented by a vector diagram (as shown in Figure 8.17), where vector $\overrightarrow{AB}$ is called the **velocity vector,** and the magnitude $|V|$ is the *speed* of the plane.

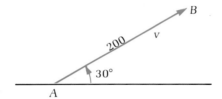

FIGURE 8.17

As an illustration of a **force vector,** consider having to lift a 10-pound mass. We would have to pull directly upward with a force of 10 pounds. This would be represented by a vertical vector, as shown in Figure 8.18.

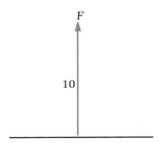

FIGURE 8.18

Definition 8.5.2	A *zero vector*, denoted by $\vec{0}$, is a vector of zero length. Its initial point and terminal point coincide.

Definition 8.5.3	A *unit vector* is a vector of magnitude 1.

Proposition 8.5.1

Two vectors $\overrightarrow{AB}$ and $\overrightarrow{CD}$ are equal if and only if they have the same direction and same length regardless of the location of their initial points. Consequently, parallel vectors of the same length and direction are equal (see Figure 8.19).

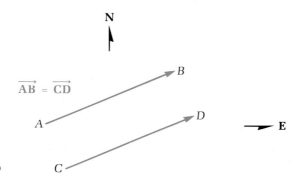

FIGURE 8.19

Note that the vectors are free to move from one position to another as long as they do not change length or direction. These are referred to as **free vectors.**

In terms of coordinates, vector $\overrightarrow{AB}$ may be moved to a new location on the plane by changing the coordinates of the initial and terminal points, as shown in Figure 8.20.

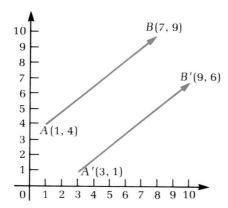

FIGURE 8.20

Notice that the change in x and the change in y are the same for both $\overrightarrow{AB}$ and $\overrightarrow{A'B'}$. To distinguish points in the plane from vectors, we use the following notation: $\langle 7-1, 9-4 \rangle$ is the same as $\langle 9-3, 6-1 \rangle$, or $\langle 6, 5 \rangle$ in both cases.

Proposition 8.5.2

If $\overrightarrow{OA}$ and $\overrightarrow{OB}$ represent two vectors, the *sum*, or *resultant*, of these vectors is represented in both direction and magnitude by $\overrightarrow{OC}$, the diagonal of the parallelogram of which $\overrightarrow{OA}$ and $\overrightarrow{OB}$ are adjacent sides. And $\overrightarrow{OA}$ and $\overrightarrow{OB}$ are called the *components* of $\overrightarrow{OC}$ (see Figure 8.21).

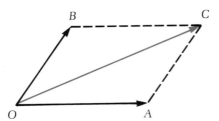

FIGURE 8.21

If $\overrightarrow{OA}$ and $\overrightarrow{OB}$ are perpendicular to each other and if θ is the angle between $\overrightarrow{OC}$ and $\overrightarrow{OA}$ (see Figure 8.22), then

$$|\overrightarrow{OA}| = |\overrightarrow{OC}| \cos \theta \qquad \text{and} \qquad |\overrightarrow{OB}| = |\overrightarrow{OC}| \sin \theta.$$

Why?

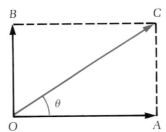

FIGURE 8.22

Example 8.17

A force $\overrightarrow{AB}$ of magnitude 6 and a force $\overrightarrow{AC}$ of magnitude 4 act at a point A. If $\overrightarrow{AB}$ is directed due east and $\overrightarrow{AC}$ has the direction N30°E, find the magnitude and direction of the resultant of these forces.

Solution: Draw a sketch to depict the given vectors as sides of a parallelogram, and draw the resultant $\overrightarrow{AD}$, as shown in Figure 8.23. From geometry, we know that angle ABD is 120°. Using the law of cosines with triangle ABD, we have

$$|\overrightarrow{AD}|^2 = |\overrightarrow{AB}|^2 + |\overrightarrow{BD}|^2 - 2|\overrightarrow{AB}|\,|\overrightarrow{BD}| \cos B$$
$$= 6^2 + 4^2 - 2(6)(4) \cos 120°$$
$$= 36 + 16 - 48(-\tfrac{1}{2})$$
$$= 36 + 16 + 24 = 76$$
$$|\overrightarrow{AD}| = \sqrt{76} = 8.72 \text{ approximately.}$$

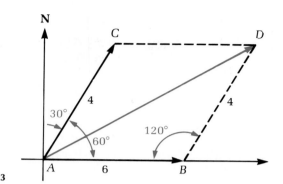

FIGURE 8.23

The direction is determined by using the law of sines:

$$\frac{AD}{\sin \angle DBA} = \frac{BD}{\sin \angle DAB}$$

$$\frac{8.72}{0.866} = \frac{4}{\sin \angle DAB}$$

$$\sin \angle DAB = \frac{4(0.866)}{8.72} = 0.3972$$

$$\angle DAB = 23°20' \text{ to the nearest } 10'.$$

Example 8.18

Find the force necessary to push a 10-pound ball up a plane that is inclined 20° to the horizontal.

Solution: The mass or force $\overrightarrow{OC}$ of the ball acts vertically downward. This force has two components: $\overrightarrow{OA}$, parallel to the inclined plane, and $\overrightarrow{OB}$, at right angles to the first (see Figure 8.24). The force $\overrightarrow{OA}$ is the one called for in this problem. By similar triangles, we know that angle AOC is 70°. Using Proposition 8.5.2,

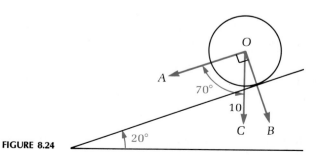

FIGURE 8.24

we get

$$|\overrightarrow{OA}| = |\overrightarrow{OC}| \cos \theta$$
$$= 10 \cos 70°$$
$$= 10(0.342)$$
$$= 3.42 \text{ pounds approximately.}$$

Example 8.19

An airplane is flying at a rate of 300 kilometers per hour in a direction of 320°. Find the westerly and northerly components of its velocity vector.

Solution: First make a sketch showing the velocity vector, and draw a parallelogram showing the desired vectors as sides of the parallelogram whose diagonal is the velocity vector (Figure 8.25).

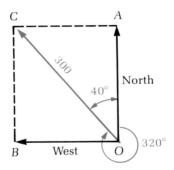

FIGURE 8.25

A plane's direction is measured in a clockwise direction from the north, just as a ship's is. Therefore angle COA equals $360° - 320°$, or $40°$. If we use Proposition 8.5.2 and triangle OCA, the northerly and westerly components can be found by

$$|\overrightarrow{OA}| = 300 \cos 40° = 300(0.7660)$$
$$= 229.8 \text{ kilometers per hour}$$

$$|\overrightarrow{OB}| = |\overrightarrow{CA}| = 300 \sin 40° = 300(0.6428)$$
$$= 192.8 \text{ kilometers per hour.}$$

Example 8.20

The components of a given vector are found to be 20 in a westerly direction and 15 in a southerly direction. Find the vector.

Solution: Draw a sketch and label the given information as shown in Figure 8.26. From the drawing, we can see that the $\tan \theta = \frac{20}{15} = 1.3333$. Therefore

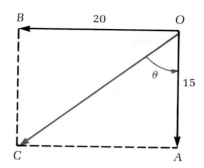

FIGURE 8.26

$\theta = 53°10'$ to the nearest $10'$. Since $\sin \theta = 20/|\overrightarrow{OC}|$,

$$|\overrightarrow{OC}| = \frac{20}{\sin \theta} = \frac{20}{0.8004}$$

$$= 25 \text{ approximately.}$$

Therefore, $\overrightarrow{OC}$ has a magnitude of 25 and a direction $S53°10'W$.

Vectors on a Coordinate System

A vector is said to be in **standard position** on a coordinate system if the tail of the arrow representing the vector is placed at the origin. Thus the coordinates of the point at the head of the arrow identify the vector (see Figure 8.27).

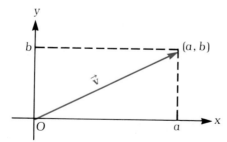

FIGURE 8.27

Notice that the magnitude of the vector $\vec{v}$, or $\langle a, b \rangle$, is simply the distance from the origin to the point (a, b). Using the Pythagorean theorem, we may formulate the following proposition:

Proposition 8.5.3

If $\vec{v} = \langle a, b \rangle$, then $|\vec{v}| = \sqrt{a^2 + b^2}$.

Example 8.21   Given $\vec{u} = \langle 2, 3 \rangle$, $\vec{v} = \langle -1, 5 \rangle$, and $\vec{w} = \langle 0, 0 \rangle$, find **(a)** $|\vec{u}|$, **(b)** $|\vec{v}|$, and **(c)** $|\vec{w}|$.

Solution: Using Proposition 8.5.3, we get

(a) $|\vec{u}| = \sqrt{2^2 + 3^2} = \sqrt{13} = 3.61$ approximately,

(b) $|\vec{v}| = \sqrt{(-1)^2 + 5^2} = \sqrt{26} = 5.10$ approximately,

(c) $|\vec{w}| = \sqrt{0^2 + 0^2} = \sqrt{0} = 0$.

It is often convenient to consider a unit vector in the same direction as a given vector.

Proposition 8.5.4   A unit vector in the same direction as vector $\vec{v} = \langle x, y \rangle$ can be expressed as

$$\left\langle \frac{x}{|\vec{v}|}, \frac{y}{|\vec{v}|} \right\rangle = \left\langle \frac{x}{\sqrt{x^2 + y^2}}, \frac{y}{\sqrt{x^2 + y^2}} \right\rangle.$$

Example 8.22   Given $\vec{v} = \langle 4, 3 \rangle$, find a unit vector in the same direction as $\vec{v}$.

Solution: Using Proposition 8.5.4, we find that

$$|\vec{v}| = \sqrt{16 + 9} = \sqrt{25} = 5.$$

Therefore the unit vector is $\langle \frac{4}{5}, \frac{3}{5} \rangle$.

If vectors are given in standard position form, then their sum can be found by adding like components (see Figure 8.28).

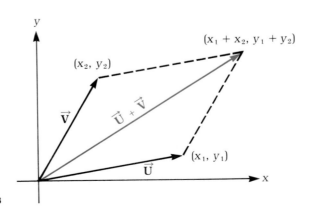

FIGURE 8.28

Definition 8.5.4

If $\vec{U} = \langle x_1, y_1 \rangle$ and $\vec{V} = \langle x_2, y_2 \rangle$, then

$$\vec{U} + \vec{V} = \langle x_1 + x_2, y_1 + y_2 \rangle \quad \text{and} \quad \vec{U} - \vec{V} = \vec{U} + (-\vec{V}).$$

Example 8.23

Given $\vec{U} = \langle -2, 3 \rangle$ and $\vec{V} = \langle 4, -7 \rangle$, find $\vec{U} + \vec{V}$ and $\vec{U} - \vec{V}$.

Solution:

$$\vec{U} + \vec{V} = \langle -2 + 4, 3 + (-7) \rangle = \langle 2, -4 \rangle,$$
$$\vec{U} - \vec{V} = \langle -2 - 4, 3 - (-7) \rangle = \langle -6, 10 \rangle.$$

When vectors are placed head to tail, the relations shown in Figure 8.29 hold.

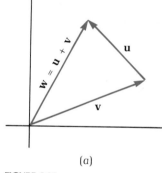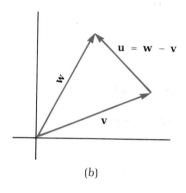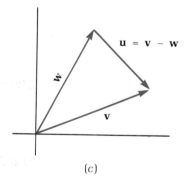

(a)   (b)   (c)

FIGURE 8.29

Polar Form of Vectors

polar form When a vector is specified by giving its length and direction, it is said to be in **polar form.** The most common notation is (ρ, θ), where ρ represents the length and θ is the angle. An example of this is $(25, 150°)$, where the angle is measured in a counterclockwise direction from the positive x axis (see Figure 8.30).

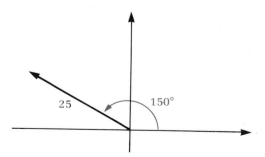

FIGURE 8.30

Example 8.24

Find the polar form of the vector $\langle 6, -4 \rangle$.

Solution: Sketch the vector on the coordinate system as shown in Figure 8.31. The related angle α can be found by the tangent relationship:

$$\tan \alpha = -4/6 = -0.666\overline{7}$$
$$= -33°40' \qquad \text{to the nearest } 10'.$$

The length of the vector is found by the Pythagorean relation:

$$|\vec{v}| = \sqrt{6^2 + (-4)^2} = \sqrt{52} = 7.21 \text{ approximately.}$$

The polar form of the vector is $(7.21, 326°20')$, or $(7.21, -33°40')$.

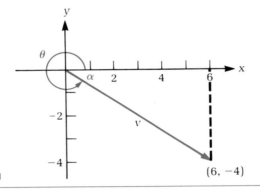

FIGURE 8.31

Example 8.25

Change the vector $(15, 130°)$ from polar to rectangular form.

Solution: Make a sketch of the vector as in Figure 8.32. Then

$$x = 15 \cos 130° = 15(-0.6428) = -9.64 \text{ approximately}$$
$$y = 15 \sin 130° = 15(0.7660) = 11.49 \text{ approximately}$$
$$\langle x, y \rangle = \langle -9.64, 11.49 \rangle.$$

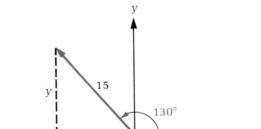

FIGURE 8.32

Thus a vector may be expressed in **rectangular form** by using the x and y coordinates of its endpoint, or in polar form by referring to its length and direction. When a vector is given in polar form, it is very convenient to plot it on polar graph paper (see Figure 8.33). The origin is called the **pole,** and the positive *polar axis* x axis is called the **polar axis.**

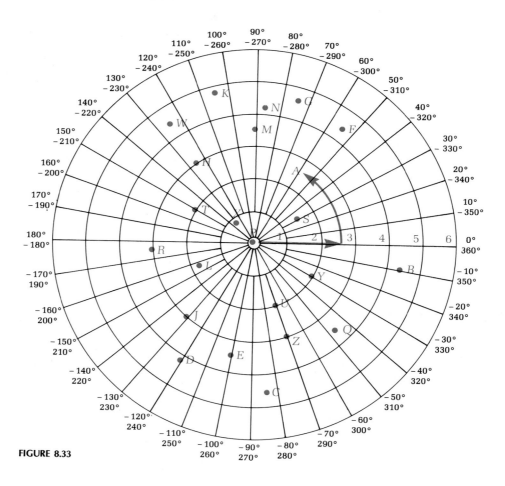

FIGURE 8.33

Example 8.26

Plot the point (2.6, 55°) on polar graph paper.

Solution: Refer to Figure 8.33. Starting at the pole, move along the 0° line 2.6 units. Then move in a counterclockwise direction to a point halfway between the 50° line and the 60° line. This is the point labeled A in the figure.

Example 8.27	Find the coordinates of the point B in Figure 8.33.
	Solution: The magnitude of B is 4.4, and the direction is $350°$. Therefore, $B = (4.4, 350°)$.

EXERCISES

1. A vertical force of 38 kilograms and a horizontal force of 17 kilograms act on a body. Find the magnitude and direction of the resultant.

2. A body is acted on by a horizontal force of 100 kilograms and a downward force of 175 kilograms. Find the direction and magnitude of the resultant.

3. Find the horizontal and vertical components of a force of 980 kilograms inclined at $73°20'$ to the horizontal.

4. Find the horizontal and vertical components of a force of 625 kilograms inclined at $21°$ to the horizontal.

5. A ship travels due east at 20 miles per hour. A man walks directly across the deck at 4 miles per hour. Find the man's direction and velocity relative to the earth.

6. A duck swimming at 5 miles per hour in still water swims directly across a current of 8 miles per hour. Find the actual direction and velocity of the duck.

7. An automobile is pulled by cable up an inclined plane to the second floor of a repair shop. If the car weighs 2400 pounds and the slope is $36°$ from the horizontal, what is the tension in the cable?

8. A mass of 385 kilograms rests on an inclined plane. A force of 238 kilograms is required to keep it from sliding down. What is the angle of inclination of the plane?

9. A guy wire runs from the top of a telephone pole 38 feet high to a point on the ground 23 feet from the pole. The tension in the guy wire is 164 pounds. Find the horizontal component.

10. A horizontal wire pulls with a force of 250 kilograms on the top of a pole 32 meters high. Its pull is opposed by that of a guy wire attached to the ground at a point 15 meters from the pole in the opposite direction. Find the tension in the guy wire if its pull just counteracts that of the horizontal wire.

11. Three competing yachts (A, B, C) in a Bermuda race are becalmed at sea. Using directional radios with a range of 25 kilometers, A and B are in radio contact, and A is 12 kilometers due north of B. Yachts A and C are also in radio contact, and C is 5 kilometers northeast of A. In which direction should B point the directional antenna to try to get in radio contact with C? Will C be within radio range of B?

12. Avon is 30 kilometers due north of Hamden and 20 kilometers northwest of Middletown. What is the relative position of Middletown to Hamden?

13. Forces of 450 and 300 kilograms act on an object. The angle between the forces is $40°$. Find the resultant force and the angle that it makes with the smaller force.

14. Forces of 25 and 40 kilograms act on an object. Determine the resultant force if the angle between the forces is $125°$. Give the angle the resultant force makes with the larger force.

15. A box is pulled with a force of 75 pounds by a rope that makes an angle of $25°$ with the horizontal. Find the vertical and horizontal components.

16. A wagon is being pulled with a force of 20 pounds. The handle of the wagon makes an angle of $45°$ with the horizontal. Find the horizontal and vertical components of the force.

Convert vectors from rectangular to polar coordinates in Exercises 17 to 26.

17. $\langle 1, 4 \rangle$ 18. $\langle 0, 3 \rangle$ 19. $\langle -3, -4 \rangle$ 20. $\langle -1, \sqrt{2} \rangle$

21. $\langle -5, 0 \rangle$ **22.** $\langle -3, -3\sqrt{3} \rangle$ **23.** $\langle 4, 4 \rangle$ **24.** $\langle 2, 0 \rangle$
25. $\langle \sqrt{3}, 1 \rangle$ **26.** $\langle -\sqrt{3}, 1 \rangle$

Convert from polar to rectangular coordinates in Exercises 27 to 36.
27. $(3, -10°)$ **28.** $(2, -20°)$ **29.** $(4, 30°)$ **30.** $(5, 45°)$
31. $(2, 3\pi/4)$ **32.** $(1, 5\pi/6)$ **33.** $(5, 60°)$ **34.** $(-3, 30°)$
35. $(0, 15°)$ **36.** $(4, 120°)$

Use polar graph paper to plot the points in Exercises 37 to 46.
37. $(2, 75°)$ **38.** $(5, 45°)$ **39.** $(3.5, 50°)$ **40.** $(1.5, 80°)$
41. $(2.25, -45°)$ **42.** $(2.25, -60°)$ **43.** $(4, 160°)$ **44.** $(3, 135°)$
45. $(1, 210°)$ **46.** $(2, 285°)$

Determine the polar coordinates of the points in Exercises 47 to 66, located in Figure 8.33.
47. C **48.** D **49.** E **50.** F
51. G **52.** H **53.** J **54.** K
55. L **56.** M **57.** N **58.** Q
59. R **60.** S **61.** T **62.** U
63. W **64.** X **65.** Y **66.** Z

Find the magnitude of each of the vectors given in standard position in Exercises 67 to 72.
67. $\langle 2, 5 \rangle$ **68.** $\langle 0, 3 \rangle$ **69.** $\langle -1, 0 \rangle$
70. $\langle -2, \sqrt{5} \rangle$ **71.** $\langle -4, 3 \rangle$ **72.** $\langle 6, 6 \rangle$

Given the vectors $\vec{U}$ and $\vec{V}$, find $\vec{U} + \vec{V}$ and $\vec{U} - \vec{V}$ in Exercises 73 to 77.
73. $\vec{U} = \langle 3, 4 \rangle, \vec{V} = \langle 5, 12 \rangle$ **74.** $\vec{U} = \langle -2, 10 \rangle, \vec{V} = \langle -1, -5 \rangle$ **75.** $\vec{U} = \langle 0, 2 \rangle, \vec{V} = \langle 4, 4 \rangle$
76. $\vec{U} = \langle -2, -5 \rangle, \vec{V} = \langle 3, -1 \rangle$ **77.** $\vec{U} = \langle 1, -1 \rangle, \vec{V} = \langle 3, -3 \rangle$

Find a unit vector in the same direction as the given vector in Exercises 78 to 81.
78. $\langle 3, 4 \rangle$ **79.** $\langle 5, -12 \rangle$ **80.** $\langle -15, -8 \rangle$ **81.** $\langle 1, 1 \rangle$

8.6

TRIGONOMETRIC FORM OF COMPLEX NUMBERS

In Section 1.12, we determined how to add, subtract, multiply, and divide complex numbers algebraically. To review, recall that a complex number z may be written as $a + bi$, where a and b are real numbers and i is the imaginary number defined by the equations $i^2 = -1$ and $i = \sqrt{-1}$.

Furthermore, a complex number may be represented geometrically. Since a complex number is determined by a pair of real numbers (a, b), we can associate (a, b) with a point of a plane. Specifically, we can represent complex numbers as points of a rectangular coordinate system known as the **complex plane** (see Figure 8.34).

complex plane

The trigonometric functions provide us with an alternate geometric representation of complex numbers. Let us now denote the distance between the

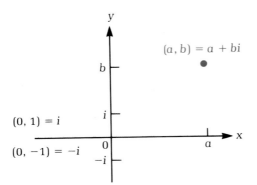

FIGURE 8.34

origin and the point that represents z as r, and let θ be an angle in standard position whose terminal side contains the point z. Note that $r = 0$ only when $z = 0$ (see Figure 8.35).

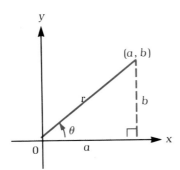

FIGURE 8.35

The point $z = a + bi$ determines a right triangle, as shown. This is comparable to the reference triangle associated with the unit circle, except that the radius does <u>not</u> have to equal 1. By the Pythagorean theorem, $r^2 = a^2 + b^2$ and $r = \sqrt{a^2 + b^2}$. Since the radius is nonnegative, we denote the distance r as *modulus* $|z|$, the **absolute value** of z or, equivalently, the **modulus** of z:

$$r = |z| = |a + bi| = \sqrt{a^2 + b^2}. \qquad (8.7)$$

The complex number $a + bi$ may also be represented in terms of the angle θ. Trigonometrically,

$$a = r \cos \theta \qquad \text{and} \qquad b = r \sin \theta.$$

Consequently,

$$z = a + bi = r \cos \theta + i(r \sin \theta),$$

or

$$z = r(\cos \theta + i \sin \theta).$$ **(8.8)**

This form of the complex number, (8.8), is also called the **trigonometric,** *argument* or **polar form** of z. The angle θ is called the **argument** of z.

Example 8.28

Express the following complex numbers in trigonometric form with $0 \le \theta < 2\pi$:

(a) $-1 + i$, **(b)** -1, **(c)** $-2 - 2i$, **(d)** $-i$, **(e)** $\sqrt{3} - i$.

Solution: For each complex number, plot the point in the complex plane, determine the value of r, find θ, and then express the number in form (8.8).

(a) $\qquad\qquad z = -1 + i \qquad$ (see Figure 8.36)

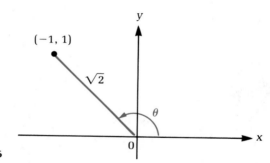

FIGURE 8.36

$$r = \sqrt{(-1)^2 + 1^2} = \sqrt{2}$$

$$\theta = 180° - 45° = 135° \quad \text{or} \quad \frac{3\pi}{4}$$

$$z = -1 + i = \sqrt{2}(\cos 135° + i \sin 135°)$$

or

$$z = -1 + i = \sqrt{2}\left(\cos \frac{3\pi}{4} + i \sin \frac{3\pi}{4}\right).$$

(b) $\qquad\qquad z = -1 \qquad$ (see Figure 8.37)

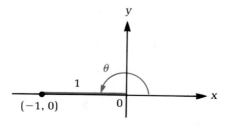

FIGURE 8.37

$$r = \sqrt{(-1)^2 + 0^2} = 1$$

$$\theta = 180° \quad \text{or} \quad \pi$$

$$z = -1 = \cos 180° + i \sin 180°$$

or

$$z = -1 = \cos \pi + i \sin \pi.$$

(c) $\qquad z = -2 - 2i \qquad$ (see Figure 8.38)

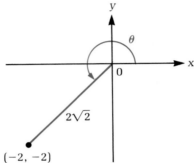

FIGURE 8.38 $(-2, -2)$

$$r = \sqrt{(-2)^2 + (-2)^2} = \sqrt{8} = 2\sqrt{2}$$

$$\theta = 180° + 45° = 225° \quad \text{or} \quad \frac{5\pi}{4}$$

$$z = -2 - 2i = 2\sqrt{2}(\cos 225° + i \sin 225°)$$

or

$$z = -2 - 2i = 2\sqrt{2}\left(\cos \frac{5\pi}{4} + i \sin \frac{5\pi}{4}\right).$$

(d) $\qquad z = -i \qquad$ (see Figure 8.39)

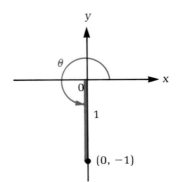

FIGURE 8.39

$$r = \sqrt{0^2 + (-1)^2} = 1$$

$$\theta = 270° \quad \text{or} \quad \frac{3\pi}{2}$$

$$z = -i = \cos 270° + i \sin 270°$$

or

$$z = -i = \cos \frac{3\pi}{2} + i \sin \frac{3\pi}{2}.$$

(e) $\quad z = \sqrt{3} - i \quad$ (see Figure 8.40)

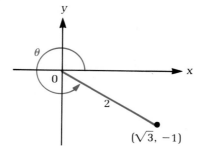

FIGURE 8.40

$$r = \sqrt{(\sqrt{3})^2 + (-1)^2} = 2$$

$$\theta = 360° - 30° = 330° \quad \text{or} \quad \frac{11\pi}{6}$$

$$z = \sqrt{3} - i = 2(\cos 330° + i \sin 330°)$$

or

$$z = \sqrt{3} - i = 2\left(\cos \frac{11\pi}{6} + i \sin \frac{11\pi}{6}\right).$$

Example 8.29

Use Appendix Table 4 or a calculator to convert the complex number $z = 3(\cos 20° + i \sin 20°)$ from trigonometric form to the rectangular form $a + bi$.

Solution:

$$3(\cos 20° + i \sin 20°) = 3(0.9397 + 0.3420i)$$
$$= 2.8191 + 1.026i.$$

It is particularly advantageous to write complex numbers in trigonometric form when we want to multiply or divide them. Recall from Section 1.12 that if $z_1 = a + bi$ and $z_2 = c + di$, then

$$z_1 \cdot z_2 = (ac - bd) + (ad + bc)i.$$

If we let

$$z_1 = r_1(\cos \theta_1 + i \sin \theta_1)$$

and

$$z_2 = r_2(\cos \theta_2 + i \sin \theta_2),$$

then

$$
\begin{aligned}
z_1 \cdot z_2 &= r_1(\cos \theta_1 + i \sin \theta_1) \cdot r_2(\cos \theta_2 + i \sin \theta_2) \\
&= r_1 r_2 [(\cos \theta_1 \cos \theta_2 - \sin \theta_1 \sin \theta_2) \\
&\quad + i(\sin \theta_1 \cos \theta_2 + \cos \theta_1 \sin \theta_2)]. \qquad \text{(multiplying)}
\end{aligned}
$$

Then by using (7.20) and (7.21) we get

$$z_1 \cdot z_2 = r_1 r_2 [\cos (\theta_1 + \theta_2) + i \sin (\theta_1 + \theta_2)] \qquad \textbf{(8.9)}$$

Formula (8.9) states that the product of two complex numbers is a complex number whose modulus is the product of the moduli of the factors and whose argument is the sum of the arguments of the factors. Similarly, a formula for the quotient of two complex numbers can be developed. Its proof is left as an exercise:

$$\frac{z_1}{z_2} = \frac{r_1}{r_2} [\cos (\theta_1 - \theta_2) + i \sin (\theta_1 - \theta_2)]. \qquad \textbf{(8.10)}$$

Example 8.30

Find **(a)** the product and **(b)** the quotient of the complex numbers

$$z_1 = 5 \left(\cos \frac{3\pi}{4} + i \sin \frac{3\pi}{4} \right)$$

and

$$z_2 = 3 \left(\cos \frac{5\pi}{3} + i \sin \frac{5\pi}{3} \right).$$

Solution:

(a) Using formula (8.9), we get

$$
\begin{aligned}
z_1 \cdot z_2 &= 15 \left[\cos \left(\frac{3\pi}{4} + \frac{5\pi}{3} \right) + i \sin \left(\frac{3\pi}{4} + \frac{5\pi}{4} \right) \right] \\
&= 15 \left(\cos \frac{29\pi}{12} + i \sin \frac{29\pi}{12} \right).
\end{aligned}
$$

principal argument

Since $29\pi/12 = 5\pi/12 + 2\pi$, we may consider the arguments to be $5\pi/12$. In fact, whenever we represent the radian measure of an argument as an angle between $-\pi$ and π, it is called the **principal argument**. So

$$z_1 \cdot z_2 = 15\left(\cos \frac{5\pi}{12} + i \sin \frac{5\pi}{12}\right).$$

(b) Using formula (8.10), we get

$$\frac{z_1}{z_2} = \frac{5}{3}\left[\cos\left(\frac{3\pi}{4} - \frac{5\pi}{3}\right) + i \sin\left(\frac{3\pi}{4} - \frac{5\pi}{3}\right)\right]$$

$$= \frac{5}{3}\left[\cos\left(\frac{-11\pi}{12}\right) + i \sin\left(\frac{-11\pi}{12}\right)\right].$$

The argument $-11\pi/12$ is equivalent to the positive angle $13\pi/12$, and the quotient may also be written as

$$\frac{z_1}{z_2} = \frac{5}{3}\left(\cos \frac{13\pi}{12} + i \sin \frac{13\pi}{12}\right).$$

Example 8.31

Find the reciprocal of the complex number $z = r(\cos \theta + i \sin \theta)$.

Solution: Since the number 1 can be written in trigonometric form as $1(\cos 0° + i \sin 0°)$,

$$\frac{1}{z} = \frac{1(\cos 0° + i \sin 0°)}{r(\cos \theta + i \sin \theta)}$$

$$= \frac{1}{r}\left[\cos(0° - \theta) + i \sin(0° - \theta)\right]$$

$$= \frac{1}{r}(\cos \theta - i \sin \theta). \qquad [\text{using } \cos(-\theta) = \cos \theta \text{ and } \sin(-\theta) = -\sin \theta]$$

Example 8.32

If $z = r(\cos \theta + i \sin \theta)$, express z^2 in trigonometric form.

Solution:

$$z^2 = r(\cos \theta + i \sin \theta) \cdot r(\cos \theta + i \sin \theta)$$
$$= r^2(\cos 2\theta + i \sin 2\theta). \qquad [\text{using (8.9)}]$$

If we use the result of Example 8.32, we can find an expression for z^3:

$$z^3 = z \cdot z^2$$
$$= r(\cos \theta + i \sin \theta) \cdot r^2(\cos 2\theta + i \sin 2\theta)$$
$$= r^3(\cos 3\theta + i \sin 3\theta). \qquad [\text{using (8.9)}]$$

The following theorem generalizes these results to any positive integral power n. The theorem's proof by mathematical induction is omitted.

Proposition 8.6.1
DeMoivre's Theorem

For every positive integer n,

$$z^n = r^n(\cos n\theta + i \sin n\theta). \tag{8.11}$$

Example 8.33

Use DeMoivre's theorem to evaluate $(\sqrt{3} + i)^{15}$.

Solution: First graph the complex number, and express it in trigonometric form (see Figure 8.41):

$$\sqrt{3} + i = 2(\cos 30° + i \sin 30°)$$

$$= 2\left(\cos \frac{\pi}{6} + i \sin \frac{\pi}{6}\right)$$

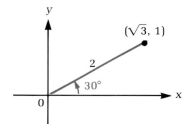

FIGURE 8.41

so

$$(\sqrt{3} + i)^{15} = \left[2\left(\cos \frac{\pi}{6} + i \sin \frac{\pi}{6}\right)\right]^{15}$$

$$= 2^{15}\left[\cos\left(15 \cdot \frac{\pi}{6}\right) + i \sin\left(15 \cdot \frac{\pi}{6}\right)\right]$$

$$= 2^{15}\left(\cos \frac{5\pi}{2} + i \sin \frac{5\pi}{2}\right).$$

Since $5\pi/2 = \pi/2 + 2\pi$, the principal argument is $\pi/2$. Thus

$$(\sqrt{3} + i)^{15} = 2^{15}\left(\cos \frac{\pi}{2} + i \sin \frac{\pi}{2}\right)$$

$$= 2^{15}[0 + i(1)]$$

$$= 2^{15}i.$$

Example 8.34 | Find $(-1 + i\sqrt{3})^3$.

Solution: First graph the complex number, and express it in trigonometric form (see Figure 8.42):

$$-1 + i\sqrt{3} = 2(\cos 120° + i \sin 120°)$$

$$= 2\left(\cos \frac{2\pi}{3} + i \sin \frac{2\pi}{3}\right)$$

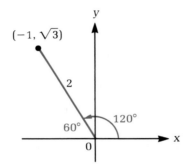

FIGURE 8.42

SO

$$(-1 + i\sqrt{3})^3 = 2^3\left[\cos\left(3 \cdot \frac{2\pi}{3}\right) + i \sin\left(3 \cdot \frac{2\pi}{3}\right)\right]$$

$$= 8(\cos 2\pi + i \sin 2\pi)$$
$$= 8[1 + i(0)]$$
$$= 8.$$

We may conclude from this example that $-1 + i\sqrt{3}$ is a cube root of 8. Recall from Section 4.3 that a polynomial of degree $n \geq 1$ has exactly n zeros. Furthermore, recall that if a complex number z is a zero of a polynomial, then its conjugate $\bar{z}$ is also a zero. Hence $-1 - i\sqrt{3}$ must also be a cube root of 8. The third cube root is the familiar real number 2.

Proposition 8.6.2 | For any integer $n \geq 2$, every complex number has n distinct roots.

DeMoivre's theorem provides the basis for finding roots of complex numbers. Let n be a positive integer and z be a complex number. A complex number v is the nth root of z if $v^n = z$. Writing v and z in trigonometric form,

$$z = r(\cos \theta + i \sin \theta)$$

and
$$v = s(\cos \phi + i \sin \phi),$$
we have
$$v^n = s^n(\cos n\phi + i \sin n\phi) = r(\cos \theta + i \sin \theta).$$
These two complex numbers are equal only if their absolute values are equal.
$$s^n = r$$
and since s and r are nonnegative,
$$s = \sqrt[n]{r}.$$
Also,
$$\cos n\phi = \cos \theta \quad \text{and} \quad \sin n\phi = \sin \theta.$$
These equations are equal if and only if $n\phi$ and θ differ by a multiple of 360°, that is,
$$n\phi = \theta + k360° \quad \text{where } k \text{ is an integer}$$
$$\phi = \frac{\theta + k360°}{n} \quad \text{or} \quad \phi = \frac{\theta + 2k\pi}{n}.$$

Proposition 8.6.3

For any nonzero complex number $z = r(\cos \theta + i \sin \theta)$, the nth roots of z are given by

$$\sqrt[n]{z} = \sqrt[n]{r}\left(\cos \frac{\theta + 2k\pi}{n} + i \sin \frac{\theta + 2k\pi}{n}\right) \tag{8.12}$$

for $k = 0, 1, 2, 3, \ldots, n - 1$.

Example 8.35

Find the five fifth roots of $32i$.

Solution: First convert $32i$ to trigonometric form:
$$32i = 32(\cos 90° + i \sin 90°)$$
$$= 32\left(\cos \frac{\pi}{2} + i \sin \frac{\pi}{2}\right).$$

Find the five roots by using formula (8.12) with $n = 5$, $r = 32$, and $\theta = \frac{\pi}{2}$ and having k take on the values 0, 1, 2, 3, and 4:

$$\sqrt[5]{32i} = 32^{1/5}\left[\cos\left(\frac{\frac{\pi}{2} + 2k\pi}{5}\right) + i \sin\left(\frac{\frac{\pi}{2} + 2k\pi}{5}\right)\right]$$
$$= 2\left[\cos\left(\frac{\pi}{10} + \frac{2k\pi}{5}\right) + i \sin\left(\frac{\pi}{10} + \frac{2k\pi}{5}\right)\right].$$

$k = 0$:

$$\sqrt[5]{32i} = 2\left(\cos\frac{\pi}{10} + i\sin\frac{\pi}{10}\right).$$

$k = 1$:

$$\sqrt[5]{32i} = 2\left[\cos\left(\frac{\pi}{10} + \frac{2\pi}{5}\right) + i\sin\left(\frac{\pi}{10} + \frac{2\pi}{5}\right)\right]$$

$$= 2\left(\cos\frac{\pi}{2} + i\sin\frac{\pi}{2}\right)$$

$$= 2[0 + i(1)]$$

$$= 2i.$$

$k = 2$:

$$\sqrt[5]{32i} = 2\left[\cos\left(\frac{\pi}{10} + \frac{4\pi}{5}\right) + i\sin\left(\frac{\pi}{10} + \frac{4\pi}{5}\right)\right]$$

$$= 2\left(\cos\frac{9\pi}{10} + i\sin\frac{9\pi}{10}\right).$$

$k = 3$:

$$\sqrt[5]{32i} = 2\left[\cos\left(\frac{\pi}{10} + \frac{6\pi}{5}\right) + i\sin\left(\frac{\pi}{10} + \frac{6\pi}{5}\right)\right]$$

$$= 2\left(\cos\frac{13\pi}{10} + i\sin\frac{13\pi}{10}\right).$$

$k = 4$:

$$\sqrt[5]{32i} = 2\left[\cos\left(\frac{\pi}{10} + \frac{8\pi}{5}\right) + i\sin\left(\frac{\pi}{10} + \frac{8\pi}{5}\right)\right]$$

$$= 2\left(\cos\frac{17\pi}{10} + i\sin\frac{17\pi}{10}\right).$$

Using a hand-held calculator or Appendix Table 4, we get the following approximate roots:

$$1.90 + 0.62i$$
$$2i$$
$$-1.90 + 0.62i$$
$$-1.18 - 1.62i.$$
$$1.18 - 1.62i.$$

Example 8.36   Find the square roots of $-1 + \sqrt{3}i$.

Solution: First graph $-1 + \sqrt{3}i$ and change it to trigonometric form (see Figure 8.43):

$$-1 + \sqrt{3}i = 2(\cos 120° + i \sin 120°)$$

$$= 2\left(\cos \frac{2\pi}{3} + i \sin \frac{2\pi}{3}\right).$$

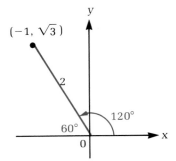

FIGURE 8.43

Using (8.12) with $n = 2$, $\theta = \dfrac{2\pi}{3}$, and $k = 0, 1$, we find

$$\sqrt{-1 + \sqrt{3}i} = \sqrt{2}\left[\cos\left(\frac{\frac{2\pi}{3} + 2k\pi}{2}\right) + i \sin\left(\frac{\frac{2\pi}{3} + 2k\pi}{2}\right)\right].$$

$k = 0$:

$$\sqrt{-1 + \sqrt{3}i} = \sqrt{2}\left[\cos\left(\frac{\frac{2\pi}{3}}{2}\right) + i \sin\left(\frac{\frac{2\pi}{3}}{2}\right)\right]$$

$$= \sqrt{2}\left(\cos \frac{\pi}{3} + i \sin \frac{\pi}{3}\right)$$

$$= \sqrt{2}\left(\frac{1}{2} + \frac{\sqrt{3}i}{2}\right)$$

$$= \frac{\sqrt{2}}{2}(1 + \sqrt{3}i).$$

$k = 1$:

$$\sqrt{-1+\sqrt{3}i} = \sqrt{2}\left[\cos\left(\frac{\frac{2\pi}{3}+2\pi}{2}\right) + i\sin\left(\frac{\frac{2\pi}{3}+2\pi}{2}\right)\right]$$

$$= \sqrt{2}\left[\cos\left(\frac{\frac{8\pi}{3}}{2}\right) + i\sin\left(\frac{\frac{8\pi}{3}}{2}\right)\right]$$

$$= \sqrt{2}\left(\cos\frac{4\pi}{3} + i\sin\frac{4\pi}{3}\right)$$

$$= \sqrt{2}\left(\frac{-1}{2} - \frac{\sqrt{3}i}{2}\right)$$

$$= -\frac{\sqrt{2}}{2}(1+\sqrt{3}i).$$

Therefore the square roots of $-1+\sqrt{3}i$ are

$$\frac{\sqrt{2}}{2}(1+\sqrt{3}i) \qquad \text{and} \qquad \frac{-\sqrt{2}}{2}(1+\sqrt{3}i).$$

EXERCISES

Express the complex numbers in Exercises 1 to 20 in trigonometric form with $0 \le \theta \le 2\pi$.

1. $1 - i$ 2. $1 - 2i$ 3. -1 4. -6

5. $6i$ 6. i 7. $2 + 2i$ 8. $-2 + 2i$

9. 15 10. 10 11. $\sqrt{3} + i$ 12. $-2 - 2\sqrt{3}i$

13. $-20i$ 14. $-5i$ 15. $-5 + 12i$ 16. $4 - 3i$

17. $-1 - \sqrt{3}i$ 18. $-4\sqrt{3} + 4i$ 19. $-5 - 5\sqrt{3}i$ 20. $-5\sqrt{3} + 5i$

Use Appendix Table 4 or a hand-held calculator to write the complex numbers in Exercises 21 to 26 in rectangular form $(a + bi)$.

21. $2[\cos (5\pi/6) + i\sin (5\pi/6)]$ 22. $3(\cos \pi + i\sin \pi)$ 23. $5[\cos (\pi/4) + i\sin (\pi/4)]$

24. $6[\cos (7\pi/4) + i\sin (7\pi/4)]$ 25. $4(\cos 55° + i\sin 55°)$ 26. $\sqrt{29}(\cos 200° + i\sin 200°)$

In Exercises 27 to 34, find $z_1 \cdot z_2$ and z_1/z_2 by changing to trigonometric form and then using (8.9) and (8.10), respectively. Check by using the methods of Section 1.12.

27. $z_1 = 1 + i, z_2 = 1 - i$ 28. $z_1 = -1 - i, z_2 = -1 + i$ 29. $z_1 = \sqrt{3} + i, z_2 = -\sqrt{3} + i$

30. $z_1 = -2 - 2\sqrt{3}i, z_2 = 3i$ 31. $z_1 = 2 + i, z_2 = 3$ 32. $z_1 = 3, z_2 = 5 - 2i$

33. $z_1 = 4i, z_2 = -3i$ 34. $z_1 = -3, z_2 = -5$

Use DeMoivre's theorem to calculate the numbers in Exercises 35 to 44. Express the results in the form $a + bi$ correct to three figures.

35. $(1 + i)^8$

36. $(3 + 3i)^5$

37. $(\sqrt{3} + i)^{12}$

38. $(1 + \sqrt{3}i)^6$

39. $(0.8912 + 0.5736i)^9$

40. $(2 + 3i)^{-3}$

41. $(2 - 5i)^{-2}$

42. $(3 - i)^6$

43. $\dfrac{60}{(1 + 2i)^5}$

44. $\dfrac{32}{(1 + i)^6}$

Find all indicated roots in Exercises 45 to 54. Use the Appendix tables or a hand-held calculator if necessary.

45. $\sqrt{2 + 2\sqrt{3}i}$

46. $\sqrt[3]{1}$

47. $\sqrt{i}$

48. $\sqrt[5]{16 - 16\sqrt{3}i}$

49. $\sqrt[3]{-i}$

50. $\sqrt[6]{-\sqrt{3} + i}$

51. $\sqrt[5]{1 + \sqrt{3}i}$

52. $\sqrt[4]{1 + i}$

53. $\sqrt[4]{1}$

54. $\sqrt[3]{-27}$

KEY WORDS

right triangle	*displacement*	*resultant*
angle of elevation	*velocity*	*components*
angle of depression	*force*	*standard position*
course	*initial point*	*polar form*
bearing	*terminal point*	*rectangular form*
oblique triangle	*magnitude*	*pole*
AAS	*direction*	*polar axis*
SSA	*displacement vector*	*complex plane*
SAS	*velocity vector*	*absolute value*
SSS	*speed*	*modulus*
law of sines	*force vector*	*trigonometric form*
law of cosines	*zero vector*	*argument*
area of triangular region	*unit vector*	*principal argument*
vector quantities	*free vector*	*DeMoivre's theorem*
vectors	*sum*	*complex roots*

KEY FORMULAS

- Law of sines:

$$\frac{a}{\sin A} = \frac{b}{\sin B} = \frac{c}{\sin C}.$$

- Law of cosines:

$$a^2 = b^2 + c^2 - 2bc \cos A,$$

$$b^2 = a^2 + c^2 - 2ac \cos B,$$

$$c^2 = a^2 + b^2 - 2ab \cos C.$$

- Triangle area:

$$\text{Area} = \tfrac{1}{2}bc \sin A = \tfrac{1}{2}ac \sin B = \tfrac{1}{2}ab \sin C,$$

$$\text{Area} = \frac{1}{2}\frac{b^2 \sin A \sin C}{\sin B},$$

$$\text{Area} = \sqrt{s(s - a)(s - b)(s - c)},$$

$$\text{where } s = \tfrac{1}{2}(a + b + c).$$

- Complex numbers: If $z_1 = r_1(\cos \theta_1 + i \sin \theta_1)$ and $z_2 = r_2(\cos \theta_2 + i \sin \theta_2)$, then

$$z_1 z_2 = r_1 r_2 [\cos (\theta_1 + \theta_2) + i \sin (\theta_1 + \theta_2)]$$

$$\frac{z_1}{z_2} = \frac{r_1}{r_2} [\cos (\theta_1 - \theta_2) + i \sin (\theta_1 - \theta_2)]$$

- Complex numbers—powers and roots: If $z = a + bi$, then

$$z^n = r^n(\cos n\theta + i \sin n\theta)$$

$$\sqrt[n]{z} = \sqrt[n]{r} \left(\cos \frac{\theta + 2k\pi}{n} + i \sin \frac{\theta + 2k\pi}{n} \right)$$

for $k = 0, 1, 2, 3, \ldots, n - 1$.

EXERCISES FOR REVIEW

Solve the triangles in Exercises 1 to 6.

1. $a = 8$, $B = 30°$, $C = 56°$
2. $c = 11$, $A = 39°$, $C = 53°$
3. $b = 20$, $c = 30$, $A = 40°$
4. $a = 12$, $c = 10$, $C = 49°$
5. $a = 21$, $c = 21$, $A = 35°$
6. $a = 918$, $b = 746$, $B = 32°20'$
7. Given $a = 7$, $b = 5$, and $C = 30°$, find c.
8. Given $a = 5$, $b = 3$, and $c = 6$, find B.
9. Given $a = 7$, $c = 9$, and $B = 61°$, find b.
10. A toboggan slide in a playground is 28 feet long and makes an angle of $39°$ with the ground. Its top is reached by a ladder 18 feet long. How steep is the ladder?
11. A ship leaves Montauk Point and sails 54 nautical miles on a heading of $135°$ and then sails 20 nautical miles on a heading of $240°$. What course should it sail to return to its starting point, and how many nautical miles must it go on this course?
12. Find the measure of the smallest angle of a triangle whose sides measure 14.2, 17.6, and 23.2 inches.

Find the area of the triangles in Exercises 13 to 18.

13. $a = 12$, $c = 8$, and $B = 60°$.
14. $a = 10$, $B = 40°$, and $C = 120°$.
15. $a = 13$, $b = 7$, and $c = 8$.
16. $b = 21.48$, $c = 38.69$, and $A = 111°20'$.
17. $a = 468.3$, $c = 372.9$, and $A = 41°50'$.
18. $c = 429.7$, $A = 73°10'$, and $B = 55°30'$.
19. A vertical force of 42 pounds and a horizontal force of 18 pounds act on an object. Find the magnitude and direction of the resultant force.
20. Find the horizontal and vertical components of a force of 100 pounds inclined at an angle of $75°$ to the horizontal.
21. A 1000-pound object is pulled up an inclined plane of slope $15°$ to the horizontal. What is the tension in the cable?
22. Forces of 250 and 360 pounds act on an object. The angle between the forces is $110°$. Find the resultant and the angle it makes with the smaller force.
23. A vacuum cleaner is pushed with a force of 25 pounds at an angle of $45°$. Find the vertical and horizontal components.
24. A mass of 450 pounds rests on an inclined plane. A force of 300 pounds is required to keep it from sliding down. What is the angle of inclination of the plane?

Convert the vectors in Exercises 25 to 28 from rectangular to polar coordinates.

25. $\langle 3, 3 \rangle$ **26.** $\langle 0, 2 \rangle$ **27.** $\langle -2, 1 \rangle$ **28.** $\langle 5, -\sqrt{3} \rangle$

Convert the vectors in Exercises 29 to 32 from polar to rectangular coordinates.

29. $(1, 30°)$ **30.** $(2, 60°)$ **31.** $(3, 140°)$ **32.** $(4, \pi/2)$

Use polar graph paper to plot the points in Exercises 33 to 36.

33. $(1, 270°)$ **34.** $(2.5, 75°)$ **35.** $(0.5, 135°)$ **36.** $(3, 300°)$

Find the magnitude of the vectors given in standard position in Exercises 37 and 38.

37. $\langle 1, 4 \rangle$ **38.** $\langle -2, 3 \rangle$

Given the vectors $\vec{U}$ and $\vec{V}$, find $\vec{U} + \vec{V}$ and $\vec{U} - \vec{V}$ in Exercises 39 and 40.

39. $\vec{U} = \langle 2, 7 \rangle, \vec{V} = \langle 8, -8 \rangle$ **40.** $\vec{U} = \langle 0, -3 \rangle, \vec{V} = \langle -1, -2 \rangle$

Find a unit vector in the same direction as the given vector in Exercises 41 and 42.

41. $\langle 6, 5 \rangle$ **42.** $\langle -7, 11 \rangle$

Express each of the complex numbers in Exercises 43 to 46 in trigonometric form, where $0 \le \theta \le 2\pi$.

43. $2 - i$ **44.** $-3 - 3i$ **45.** $\sqrt{3} - i$ **46.** $4 + 3i$

Use Appendix Table 4 or a hand-held calculator to write the complex numbers in Exercises 47 to 50 in rectangular form ($a + bi$).

47. $8(\cos 75° + i \sin 75°)$ **48.** $3(\cos 315° + i \sin 315°)$

49. $2[\cos (3\pi/4) + i \sin (3\pi/4)]$ **50.** $5[\cos (7\pi/6) + i \sin (7\pi/6)]$

Find $z_1 \cdot z_2$ and z_1/z_2 in Exercises 51 and 52.

51. $z_1 = 1 - i, z_2 = 5$ **52.** $z_1 = 2 - i, z_2 = 5 + 2i$

Use DeMoivre's theorem to calculate the numbers in Exercises 53 and 54. Express the results in the form $a + bi$ correct to three places.

53. $(1 - i)^6$ **54.** $(2 - 3i)^3$

Find all the indicated roots in Exercises 55 and 56. Use Appendix tables or a hand-held calculator.

55. Find the square roots of $2 - 2\sqrt{3}i$.   **56.** Find the fifth roots of -243.

CHAPTER TEST

1. Solve the triangles:
   (a) $b = 18.6, c = 27.2, B = 41.6°$
   (b) $a = 12, b = 16, c = 110°$

2. Express each complex number in trigonometric form, where $0 \le \theta < 2\pi$.
   (a) $-1 + 2i$,   (b) $\sqrt{3} - 2i$

3. Use Appendix Table 4 or a hand-held calculator to write the complex number in rectangular form ($a + bi$).

$$z = 2\left(\cos \frac{7\pi}{12} + i \sin \frac{7\pi}{12} \right)$$

4. Find **(a)** $z_1 z_2$ and **(b)** z_1/z_2 given $z_1 = 2 - i$ and $z_2 = 4 + 3i$.

5. Use DeMoivre's theorem to calculate $(2 + 3i)^3$.

6. Find the square roots of $-3 - 3i$.

7. A triangular corner lot has frontages of 52 and 78 meters with the rear line equal to 105 meters. Find the area of the lot.

8. Find the minimum force necessary to keep a 2400-pound car from rolling down a hill which is inclined at 5° with the horizontal.

9. The Eiffel Tower is 300 meters high. The top of the Arch of Triumph is 54 meters higher than the base of the tower. From the top of the arch the tower subtends an angle of 10°10'. Find the horizontal distance from the arch to the tower.

10. Two lifeguards, stationed 100 yards apart, spot a swimmer calling for help. Which lifeguard will reach the swimmer first (assuming they travel at the same rate) if the angles are as shown. What distance did each lifeguard swim?

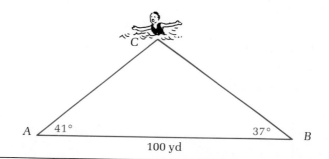

SYSTEMS OF LINEAR EQUATIONS AND LINEAR INEQUALITIES

9

INTRODUCTION

In this chapter, we are concerned exclusively with *systems of linear equations and of linear inequalities*, important topics in mathematics with applications in many areas. A high school algebra course introduces the process of solving two linear equations in two variables by using graphs, elimination, or substitution techniques. We review these relatively simple but powerful methods. Then we examine the solutions of n linear equations in n variables, $n \geq 2$, using matrices. Finally, we discuss linear programming as an application of linear inequalities.

9.1   SYSTEMS OF LINEAR EQUATIONS IN TWO VARIABLES

Many applied problems involve conditions that must all be satisfied at the same time. These are written in the form of simultaneous linear equations in two variables. Suppose, for example, that the sum of two numbers is 52. If one of the numbers is 6 more than the second number, can you determine the two numbers? Perhaps, after some trial and error, you come to the conclusion that the two numbers are 23 and 29, since their sum is 52, while their difference is 6. How did you arrive at this solution? Did you use some technique that you learned in high school algebra, or was it a good guess? Suppose that 720 teachers from secondary schools attended a professional convention in Boston last summer. The number of male teachers is 90 less than twice the number of female teachers. Can you determine the number of male and female teachers who attended the convention? If you have organized your thinking and found that 450 male teachers and 270 female teachers attended the convention, you have some insight into solving similar problems.

But many times we are faced with problems too complicated to solve by trial and error without using any mathematical model. Consider, for example, that the manager of a grocery store wishes to mix peanuts that sell for $1.80 a pound with pecans that sell for $3.00 a pound. Can you determine the number of pounds of peanuts that should be mixed with 20 pounds of pecans to produce a mixture selling for $2.20 a pound? As another illustration, suppose that solution A is 30 percent alcohol and 70 percent water while solution B is 45 percent alcohol and 55 percent water. How much of solution A must be mixed with 30 liters of solution B to produce another solution which is 35 percent alcohol? Obviously, guesswork is not helpful in problems of this nature, and we must seek a systematic procedure that expresses the relevant information in linear equations. These equations may then be solved by graphing, substitution, or elimination methods. We may also use matrices to solve these linear equations.

Recall from Chapter 3 that $ax + by + c = 0$ is a linear equation that describes a particular kind of relationship between two unknown quantities. For example, the equation $2x + 3y = 18$ is a linear equation in two variables, x and y. A **solution** of this equation is an ordered pair of numbers (x, y) that, when substituted for x and y in this equation, converts it to a true statement. The pair $(6, 2)$, for example, is one solution, since $2(6) + 3(2) = 18$. Similarly, many other ordered pairs such as $(0, 6)$ $(3, 4)$, $(9, 0)$, $(12, -2)$, and $(15, -4)$ are all solutions of the linear equation $2x + 3y = 18$. In fact, a linear equation in more than one unknown has an infinite number of solutions.

Consider the following pair of linear equations:

$$2x + 3y = 18$$
$$3x - 2y = 1$$

$$(9.1)$$

We have two equations in two unknowns, and we wish to determine the solution set. Adding 2 times the first equation to 3 times the second equation eliminates y, and we have

$$\begin{array}{r} 4x + 6y = 36 \\ 9x - 6y = 3 \\ \hline 13x = 39 \\ x = 3. \end{array}$$

Substituting this value of x in either of the two original equations, we obtain $y = 4$. Thus the solution to the system of linear equations (9.1) that satisfies both equations is the ordered pair $(3, 4)$. Had we wanted to eliminate x instead of y, we would have multiplied the first equation by 3 and second equation by 2. Thus

$$6x + 9y = 54$$
$$6x - 4y = 2.$$

Then subtracting the second equation from the first equation, we would have obtained

$$\begin{array}{r} 6x + 9y = 54 \\ -6x + 4y = -2 \\ \hline 13y = 52 \\ y = 4. \end{array}$$

method of elimination
We get $y = 4$ as before. Note that the **method of elimination** which we have used here assumes that we can multiply "equals by equals," add "equals to equals," and subtract "equals from equals" without changing equalities.

substitution method
Another approach for solving simultaneous linear equations is the **substitution method.** We can solve each of the two linear equations in (9.1) for x. The first equation, $2x + 3y = 18$, gives

$$x = \frac{18 - 3y}{2},$$

and the second equation, $3x - 2y = 1$, yields

$$x = \frac{1 + 2y}{3}.$$

Equating these two expressions, we have

$$\frac{18 - 3y}{2} = \frac{1 + 2y}{3}$$

$$54 - 9y = 2 + 4y$$

$$-13y = -52$$

$$y = 4.$$

The corresponding value of $x = 3$ is obtained by substituting $y = 4$ in either of the two equations. If we had wanted to solve each of the original equations for y instead of x, we would have obtained

$$y = \frac{18 - 2x}{3} \qquad \text{and} \qquad y = \frac{3x - 1}{2}.$$

Equating these expressions, we obtain

$$\frac{18 - 2x}{3} = \frac{3x - 1}{2}$$

$$36 - 4x = 9x - 3$$

$$-13x = -39$$

$$x = 3.$$

We substitute the value of x in either of the two equations and obtain $y = 4$, as before.

Another method of finding the solution set to simultaneous linear equations is to use graphs (see Figure 9.1). To represent a linear equation graphically, we

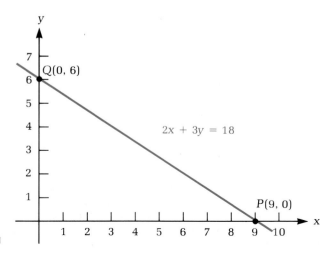

FIGURE 9.1

choose two distinct two points whose coordinates satisfy the linear equation. Consider again the equations in (9.1). The points $P(9, 0)$ and $Q(0, 6)$ lie in the line $2x + 3y = 18$; the straight line joining these points represents the graph of this linear equation.

The second equation, $3x - 2y = 1$, like the first, is satisfied by infinitely many points. Some of the points are $A(1, 1)$, $B(3, 4)$, $C(5, 7)$, and $D(7, 10)$. The line joining any two of these points determines the graph of the linear equation $3x - 2y = 1$ (see Figure 9.2).

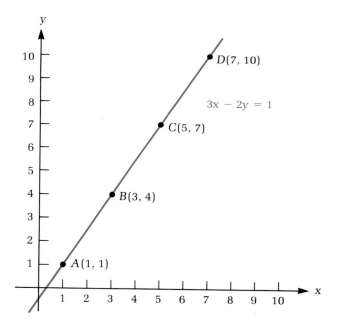

FIGURE 9.2

Our interest is to find ordered pairs that satisfy both the above equations simultaneously. Consequently, the solution set of (9.1) is the intersection of the solution sets of the individual equations. Having sketched the graphs of the equations, we simply read off the coordinates of the point of intersection of the two straight lines. The solution set consists of the ordered pair $(3, 4)$, as shown in Figure 9.3.

The graphical method we have used to find the point of intersection of two lines has serious practical limitations, since we can read off only an **approximation** of the solution from the graph. Unless there are reasons to the contrary, we suggest that you avoid the use of graphical methods for solving linear equations.

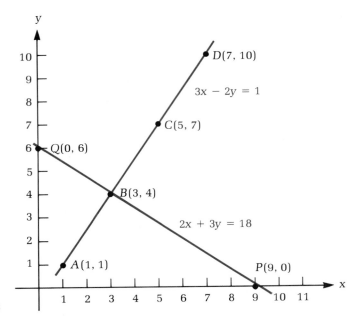

FIGURE 9.3

The graphical representation of a system of two linear equations in two variables may result in one of the three possibilities:

1. The lines intersect in exactly one point. The coordinates of the point of intersection comprise the solution set. The system of equations is said to be **independent.**

independent system

2. The lines are parallel. The solution set in this case is the empty set. The system of equations is said to be **inconsistent.**

inconsistent system

3. The lines coincide. In this case, the infinite number of solutions of either equation is the solution set. The system of equations is said to be **dependent.**

dependent system

We have provided an illustration of the first case in Figure 9.3. The remaining possibilities are illustrated in Examples 9.1 and 9.2.

Example 9.1 | Solve the following system of linear equations graphically:

$$2x + 3y = 12$$

$$6x + 9y = 54.$$

Solution: The ordered pairs (6, 0), (3, 2), and (0, 4) are a few of the solutions of the linear equation $2x + 3y = 12$. Some of the ordered pairs that satisfy the linear equation $6x + 9y = 54$ are $(9, 0)$, $(3, 4)$, $(6, 2)$, and $(0, 6)$. The graph of these equations is shown in Figure 9.4. Note that the linear equation $2x + 3y = 12$, when expressed in slope-intercept form (3.2), yields

$$y = -\frac{2}{3}x + 4.$$

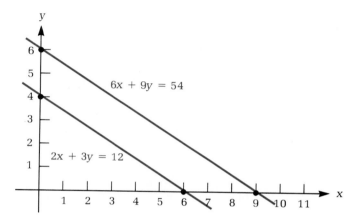

FIGURE 9.4

This implies that the slope of the line is $m_1 = -\frac{2}{3}$, and the y intercept is 4. Similarly, solving $6x + 9y = 54$, we have

$$y = -\frac{2}{3}x + 6.$$

Thus $m_2 = -\frac{2}{3}$, and the y intercept is 6. Since the slopes m_1 and m_2 are equal and the y intercepts are unequal, we conclude that $2x + 3y = 12$ and $6x + 9y = 54$ are distinct parallel lines and so have no common point of intersection. The solution set is the empty set.

Example 9.2

Solve the following system of linear equations graphically:

$$x - 2y = 4$$
$$4x - 8y = 16.$$

Solution: Note that $x - 2y = 4$ and $4x - 8y = 16$ are equations of the same straight line. (Why?) The ordered pairs $(0, -2)$, $(2, -1)$, $(4, 0)$, and $(8, 2)$ and infinitely many more points that satisfy the first equation $x - 2y = 4$ also satisfy the equation $4x - 8y = 16$. Hence the lines coincide (see Figure 9.5). Since

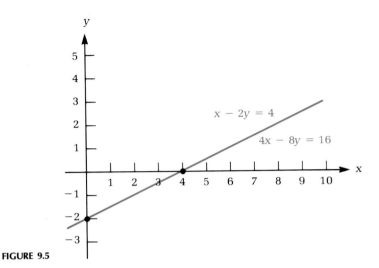

FIGURE 9.5

every ordered pair that is a solution of either linear equation is also a solution of the other linear equation, it follows that the system has an infinite number of solutions.

Let us now examine how the method of elimination is used in solving linear equations that are not independent.

Example 9.3

Solve the following system of linear equations:

$$3x + 4y = 12$$
$$9x + 12y = 25.$$

Solution: Adding (-3) times the first equation to the second equation eliminates both x and y, and we obtain

$$-9x - 12y = -36$$
$$\underline{9x + 12y = 25}$$
$$0 = -11,$$

which is a false statement. The system of equations is *inconsistent* and therefore has no solution.

Also, we may observe that $3x + 4y = 12$ and $9x + 12y = 25$ are distinct parallel lines. Therefore, they have no common point of intersection, and the solution set is the empty set.

Example 9.4

Solve this system of linear equations:

$$x - 3y = 6$$
$$2x - 6y = 12.$$

Solution: Adding -2 times the first equation to the second equation, we get

$$-2x + 6y = -12$$
$$2x - 6y = 12$$
$$\overline{\,0 = 0,}$$

which is true. This means that $x - 3y = 6$ and $2x - 6y = 12$ are equations of the same straight line. Therefore their graphs coincide, and every solution of either equation is also a solution of the other equation. The system of equations is dependent and has an infinite number of solutions.

The following examples are practical problems that can be translated to two equations in two unknowns.

Example 9.5

Anita deposited $9000 in a savings bank, some at an annual rate of 5.5 percent interest and the remainder at an annual rate of 6 percent. At the end of the year, she collected the interest, which totaled $520. Determine how much money was invested at each rate.

Solution: Let x and y represent the amounts invested at 5.5 and 6 percent, respectively. Then $0.055x$ is the interest she received from one account, and $0.06y$ is the interest she received from the other account. The system of equations is

$$x + y = 9000$$
$$(0.055)x + (0.06)y = 520,$$

which may be expressed in a simplified form as follows:

$$x + y = 9000$$
$$11x + 12y = 104{,}000.$$

Solving this system of two equations in two unknowns, we have

$$x = 4000 \quad \text{and} \quad y = 5000.$$

Thus $4000 was invested at 5.5 percent, and $5000 was invested at 6 percent.

Example 9.6   The manager of a grocery store wishes to mix peanuts that sell for $1.80 a pound with pecans that sell for $3.00 a pound. How many pounds of peanuts should be mixed with 20 pounds of pecans so that the mixture will sell for $2.20 a pound?

Solution: Let x and y represent the number of pounds of peanuts and of the mixture, respectively. The linear equation involving the total number of pounds is

$$y = x + 20.$$

Since x pounds of peanuts sell for $1.80 a pound, 20 pounds of pecans sell for $3.00 a pound, and y pounds of mixture sell for $2.20 a pound, we have the total-cost equation

$$(2.20)y = (1.80)x + 20(3.00).$$

Thus we have the following system of linear equations to solve:

$$y = x + 20$$
$$22y = 18x + 600.$$

Solving this system of two equations in two unknowns, we have

$$x = 40 \quad \text{and} \quad y = 60.$$

Hence the manager should mix 40 pounds of peanuts with 20 pounds of pecans.

EXERCISES

Determine the solution set for the systems of linear equations in Exercises 1 to 8.

1. $x + y = 1$
   $x - y = 1$

2. $x + 2y = 5$
   $2x + 3y = 8$

3. $2x - 3y = 13$
   $4x + y = 5$

4. $2x - y = 4$
   $5x - y = 13$

5. $x + y = 5$
   $2x + 3y = 12$

6. $x - 3y = 7$
   $2x + y = 7$

7. $3x - 2y = 6$
   $x + y = 4$

8. $2x - 3y = 5$
   $3x - 2y = -4$

Determine whether the systems of linear equations in Exercises 9 to 14 are independent, dependent, or inconsistent.

9. $x + y = 3$
   $3x + 3y = 12$

10. $3x - y = 4$
    $3x - y = 8$

11. $3x - y = 8$
    $2x + 3y = 9$

12. $x - y = 0$
    $x + 3y = 8$

13. $x + y = 3$
    $2x + 2y = 6$

14. $x - 2y = 0$
    $3x + 2y = 8$

Solve graphically each system of linear equations in Exercises 15 to 20.

15. $x - y = 2$
    $2x + 3y = 9$

16. $2x + y = 5$
    $x + 2y = 4$

17. $x - y = 0$
    $x + 3y = 8$

18. $x + 2y = 2$
    $x - 2y = 2$

19. $x + 3y = 9$
    $3x + 2y = 13$

20. $x - y = -1$
    $3x + 2y = 12$

21. The sum of two numbers is 78. If one number is 12 more than the second number, find the two numbers.

22. Four shirts and three ties cost $84 in a department store, whereas five shirts and two ties cost $91. Determine the price of each shirt and each tie.

23. At a theater performance, children were charged $0.75, whereas adults paid $1.25 for an admission ticket. There were 600 fewer children's tickets sold than adult tickets. If the total receipts for the performance were $3250, how many of each type of admission ticket were sold?

24. Adolfo deposited $10,000 in a savings bank, some at an annual rate of 5.5 percent and the remainder at an annual rate of 6 percent. The total interest collected at the end of 1 year was $570. How much money was deposited at 5.5 and at 6 percent?

25. A total of $12,000 is invested in two savings plans. One savings plan has a 5.5 percent annual rate of interest, and the other has a 7 percent rate. If a total of $765 is collected as interest at the end of 1 year, how much money was invested at 5.5 and at 7 percent?

26. The manager of a grocery store wishes to mix dried fruits A that sell for $5.76 a kilogram with dried fruits B that sell for $7.86 a kilogram. How many kilograms of dried fruits A should be mixed with 20 kilograms of dried fruits B to produce a mixture selling for $6.81 a kilogram?

27. How many kilograms of nuts selling for $5.76 a kilogram should be mixed with 10 kilograms of nuts selling for $7.80 a kilogram so as to produce a mixture selling for $6.78 a kilogram?

28. A grocer wants to market a new brand of mixed nuts by mixing cashews that sell for $7 a pound and peanuts that sell for $2 a pound. How many pounds of peanuts should be mixed with 15 pounds of cashews to produce a mixture selling for $3 a pound?

29. Solution A is 26 percent alcohol and 74 percent water. Solution B is 42 percent alcohol and 58 percent water. How much of each solution must be mixed to make 120 quarts of a solution that is 30 percent alcohol?

30. Solution A is 40 percent alcohol and 60 percent water. Solution B is 30 percent alcohol and 70 percent water. How much of each solution must be mixed to make 100 liters of a solution that is 32 percent alcohol?

31. A dairy produces milk that is 3 percent butterfat and cream that is 30 percent butterfat. How many gallons of milk should be mixed with 15 gallons of cream to obtain a mixture that is 9 percent butterfat?

32. A dairy has 45 gallons of milk that is 4 percent butterfat and 20 gallons of cream that is 24 percent butterfat. How many gallons of milk should be mixed with the cream to obtain "half and half," that is 12 percent butterfat? (Note that "half and half" is not strictly equal parts of milk and cream.)

9.2

SYSTEMS OF LINEAR EQUATIONS IN THREE VARIABLES

The method of elimination can also be used to solve a system of three linear equations in three variables, four linear equations in four variables, and so on. We simply eliminate one variable at a time, and at each step we reduce by 1 the number of variables as well as the number of equations. To illustrate how it works for a system of three linear equations in three variables, we consider the following examples.

Example 9.7

Solve the following system of linear equations:

$$x + 2y + z = 4$$
$$2x + 3y + 4z = 19$$
$$3x - y + z = 9.$$

(9.2)

Solution: First, we eliminate one of the variables and reduce the problem to solving a system of two linear equations in two variables. We choose two pairs of equations and eliminate the *same* variable from each pair. To eliminate x from the first two equations, for instance, we add (-2) times the first equation to the second equation, and we have

$$
\begin{array}{rcr}
-2x - 4y - 2z &=& -8 \\
2x + 3y + 4z &=& 19 \\
\hline
-y + 2z &=& 11.
\end{array}
$$

Similarly, to eliminate x from the first and third equations, we add (-3) times the first equation to the third equation, and we have

$$
\begin{array}{rcr}
-3x - 6y - 3z &=& -12 \\
3x - y + z &=& 9 \\
\hline
-7y - 2z &=& -3.
\end{array}
$$

This leaves us with the following system of two linear equations in two variables y and z:

$$
\begin{aligned}
-y + 2z &= 11 \\
-7y - 2z &= -3.
\end{aligned}
\tag{9.3}
$$

We continue the method of elimination. Adding these equations, we eliminate z. Thus, we have

$$
\begin{array}{rcr}
-y + 2z &=& 11 \\
-7y - 2z &=& -3 \\
\hline
-8y &=& 8 \\
y &=& -1.
\end{array}
$$

Substituting this value of y in either of the two equations in (9.3), we obtain $z = 5$. Finally, if we substitute $y = -1$ and $z = 5$ into any of the original equations (9.2), we get $x = 1$. The solution of the system of linear equations (9.2) is

$$
x = 1, \qquad y = -1, \qquad \text{and} \qquad z = 5.
$$

When we deal with a system of three linear equations in three variables, sometimes there are no solutions or there are an infinite number of solutions. Consider, for instance, Example 9.8.

Example 9.8

Solve the following system of linear equations:

$$
\begin{aligned}
x + 3y - 4z &= 5 \\
2x - y + 5z &= 7 \\
7x + 7y - 2z &= 31.
\end{aligned}
\tag{9.4}
$$

Solution: Adding the first equation to 3 times the second equation eliminates the variable y, and we have

$$\begin{array}{rcl} x + 3y - 4z &=& 5 \\ 6x - 3y + 15z &=& 21 \\ \hline 7x \qquad + 11z &=& 26. \end{array}$$

Similarly, to eliminate y from the second and third equations, we add 7 times the second equation to the third equation, thus obtaining

$$\begin{array}{rcl} 14x - 7y + 35z &=& 49 \\ 7x + 7y - 2z &=& 31 \\ \hline 21x \qquad + 33z &=& 80. \end{array}$$

This leaves us with the following system of linear equations in two variables x and z:

$$7x + 11z = 26$$

$$21x + 33z = 80. \tag{9.5}$$

Adding -3 times the first equation to the second equation, we have

$$\begin{array}{rcl} -21x - 33z &=& -78 \\ 21x + 33z &=& 80 \\ \hline 0 &=& 2, \end{array}$$

which is a *contradiction*. Thus there are no values of x and z for which either of the equations in (9.5) holds. Consequently, there are no values of x, y, and z for which the system of linear equations (9.4) holds. So we conclude that the system of equations has *no solution*.

Let us now look at an example in which the system of linear equations has an *infinite number of solutions*.

Example 9.9 Solve the following system of linear equations:

$$\begin{array}{rcl} x - y + z &=& 5 \\ 3x + 5y - 13z &=& -9 \\ 5x + 3y - 11z &=& 1. \end{array}$$

Solution: Adding 5 times the first equation to the second equation eliminates y, and we have

$$\begin{array}{rcl} 5x - 5y + 5z &=& 25 \\ 3x + 5y - 13z &=& -9 \\ \hline 8x \qquad - 8z &=& 16, \end{array}$$

or

$$x - z = 2$$
$$x = 2 + z.$$

Adding 3 times the first equation to the third equation, we get

$$\begin{array}{r} 3x - 3y + 3z = 15 \\ 5x + 3y - 11z = 1 \\ \hline 8x - 8z = 16, \end{array}$$

which agains leads us to the same equation:

$$x = 2 + z.$$

Substituting $x = 2 + z$ in any of the original equations and solving for y, we have

$$y = -3 + 2z.$$

Thus, we have

$$x = 2 + z \qquad \text{and} \qquad y = -3 + 2z.$$

Clearly, z can take any arbitrary value. Thus, if $z = c$, then

$$x = 2 + c \qquad \text{and} \qquad y = -3 + 2c.$$

We leave it for the reader to verify, by substituting

$$x = 2 + c, \qquad y = -3 + 2c, \qquad z = c$$

in the original set of equations, that is, in fact, a solution of the three linear equations in three variables.

    Note that for each value of c we have a solution. Because c can assume any value, we conclude that the system has an infinite number of solutions.

    In these examples we started by eliminating a variable from any pair of equations. A more systematic way of applying the method of elimination to the solution of a system of n linear equations in n variables, $n > 2$, is to eliminate the first variable from the second equation, the first two variables from the third equation, the first three variables from the fourth equation, and so on. When we deal with four linear equations in four variables, this method of elimination works in a manner so that the value of fourth variable can be read off the fourth equation. And by working backward, substitution of the value of the fourth variable into the third equation gives the value of the third variable, substitution of the values of third and fourth variables into the second equation gives the value of the second variable, and so on. Generally speaking, the method of elimination we have used in the examples in this section becomes rather tedious for more than three variables, and it is preferable to use the alternative technique which we discuss later in this chapter. As we shall see, these alternative methods have the added advantage that they can be readily programmed for use by computers.

Geometrically speaking, a linear equation in three variables

$$ax + by + cz = d$$

represents an equation of a plane in three dimensions. If there are two planes, they intersect in a common line or are coincident or are parallel. If the planes are parallel, the system does not have any solution; but if the planes coincide, then the points (x, y, z) that lie on one plane are also points on the other plane, and the system has an infinite number of solutions. If three planes intersect at a single point, the coordinates (x, y, z) of that point determine the solution for the corresponding system of linear equations. If there are more than three planes, the system will have a unique solution if and only if all the planes pass through a common point of intersection. The coordinates of this point constitute the solution for the system of equations.

EXERCISES

Solve the systems of linear equations in Exercises 1 to 12.

1.
$$x + y + z = 8$$
$$x - y + z = 2$$
$$x + y - z = 12$$

2.
$$x + y + 5z = 8$$
$$2x + y - z = 3$$
$$3x + 2y + 5z = 12$$

3.
$$x + y - z = 4$$
$$2x + y + 3z = 10$$
$$3x + 2y - 2z = 10$$

4.
$$x + 2y - z = 6$$
$$3x - y + 2z = 9$$
$$5x + 3y = 21$$

5.
$$x + 2y + 3z = 6$$
$$4x + 8y + 3z = 15$$
$$2x + 4y - z = 5$$

6.
$$x + 2y + z = 6$$
$$3x - y + z = 2$$
$$2x + 3y + 4z = 12$$

7.
$$x + y + z = 3$$
$$2x + 4y - 3z = -6$$
$$4x + 3y - 5z = -5$$

8.
$$x + y + z = 2$$
$$x + 3y + 2z = 1$$
$$2x + y - z = 2$$

9.
$$x - y + 2z = 2$$
$$4x - 7y + 2z = -4$$
$$2x - 3y + 2z = 1$$

10.
$$x - 3y + 5z = 5$$
$$2x + 2y - 3z = 3$$
$$3x - y + 2z = 12$$

11.
$$x - 3y + 5z = 5$$
$$2x + 2y - 3z = 3$$
$$3x - y + 4z = 14$$

12.
$$x - 2y + z = 8$$
$$2x - 3y - 5z = 7$$
$$3x + 5y + 7z = 6$$

9.3

MATRICES: AN INTRODUCTION

Since the introduction of high-speed electronic computers, matrices have found increasing applications in biology, sociology, economics, engineering, physics, psychology, statistics, and many other areas of applied mathematics. This section is devoted to matrix algebra, namely, to the rules according to which matrices are combined. As we shall see, matrices can be added or subtracted. We can also multiply a matrix by a real number. Section 9.4 is devoted to the rules according to which some matrices can be multiplied. Then in Section 9.5, we examine the solutions of systems of linear equations using matrices.

Definition 9.3.1 A *matrix* is a rectangular array of numbers arranged in rows and columns.

In other words, a matrix is a set of numbers arranged in a pattern that suggests the geometric form of a rectangle. Some simple examples are

$$[2 \quad 1 \quad 3 \quad 4], \quad \begin{bmatrix} 3 & 1 & 4 \\ 5 & 3 & 6 \end{bmatrix}, \quad \begin{bmatrix} 1 & 2 \\ 3 & 4 \\ 5 & 6 \end{bmatrix}, \quad \begin{bmatrix} -1 & 4 \\ 2 & -3 \end{bmatrix}, \quad \begin{bmatrix} 8 \\ 1 \\ 3 \\ -1 \end{bmatrix}.$$

It is sometimes necessary to refer to a general matrix; in that event, it is assumed that the matrix (call it A) consists of $m \cdot n$ numbers arranged in m rows and n columns. The individual numbers in the matrix A are denoted by a_{ij}, where i denotes the row in which the element is located and j refers to the column. For example, in the matrix A, a_{12} is the element in the first row and the second column, a_{21} is the element in the second row and the first column, a_{33} is the element in the third row and the third column, a_{32} is the element in the third row and the second column, and so on. Thus

$$A = \begin{bmatrix} a_{11} & a_{12} & a_{13} & \cdots & a_{1n} \\ a_{21} & a_{22} & a_{23} & \cdots & a_{2n} \\ a_{31} & a_{32} & a_{33} & \cdots & a_{3n} \\ \cdots & \cdots & \cdots & \cdots & \cdots \\ a_{m1} & a_{m2} & a_{m3} & \cdots & a_{mn} \end{bmatrix}.$$

Definition 9.3.2 A matrix with m rows and n columns is called an $m \times n$ matrix, read "m by n matrix."

Definition 9.3.3 The *dimensions of a matrix* are given in the form $m \times n$, where m represents the number of rows and n represents the number of columns.

Thus the matrix

$$\begin{bmatrix} 1 & 6 & 7 \\ 0 & -3 & 9 \end{bmatrix}$$

with two rows and three columns has dimensions 2×3. Note that in stating the dimensions of a matrix, we always list the number of rows *first* and the number of columns *second*.

Definition 9.3.4 A matrix of dimensions $m \times n$ is called a *zero matrix* if all its elements are zero.

The matrix

$$A = \begin{bmatrix} 0 & 0 & 0 \\ 0 & 0 & 0 \end{bmatrix}$$

is a zero matrix of dimensions 2×3.

Definition 9.3.5 | A matrix with n rows and n columns is called a *square matrix of order n.*

Thus, for example,

$$B = \begin{bmatrix} b_{11} & b_{12} & b_{13} \\ b_{21} & b_{22} & b_{23} \\ b_{31} & b_{32} & b_{33} \end{bmatrix}$$

main diagonal   is a square matrix of order 3. The elements b_{11}, b_{22}, and b_{33} constitute the **main diagonal** of matrix B.

Definition 9.3.6 | A square matrix of order n, that has ones along the main diagonal and zeros elsewhere, is called the *identity matrix* and is denoted by I.

Thus

$$I = \begin{bmatrix} 1 & 0 & 0 \\ 0 & 1 & 0 \\ 0 & 0 & 1 \end{bmatrix}$$

is an identity matrix of order 3.

Definition 9.3.7 | Two matrices are said to be *equal* if and only if they have the same dimensions and their corresponding elements are equal.

Example 9.10 | Given that

$$\begin{bmatrix} x & 4 & 3 \\ 6 & y & 7 \end{bmatrix} = \begin{bmatrix} 7 & 4 & z \\ 6 & 0 & 7 \end{bmatrix},$$

then $x = 7$, $y = 0$, and $z = 3$.

Example 9.11 | Consider the matrices

$$A = \begin{bmatrix} 1 & 2 & -1 \\ 2 & 3 & 0 \end{bmatrix}, \qquad B = \begin{bmatrix} 1 & 2 & -1 \\ 2 & 3 & 4 \end{bmatrix}.$$

Here $A \neq B$ because there are corresponding elements that are not equal. The matrices

$$C = \begin{bmatrix} 0 & 1 \\ 2 & 5 \end{bmatrix} \quad \text{and} \quad D = \begin{bmatrix} 0 & 1 & 6 \\ 2 & 5 & 1 \end{bmatrix}$$

are not equal because they are of different dimensions.

Addition and Subtraction of Matrices

Definition 9.3.8

If A and B are two matrices, each of dimension $m \times n$, then their sum (or difference) is defined to be another matrix C also of dimension $m \times n$ such that every element of C is the sum (or difference) of the corresponding elements of matrices A and B.

Thus two matrices may be added or subtracted if and only if they have the same dimensions.

Example 9.12

(a) $[4 \quad 3 \quad 1] + [5 \quad 1 \quad 4] = [4+5 \quad 3+1 \quad 1+4] = [9 \quad 4 \quad 5]$

(b) $[4 \quad 3 \quad 1] - [5 \quad 1 \quad 4] = [4-5 \quad 3-1 \quad 1-4]$
$$= [-1 \quad 2 \quad -3]$$

(c) $\begin{bmatrix} 3 \\ 1 \\ 5 \end{bmatrix} + \begin{bmatrix} -4 \\ 2 \\ 1 \end{bmatrix} = \begin{bmatrix} 3-4 \\ 1+2 \\ 5+1 \end{bmatrix} = \begin{bmatrix} -1 \\ 3 \\ 6 \end{bmatrix}$

(d) $\begin{bmatrix} 3 \\ 1 \\ 5 \end{bmatrix} - \begin{bmatrix} -4 \\ 2 \\ 1 \end{bmatrix} = \begin{bmatrix} 3-(-4) \\ 1-2 \\ 5-1 \end{bmatrix} = \begin{bmatrix} 7 \\ -1 \\ 4 \end{bmatrix}$

Example 9.13

If

$$A = \begin{bmatrix} 8 & 4 & 2 \\ 9 & 5 & 3 \end{bmatrix} \quad \text{and} \quad B = \begin{bmatrix} 5 & 1 & 0 \\ 3 & 2 & 4 \end{bmatrix},$$

then A and B have same dimensions. And

$$A + B = \begin{bmatrix} 8+5 & 4+1 & 2+0 \\ 9+3 & 5+2 & 3+4 \end{bmatrix} = \begin{bmatrix} 13 & 5 & 2 \\ 12 & 7 & 7 \end{bmatrix}$$

and

$$A - B = \begin{bmatrix} 8-5 & 4-1 & 2-0 \\ 9-3 & 5-2 & 3-4 \end{bmatrix} = \begin{bmatrix} 3 & 3 & 2 \\ 6 & 3 & -1 \end{bmatrix}.$$

To understand this concept more clearly, consider Example 9.14.

Example 9.14

The distribution of machines shipped from two plants P_1 and P_2 to four warehouses W_1, W_2, W_3, and W_4 during January 1984 is as follows:

$$
\begin{array}{c}
\text{WAREHOUSE} \\
\begin{array}{cc}
\text{PLANT} & \begin{array}{cccc} W_1 & W_2 & W_3 & W_4 \end{array} \\
\begin{array}{c} P_1 \\ P_2 \end{array} & \begin{bmatrix} 20 & 26 & 32 & 41 \\ 33 & 29 & 36 & 18 \end{bmatrix}
\end{array}
\end{array}
$$

The distribution of machines from the plants to the warehouses during February 1984 is as follows:

$$
\begin{array}{cc}
 & \begin{array}{cccc} W_1 & W_2 & W_3 & W_4 \end{array} \\
\begin{array}{c} P_1 \\ P_2 \end{array} & \begin{bmatrix} 31 & 17 & 36 & 39 \\ 24 & 21 & 31 & 25 \end{bmatrix}
\end{array}
$$

The combined shipment of machines to warehouses over these 2 months is the sum of the corresponding elements in the above matrices; that is,

$$
\begin{array}{cc}
 & \begin{array}{cccc} W_1 & W_2 & W_3 & W_4 \end{array} \\
\begin{array}{c} P_1 \\ P_2 \end{array} & \begin{bmatrix} 20+31 & 26+17 & 32+36 & 41+39 \\ 33+24 & 29+21 & 36+31 & 18+25 \end{bmatrix}
\end{array}
$$

$$
= \begin{array}{cc}
 & \begin{array}{cccc} W_1 & W_2 & W_3 & W_4 \end{array} \\
 & \begin{bmatrix} 51 & 43 & 68 & 80 \\ 57 & 50 & 67 & 43 \end{bmatrix}
\end{array}.
$$

Definition 9.3.9

Let A be an $m \times n$ matrix and k be a real number. Then kA is another $m \times n$ matrix B such that

$$b_{ij} = k \cdot a_{ij} \qquad \text{for every } (i, j).$$

Thus if

$$A = \begin{bmatrix} 2 & 3 \\ 1 & 5 \end{bmatrix},$$

then

$$A + A = \begin{bmatrix} 2 & 3 \\ 1 & 5 \end{bmatrix} + \begin{bmatrix} 2 & 3 \\ 1 & 5 \end{bmatrix} = \begin{bmatrix} 2+2 & 3+3 \\ 1+1 & 5+5 \end{bmatrix}$$

$$= \begin{bmatrix} 4 & 6 \\ 2 & 10 \end{bmatrix} = 2A$$

Example 9.15

If
$$A = \begin{bmatrix} 4 & -2 & 5 \\ -3 & 1 & 6 \end{bmatrix},$$

then
$$(-1)A = \begin{bmatrix} -4 & 2 & -5 \\ 3 & -1 & -6 \end{bmatrix}$$

and
$$3A = \begin{bmatrix} 12 & -6 & 15 \\ -9 & 3 & 18 \end{bmatrix}.$$

In general, if k is any real number, then
$$kA = \begin{bmatrix} 4k & -2k & 5k \\ -3k & k & 6k \end{bmatrix}.$$

Example 9.16

If
$$A = \begin{bmatrix} 3 & 2 \\ 4 & 1 \end{bmatrix}, \quad B = \begin{bmatrix} 5 & 6 \\ 7 & 8 \end{bmatrix}, \quad \text{and} \quad C = \begin{bmatrix} -1 & 1 \\ 2 & -3 \end{bmatrix},$$

then
$$3A + 2B = \begin{bmatrix} 9 & 6 \\ 12 & 3 \end{bmatrix} + \begin{bmatrix} 10 & 12 \\ 14 & 16 \end{bmatrix} = \begin{bmatrix} 19 & 18 \\ 26 & 19 \end{bmatrix}$$

and
$$3B + 4C = \begin{bmatrix} 15 & 18 \\ 21 & 24 \end{bmatrix} + \begin{bmatrix} -4 & 4 \\ 8 & -12 \end{bmatrix} = \begin{bmatrix} 11 & 22 \\ 29 & 12 \end{bmatrix}.$$

EXERCISES

Give the dimensions of the matrices in Exercises 1 to 6.

1. $\begin{bmatrix} 2 & 5 & 6 \\ 3 & 0 & 4 \end{bmatrix}$

2. $\begin{bmatrix} 1 & 2 & 3 & 4 \\ 2 & 4 & 6 & 19 \\ 5 & -1 & 12 & 11 \end{bmatrix}$

3. $\begin{bmatrix} 4 & 5 & -3 \\ 3 & -1 & 10 \\ 2 & 5 & 8 \end{bmatrix}$

4. $\begin{bmatrix} 1 & 5 & 6 & 8 & 9 & 0 \\ 4 & -2 & 7 & 3 & 1 & 3 \end{bmatrix}$

5. $\begin{bmatrix} 3 & 4 & 6 \end{bmatrix}$

6. $\begin{bmatrix} 6 & 8 \\ 2 & 1 \end{bmatrix}$

7. Given matrix A, find the values of elements a_{12}, a_{21}, a_{32}, and a_{23}:

$$A = \begin{bmatrix} 5 & -4 & 0 \\ 1 & 3 & 2 \\ 6 & -2 & 4 \end{bmatrix}.$$

8. Given matrix B, find the values of elements b_{13}, b_{32}, b_{34}, and b_{21}:

$$B = \begin{bmatrix} 10 & 9 & -2 & 12 \\ 3 & 7 & 1 & 4 \\ 2 & 6 & 8 & -6 \end{bmatrix}.$$

9. Given that

$$\begin{bmatrix} -2 & x & 4 & 0 \\ 3 & 2 & y & 3 \end{bmatrix} = \begin{bmatrix} -2 & 3 & z & 0 \\ 3 & 2 & 1 & u \end{bmatrix},$$

find the values of x, y, z, and u.

10. Given that

$$\begin{bmatrix} x+1 & 3 & -1 \\ 4 & -1 & 0 \\ 2 & z+3 & -3 \end{bmatrix} = \begin{bmatrix} 5 & 3 & -1 \\ 4 & y+2 & 0 \\ 2 & 0 & -3 \end{bmatrix},$$

find the values of x, y, and z.

Find x, y, and z in Exercises 11 to 13.

11. $[1 \quad 2 \quad 3] + [x \quad y \quad z] = [4 \quad 5 \quad 10]$

12. $[-1 \quad 2 \quad 3] + 3[x \quad y \quad z] = [8 \quad 17 \quad 6]$

13. $\begin{bmatrix} -1 \\ 2 \\ 3 \end{bmatrix} + 2\begin{bmatrix} x \\ y \\ z \end{bmatrix} = \begin{bmatrix} 4 \\ -3 \\ 5 \end{bmatrix}$

Find $A + B$ and $A - B$, if possible, in Exercises 14 to 26.

14. $A = [4 \quad 2 \quad 1 \quad 6]$, $B = [5 \quad 4 \quad 3 \quad 2]$

15. $A = [2 \quad 3 \quad 4 \quad 1 \quad 7]$, $B = [-2 \quad -1 \quad 0 \quad 3 \quad -2]$

16. $A = [5 \quad 6 \quad 8 \quad 2]$, $B = [2 \quad 8 \quad 9]$

17. $A = \begin{bmatrix} 2 \\ 3 \\ 1 \\ -2 \end{bmatrix}$, $B = \begin{bmatrix} -1 \\ 1 \\ 4 \\ 3 \end{bmatrix}$

18. $A = \begin{bmatrix} 5 \\ 8 \\ 9 \\ 10 \end{bmatrix}$, $B = \begin{bmatrix} 6 \\ 0 \\ 2 \end{bmatrix}$

19. $A = \begin{bmatrix} 0 \\ 5 \\ 7 \end{bmatrix}$, $B = \begin{bmatrix} 4 \\ -1 \\ -3 \end{bmatrix}$

20. $A = [2 \quad 3 \quad 5]$, $B = \begin{bmatrix} 9 \\ -1 \\ 4 \end{bmatrix}$

21. $A = \begin{bmatrix} 1 & 2 \\ 3 & 4 \end{bmatrix}$, $B = \begin{bmatrix} 2 & -1 \\ -1 & 3 \end{bmatrix}$

22. $A = \begin{bmatrix} 3 & 4 \\ 1 & 2 \end{bmatrix}$, $B = \begin{bmatrix} 2 & 3 & 5 \\ 1 & 5 & 8 \end{bmatrix}$

23. $A = \begin{bmatrix} 1 & 2 & 3 \\ 5 & 1 & 4 \end{bmatrix}, B = \begin{bmatrix} 2 & 4 & 5 & 7 \\ 1 & 5 & 2 & 1 \end{bmatrix}$

24. $A = \begin{bmatrix} 1 & 2 & 3 \\ 3 & 1 & 2 \end{bmatrix}, B = \begin{bmatrix} -2 & -1 & -4 \\ 1 & 3 & 2 \end{bmatrix}$

25. $A = \begin{bmatrix} 2 & 3 & 1 \\ 0 & 1 & 8 \end{bmatrix}, B = \begin{bmatrix} 1 & 2 \\ 4 & 6 \\ 7 & 1 \end{bmatrix}$

26. $A = \begin{bmatrix} 2 & -1 & 1 \\ 2 & 3 & 4 \\ 5 & -1 & 1 \end{bmatrix}, B = \begin{bmatrix} 4 & -2 & 3 \\ 5 & -1 & 2 \\ 2 & 1 & -3 \end{bmatrix}$

Given that

$$A = \begin{bmatrix} 5 & 8 \\ 7 & 9 \end{bmatrix}, \quad B = \begin{bmatrix} 1 & 2 \\ 3 & 4 \end{bmatrix}, \quad \text{and} \quad C = \begin{bmatrix} 7 & 5 \\ 6 & 9 \end{bmatrix},$$

perform the indicated operations in Exercises 27 to 32.

27. $2A + 3B$

28. $4A + 5B + 2C$

29. $A + 2B + 3C$

30. $3A - 2B + C$

31. $A + (B + C)$

32. $(A + B) + C$

9.4

MULTIPLICATION OF MATRICES

We observed that matrices can be added or subtracted if and only if they have the same dimensions. We also examined the multiplication of a matrix by a real number. Continuing the line of reasoning of matrix addition or subtraction, we might be led to define the product of two $m \times n$ matrices as the matrix obtained by pairwise multiplying their respective elements. This would be an exercise in futility because when mathematicians talk about the product of two matrices, they are actually referring to a different and much more useful concept. Since the definition of matrix multiplication is fairly complicated, we illustrate it first in connection with a $1 \times n$ matrix (also called a **row vector**) and an $n \times 1$ matrix (called a **column vector**).

Suppose that the business manager of a baseball team reports that 1200 people bought box seats for a big game, 5600 bought reserved seats, and 2800 bought general admission tickets. These figures may be represented by a *row vector* (1×3 matrix) as follows:

$$A = [1200 \quad 5600 \quad 2800].$$

Suppose further that a box seat costs $8.00, a reserved seat costs $5.00, and a general admission ticket costs $2.50. This information can be represented in a *column vector* (3×1 matrix):

$$B = \begin{bmatrix} 8.00 \\ 5.00 \\ 2.50 \end{bmatrix}.$$

What are the receipts for this game?

We must multiply the number of tickets sold by the corresponding price for each ticket and then add the amounts to get the total receipts. Clearly, the multiplication should have the form

$$AB = \begin{bmatrix} 1200 & 5600 & 2800 \end{bmatrix} \begin{bmatrix} 8.00 \\ 5.00 \\ 2.50 \end{bmatrix}$$

$$= (1200)(8.00) + (5600)(5.00) + (2800)(2.50)$$
$$= 9600 + 28,000 + 7000$$
$$= 44,600.$$

Therefore the total receipts are $44,600.

Note that this total is the sum of the products obtained by multiplying the *first* element of the row vector by the *first* element of the column vector, the *second* element of the row vector by the *second* element of the column vector, and the *third* element of the row vector by the *third* element of the column vector. Since sums of products like this arise in many applications, we *define* the product of a row vector and a column vector if and only if both row and column vectors have the same number of elements. Aside from the fact that this kind of vector product has important applications, it also defines the product of two matrices in the following way:

Definition 9.4.1

Let A be an $m \times n$ matrix, and let B be an $n \times p$ matrix. The product AB is the $m \times p$ matrix whose element in the ith row and jth column is the sum of the products of the elements in the ith row of A and the corresponding elements in the jth column of B.

In other words,

$$\begin{bmatrix} a_{11} & a_{12} & \cdots & a_{1n} \\ a_{21} & a_{22} & \cdots & a_{2n} \\ \hdotsfor{4} \\ \boxed{a_{i1} \quad a_{i2} \quad \cdots \quad a_{in}} \\ \hdotsfor{4} \\ a_{m1} & a_{m2} & \cdots & a_{mn} \end{bmatrix} \times \begin{bmatrix} b_{11} & b_{12} & \cdots & b_{ij} & \cdots & b_{1p} \\ b_{21} & b_{22} & \cdots & b_{2j} & \cdots & b_{2p} \\ \hdotsfor{6} \\ b_{n1} & b_{n2} & \cdots & b_{nj} & \cdots & b_{np} \end{bmatrix}$$

$$= \begin{bmatrix} c_{11} & c_{12} & \cdots & c_{ij} & \cdots & c_{1p} \\ c_{21} & c_{22} & \cdots & c_{2j} & \cdots & c_{2p} \\ & & & & & \\ c_{i1} & c_{i2} & \cdots & \boxed{c_{ij}} & \cdots & c_{1p} \\ & & & & & \\ c_{m1} & c_{m2} & \cdots & c_{mj} & \cdots & c_{mp} \end{bmatrix},$$

where

$$c_{ij} = [a_{i1} \quad a_{i2} \quad \cdots \quad a_{in}]\begin{bmatrix} b_{1j} \\ b_{2j} \\ \vdots \\ b_{nj} \end{bmatrix} = a_{i1}b_{1j} + a_{i2}b_{2j} + \cdots + a_{in}b_{nj}.$$

If $i = 1$ and $j = 1$, then

$$c_{11} = a_{11}b_{11} + a_{12}b_{21} + \cdots + a_{1n}b_{n1}.$$

Note that this is the direct result of multiplying the elements in the *first* row of matrix A by the corresponding elements in the *first* column of matrix B.
    If $i = 2$ and $j = 3$, then

$$c_{23} = a_{21}b_{13} + a_{22}b_{23} + \cdots + a_{2n}b_{n3}.$$

Similarly, this is the sum of the products of the elements in the *second* row of matrix A with the corresponding elements of the *third* column of matrix B.
    Observe that the product matrix AB is defined if and only if the number of columns of the first matrix A equals the number of rows of the second matrix B. Further, the product matrix AB has the same number of rows as matrix A and the same number of columns as matrix B. Thus

$$A_{(m \times n)} \qquad\qquad B_{(n \times p)} = (AB)_{(m \times p)}$$

must be equal

dimensions of AB

Example 9.17

Let

$$A = \begin{bmatrix} 2 & 3 & 1 \\ 4 & 6 & 7 \end{bmatrix}_{(2 \times 3)} \qquad\qquad B = \begin{bmatrix} 4 & 5 \\ 3 & 1 \\ 1 & 2 \end{bmatrix}_{(3 \times 2)}$$

Do they match?

If so, then the dimensions of AB are 2×2.

Matrix AB is defined since the number of columns of matrix A is the same as the number of rows of matrix B. The product matrix $C = AB$ will have two rows and two columns. Thus

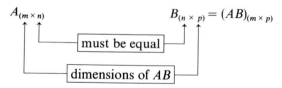

$$\begin{bmatrix} 2 & 3 & 1 \\ 4 & 6 & 7 \end{bmatrix} \begin{bmatrix} 4 & 5 \\ 3 & 1 \\ 1 & 2 \end{bmatrix} = \begin{bmatrix} c_{11} & c_{12} \\ c_{21} & c_{22} \end{bmatrix}.$$

To obtain c_{11}, we multiply the elements in the first row of matrix A by the corresponding elements in the first column of matrix B and then add. Thus

$$c_{11} = \begin{bmatrix} 2 & 3 & 1 \end{bmatrix} \begin{bmatrix} 4 \\ 3 \\ 1 \end{bmatrix} = 2(4) + 3(3) + 1(1) = 18.$$

Similarly,

$$c_{12} = \begin{bmatrix} 2 & 3 & 1 \end{bmatrix} \begin{bmatrix} 5 \\ 1 \\ 2 \end{bmatrix} = 2(5) + 3(1) + 1(2) = 15$$

$$c_{21} = \begin{bmatrix} 4 & 6 & 7 \end{bmatrix} \begin{bmatrix} 4 \\ 3 \\ 1 \end{bmatrix} = 4(4) + 6(3) + 7(1) = 41$$

$$c_{22} = \begin{bmatrix} 4 & 6 & 7 \end{bmatrix} \begin{bmatrix} 5 \\ 1 \\ 2 \end{bmatrix} = 4(5) + 6(1) + 7(2) = 40.$$

Thus

$$C = AB = \begin{bmatrix} 18 & 15 \\ 41 & 40 \end{bmatrix}.$$

Properties of Matrix Multiplication

We recall that the product AB of two matrices A and B is defined if and only if the number of columns of matrix A equals the number of rows of matrix B. The product matrix AB has the same number of rows as A and the same number of columns as B. Thus if

$$A = \begin{bmatrix} 1 & 2 & 4 \\ 2 & 3 & 6 \end{bmatrix}_{(2 \times 3)} \quad \text{and} \quad B = \begin{bmatrix} 5 & 7 & 4 \\ 4 & 6 & 0 \\ 4 & 2 & 1 \end{bmatrix}_{(3 \times 3)}$$

then the product AB is defined, since A has three columns and B has three rows. The resulting matrix AB has two rows and three columns. To clarify the order of multiplication, we say that B is premultiplied by A and that A is postmultiplied by B.

The product BA, however, is not defined in the above example, since the number of columns in B is not equal to the number of rows in A. Even if products AB and BA are defined, AB does not, in general, equal BA.

Example 9.18

Let

$$A = \begin{bmatrix} 1 & 2 & 3 \\ 2 & 3 & 5 \end{bmatrix}_{(2 \times 3)} \quad \text{and} \quad B = \begin{bmatrix} 4 & 2 & 3 \\ 2 & 1 & 1 \\ 1 & 2 & 0 \end{bmatrix}_{(3 \times 3)}.$$

Then

$$AB = \overrightarrow{\begin{bmatrix} 1 & 2 & 3 \\ 2 & 3 & 5 \end{bmatrix}} \downarrow \begin{bmatrix} 4 & 2 & 3 \\ 2 & 1 & 1 \\ 1 & 2 & 0 \end{bmatrix}$$

$$= \begin{bmatrix} 1(4) + 2(2) + 3(1) & 1(2) + 2(1) + 3(2) & 1(3) + 2(1) + 3(0) \\ 2(4) + 3(2) + 5(1) & 2(2) + 3(1) + 5(2) & 2(3) + 3(1) + 5(0) \end{bmatrix}$$

$$= \begin{bmatrix} 11 & 10 & 5 \\ 19 & 17 & 9 \end{bmatrix}.$$

The product BA is not defined, since the number of columns of B is not equal to the number of rows of A.

Example 9.19

Let

$$A = \begin{bmatrix} 1 & 2 & 3 \\ 2 & 3 & 1 \end{bmatrix} \quad \text{and} \quad B = \begin{bmatrix} 4 & 5 \\ 2 & 1 \\ 1 & 3 \end{bmatrix}.$$

Then

$$AB = \begin{bmatrix} 1 & 2 & 3 \\ 2 & 3 & 1 \end{bmatrix} \begin{bmatrix} 4 & 5 \\ 2 & 1 \\ 1 & 3 \end{bmatrix}$$

$$= \begin{bmatrix} 1(4) + 2(2) + 3(1) & 1(5) + 2(1) + 3(3) \\ 2(4) + 3(2) + 1(1) & 2(5) + 3(1) + 1(3) \end{bmatrix} = \begin{bmatrix} 11 & 16 \\ 15 & 16 \end{bmatrix}$$

and

$$BA = \begin{bmatrix} 4 & 5 \\ 2 & 1 \\ 1 & 3 \end{bmatrix} \begin{bmatrix} 1 & 2 & 3 \\ 2 & 3 & 1 \end{bmatrix}$$

$$= \begin{bmatrix} 4(1) + 5(2) & 4(2) + 5(3) & 4(3) + 5(1) \\ 2(1) + 1(2) & 2(2) + 1(3) & 2(3) + 1(1) \\ 1(1) + 3(2) & 1(2) + 3(3) & 1(3) + 3(1) \end{bmatrix}$$

$$= \begin{bmatrix} 14 & 23 & 17 \\ 4 & 7 & 7 \\ 7 & 11 & 6 \end{bmatrix}.$$

Note that AB and BA are both defined, but AB is 2×2 while BA is 3×3.

Example 9.20

Let

$$A = \begin{bmatrix} 3 & 4 \\ -1 & 2 \end{bmatrix} \quad \text{and} \quad B = \begin{bmatrix} 2 & 1 \\ 3 & 2 \end{bmatrix}.$$

Then

$$AB = \begin{bmatrix} 3 & 4 \\ -1 & 2 \end{bmatrix}\begin{bmatrix} 2 & 1 \\ 3 & 2 \end{bmatrix} = \begin{bmatrix} 3(2) + 4(3) & 3(1) + 4(2) \\ -1(2) + 2(3) & -1(1) + 2(2) \end{bmatrix}$$

$$= \begin{bmatrix} 18 & 11 \\ 4 & 3 \end{bmatrix},$$

whereas

$$BA = \begin{bmatrix} 2 & 1 \\ 3 & 2 \end{bmatrix}\begin{bmatrix} 3 & 4 \\ -1 & 2 \end{bmatrix} = \begin{bmatrix} 2(3) + 1(-1) & 2(4) + 1(2) \\ 3(3) + 2(-1) & 3(4) + 2(2) \end{bmatrix}$$

$$= \begin{bmatrix} 5 & 10 \\ 7 & 16 \end{bmatrix}.$$

Thus AB and BA are both defined and are of same dimensions, but they are not equal.

We have shown that the commutative law, $AB = BA$, does not hold, in general. However, the associative and distributive laws do apply for matrix multiplication. That is, if A, B, and C are three matrices whose dimensions are properly related, then

$$A(BC) = (AB)C$$

$$A(B + C) = AB + AC$$

$$(B + C)A = BA + CA.$$

We leave it for the reader to verify these properties (see Exercises 33 and 34).

EXERCISES

For the matrices in Exercises 1 to 5, determine the dimensions of the product matrix AB.

1. A is a 3×3 matrix and B is a 3×2 matrix.   **2.** A is a 5×3 matrix and B is a 4×3 matrix.

3. A is a 2×5 matrix and B is a 5×4 matrix.   **4.** A is a 4×7 matrix and B is a 7×5 matrix.

5. A is a 3×2 matrix and B is a 2×6 matrix.

For the pairs of matrices in Exercises 6 to 12, determine whether it is possible to compute AB, BA, both, or neither.

6. A is a 3×5 matrix and B is a 5×2 matrix.   **7.** A is a 2×3 matrix and B is a 3×2 matrix.

8. A is a 3×4 matrix and B is a 4×3 matrix.   **9.** A is a 4×5 matrix and B is a 4×2 matrix.

10. A is a 5×7 matrix and B is a 7×5 matrix.

11. A is a 3×2 matrix and B is a 2×3 matrix.

12. A is a 3×1 matrix and B is a 1×3 matrix.

Find AB and BA in Exercises 13 to 16.

13. $A = \begin{bmatrix} 3 & 4 \\ 2 & 3 \end{bmatrix}$, $B = \begin{bmatrix} 5 & 1 \\ 1 & 2 \end{bmatrix}$

14. $A = \begin{bmatrix} 6 & 1 \\ 2 & 5 \end{bmatrix}$, $B = \begin{bmatrix} 3 & 7 \\ 4 & -2 \end{bmatrix}$

15. $A = \begin{bmatrix} 1 & 2 \\ 4 & 5 \end{bmatrix}$, $B = \begin{bmatrix} 3 & 1 \\ 2 & 7 \end{bmatrix}$

16. $A = \begin{bmatrix} 1 & 2 & 5 \\ 2 & 3 & 9 \end{bmatrix}$, $B = \begin{bmatrix} 5 & 7 \\ 4 & -1 \\ -3 & -2 \end{bmatrix}$

Given that

$$A = \begin{bmatrix} 1 & 5 & 3 \\ 7 & 4 & 6 \end{bmatrix}, \qquad B = \begin{bmatrix} 2 & 3 & 4 \\ 3 & 5 & 7 \\ -1 & 5 & 3 \end{bmatrix}, \qquad \text{and} \qquad C = \begin{bmatrix} 3 & 1 & 2 \\ 1 & 4 & 3 \\ 2 & 3 & 5 \end{bmatrix},$$

perform the indicated operations in Exercises 17 to 20.

17. AB **18.** AC **19.** BC **20.** CB

Given that

$$A = \begin{bmatrix} 1 & 2 \\ 5 & 4 \end{bmatrix}, \qquad B = \begin{bmatrix} 2 & 7 \\ 3 & 5 \end{bmatrix}, \qquad \text{and} \qquad C = \begin{bmatrix} 5 & -2 \\ 2 & 4 \end{bmatrix},$$

perform the indicated operations in Exercises 21 to 32.

21. AB **22.** BA **23.** AC **24.** BC

25. CA **26.** CB **27.** $A(BC)$ **28.** $(AB)C$

29. $A(B + C)$ **30.** $(B + C)A$ **31.** $B(A + C)$ **32.** $(A + C)B$

33. Given that

$$A = \begin{bmatrix} 3 & 1 & 2 \\ 2 & 3 & 4 \end{bmatrix}, \qquad B = \begin{bmatrix} 1 & 3 \\ 2 & 1 \\ 4 & 5 \end{bmatrix}, \qquad \text{and} \qquad C = \begin{bmatrix} 0 & 3 \\ 2 & 4 \\ 3 & 5 \end{bmatrix},$$

show that $A(B + C) = AB + AC$.

34. Given that

$$A = \begin{bmatrix} 2 & 0 & -1 \\ -1 & -2 & 2 \end{bmatrix}, \qquad B = \begin{bmatrix} -1 & 0 & -1 \\ 2 & -3 & -1 \\ 3 & 0 & -1 \end{bmatrix}, \qquad \text{and} \qquad C = \begin{bmatrix} 1 \\ 1 \\ -2 \end{bmatrix},$$

show that $A(BC) = (AB)C$.

35. Suppose that

$$A = \begin{bmatrix} 2 & 6 \\ 3 & 9 \end{bmatrix} \qquad \text{and} \qquad B = \begin{bmatrix} -6 & 15 \\ 2 & -5 \end{bmatrix}$$

Is AB a zero matrix? If so, can you conclude that if $AB = 0$, then either A or B or both matricess are zero?

<table>
<tr><td>**9.5**</td><td>**SOLUTIONS OF LINEAR EQUATIONS: GAUSS-JORDAN ELIMINATION METHOD**</td></tr>
</table>

Gauss-Jordan
elimination
method

We introduced the *method of elimination* for solving systems of linear equations in Sections 9.1 and 9.2. We now suggest that this powerful method leads to an expedient means of solving a large system of linear equations. This method not only is used for electronic machine calculations but also is fundamental to the understanding of the basic concepts involved. Consider, for example, the following set of equations:

$$2x + 3y + 4z = 12$$
$$x + 2y + z = 6 \tag{9.6}$$
$$3x - y + z = 2.$$

Our objective for this method of elimination is first to transform this set of linear equations to a triangular system

$$x + b_{12}y + b_{13}z = c_1$$
$$y + b_{23}z = c_2$$
$$z = c_3$$

and then to reduce it to the form

$$x + 0 \cdot y + 0 \cdot z = c_1^*$$
$$0 \cdot x + y + 0 \cdot z = c_2^*$$
$$0 \cdot x + 0 \cdot y + z = c_3$$

so that $x = c_1^*$, $y = c_2^*$, and $z = c_3$ is the obvious solution to the original set of linear equations (9.6).

An interchange in the first two equations in (9.6) yields

$$x + 2y + z = 6$$
$$2x + 3y + 4z = 12 \tag{9.7}$$
$$3x - y + z = 2.$$

The first equation in (9.7) has 1 as its leading coefficient and can therefore be used to eliminate the variable x from the second and third equations in (9.7). Subtracting 2 times the first equation from the second and 3 times the first equation from the third, we obtain

$$x + 2y + z = 6$$
$$-y + 2z = 0 \tag{9.8}$$
$$-7y - 2z = -16.$$

Now we multiply the second equation in (9.8) by -1 so as to have 1 as its leading coefficient. This yields

$$\begin{aligned} x + 2y + z &= 6 \\ y - 2z &= 0 \\ -7y - 2z &= -16. \end{aligned} \tag{9.9}$$

Adding 7 times the second equation in (9.9) to the third equation results in the elimination of y from the third equation, and we have

$$\begin{aligned} x + 2y + z &= 6 \\ y - 2z &= 0 \\ -16z &= -16. \end{aligned} \tag{9.10}$$

We now divide the third equation in (9.10) by the coefficient of z, and our system is in triangular form:

$$\begin{aligned} x + 2y + z &= 6 \\ y - 2z &= 0 \\ z &= 1. \end{aligned} \tag{9.11}$$

Recall that our next objective is to reduce each equation to just one variable. To eliminate z from the first two equations, we add 2 times the third equation to the second equation and -1 times the third equation to the first equation. Thus we obtain

$$\begin{aligned} x + 2y &= 5 \\ y &= 2 \\ z &= 1. \end{aligned} \tag{9.12}$$

Adding -2 times the second equation in (9.12) to the first equation, we get

$$\begin{aligned} x &= 1 \\ y &= 2 \\ z &= 1. \end{aligned} \tag{9.13}$$

We leave it for the reader to verify that $x = 1$, $y = 2$, and $z = 1$ is the solution set to the original set of linear equations (9.6).

| Definition 9.5.1 | Two systems of linear equations in n variables are said to be *equivalent* if and only if every solution of one system is also a solution of the other. |

The elimination process reduces a given set of linear equations to an equivalent set of equations from which the solutions can be easily read. Thus the system of equations (9.6) and (9.8) are equivalent. Similarly, (9.8), (9.9), (9.10), and (9.11) are also equivalent, as are (9.12) and (9.13). This implies that the original set of equations (9.6) is equivalent to (9.13).

Observe that in finding the solution to the set of linear equations (9.6) we used operations of the following form:

1. The multiplication of an equation of a system by a constant
2. The interchange of any two equations in the system
3. The addition of an arbitrary multiple of one equation to another equation in the system

It is easy to show that the above operations preserve the solution set of the system of linear equations.

In reducing the given system of linear equations to equivalent systems, we actually operate on the coefficients and the corresponding constants. The variables simply serve to keep their coefficients properly aligned in columns.

The system of linear equations (9.6) can be written in matrix form as

$$\begin{bmatrix} 2 & 3 & 4 \\ 1 & 2 & 1 \\ 3 & -1 & 1 \end{bmatrix} \begin{bmatrix} x \\ y \\ z \end{bmatrix} = \begin{bmatrix} 12 \\ 6 \\ 2 \end{bmatrix},$$

or more precisely as

$$AX = B,$$

where A is the **coefficient matrix** and B is the **vector of constants.**

Definition 9.5.2

The *augmented matrix* for the system $AX = B$ is the matrix $[A:B]$ formed by adjoining the column vector B of constants to the right of the coefficient matrix A.

Thus the augmented matrix of the system of equations (9.6) is

$$\begin{bmatrix} 2 & 3 & 4 & | & 12 \\ 1 & 2 & 1 & | & 6 \\ 3 & -1 & 1 & | & 2 \end{bmatrix}.$$

Each row of the augmented matrix contains the coefficients of the unknowns and the constant term of the corresponding equation.

The technique of solving linear equations by using matrices involves first obtaining a 1 in the a_{11} position, either by dividing the first row by its leading term or by interchanging the first row with another row having 1 as its leading term. Suitable multiples of the new first row are then added to the remaining rows to obtain zeros in the a_{21} and a_{31} positions. The process is repeated by selecting a new row among the altered rows, dividing it by its leading nonzero term, if necessary, to locate a 1 in the a_{22} position. The new second row thus obtained is then used to obtain a zero in the a_{32} position. The process is continued until we obtain an equivalent augmented matrix in the triangular form as follows:

$$[A^*:B^*] = \begin{bmatrix} 1 & b_{12} & b_{13} & c_1 \\ 0 & 1 & b_{23} & c_2 \\ 0 & 0 & 1 & c_3 \end{bmatrix}.$$

Adding suitable multiples of the last row to the remaining rows yields zeros in the b_{13} and b_{23} positions. Again adding a suitable multiple of the second row to the first row places a zero in the b_{12} position. Thus we have an identity matrix on the left augmented by the solution vector on the right as follows:

$$\begin{bmatrix} 1 & 0 & 0 & c_1^* \\ 0 & 1 & 0 & c_2^* \\ 0 & 0 & 1 & c_3 \end{bmatrix}.$$

Obviously, $x = c_1^*$, $y = c_2^*$, and $z = c_3$ is the solution to the original set of linear equations (9.6). This process of obtaining solutions is called the **Gauss-Jordan elimination method.**

The operations we performed on the system of linear equations (9.6) correspond exactly to those we perform now on the rows of the augmented matrix $[A:B]$. The augmented matrix is

ROW	x	y	z	CONSTANT	
R_1	2	3	4	12	
R_2	1	2	1	6	**(9.6a)**
R_3	3	-1	1	2	

Since 1 happens to be in the a_{21} position, we begin by interchanging the first two rows of the above matrix so as to obtain 1 in the a_{11} position:

ROW OPERATION	x	y	z	CONSTANT	
$R_1' = R_2$	1	2	1	6	
$R_2' = R_1$	2	3	4	12	**(9.7a)**
R_3	3	-1	1	2	

Next, we want to obtain zeros in the a_{21} and a_{31} positions. This can be accomplished simply by multiplying the elements of the new first row R_1' by -2 and -3 and then adding the results to the corresponding elements of the new second row R_2' and the third row R_3, respectively:

ROW OPERATION	x	y	z	CONSTANT	
R_1'	1	2	1	6	
$R_2'' = R_2' + (-2)R_1'$	0	-1	2	0	**(9.8a)**
$R_3' = R_3 + (-3)R_1'$	0	-7	-2	-16	

Now we want to get 1 in the a_{22} position. This can be accomplished by multiplying the elements of the second row R_2'' by -1. We obtain

$$
\begin{array}{cc}
\text{ROW OPERATION} & \begin{array}{cccc} x & y & z & \text{CONSTANT} \end{array} \\
\begin{array}{c} R_1' \\ R_2''' = (-1)R_2'' \\ R_3' \end{array} &
\left[\begin{array}{ccc|c}
1 & 2 & 1 & 6 \\
0 & 1 & -2 & 0 \\
0 & -7 & -2 & -16
\end{array}\right].
\end{array}
\tag{9.9a}
$$

To obtain 0 in the a_{32} position, we multiply the elements in the second row R_2''' by 7 and add the results to the corresponding elements of the third row R_3':

$$
\begin{array}{cc}
\text{ROW OPERATION} & \begin{array}{cccc} x & y & z & \text{CONSTANT} \end{array} \\
\begin{array}{c} R_1' \\ R_2''' \\ R_3'' = R_3' + 7R_2''' \end{array} &
\left[\begin{array}{ccc|c}
1 & 2 & 1 & 6 \\
0 & 1 & -2 & 0 \\
0 & 0 & -16 & -16
\end{array}\right].
\end{array}
\tag{9.10a}
$$

If we divide the elements of the third row R_3'' by -16, the above matrix becomes

$$
\begin{array}{cc}
\text{ROW OPERATION} & \begin{array}{cccc} x & y & z & \text{CONSTANT} \end{array} \\
\begin{array}{c} R_1' \\ R_2''' \\ R_3''' = (-\frac{1}{16})R_3'' \end{array} &
\left[\begin{array}{ccc|c}
1 & 2 & 1 & 6 \\
0 & 1 & -2 & 0 \\
0 & 0 & 1 & 1
\end{array}\right].
\end{array}
\tag{9.11a}
$$

Now we use the last row R_3''' to obtain zeros in the a_{13} and a_{23} positions. Adding -1 and 2 times the elements of R_3''' to the corresponding elements of the first and second rows of the preceding matrix yields

$$
\begin{array}{cc}
\text{ROW OPERATION} & \begin{array}{cccc} x & y & z & \text{CONSTANT} \end{array} \\
\begin{array}{c} R_1'' = R_1' + (-1)R_3''' \\ R_2^{\text{IV}} = R_2''' + 2R_3''' \\ R_3''' \end{array} &
\left[\begin{array}{ccc|c}
1 & 2 & 0 & 5 \\
0 & 1 & 0 & 2 \\
0 & 0 & 1 & 1
\end{array}\right].
\end{array}
\tag{9.12a}
$$

Finally, we multiply the elements of the second row R_2^{IV} in the preceding matrix by -2 and add the results to the corresponding elements of the first row R_1''. The resulting matrix is

$$
\begin{array}{cc}
\text{ROW OPERATION} & \begin{array}{cccc} x & y & z & \text{CONSTANT} \end{array} \\
\begin{array}{c} R_1''' = R_1'' + (-2)R_2^{\text{IV}} \\ R_2^{\text{IV}} \\ R_3''' \end{array} &
\left[\begin{array}{ccc|c}
1 & 0 & 0 & 1 \\
0 & 1 & 0 & 2 \\
0 & 0 & 1 & 1
\end{array}\right].
\end{array}
\tag{9.13a}
$$

Thus $x = 1$, $y = 2$, and $z = 1$ is the solution to the system of linear equations.

The sequence of operations used to obtain an identity matrix on the left and the solution vector on the right is by no means unique. Whatever the sequence of operations, each row in the final matrix eventually involves one unknown with a 1 as its coefficient so that the corresponding constant term is the solution for that unknown.

We have investigated so far the Gauss-Jordan elimination method for solving n linear equations in n variables where the system of linear equations possesses a solution and the solution is unique. We now consider systems of linear equations in which it is not possible to obtain an identity matrix augmented by a unique solution vector. In some cases, no solution for the system exists. In others, there are infinitely many solutions to the system.

Example 9.21

Solve the following set of linear equations:

$$x + 2y + 3z = 6$$

$$2x + 4y - z = 5$$

$$4x + 8y + 3z = 15.$$

Solution: The augmented matrix, in this case, is

	x	y	z	CONSTANT
R_1	1	2	3	6
R_2	2	4	-1	5
R_3	4	8	3	15

Since 1 happens to be in the a_{11} position, we perform the following operations:

1. $R_2' = R_2 + (-2)R_1.$
2. $R_3' = R_3 + (-4)R_1.$

These operations transform the preceding matrix to

	x	y	z	CONSTANT
R_1	1	2	3	6
R_2'	0	0	-7	-7
R_3'	0	0	-9	-9

To obtain 1s in the a_{23} and a_{33} positions, we divide the second and third rows by (-7) and (-9), respectively. In other words, we perform the following operations:

1. $R_2'' = \left(-\dfrac{1}{7}\right)R_2'.$

2. $R_3'' = \left(-\dfrac{1}{9}\right)R_3'.$

These operations transform the preceding matrix to

$$
\begin{array}{c}
\\
R_1 \\
R_2'' = \left(-\dfrac{1}{7}\right)R_2' \\
R_3'' = \left(-\dfrac{1}{9}\right)R_3'
\end{array}
\begin{array}{cccc}
x & y & z & \text{CONSTANT} \\
\left[\begin{array}{ccc|c}
1 & 2 & 3 & 6 \\
0 & 0 & 1 & 1 \\
0 & 0 & 1 & 1
\end{array}\right].
\end{array}
$$

To obtain zeros in the a_{13} and a_{23} positions, we perform the following operations:

1. $R_1' = R_1 + (-3)R_3''.$
2. $R_2''' = R_2'' + (-1)R_3''.$

These transform the matrix to

$$
\begin{array}{c}
R_1' \\
R_2''' \\
R_3''
\end{array}
\begin{array}{cccc}
x & y & z & \text{CONSTANT} \\
\left[\begin{array}{ccc|c}
1 & 2 & 0 & 3 \\
0 & 0 & 0 & 0 \\
0 & 0 & 1 & 1
\end{array}\right].
\end{array}
$$

The last matrix corresponds to the linear equations

$$x + 2y = 3$$
$$z = 1.$$

The first equation implies that $x = 3 - 2y$. Clearly, y can take any arbitrary value. Thus if $y = c$, then

$$x = 3 - 2c, \qquad y = c, \qquad \text{and} \qquad z = 1.$$

We leave it for the reader to verify by substituting these values into the system that $x = 3 - 2c$, $y = c$, and $z = 1$ is a solution of three linear equations in three unknowns. Note that for each value of c, we obtain a solution. Because c can assume any value, we conclude that the system is consistent and has an infinite number of solutions.

Another special case is illustrated by Example 9.22.

Example 9.22

Solve the following set of linear equations:

$$
\begin{aligned}
x - y + 2z &= 2 \\
-2x + 3y - 2z &= -1 \\
4x - 7y + 2z &= -4.
\end{aligned}
$$

Solution: The augmented matrix is

$$
\begin{array}{cccc}
 & x & y & z & \text{CONSTANT} \\
R_1 \\
R_2 \\
R_3
\end{array}
\left[
\begin{array}{ccc|c}
1 & -1 & 2 & 2 \\
-2 & 3 & -2 & -1 \\
4 & -7 & 2 & -4
\end{array}
\right].
$$

We perform the following operations:

1. $R_2' = R_2 + 2R_1$.
2. $R_3' = R_3 + (-4)R_1$.

They transform the preceding matrix to

$$
\begin{array}{cccc}
 & x & y & z & \text{CONSTANT} \\
R_1 \\
R_2' \\
R_3'
\end{array}
\left[
\begin{array}{ccc|c}
1 & -1 & 2 & 2 \\
0 & 1 & 2 & 3 \\
0 & -3 & -6 & -12
\end{array}
\right].
$$

To obtain zero in the a_{32} position, we perform the following operation:

$$R_3'' = R_3' + 3R_2',$$

which transforms the matrix to

$$
\begin{array}{cccc}
 & x & y & z & \text{CONSTANT} \\
R_1 \\
R_2' \\
R_3''
\end{array}
\left[
\begin{array}{ccc|c}
1 & -1 & 2 & 2 \\
0 & 1 & 2 & 3 \\
0 & 0 & 0 & -3
\end{array}
\right].
$$

The last row determines the equation

$$0 \cdot x + 0 \cdot y + 0 \cdot z = -3.$$

Because there are no values of x, y, and z for which the equation holds, we conclude that the system is inconsistent and has no solutions.

EXERCISES

Using the Gauss-Jordan elimination method, solve the systems of linear equations in Exercises 1 to 18.

1. $\begin{aligned} x + 2y &= 5 \\ 2x + 3y &= 8 \end{aligned}$

2. $\begin{aligned} x + y &= 5 \\ 2x + 3y &= 12 \end{aligned}$

3. $\begin{aligned} 2x + 3y &= 13 \\ x + 2y &= 8 \end{aligned}$

4. $\begin{aligned} 3x + 4y &= 25 \\ -2x + 5y &= 14 \end{aligned}$

5. $\begin{aligned} 3x + 2y &= 12 \\ 2x - y &= 1 \end{aligned}$

6. $\begin{aligned} x - 2y &= 7 \\ 4x - 3y &= 18 \end{aligned}$

7. $\begin{aligned} x - 3y + z &= -6 \\ 2x + 3y + 3z &= 5 \\ 3x - y + z &= 0 \end{aligned}$

8. $\begin{aligned} x - y - z &= 2 \\ 2x + 4y - 3z &= -4 \\ -2x + 3y + 2z &= -5 \end{aligned}$

9. $\begin{aligned} x - 2y + z &= 11 \\ 2x - 4y - 3z &= 7 \\ 3x + 5y + 6z &= 9 \end{aligned}$

10. $\begin{aligned} 3x + 4y - z &= 8 \\ 2x - 5y + 3z &= 1 \\ x + y + z &= 6 \end{aligned}$

11. $\begin{aligned} x + 2y + z &= 5 \\ x - y + 2z &= 12 \\ 2x + 3y - z &= -1 \end{aligned}$

12. $\begin{aligned} 2x + 2y - z &= -1 \\ 3x + 2y - 2z &= -2 \\ x - y + z &= 6 \end{aligned}$

13. $\begin{aligned} x + y + 5z &= 8 \\ 2x + y - z &= 3 \\ 3x + 2y + 5z &= 12 \end{aligned}$

14. $\begin{aligned} 2x - 2y + z &= 17 \\ x - y - z &= -2 \\ x - 5y - z &= 6 \end{aligned}$

15. $\begin{aligned} x - 2y + 3z &= 9 \\ 3x - y - 2z &= 0 \\ 6x + 3y - z &= 1 \end{aligned}$

16. $\begin{aligned} x + 2y + 3z &= 6 \\ 2x - 2y + 5z &= 5 \\ 4x - y - 3z &= 0 \end{aligned}$

17. $\begin{aligned} x + y - z &= 0 \\ 2x + y + 3z &= 9 \\ x + 3y + z &= 6 \end{aligned}$

18. $\begin{aligned} x + 2y + 3z &= 14 \\ -x + y + 2z &= 7 \\ 3x - y + 5z &= 16 \end{aligned}$

Find the solution or solutions, if they exist, for the systems of linear equations in Exercises 19 to 30.

19. $\begin{aligned} x + y &= 3 \\ 2x + 2y &= 5 \end{aligned}$

20. $\begin{aligned} x + y &= 4 \\ 3x + 3y &= 12 \end{aligned}$

21. $\begin{aligned} x - 3y &= 6 \\ 2x - 6y &= 12 \end{aligned}$

22. $\begin{aligned} 2x - y &= 1 \\ 3x + 2y &= 12 \end{aligned}$

23. $\begin{aligned} x + y - z &= 4 \\ 2x + y + 3z &= 10 \\ 3x + 2y - 2z &= 10 \end{aligned}$

24. $\begin{aligned} x + y - z &= 5 \\ 2x + y + 3z &= 3 \\ 5x + 4y &= 18 \end{aligned}$

25. $\begin{aligned} x - y + z &= 5 \\ 3x + 5y - 13z &= -9 \\ 5x + 3y - 11z &= 1 \end{aligned}$

26. $\begin{aligned} x + 2y - 3z &= 7 \\ 2x - y + z &= 5 \\ 3x + y - 2z &= 12 \end{aligned}$

27. $\begin{aligned} x + 2y - z &= 6 \\ 3x - y + 2z &= 9 \\ 5x + 3y &= 21 \end{aligned}$

28. $\begin{aligned} -x + 2y + z &= 2 \\ 4x - y + 6z &= -4 \\ 2x + 3y + 8z &= 2 \end{aligned}$

29. $\begin{aligned} x - y + z &= 3 \\ 2x + 3y - z &= -5 \\ 4x + y + z &= 1 \end{aligned}$

30. $\begin{aligned} x + 2y - z &= 5 \\ 2x + 4y + 4z &= 22 \\ 4x + 8y + 3z &= 34 \end{aligned}$

9.6

SOLUTIONS OF LINEAR EQUATIONS USING MATRIX INVERSES

In elementary algebra, the equation

$$ax = b \qquad (a \neq 0)$$

can be solved for the unknown x by dividing both sides of the equation by a, thus obtaining

$$x = \frac{b}{a}.$$

Alternatively, we can multiply both sides of the equation by the multiplicative inverse of a, namely $\frac{1}{a}$, or a^{-1}. This yields

$$a^{-1}(ax) = a^{-1}b$$
$$(a^{-1}a)x = a^{-1}b$$
$$x = a^{-1}b \qquad (a^{-1}a = 1)$$

as the solution to the equation $ax = b$. The concept of multiplicative inverse can be extended to matrix multiplication and then used effectively for solving a set of n linear equations in n unknowns.

Definition 9.6.1

If A is a square matrix of order n and B is another square matrix, also of order n, such that

$$AB = BA = I,$$

then B is called an *inverse* of matrix A.

Example 9.23

Let

$$A = \begin{bmatrix} 9 & 7 \\ 5 & 4 \end{bmatrix} \quad \text{and} \quad B = \begin{bmatrix} 4 & -7 \\ -5 & 9 \end{bmatrix}.$$

Then

$$AB = \begin{bmatrix} 9 & 7 \\ 5 & 4 \end{bmatrix}\begin{bmatrix} 4 & -7 \\ -5 & 9 \end{bmatrix} = \begin{bmatrix} 1 & 0 \\ 0 & 1 \end{bmatrix}$$

$$BA = \begin{bmatrix} 4 & -7 \\ -5 & 9 \end{bmatrix}\begin{bmatrix} 9 & 7 \\ 5 & 4 \end{bmatrix} = \begin{bmatrix} 1 & 0 \\ 0 & 1 \end{bmatrix}.$$

Thus B is an inverse of matrix A.

We do not wish to leave an impression that all nonzero square matrices have inverses. Consider the matrix

$$A = \begin{bmatrix} 1 & 0 \\ 0 & 0 \end{bmatrix}$$

and suppose that

$$B = \begin{bmatrix} a & b \\ c & d \end{bmatrix}$$

is an inverse of A. Then $AB = I$. That is,

$$\begin{bmatrix} 1 & 0 \\ 0 & 0 \end{bmatrix}\begin{bmatrix} a & b \\ c & d \end{bmatrix} = \begin{bmatrix} 1 & 0 \\ 0 & 1 \end{bmatrix}$$

$$\begin{bmatrix} a & b \\ 0 & \boxed{0} \end{bmatrix} = \begin{bmatrix} 1 & 0 \\ 0 & \boxed{1} \end{bmatrix}.$$

Since these matrices are equal, their elements must equate. Specifically,

$$0 = 1,$$

which is false. Hence we conclude that matrix A does not have an inverse.

Proposition 9.6.1

The inverse of a square matrix A, if it exists, is unique.

Proof: Let A be a matrix with A^{-1} as its inverse. Suppose that B is another matrix, which is also an inverse of A. Then

$$AA^{-1} = A^{-1}A = I$$

and

$$AB = BA = I.$$

Multiplying both sides of $AB = I$ by A^{-1}, we have

$$A^{-1}(AB) = A^{-1}I = A^{-1}.$$

Also,

$$A^{-1}(AB) = (A^{-1}A)B = IB = B.$$

Hence

$$B = A^{-1}.$$

In view of the above result, we denote the inverse of a matrix A by the symbol A^{-1}. Note that in matrix theory, it is not acceptable to use the symbol $\frac{1}{A}$ in place of A^{-1}.

We now describe a technique for computing A^{-1} of a square matrix A whenever it exists. The Gauss-Jordan elimination procedure that we have used in solving a system of linear equations can also be applied to compute the inverse of the coefficient matrix.

To compute A^{-1} by this method, we set up the equations

$$AX = E_1, \qquad AX = E_2, \qquad \ldots, \qquad AX = E_n,$$

where E_i is a column vector with a 1 in the ith row and 0s elsewhere. Then the augmented matrix is the $n \times 2n$ matrix

$$[A:I_n],$$

where the $n \times n$ identity matrix I_n appears to the right of matrix A. If A^{-1} exists, we can reduce $[A:I_n]$ to $[I_n:A^{-1}]$ by repeatedly using elementary row operations. If A does not have an inverse, then it is impossible to reduce $[A:I_n]$ into the form where I_n is the $n \times n$ identity matrix appearing on the left.

Example 9.24

Find the inverse of

$$A = \begin{bmatrix} 1 & 3 \\ 2 & 7 \end{bmatrix}.$$

Solution: The elementary operations are performed in the order specified in Table 9.1.

TABLE 9.1

Row Operation	a_1	a_2	E_1	E_2
R_1	1	3	1	0
R_2	2	7	0	1
R_1	1	3	1	0
$R'_2 = R_2 + (-2)R_1$	0	1	-2	1
$R'_1 = R_1 + (-3)R'_2$	1	0	7	-3
R'_2	0	1	-2	1

The augmented matrix $[A:I_2]$ is reduced to $[I_2:A^{-1}]$, where

$$A^{-1} = \begin{bmatrix} 7 & -3 \\ -2 & 1 \end{bmatrix}.$$

We now illustrate briefly how A^{-1} is used to solve a system of n linear equations in n unknowns. Consider, for instance, Example 9.25.

Example 9.25

Solve the following system of linear equations:

$$\begin{aligned} 2x + 3y &= 5 \\ -5x - 2y &= 4. \end{aligned} \qquad (9.14)$$

Solution: In matrix notation, we have

$$\begin{bmatrix} 2 & 3 \\ -5 & -2 \end{bmatrix} \begin{bmatrix} x \\ y \end{bmatrix} = \begin{bmatrix} 5 \\ 4 \end{bmatrix},$$

or

$$AX = B, \qquad (9.15)$$

$$A = \begin{bmatrix} 2 & 3 \\ -5 & -2 \end{bmatrix}, \qquad X = \begin{bmatrix} x \\ y \end{bmatrix}, \qquad \text{and} \qquad B = \begin{bmatrix} 5 \\ 4 \end{bmatrix}.$$

Assuming that we can find A^{-1} such that $AA^{-1} = A^{-1}A = I_2$, we multiply both sides of (9.15) by A^{-1} and obtain

$$A^{-1}(AX) = A^{-1}B. \qquad (9.16)$$

Using the associative property of matrix multiplication, we have

$$(A^{-1}A)X = A^{-1}B. \qquad (9.17)$$

Since $A^{-1}A = I_2$, we obtain

$$X = A^{-1}B. \qquad (9.18)$$

In this example,

$$A^{-1} = \frac{1}{11} \begin{bmatrix} -2 & -3 \\ 5 & 2 \end{bmatrix},$$

since $A^{-1}A = AA^{-1} = I_2$. Thus

$$X = A^{-1}B = \frac{1}{11} \begin{bmatrix} -2 & -3 \\ 5 & 2 \end{bmatrix} \begin{bmatrix} 5 \\ 4 \end{bmatrix} = \begin{bmatrix} -2 \\ 3 \end{bmatrix},$$

and $x = -2$, $y = 3$ is the solution to the linear equations.

Example 9.26

Find the inverse of

$$A = \begin{bmatrix} 1 & 1 & 1 \\ 3 & 4 & -1 \\ 2 & -5 & 3 \end{bmatrix}$$

and use A^{-1} to solve the following system of linear equations:

$$x + y + z = 9$$
$$3x + 4y - z = 13$$
$$2x - 5y + 3z = 8.$$

Solution: The elementary operations are performed in the order specified in Table 9.2.

TABLE 9.2

Row Operation	a_1	a_2	a_3	E_1	E_2	E_3
R_1	1	1	1	1	0	0
R_2	3	4	-1	0	1	0
R_3	2	-5	3	0	0	1
R_1	1	1	1	1	0	0
$R_2' = R_2 + (-3)R_1$	0	1	-4	-3	1	0
$R_3' = R_3 + (-2)R_1$	0	-7	1	-2	0	1
$R_1' = R_1 + (-1)R_2'$	1	0	5	4	-1	0
R_2'	0	1	-4	-3	1	0
$R_3'' = \left(-\dfrac{1}{27}\right)(R_3' + 7R_2')$	0	0	1	$\frac{23}{27}$	$-\frac{7}{27}$	$-\frac{1}{27}$
$R_1'' = R_1' + (-5)R_3''$	1	0	0	$-\frac{7}{27}$	$\frac{8}{27}$	$\frac{5}{27}$
$R_2'' = R_2' + 4R_3''$	0	1	0	$\frac{11}{27}$	$-\frac{1}{27}$	$-\frac{4}{27}$
R_3''	0	0	1	$\frac{23}{27}$	$-\frac{7}{27}$	$-\frac{1}{27}$

Therefore

$$A^{-1} = \frac{1}{27} \begin{bmatrix} -7 & 8 & 5 \\ 11 & -1 & -4 \\ 23 & -7 & -1 \end{bmatrix}$$

and

$$\begin{bmatrix} x \\ y \\ z \end{bmatrix} = A^{-1} \begin{bmatrix} 9 \\ 13 \\ 8 \end{bmatrix} = \frac{1}{27} \begin{bmatrix} -7 & 8 & 5 \\ 11 & -1 & -4 \\ 23 & -7 & -1 \end{bmatrix} \cdot \begin{bmatrix} 9 \\ 13 \\ 8 \end{bmatrix} = \begin{bmatrix} 3 \\ 2 \\ 4 \end{bmatrix}.$$

Hence $x = 3$, $y = 2$, $z = 4$ is the solution to the system of linear equations.

Frequently, we wish to solve simultaneously several systems of linear equations

$$AX = B_1, \qquad AX = B_2, \qquad \ldots, \qquad AX = B_n,$$

all of which have the same coefficient matrix A. If A^{-1} exists, then we can multiply the augmented matrix

$$[A:B_1, B_2, \ldots, B_n]$$

by A^{-1} to obtain

$$[I_n:A^{-1}B_1, A^{-1}B_2, \ldots, A^{-1}B_n].$$

Note that $A^{-1}B_1$, $A^{-1}B_2, \ldots, A^{-1}B_n$ are solutions to the systems of linear equations

$$AX = B_1, \qquad AX = B_2, \qquad \ldots, \qquad AX = B_n,$$

respectively.

Example 9.27   Find the inverse of

$$A = \begin{bmatrix} 2 & 3 & -1 \\ 1 & 2 & 1 \\ -1 & -1 & 3 \end{bmatrix}$$

and use it to solve the following systems of linear equations:

(a) $2x + 3y - z = 4$
    $x + 2y + z = 7$
    $-x - y + 3z = 7$

(b) $2x + 3y - z = 1$
    $x + 2y + z = 4$
    $-x - y + 3z = 5$

(c) $2x + 3y - z = 26$
$x + 2y + z = 11$
$-x - y + 3z = -18.$

Solution: We leave it for the reader to verify that

$$A^{-1} = \begin{bmatrix} 7 & -8 & 5 \\ -4 & 5 & -3 \\ 1 & -1 & 1 \end{bmatrix}.$$

The linear equations to be solved are

$$AX = B_1, \qquad AX = B_2, \qquad \text{and} \qquad AX = B_3,$$

where

$$A = \begin{bmatrix} 2 & 3 & -1 \\ 1 & 2 & 1 \\ -1 & -1 & 3 \end{bmatrix}, \qquad X = \begin{bmatrix} x \\ y \\ z \end{bmatrix}, \qquad B_1 = \begin{bmatrix} 4 \\ 7 \\ 7 \end{bmatrix},$$

$$B_2 = \begin{bmatrix} 1 \\ 4 \\ 5 \end{bmatrix}, \qquad B_3 = \begin{bmatrix} 26 \\ 11 \\ -18 \end{bmatrix}.$$

The augmented matrix associated with the system is

$$[A:B_1, B_2, B_3] = \begin{bmatrix} 2 & 3 & -1 & | & 4 & 1 & 26 \\ 1 & 2 & 1 & | & 7 & 4 & 11 \\ -1 & -1 & 3 & | & 7 & 5 & -18 \end{bmatrix}.$$

Now we multiply the augmented matrix $[A:B_1, B_2, B_3]$ by A^{-1} and obtain

$$\begin{bmatrix} 7 & -8 & 5 \\ -4 & 5 & -3 \\ 1 & -1 & 1 \end{bmatrix} \begin{bmatrix} 2 & 3 & -1 & | & 4 & 1 & 26 \\ 1 & 2 & 1 & | & 7 & 4 & 11 \\ -1 & -1 & 3 & | & 7 & 5 & -18 \end{bmatrix}$$

$$= \begin{bmatrix} 1 & 0 & 0 & | & 7 & 0 & 4 \\ 0 & 1 & 0 & | & -2 & 1 & 5 \\ 0 & 0 & 1 & | & 4 & 2 & -3 \end{bmatrix}.$$

Hence $x = 7$, $y = -2$, $z = 4$ is the solution to the system of equations in (a); $x = 0$, $y = 1$, $z = 2$ is the solution to the linear equations in (b); and $x = 4$, $y = 5$, $z = -3$ is the solution to the system of linear equations in (c).

A great deal of time is saved if A^{-1} is known, because then we have only to compute $A^{-1} B$ for the new column vector B. This is an advantage over the elimination method discussed earlier.

EXERCISES

Find, if possible, the inverse of each matrix in Exercise 1 to 4.

1. $\begin{bmatrix} 3 & 2 \\ 5 & 3 \end{bmatrix}$
   2. $\begin{bmatrix} 6 & 5 \\ 4 & 3 \end{bmatrix}$
   3. $\begin{bmatrix} 12 & 11 \\ 13 & 12 \end{bmatrix}$
   4. $\begin{bmatrix} -8 & 10 \\ 6 & -7 \end{bmatrix}$

Using the Gauss-Jordan elimination method, find the inverse of each matrix in Exercises 5 to 12.

5. $\begin{bmatrix} 2 & -1 & 2 \\ 4 & 1 & 2 \\ 8 & -1 & 1 \end{bmatrix}$
   6. $\begin{bmatrix} 1 & 2 & 1 \\ 2 & 3 & 4 \\ 3 & -1 & 1 \end{bmatrix}$
   7. $\begin{bmatrix} 1 & -3 & 1 \\ 2 & 3 & 3 \\ 3 & -1 & 1 \end{bmatrix}$

8. $\begin{bmatrix} 2 & 2 & -1 \\ 3 & 2 & -2 \\ 1 & -1 & 1 \end{bmatrix}$
   9. $\begin{bmatrix} 1 & 2 & 1 \\ 1 & -1 & 2 \\ 2 & 3 & -1 \end{bmatrix}$
   10. $\begin{bmatrix} 3 & 1 & 0 \\ 1 & -1 & 2 \\ 1 & 1 & 1 \end{bmatrix}$

11. $\begin{bmatrix} 3 & 4 & -1 \\ 2 & -5 & 3 \\ 1 & 1 & 1 \end{bmatrix}$
   12. $\begin{bmatrix} 2 & 3 & -1 \\ 1 & 2 & 1 \\ -1 & -1 & 3 \end{bmatrix}$

13. Find the inverse of the matrix

$$A = \begin{bmatrix} 4 & -1 \\ 2 & 1 \end{bmatrix}$$

and use A^{-1} to solve the linear equations

$$4x - y = 17$$
$$2x + y = 7.$$

14. Find the inverse of

$$A = \begin{bmatrix} 5 & 3 \\ 3 & 2 \end{bmatrix}$$

and use A^{-1} to solve the following systems of linear equations:

(a) $5x + 3y = 8$ (b) $5x + 3y = 2$ (c) $5x + 3y = 13$ (d) $5x + 3y = 11$
    $3x + 2y = 5$ $3x + 2y = 1$ $3x + 2y = 8$ $3x + 2y = 7$

15. Find the inverse of

$$A = \begin{bmatrix} 3 & 2 \\ 4 & 1 \end{bmatrix}$$

and use A^{-1} to solve the following systems of linear equations:

(a) $3x + 2y = 12$ (b) $3x + 2y = 6$ (c) $3x + 2y = 11$ (d) $3x + 2y = 1$
    $4x + y = 11$ $4x + y = 13$ $4x + y = -2$ $4x + y = 3$

16. Find the inverse of

$$A = \begin{bmatrix} 1 & 1 & 1 \\ 2 & 5 & -3 \\ 3 & 4 & -7 \end{bmatrix}$$

and use A^{-1} to solve the following systems of linear equations:

(a)
$$x + y + z = 3$$
$$2x + 5y - 3z = 4$$
$$3x + 4y - 7z = 0$$

(b)
$$x + y + z = 6$$
$$2x + 5y - 3z = 3$$
$$3x + 4y - 7z = -10$$

(c)
$$x + y + z = -1$$
$$2x + 5y - 3z = 13$$
$$3x + 4y - 7z = 27$$

(d)
$$x + y + z = 6$$
$$2x + 5y - 3z = 13$$
$$3x + 4y - 7z = 10$$

17. Find the inverse of

$$A = \begin{bmatrix} 1 & 1 & 5 \\ 2 & 1 & -1 \\ 3 & 2 & 5 \end{bmatrix}$$

and use A^{-1} to solve the following systems of linear equations:

(a)
$$x + y + 5z = 0$$
$$2x + y - z = 8$$
$$3x + 2y + 5z = 7$$

(b)
$$x + y + 5z = 26$$
$$2x + y - z = -1$$
$$3x + 2y + 5z = 30$$

(c)
$$x + y + 5z = 41$$
$$2x + y - z = 1$$
$$3x + 2y + 5z = 49.$$

18. A department store has 90 men's suits of three different types that must be sold this spring. If the type 1 suit sells for $50, the type 2 suit for $60, and the type 3 suit for $75, the net sales amounts to $5800; but if these suits are sold for $45, $55, and $60, respectively, then the total sales revenue is only $4950. Using the inverse of the coefficient matrix, determine the number of suits each type the store has in stock.

19. An airline has three types of cargo planes, 1, 2, and 3. They can carry three types of equipment as shown in the following matrix:

PLANES

$$\begin{array}{c} \\ A \\ \text{EQUIPMENT} \quad B \\ C \end{array} \begin{array}{ccc} 1 & 2 & 3 \\ \begin{bmatrix} 8 & 6 & 10 \\ 5 & 4 & 2 \\ 3 & 6 & 4 \end{bmatrix} \end{array}$$

Using the inverse of the matrix, determine the number of planes of each type required to deliver 46 pieces of equipment A, 25 pieces of equipment B, and 25 pieces of equipment C.

9.7

SYSTEMS OF LINEAR INEQUALITIES IN TWO VARIABLES

There are many problems in education, marketing, and management that call for several conditions expressed in the form of linear inequalities to be satisfied simultaneously. The funds allocated for faculty salaries limit the number of new teachers that can be hired, the size of the parking lot limits the number of automobiles that can be parked at a given time, the production capacity of a factory limits the quantity of goods that can be shipped to warehouses, and so on.

Linear equalities are of the form

$$ax + by = c,$$

linear equality where a, b, and c are real numbers and not both a and b are zero. If the equality sign is replaced by $<$, $\leq$, $>$, or $\geq$, the resulting inequality is called a **linear inequality.** Thus

$$3x + 4y \geq 12 \qquad \text{and} \qquad 2x - 3y \leq 6$$

are examples of **linear inequalities in two variables** x and y.

As shown in Figure 9.6, a line divides a plane into three parts: the line itself and two half-planes on either side of the line. The line does not belong to either of the half-planes, and half-planes I and II are themselves disjointed. Just as the points on the line can be described by means of linear equations, the points in the half-planes can be described by linear inequalities.

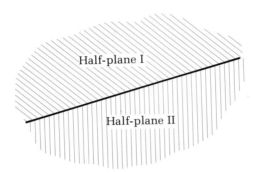

FIGURE 9.6

To plot the graph of a linear inequality in two unknowns, we first graph the line by choosing two points that satisfy the linear equation and connecting them with a straight line. To plot $3x + 4y \geq 12$, for example, we need to determine which of the two half-planes satisfies the linear inequality $3x + 4y > 12$. The proper half-plane can be determined by choosing an arbitrary point in one of the half-planes. If the point's coordinates satisfy the inequality, then the half-plane that contains that point is the graph of the inequality. If the coordinates of the point do not satisfy the inequality, then the other half-plane is the desired graph. The origin $(0, 0)$ is often a good choice for the test because the resulting arithmetic is obviously simplified. [If the point $(0, 0)$ lies on the line, then another point must be chosen.]

Thus if we substitute $x = 0$, $y = 0$ into the linear inequality $3x + 4y > 12$, we obtain

$$3(0) + 4(0) > 12$$

$$0 > 12,$$

which is a false statement. Thus the point $(0, 0)$ is not a part of the half-plane determined by $3x + 4y > 12$. The shaded region in Figure 9.7 together with the line $3x + 4y = 12$ is the desired graph of the linear inequality $3x + 4y \geq 12$.

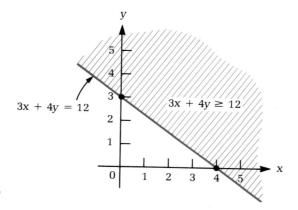

FIGURE 9.7

Example 9.28

Sketch the graph of the linear inequality

$$x + 2y > 4.$$

Solution: The region we want to sketch is one of the half-planes determined by the linear equality $x + 2y = 4$ [see Figure 9.8(a)]. We choose an arbitrary point, say, the origin $(0, 0)$, as a testing point. Substituting $x = 0$, $y = 0$ in the inequality $x + 2y > 4$, we obtain

$$0 + 2(0) > 4$$

$$0 > 4,$$

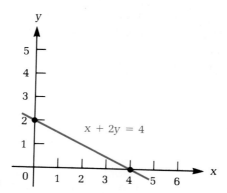

FIGURE 9.8(a)

which is false. This implies that the origin is not located in the desired half-plane. You can verify that (1, 2), (2, 3), and (3, 1) lie in the desired half-plane of $x + 2y > 4$ [see Figure 9.8(b)]. The line representing the equation is dotted, indicating that it is not a part of the graph of the inequality.

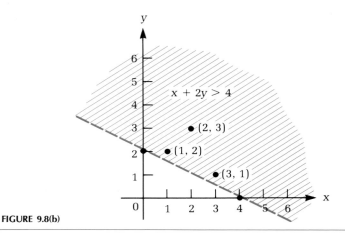

FIGURE 9.8(b)

We now consider **systems of linear inequalities** in two variables x and y, such as

$$
\begin{array}{ccc}
\begin{array}{c} x \geq 0 \\ y \geq 0 \end{array}
& \text{or}
& \begin{array}{c} 2x + y \leq 2 \\ 2x - 3y \geq 6 \end{array}
& \text{or}
& \begin{array}{c} 4x + 2y \leq 48 \\ 2x + 3y \leq 40 \\ x \geq 0 \\ y \geq 0 \end{array}
\end{array}
$$

To solve one of these systems, we must determine what values of x and y satisfy all the linear inequalities simultaneously. The set of ordered pairs (x, y) so obtained is called the **solution set of the system of linear inequalities.**

Example 9.29   Determine the region satisfying the linear inequalities

$$x \geq 0$$
$$y \geq 0.$$

Solution: The linear inequality $y \geq 0$ corresponds to the upper half of the xy plane and includes the line $y = 0$ (see Figure 9.9). Similarly, the inequality $x \geq 0$ refers to the right half-plane of the xy plane, and this region includes the line $x = 0$ (see Figure 9.10). Thus the linear inequalities $x \geq 0$ and $y \geq 0$ together refer to the first quadrant, including the axes (see Figure 9.11).

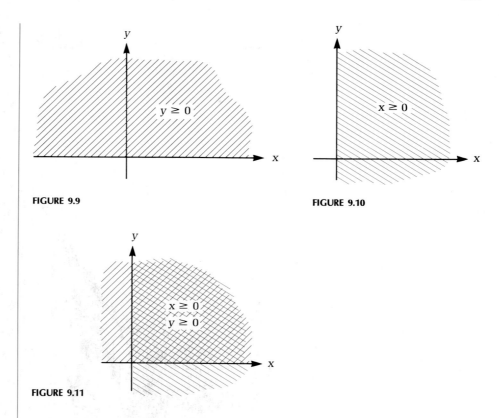

FIGURE 9.9

FIGURE 9.10

FIGURE 9.11

Notice that the area shaded in black refers to the inequality $y \geq 0$, the area shaded in color corresponds to the inequality $x \geq 0$, and the crosshatched area represents $x \geq 0$ and $y \geq 0$.

These inequalities play a very important role in essentially all linear programming problems, which we consider in the next section.

Example 9.30

Determine the region satisfying the linear inequalities

$$2x + y \leq 2$$
$$2x - 3y \geq 6.$$

Solution: First we draw the graph of the linear equality $2x + y = 2$. Using $(0, 0)$ as a test point, we determine that the lower half-plane corresponds to the linear inequality $2x + y < 2$. This region shaded in black together with the line $2x + y = 2$ represents the graph of $2x + y \leq 2$ (see Figure 9.12). Similarly, we graph the linear inequality $2x - 3y \geq 6$. The region shaded in color together with the line $2x - 3y = 6$ refers to the graph of $2x - 3y \geq 6$ (see Figure 9.13). The solution set of the inequalities $2x + y \leq 2$ and $2x - 3y \geq 6$ is the intersection of the solution sets of the two inequalities and is represented by the cross-hatched region (see Figure 9.14). You should check a sample point in this region to be convinced that the region satisfies the linear inequalities simultaneously.

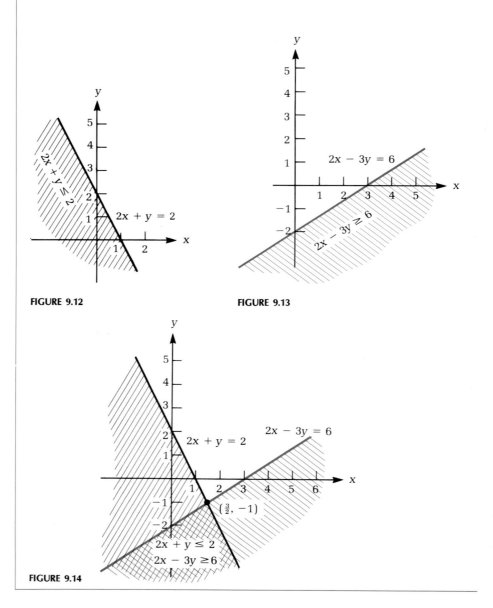

FIGURE 9.12

FIGURE 9.13

FIGURE 9.14

Example 9.31

Determine the region satisfying the inequalities

$$5x + 3y \leq 150$$

$$x + y \leq 40$$

$$x \geq 0$$

$$y \geq 0.$$

Solution: This is a system of four linear inequalities in two unknowns, x and y. We found in Example 9.29 that the linear inequalities $x \geq 0$ and $y \geq 0$ refer to the right half and the upper half, respectively, of the xy plane, with their intersection being the first quadrant, including the axes. As far as the other inequalities are concerned, we simply graph the lines $5x + 3y = 150$ and $x + y = 40$ and find their respective half-planes for $5x + 3y \leq 150$ and $x + y \leq 40$. The solution set of the system is depicted by the intersection of these four linear inequalities (see Figure 9.15).

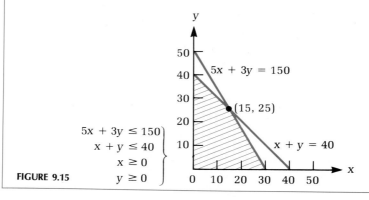

FIGURE 9.15

Example 9.32

Determine the region satisfying the inequalities

$$2x + y \geq 100$$

$$x + y \geq 80$$

$$x \geq 0$$

$$y \geq 0.$$

Solution: The graph of the inequalities is shown in Figure 9.16. The solution set is represented by the shaded region, which is understood to extend indefinitely upward and to the right.

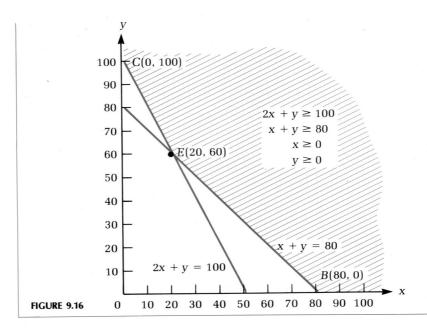

FIGURE 9.16

EXERCISES

Draw graphs of the linear inequalities in Exercises 1 to 6.

1. $x + 6y \geq 10$

2. $4x + 3y \geq 12$

3. $3x + 2y < 6$

4. $3x - 4y \leq 12$

5. $x + 3y \geq 8$

6. $8x + 7y \leq 28$

Determine the region satisfied by each system of linear inequalities in Exercises 7 to 15.

7. $x + y \leq 4$ and $2x - y \leq 4$

8. $x - y \geq 6$ and $2x + 3y \leq 5$

9. $x \geq 0, y \geq 0,$ and $3x + 4y \leq 12$

10. $x \geq 0, y \geq 0,$ and $x + 2y \leq 8$

11. $x \geq 0, y \geq 0, x + y \leq 6,$ and $3x + 8y \leq 24$

12. $x \geq 0, y \geq 0, 2x + y \leq 24,$ and $2x + 3y \leq 40$

13. $x \geq 0, y \geq 0, 2x + y \leq 100,$ and $x + y \leq 75$

14. $x \geq 0, y \geq 0, 2x + 5y \leq 30,$ and $4x + 5y \leq 40$

15. $x \geq 0, y \geq 0, x - y \geq 4,$ and $3x + 4y \geq 24$

9.8

LINEAR PROGRAMMING: AN INTRODUCTION

Business firms are continuously expanding their operations in both size and complexity, creating new uncertainties in a highly competitive market. Business executives must, therefore, continue to systematically explore such problems as how goods can be produced most economically with given material and equipment, how crude oil should be allocated among refineries to obtain the optimal production, how products can be shipped from various plants to retail outlets at a minimum cost, and so on. If both the objective and the associated constraints can be translated to the form of linear equations or linear inequalities, then a technique called **linear programming** may be used to solve such problems. The adjective **linear** describes a relationship between two or more variables; the term **programming** refers to the mathematical technique used to obtain the optimal solution. The following examples illustrate the process.

linear programming

Example 9.33

Find the maximum value of $P = 5x + 3y$ subject to the linear constraints

$$2x + 4y \leq 80$$

$$5x + 2y \leq 80$$

$$x \geq 0$$

$$y \geq 0.$$

Solution: The graph of the linear inequalities is shown in Figure 9.17. Notice that the region of feasible solutions is the shaded area bounded by the graphs of the linear equalities $2x + 4y = 80$ and $5x + 2y = 80$ and by the coordinate axes.

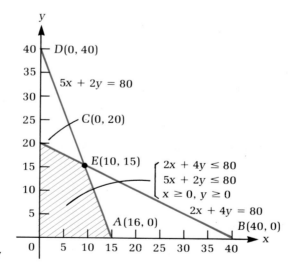

FIGURE 9.17

objective function

The problem now reduces to finding points (x, y) in the solution set at which the **objective function** $P = 5x + 3y$ attains the maximum value. However, the evaluation of P at each possible point (x, y) in the solution set would be tedious. In many cases, it would be impossible. This is where the theory of linear programming simplifies the task considerably. **It can be shown that a linear function**

takes on its maximum and minimum values at a point of intersection of the lines which determine the half-planes. This means that if there is a maximum value of P, it must occur at one of the points of intersection $O(0, 0)$, $A(16, 0)$, $E(10, 15)$, or $C(0, 20)$.

The values of $P = 5x + 3y$ at each of the points of intersection are displayed in Table 9.3. We see that the maximum value of P occurs at the point $E(10, 15)$.

TABLE 9.3	
Point (x, y)	$P = 5x + 3y$
$O(0, 0)$	$5(0) + 3(0) = 0$
$A(16, 0)$	$5(16) + 3(0) = 80$
$E(10, 15)$	$5(10) + 3(15) = 95$
$C(0, 20)$	$5(0) + 3(20) = 60$

Example 9.34

A manufacturer makes two products, each requiring time on three different machines. Each unit of product 1 requires 10 minutes on machine A, 6 minutes on machine B, and 9 minutes on machine C; each unit of product 2 requires 5 minutes on machine A, 12 minutes on machine B, and 9 minutes on machine C. Both machines A and C are available for 7.5 hours each, and machine B is available for 8 hours. The profits are \$5 and \$3 on products 1 and 2, respectively. Determine the combination of the products that will maximize profit.

Solution: The information in the problem is summarized in Table 9.4.

TABLE 9.4			
	Product 1	Product 2	Time Available (in Minutes)
Machine A	10	5	450
Machine B	6	12	480
Machine C	9	9	450
Profit	\$5	\$3	

Let x and y represent the number of units of products 1 and 2, respectively. This means that

$$x \geq 0, \qquad y \geq 0.$$

Each unit of product 1 requires 10 minutes on machine A; so x units will require $10x$ minutes. Similarly, y units of product 2 require $5y$ minutes. This means that machine A must be used for $10x + 5y$ minutes. Since machine A is available for 7.5 hours (450 minutes), we have the linear inequality

$$10x + 5y \leq 450.$$

Using the same reasoning for machines B and C, we have the linear inequalities

$$6x + 12y \leq 480$$

and

$$9x + 9y \leq 450,$$

respectively. This system of three linear inequalities together with the inequalities $x \geq 0$ and $y \geq 0$ provides the linear constraints which occur in the manufacturing process.

The region of feasible solutions is bounded by the graph of the linear equations $10x + 5y = 450$, $6x + 12y = 480$, and $9x + 9y = 450$ and by the coordinate axes, as shown by the shaded area in Figure 9.18.

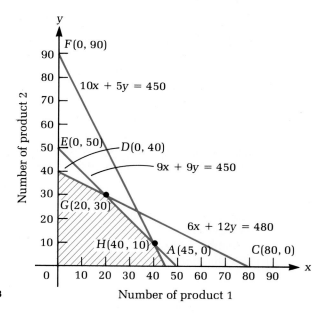

FIGURE 9.18

Since the manufacturer makes \$5 on each unit of product 1 and \$3 on each unit of product 2, the objective function to be maximized subject to the above linear constraints is

$$P = 5x + 3y.$$

As was pointed out in the solution of Example 9.33, the maximum value of P must occur at one of the points of intersection $O(0, 0)$, $A(45, 0)$, $H(40, 10)$, $G(20, 30)$, and $D(0, 40)$. The values of P at each of these points are displayed in Table 9.5. Clearly, the optimal solution is at the point $H(40, 10)$ because production of 40 units of product 1 and 10 units of product 2 will maximize the profit.

TABLE 9.5

Point (x, y)	$P = 5x + 3y$
$O(0, 0)$	$5(0) + 3(0) = 0$
$A(45, 0)$	$5(45) + 3(0) = 225$
$H(40, 10)$	$5(40) + 3(10) = 230$
$G(20, 30)$	$5(20) + 3(30) = 190$
$D(0, 40)$	$5(0) + 3(40) = 120$

Note that in Examples 9.33 and 9.34, linear programming has been used to maximize an objective function subject to certain linear constraints. The next example deals with minimization of cost.

Example 9.35

Nancy must supplement her diet with at least 80 milligrams of calcium and 17 milligrams of iron. This requirement can be met from two types of vitamin tablets, P and Q. Each vitamin P tablet contains 5 milligrams of calcium and 1 milligram of iron, whereas each vitamin Q table contains 8 milligrams of calcium and 2 milligrams of iron. The vitamin P and Q tablets cost $0.06 and $0.08 each, respectively. Determine the number of tablets of each kind that she must purchase to meet her daily requirements at a minimum cost.

Solution: Let x and y represent the number of vitamin P and Q tablets, respectively. The calcium provided by x number of P tablets and y number of Q tablets is given by $5x + 8y$; the iron contained in x number of P tablets and y number of Q tablets is given by $x + 2y$. The minimum daily requirement of calcium and iron can be expressed as

$$5x + 8y \geq 80$$

$$x + 2y \geq 17.$$

Further, x number of P tablets and y number of Q tablets cost $6x + 8y$. The problem asks for the best possible combination of P and Q tablets that would minimize the objective function

$$C = 6x + 8y$$

subject to the linear constraints

$$5x + 8y \geq 80$$

$$x + 2y \geq 17$$

$$x \geq 0$$

$$y \geq 0.$$

To determine the region of feasible solutions, we graph the linear equations, as shown in Figure 9.19. The points on line *AD* whose coordinates (x, y) satisfy the linear equality $5x + 8y = 80$ represent the combination of the two types of vitamin P and Q tablets that yields exactly 80 milligrams of calcium. Similarly, the points on line *BC* satisfying the linear equality $x + 2y = 17$ are combinations of the two types of P and Q tablets that provides 17 milligrams of iron. The area of feasible solutions consists of all points falling on or to the right of line segments *DE* and *EB*.

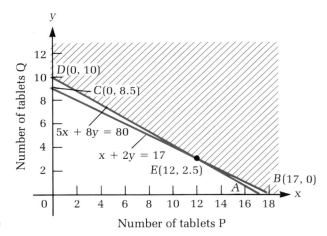

FIGURE 9.19

To determine the minimum cost combination, we test each point of intersection. Point *E* can be determined by solving the linear equations $5x + 8y = 80$ and $x + 2y = 17$. The values of the objective function $C = 6x + 8y$ at $D(0, 10)$, $E(12, 2.5)$, and $B(17, 0)$ are as follows:

Point $D(0, 10)$: $C = 6(0) + 8(10) = 80$

Point $E(12, 2.5)$: $C = 6(12) + 8(2.5) = 92$

Point $B(17, 0)$: $C = 6(17) + 8(0) = 102$.

> Points E and B need no further consideration because of their higher costs. The optimum solution is therefore at $D(0, 10)$, when $C = 80$. This means that the specified requirements of calcium and iron are met at a minimum cost of $0.80 when Nancy purchases 10 tablets of type Q.

EXERCISES

Determine graphically the values of x and y that maximize the objective function P in Exercises 1 to 6 subject to the given linear constraints.

1. $P = 8x + 7y$
$2x + 3y \le 24$
$2x + y \le 16$
$x \ge 0, y \ge 0$

2. $P = 8x + 6y$
$x + 2y \le 150$
$x + y \le 100$
$x \ge 0, y \ge 0$

3. $P = 15x + 10y$
$3x + 2y \le 80$
$2x + 3y \le 70$
$x \ge 0, y \ge 0$

4. $P = 16x + 20y$
$10x + 20y \le 360$
$20x + 10y \le 360$
$x \ge 0, y \ge 0$

5. $P = 6x + 9y$
$2x + 5y \le 50$
$x + y \le 11$
$2x + y \le 20$
$x \ge 0, y \ge 0$

6. $P = 48x + 36y$
$4x + 3y \le 120$
$4x + y \le 80$
$2x + 4y \le 120$
$x \ge 0, y \ge 0$

Determine graphically the values of x and y that minimize the objective function C in Exercises 7 to 10 subject to the given linear constraints.

7. $C = 9x + 15y$
$3x + 4y \ge 25$
$x + 3y \ge 15$
$x \ge 0, y \ge 0$

8. $C = 28x + 35y$
$2x + y \ge 110$
$2x + 3y \ge 170$
$x \ge 0, y \ge 0$

9. $C = 9x + 7y$
$20x + 15y \ge 1250$
$10x + 10y \ge 700$
$x \ge 0, y \ge 0$

10. $C = 35x + 40y$
$12x + 9y \ge 750$
$15x + 12y \ge 975$
$x \ge 0, y \ge 0$

11. A company manufactures two products. Each product must pass through two assembly points, A and B. Each unit of product 1 requires 4 hours at A and 4 hours at B. Product 2 requires 3 hours at A and 1 hour at B. There are 110 hours available at A and 90 hours available at B. The net profits on each unit of products 1 and 2 are $10 and $6, respectively. Determine the combination of products 1 and 2 that will maximize the profit.

12. The Alexander Manufacturing Company produces glass doors and windows. Each glass door requires 6 worker hours in department 1 and 2 worker hours in department 2; each glass window requires 4 worker hours in department 1 and 3 worker hours in department 2. The profits on each glass door and window are $25 and $20, respectively. There are 60 worker hours available in department 1 and 30 worker hours available in department 2. Determine the combination of glass doors and windows that will maximize the profit.

13. A manufacturer makes two kinds of radios, model 1 and model 2. The process for model 1 requires 2 hours in department A, 2 hours in department B, and 1 hour in department C. Model 2, however, requires 1.5 hours in department A, 0.5 hour in department B, and 2 hours in department C. The profits on each unit of model 1 and model 2 are $45 and $20, respectively. If there are 60 hours available in department A, 40 hours in department B, and 60 hours in department C, determine the combinations of the products that will maximize the profit.

14. An airline agrees to charter planes for a group. The group needs at least 160 first-class seats and at least 300 tourist-class seats. The airline must use at least two of its model 207 planes, which have 20 first-class seats and 30 tourist-class seats, and at least two of its model 227 planes, which have 20 first-class and 60 tourist-class seats. Each flight of model 207 plane costs the airline $1000, while each flight of model 227 plane costs of $1500. Determine the combination of model 207 and model 227 planes that the airline should use to minimize the flight costs.

15. Harold needs to prepare a breakfast that will supply him with at least 7 units of vitamin A and at least 11 units of vitamin B. This requirement can be met by two types of food. Each unit of food 1 contains 1 unit of vitamin A and 2 units of vitamin B and costs $0.08; each unit of food 2 contains 2 units of vitamin A and 1 unit of vitamin B and costs $0.12. Determine the best possible combination of the two types of food that will minimize the cost while meeting the nutritional requirements.

16. A pound of fish contains 3 units of carbohydrates, 6 units of vitamins, and 12 units of protein and costs $1.75, whereas a quart of milk contains 2 units of carbohydrates, 3 units of vitamins, and 4 units of protein and costs $0.60. If the minimum daily requirements are 12 units of carbohydrates, 15 units of vitamins, and 36 units of protein, determine the best possible combination of fish and milk in order to minimize the cost.

KEY WORDS

system of linear equations
system of linear inequalities
solution of a system of equations
method of elimination
substitution method
independent system of equations
inconsistent system of equations
dependent system of equations
matrix
m × n matrix

dimensions of a matrix
zero matrix
square matrix of order n
main diagonal
identity matrix
equal matrices
addition and subtraction of matrices
multiplication of matrices
coefficient matrix
vector of constants

equivalent systems of equations
augmented matrix
Gauss-Jordan elimination method
inverse of a square matrix
linear inequality
solution set of a system of linear inequalities
linear programming
objective function

EXERCISES FOR REVIEW

Determine the solution set for the linear equations in Exercises 1 to 4.

1. $2x + 3y = 21$
$3x + 2y = 19$

2. $5x - 3y = 10$
$2x + y = 4$

3. $2x + y = 5$
$3x + 2y = 6$

4. $2x - 3y = 4$
$5x + 4y = 1$

5. The sum of two numbers is 151, and their difference is 15. Find the numbers.

6. At an Ice Capades performance, an adult admission ticket was $3.50 and children were charged $2.00. There were 800 more adult tickets sold than children's tickets. If the total receipts for the show were $6650, how many of each type of admission ticket were sold?

7. The manager of a grocery store wishes to mix peanuts that sell for $4 a pound with cashews that sell for $7 a pound. How many pounds of peanuts should be mixed with 8 pounds of cashews to sell the mixture for $5 a pound?

8. Antifreeze A is 18 percent alcohol and 82 percent water. Antifreeze B is 10 percent alcohol and 90 percent water. How many liters of each should be mixed to get 20 liters of a mixture that is 15 percent alcohol?

Solve the system of linear equations in Exercises 9 and 10.

9. $x + 2y + z = 6$
$3x - y + z = 2$
$2x + 3y + 4z = 12$

10. $x + y + z = 3$
$2x + 4y - 3z = -6$
$4x + 3y - 5z = -5$

11. Given that

$$\begin{bmatrix} x & 0 & -1 \\ 1 & 4 & 2 \\ 3 & y & 5 \end{bmatrix} = \begin{bmatrix} 2 & 0 & -1 \\ 1 & 4 & 2 \\ 3 & 6 & z \end{bmatrix},$$

find the values of x, y, and z.

Given that

$$A = \begin{bmatrix} 2 & 4 & 1 \\ 5 & 0 & -3 \end{bmatrix} \quad \text{and} \quad B = \begin{bmatrix} 4 & 1 & -2 \\ 3 & 1 & 4 \end{bmatrix},$$

perform the indicated operations in Exercises 12 to 14.

12. $A + B$ **13.** $A - B$ **14.** $3A + 2B$

Given that

$$A = \begin{bmatrix} 2 & 1 & -3 \\ 5 & 2 & 4 \end{bmatrix}, \quad B = \begin{bmatrix} 2 & 3 & 4 \\ 4 & 1 & 3 \\ 3 & 2 & 1 \end{bmatrix}, \quad C = \begin{bmatrix} 5 & 2 & 3 \\ 3 & 1 & 4 \\ 7 & 3 & 2 \end{bmatrix},$$

perform the indicated operations in Exercises 15 to 18.

15. AB **16.** AC **17.** BC **18.** CB

19. Given that

$$A = \begin{bmatrix} 1 & 2 & 2 \\ 3 & 4 & 6 \end{bmatrix}, \quad B = \begin{bmatrix} 2 & 4 & 1 \\ 4 & 1 & 3 \\ 1 & -1 & 2 \end{bmatrix}, \quad C = \begin{bmatrix} -1 & 3 \\ 3 & 3 \\ 5 & 1 \end{bmatrix},$$

show that $A(BC) = (AB)C$.

20. Given that

$$A = \begin{bmatrix} 2 & 1 \\ 3 & 5 \end{bmatrix}, \quad B = \begin{bmatrix} 4 & 7 \\ -2 & 3 \end{bmatrix}, \quad C = \begin{bmatrix} 5 & 1 \\ 2 & 4 \end{bmatrix},$$

show that $A(B + C) = AB + AC$.

Using the Gauss-Jordan elimination method, solve the systems of linear equations in Exercises 21 to 24.

21. $\begin{aligned} x + 2y &= 4 \\ 2x + 3y &= 7 \end{aligned}$ **22.** $\begin{aligned} 2x - y &= 3 \\ 3x + 2y &= 8 \end{aligned}$ **23.** $\begin{aligned} x + 2y + 3z &= 5 \\ 3x - 5y + 4z &= 16 \\ 5x + 8y + 11z &= 19 \end{aligned}$ **24.** $\begin{aligned} x + 2y + z &= 4 \\ 2x + 3y + 4z &= 19 \\ 3x - y + z &= 9 \end{aligned}$

25. Find the inverse of

$$A = \begin{bmatrix} 2 & 5 \\ 5 & 13 \end{bmatrix}$$

and use it to solve the following systems of linear equations:

(a) $\begin{aligned} 2x + 5y &= 19 \\ 5x + 13y &= 49 \end{aligned}$ **(b)** $\begin{aligned} 2x + 5y &= 4 \\ 5x + 13y &= 11. \end{aligned}$

26. Find the inverse of

$$A = \begin{bmatrix} 1 & 1 & 5 \\ 2 & 1 & -1 \\ 3 & 2 & 5 \end{bmatrix}$$

and use A^{-1} to solve the following system of linear equations:

$$x + y + 5z = 7$$
$$2x + y - z = 1$$
$$3x + 2y + 5z = 9.$$

Determine the region satisfied by the system of inequalities in Exercises 27 to 30.

27. $2x + 3y \le 40$
$4x + 2y \le 48$
$x \ge 0, y \ge 0$

28. $x - y \ge 1$
$x + y \ge 5$
$x < 7$

29. $2x + 5y \le 50$
$x + y \le 11$
$2x + y \le 20$
$x \ge 0, y \ge 0$

30. $20x + 15y \ge 1250$
$10x + 10y \ge 700$
$x \ge 0, y \ge 0$

Determine graphically the values of x and y that maximize the objective function P in Exercises 31 and 32 subject to the given linear constraints.

31. $P = 6x + 4y$
$x + 2y \le 240$
$3x + 2y \le 300$
$x \ge 0, y \ge 0$

32. $P = 4x + 5y$
$2x + 3y \le 24$
$2x + y \le 16$
$x \ge 0, y \ge 0$

Determine graphically the values of x and y that minimize the objective function C in Exercises 33 and 34 subject to the given linear constraints.

33. $C = 3x + 4y$
$x \le 4, y \le 3$
$x + y \ge 6$
$x \ge 0, y \ge 0$

34. $C = 6x + 7y$
$3x + 9y \ge 36$
$6x + 2y \ge 24$
$2x + 2y \ge 16$
$x \ge 0, y \ge 0$

35. A company manufactures two kinds of wall clocks, a battery-operated model and an electric model. The company can make up to 500 electric clocks a day and up to 400 battery-operated clocks a day, but it can only make a total of 600 clocks a day. It takes 2 hours to make the electric model and 3 hours to make the battery-operated model. The company's workforce can provide 1400 hours a day. If the company makes a profit of $3 on each electric model and $4 on each battery-operated model, determine the combination of electric and battery-operated clocks that will maximize profit.

36. A small-scale manufacturer has production facilities for producing two different products. Each product requires three different operations: grinding, assembling, and testing. Product 1 requires 30 minutes for grinding, 40 minutes for assembling, and 20 minutes for testing; product 2 requires 15, 80, and 90 minutes for grinding, assembling, and testing, respectively. The production run calls for at least 15 hours of grinding time, at least 40 hours of assembling time, and at least 30 hours of testing time. If product 1 costs $10 and product 2 costs $15 to manufacture, determine the number of units of each product that the firm should produce to minimize the cost of operations.

37. Leslie needs a diet that will supply her with a least 1400 calories, 165 units of vitamin B, and 240 units of calcium. This requirement can be met from two types of food. Each unit of food 1 contains 35 calories, 5 units of vitamin B, and 5 units of calcium and costs $0.05; each unit of food 2 contains 20 calories, 5 units of vitamin B, and 4 units of calcium and costs $0.04. Determine the combination of the two types of food that Leslie should buy to meet her daily requirements at a minimum cost.

38. A factory uses two kinds of petroleum products in its manufacturing process, regular (R) and low sulfur (L). Each gallon of R used emits 0.03 pound of sulfur dioxide and 0.01 pound of lead pollutants, while each gallon of L used emits 0.01 pound of sulfur dioxide and 0.01 pound of lead pollutants. Each gallon of R costs $0.50, and each gallon of L costs $0.60. The factory needs to use at least 100 gallons of petroleum products each day. Federal regulations allow emission of no more than 6 pounds of sulfur dioxide and no more than 4 pounds of lead pollutants. Determine the number of gallons of each type of product the factory should use to minimize its cost.

CHAPTER TEST

Given that

$$A = \begin{bmatrix} 3 & 4 & 6 \\ -1 & 2 & 8 \end{bmatrix} \quad \text{and} \quad B = \begin{bmatrix} -1 & 5 & 8 \\ 2 & -3 & 10 \end{bmatrix},$$

perform the indicated operations in Exercises 1 to 3.

1. $A + B$ **2.** $A - B$ **3.** $3A + 5B$

Given that

$$A = \begin{bmatrix} 5 & 7 \\ -1 & 6 \end{bmatrix}, \quad B = \begin{bmatrix} 4 & 6 \\ 2 & -1 \end{bmatrix}, \quad \text{and} \quad C = \begin{bmatrix} 5 & -1 \\ 3 & 4 \end{bmatrix},$$

perform the indicated operations in Exercises 4 to 12.

4. AB **5.** BA **6.** AC

7. $AB + AC$ **8.** $A(B + C)$ **9.** BC

10. $A(BC)$ **11.** $(AB)C$ **12.** $BA + BC$

Determine the solution set of the following linear systems of equations:

13. $2x + y = 5$
$3x + 2y = 6$

14. $\begin{aligned} x + 2y + z &= 6 \\ 2x + 3y + 4z &= 12 \\ 3x - y + z &= 2 \end{aligned}$

15. Using the Gauss-Jordan elimination method, solve the following system of linear equations:

$$x + 2y = 8$$
$$2x + 3y = 13.$$

16. Find the inverse of

$$A = \begin{bmatrix} 5 & 3 \\ 3 & 2 \end{bmatrix},$$

and use A^{-1} to solve the following system of linear equations:

$$5x + 3y = 8$$
$$3x + 2y = 5.$$

17. Solve the graph of the linear inequality $2x - 3y \geq 6$.

18. Determine graphically the values of x and y that maximize $P = 6x + 4y$ subject to the following linear constraints:

$$x + 2y \leq 240$$

$$3x + 2y \leq 300$$

$$x \geq 0, \qquad y \geq 0.$$

19. Determine graphically the values of x and y that minimize $C = 3x + 4y$ subject to the following linear constraints:

$$x \leq 4, \qquad y \leq 3, \qquad x + y \geq 6, \qquad x \geq 0, \qquad y \geq 0.$$

20. A pound of fish contains 3 units of carbohydrates, 6 units of vitamins, and 12 units of protein and costs $1.75, whereas a quart of milk contains 2 units of carbohydrates, 3 units of vitamins, and 4 units of proteins and costs $0.60. If the minimum daily requirements are 12 units of carbohydrates, 15 units of vitamins, and 36 units of protein, determine the best possible combination of fish and milk to minimize the cost.

SEQUENCES AND SERIES

10

INTRODUCTION

The study of sequences is the study of another type of function. Traditionally, sequences have played an important role in mathematics and have numerous applications in other fields. Suppose, for example, that the number of bacteria present in a certain culture is growing in such a way that the number is 1.35 times as large as it was 1 hour ago. Given that the number of bacteria is 1000 at 1:00 P.M., how many would there be at 5:00 P.M.? Or suppose that the annual depreciation of a certain automobile is 20 percent of its value at the beginning of the year. If the original cost of the automobile is $14,000, what will be its resale value after 3 years? Sequences deal with questions of this type.

In this chapter, we study two types of sequences—arithmetic and geometric—as well as the sums of such sequences, known as **series.** We examine the behavior of an infinite sequence as the number of terms gets larger and larger, and we look at the sums of infinite geometric sequences. Finally, we consider mathematical induction.

10.1   SEQUENCES AND SERIES

infinite sequence   An **infinite sequence** is a real-valued function whose domain consists of positive integers. Intuitively, in infinite sequence is an ordered set of numbers such as

$$4, 7, 10, 13, 16, \ldots.$$

We label each number of the sequence by its position, using a variable such as a. In other words,

$$a_1 = 4, \qquad a_2 = 7, \qquad a_3 = 10, \qquad a_4 = 13, \ldots.$$

The subscripts of the symbols $a_1, a_2, a_3, \ldots$ are called **indices;** they locate the numbers in the sequence. The numbers $a_1, a_2, a_3, \ldots$ are called the *terms* of the sequence $\{a_n\}$, and a_n is called the **general term.**

Definition 10.1.1   An infinite sequence, $\{a_n\}$, is a *real-valued function* that has the set of positive integers as its domain.

Example 10.1   Find the first six terms of the sequence defined by

$$a_n = n^2 + 3.$$

Solution: If we set $n = 1, 2, 3, 4, 5,$ and 6 in the general term

$$a_n = n^2 + 3,$$

we have

$$a_1 = 1^2 + 3 = 4 \qquad a_2 = 2^2 + 3 = 7 \qquad a_3 = 3^2 + 3 = 12$$

$$a_4 = 4^2 + 3 = 19 \qquad a_5 = 5^2 + 3 = 28 \qquad a_6 = 6^2 + 3 = 39.$$

Thus the first six terms of the sequence are

$$4, 7, 12, 19, 28, 39.$$

Example 10.2

Find the first nine terms of the sequence defined by

$$a_n = \frac{n}{n+1}.$$

Solution: We substitute successively $n = 1, 2, 3, 4, 5, 6, 7, 8,$ and 9 in the general term a_n. We compute the first nine terms of the sequence as follows:

$$a_1 = \frac{1}{1+1} = \frac{1}{2} \qquad a_2 = \frac{2}{2+1} = \frac{2}{3} \qquad a_3 = \frac{3}{3+1} = \frac{3}{4}$$

$$a_4 = \frac{4}{4+1} = \frac{4}{5} \qquad a_5 = \frac{5}{5+1} = \frac{5}{6} \qquad a_6 = \frac{6}{6+1} = \frac{6}{7}$$

$$a_7 = \frac{7}{7+1} = \frac{7}{8} \qquad a_8 = \frac{8}{8+1} = \frac{8}{9} \qquad a_9 = \frac{9}{9+1} = \frac{9}{10}.$$

Hence the first nine terms of the sequence are

$$\frac{1}{2}, \frac{2}{3}, \frac{3}{4}, \frac{4}{5}, \frac{5}{6}, \frac{6}{7}, \frac{7}{8}, \frac{8}{9}, \frac{9}{10}.$$

Another equally acceptable way to specify a sequence is to list the first few terms as long as the general term of the sequence can be determined. Consider, for example, the first five terms

$$a_1 = \frac{1}{2}, \qquad a_2 = \frac{1}{4}, \qquad a_3 = \frac{1}{8}, \qquad a_4 = \frac{1}{16}, \qquad a_5 = \frac{1}{32}.$$

Does it seem reasonable to guess that $a_6 = \frac{1}{64}$? If that number comes to your mind, you have seen the pattern. Can you now determine the general term a_n? Note that

$$a_1 = \left(\frac{1}{2}\right)^1, \qquad a_2 = \left(\frac{1}{2}\right)^2, \qquad a_3 = \left(\frac{1}{2}\right)^3, \qquad a_4 = \left(\frac{1}{2}\right)^4, \qquad a_5 = \left(\frac{1}{2}\right)^5.$$

Observe that the power of $\left(\frac{1}{2}\right)$ in each of the terms is the same as the index of the symbol $a_1, a_2, a_3, \ldots$. Thus

$$a_n = \left(\frac{1}{2}\right)^n.$$

Once the general term a_n is known, we can find any term in the sequence $\{a_n\}$ for a given value of n. For example,

$$a_9 = \left(\frac{1}{2}\right)^9 = \frac{1}{512}, \qquad a_{10} = \left(\frac{1}{2}\right)^{10} = \frac{1}{1024},$$

and so on.

Example 10.3

Determine the general term a_n in the sequence

$$\frac{1}{2}, \frac{1}{4}, \frac{1}{6}, \frac{1}{8}, \frac{1}{10}, \ldots$$

Solution: Observe that

$$a_1 = \frac{1}{2} = \frac{1}{2\cdot 1} \qquad a_2 = \frac{1}{4} = \frac{1}{2\cdot 2} \qquad a_3 = \frac{1}{6} = \frac{1}{2\cdot 3}$$

$$a_4 = \frac{1}{8} = \frac{1}{2\cdot 4} \qquad a_5 = \frac{1}{10} = \frac{1}{2\cdot 5},$$

and so on. Hence the general term of the sequence is

$$a_n = \frac{1}{2n}.$$

Sigma Notation

The symbol $\sum$, the Greek letter sigma, is used to indicate a sum, for instance,

$$\sum_{i=1}^{5} i^2,$$

where following the $\sum$ is a term, i^2. The **index of summation** i takes on successive integral values to form the different terms of the sum. The equation below the sigma, $i = 1$, indicates the index of the summation and the first integral value of the index. The number above $\sum$ represents the last integral value of the index.

Example 10.4

Find

$$\sum_{i=1}^{5} i^2.$$

Solution: To evaluate the sum, we substitute successively $i = 1, 2, 3, 4,$ and 5 in the term i^2 and then add the resulting terms. Thus,

$$\sum_{i=1}^{5} i^2 = 1^2 + 2^2 + 3^2 + 4^2 + 5^2 = 1 + 4 + 9 + 16 + 25 = 55.$$

Example 10.5

The first six terms in the sequence defined by

$$a_n = 3n + 2$$

are 5, 8, 11, 14, 17, and 20. The sum of these terms is represented by

$$\sum_{n=1}^{6} (3n + 2) = 5 + 8 + 11 + 14 + 17 + 20 = 75.$$

Partial Sum of Sequences

With each sequence $\{a_n\}$, we can associate another sequence $\{S_n\}$ defined by the equation

$$S_n = \sum_{j=1}^{n} a_j$$
$$= a_1 + a_2 + a_3 + \cdots + a_n.$$

sequence of partial sums

This sequence is called the **sequence of partial sums** of the sequence $\{a_n\}$. Evidently,

$$S_1 = a_1 \qquad \text{(first term of sequence)}$$
$$S_2 = a_1 + a_2 \qquad \text{(sum of first two terms)}$$
$$S_3 = a_1 + a_2 + a_3 \qquad \text{(sum of first three terms)}$$
$$\vdots$$
$$S_n = a_1 + a_2 + a_3 + \cdots + a_n. \qquad \text{(sum of first } n \text{ terms)}$$

Example 10.6

Find S_1, S_2, S_3, S_4, S_5, and S_6 for the following sequences:

(a) $a_n = 3n + 2$

(b) $a_n = \left(\dfrac{1}{2}\right)^{n-1}$.

Solution:

(a) Let $n = 1, 2, 3, 4, 5,$ and 6. Then

$$a_1 = 5, \qquad a_2 = 8, \qquad a_3 = 11, \qquad a_4 = 14, \qquad a_5 = 17, \qquad a_6 = 20.$$

Thus

$$S_1 = a_1 = 5, \qquad S_2 = a_1 + a_2 = 5 + 8 = 13,$$
$$S_3 = a_1 + a_2 + a_3 = S_2 + a_3 = 13 + 11 = 24.$$

Using the relation

$$S_n = S_{n-1} + a_n \qquad (n > 1),$$

we obtain

$$S_4 = S_3 + a_4 = 24 + 14 = 38$$
$$S_5 = S_4 + a_5 = 38 + 17 = 55$$
$$S_6 = S_5 + a_6 = 55 + 20 = 75.$$

(b) In the sequence $a_n = \left(\dfrac{1}{2}\right)^{n-1}$, we have

$$S_1 = 1$$

$$S_2 = 1 + \frac{1}{2} = \frac{3}{2}$$

$$S_3 = S_2 + a_3 = \frac{3}{2} + \frac{1}{4} = \frac{7}{4}$$

$$S_4 = S_3 + a_4 = \frac{7}{4} + \frac{1}{8} = \frac{15}{8}$$

$$S_5 = S_4 + a_5 = \frac{15}{8} + \frac{1}{16} = \frac{31}{16}$$

$$S_6 = S_5 + a_6 = \frac{31}{16} + \frac{1}{32} = \frac{63}{32}.$$

EXERCISES

Find the first five terms in Exercise 1 to 26.

1. $a_n = 2n$

2. $a_n = 2n + 5$

3. $a_n = 3n + 1$

4. $a_n = 3n - 2$

5. $a_n = 4n - 5$

6. $a_n = 5n - 4$

7. $a_n = n^2$

8. $a_n = 2n^2 - 1$

9. $a_n = (n - 1)(n + 1)$

10. $a_n = n(n + 2)$

11. $a_n = 2^n$

12. $a_n = 2^n - 1$

13. $a_n = 3^n$

14. $a_n = 3^{n-1}$

15. $a_n = \left(\dfrac{1}{2}\right)^{n-1}$

16. $a_n = \left(\dfrac{1}{3}\right)^{n+1}$

17. $a_n = (-1)^n$

18. $a_n = \dfrac{(-1)^n}{n + 1}$

19. $a_n = \dfrac{n(n + 1)}{2}$

20. $a_n = \dfrac{2n + 1}{3n + 2}$

21. $a_n = \dfrac{n}{n^2 + 1}$

22. $a_n = \dfrac{n^2 + 1}{n^2 + 2}$

23. $a_n = \dfrac{n(n + 1)(2n + 1)}{6}$

24. $a_n = (-1)^n + 1$

25. $a_n = 1 - (-1)^n$ **26.** $a_n = \dfrac{(-1)^n(n+1)}{n+2}$

Determine the general term a_n in Exercises 27 to 40.

27. $2, 5, 8, 11, 14, \ldots, a_n, \ldots$

28. $4, 7, 10, 13, 16, \ldots, a_n, \ldots$

29. $3, 5, 7, 9, 11, \ldots, a_n, \ldots$

30. $5, 10, 15, 20, 25, \ldots, a_n, \ldots$

31. $2, 4, 8, 16, 32, \ldots, a_n, \ldots$

32. $3, 9, 27, 81, 243, \ldots, a_n, \ldots$

33. $1, -1, 1, -1, 1, \ldots, a_n, \ldots$

34. $4, 16, 64, 256, 1024, \ldots, a_n, \ldots$

35. $1, \dfrac{1}{2}, \dfrac{1}{4}, \dfrac{1}{8}, \dfrac{1}{16}, \ldots, a_n, \ldots$

36. $1, \dfrac{1}{3}, \dfrac{1}{9}, \dfrac{1}{27}, \dfrac{1}{81}, \ldots, a_n, \ldots$

37. $2, \dfrac{3}{2}, \dfrac{4}{3}, \dfrac{5}{4}, \dfrac{6}{5}, \ldots, a_n, \ldots$

38. $\dfrac{1}{3}, \dfrac{1}{6}, \dfrac{1}{9}, \dfrac{1}{12}, \dfrac{1}{15}, \ldots, a_n, \ldots$

39. $\dfrac{-1}{2}, \dfrac{1}{4}, \dfrac{-1}{8}, \dfrac{1}{16}, \dfrac{-1}{32}, \ldots, a_n, \ldots$

40. $\dfrac{1}{2}, \dfrac{2}{3}, \dfrac{3}{4}, \dfrac{4}{5}, \dfrac{6}{5}, \ldots, a_n, \ldots$

Evaluate the sums in Exercises 41 to 50.

41. $\displaystyle\sum_{i=1}^{4} i$

42. $\displaystyle\sum_{i=1}^{5} (i+2)$

43. $\displaystyle\sum_{i=1}^{5} (2i+1)$

44. $\displaystyle\sum_{i=1}^{6} (3i-1)$

45. $\displaystyle\sum_{i=1}^{6} i^2$

46. $\displaystyle\sum_{i=1}^{6} (i+1)^2$

47. $\displaystyle\sum_{j=1}^{5} (j-1)(j+1)$

48. $\displaystyle\sum_{j=1}^{6} j(j+3)$

49. $\displaystyle\sum_{j=1}^{6} (j+4)(j+5)$

50. $\displaystyle\sum_{j=1}^{5} (j^2+1)$

Find the partial sums $S_1, S_2, S_3, S_4,$ and S_5 in Exercises 51 to 66.

51. $a_n = 2n+1$

52. $a_n = 3n-1$

53. $a_n = 4n+3$

54. $a_n = 5n+2$

55. $a_n = n^2$

56. $a_n = n(n+1)$

57. $a_n = (n-1)(n+1)$

58. $a_n = (n+1)(n+2)$

59. $a_n = (-1)^n$

60. $a_n = 2^n$

61. $a_n = 3^{n-1}$

62. $a_n = \dfrac{1}{n}$

63. $a_n = \dfrac{n-1}{n}$

64. $a_n = \dfrac{n+1}{2}$

65. $a_n = \left(\dfrac{1}{2}\right)^n$

66. $a_n = \left(\dfrac{1}{3}\right)^{n-1}$

10.2

ARITHMETIC SEQUENCES

Consider the sequence

$$4, 7, 10, 13, 16, 19, \ldots.$$

Each term in the sequence after the first is obtained by adding 3 to the preceding term. The first few terms of the sequence are

$$a_1 = 4, \qquad a_2 = 4 + 3 = 7, \qquad a_3 = 7 + 3 = 10,$$

and so on.

Definition 10.2.1

> A sequence in which the difference between any term and the succeeding term is constant is called an *arithmetic sequence*, or *arithmetic progression*. The constant difference is called the *common difference*.

An arithmetic sequence is completely determined if the first term a_1 and the common difference d are known. In general, an arithmetic sequence is of the form

$$a_1, a_2, a_3, a_4, \ldots,$$

where

$$a_2 = a_1 + d$$

$$a_3 = a_2 + d = (a_1 + d) + d = a_1 + 2d$$

$$a_4 = a_3 + d = (a_1 + 2d) + d = a_1 + 3d.$$

Note that the coefficient of d in any term is 1 less than the number of the term in the sequence. Thus the general term a_n in an arithmetic sequence is

$$a_n = a_1 + (n - 1)d, \qquad\qquad \textbf{(10.1)}$$

where a_1 is the first term in the sequence and d is the common difference.

Example 10.7

Find the eleventh term of the sequence

$$5, 9, 13, 17, 21, \ldots.$$

Solution: Here $a_1 = 5$, $d = 4$, and $n = 11$. Using (10.1), we obtain

$$a_{11} = 5 + (11 - 1)4 = 45.$$

Example 10.8

Find the 23d term in an arithmetic sequence whose 4th and 9th terms are 16 and 31, respectively.

Solution: If a_1 is the first term and d is the common difference, then

$$a_4 = a_1 + 3d = 16$$

$$a_9 = a_1 + 8d = 31.$$

Subtracting the first equation from the second, we get $5d = 15$, or $d = 3$. Substituting $d = 3$ in the equation $a_1 + 3d = 16$, we have $a_1 = 7$. Hence

$$a_{23} = a_1 + 22d$$
$$= 7 + (22)(3) = 73.$$

Example 10.9

Determine three consecutive numbers in an arithmetic sequence such that their sum is 21 and their product is 315.

Solution: Let $a - d$, a, and $a + d$ be three consecutive numbers in an arithmetic sequence. Then

$$(a - d) + a + (a + d) = 21$$
$$3a = 21$$
$$a = 7.$$

Also,

$$a(a - d)(a + d) = 315$$
$$a(a^2 - d^2) = 315.$$

Since $a = 7$, we have

$$7(49 - d^2) = 315.$$

Solving this equation, we obtain $d = \pm 2$. Thus the numbers are

$$5, 7, 9 \quad \text{or} \quad 9, 7, 5.$$

Proposition 10.2.1

If $a_1, a_2, a_3, \ldots, a_n, \ldots$ is an arithmetic sequence, where d is the common difference, then the sum of the first n terms is given by

$$S_n = \frac{n}{2}(a_1 + a_n). \tag{10.2}$$

Proof: Observe that each term in an arithmetic sequence is obtained by adding the common difference d to the term that precedes it or by subtracting d from the term that follows it. Thus

$$S_n = a_1 + (a_1 + d) + (a_1 + 2d) + \cdots + (a_n - d) + a_n.$$

Writing S_n in reverse order, we have

$$S_n = a_n + (a_n - d) + (a_n - 2d) + \cdots + (a_1 + d) + a_1.$$

Adding the above statements and rearranging terms, we have

$$2S_n = (a_1 + a_n) + (a_1 + a_n) + \cdots + (a_1 + a_n)$$
$$= n(a_1 + a_n),$$

or

$$S_n = \frac{n}{2}(a_1 + a_n).$$

Let us look at an example.

Example 10.10

There are 60 terms in an arithmetic sequence, of which the first term is 9 and the last term is 127. Find S_{60}.

Solution: Here

$$a_1 = 9, \qquad a_n = 127, \qquad \text{and} \qquad n = 60.$$

Using (10.2), we have

$$S_{60} = \frac{60}{2}(9 + 127)$$

$$= 4080.$$

Equation (10.2) is useful if the first term a_1, the last term a_n, and the number of terms n are known. If the last term a_n is *not* known, then we use the following proposition.

Proposition 10.2.2

The sum of the first n terms of an arithmetic sequence with a_1 as the first term and d the common difference is given by

$$S_n = \frac{n}{2}[2a_1 + (n-1)d]. \qquad\qquad (10.3)$$

Proof: Substituting

$$a_n = a_1 + (n-1)d$$

in (10.2), we obtain

$$S_n = \frac{n}{2}[a_1 + a_1 + (n-1)d]$$

$$= \frac{n}{2}[2a_1 + (n-1)d],$$

where a_1 is the first term and d is the common difference.

Example 10.11

Find the sum of the first 80 odd numbers.

Solution: The problem is to find the sum of the first 80 terms of the arithmetic sequence

$$1, 3, 5, 7, \ldots .$$

Clearly, $a_1 = 1$, $d = 2$, and $n = 80$. Using (10.3), we have

$$S_{80} = \frac{80}{2}[2(1) + (80 - 1)(2)]$$

$$= 6400.$$

Example 10.12

A saleswoman in a shoe store sold seven pairs of shoes the day she was hired; each day thereafter, she sold four more pairs of shoes than she had sold the day before. Determine her sale on the 20th day. What are her total sales after 20 days?

Solution: The number of pairs of shoes sold each day constitute an arithmetic sequence with first term $a_1 = 7$ and a common difference of $d = 4$. Using (10.1), we obtain

$$a_{20} = 7 + (20 - 1)4 = 83.$$

Since the last term is known, we may use (10.2) and obtain

$$S_{20} = \frac{20}{2}(7 + 83) = 900.$$

Therefore she sold 83 pairs on her 20th day and a total of 900 after her first 20 days.

Example 10.13

Mario deposited $1000 in his checking account at the beginning of the semester. He withdrew $25 the first week, $30 the second week, $35 the third week, and so on until his money ran out. How much money is in his checking account after 12 weeks? How long will the initial deposit of $1000 last?

Solution: If we use (10.3) with $a_1 = 25$, $d = 5$, and $n = 12$, the total withdrawal during the first 12 weeks is given by

$$S_{12} = \frac{12}{2}[2(25) + (12 - 1)(5)]$$

$$= 630.$$

Thus the amount on deposit is $1000 - $630 = $370.
   To determine how long the deposit of $1000 will suffice, suppose that n is the number of weeks the money will last. Then $S_n = 1000$, and again by (10.3) we have

$$1000 = \frac{n}{2}[2(25) + (n - 1)(5)]$$

or

$$\frac{n}{2}(5n + 45) = 1000$$

$$n(n + 9) = 400$$

$$n^2 + 9n - 400 = 0$$

$$(n + 25)(n - 16) = 0.$$

It follows that $n = 16$ or $n = -25$. Since n must be positive, we conclude that $n = 16$. Therefore, his money will last 16 weeks.

EXERCISES

Find the 10th term for the sequences in Exercises 1 to 10.

1. 4, 7, 10, 13, . . .   **2.** 5, 9, 13, 17, . . .   **3.** 3, 7, 11, 15, . . .   **4.** 15, 13, 11, 9, . . .

5. $-8, -6, -4, -2, \ldots$ **6.** $-6, -1, 4, 9, \ldots$ **7.** 8, 5, 2, -1, . . .   **8.** 14, 12, 10, 8, . . .

9. 27, 22, 17, 12, . . .   **10.** 18, 12, 6, 0, . . .

Find a_{15} and S_{15} for the arithmetic sequences in Exercises 11 to 18.

11. 8, 11, 14, 17, . . .   **12.** 5, 7, 9, 11, . . .   **13.** 5, 9, 13, 17, . . .   **14.** 12, 10, 8, 6, . . .

15. 15, 12, 9, 6, . . .   **16.** 16, 11, 6, 1, . . .   **17.** 15, 11, 7, 3, . . .   **18.** 21, 17, 13, 9, . . .

Find the sum of the arithmetic sequences in Exercises 19 to 22.

19. The first 100 positive integers.   **20.** The first 60 odd positive integers.

21. All even numbers from 2 through 100.   **22.** All odd numbers from 13 through 39, inclusive.

23. Insert five terms between 3 and 15 to obtain an arithmetic sequence with seven numbers.

24. Insert seven terms between 5 and 33 to obtain an arithmetic sequence with nine terms.

25. The 3rd and 16th terms of an arithmetic sequence are 13 and 78, respectively. Determine the first term and the common difference. Determine also a_{20} and S_{20}.

26. There are 30 terms in an arithmetic sequence of which the first is 6 and the last is 93. Determine S_{30} and S_{40}.

27. The sum of three consecutive terms in an arithmetic sequence is 24, and their product is 480. Find the numbers.

28. The sum of three consecutive terms in an arithmetic sequence is 15, and the sum of their squares is 83. Find the numbers.

29. Jesse runs 1 mile the first day and increases her daily run by 0.1 mile every day. How many miles will she run on the fifth day? How far has she run from the beginning to the end of the 15th day?

30. Bruce starts a Christmas club by depositing $0.25 the first week, and in each subsequent week he deposits $0.10 more than the week before. How much money will he have in his account after 50 weeks?

31. Scott starts a savings account by depositing $100. Each month thereafter he deposits $25 more than the month before. What are his total savings (without interest) at the end of 2 years?

32. A salesman in a department store had sales of $150 his first day on the job. On each subsequent working day, he had sales of $15 more than he had the day before. Determine his total sales at the end of 15 working days.

33. Ada saved $300 the first year she was hired. Each year thereafter she saved $100 more than the year before. What are her total savings (without interest) at the end of 12 years?

34. Peter is hired with an initial salary of $12,000 in a school system. If his salary increases $600 annually, how many years will it take him to earn $15,000 annually?

35. A drilling company charges a flat $50 setup charge plus $3 for the first foot of a well, $5 for the second foot, $7 for the third foot, and so on. What will the cost be to drill a well 150 feet deep?

36. A stack of poles has 15 in the bottom row, 14 in the second row, 13 in the third row, and so on. Determine the total number of poles if there is only one pole in the top row.

37. Judy stacks 25 blocks in the bottom row, 23 in the second row, 21 in the third row, and so on. Determine the total number of blocks if there are only three blocks in the top row.

38. Leon stacks 20 blocks in the bottom row, 19 in the second row, 18 in the third row, and so on. Determine the number of rows if he has a stack of 155 blocks.

39. Elaine's father has deposited $4050 in her checking account at the beginning of her school year. She withdraws $25 the first week, $30 the second week, $35 the third week, and so on until her money runs out.
    (a) How much money does she withdraw from her account in the 10th week?
    (b) How much money is in her checking account after 18 weeks?
    (c) What is the balance in her account at the end of 36 weeks?

40. Show that the sum of the first n positive integers is $\dfrac{n(n+1)}{2}$.

41. Show that the sum of the first n odd positive integers is n^2.

10.3 GEOMETRIC SEQUENCES

Consider the sequences

(a) 1, 2, 4, 8, 16, . . .

(b) 3, 9, 27, 81, 243, . . .

(c) $2, -\dfrac{1}{2}, \dfrac{1}{8}, -\dfrac{1}{32}, \dfrac{1}{128}, \ldots$

Each term after the first is the product of the preceding term and a constant. This constant is 2 in the first sequence, 3 in the second sequence, and $-\frac{1}{4}$ in the third sequence.

Definition 10.3.1

A sequence in which the ratio of any term to the preceding term is constant is called a **geometric sequence**, or **geometric progression**. The constant is called the **common ratio**.

Thus if a_1 is the first number and r is the common ratio, then

$$a_2 = a_1 r,$$

$$a_3 = a_2 r = (a_1 r)r = a_1 r^2,$$

$$a_4 = a_3 r = (a_1 r^2)r = a_1 r^3,$$

and so on. Observe that the power of r is 1 less than the number of terms in the geometric sequence. Thus the general term a_n in a geometric sequence is given by

$$a_n = a_1 r^{n-1}.$$ (10.4)

Example 10.14   Find the sixth term of a geometric sequence whose first term is 4 and whose common ratio is 3.

Solution: Here $a_1 = 4$, $r = 3$, and $n = 6$. Thus

$$
\begin{aligned}
a_6 &= a_1 r^5 \\
&= (4)(3^5) \\
&= (4)(243) \\
&= 972.
\end{aligned}
$$

Example 10.15   Find the 11th term in a sequence whose 3rd term is 12 and whose 8th term is 384.

Solution: We have

$$a_3 = a_1 r^2 = 12 \qquad \text{and} \qquad a_8 = a_1 r^7 = 384.$$

Thus

$$\frac{a_1 r^7}{a_1 r^2} = \frac{384}{12} = 32,$$

that is,

$$r^5 = 32 \qquad \text{or} \qquad r = 2.$$

Since

$$a_1 r^2 = 12 \qquad \text{and} \qquad r = 2,$$

it follows that

$$a_1 = 3.$$

Therefore

$$a_{11} = a_1 r^{10} = 3(2^{10}) = 3(1024) = 3072.$$

Proposition 10.3.1

The sum of the first n terms in a geometric sequence with a_1 as the first term and r the common ratio is given by

$$S_n = \begin{cases} \dfrac{a_1(1 - r^n)}{(1 - r)} & r \neq 1 \\ na_1 & r = 1. \end{cases} \qquad (10.5)$$

Proof: Let

$$S_n = a_1 + a_1 r + a_1 r^2 + \cdots + a_1 r^{n-2} + a_1 r^{n-1}. \qquad (10.6)$$

Multiplying this equation by r, we obtain

$$rS_n = a_1 r + a_1 r^2 + a_1 r^3 + \cdots + a_1 r^{n-1} + a_1 r^n. \qquad (10.7)$$

Subtracting (10.7) from (10.6), we have

$$S_n(1 - r) = a_1(1 - r^n),$$

or

$$S_n = \frac{a_1(1 - r^n)}{(1 - r)} \qquad (r \neq 1).$$

If $r = 1$, then from (10.6), we have

$$S_n = a_1 + a_1 + a_1 + \cdots + a_1$$
$$= na_1.$$

Example 10.16

Find the sum of the first six terms of the geometric sequence

$$4, 12, 36, 108, \ldots .$$

Solution: Since $a_1 = 4$, $r = 3$, and $n = 6$, the required sum is

$$S_6 = \frac{4(1 - 3^6)}{(1 - 3)}$$

$$= 1456.$$

Example 10.17

The profit of a department store has shown an annual increase of 6 percent. Assuming that the current market trends continue, what will the store's annual profit be in the fifth year, given that the first year's profit was $25,000? Determine also the total profit for the first 5 years.

Solution: Here $a_1 = \$25,000$ and $n = 5$. Since the profit increases at an annual rate of 6 percent, $r = 1.06$. Thus the profit in the fifth year is given by

$$a_5 = a_1 r^4 = 25,000(1.06)^4$$
$$= 25,000(1.262477)$$
$$= 31,561.92.$$

That is, the profit in the fifth year is $31,561.92. The total profit for the first 5 years is

$$S_5 = \frac{25,000[1 - (1.06)^5]}{1 - 1.06}$$

$$= \frac{25,000(1 - 1.338226)}{1 - 1.06}$$

$$= \frac{25,000(-0.338226)}{-0.06}$$

$$= 140,927.50.$$

Therefore, the profit for the first 5 years is $140,927.50.

EXERCISES

Find the eighth term for the geometric sequences in Exercises 1 to 10.

1. $2, 6, 18, 54, \ldots$

2. $4, 8, 16, 32, \ldots$

3. $-1, 3, -9, 27, \ldots$

4. $-2, 8, -32, 128, \ldots$

5. $1, 4, 16, 64, \ldots$

6. $3, 9, 27, 81, \ldots$

7. $1, \dfrac{1}{2}, \dfrac{1}{4}, \dfrac{1}{8}, \ldots$

8. $1, \dfrac{1}{3}, \dfrac{1}{9}, \dfrac{1}{27}, \ldots$

9. $1, \dfrac{2}{3}, \dfrac{4}{9}, \dfrac{8}{27}, \ldots$

10. $1, \dfrac{3}{4}, \dfrac{9}{16}, \dfrac{27}{64}, \ldots$

Find the sixth term of the geometric sequence whose first term a_1 and common ratio r are given in Exercises 11 to 16.

11. $a_1 = 3, r = 2$

12. $a_1 = 2, r = 3$

13. $a_1 = 4, r = 3$

14. $a_1 = 5, r = 4$

15. $a_1 = 27, r = \dfrac{1}{3}$

16. $a_1 = 3, r = -\dfrac{1}{3}$

Find the sum of the first nine terms in the geometric sequences in Exercises 17 to 24.

17. $2, 4, 8, 16, \ldots$

18. $1, 3, 9, 27, \ldots$

19. $2, 10, 50, 250, \ldots$

20. $16, 8, 4, 2, \ldots$

21. $64, 16, 4, 1, \ldots$

22. $8, 4, 2, 1, \ldots$

23. $6, 3, \dfrac{3}{2}, \dfrac{3}{4}, \ldots$

24. $9, 3, 1, \dfrac{1}{3}, \ldots$

Find a_{10} and S_{10} in the geometric sequences in Exercises 25 to 28, given the first term a_1 and the common ratio r.

25. $a_1 = 3, r = 2$ **26.** $a_1 = 2, r = 3$ **27.** $a_1 = 27, r = \dfrac{1}{3}$ **28.** $a_1 = 128, r = \dfrac{1}{2}$

29. Determine the geometric sequence whose third term is 12 and whose sixth term is 96.

30. Determine the geometric sequence whose fifth term is 80 and whose eighth term is 640.

31. A department store offered Anita a job with a starting salary of $15,000 and a 9 percent annual increase for the next 10 years. How much money will she be making in her eighth year of service with the store?

32. The annual depreciation of a certain typewriter is 30 percent of its value at the beginning of the year. If the new typewriter costs $750, what will its resale value be in the sixth year of its use?

33. The annual depreciation of a certain automobile is 25 percent of its value at the beginning of each year. If the original cost of the automobile is $8000, what will the resale value of the car be after 5 years?

34. A refrigerator costs $1200 and has an annual depreciation of 40 percent. What will its resale value be in the fourth year of its use?

35. A tennis ball falling from a height of 64 feet rebounds one-fourth of its preceding height after each fall. What distance has the ball traveled when it hits the ground for the sixth time?

36. A ball thrown vertically into the air 60 feet falls and rebounds 40 feet. It falls again and rebounds $26\dfrac{2}{3}$ feet, and so on. Find the total distance traveled by the ball when it hits the ground for the 10th time.

37. The profit of a grocery store has shown an annual increase of 5 percent. Assuming that current market trends continue, what will the store's annual profit be in the third year, given that the first year's profit was $20,000? Determine also the total profit for the first 5 years.

38. The expenditure of the Saunders Smelting Company for control of air pollution was $15,000 in 1980. Assuming that the expenditure increases 8 percent annually, what will the company's annual expenditure be in 1986? Determine also the total amount expended from 1980 to 1986.

10.4

LIMIT OF A SEQUENCE

In the preceding two sections, we discussed arithmetic and geometric sequences. Now we examine intuitively the behavior of the general term a_n of an infinite sequence when n becomes large.

Example 10.18

Consider the infinite geometric sequence

$$\frac{1}{2}, \frac{1}{4}, \frac{1}{8}, \frac{1}{16}, \ldots, \left(\frac{1}{2}\right)^n, \ldots$$

In this sequence,

$$a_1 = \frac{1}{2}, \qquad a_2 = \frac{1}{4}, \qquad a_3 = \frac{1}{8}, \qquad \ldots, \qquad a_n = \left(\frac{1}{2}\right)^n, \ldots$$

Since a sequence is a special type of function, we can graph it in the usual coordinate system, as shown in Figure 10.1.

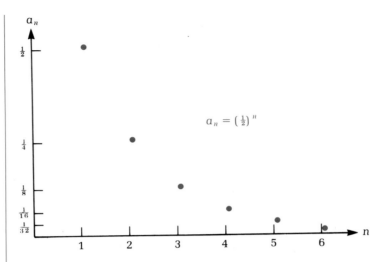

FIGURE 10.1

Note that as the successive terms in the sequence get smaller and smaller, they come closer and closer to zero despite the fact that no term in the sequence is actually zero. Accordingly, we write

$$\left(\frac{1}{2}\right)^n \to 0 \qquad \text{or} \qquad \lim_{n \to \infty} \left(\frac{1}{2}\right)^n = 0$$

and say that as n gets larger and larger or as n approaches infinity, the sequence $\left(\frac{1}{2}\right)^n$ *converges* to 0 or has *limit* 0.

Example 10.19

Consider the infinite sequence

$$1, \frac{1}{2}, \frac{1}{3}, \frac{1}{4}, \ldots, \frac{1}{n}, \ldots$$

Here

$$a_1 = 1, \qquad a_2 = \frac{1}{2}, \qquad a_3 = \frac{1}{3}, \qquad \ldots, \qquad a_n = \frac{1}{n}, \ldots$$

Again, as n gets larger and larger, the terms of the sequence come closer and closer to zero (yet no term in the sequence is actually zero). Intuitively, it appears that the terms of the sequence converge to zero as n approaches infinity. Thus

$$\lim_{n \to \infty} \left(\frac{1}{n}\right) = 0.$$

Example 10.20

Consider the infinite sequence

$$3, \frac{7}{2}, \frac{11}{3}, \frac{15}{4}, \ldots, \frac{4n-1}{n}, \ldots$$

Here

$$a_1 = 3, \quad a_2 = \frac{7}{2}, \quad a_3 = \frac{11}{3}, \quad a_4 = \frac{15}{4}, \quad a_5 = \frac{19}{5}, \ldots,$$

$$a_{10} = \frac{39}{10}, \quad \ldots, \quad a_{50} = \frac{199}{50}, \quad \ldots, \quad a_{100} = \frac{399}{100}, \quad \ldots,$$

$$a_{1000} = \frac{3999}{1000}, \ldots$$

The pattern shows the general term a_n to be $\dfrac{4n-1}{n}$. As n takes on larger and larger values, the terms of the sequence get closer and closer to 4. Thus

$$\lim_{n \to \infty} \left(\frac{4n-1}{n} \right) = 4.$$

Example 10.21

Consider the infinite sequence with the general term

$$a_n = \frac{3n+1}{2n-1}.$$

Let us see whether this infinite sequence has a limit when $n \to \infty$. Substituting $n = 1, 2, 3, \ldots$, we see that

$$a_1 = 4, \quad a_2 = \frac{7}{3} = 2.333, \quad a_3 = 2,$$

$$a_4 = \frac{13}{7} = 1.857, \quad a_5 = \frac{16}{9} = 1.777, \quad a_6 = \frac{19}{11} = 1.727, \quad \ldots,$$

$$a_{13} = \frac{8}{5} = 1.6, \quad \ldots, \quad a_{51} = \frac{154}{101} = 1.525, \quad \ldots,$$

$$a_{101} = \frac{304}{201} = 1.512, \quad \ldots, \quad a_{501} = \frac{1054}{1001} = 1.502, \quad \ldots.$$

Figure 10.2 represents the graph of the first six terms of this sequence. Note that as n becomes larger and larger, the terms of this sequence come closer and closer to a particular number. This observation suggests that a limit may exist.

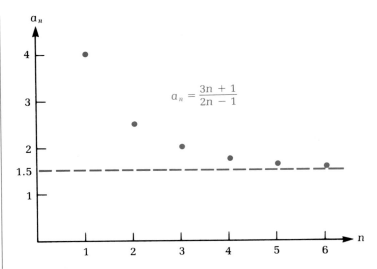

FIGURE 10.2

Can you guess this number? If not, let us look at the problem a little differently. Since $n \neq 0$, we can divide both numerator and denominator of the general term a_n by n. Thus

$$\frac{3n + 1}{2n - 1} = \frac{3 + 1/n}{2 - 1/n}.$$

We found in Example 10.19 that $\lim_{n \to \infty} \dfrac{1}{n} = 0$; using this fact, we observe that

$$\lim_{n \to \infty} \left(3 + \frac{1}{n}\right) = 3 \qquad \text{and} \qquad \lim_{n \to \infty} \left(2 - \frac{1}{n}\right) = 2.$$

Thus it is reasonable to suggest that

$$\lim_{n \to \infty} \frac{3n + 1}{2n - 1} = \lim_{n \to \infty} \frac{3 + 1/n}{2 - 1/n} = \frac{3}{2}.$$

limit of a sequence

These considerations lead to the following definition of the limit of a sequence.

Definition 10.4.1

The infinite sequence $\{a_n\}$ converges to the number L, or has limit L, if and only if the absolute value of the difference between the nth term in the sequence and the number L is as small as we please for sufficiently large n.

We do not wish to leave the impression that all sequences approach a particular number for sufficiently large n. We illustrate this in the following examples.

Example 10.22

Consider the infinite sequence

$$3, 5, 7, 9, \ldots, 2n + 1, \ldots.$$

divergent sequence

As n increases, the terms of the sequence $a_n = 2n + 1$ get larger and larger but do not approach any particular number. Intuitively, the sequence does *not* have a limit. Accordingly, we say that the sequence $a_n = 2n + 1$ *diverges*.

Example 10.23

Consider the infinite sequence

$$-1, 1, -1, 1, \ldots, (-1)^n, \ldots.$$

Apparently,

$$a_n = \begin{cases} 1 & \text{if } n \text{ is even} \\ -1 & \text{if } n \text{ is odd.} \end{cases}$$

This sequence does not tend to a limit, since there are two numbers, 1 and -1, to which the terms of the sequence approach.

Example 10.24

Consider the infinite sequence

$$\frac{1}{2}, -\frac{2}{3}, \frac{3}{4}, -\frac{4}{5}, \ldots, \frac{(-1)^{n+1}n}{n+1}, \ldots.$$

This sequence does not converge. (Why?) Observe that as n increases, the nth term in the sequence oscillates; the terms $a_1, a_3, a_5, \ldots$ approach 1, whereas the terms $a_2, a_4, a_6, \ldots$ approach -1. Since there are two numbers that the terms of the sequence approach, we conclude that the sequence diverges.

Examples 10.23 and 10.24 lead us to Proposition 10.4.1, which we state without proof.

Proposition 10.4.1

The limit of an infinite sequence, if it exists, is unique.

Limits of sequences have properties that may be used to evaluate the limit of a combination of many sequences. If $\{a_n\}$ and $\{b_n\}$ are two infinite sequences and c is a constant, we can form new sequences $\{ca_n\}$, $\{a_n \pm b_n\}$, and $\{a_n b_n\}$. The sequence $\left\{\dfrac{a_n}{b_n}\right\}$ can be formed if and only if $b_n \neq 0$ for each n.

We now state the following propositions (without proof). Note that in each case L, L_1, and L_2 are finite.

Proposition 10.4.2

$\lim\limits_{n \to \infty} c = c$, where c is constant.

Proposition 10.4.3

If $\lim\limits_{n \to \infty} a_n = L$, then

$$\lim\limits_{n \to \infty} ca_n = c \lim\limits_{n \to \infty} a_n = cL,$$

where c is a constant.

Proposition 10.4.4

If $\lim\limits_{n \to \infty} a_n = L_1$ and $\lim\limits_{n \to \infty} b_n = L_2$, then

$$\lim\limits_{n \to \infty} (a_n \pm b_n) = \lim\limits_{n \to \infty} a_n \pm \lim\limits_{n \to \infty} b_n = L_1 \pm L_2.$$

Proposition 10.4.5

If $\lim\limits_{n \to \infty} a_n = L_1$ and $\lim\limits_{n \to \infty} b_n = L_2$, then

$$\lim\limits_{n \to \infty} a_n b_n = \left(\lim\limits_{n \to \infty} a_n \right)\left(\lim\limits_{n \to \infty} b_n \right) = L_1 L_2.$$

Proposition 10.4.6

If $\lim\limits_{n \to \infty} a_n = L_1$ and $\lim\limits_{n \to \infty} b_n = L_2$, then

$$\lim\limits_{n \to \infty} \frac{a_n}{b_n} = \frac{\lim\limits_{n \to \infty} a_n}{\lim\limits_{n \to \infty} b_n} = \frac{L_1}{L_2},$$

where $L_2 \neq 0$ and $b_n \neq 0$ for each n.

In other words, the limit of the sum (or difference), product, and quotient of two sequences equals the sum (or difference), product, and quotient, respectively, of their limits.

To illustrate these propositions, consider the following examples.

Example 10.25

Evaluate $\lim\limits_{n \to \infty} a_n$, where

$$a_n = \frac{n + 1}{2n}.$$

Solution: Since

$$\frac{n+1}{2n} = \frac{1}{2}\left(1 + \frac{1}{n}\right),$$

it is easy to see that

$$\lim_{n \to \infty} a_n = \lim_{n \to \infty} \frac{n+1}{2n}$$

$$= \frac{1}{2} \lim_{n \to \infty} \left(1 + \frac{1}{n}\right) \qquad \text{(Proposition (10.4.3)}$$

$$= \frac{1}{2}\left(\lim_{n \to \infty} 1 + \lim_{n \to \infty} \frac{1}{n}\right) \qquad \text{(Proposition 10.4.4)}$$

$$= \frac{1}{2}(1 + 0) \qquad \text{since } \lim_{n \to \infty} \frac{1}{n} = 0 \qquad \text{(Proposition 10.4.2)}$$

$$= \frac{1}{2}.$$

Example 10.26

Evaluate $\lim_{n \to \infty} b_n$, where

$$b_n = \frac{3n - 2}{2n + 1}.$$

Solution: Note that since

$$\frac{3n - 2}{2n + 1} = \frac{3 - 2/n}{2 + 1/n},$$

we have

$$\lim_{n \to \infty} b_n = \lim_{n \to \infty} \frac{3n - 2}{2n + 2} = \lim_{n \to \infty} \frac{3 - 2/n}{2 + 1/n}.$$

Dealing with numerator and denominator separately, we have

$$\lim_{n \to \infty} \left(3 - \frac{2}{n}\right) = \lim_{n \to \infty} 3 - \lim_{n \to \infty} \frac{2}{n} \qquad \text{(Proposition 10.4.4)}$$

$$= \lim_{n \to \infty} 3 - 2 \lim_{n \to \infty} \frac{1}{n} \qquad \text{(Proposition 10.4.3)}$$

$$= 3 - 2(0) \qquad \text{since } \lim_{n \to \infty} \frac{1}{n} = 0 \qquad \text{(Proposition 10.4.2)}$$

$$= 3.$$

Similarly,

$$\lim_{n \to \infty} \left(2 + \frac{1}{n} \right) = \lim_{n \to \infty} 2 + \lim_{n \to \infty} \frac{1}{n} \qquad \text{(Proposition 10.4.4)}$$

$$= 2 + 0 \qquad \text{since } \lim_{n \to \infty} \frac{1}{n} = 0 \qquad \text{(Proposition 10.4.2)}$$

$$= 2.$$

Thus

$$\lim_{n \to \infty} \frac{3 - 2/n}{2 + 1/n} = \frac{\displaystyle\lim_{n \to \infty} (3 - 2/n)}{\displaystyle\lim_{n \to \infty} (2 + 1/n)} \qquad \text{(Proposition 10.4.6)}$$

$$= \frac{3}{2}.$$

Example 10.27   Let

$$a_n = \frac{n + 1}{2n} \qquad \text{and} \qquad b_n = \frac{3n - 2}{2n + 3}.$$

Verify that

$$\lim_{n \to \infty} a_n b_n = \left(\lim_{n \to \infty} a_n \right) \left(\lim_{n \to \infty} b_n \right).$$

Solution: We have established in Examples 10.25 and 10.26, respectively, that

$$\lim_{n \to \infty} \frac{n + 1}{2n} = \frac{1}{2} \qquad \text{and} \qquad \lim_{n \to \infty} \frac{3n - 2}{2n + 3} = \frac{3}{2}.$$

Now

$$a_n b_n = \left(\frac{n + 1}{2n} \right) \left(\frac{3n - 2}{2n + 3} \right)$$

$$= \frac{3n^2 + n - 2}{4n^2 + 6n}$$

$$= \frac{3 + 1/n - 2/n^2}{4 + 6/n}.$$

Since

$$\lim_{n \to \infty} \frac{1}{n} = 0$$

and

$$\lim_{n \to \infty} \frac{1}{n^2} = \lim_{n \to \infty} \left(\frac{1}{n} \cdot \frac{1}{n} \right) = \left(\lim_{n \to \infty} \frac{1}{n} \right) \left(\lim_{n \to \infty} \frac{1}{n} \right) = 0 \cdot 0 = 0,$$

it is easy to see that

$$\lim_{n \to \infty} a_n b_n = \frac{\lim\limits_{n \to \infty} (3 + 1/n - 2/n^2)}{\lim\limits_{n \to \infty} (4 + 6/n)} \qquad \text{(Proposition 10.4.6)}$$

$$= \frac{3 + 0 - 0}{4 + 0}$$

$$= \frac{3}{4}$$

$$= \frac{1}{2} \cdot \frac{3}{2}$$

$$= \left(\lim_{n \to \infty} a_n \right) \left(\lim_{n \to \infty} b_n \right).$$

Note that this illustrates Proposition 10.4.5.

EXERCISES

In Exercises 1 to 26, the nth term of a sequence $\{a_n\}$ is given. Evaluate $\lim\limits_{n \to \infty} a_n$ if it exists.

1. $a_n = \dfrac{2n}{n + 1}$

2. $a_n = \dfrac{2n - 1}{2n + 1}$

3. $a_n = \dfrac{2n + 1}{3n + 1}$

4. $a_n = \dfrac{2n + 3}{3n + 4}$

5. $a_n = \dfrac{4n + 1}{2n + 1}$

6. $a_n = \dfrac{5n + 2}{3n + 1}$

7. $a_n = \dfrac{2n}{n^2 + 1}$

8. $a_n = \dfrac{n^2 + 1}{2n^2 + 3}$

9. $a_n = \dfrac{2n - 1}{n^2 + n + 2}$

10. $a_n = \dfrac{5 - 4n^2}{8 + 5n^2}$

11. $a_n = \dfrac{n^2 - 1}{n + 1}$

12. $a_n = \dfrac{3n - 1}{n^2}$

13. $a_n = \dfrac{(-1)^n}{n + 2}$

14. $a_n = \dfrac{(-1)^n(n + 1)}{n + 2}$

15. $a_n = \dfrac{(-1)^n(n + 1)}{2n}$

16. $a_n = n^n$

17. $a_n = \dfrac{(-1)^n(2n + 1)}{2n + 3}$

18. $a_n = \dfrac{(-1)^n(3n + 2)}{4n + 5}$

19. $a_n = \dfrac{n^2 + n + 1}{2n^2 + 3n + 4}$

20. $a_n = \dfrac{2n^2 + 3n + 1}{3n^2 + 4n + 1}$

21. $a_n = \dfrac{n^3 + n^2 + 1}{2n^2 + 3}$

22. $a_n = \dfrac{3n^3 + 4n^2 + 5}{5n^3 + 2n^2 + 7}$

23. $a_n = \dfrac{2n^3 + 3n^2 + n + 1}{n^3 + n^2 + n + 1}$

24. $a_n = \dfrac{2n^3 + 3n^2 + 1}{3n^2 + 1}$

25. $a_n = \dfrac{2^n - 1}{2^n}$

26. $a_n = \dfrac{3^n - 1}{3^n}$

27. Given two infinite sequences

$$a_n = \frac{2n + 1}{3n + 2} \quad \text{and} \quad b_n = \frac{n + 1}{2n - 1},$$

verify that

(a) $\displaystyle\lim_{n \to \infty} a_n b_n = \left(\lim_{n \to \infty} a_n\right)\left(\lim_{n \to \infty} b_n\right)$

(b) $\displaystyle\lim_{n \to \infty} \frac{a_n}{b_n} = \frac{\displaystyle\lim_{n \to \infty} a_n}{\displaystyle\lim_{n \to \infty} b_n} \qquad (b_n \neq 0)$

(c) $\displaystyle\lim_{n \to \infty} (a_n + b_n) = \lim_{n \to \infty} a_n + \lim_{n \to \infty} b_n.$

28. Given two infinite sequences

$$a_n = \frac{n + 1}{2n + 1} \quad \text{and} \quad b_n = \frac{2n + 3}{3n + 4},$$

verify that

(a) $\displaystyle\lim_{n \to \infty} (a_n + b_n) = \lim_{n \to \infty} a_n + \lim_{n \to \infty} b_n$

(b) $\displaystyle\lim_{n \to \infty} a_n b_n = \left(\lim_{n \to \infty} a_n\right)\left(\lim_{n \to \infty} b_n\right)$

(c) $\displaystyle\lim_{n \to \infty} \frac{a_n}{b_n} = \frac{\displaystyle\lim_{n \to \infty} a_n}{\displaystyle\lim_{n \to \infty} b_n} \qquad (b_n \neq 0).$

10.5

SUMS OF INFINITE GEOMETRIC SEQUENCES

Recall from Proposition 10.3.1 that the sum S_n of the first n terms of a geometric sequence is given by

$$S_n = \frac{a_1(1 - r^n)}{1 - r} \qquad (r \neq 1),$$

where a_1 is the first term in the sequence and r is the common ratio. In this section, we study

$$S = \lim_{n \to \infty} S_n$$

when $|r| < 1$.

Let us begin by finding the sum of the infinite geometric sequence

$$1, \frac{1}{2}, \frac{1}{4}, \frac{1}{8}, \frac{1}{16}, \frac{1}{32}, \cdots$$

if it exists.

Example 10.28 | Find

$$1 + \frac{1}{2} + \frac{1}{4} + \frac{1}{8} + \frac{1}{16} + \frac{1}{32} + \cdots.$$

Solution: Notice that

$$S_1 = 1$$

$$S_2 = 1 + \frac{1}{2} = \frac{3}{2}$$

$$S_3 = S_2 + a_3 = \frac{3}{2} + \frac{1}{4} = \frac{7}{4}$$

$$S_4 = S_3 + a_4 = \frac{7}{4} + \frac{1}{8} = \frac{15}{8}$$

$$S_5 = S_4 + a_5 = \frac{15}{8} + \frac{1}{16} = \frac{31}{16}$$

$$S_6 = S_5 + a_5 = \frac{31}{16} + \frac{1}{32} = \frac{63}{32}$$

and, in general,

$$S_n = \frac{a_1(1 - r^n)}{1 - r} = \frac{1\left[1 - \left(\frac{1}{2}\right)^n\right]}{1 - \frac{1}{2}}.$$

Since

$$\lim_{n \to \infty} \left(\frac{1}{2}\right)^n = 0, \qquad \text{(Example 10.18)}$$

we have

$$\lim_{n \to \infty} S_n = \frac{1(1 - 0)}{\left(\frac{1}{2}\right)} = 2.$$

Hence

$$1 + \frac{1}{2} + \frac{1}{4} + \frac{1}{8} + \cdots = 2.$$

In general,

$$S_n = \frac{a_1(1 - r^n)}{1 - r} \qquad (r \neq 1)$$

$$= \frac{a_1}{1 - r} - \frac{a_1 r^n}{1 - r},$$

where a_1 is the first term, r is the common ratio, and n is a positive integer. For $|r| < 1$, r^n decreases as n increases; and by taking n sufficiently large, we can make r^n as close to zero as we wish. In other words,

$$\lim_{n \to \infty} r^n = 0 \qquad (|r| < 1),$$

and we have

$$S = \lim_{n \to \infty} S_n = \frac{a_1}{1 - r} \qquad (|r| < 1).$$

This leads us to the following result.

Proposition 10.5.1

If $|r| < 1$, then the sum of the infinite geometric sequence with a_1 as the first term and r as the common ratio is

$$S = \sum_{i=1}^{\infty} a_1 r^{i-1} = a_1 + a_1 r + a_1 r^2 + \cdots + a_1 r^{n-1} + \cdots,$$

which reduces to

$$S = \frac{a_1}{1 - r}. \tag{10.8}$$

Example 10.29

Find

$$1 + \frac{2}{5} + \frac{4}{25} + \frac{8}{125} + \cdots.$$

Solution: Here $a_1 = 1$ and $r = \frac{2}{5}$. Since $|r| < 1$, the sum exists and is given by

$$S = \frac{a_1}{1 - r} = \frac{1}{1 - \frac{2}{5}} = \frac{5}{3}.$$

Example 10.30

Express $0.5555 \ldots$ in the form $\frac{a}{b}$, where a and b are integers and $b \neq 0$.

Solution: First, write $0.5555 \ldots$ as

$$0.5555 \ldots = 0.5 + 0.05 + 0.005 + 0.0005 + \cdots,$$

which is the sum of the terms of an infinite geometric sequence with

$$a_1 = 0.5 \qquad \text{and} \qquad r = 0.1.$$

Since $|r| < 1$, the sum exists and is given by

$$S = \frac{a_1}{1 - r} = \frac{0.5}{1 - 0.1} = \frac{5}{9}.$$

Example 10.31

Express $3.434343\ldots$ in the form $\dfrac{a}{b}$, where a and b are integers and $b \neq 0$.

Solution: Note that

$$3.434343\ldots = 3 + (0.43 + 0.0043 + 0.000043 + \cdots).$$

The sum in parentheses is an infinite geometric sequence with

$$a_1 = 0.43 \qquad \text{and} \qquad r = 0.01.$$

Thus

$$\frac{a_1}{1-r} = \frac{0.43}{1-0.01} = \frac{43}{99}.$$

Consequently,

$$3.434343\ldots = 3 + \frac{43}{99} = \frac{340}{99}.$$

EXERCISES

Find the sums of the infinite geometric sequences in Exercises 1 to 10.

1. $1 + \dfrac{1}{4} + \dfrac{1}{16} + \dfrac{1}{64} + \cdots$

2. $1 + \dfrac{1}{5} + \dfrac{1}{25} + \dfrac{1}{125} + \cdots$

3. $1 - \dfrac{1}{2} + \dfrac{1}{4} - \dfrac{1}{8} + \cdots$

4. $1 - \dfrac{1}{3} + \dfrac{1}{9} - \dfrac{1}{27} + \cdots$

5. $2 + \dfrac{4}{3} + \dfrac{8}{9} + \dfrac{16}{27} + \cdots$

6. $2 - \dfrac{4}{3} + \dfrac{8}{9} - \dfrac{16}{27} + \cdots$

7. $\dfrac{1}{3} - \dfrac{2}{9} + \dfrac{4}{27} - \dfrac{8}{81} + \cdots$

8. $1 - \dfrac{1}{6} + \dfrac{1}{36} - \dfrac{1}{216} + \cdots$

9. $1 + 0.1 + 0.01 + 0.001 + \cdots$

10. $1 + 0.9 + 0.81 + 0.729 + \cdots$

Express the repeating decimals in Exercises 11 to 30 in the form a/b, where a and b are integers and $b \neq 0$.

11. $0.1111\ldots$
12. $0.4444\ldots$
13. $0.7777\ldots$
14. $0.9999\ldots$
15. $0.121212\ldots$
16. $0.212121\ldots$
17. $0.101101101\ldots$
18. $0.123123123\ldots$
19. $0.212212212\ldots$
20. $0.109109109\ldots$
21. $0.323323323\ldots$
22. $0.425425425\ldots$
23. $0.271271271\ldots$
24. $2.142142142\ldots$
25. $3.006006006\ldots$
26. $1.009009009\ldots$
27. $3.03030303\ldots$
28. $2.678678678\ldots$
29. $1.456456456\ldots$
30. $4.789789789\ldots$

10.6

MATHEMATICAL INDUCTION

Suppose we assert that the formula

$$1 + 2 + 3 + 4 + \cdots + n = \frac{n(n+1)}{2}$$

holds for all positive integers. We can verify that this assertion is true by actual substitution of $n = 1$, $n = 2$, $n = 3$, $n = 4, \ldots$, thus obtaining

	left-hand side	right-hand-side
$n = 1$	1	$\dfrac{1(1 + 1)}{2} = 1$
$n = 2$	$1 + 2 = 3$	$\dfrac{2(2 + 1)}{2} = 3$
$n = 3$	$1 + 2 + 3 = 6$	$\dfrac{3(3 + 1)}{2} = 6$
$n = 4$	$1 + 2 + 3 + 4 = 10$	$\dfrac{4(4 + 1)}{2} = 10$

and so on. However, it is impossible to verify the above formula for all positive integers in this manner. We shall therefore develop a method of proof called *mathematical induction*, **mathematical induction,** which is essentially based on the following property of the set N of positive integers.

Axiom of Mathematical Induction

If a subset S of N had the properties that

(i) $1 \in S$ and

(ii) $k \in S$ implies that $(k + 1) \in S$,

then $S = N$.

These properties form the basis of the following proposition which we state without proof.

Proposition 10.6.1

Let $p(n)$ be a statement, or an equation, that involves a positive integer n. Then $p(n)$ is true for every integer n

(i)   if $p(1)$ is true, and

(ii)   for a positive integer k, if $p(k)$ is true, then $p(k + 1)$ is also true.

Example 10.32

Prove that
$$1 + 3 + 5 + \cdots + (2n - 1) = n^2.$$

Solution: Let $p(n)$ denote the given statement for all $n \in N$.

(i)   Substituting $n = 1$ in $p(n)$, we have 1 on the left-hand side and $1^2 = 1$ on the right-hand side. Thus $p(1)$ is true.

(ii)   Assume that $p(k)$ is true; that is, assume the validity of

$$p(k): \quad 1 + 3 + 5 + \cdots + (2k - 1) = k^2.$$

We need to show that

$$p(k + 1):\ 1 + 3 + 5 + \cdots + [2(k + 1) - 1]$$
$$= 1 + 3 + 5 + \cdots + (2k + 1)$$
$$= (k + 1)^2$$

induction hypothesis

is true. Our *induction hypothesis* is that

$$1 + 3 + 5 + \cdots + (2k - 1) = k^2.$$

Adding $2k + 1$ to both sides of the equation, we have

$$1 + 3 + 5 + \cdots + (2k - 1) + (2k + 1) = k^2 + 2k + 1$$
$$= (k + 1)^2,$$

which establishes the second step necessary to complete the proof.

Since $p(1)$ is true and $p(k)$ implies that $p(k + 1)$ is true, we conclude that $p(n)$ is true for all $n \in N$.

Example 10.33

Prove that $2^{2n} - 1$ is divisible by 3.

Solution: Let $p(n)$ denote the statement that $2^{2n} - 1$ is divisible by 3.

(i)   Let $n = 1$. Then

$$2^{2 \cdot 1} - 1 = 2^2 - 1 = 4 - 1 = 3,$$

which is certainly divisible by 3. This means that $p(1)$ is true.

(ii)   Assume the validity of

$$p(k):\ 2^{2k} - 1 \text{ is divisible by 3.}$$

We must show that

$$p(k + 1):\ 2^{2(k + 1)} - 1 = 2^{2k + 2} - 1$$

is also divisible by 3. Our hypothesis that $p(k)$ is true implies the existence of a positive integer m such that

$$2^{2k} - 1 = 3m$$

$$2^{2k} = 3m + 1.$$

Multiplying both sides of the last equation by 2^2, we obtain

$$2^{2k + 2} = 2^2(3m + 1)$$
$$= 4(3m + 1)$$
$$= 12m + 4$$
$$= 3(4m + 1) + 1.$$

That is,

$$2^{2k + 2} - 1 = 3(4m + 1),$$

which implies that $2^{2k+2} - 1$ is divisible by 3. This establishes the fact that $k + 1 \in S$. Hence $p(n)$ is true for all $n \in N$.

Example 10.34

Prove that $3^n > 2n$.

Solution: Let $p(n)$ denote the given statement for all positive integers.

(i)   Let $n = 1$. Then

$$p(1): \ 3 > 2,$$

which is certainly a true statement. Thus $p(1)$ holds.

(ii)   Assume the validity of

$$p(k): \ 3^k > 2k.$$

We must show that

$$p(k + 1): \ 3^{k+1} > 2k + 2.$$

The induction hypothesis is

$$3^k > 2k.$$

Multiplying both sides of this equation by 3, we obtain

$$3^{k+1} > 3(2k) = 6k.$$

Since $6k > 2k + 2$ for all $k \in N$ (why?), we have

$$3^{k+1} > 6k > 2k + 2.$$

This means that $p(k + 1)$ is true. By the principal of mathematical induction, the assertion is true for all positive integers.

EXERCISES

Using mathematical induction, prove the statements in Exercises 1 to 22 for all positive integers.

1. $2 + 4 + 6 + \cdots + 2n = n(n + 1)$

2. $4 + 8 + 12 + \cdots + 4n = 2n(n + 1)$

3. $2 + 6 + 10 + \cdots + (4n - 2) = 2n^2$

4. $1 + 4 + 7 + \cdots + (3n - 2) = \dfrac{n(3n - 1)}{2}$

5. $1 + 2 + 2^2 + 2^3 + \cdots + 2^{n-1} = 2^n - 1$

6. $1 + 3 + 3^2 + 3^3 + \cdots + 3^{n-1} = \dfrac{3^n - 1}{2}$

7. $1^2 + 2^2 + 3^2 + \cdots + n^2 = \dfrac{n(n + 1)(2n + 1)}{6}$

8. $1^2 + 3^2 + 5^2 + \cdots + (2n - 1)^2 = \dfrac{n(2n - 1)(2n + 1)}{3}$

9. $1^3 + 2^3 + 3^3 + \cdots + n^3 = \left[\dfrac{n(n+1)}{2}\right]^2$

10. $1^3 + 3^3 + 5^3 + \cdots + (2n-1)^3 = n^2(2n^2 - 1)$

11. $3 + 3^2 + 3^3 + \cdots + 3^n = \dfrac{3(3^n - 1)}{2}$

12. $4 + 4^2 + 4^3 + \cdots + 4^n = \dfrac{4(4^n - 1)}{3}$

13. $(1)(2) + (2)(3) + (3)(4) + \cdots + n(n+1) = \dfrac{n(n+1)(n+2)}{3}$

14. $(1)(3) + (2)(4) + (3)(5) + \cdots + n(n+2) = \dfrac{n(n+1)(2n+7)}{6}$

15. $\dfrac{1}{(1)(2)} + \dfrac{1}{(2)(3)} + \dfrac{1}{(3)(4)} + \cdots + \dfrac{1}{n(n+1)} = \dfrac{n}{n+1}$

16. $n(n+1)$ is divisible by 2

17. $n(n^2 + 2)$ is divisible by 3

18. $3^{2n} - 1$ is divisible by 8

19. $2^{4n} - 1$ is divisible by 15

20. $a^n - b^n$ is divisible by $a - b$

21. $2^n \geq 2n$

22. $5^n \geq 4n + 1$

KEY WORDS

infinite sequence
indices
terms of the sequence
general term
sigma notation
index of summation
series
sequence of partial sums

arithmetic sequence
arithmetic progression
common difference
geometric sequence
geometric progression
common ratio
limit of a sequence
convergent sequence

divergent sequence
sum of infinite geometric sequences
mathematical induction
axiom of mathematical induction
induction hypothesis

KEY FORMULAS

- The general term a_n in an arithmetic sequence is
$$a_n = a_1 + (n-1)d,$$
where a_1 is the first term and d is the common difference.

- The sum of the first n terms of an arithmetic sequence with a_1 as the first term and a_n as the general term is
$$S_n = \frac{n}{2}(a_1 + a_n).$$

- The sum of the first n terms of an arithmetic sequence with a_1 as the first term and d the common difference is
$$S_n = \frac{n}{2}[2a_1 + (n-1)d].$$

- The general term a_n in a geometric sequence is
$$a_n = a_1 r^{n-1},$$
where a_1 is the first term and r is the common ratio.

- The sum of the first n terms in a geometric sequence with a_1 as the first term and r the common ratio is

$$S_n = \begin{cases} \dfrac{a_1(1 - r^n)}{1 - r} & r \neq 1 \\ na_1 & r = 1. \end{cases}$$

- If $|r| < 1$, then the sum of the infinite geometric sequence with a_1 as the first term and r the common ratio is

$$S = \frac{a_1}{1 - r}.$$

EXERCISES FOR REVIEW

Find the first five terms for the sequences in Exercises 1 to 4.

1. $a_n = 3n + 2$
2. $a_n = n^2 + n + 1$
3. $a_n = 2^n + 1$
4. $a_n = \left(\dfrac{1}{3}\right)^n$

Determine the general term a_n for the sequences in Exercises 5 to 10.

5. $3, 6, 9, 12, 15, \ldots$
6. $1, 2, 4, 8, 16, \ldots$
7. $2, 7, 12, 17, 22, \ldots$
8. $3, 7, 11, 15, 19, \ldots$
9. $3, 9, 27, 81, 243, \ldots$
10. $1, -1, 1, -1, 1, \ldots$

Evaluate the sums in Exercises 11 to 16.

11. $\sum_{i=1}^{5} (i + 1)$
12. $\sum_{i=1}^{5} 2i$
13. $\sum_{i=1}^{4} (i^2 + 1)$
14. $\sum_{j=1}^{6} (2^j - 1)$
15. $\sum_{i=1}^{8} (2i + 1)$
16. $\sum_{k=1}^{10} (5k + 3)$

Find the partial sums S_1, S_2, S_3, and S_4 in Exercises 17 to 20.

17. $a_n = 4n + 3$
18. $a_n = n(n + 1)$
19. $a_n = 2^n$
20. $a_n = \left(\dfrac{1}{2}\right)^n$

Find a_8 and S_8 for the arithmetic sequences in Exercises 21 to 24.

21. $3, 5, 7, 9, \ldots$
22. $2, 5, 8, 11, \ldots$
23. $1, 6, 11, 16, \ldots$
24. $13, 9, 5, 1, \ldots$

Find S_{20} in Exercises 25 to 28.

25. $2, 6, 18, 54, \ldots$
26. $1, 2, 4, 8, \ldots$
27. $1, 5, 25, 125, \ldots$
28. $16, 8, 4, 2, \ldots$

Find a_{12} and S_{12} in Exercises 29 and 30, given the first term a_1 and the common ratio r.

29. $a_1 = 2, r = 3$
30. $a_1 = 5, r = 2$

31. The sums of three consecutive numbers in an arithmetic sequence is 21, and their product is 336. Find the numbers.

32. Jovita is hired at an initial salary of $15,000 in a consulting firm. If her salary increases by $1200 annually, how many years will it take her to earn $24,600 anually?

33. Determine the geometric sequence whose fifth term is 80 and whose eighth term is 640.

34. The annual depreciation of a certain typewriter is 25 percent of its value at the beginning of the year. If the new typewriter costs $1200, what will its resale value be in the fourth year of its use?

In Exercises 35 to 40, the nth term of a sequence $\{a_n\}$ is given. Evaluate $\lim_{n \to \infty} a_n$ if it exists.

35. $a_n = \dfrac{3n - 2}{n}$
36. $a_n = \dfrac{3n}{2n^2 + 1}$
37. $a_n = \dfrac{2n^2 + n + 1}{3n^2 - 4n + 2}$

38. $a_n = \dfrac{(-1)^n}{n+1}$ **39.** $a_n = \dfrac{2^n - 1}{2^n}$ **40.** $a_n = \dfrac{(-1)^n(n+1)}{3n}$

Find the sums of the infinite sequences in Exercises 41 and 42.

41. $2 + \dfrac{4}{3} + \dfrac{8}{9} + \dfrac{16}{27} + \cdots$

42. $1 + 0.1 + 0.01 + 0.001 + \cdots$

Express the repeating decimals in Exercises 43 to 46 in the form $\dfrac{a}{b}$, where a and b are integers and $b \neq 0$.

43. $0.212121\ldots$ **44.** $0.105105105\ldots$ **45.** $1.271271271\ldots$ **·46.** $2.636363\ldots$

Using mathematical induction, prove the statements in Exercises 47 and 48.

47. $1 + 2 + 3 + \cdots + n = \dfrac{n(n+1)}{2}$

48. $1^2 + 2^2 + 3^2 + \cdots + n^2 = \dfrac{n(n+1)(2n+1)}{6}$

CHAPTER TEST

1. Find the first five terms of the sequence $a_n = 3n + 1$.

2. Find the 75th term in the arithmetic sequence $2, 5, 8, 11, \ldots$.

3. The sum of three consecutive terms in an arithmetic sequence is 18, and their product is 210. Find the numbers.

4. Find the sum of the first 45 odd numbers.

5. Determine the geometric sequence whose third term is 12 and sixth term is 96.

6. Find a_{10} and S_{10} in the geometric sequence, $3, 6, 12, 24, \ldots$.

7. Find the sum of the infinite geometric sequence

$$1 - 0.5 + 0.25 - 0.125 + 0.0625 - \cdots.$$

8. Express the repeating decimal

$$0.2121212121\ldots$$

in the form a/b, where a and b are integers and $b \neq 0$.

9. A stack of poles has 15 in the bottom row, 14 in the second row, 13 in the third row, and so on. Determine the total number of poles if there is only one pole in the top row.

10. The annual depreciation of a certain personal computer is 30 percent of its value at the beginning of each year. If the new personal computer costs \$2500, what will be its resale value at the beginning of the fourth year of its use?

11. Evaluate $\lim\limits_{n \to \infty} a_n$ (if it exists) where

$$a_n = \frac{(-1)^n \cdot n}{n+1}.$$

12. Use mathematical induction to prove that

$$2 + 4 + 6 + 8 + \cdots + 2n = n(n+1).$$

ADDITIONAL TOPICS
IN ALGEBRA

11

CIRCLES

Definition 11.1.1

> A **circle** is the collection of all points in the plane that are at a fixed distance from a fixed point C. The fixed point is called the *center*, and the fixed distance is called the **radius** of the circle.

To find the equation of the circle, we let $C(h, k)$ be the center of the circle and r be its radius. If $P(x, y)$ is an arbitrary point in the plane as shown in Figure 11.1 then P lies on the circle if and only if

$$d_{CP} = r.$$

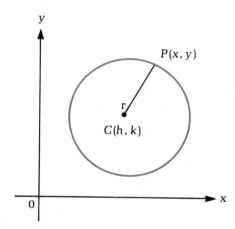

FIGURE 11.1

Using the distance formula (Proposition 3.1.1), we get

$$\sqrt{(x - h)^2 + (y - k)^2} = r.$$

Squaring both sides of this expression, we obtain

$$(x - h)^2 + (y - k)^2 = r^2, \tag{11.1}$$

which is the **standard equation of a circle** with its center at the point $C(h, k)$ and radius r.

In particular, if $h = 0$ and $k = 0$, then (11.1) reduces to

$$x^2 + y^2 = r^2, \qquad \text{(11.2)}$$

which is the equation of the circle of radius r with its center at the origin.

Example 11.1   Find an equation of the circle that has center C at the origin and radius $r = 5$. Graph the circle.

Solution: Substituting $r = 5$ in (11.2), we have

$$x^2 + y^2 = 25.$$

The graph of the circle is shown in Figure 11.2.

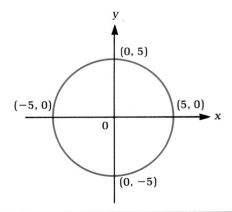

FIGURE 11.2

Example 11.2   Find an equation of the circle with center $C(3, 4)$ and radius $r = 6$.

Solution: Substituting $h = 3$, $k = 4$, and $r = 6$ in (11.1), we get

$$(x - 3)^2 + (y - 4)^2 = 36,$$

which on simplification yields

$$x^2 + y^2 - 6x - 8y - 11 = 0.$$

Example 11.3   Find the equation of the circle having as its diameter the line segment joining the points $(3, 7)$ and $(-5, 1)$.

Solution: From plane geometry, we know that the center of the circle is the middle point of the diameter. The coordinates of the center, by Proposition 3.1.2, are

$$\left(\frac{3-5}{2}, \frac{7+1}{2}\right) = (-1, 4).$$

Using the distance formula, we see that the length of the diameter is

$$\sqrt{[3-(-5)]^2 + (7-1)^2} = \sqrt{8^2 + 6^2}$$
$$= \sqrt{64 + 36} = \sqrt{100} = 10.$$

Hence the radius of the circle $r = 5$. The equation of the circle with $(-1, 4)$ as its center and $r = 5$ is

$$(x + 1)^2 + (y - 4)^2 = 25,$$

which on simplification yields

$$x^2 + y^2 + 2x - 8y - 8 = 0.$$

Example 11.4

Show that the equation

$$x^2 + y^2 - 10x + 4y + 4 = 0$$

represents a circle. Find its center and radius. Graph the circle.

Solution: To graph the circle, we must find its center and its radius. We re-arrange the given equation in the form

$$(x^2 - 10x \quad) + (y^2 + 4y \quad) = -4$$

and then complete the squares within each set of parentheses. To complete the squares of the expression $x^2 + ax$ in the first set of parentheses, we add $\left(\frac{a}{2}\right)^2$, the square of one-half the coefficient of x, to both sides of the equation. Similarly, for the expression $y^2 + by$ in the second set of parentheses, we add $\left(\frac{b}{2}\right)^2$ to both sides of the equation. Thus we obtain

$$(x^2 - 10x + 25) + (y^2 + 4y + 4) = -4 + 25 + 4,$$

or

$$(x - 5)^2 + (y + 2)^2 = 25,$$

which is of the form (11.1). Hence, the equation

$$x^2 + y^2 - 10x + 4y + 4 = 0$$

represents a circle. The center of the circle is $(5, -2)$, and its radius $r = 5$. The graph of the circle is shown in Figure 11.3.

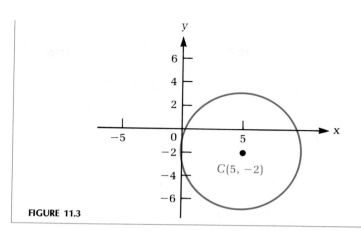

FIGURE 11.3

Example 11.5

Graph

$$y = \sqrt{9 - x^2}.$$

Solution: Squaring both sides of the equation, we get

$$y^2 = 9 - x^2,$$

or

$$x^2 + y^2 = 9,$$

which is an equation of a circle having its center at the origin and radius $r = 3$. Since in the equation $y = \sqrt{9 - x^2}$, y takes only nonnegative values, we conclude that the graph is an upper *semicircle*, as shown in Figure 11.4.

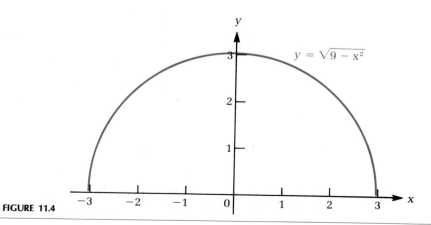

FIGURE 11.4

EXERCISES

In Exercises 1 to 10, find the equation of the circle having the given point C as its center and r as its radius.

1. $C(0, 0)$, $r = 4$ **2.** $C(0, 0)$, $r = 5$ **3.** $C(1, 2)$, $r = 5$ **4.** $C(4, 2)$, $r = 6$

5. $C(-4, 1)$, $r = 8$ **6.** $C(5, -3)$, $r = 10$ **7.** $C(-3, -4)$, $r = 7$ **8.** $C(3, 0)$, $r = 5$

9. $C(0, -5)$, $r = 12$ **10.** $C(-5, -12)$, $r = 13$

In Exercises 11 to 20, find the center and radius of the circle.

11. $x^2 + y^2 = 64$ **12.** $(x - 2)^2 + (y - 3)^2 = 36$ **13.** $(x - 1)^2 + (y + 1)^2 = 25$

14. $(x - 3)^2 + (y - 4)^2 = 49$ **15.** $x^2 + y^2 - 6x - 8y - 56 = 0$ **16.** $x^2 + y^2 + 6x - 10y + 9 = 0$

17. $x^2 + y^2 + 6x - 12y + 20 = 0$ **18.** $x^2 + y^2 + 4x - 6y - 1 = 0$ **19.** $x^2 + y^2 + 14x + 8y + 29 = 0$

20. $x^2 + y^2 + 8x - 12y + 52 = 0$

21. Find the equation of the circle having as its diameter the line segment joining the points $(4, 1)$ and $(6, 5)$.

22. Find the equation of the circle having as its diameter the line segment joining the points $(5, -5)$ and $(7, 3)$.

Graph the equations in Exercises 23 to 28.

23. $x^2 + y^2 = 16$ **24.** $x^2 + y^2 + 12x - 2y - 27 = 0$ **25.** $x^2 + y^2 - 6x = 0$

26. $x^2 + y^2 - 4x + 2y - 31 = 0$ **27.** $y = \sqrt{16 - x^2}$ **28.** $y = -\sqrt{25 - x^2}$

11.2

CONIC SECTIONS

conic

Consider a double-napped *right circular cone*, as shown in Figure 11.5. The curve formed by intersecting or cutting a double-napped right circular cone with a plane is called a **conic section,** or simply a **conic.** If the cutting plane does not pass through the vertex of the cone, the conic belongs to one of four types:

1. If the cutting plane cuts through one nappe and is perpendicular to the axis of the cone, as shown in Figure 11.6(*a*), the conic is called a **circle.**

2. If the plane cuts through one nappe and is not perpendicular to the axis of the cone, as shown in Figure 11.6(*b*), the conic is called an **ellipse.**

ellipse

3. If the cutting plane is parallel to one side of the cone, as shown in Figure 11.6(*c*), the conic is called a **parabola.**

4. If the cutting plane is parallel to the axis of the cone and cuts through both nappes of the cone, as shown in Figure 11.6(*d*), the conic is called a **hyperbola.**

The conic sections were studied by the ancient Greek mathematicians, who used methods very similar to modern elementary geometry. The circle, which is the simplest conic section, is discussed in Section 11.1. We now investigate parabolas, ellipses, and hyperbolas.

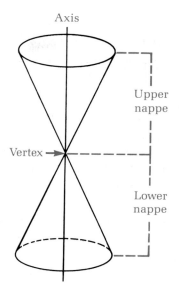

Axis

Upper
nappe

Vertex →

Lower
nappe

FIGURE 11.5

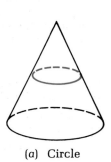

(*a*) Circle

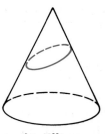

(*b*) Ellipse

(*c*) Parabola

(*d*) Hyperbola

FIGURE 11.6

Parabolas

Recall from Section 3.6 that the graph of

$$y = ax^2 + bx + c, \tag{11.3}$$

where a, b, and c are real numbers and $a \neq 0$, is a parabola that opens upward if $a > 0$ and downward if $a < 0$. By interchanging the variables x and y, we obtain

$$x = ay^2 + by + c. \tag{11.4}$$

The graph of this equation is also a parabola, which opens to the right if $a > 0$ and to the left if $a < 0$. If $b = c = 0$, then $x = ay^2$. Figures 11.7 and 11.8 show the graph of $x = ay^2$ for $a = 1$ and $a = -1$, respectively. Observe that the parabolas have their vertices at the origin and are symmetric about the x axis.

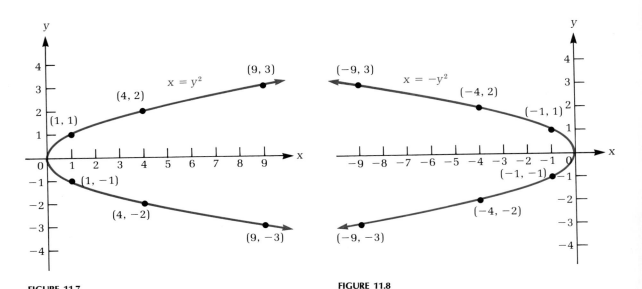

FIGURE 11.7   FIGURE 11.8

If $b = 0$ and $c \neq 0$, then

$$x = ay^2 + c,$$

and its graph is obtained by shifting the graph of $x = ay^2$ some $|c|$ units to the right if $c > 0$ and to the left if $c < 0$.

Remember that we expressed

$$y = ax^2 + bx + c \qquad (a \neq 0)$$

in the form

$$y = a(x - h)^2 + k$$

where

$$h = -\frac{b}{2a} \qquad \text{and} \qquad k = \frac{4ac - b^2}{4a},$$

and thus we identified the axis of symmetry $x = h$ and the vertex (h, k) of the parabola. In a similar manner, we may express the equation

$$x = ay^2 + by + c$$

in the form

$$x = a(y - k)^2 + h \qquad \qquad \textbf{(11.5)}$$

where

$$k = -\frac{b}{2a} \qquad \text{and} \qquad h = \frac{4ac - b^2}{4a}.$$

The graph of (11.5) is a parabola that opens to the right if $a > 0$ and to the left if $a < 0$. The line $y = k$ is the axis of symmetry for the parabola with vertex (h, k).

Example 11.6

Graph

$$x = 3y^2 - 6y + 1.$$

Solution: Comparing this equation with the general form (11.4), we see that $a = 3$, $b = -6$, and $c = 1$. Since $a > 0$, the parabola opens to the right. Further,

$$h = \frac{4ac - b^2}{4a} = -2 \qquad \text{and} \qquad k = -\frac{b}{2a} = 1.$$

Thus the vertex of the parabola is $(-2, 1)$. The line $y = 1$, which passes through the vertex, is the axis for the parabola. By plotting a few additional points, we get the graph shown in Figure 11.9.

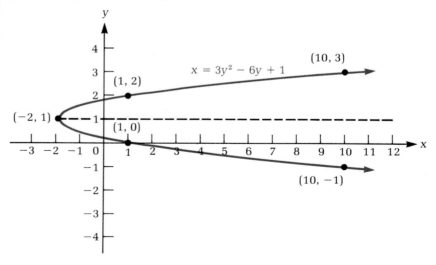

FIGURE 11.9

We now define a parabola geometrically and use the definition to develop a standard form for the equation of a parabola.

Definition 11.2.1

> A *parabola* is the collection of all points in the plane that are equidistant from a fixed point and from a fixed line.

directrix   The fixed point F is called the *focus*, and the fixed line MN is the *directrix* for the parabola (see Figure 11.10). The line passing through F and perpendicular to MN is the axis of symmetry for the parabola. The midpoint V between the focus and the directrix is the vertex for the parabola.

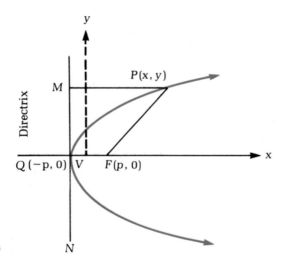

FIGURE 11.10

To obtain the standard equation of a parabola, we take as the axis for the parabola the x axis and as vertex V the origin. Let Q be the point of intersection of the directrix with the x axis. Let $d_{QF} = 2p$ be the distance between Q and F such that $d_{QV} = d_{VF} = p$. Then the coordinates of F and Q are $(p, 0)$ and $(-p, 0)$, respectively. The equation of the directrix is $x = -p$. Let $P(x, y)$ be any point on the parabola. Then by Definition 11.2.1,

$$d_{MP} = d_{PF}.$$

Observe from Figure 11.10 that

$$d_{MP} = x + p.$$

From the distance formula,

$$d_{PF} = \sqrt{(x - p)^2 + (y - 0)^2}.$$

Since $d_{MP} = d_{PF}$, we have

$$\sqrt{(x - p)^2 + y^2} = x + p.$$

Squaring both sides, we get

$$(x - p)^2 + y^2 = (x + p)^2,$$

which simplifies to

$$y^2 = 4px. \tag{11.6}$$

Equation (11.6) is the **standard equation of the parabola** when the axis of symmetry for the parabola is the x axis and the vertex is the origin. If $p > 0$, the parabola opens to the right; if $p < 0$, the parabola opens to the left.

If the axis of symmetry for the parabola is the y axis and its vertex is the origin, then the standard equation of the parabola is

$$x^2 = 4py. \tag{11.7}$$

The parabola opens upward or downward depending on whether p is positive or negative, respectively.

Example 11.7

Find the vertex, focus, and directrix of the parabola

$$y^2 = 16x.$$

Solution: Since the equation is in standard form (11.6) with $p = 4$, the coordinates of the vertex are $(0, 0)$, and the coordinates of the focus are $(4, 0)$. The equation for the axis of the parabola is $y = 0$; the equation of the directrix is $x = -4$.

Example 11.8

Find the equation of the parabola that is symmetric with respect to the y axis, passes through the point $(3, -2)$, and has as its vertex the point $(0, 0)$.

Solution: Since the vertex is $(0, 0)$ and the parabola is symmetric with respect to the y axis, the equation is of the form (11.7). To find p, we substitute $x = 3$ and $y = -2$ in (11.7), obtaining

$$9 = -8p$$

$$p = -\frac{9}{8}.$$

Thus the equation is

$$x^2 = 4\left(-\frac{9}{8}\right)y$$

$$x^2 = -\frac{9}{2}\,y.$$

Next, we discuss the algebraic representation of an ellipse.

The Ellipse

Definition 11.2.2

> An *ellipse* is the collection of all points in the plane such that the sum of their distances from two fixed points is constant.

foci The fixed points are called *foci* for the ellipse, and the middle point of the line joining the foci is called the *center* for the ellipse.

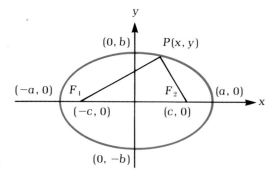

FIGURE 11.11

To determine the standard equation of an ellipse, we let $F_1(-c, 0)$ and $F_2(c, 0)$ be the coordinates of the foci, as shown in Figure 11.11. Let $P(x, y)$ be an arbitrary point on the ellipse. Then, by Definition 11.2.2,

$$d_{PF_1} + d_{PF_2} = 2a. \qquad \text{(constant)} \qquad\qquad \textbf{(11.8)}$$

Using the distance formula, we get

$$d_{PF_1} = \sqrt{[x - (-c)]^2 + (y - 0)^2} = \sqrt{(x + c)^2 + y^2}$$

$$d_{PF_2} = \sqrt{(x - c)^2 + (y - 0)^2} = \sqrt{(x - c)^2 + y^2}$$

so that

$$d_{PF_1} + d_{PF_2} = \sqrt{(x + c)^2 + y^2} + \sqrt{(x - c)^2 + y^2} = 2a$$

or

$$\sqrt{(x + c)^2 + y^2} = 2a - \sqrt{(x - c)^2 + y^2}.$$

Squaring both sides, we obtain

$$(x^2 + 2cx + c^2) + y^2 = 4a^2 + (x^2 - 2cx + c^2 + y^2) - 4a\sqrt{(x - c)^2 + y^2},$$

which on simplification yields

$$a^2 - cx = a\sqrt{(x - c)^2 + y^2}.$$

Squaring again and rearranging terms, we have

$$(a^2 - c^2)x^2 + a^2y^2 = a^2(a^2 - c^2). \tag{11.9}$$

Since $a > c$, it follows that $a^2 > c^2$ and $a^2 - c^2 > 0$. Dividing (11.9) by $a^2(a^2 - c^2)$, we obtain

$$\frac{x^2}{a^2} + \frac{y^2}{b^2} = 1, \tag{11.10}$$

where $b^2 = a^2 - c^2$. Equation (11.10) is the **standard equation of the ellipse** having foci $(-c, 0)$ and $(c, 0)$.

vertices
major axis

minor axis

The points $(a, 0)$ and $(-a, 0)$ are called the **vertices** of the ellipse. The line segment joining the vertices is called the **major axis,** and its length is $2a$. The line segment joining the points $(0, b)$ and $(0, -b)$ on the y axis is called the **minor axis,** and its length is $2b$. Note that the middle point of both the major and minor axes is the center for the ellipse.

Example 11.9

Graph the ellipse

$$\frac{x^2}{25} + \frac{y^2}{16} = 1$$

and find the coordinates of the foci.

Solution: Comparing the equation with (11.10), we observe that

$$a^2 = 25 \quad \text{and} \quad b^2 = 16.$$

The vertices for the ellipse are $(5, 0)$ and $(-5, 0)$; the line segment joining these two points is the major axis, and the line segment joining $(0, 4)$ and $(0, -4)$ is the minor axis.

Since $b^2 = a^2 - c^2$, it follows that

$$c^2 = a^2 - b^2 = 25 - 16 = 9.$$

The foci for the ellipse are $(-3, 0)$ and $(3, 0)$. The graph of this ellipse is shown in Figure 11.12.

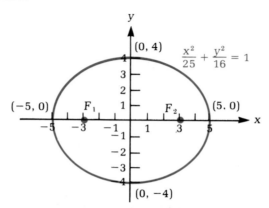

FIGURE 11.12

When the foci are on the y axis instead of the x axis with the center still at the origin, the equation of the ellipse is

$$\frac{x^2}{b^2} + \frac{y^2}{a^2} = 1. \tag{11.11}$$

The points $(0, a)$ and $(0, -a)$ are the vertices of the ellipse. The line joining these vertices on the y axis is now the major axis, and the line joining $(b, 0)$ and $(-b, 0)$ on the x axis is the minor axis.

Example 11.10

Determine the equation of the ellipse given that the foci are $(0, 3)$ and $(0, -3)$ and the length of the major axis is 8. Sketch the graph of the ellipse.

Solution: The points $(0, a)$ and $(0, -a)$ are the vertices of the ellipse. Therefore the length of the major axis is $2a$, and

$$2a = 8$$

$$a = 4.$$

Since $a = 4$ and $c = 3$, it follows that

$$b^2 = a^2 - c^2 = 16 - 9 = 7.$$

Substituting $a^2 = 16$ and $b^2 = 7$ in equation (11.11), we have

$$\frac{x^2}{7} + \frac{y^2}{16} = 1$$

as the equation for the ellipse. Its graph is displayed in Figure 11.13.

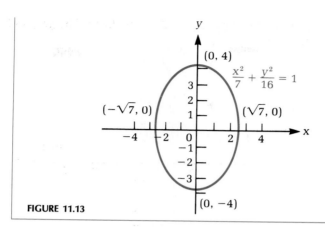

FIGURE 11.13

The hyperbola is the last of the conic sections we have left to explore.

The Hyperbola

Definition 11.2.3

A *hyperbola* is the collection of all points in the plane such that the absolute value of the difference of the distances from any point to two fixed points is constant.

The two fixed points are called the *foci* of the hyperbola. As in the case of the ellipse, let $F_1(-c, 0)$ and $F_2(c, 0)$ be the coordinates of the foci, as shown in Figure 11.14. Let $P(x, y)$ be a point in the plane such that the absolute value of

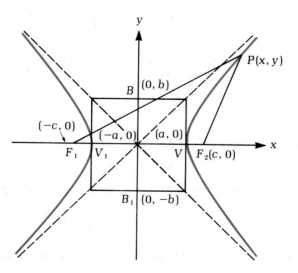

FIGURE 11.14

the difference of its distance from F_1 and F_2 is constant, say, $2a$. In other words,

$$|d_{PF_1} - d_{PF_2}| = 2a.$$

We have two cases to consider.

Case 1

If $d_{PF_1} > d_{PF_2}$, then $d_{PF_1} - d_{PF_2} = 2a$. Using the distance formula, we have

$$\sqrt{(x + c)^2 + y^2} - \sqrt{(x - c)^2 + y^2} = 2a.$$

Case 2

If $d_{PF_1} < d_{PF_2}$, then $d_{PF_1} - d_{PF_2} = -2a$, and we have

$$\sqrt{(x + c)^2 + y^2} - \sqrt{(x - c)^2 + y^2} = -2a.$$

In either case, if we square twice and rearrange the terms, we obtain

$$(a^2 - c^2)x^2 + a^2y^2 = a^2(a^2 - c^2). \tag{11.12}$$

This is the same as equation (11.9), which we derived for an ellipse. In the case of a hyperbola, $a < c$, which means that $a^2 < c^2$ and hence $a^2 - c^2 < 0$. Dividing (11.12) by $a^2(a^2 - c^2)$, we obtain

$$\frac{x^2}{a^2} - \frac{y^2}{b^2} = 1, \tag{11.13}$$

where $b^2 = c^2 - a^2$.

Equation (11.13) is the **standard form of the equation of the hyperbola** having foci $(\pm c, 0)$ on the x axis. If the hyperbola is such that its foci are $(0, \pm c)$ on the y axis, then the equation of the curve in its standard form is

$$\frac{y^2}{a^2} - \frac{x^2}{b^2} = 1. \tag{11.14}$$

The points $(a, 0)$ and $(-a, 0)$ in Figure 11.14 are the vertices for the equation of hyperbola (11.13). The line segment joining the vertices is called the ***transverse axis***; its length is $2a$. The line segment joining $B_1(0, -b)$ and $B(0, b)$ ***conjugate axis*** is called the **conjugate axis.** The dashed lines in the figure, which are obtained by extending the diagonals of the rectangle with sides of length $2a$ and $2b$, are ***asymptotes*** the **asymptotes** of the hyperbola. The equations of the asymptotes are given by

$$\frac{x^2}{a^2} - \frac{y^2}{b^2} = 0$$

or

$$y^2 = \frac{b^2}{a^2}x^2.$$

Hence

$$y = \pm \frac{b}{a} x. \tag{11.15}$$

Example 11.11

Determine the vertices, foci, and equations of the asymptotes for the hyperbola
$$16x^2 - 9y^2 = 144.$$

Solution: Dividing both sides of the equation by 144, we get
$$\frac{x^2}{9} - \frac{y^2}{16} = 1,$$

which is the standard equation of the hyperbola having its center at the origin and its foci on the x axis. Comparing this equation with (11.13), we see that

$$a = 3, \qquad b = 4, \qquad \text{and} \qquad c = \sqrt{a^2 + b^2} = \sqrt{9 + 16} = 5.$$

Thus the hyperbola has vertices $(-3, 0)$ and $(3, 0)$ and foci $(-5, 0)$ and $(5, 0)$. The conjugate axis extends from $(0, -4)$ to $(0, 4)$. The equations of the asymptotes are

$$y = -\tfrac{4}{3}x \qquad \text{and} \qquad y = \tfrac{4}{3}x.$$

Example 11.12

The vertices for the hyperbola are $(0, \pm 6)$, and the equations of the asymptotes are $2y = \pm 3x$. Find the coordinates of foci and the equation of the hyperbola.

Solution: Since the vertices are on the y axis, the standard equation of the hyperbola is of the form

$$\frac{y^2}{a^2} - \frac{x^2}{b^2} = 1.$$

The equation of the asymptotes is $2y = \pm 3x$, which in the standard form is

$$y = \pm \frac{a}{b} x = \pm \frac{3}{2} x.$$

Observe that $a = 6$, so

$$\frac{6}{b} = \frac{3}{2}.$$

Hence

$$b = 4 \qquad \text{and} \qquad c = \sqrt{a^2 + b^2} = \sqrt{36 + 16}$$
$$= \sqrt{52} = 2\sqrt{13}.$$

The coordinates of the foci are $(0, \pm 2\sqrt{13})$, and the equation of the hyperbola is

$$\frac{y^2}{36} - \frac{x^2}{16} = 1.$$

EXERCISES

Graph the parabolas in Exercises 1 to 14.

1. $y^2 = 2x$
4. $y^2 = 12x$

2. $y^2 = \frac{3}{2}x$
5. $y^2 = -\frac{1}{2}x$

3. $y^2 = 4x$
6. $y^2 = -5x$

7. $x^2 = 4y$

8. $x^2 = \frac{7}{2}y$

9. $x^2 = \frac{y}{2}$

10. $x^2 = -5y$
13. $x = y^2 - 6y + 9$

11. $x = 3y^2 + 8y - 3$
14. $x = 2y^2 + 4y + 3$

12. $x = -2y^2 + 2y - 3$

Find the vertex, focus, and directrix of the parabolas in Exercises 15 to 20.

15. $y^2 = 40x$

16. $y^2 = 4x$

17. $y^2 = -\frac{7}{3}x$

18. $x^2 = -\frac{3}{2}y$

19. $x^2 = 10y$

20. $x^2 = \frac{1}{4}y$

In Exercises 21 to 26, find the equation of the parabola having its vertex at the origin, given that

21. its focus is $(6, 0)$.
23. its directrix is $y = 3$.
25. it passes through the point $(4, 8)$, and the axis of symmetry is the x axis.
26. it passes through the point $(-2, -4)$, and the axis of symmetry is the y axis.

22. its focus is $(0, 4)$.
24. its directrix is $x + 2 = 0$.

Graph the ellipses in Exercises 27 to 32.

27. $\frac{x^2}{16} + \frac{y^2}{9} = 1$
30. $25x^2 + 16y^2 = 400$

28. $\frac{x^2}{9} + \frac{y^2}{4} = 1$
31. $25x^2 + 36y^2 = 900$

29. $9x^2 + 4y^2 = 36$
32. $16x^2 + 9y^2 = 144$

Determine the equation of the ellipses in Exercises 33 to 36, and sketch their graphs.

33. Foci $(2, 0)$ and $(-2, 0)$; length of major axis is 8.
35. Vertices $(5, 0)$ and $(-5, 0)$; foci $(3, 0)$ and $(-3, 0)$.

34. Foci $(0, 2)$ and $(0, -2)$; length of major axis is 6.
36. Vertices $(0, 6)$ and $(0, -6)$, foci $(0, 4)$ and $(0, -4)$.

Graph the hyperbolas in Exercises 37 to 42.

37. $x^2 - 4y^2 = 1$
40. $16x^2 - 25y^2 = 400$

38. $4x^2 - y^2 = 1$
41. $9y^2 - 25x^2 = 225$

39. $4x^2 - 9y^2 = 36$
42. $9y^2 - 4x^2 = 36$

Determine the equations of the hyperbolas in Exercises 43 to 50 in the standard form (11.13) or (11.14).

43. Vertices (± 2, 0); foci (± 3, 0).

44. Vertices (± 4, 0); foci (± 5, 0).

45. Vertices (0, ± 3); foci (0, ± 5).

46. Vertices (0, ± 5); asymptotes $y = \pm\dfrac{5}{4}x$.

47. Vertices (± 3, 0); asymptotes $y = \pm 2x$.

48. Foci (0, ± 6); length of conjugate axis is $4\sqrt{5}$.

49. Foci (± 5, 0); asymptotes $y = \pm\dfrac{x}{2}$.

50. Ends of conjugate axis (± 4, 0); asymptotes $y = \pm\dfrac{3}{2}x$.

11.3

FUNDAMENTAL PRINCIPLE OF COUNTING

Suppose that four members A, B, C, and D are eligible for appointment to the offices of president and secretary of a club. In how many ways can these two positions be filled, assuming that one person cannot hold both positions? Can you list all the possibilities? To handle problems like this one, it helps to have a tree diagram such as in Figure 11.15.

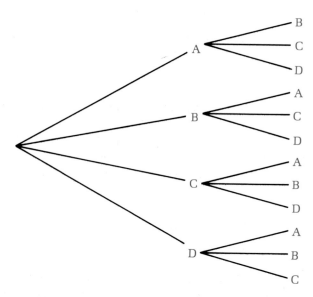

FIGURE 11.15

Note that the tree diagram takes *order* into account. Thus AB and BA count as two distinct arrangements; the first one refers to the possibility that A is the president and B is the secretary, while the second one denotes the case in which B is the president and A is the secretary. *Order is the essence of such arrangements, and any change in order yields a completely different arrangement.*

The diagram shows that there are four branches corresponding to the appointment of A, B, C, and D as president. After this appointment, three candidates are left, and any can be chosen as the secretary. For instance, if A is appointed as president, there are three remaining candidates B, C, and D for the secretary. Since there are four possibilities for president and corresponding to each of the four possibilities there are three possibilities for secretary, in all there are $4 \times 3 = 12$ different ways the two positions can be filled. Looking at the tree, we see that it has 12 different paths along the branches corresponding to the 12 different ways in which the president and the secretary can be appointed.

As another illustration, suppose that a commuter train to New York has six coaches. In how many ways can Mr. Wong and Ms. Bowman be assigned to a coach so that they do not ride in the same coach? Apparently, Mr. Wong can be assigned to any one of the six coaches, and after he is seated, Ms. Bowman can go in any of the five remaining coaches. Thus there are $6 \times 5 = 30$ different possibilities. These examples illustrate the **fundamental principle of counting.**

fundamental principle of counting

> If an operation consists of two separate steps, of which the first can be performed in n different ways, and corresponding to each of these n ways there are p ways of performing the second step, then the entire operation can be completed in $n \times p$ different ways.

Example 11.13

A bus offers six routes between New York and Chicago. In how many ways can a visitor from New York go to Chicago and return by a different route?

Solution: There are two steps: choosing a route from New York to Chicago and choosing a return trip from Chicago to New York. Since a visitor may use any of the six routes from New York to Chicago, the first step can be performed in 6 ways. After a route to Chicago is selected, five routes are left and any one of them can be used for the return trip. The second step can be performed in 5 ways. Thus, there are $6 \times 5 = 30$ different ways.

Example 11.14

Mike wants to buy one of the seven selected gifts from a catalog and have it sent by United Parcel Post, Federal Express, regular parcel post, or first-class mail. In how many different ways can he complete the transaction?

Solution: Two steps in the transaction are choosing a gift and selecting a method of sending it. Since seven different gifts are available, the first step can be performed in 7 ways. The second step can be completed in 4 ways corresponding to four methods of sending the gift. By the fundamental principle of counting, there are $7 \times 4 = 28$ different ways in which the transaction can take place.

We now generalize the fundamental principle of counting to apply to operations involving more than two steps.

> If an operation consists of m separate steps, of which the first can be performed in n_1 ways, the second can be performed in n_2 ways, the third can be performed in n_3 ways, and so on for the m steps, then the entire operation can be performed in
>
> $$n_1 \times n_2 \times n_3 \times \cdots \times n_m$$
>
> ways.

Example 11.15

A restaurant menu offers a choice of four soups, five meat dishes, six desserts, and a choice of coffee, tea, milk, or a soft drink. In how many ways can someone order a meal consisting of a soup, meat dish, dessert, and beverage?

Solution: The problem requires the filling of four places shown:

Soup	Meat dish	Dessert	Beverage
☐	☐	☐	☐

There are 4 ways of filling the first place because four soups are available. Then there are 5 ways of filling the second place, corresponding to the five meat dishes. The first two places can be filled in $4 \times 5 = 20$ ways. The third place can be filled in 6 ways, corresponding to the six choices of dessert. Thus, the first three places can be filled in $4 \times 5 \times 6 = 120$ ways. There are 4 ways to fill the fourth place, corresponding to the four choices of beverage. By the fundamental principle of counting, there are

$$4 \times 5 \times 6 \times 4 = 480$$

ways of ordering a meal consisting of soup, a meat dish, dessert, and beverage.

Example 11.16

In a certain state, automobile license plates start with one letter of the alphabet followed by five digits (with repetitions permitted). Assuming that all 26 letters of the alphabet and all digits, 0 through 9, are used, how many different license plates can be made?

Solution: There are six places to fill:

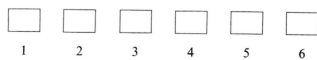

There are 26 letters in the alphabet, and any can be placed in the first place. Each of the next five places can be filled in 10 different ways. By the fundamental principle of counting, the number of possible license plates is

$$26 \times 10 \times 10 \times 10 \times 10 \times 10 = 2,600,000.$$

Example 11.17   In how many ways can six people be assigned as managers to the six branches of a supermarket chain?

Solution: The problem involves the filling of six places. Any of the six people can be assigned to the first branch, leaving five persons. There are then 5 ways of assigning the second person to the next store. The first two places can therefore be filled in $6 \times 5 = 30$ ways. Now there are four persons yet to be assigned, and any can be assigned to the third store. Thus the first three persons can be placed in $6 \times 5 \times 4 = 120$ different ways. Using this reasoning, we see that all six persons can be assigned to the six branches in

$$6 \times 5 \times 4 \times 3 \times 2 \times 1 = 720$$

ways.

EXERCISES

1. A girl has 8 skirts and 6 blouses. How many different skirt-blouse combinations can she wear?

2. There are 4 different routes from city A to city B. In how many ways can a person make a round trip without taking the same route both ways?

3. A club has 4 candidates for president, 6 for vice-president, and 5 for treasurer. How many different executive councils could be elected?

4. A manufacturer offers a basic kitchen cabinet with 5 styles of hinges, 6 styles of doorknobs, and 8 wood finishes. How many different styles of kitchen cabinets are available?

5. A car manufacturer offers 15 color choices, 8 choices of body style, and 10 choices of upholstery. How many cars are required for a complete display of every possible combination?

6. The owner of a sailboat received six different signal flags when she purchased the boat. If the order in which the flags are arranged determines the signal being sent, how many three-flag signals are possible?

7. In how many ways can 6 executives be seated in a conference room with 8 chairs, assuming that chairs are placed in a row?

8. In how many ways can 5 students be seated in a classroom with 12 desks, assuming that chairs are placed in a row?

9. Five finalists in the Miss U.S.A. contest are Miss California, Miss New York, Miss Ohio, Miss Rhode Island, and Miss Texas. In how many ways can the judges choose the winner, first runner-up, and second runner-up?

10. In how many ways can an accounting book, a biology book, a chemistry book, and a mathematics book be arranged on a shelf?

11. Six candidates for a managerial position of a department store are ranked according to the weighted average of several criteria. How many different rankings are possible if no two candidates receive the same rank?

12. A department store manager wants to display 7 brands of a product along one shelf of an aisle. In how many ways can he arrange these brands?

13. How many 5-digit numerals can be formed by using the digits 2, 3, 4, 5, and 6 with no digits being repeated? How many of these are greater than 40,000? How many of these numerals are less than 30,000?

14. How many 5-digit numerals can be formed by using 4, 5, 6, 7, and 8 with no digit being repeated? How many of these are multiples of 5?

15. Eight guests have been invited to sit in the 8 seats at the head banquet table. How many different seating arrangements are possible if all guests accept the invitation?

16. In how many ways can 4 boys and 4 girls be seated in a row of 8 chairs if
   (a) the seats are not assigned?
   (b) girls and boys have to be seated alternately?
   (c) the first and the last chair are to be occupied by girls only?

17. A telephone number consists of a sequence of 7 digits (with repetitions permitted) selected from the numbers 0 through 9. How many different telephone numbers are possible if the first digit cannot be 0 or 1?

18. An automobile license plate contains 3 letters followed by 3 numerals (with repetitions permitted). A typical license plate is CZZ883. Assume that all 26 letters of the alphabet and all numerals, 0 through 9, are used. How many license plates can be made?

11.4

PERMUTATIONS

In this section, we show that the fundamental principle of counting provides a general method for finding the number of distinct arrangements of n different things taken r at a time. For many types of problems, this approach can be shortened by means of convenient symbols and formulas which we now introduce.

Definition 11.4.1

> The product of the first n positive integers is called n *factorial* and is denoted by $n!$. If $n = 0$, then we define $0! = 1$.

Thus

$$n! = n(n - 1)(n - 2)(n - 3) \cdots 3 \cdot 2 \cdot 1$$
$$= 1 \cdot 2 \cdot 3 \cdots (n - 2)(n - 1)n.$$

In particular,

$$1! = 1$$

$$2! = 2 \times 1 = 2$$

$$3! = 3 \times 2 \times 1 = 6$$

$$4! = 4 \times 3 \times 2 \times 1 = 24$$

$$5! = 5 \times 4 \times 3 \times 2 \times 1 = 120.$$

Factorial notation is very useful for representing large numbers. Note that

$$5! = 5 \times 4!, \qquad 6! = 6 \times 5!, \qquad \ldots, \qquad 100! = 100 \times 99!.$$

In general,

$$n! = n \times (n - 1)!.$$

Arithmetic involving factorials can be carried out easily if we are careful in using the factorial as it is defined.

Example 11.18

Compute the following:

(a) $7! - 4!$ **(b)** $(7 - 4)!$

Solution:

(a) $7! = 7 \times 6 \times 5 \times 4 \times 3 \times 2 \times 1 = 5040$

   $4! = 4 \times 3 \times 2 \times 1 = 24$

$$7! - 4! = 5040 - 24 = 5016$$

(b) $(7 - 4)! = 3! = 3 \times 2 \times 1 = 6$

   Observe that

$$(7 - 4)! \neq 7! - 4!$$

and in general

$$(n - r)! \neq n! - r!,$$

where n and r are positive integers.

Example 11.19

Compute $\dfrac{8!}{4!}$.

Solution: Note that

$$\frac{8!}{4!} \neq 2!,$$

and we must proceed as follows:

$$8! = 8 \times 7 \times 6 \times 5 \times 4 \times 3 \times 2 \times 1 = 40{,}320$$

$$4! = 4 \times 3 \times 2 \times 1 = 24$$

Thus,

$$\frac{8!}{4!} = \frac{40{,}320}{24} = 1680.$$

The arithmetic is simpler if we do not multiply the numbers before dividing, that is, if we write

$$\frac{8!}{4!} = \frac{8 \times 7 \times 6 \times 5 \times \cancel{4} \times \cancel{3} \times \cancel{2} \times \cancel{1}}{\cancel{4} \times \cancel{3} \times \cancel{2} \times \cancel{1}}$$

$$= 8 \times 7 \times 6 \times 5$$

$$= 1680$$

We could also write

$$8! = 8 \times 7 \times 6 \times 5 \times (4!).$$

Then

$$\frac{8!}{4!} = \frac{8 \times 7 \times 6 \times 5 \times (4!)}{4!}$$

$$= 8 \times 7 \times 6 \times 5$$

$$= 1680,$$

as before. In latter form, 4! can be divided out of both the numerator and the denominator of the fraction.

Definition 11.4.2

Any arrangement of r objects taken from a set of n objects is called a *permutation* of n objects taken r at a time. For the sake of brevity, the number of possible arrangements is denoted by $P(n, r)$, $r \leq n$.

Suppose that 8 members

$$\{A, B, C, D, E, F, G, H\}$$

are eligible for appointment to the office of president, vice-president, and secretary of a club. We say that this is a permutation of 8 distinct objects (members of the club) taken 3 at a time (offices to be filled). We denote this by

$$P(8, 3).$$

To compute $P(8, 3)$, we observe that any of the 8 persons can be appointed as president and any of the remaining 7 can be offered the position of vice-president. This leaves 6 persons, and any can be appointed as secretary. By the fundamental principle of counting,

$$P(8, 3) = \underbrace{8 \times 7 \times 6}_{\substack{\text{three} \\ \text{factors}}} = 336.$$

Suppose we had 10 members of the club, and we had to elect a president, vice-president, secretary, and treasurer. Then we would have

$$P(10, 4) = \underbrace{10 \times 9 \times 8 \times 7}_{\text{four factors}} = 5040.$$

How many selections would be possible if 12 members of the club were eligible for election and we had to elect 5 officers? In that case we would have

$$P(12, 5) = \underbrace{12 \times 11 \times 10 \times 9 \times 8}_{\text{five factors}} = 95{,}040.$$

As we progress, we see some patterns:

1. The number of factors in each product is the same as the number of positions.
2. The first number in the product is the same as the number of persons available. This number is multiplied by a sequence of numbers that decrease by 1 until we have as many numbers as there are positions to be filled.
3. The last number in the product is 1 more than the difference between the number of persons and the number of positions to be filled.

Proceeding along these lines, we see that

$$P(n, r) = \underbrace{n(n - 1)(n - 2) \cdots (n - r + 1)}_{r \text{ factors}}.$$

To express $P(n, r)$ in terms of factorials, we multiply the numerator and the denominator of

$$\frac{n(n - 1)(n - 2) \cdots (n - r + 1)}{1}$$

by $(n - r)(n - r - 1) \cdots 2 \cdot 1$. Hence,

$$P(n, r) = \frac{n(n - 1)(n - 2) \cdots (n - r + 1)(n - r)(n - r - 1) \cdots 2 \cdot 1}{(n - r)(n - r - 1) \cdots 2 \cdot 1}.$$

The numerator is the product of the integers from n to 1, while the denominator is the product of first $n - r$ integers, so that

$$P(n, r) = \frac{n!}{(n - r)!}.$$

Thus, we have shown that

$$P(n, r) = n(n - 1)(n - 2) \cdots (n - r + 1) = \frac{n!}{(n - r)!}. \qquad \textbf{(11.16)}$$

If $r = n$, then $n - r = 0$ and $0! = 1$. Thus (11.16) reduces to

$$P(n, n) = n! \qquad \text{(11.17)}$$

Example 11.20

(a) $P(5, 2) = \dfrac{5!}{(5 - 2)!} = \dfrac{5!}{3!} = 5 \times 4 = 20$

(b) $P(6, 3) = \dfrac{6!}{(6 - 3)!} = \dfrac{6!}{3!} = 6 \times 5 \times 4 = 120$

(c) $P(9, 4) = \dfrac{9!}{(9 - 4)!} = \dfrac{9!}{5!} = 9 \times 8 \times 7 \times 6 = 3024$

(d) $P(8, 8) = 8! = 8 \times 7 \times 6 \times 5 \times 4 \times 3 \times 2 \times 1 = 40{,}320$

(e) $P(52, 52) = 52! = 8.065817512 \times 10^{67}$, which is a very large number.

Example 11.21

How many "words" of 4 letters can be formed from the letters a, b, c, d, e, and f if each letter is used only once?

Solution: Since these letters in different orders constitute different words, the result is the number of permutations of 6 objects taken 4 at a time. That is,

$$P(6, 4) = \dfrac{6!}{(6 - 4)!} = \dfrac{6!}{2!} = 6 \times 5 \times 4 \times 3 = 360 \text{ words.}$$

Example 11.22

What is the possible number of ways in which 6 students can be seated in a classroom with 12 desks?

Solution: There are 6 seats to fill and 12 desks from which to choose. The result is the number of permutations of 12 distinct objects taken 6 at a time. Thus

$$P(12, 6) = \dfrac{12!}{(12 - 6)!} = \dfrac{12!}{6!} = 12 \times 11 \times 10 \times 9 \times 8 \times 7$$
$$= 665{,}280.$$

Example 11.23

A dormitory has 9 floors. In how many ways can 4 students reside if they must live on different floors?

Solution: Of 9 available floors, 4 students need to occupy 4 floors. This corresponds to a selection of 4 objects from a set of 9 objects, which can be accomplished in

$$P(9, 4) = \dfrac{9!}{(9 - 4)!} = \dfrac{9!}{5!} = 9 \times 8 \times 7 \times 6 = 3024 \text{ ways.}$$

Example 11.24

In how many ways can nine employees stand in a line to receive their paychecks at the payroll counter?

Solution: This corresponds to the case where $r = n$. Thus, the number of permutations of a set of 9 objects, taken all together, is

$$P(9, 9) = 9! = 362,880.$$

Example 11.25

In an inn, 6 rooms in a row are to be assigned at random to 6 guests, two of whom are friends. What is the number of possible arrangements so that two friends are assigned rooms side by side?

Solution: Let A, B, C, D, E, and F represent 6 guests. Suppose that D and E are the 2 friends who would like to have rooms side by side. Since D and E are to appear together, we may count them as one entity. Then the number of permutations of 5 different letters

$$A, B, C, DE, F$$

taken all together is $5! = 120$. But every such arrangement gives $2! = 2$ separate arrangements when D and E are arranged among themselves as DE and ED. Thus, the number of possible arrangements is

$$(5!)(2!) = 120 \times 2 = 240.$$

Example 11.26

Five married couples have bought 10 seats for a Broadway show. What is the number of seating arrangements if all the men are to sit together and all the women are to sit together?

Solution: To begin, men occupy either the first 5 seats or the last 5 seats. This leaves the remaining seats to be occupied by women as follows:

	Men					Women				
	1	2	3	4	5	6	7	8	9	10

	Women					Men				
	1	2	3	4	5	6	7	8	9	10

The number of arrangements in which 5 men can be seated is $5! = 120$. Similarly, 5 women can be seated in $5! = 120$ ways. Thus, the number of possible arrangements is

$$2(5!)(5!) = 2 \times 120 \times 120 = 28,800.$$

EXERCISES

In Exercises 1 to 16, evaluate the expression.

1. $6! - 4!$ **2.** $(6 - 4)!$ **3.** $8! - 5!$ **4.** $(8 - 5)!$

5. $(10 - 4)!$ **6.** $10! - 4!$ **7.** $\dfrac{10!}{4!}$ **8.** $\dfrac{12!}{5!}$

9. $2(7!)$ **10.** $8! - 2(7!)$ **11.** $\dfrac{8!}{3!5!}$ **12.** $\dfrac{10!}{4!6!}$

13. $\dfrac{11!}{4!7!}$ **14.** $\dfrac{13!}{5!6!2!}$ **15.** $\dfrac{15!}{4!5!6!}$ **16.** $\dfrac{12!}{(4!)^3}$

In Exercises 17 to 36, evaluate the term.

17. $P(8, 1)$ **18.** $P(8, 2)$ **19.** $P(8, 3)$ **20.** $P(8, 4)$

21. $P(9, 3)$ **22.** $P(9, 5)$ **23.** $P(10, 4)$ **24.** $P(10, 6)$

25. $P(12, 4)$ **26.** $P(12, 5)$ **27.** $P(14, 3)$ **28.** $P(16, 2)$

29. $P(40, 2)$ **30.** $P(52, 1)$ **31.** $P(52, 3)$ **32.** $P(100, 2)$

33. $P(n, 1)$ **34.** $P(n, 2)$ **35.** $P(n, 3)$ **36.** $P(n, 5)$

In Exercises 37 to 44, determine the number of nonsense words that can be formed from the given word.

37. time   **38.** globe   **39.** single   **40.** travel

41. second   **42.** holiday   **43.** counter   **44.** seating

45. How many nonsense words can be formed from the letters of the word *payment* taken
     (a) 3 at a time?   **(b)** 4 at a time?   **(c)** 5 at a time?   **(d)** 6 at a time?

46. How many 4-letter code words can be made from the 26 letters of the alphabet if no letter is repeated?

47. How many 4-digit numerals can be formed from the digits 3, 4, 5, 6, 7, and 8 if no digit is repeated?

48. How many license plates bearing 4 letters can be made from the letters of the alphabet if repetition of letters is not permitted?

49. What is the total number of ways in which 5 students can be seated in a classroom with 10 desks?

50. A commuter train has 6 coaches. In how many ways can 4 people travel if they must ride in different coaches?

51. In how many ways can the letters of the word *spectrum* be arranged so that *t* and *r* are always together?

52. In how many ways can the letters of the word *matrix* be arranged so that *m* and *a* are always together?

53. In how many ways can the letters of the word *social* be arranged? How many of these arrangements begin with *s*? How many of these arrangements begin with *s* and end in *l*? How many of these arrangements have *o* and *c* together?

54. In an inn, 8 rooms in a row are to be assigned at random to 8 guests, 2 of whom are from the same town. What is the number of possible arrangements so that the 2 guests from the same town are assigned rooms side by side?

55. There are 6 guests to be seated at the head banquet table. How many different seating arrangements are possible if the invitation is accepted by all the guests?

56. A state report on crime consists of nine volumes numbered 1 to 9. In how many ways can these volumes be placed on a shelf?

57. Seven speakers have been scheduled to address a convention. In how many different ways can their speaking assignments be arranged?

58. Eight horses are entered in a race. In how many ways can the horses finish?

59. Eight students are to take an examination, 2 in mathematics and the remaining 6 in different subjects. In how many ways can they be seated in a row so that the students taking the mathematics examination do not sit next to each other?

60. Six couples have bought 12 seats in a row for a certain show. In how many ways can these 12 persons be seated? What is the number of seating arrangements if all the women are to sit together and all the men are to sit together?

11.5 COMBINATIONS

In previous sections, we obtained formulas that enabled us to count the number of ways in which r objects can be arranged in a row from a set of n different objects. Many problems require us to select r objects from a set of n objects without any regard to *order*. Such problems involve combinations. For example, a hand of cards consisting of jack, queen, king, ace, and 10 of hearts is the same as one that is dealt in the order: queen, jack, ace, king, and 10 of hearts. Again, choosing an executive committee of three members, "Susan, Elaine, and Kerry," is the same committee as "Kerry, Susan, and Elaine." The order in which the members of the committee have been selected is of no significance whatsoever. Thus the fundamental difference between a permutation and a combination lies in the fact that the *order of the arrangement is not relevant in a combination, while in a permutation problem the order is essential.* To illustrate this point, consider the digits 1, 2, 3, and 4 taken 3 at a time without repetition. An application of the fundamental principle of counting results in 24 different arrangements. Those involving only 1, 2, and 3, are

$$(1, 2, 3), \quad (1, 3, 2), \quad (2, 1, 3), \quad (2, 3, 1), \quad (3, 1, 2), \quad (3, 2, 1).$$

These permutations constitute the same combination. Similarly, we can take a subset consisting of three digits 1, 2, and 4 and permute its elements in six ways, yielding

$$(1, 2, 4), \quad (1, 4, 2), \quad (2, 1, 4), \quad (2, 4, 1), \quad (4, 1, 2), \quad (4, 2, 1).$$

Or suppose that a club of six persons decides to form an executive council consisting of any three members, and we are interested in finding the number of executive councils that can be formed. If the six persons are identified by the letters A, B, C, D, E, and F, then there are 20 possible combinations—each consisting of three members:

ABC,	ABD,	ABE,	ABF,	ACD,	ACE,	ACF,
ADE,	ADF,	AEF,	BCD,	BCE,	BCF,	BDE,
BDF,	BEF,	CDE,	CDF,	CEF,	DEF.	

Had we wished to consider the order in which three members could be selected to form an executive council, the answer would be

$$P(6, 3) = 6 \times 5 \times 4 = 120.$$

Each combination of three persons would then be counted six times and thus constitute six different permutations. Since this argument is valid for any three members constituting any executive council, the total number of ways in which three members can be selected is given by

$$\frac{P(6, 3)}{6} = \frac{120}{6} = 20.$$

These examples underscore the important distinction between a permutation and a combination. **In a permutation, order is taken into consideration, while in combination problems, the order is of no significance.**

It is usually neither desirable nor feasible to list all the possible combinations and then count them in the usual manner. The list may simply get too long. In the above example, if there had been 25 members in the club, with five to be selected, the list would have had 53,130 different possible combinations!

Given a counting problem, we must first decide from the nature of the problem whether it involves permutation formulas or combination techniques. The decision depends on the answer to the question: *Is order important in the problem?* For example, if we are to arrange four books on a shelf, it is natural to regard ABCD and ACDB as two different arrangements, and we must take order into consideration. But if we are selecting four books to take along with us on vacation, order hardly matters, and combinations are involved.

Definition 11.5.1

Any collection of r objects selected with complete disregard to their order from a set of n different objects is called a *combination* of the n objects taken r at a time. Symbolically, the total number of such possible combinations is denoted by

$$\binom{n}{r}, \quad \text{where } r \leq n.$$

With this notation in mind, the formula for $\binom{n}{r}$ is

$$\binom{n}{r} = \frac{P(n, r)}{r!}. \tag{11.18}$$

Applying (11.16), we have

$$\binom{n}{r} = \frac{n!}{r!(n - r)!}. \tag{11.19}$$

If $r = 0$ and $0! = 1$, then substituting $r = 0$ in (11.19), we get

$$\binom{n}{0} = \frac{n!}{0!(n - 0)!} = \frac{n!}{n!} = 1. \tag{11.20}$$

If $r = n$, then $n - r = 0$ and $0! = 1$. Then (11.19) reduces to

$$\binom{n}{n} = \frac{n!}{n!0!} = \frac{n!}{n!} = 1. \tag{11.21}$$

Example 11.27

(a) $\binom{5}{2} = \frac{P(5, 2)}{2!} = \frac{5!}{2!3!} = \frac{5 \times 4}{2 \times 1} = 10$

(b) $\binom{7}{3} = \frac{P(7, 3)}{3!} = \frac{7!}{3!4!} = \frac{7 \times 6 \times 5}{3 \times 2 \times 1} = 35$

(c) $\binom{9}{4} = \frac{P(9, 4)}{4!} = \frac{9!}{4!5!} = \frac{9 \times 8 \times 7 \times 6}{4 \times 3 \times 2 \times 1} = 126$

(d) $\binom{12}{5} = \frac{P(12, 5)}{5!} = \frac{12!}{5!7!} = \frac{12 \times 11 \times 10 \times 9 \times 8}{5 \times 4 \times 3 \times 2 \times 1} = 792$

(e) $\binom{8}{0} = 1$

(f) $\binom{8}{8} = 1.$

Example 11.28

In how many ways can a congressional subcommittee of 6 be chosen from a committee of 12 members?

Solution: The number of possibilities of selecting 6 members from a group of 12 is given by

$$\binom{12}{6} = \frac{P(12, 6)}{6!} = \frac{12!}{6!6!}$$

$$= \frac{12 \times 11 \times 10 \times 9 \times 8 \times 7}{6 \times 5 \times 4 \times 3 \times 2 \times 1} = 924.$$

Example 11.29

In how many ways can a poker hand of 5 cards be dealt from a standard bridge deck of 52 cards?

Solution: The selection of 5 cards can be made in

$$\binom{52}{5} = \frac{P(52, 5)}{5!} = \frac{52!}{5!47!}$$

$$= \frac{52 \times 51 \times 50 \times 49 \times 48}{5 \times 4 \times 3 \times 2 \times 1}$$

$$= 2,598,960 \text{ ways.}$$

Example 11.30

A committee of 3 labor and 4 management personnel is to be selected from 6 labor and 8 management representatives. In how many ways can the committee be formed?

Solution: The labor representation can be accomplished in

$$\binom{6}{3} = \frac{6!}{3!3!} = 20 \text{ ways.}$$

The management personnel can be selected in

$$\binom{8}{4} = \frac{8!}{4!4!} = 70 \text{ ways.}$$

An application of the fundamental principle of counting yields

$$20 \times 70 = 1400 \text{ ways.}$$

Example 11.31

In how many ways can an employer select 10 new employees from a group of 12 applicants if one particular applicant must be **(a)** selected? **(b)** rejected?

Solution: The number of combinations of 12 applicants taken 10 at a time is

$$\binom{12}{10} = \frac{12!}{10!2!} = 66.$$

(a) If one particular applicant must be selected, we need to select 9 more of the remaining 11. This can be accomplished in

$$\binom{11}{9} = \frac{11!}{9!2!} = 55 \text{ ways.}$$

(b) If one applicant is not to be selected, the employer needs to choose 10 applicants from the remaining group of 11. This can be done in

$$\binom{11}{10} = \frac{11!}{10!1!} = 11 \text{ ways.}$$

Example 11.32

In how many ways can 13 detectives be assigned to two groups containing 6 and 7 detectives?

Solution: Each selection of 6 detectives for the first group leaves the remaining 7 detectives to be assigned to the second group. Hence the required number of combinations is given by

$$\binom{13}{6} = \frac{13!}{6!7!} = 1716 \text{ ways.}$$

Example 11.33

In how many ways can 18 salespeople be assigned to three groups containing 5, 6, and 7 salespeople?

Solution: For the first group of 5, there are 18 persons available, so that there are

$$\binom{18}{5} \text{ ways}$$

in which 5 people can be assigned to the first group. This leaves 13 persons available from which to select 6 persons to go in the second group; consequently, 6 salespeople for the second group can be selected in

$$\binom{13}{6} \text{ ways.}$$

This leaves 7 persons to go in the third group. By multiplying these expressions, we obtain the number of ways in which 18 salespeople can be assigned to three distinct groups. Thus

$$\binom{18}{5}\binom{13}{6} = \frac{18!}{5!13!}\frac{13!}{6!7!} = \frac{18!}{5!6!7!} = 14,702,688.$$

Pascal's Triangle

The number of combinations of n objects taken r at a time, which we have denoted by $\binom{n}{r}$ in this section, plays an important role in probability, statistics, and other branches of mathematics. One method of determining these numbers originated with the French mathematician Blaise Pascal (1623–1662). He presented the following arrangement:

				1		1				$n = 1,$	$r = 0, 1.$
			1		2		1			$n = 2,$	$r = 0, 1, 2.$
		1		3		3		1		$n = 3,$	$r = 0, 1, 2, 3.$
	1		4		6		4		1	$n = 4,$	$r = 0, 1, 2, 3, 4.$
1		5		10		10		5		1	$n = 5,$	$r = 0, 1, 2, 3, 4, 5.$
1	6		15		20		15		6	1	$n = 6,$	$r = 0, 1, \ldots, 6.$

As we continue, the following patterns begin to emerge:

1. Each row begins with a 1 and ends with a 1.
2. Any other term in a row is the sum of the two terms in the row above it.

For example, the numbers in the fourth row are

$$1, \quad 1+3, \quad 3+3, \quad 3+1, \quad 1$$

that is,

$$1, \quad 4, \quad 6, \quad 4, \quad 1.$$

The numbers in the fifth row are

$$1, \quad 1+4, \quad 4+6, \quad 6+4, \quad 4+1, \quad 1$$

or

$$1, \quad 5, \quad 10, \quad 10, \quad 5, \quad 1$$

and so on. The property of the numbers $\binom{n}{r}$ on which *Pascal's triangle* is based is

$$\binom{n}{r} + \binom{n}{r-1} = \binom{n+1}{r}. \tag{11.22}$$

We leave this as an exercise for the reader. (See Exercise 28.)

We also observe that there is a symmetry about the middle term (for n even) or middle pair (for n odd). Precisely, for any positive integer n,

$$\binom{n}{r} = \binom{n}{n-r}, \quad r = 0, 1, 2, \ldots, n. \tag{11.23}$$

We suggest that you use (11.19) to show that each side of the equation is equal to

$$\frac{n!}{r!(n-r)!}.$$

EXERCISES

Evaluate the expressions in Exercises 1 to 5.

1. $\binom{8}{3}$ 2. $\binom{10}{4}$ 3. $\binom{20}{12}$ 4. $\binom{50}{49}$ 5. $\binom{100}{98}$

6. How many committees of 6 representatives can be formed from a group of 12 persons?

7. An electronic circuit may fail at 7 stages. If it fails at exactly 3 stages, in how many ways can this happen?

8. In how many ways can you answer a 16-question true-false examination if you make the same number of answers true as you make false?

9. The Boston Red Sox have 25 players on the team. Assume that each player can play any position. In how many ways can the manager choose 9 players to play in a game?

10. A contractor needs 6 bricklayers, and 14 apply for the job. In how many ways can the contractor select 6 bricklayers?

11. Ms. Hawkes plans to accompany her friend on a sales trip to Chicago. She has 4 coats and 6 dresses in her wardrobe. In how many ways can she choose 3 dresses and 3 coats for the trip?

12. A certain class has 8 men and 4 women.
    (a) In how many ways can the teacher select a committee of 6 students?
    (b) How many of these committees have 2 women?

13. In how many ways can a congressional committee of 5 be chosen from 5 Republicans and 5 Democrats if
    (a) all are equally eligible?
    (b) the committee must consist of three Democrats and two Republicans?

14. Among the eight nominees for two vacancies on a school board are 4 women and 4 men. In how many ways can these vacancies be filled with
    (a) any 2 of the 8 nominees?
    (b) any 2 of the female nominees?
    (c) 1 of the female nominees and 1 of the male nominees?

15. A shipment of 15 alarm clocks contains 1 that is defective. In how many ways can an inspector choose 3 of the alarm clocks for inspection so that
    (a) the defective alarm clock is not included?
    (b) the defective alarm clock is included?

16. Ten managers apply for promotion to district sales representatives from which five are to be chosen.
    (a) In how many ways can these promotions be decided?
    (b) How many of these choices will include a particular candidate?

17. A group of 15 workers decides to send a delegation of 5 to the management to discuss their grievances.
    (a) How many delegations are possible?
    (b) How many of these delegations will include a particular worker?
    (c) How many of these delegations will exclude a particular worker?

18. From a regular deck of 52 cards, 4 are to be selected. How many different selections are possible if the cards
    (a) may be of any suit and any denomination?
    (b) must include exactly 3 spades?
    (c) must be of the same suit?

19. A poker hand is a set of 5 cards selected from a standard deck of 52 cards. How many poker hands contain
    (a) a full house (three cards of one face value and a pair of cards of another face value)?
    (b) three-of-a-kind (three cards of one face value and two different cards)?
    (c) four-of-a-kind (four cards of one face value and one different card)?
    (d) a royal flush (ten, jack, queen, king, and ace of the same suit)?
    (e) a straight flush (five cards in sequence and of the same suit but not a royal flush)?

20. Ten persons are going on a field trip for a history course in 3 cars that will hold 2, 3, and 5 persons. In how many ways could they go on the trip?

21. There are 3 offices available for a staff of 12. The first office can accommodate 3 persons, the second and third offices can take 4 and 5 persons, respectively. How many different arrangements are possible?

22. In how many ways can a coach select 9 baseball players from a group of 14 players if a particular player must be
    (a) included?
    (b) excluded?

23. From a company of 12 soldiers, a squad of 4 is chosen each night. For how many nights could a squad go on duty without two of the squads being identical? In how many of these squads would a particular soldier be included?

24. In how many ways can 5 red balls be drawn from an urn if
    (a) the urn contains only 7 red balls?
    (b) the urn contains 8 red balls?
    (c) the urn contains 10 red, 3 white, and 4 green balls?

25. A baseball squad has 6 outfielders, 7 infielders, 5 pitchers, and 2 catchers. In how many different ways can a team of 9 players be chosen?

26. A committee to consider labor-management problems is composed of 6 outside arbitrators, 7 employees, and 3 employers. The committee is to be selected from 12 arbitrators, 50 employees, and 5 employers. How many committees are possible?

27. Use Pascal's triangle to calculate the values of

$$\binom{7}{r} \qquad \text{for } r = 0, 1, \ldots, 7.$$

28. Using only equation (11.19), show that

$$\binom{n}{r} + \binom{n}{r-1} = \binom{n+1}{r}$$

11.6 BINOMIAL THEOREM

The quantity

$$\binom{n}{r}$$

binomial coefficient   is called a **binomial coefficient** because of the fundamental role it plays in the formulation of the binomial theorem. Expansions of integral powers of $(a + b)^n$ for $n = 0, 1, 2, \ldots$ occur frequently in algebra and are beginning to appear in almost all phases of mathematics. Here we undertake a systematic development of a formula that produces such expansions.

The following identities could be established by direct multiplication:

$$(a + b)^0 = 1$$

$$(a + b)^1 = a + b$$

$$(a + b)^2 = a^2 + 2ab + b^2$$

$$(a + b)^3 = a^3 + 3a^2b + 3ab^2 + b^3$$

$$(a + b)^4 = a^4 + 4a^3b + 6a^2b^2 + 4ab^3 + b^4$$

$$(a + b)^5 = a^5 + 5a^4b + 10a^3b^2 + 10a^2b^3 + 5ab^4 + b^5$$

$$(a + b)^6 = a^6 + 6a^5b + 15a^4b^2 + 20a^3b^3 + 15a^2b^4 + 6ab^5 + b^6.$$

As we progress, this process of direct multiplication soon becomes cumbersome and tedious. We begin to wonder whether there is a better method to arrive at the same results. Note that as we continue expanding larger and larger powers of $a + b$, several patterns emerge, leading to a part of the solution. Can you observe any?

The following are some of the patterns:

1. The coefficients of the first and last terms are both 1.

2. There are $n + 1$ terms in the expansion of $(a + b)^n$.

3. The exponents of a start with n and then decrease by 1 until the exponent of a has decreased to 0 in the last term. The exponent of b is 0 in the first term and then increases by 1 until the exponent of b is n in the last term.

4. The sum of exponents of a and b in a given term is always n.

5. There is a symmetry about the middle term (for n even) or middle pair (for n odd).

Proceeding along these lines, we see that the $n + 1$ terms in the expansion of $(a + b)^n$, without their coefficients, are

$$a^n, \qquad a^{n-1}b, \qquad a^{n-2}b^2, \qquad a^{n-3}b^3, \qquad \ldots, \qquad a^{n-r}b^r, \qquad \ldots, \qquad b^n.$$

Now there remains the question of determining the coefficients of these terms. Why, in the expansion of $(a + b)^3$, do the terms $3a^2b$ and $3ab^2$ exist? To handle such questions, let us examine the actual multiplication of $(a + b)^3$. We have

$$\begin{aligned}
(a + b)^3 &= (a + b)(a + b)(a + b) \\
&= a \cdot a \cdot a + a \cdot a \cdot b + a \cdot b \cdot a + a \cdot b \cdot b + b \cdot a \cdot a \\
&\quad + b \cdot a \cdot b + b \cdot b \cdot a + b \cdot b \cdot b \qquad \text{(8 terms)} \\
&= a^3 + 3a^2b + 3ab^2 + b^3.
\end{aligned}$$

In the eight terms involved in the expansion of $(a + b)^3$, there are three a^2b terms corresponding to the different ways of selecting two a's and one b, one from each of the factors $a + b$. The coefficient of a^3 is 1 because there is only one way in which three a's can be selected, one from each factor $a + b$. The coefficient of ab^2 is 3 because there are three ways of selecting one a and two b's, one from each factor $a + b$. The coefficient of b^3 is 1 because there is only one way of selecting three b's, one from each factor $a + b$. Let us try to obtain the expansion of $(a + b)^4$. The terms in the expansion must be of the form

$$a^4, \qquad a^3b, \qquad a^2b^2, \qquad ab^3, \qquad b^4,$$

and we need to determine how many there are of each type. Each term is obtained by selecting exactly one letter from each factor $a + b$. As far as a^4 is concerned, we select no b and four a's, one from each factor $a + b$. Since this can be done in only $\binom{4}{0}$, or 1, way, a^4 occurs exactly once in the expansion, and its coefficient is 1. To get the second term in the expansion, we select three a's and one b. This can be done in $\binom{4}{1}$, or 4, ways; hence the coefficient of a^3b is 4.

Similarly, the coefficient of a^2b^2 is the number of ways we can select two a's and two b's; namely, $\binom{4}{2}$, or 6, ways. The term ab^3 can be obtained in $\binom{4}{3}$, or 4, ways, and the last term b^4 can be obtained in only 1 way. Proceeding along these lines, we obtain

$$(a + b)^4 = \binom{4}{0}a^4 + \binom{4}{1}a^3b + \binom{4}{2}a^2b^2 + \binom{4}{3}ab^3 + \binom{4}{4}b^4.$$

If we apply the same reasoning to the expansion of $(a + b)^n$, where n is a positive integer, we obtain the following result.

Proposition 11.6.1

Binomial Theorem

If n is a positive integer, then

$$(a + b)^n = \binom{n}{0}a^n + \binom{n}{1}a^{n-1}b + \binom{n}{2}a^{n-2}b^2$$

$$+ \cdots + \binom{n}{r}a^{n-r}b^r + \cdots + \binom{n}{n}b^n.$$

Note that the $n + 1$ terms in the expansion of $(a + b)^n$, without their coefficients, are

$$a^n, \quad a^{n-1}b, \quad a^{n-2}b^2, \quad a^{n-3}b^3, \quad \ldots, \quad a^{n-r}b^r, \quad \ldots,$$
$$a^2b^{n-2}, \quad ab^{n-1}, \quad b^n.$$

In other words, each term in the expansion is of the form

$$a^{n-r}b^r, \qquad r = 0, 1, 2, \ldots, n.$$

The coefficient of this general term is $\binom{n}{r}$, since this corresponds to the number of ways in which r b's and $(n - r)$ a's can be selected, and the complete general term is

$$\binom{n}{r}a^{n-r}b^r.$$

A summation of this general term for $r = 0, 1, 2, \ldots, n$ yields the above assertion.

Corollary 1

From Proposition 11.6.1, with $a = 1$, it follows that

$$(1 + b)^n = \binom{n}{0} + \binom{n}{1}b + \binom{n}{2}b^2 + \binom{n}{3}b^3 + \cdots + \binom{n}{r}b^r + \cdots + \binom{n}{n}b^n.$$

Corollary 2

With $a = b = 1$ in Proposition 11.6.1, it follows that

$$(1 + 1)^n = 2^n = \binom{n}{0} + \binom{n}{1} + \binom{n}{2} + \binom{n}{3} + \cdots + \binom{n}{r} + \cdots + \binom{n}{n}.$$

Example 11.34

Expand $(x + y)^6$.

Solution:

$$(x + y)^6 = \binom{6}{0}x^6 + \binom{6}{1}x^5 y + \binom{6}{2}x^4 y^2 + \binom{6}{3}x^3 y^3$$

$$+ \binom{6}{4}x^2 y^4 + \binom{6}{5}xy^5 + \binom{6}{6}y^6$$

$$= x^6 + 6x^5 y + 15x^4 y^2 + 20x^3 y^3 + 15x^2 y^4 + 6xy^5 + y^6.$$

The calculation of the coefficients is simplified by making use of the fact (11.23) that

$$\binom{n}{r} = \binom{n}{n - r}.$$

In Example 11.34 we needed only to compute up to $\binom{6}{3}$ and then recognize that

$$\binom{6}{4} = \binom{6}{2}, \qquad \binom{6}{5} = \binom{6}{1}, \qquad \text{and} \qquad \binom{6}{6} = \binom{6}{0} = 1.$$

Example 11.35

Expand $(x + 2y)^7$.

Solution:

$$(x + 2y)^7 = \binom{7}{0}x^7 + \binom{7}{1}x^6(2y) + \binom{7}{2}x^5(2y)^2 + \binom{7}{3}x^4(2y)^3$$

$$+ \binom{7}{4}x^3(2y)^4 + \binom{7}{5}x^2(2y)^5 + \binom{7}{6}x(2y)^6 + \binom{7}{7}(2y)^7$$

$$= x^7 + 14x^6 y + 84x^5 y^2 + 280x^4 y^3 + 560x^3 y^4$$

$$+ 672x^2 y^5 + 448xy^6 + 128y^7.$$

Example 11.36

Expand $(1 + 2x)^6$.

Solution: Letting $n = 6$ and $b = 2x$ in Corollary 1, we have

$$(1 + 2x)^6 = \binom{6}{0} + \binom{6}{1}2x + \binom{6}{2}(2x)^2 + \binom{6}{3}(2x)^3$$

$$+ \binom{6}{4}(2x)^4 + \binom{6}{5}(2x)^5 + \binom{6}{6}(2x)^6$$

$$= 1 + 12x + 60x^2 + 160x^3 + 240x^4 + 192x^5 + 64x^6.$$

Example 11.37

Expand $(1 - 3x)^4$.

Solution: With $n = 4$ and $b = -3x$ in Corollary 1, we have

$$(1 - 3x)^4 = [1 + (-3x)]^4$$

$$= \binom{4}{0} + \binom{4}{1}(-3x) + \binom{4}{2}(-3x)^2 + \binom{4}{3}(-3x)^3 + \binom{4}{4}(-3x)^4$$

$$= 1 - 12x + 54x^2 - 108x^3 + 81x^4.$$

Example 11.38

Expand $(2x - 3y)^5$.

Solution: The expansion of $(2x - 3y)^5$ can be expressed as $[2x + (-3y)]^5$. Then

$$[2x + (-3y)]^5 = \binom{5}{0}(2x)^5 + \binom{5}{1}(2x)^4(-3y) + \binom{5}{2}(2x)^3(-3y)^2$$

$$+ \binom{5}{3}(2x)^2(-3y)^3 + \binom{5}{4}(2x)(-3y)^4 + \binom{5}{5}(-3y)^5$$

$$= 32x^5 - 240x^4y + 720x^3y^2 - 1080x^2y^3 + 810xy^4 - 243y^5.$$

Example 11.39

Using the binomial theorem, find the numerical value of $(1.06)^5$.

Solution: Let $n = 5$ and $b = 0.06$ in Corollary 1. It follows that

$$(1 + 0.06)^5 = \binom{5}{0} + \binom{5}{1}0.06 + \binom{5}{2}(0.06)^2 + \binom{5}{3}(0.06)^3$$

$$+ \binom{5}{4}(0.06)^4 + \binom{5}{5}(0.06)^5$$

$$= 1 + 5(0.06) + 10(0.06)^2 + 10(0.06)^3 + 5(0.06)^4 + (0.06)^5$$

$$= 1 + 0.30 + 0.036 + 0.00216 + \cdots$$

$$\approx 1.3382.$$

This example may serve as a practical illustration of the binomial theorem. Consider, for instance, an investment of $1 in a savings bank that pays 6 percent interest compounded annually. How much will this investment be worth in 5 years? The answer is generally available in the appropriate tables, but in case such tables are not readily available, we can always obtain an approximate answer by using the first few terms in the binomial expansion. Using the first four terms in the expansion, we can say that the investment of $1 will be worth $1.3382 in 5 years.

Example 11.40

Without doing the actual expansion, find the ninth term in the expansion of $(2x + y)^{14}$.

Solution: The general term, the $(r + 1)$st term in the expansion of the binomial theorem, is given by

$$T_{r+1} = \binom{n}{r} a^{n-r} b^r.$$

Accordingly, we have

$$T_9 = \binom{14}{8} (2x)^6 y^8$$

$$= \binom{14}{8} (64x^6) y^8$$

$$= 192{,}192 x^6 y^8.$$

Example 11.41

Without expanding, find the middle term in the expansion of $\left(x - \dfrac{1}{x} \right)^{12}$.

Solution: Since $n = 12$, it follows that there are 13 terms in the expansion of $\left(x - \dfrac{1}{x} \right)^{12}$. Accordingly, the seventh term represents the middle term. Thus

$$T_7 = \binom{12}{6} x^6 \left(-\frac{1}{x} \right)^6 = 924 x^6 (-1)^6 x^{-6} = 924.$$

Example 11.42

Find the two middle terms in the expansion of $\left(2a - \dfrac{a^2}{4} \right)^9$.

Solution: With $n = 9$, there are 10 terms in this expansion, and the two middle terms are the fifth and sixth, given by T_5 and T_6:

$$T_5 = \binom{9}{4} (2a)^5 \left(-\frac{a^2}{4} \right)^4 = 126(32a^5) \left(\frac{a^8}{256} \right) = \frac{63}{4} a^{13}$$

$$T_6 = \binom{9}{5} (2a)^4 \left(-\frac{a^2}{4} \right)^5 = 126(16a^4) \left(-\frac{a^{10}}{1024} \right) = -\frac{63}{32} a^{14}.$$

Example 11.43

Find the term independent of x in the expansion of $\left(2x^2 - \dfrac{1}{x} \right)^{12}$.

Solution: We have

$$T_{r+1} = \binom{12}{r}(2x^2)^{12-r}\left(-\frac{1}{x}\right)^r$$

$$= \binom{12}{r}2^{12-r}x^{24-2r}(-1)^r x^{-r}$$

$$= \binom{12}{r}(-1)^r 2^{12-r}x^{24-3r}.$$

It is evident that T_{r+1} will be independent of x if $24 - 3r = 0$, that is, $r = 8$. Thus T_9 is the term independent of x and is given by

$$\binom{12}{8}(-1)^8 2^4 = \binom{12}{4}2^4 = 7920.$$

Example 11.44

Without expanding, find the term of $(2x + 3y)^6$ involving $x^2 y^4$.

Solution: The general term for this expression is given by

$$T_{r+1} = \binom{6}{r}(2x)^{6-r}(3y)^r.$$

Since the exponent of y is r, then $r = 4$. Hence the term involving $x^2 y^4$ is given by

$$T_5 = \binom{6}{4}(2x)^2(3y)^4 = 15(4x^2)(81y^4)$$

$$= 4860x^2 y^4.$$

EXERCISES

Expand the expressions in Exercises 1 to 10.

1. $(x + y)^7$

2. $(2x + y)^4$

3. $(x + 2y)^3$

4. $(x - 2y)^5$

5. $(2x + 3y)^5$

6. $(3x - 2y)^6$

7. $(1 - x)^8$

8. $\left(x + \dfrac{1}{x}\right)^4$

9. $\left(2x - \dfrac{1}{x}\right)^6$

10. $(1 + 3x)^5$

Evaluate the expressions in Exercises 11 to 15, without doing actual computations.

11. $\binom{4}{0} + \binom{4}{1} + \binom{4}{2} + \binom{4}{3} + \binom{4}{4}$

12. $\dbinom{5}{0} + \dbinom{5}{1} + \dbinom{5}{2} + \dbinom{5}{3} + \dbinom{5}{4} + \dbinom{5}{5}$

13. $\dbinom{6}{0} + \dbinom{6}{1} + \dbinom{6}{2} + \dbinom{6}{3} + \dbinom{6}{4} + \dbinom{6}{5} + \dbinom{6}{6}$

14. $\dbinom{7}{0} + \dbinom{7}{1} + \dbinom{7}{2} + \dbinom{7}{3} + \dbinom{7}{4} + \dbinom{7}{5} + \dbinom{7}{6} + \dbinom{7}{7}$

15. $\dbinom{n}{0} + \dbinom{n}{1} + \dbinom{n}{2} + \dbinom{n}{3} + \dbinom{n}{4} + \dbinom{n}{5} + \dbinom{n}{6} + \dbinom{n}{7} + \cdots + \dbinom{n}{n}$

Using the binomial theorem, find approximations for the expressions in Exercises 16 to 23.

16. $(1.05)^6$ 17. $(1.07)^4$ 18. $(0.98)^6$ 19. $(0.99)^5$

20. $(1.04)^5$ 21. $(1.006)^4$ 22. $(1.08)^{10}$ 23. $(1.09)^{12}$

24. Find in simplified form
   (a) the fourth term in the expansion of $(2x - y)^9$.

   (b) the sixth term in the expansion of $\left(\dfrac{x}{3} + \dfrac{3}{x}\right)^{12}$.

   (c) the middle term in the expansion of $(2x - 3y)^{14}$.
   (d) the middle terms in the expansion of $(3x + 4y)^{11}$.

25. Find the coefficient of
   (a) x^4 in the expansion of $(x - x^{-2})^{10}$.
   (b) x^5 in the expansion of $(x + x^{-3})^{17}$.
   (c) x^n in the expansion of $(1 + x)^{2n}$.

26. Find the term independent of x in the expansion of $(x^3 - x^{-2})^{10}$. What is the sum of all terms if $x = 1$?

27. Washburn plans to deposit $5000 in a savings account that pays 12 percent interest compounded annually. How much money will be in the account in 10 years?

28. Prove that

$$\dbinom{n}{0} - \dbinom{n}{1} + \dbinom{n}{2} - \dbinom{n}{3} + \dbinom{n}{4} - \cdots + (-1)^n\dbinom{n}{n} = 0.$$

KEY WORDS

circle
center
radius
standard equation
semicircle
diameter
conic section
right circular cone
double-napped cone
upper nappe

lower nappe
focus
directrix
ellipse
foci
center of the ellipse
vertices
major axis
minor axis
hyperbola

transverse axis
conjugate axis
asymptotes
tree diagram
fundamental principle of counting
n factorial
permutation
combination
binomial coefficient
binomial theorem

KEY FORMULAS

- The standard equation of a circle with its center at the point $C(h, k)$ and radius r is

$$(x - h)^2 + (y - k)^2 = r^2.$$

- If the *axis of symmetry* of the parabola is the x axis and its *vertex* is the origin, then the standard equation of the parabola is

$$y^2 = 4px.$$

- If the *axis of symmetry* of the parabola is the y axis and its *vertex* is the origin, then the standard equation of the parabola is

$$x^2 = 4py.$$

- The standard equation of the ellipse is

$$\frac{x^2}{a^2} + \frac{y^2}{b^2} = 1.$$

The points $(a, 0)$ and $(-a, 0)$ are the *vertices* of the ellipse. The line joining the vertices is the *major axis;* the line joining the points $(0, -b)$ and $(0, b)$ on the y axis is the *minor axis.* The points $(\pm c, 0)$, where $b^2 = a^2 - c^2$, are the *foci* of the ellipse.

- The standard equation of the hyperbola having foci $(\pm c, 0)$ on the x axis is

$$\frac{x^2}{a^2} - \frac{y^2}{b^2} = 1.$$

The points $(a, 0)$ and $(-a, 0)$ are the *vertices;* the line joining the vertices is the *transverse axis,* and the line joining the points $(0, -b)$ and $(0, b)$ is the *conjugate axis.* The equation of the asymptotes is

$$y = \pm \frac{b}{a} x.$$

- The standard equation of the hyperbola having its foci $(0, \pm c)$ on the y axis is

$$\frac{y^2}{a^2} - \frac{x^2}{b^2} = 1.$$

- The number of permutations of n objects taken r at a time is

$$P(n, r) = n(n - 1)(n - 2) \cdots (n - r + 1) = \frac{n!}{(n - r)!}.$$

- The number of combinations of n objects taken r at a time is

$$\binom{n}{r} = \frac{P(n, r)}{r!} = \frac{n!}{r!(n - r)!}.$$

- $$\binom{n}{r} = \binom{n}{n - r}.$$

- *Binomial Theorem:* If n is a positive integer, then

$$(a + b)^n = \binom{n}{0} a^n + \binom{n}{1} a^{n-1}b + \binom{n}{2} a^{n-2}b^2$$

$$+ \cdots + \binom{n}{r} a^{n-r}b^r + \cdots + \binom{n}{n} b^n.$$

EXERCISES FOR REVIEW

Graph the equations in Exercises 1 to 8.

1. $x = y^2 - 4y + 4$
2. $x = y^2 + 2y - 8$
3. $x^2 + y^2 = 9$
4. $y = -\sqrt{16 - x^2}$
5. $25x^2 + 9y^2 = 225$
6. $9x^2 + 16y^2 = 144$
7. $9x^2 - 16y^2 = 144$
8. $4y^2 - 9x^2 = 36$

Find the center and radius of the circles in Exercises 9 and 10.

9. $x^2 + y^2 - 6x - 8y - 11 = 0$
10. $x^2 + y^2 + 4x + 6y - 3 = 0$

Find the equation of the parabola having the given properties in Exercises 11 to 14.

11. Focus is $(4, 0)$; vertex is $(0, 0)$.
12. Focus is $(0, 2)$; directrix is $y = -2$.

13. It passes through the point (2, 4), and the axis of symmetry is the x axis.
14. It passes through the point (2, 4), and the axis of symmetry is the y axis.

Find the equation of the ellipse having the given properties in Exercises 15 and 16.

15. Foci $(-1, 0)$ and $(1, 0)$; length of major axis is 6.
16. Vertices $(0, -5)$ and $(0, 5)$; length of minor axis is 3.

Find the equation of the hyperbola having the given properties in Exercises 17 and 18.

17. Vertices $(0, 4)$ and $(0, -4)$; equation of asymptotes is $y = \pm\frac{4}{3}x$.
18. Foci $(2, 0)$ and $(-2, 0)$; length of transverse axis is 3.
19. A restaurant offers 8 kinds of sandwiches, which it serves with coffee, tea, milk, or a soft drink. In how many ways can one order a sandwich and a beverage?
20. In a drive-in restaurant, a customer can order a hamburger rare, medium rare, medium, or well done, also with or without mustard, with or without onions, and with or without relish. In how many ways can a person order a hamburger?
21. Five persons enter a train coach in which 10 seats are vacant. In how many ways can these 5 persons be seated?
22. How many 5-letter "words" can be made from 26 letters of the alphabet if
    (a) no letter is repeated?
    (b) letters may be repeated any number of times?
23. In a city health department, there are 6 adjacent offices to be occupied by 6 nurses, A, B, C, D, E, and F. In how many different ways can these nurses be assigned to these offices?
24. There are 8 buildings on the campus of a state college. In how many ways can a prospective student visit the campus in its entirety?
25. A true-false test consists of 20 questions. In how many ways can a student check off the answers to these questions?
26. An automobile license plate contains 2 letters followed by 4 numerals (with repetitions permitted). Assuming that all 26 letters of the alphabet and all numerals, 0 through 9, are used, how many license plates can be made?
27. How many 5-digit numerals can be formed from the digits 1, 2, 3, 4, 5, and 6 if
    (a) no digit is repeated?
    (b) digits may be repeated any number of times?
28. In how many ways can a committee of 3 be chosen from a group of 10?
29. In a literature class, a student must select 4 books to read from a reading list of 9. How many choices does a student have?
30. In a retail store, there are 12 varieties of green beans. If Susan decides to sample 5 of these 12 varieties, how many samples are available to her?
31. The price of a New England tour includes 4 stopovers to be selected from among 8 cities. In how many ways can one plan such a tour if
    (a) the order of the stopovers does matter?
    (b) the order of the stopovers does not matter?
32. How many groups of 3 boys and 4 girls can be formed from a total of 6 boys and 8 girls?
33. From a committee of 5 women and 7 men, an ad hoc committee is to be formed consisting of 3 women and 3 men. In how many ways can this committee be formed?
34. A poker hand is a set of 5 cards selected from a standard deck of 52 cards. What is the number of possible poker hands that contain
    (a) one pair and three different cards?
    (b) two pairs and one different card?

35. A shipment of 20 alarm clocks contains 3 that are defective. In how many ways can an inspector choose 5 of the alarm clocks so that
  (a) no defective alarm clock is included?
  (b) two defective alarm clocks are included?
  (c) all the defective alarm clocks are included?

Expand the expressions in Exercises 36 to 39.

36. $(3x + 2y)^4$ **37.** $(4x - 3y)^5$ **38.** $(1 - 2x)^6$ **39.** $(1 + 3x)^6$

40. Without actual expansion, find the sixth term in the expansion of $(2x + 3y)^8$.

41. Without expanding, find the middle term in the expansion of $\left(x - \dfrac{1}{x}\right)^{10}$.

Using the binomial theorem, find approximations for the expressions in Exercises 42 to 45.

42. $(1.08)^5$ **43.** $(0.99)^6$ **44.** $(1.09)^6$ **45.** $(0.995)^4$

CHAPTER TEST

1. Graph the following:
  (a) $(x - 3)^2 + (y - 4)^2 = 25$ **(b)** $x = y^2 + 2y - 8$

2. Find the center and the radius of the circle

$$x^2 + y^2 - 6x - 8y - 11 = 0.$$

3. Find the equation of the parabola given that its focus is $(4, 0)$ and its vertex is $(0, 0)$.

4. Find the equation of the ellipse given that its foci are $(-1, 0)$ and $(1, 0)$ and the length of the major axis is 6.

5. Determine the vertices, foci, and equations for the asymptotes of the hyperbola $16x^2 - 9y^2 = 144$.

6. A telephone number consists of a sequence of 7 digits (with repetitions permitted) selected from the numbers 0 through 9. How many different telephone numbers are possible if the first digit cannot be 0 or 1?

7. Grades of A, B, C, D, and F are assigned to a class of 5 students. In how many ways can these students be graded if no two students receive the same grade? In how many ways can grades be assigned if only A or B is to be assigned?

8. A committee of 4 labor and 4 management personnel is to be selected from 6 labor and 8 management personnel. In how many ways can the committee be formed?

9. In how many ways can 15 salespeople be assigned to three groups consisting of 4, 5, and 6 members, respectively?

10. From a regular deck of cards, 5 are to be chosen. How many selections are possible if the cards must be the same suit?

11. Expand $(4x - 5y)^4$.

12. Without doing the actual expansion, find the sixth term in the expansion of $(2x + 3y)^8$.

APPENDIX TABLES

TABLE 1   Exponential Functions

x	e^x	e^{-x}	x	e^x	e^{-x}
0.0	1.0000	1.0000	5.0	148.51	0.00674
0.1	1.1052	0.90484	5.1	164.13	0.00609
0.2	1.2214	0.81873	5.2	181.39	0.00551
0.3	1.3499	0.74082	5.3	200.47	0.00498
0.4	1.4918	0.67032	5.4	221.55	0.00451
0.5	1.6487	0.60653	5.5	244.75	0.00408
0.6	1.8221	0.54881	5.6	270.40	0.00369
0.7	2.0138	0.49659	5.7	298.90	0.00334
0.8	2.2255	0.44933	5.8	330.31	0.00302
0.9	2.4596	0.40657	5.9	365.08	0.00274
1.0	2.7183	0.36788	6.0	403.45	0.00247
1.1	3.0042	0.33287	6.1	445.89	0.00224
1.2	3.3201	0.30119	6.2	492.77	0.00203
1.3	3.6693	0.27253	6.3	544.61	0.00184
1.4	4.0552	0.24660	6.4	601.86	0.00166
1.5	4.4817	0.22313	6.5	665.17	0.00150
1.6	4.9530	0.20190	6.6	735.15	0.00136
1.7	5.4739	0.18268	6.7	812.43	0.00123
1.8	6.0496	0.16530	6.8	897.89	0.00111
1.9	6.6859	0.14957	6.9	992.38	0.00100
2.0	7.3891	0.13534	7.0	1,096.69	0.00091
2.1	8.1662	0.12246	7.1	1,212.07	0.00082
2.2	9.0250	0.11080	7.2	1,339.57	0.00074
2.3	9.9742	0.10026	7.3	1,480.49	0.00068
2.4	11.023	0.09072	7.4	1,636.25	0.00061
2.5	12.182	0.08208	7.5	1,808.0	0.00055
2.6	13.464	0.07427	7.6	1,998.2	0.00050
2.7	14.880	0.06721	7.7	2,208.3	0.00045
2.8	16.445	0.06081	7.8	2,440.6	0.00041
2.9	18.174	0.05502	7.9	2,697.3	0.00037
3.0	20.086	0.04979	8.0	2,981.0	0.00034
3.1	22.198	0.04505	8.1	3,294.5	0.00030
3.2	24.533	0.04076	8.2	3,641.0	0.00027
3.3	27.113	0.03688	8.3	4,023.9	0.00025
3.4	29.964	0.03337	8.4	4,447.1	0.00022
3.5	33.115	0.03020	8.5	4,914.8	0.00020
3.6	36.598	0.02732	8.6	5,431.7	0.00018
3.7	40.447	0.02472	8.7	6,002.9	0.00017
3.8	44.701	0.02237	8.8	6,634.2	0.00015
3.9	49.402	0.02024	8.9	7,332.0	0.00014
4.0	54.598	0.01832	9.0	8,103.1	0.00012
4.1	60.310	0.01657	9.1	8,955.3	0.00011
4.2	66.686	0.01500	9.2	9,897.1	0.00010
4.3	73.700	0.01357	9.3	10,938	0.00009
4.4	81.451	0.01228	9.4	12,088	0.00008
4.5	90.017	0.01111	9.5	13,360	0.00007
4.6	99.484	0.01005	9.6	14,765	0.00007
4.7	109.95	0.00910	9.7	16,318	0.00006
4.8	121.51	0.00823	9.8	18,034	0.00006
4.9	134.29	0.00745	9.9	19,930	0.00005

TABLE 2 Common Logarithms

N	0	1	2	3	4	5	6	7	8	9
1.0	0.0000	0.0043	0.0086	0.0128	0.0170	0.0212	0.0253	0.0294	0.0334	0.0374
1.1	0.0414	0.0453	0.0492	0.0531	0.0569	0.0607	0.0645	0.0682	0.0719	0.0755
1.2	0.0792	0.0828	0.0864	0.0899	0.0934	0.0969	0.1004	0.1038	0.1072	0.1106
1.3	0.1139	0.1173	0.1206	0.1239	0.1271	0.1303	0.1335	0.1367	0.1399	0.1430
1.4	0.1461	0.1492	0.1523	0.1553	0.1584	0.1614	0.1644	0.1673	0.1703	0.1732
1.5	0.1761	0.1790	0.1818	0.1847	0.1875	0.1903	0.1931	0.1959	0.1987	0.2014
1.6	0.2041	0.2068	0.2095	0.2122	0.2148	0.2175	0.2201	0.2227	0.2253	0.2279
1.7	0.2304	0.2330	0.2355	0.2380	0.2405	0.2430	0.2455	0.2480	0.2504	0.2529
1.8	0.2553	0.2577	0.2601	0.2625	0.2648	0.2672	0.2695	0.2718	0.2742	0.2765
1.9	0.2788	0.2810	0.2833	0.2856	0.2878	0.2900	0.2923	0.2945	0.2967	0.2989
2.0	0.3010	0.3032	0.3054	0.3075	0.3096	0.3118	0.3139	0.3160	0.3181	0.3201
2.1	0.3222	0.3243	0.3263	0.3284	0.3304	0.3324	0.3345	0.3365	0.3385	0.3404
2.2	0.3424	0.3444	0.3464	0.3483	0.3502	0.3522	0.3541	0.3560	0.3579	0.3598
2.3	0.3617	0.3636	0.3655	0.3674	0.3692	0.3711	0.3729	0.3747	0.3766	0.3784
2.4	0.3802	0.3820	0.3838	0.3856	0.3874	0.3893	0.3909	0.3927	0.3945	0.3962
2.5	0.3979	0.3997	0.4014	0.4031	0.4048	0.4065	0.4082	0.4099	0.4116	0.4133
2.6	0.4150	0.4166	0.4183	0.4200	0.4216	0.4232	0.4249	0.4265	0.4281	0.4298
2.7	0.4314	0.4330	0.4346	0.4362	0.4378	0.4393	0.4409	0.4425	0.4440	0.4456
2.8	0.4472	0.4487	0.4502	0.4518	0.4533	0.4548	0.4564	0.4579	0.4594	0.4609
2.9	0.4624	0.4639	0.4654	0.4669	0.4683	0.4698	0.4713	0.4728	0.4742	0.4757
3.0	0.4771	0.4786	0.4800	0.4814	0.4829	0.4843	0.4857	0.4871	0.4886	0.4900
3.1	0.4914	0.4928	0.4942	0.4955	0.4969	0.4983	0.4997	0.5011	0.5024	0.5038
3.2	0.5051	0.5065	0.5079	0.5092	0.5105	0.5119	0.5132	0.5145	0.5159	0.5172
3.3	0.5185	0.5198	0.5211	0.5224	0.5237	0.5250	0.5263	0.5276	0.5289	0.5302
3.4	0.5315	0.5328	0.5340	0.5353	0.5366	0.5378	0.5391	0.5403	0.5416	0.5428
3.5	0.5441	0.5453	0.5465	0.5478	0.5490	0.5502	0.5514	0.5527	0.5539	0.5551
3.6	0.5563	0.5575	0.5587	0.5599	0.5611	0.5623	0.5635	0.5647	0.5658	0.5670
3.7	0.5682	0.5694	0.5705	0.5717	0.5729	0.5740	0.5752	0.5763	0.5775	0.5786
3.8	0.5798	0.5809	0.5821	0.5832	0.5843	0.5855	0.5866	0.5877	0.5888	0.5899
3.9	0.5911	0.5922	0.5933	0.5944	0.5955	0.5966	0.5977	0.5988	0.5999	0.6010
4.0	0.6021	0.6031	0.6042	0.6053	0.6064	0.6075	0.6085	0.6096	0.6107	0.6117
4.1	0.6128	0.6138	0.6149	0.6160	0.6170	0.6180	0.6191	0.6201	0.6212	0.6222
4.2	0.6232	0.6243	0.6253	0.6263	0.6274	0.6284	0.6294	0.6304	0.6314	0.6325
4.3	0.6335	0.6345	0.6355	0.6365	0.6375	0.6385	0.6395	0.6405	0.6415	0.6425
4.4	0.6435	0.6444	0.6454	0.6464	0.6474	0.6484	0.6493	0.6503	0.6513	0.6522
4.5	0.6532	0.6542	0.6551	0.6561	0.6571	0.6580	0.6590	0.6599	0.6609	0.6618
4.6	0.6628	0.6637	0.6646	0.6656	0.6665	0.6675	0.6684	0.6693	0.6702	0.6712
4.7	0.6721	0.6730	0.6739	0.6749	0.6758	0.6767	0.6776	0.6785	0.6794	0.6803
4.8	0.6812	0.6821	0.6830	0.6839	0.6848	0.6857	0.6866	0.6875	0.6884	0.6893
4.9	0.6902	0.6911	0.6920	0.6928	0.6937	0.6946	0.6955	0.6964	0.6972	0.6981
5.0	0.6990	0.6998	0.7007	0.7016	0.7024	0.7033	0.7042	0.7050	0.7059	0.7067
5.1	0.7076	0.7084	0.7093	0.7101	0.7110	0.7118	0.7126	0.7135	0.7143	0.7152
5.2	0.7160	0.7168	0.7177	0.7185	0.7193	0.7202	0.7210	0.7218	0.7226	0.7235
5.3	0.7243	0.7251	0.7259	0.7267	0.7275	0.7284	0.7292	0.7300	0.7308	0.7316
5.4	0.7324	0.7332	0.7340	0.7348	0.7356	0.7364	0.7372	0.7380	0.7388	0.7396

TABLE 2 (continued)

N	0	1	2	3	4	5	6	7	8	9
5.5	0.7404	0.7412	0.7419	0.7427	0.7435	0.7443	0.7451	0.7459	0.7466	0.7474
5.6	0.7482	0.7490	0.7497	0.7505	0.7513	0.7520	0.7528	0.7536	0.7543	0.7551
5.7	0.7559	0.7566	0.7574	0.7582	0.7589	0.7597	0.7604	0.7612	0.7619	0.7627
5.8	0.7634	0.7642	0.7649	0.7657	0.7664	0.7672	0.7679	0.7686	0.7694	0.7701
5.9	0.7709	0.7716	0.7723	0.7731	0.7738	0.7745	0.7752	0.7760	0.7767	0.7774
6.0	0.7782	0.7789	0.7796	0.7803	0.7810	0.7818	0.7825	0.7832	0.7839	0.7846
6.1	0.7853	0.7860	0.7868	0.7875	0.7882	0.7889	0.7896	0.7903	0.7910	0.7917
6.2	0.7924	0.7931	0.7938	0.7945	0.7952	0.7959	0.7966	0.7973	0.7980	0.7987
6.3	0.7993	0.8000	0.8007	0.8014	0.8021	0.8028	0.8035	0.8041	0.8048	0.8055
6.4	0.8062	0.8069	0.8075	0.8082	0.8089	0.8096	0.8102	0.8109	0.8116	0.8122
6.5	0.8129	0.8136	0.8142	0.8149	0.8156	0.8162	0.8169	0.8176	0.8182	0.8189
6.6	0.8195	0.8202	0.8209	0.8215	0.8222	0.8228	0.8235	0.8241	0.8248	0.8254
6.7	0.8261	0.8267	0.8274	0.8280	0.8287	0.8293	0.8299	0.8306	0.8312	0.8319
6.8	0.8325	0.8331	0.8338	0.8344	0.8351	0.8357	0.8363	0.8370	0.8376	0.8382
6.9	0.8388	0.8395	0.8401	0.8407	0.8414	0.8420	0.8426	0.8432	0.8439	0.8445
7.0	0.8451	0.8457	0.8463	0.8470	0.8476	0.8482	0.8488	0.8494	0.8500	0.8506
7.1	0.8513	0.8519	0.8525	0.8531	0.8537	0.8543	0.8549	0.8555	0.8561	0.8567
7.2	0.8573	0.8579	0.8585	0.8591	0.8597	0.8603	0.8609	0.8615	0.8621	0.8627
7.3	0.8633	0.8639	0.8645	0.8651	0.8657	0.8663	0.8669	0.8675	0.8681	0.8686
7.4	0.8692	0.8698	0.8704	0.8710	0.8716	0.8722	0.8727	0.8733	0.8739	0.8745
7.5	0.8751	0.8756	0.8762	0.8768	0.8774	0.8779	0.8785	0.8791	0.8797	0.8802
7.6	0.8808	0.8814	0.8820	0.8825	0.8831	0.8837	0.8842	0.8848	0.8854	0.8859
7.7	0.8865	0.8871	0.8876	0.8882	0.8887	0.8893	0.8899	0.8904	0.8910	0.8915
7.8	0.8921	0.8927	0.8932	0.8938	0.8943	0.8949	0.8954	0.8960	0.8965	0.8971
7.9	0.8976	0.8982	0.8987	0.8993	0.8998	0.9004	0.9009	0.9015	0.9020	0.9025
8.0	0.9031	0.9036	0.9042	0.9047	0.9053	0.9058	0.9063	0.9069	0.9074	0.9079
8.1	0.9085	0.9090	0.9096	0.9101	0.9106	0.9112	0.9117	0.9122	0.9128	0.9133
8.2	0.9138	0.9143	0.9149	0.9154	0.9159	0.9165	0.9170	0.9175	0.9180	0.9186
8.3	0.9191	0.9196	0.9201	0.9206	0.9212	0.9217	0.9222	0.9227	0.9232	0.9238
8.4	0.9243	0.9248	0.9253	0.9258	0.9263	0.9269	0.9274	0.9279	0.9284	0.9289
8.5	0.9294	0.9299	0.9304	0.9309	0.9315	0.9320	0.9325	0.9330	0.9335	0.9340
8.6	0.9345	0.9350	0.9355	0.9360	0.9365	0.9370	0.9375	0.9380	0.9385	0.9390
8.7	0.9395	0.9400	0.9405	0.9410	0.9415	0.9420	0.9425	0.9430	0.9435	0.9440
8.8	0.9445	0.9450	0.9455	0.9460	0.9465	0.9469	0.9474	0.9479	0.9484	0.9489
8.9	0.9494	0.9499	0.9504	0.9509	0.9513	0.9518	0.9523	0.9528	0.9533	0.9538
9.0	0.9542	0.9547	0.9552	0.9557	0.9562	0.9566	0.9571	0.9576	0.9581	0.9586
9.1	0.9590	0.9595	0.9600	0.9605	0.9609	0.9614	0.9619	0.9624	0.9628	0.9633
9.2	0.9638	0.9643	0.9647	0.9652	0.9657	0.9661	0.9666	0.9671	0.9675	0.9680
9.3	0.9685	0.9689	0.9694	0.9699	0.9703	0.9708	0.9713	0.9717	0.9722	0.9727
9.4	0.9731	0.9736	0.9741	0.9745	0.9750	0.9754	0.9759	0.9763	0.9768	0.9773
9.5	0.9777	0.9782	0.9786	0.9791	0.9795	0.9800	0.9805	0.9809	0.9814	0.9818
9.6	0.9823	0.9827	0.9832	0.9836	0.9841	0.9845	0.9850	0.9854	0.9859	0.9863
9.7	0.9868	0.9872	0.9877	0.9881	0.9886	0.9890	0.9891	0.9899	0.9903	0.9908
9.8	0.9912	0.9917	0.9921	0.9926	0.9930	0.9934	0.9939	0.9943	0.9948	0.9952
9.9	0.9956	0.9961	0.9965	0.9969	0.9974	0.9978	0.9983	0.9987	0.9991	0.9996

TABLE 3   Natural Logarithms

N	.00	.01	.02	.03	.04	.05	.06	.07	.08	.09
1.0	0.0000	0.0100	0.0198	0.0296	0.0392	0.0488	0.0583	0.0677	0.0770	0.0862
1.1	0.0953	0.1044	0.1133	0.1222	0.1310	0.1398	0.1484	0.1570	0.1655	0.1740
1.2	0.1823	0.1906	0.1989	0.2070	0.2151	0.2231	0.2311	0.2390	0.2469	0.2546
1.3	0.2624	0.2700	0.2776	0.2852	0.2927	0.3001	0.3075	0.3148	0.3221	0.3293
1.4	0.3365	0.3436	0.3507	0.3577	0.3646	0.3716	0.3784	0.3853	0.3920	0.3988
1.5	0.4055	0.4121	0.4187	0.4253	0.4318	0.4383	0.4447	0.4511	0.4574	0.4637
1.6	0.4700	0.4762	0.4824	0.4886	0.4937	0.5008	0.5068	0.5128	0.5188	0.5247
1.7	0.5306	0.5365	0.5423	0.5481	0.5539	0.5596	0.5653	0.5710	0.5766	0.5822
1.8	0.5878	0.5933	0.5988	0.6043	0.6098	0.6152	0.6206	0.6259	0.6313	0.6366
1.9	0.6419	0.6471	0.6523	0.6575	0.6627	0.6678	0.6729	0.6780	0.6831	0.6881
2.0	0.6931	0.6981	0.7031	0.7080	0.7129	0.7178	0.7227	0.7275	0.7324	0.7372
2.1	0.7419	0.7467	0.7514	0.7561	0.7608	0.7655	0.7701	0.7747	0.7793	0.7839
2.2	0.7885	0.7930	0.7975	0.8020	0.8065	0.8109	0.8154	0.8198	0.8242	0.8286
2.3	0.8329	0.8372	0.8416	0.8459	0.8502	0.8544	0.8587	0.8629	0.8671	0.8713
2.4	0.8755	0.8796	0.8838	0.8879	0.8920	0.8961	0.9002	0.9042	0.9083	0.9123
2.5	0.9163	0.9203	0.9243	0.9282	0.9322	0.9361	0.9400	0.9439	0.9478	0.9517
2.6	0.9555	0.9594	0.9632	0.9670	0.9708	0.9746	0.9783	0.9821	0.9858	0.9895
2.7	0.9933	0.9969	1.0006	1.0043	1.0080	1.0116	1.0152	1.0188	1.0225	1.0260
2.8	1.0296	1.0332	1.0367	1.0403	1.0438	1.0473	1.0508	1.0543	1.0578	1.0613
2.9	1.0647	1.0682	1.0716	1.0750	1.0784	1.0818	1.0852	1.0886	1.0919	1.0953
3.0	1.0986	1.1019	1.1053	1.1086	1.1119	1.1151	1.1184	1.1217	1.1249	1.1282
3.1	1.1314	1.1346	1.1378	1.1410	1.1442	1.1474	1.1506	1.1537	1.1569	1.1600
3.2	1.1632	1.1663	1.1694	1.1725	1.1756	1.1787	1.1817	1.1848	1.1878	1.1909
3.3	1.1939	1.1969	1.2000	1.2030	1.2060	1.2090	1.2119	1.2149	1.2179	1.2208
3.4	1.2238	1.2267	1.2296	1.2326	1.2355	1.2384	1.2413	1.2442	1.2470	1.2499
3.5	1.2528	1.2556	1.2585	1.2613	1.2641	1.2669	1.2698	1.2726	1.2754	1.2782
3.6	1.2809	1.2837	1.2865	1.2892	1.2920	1.2947	1.2975	1.3002	1.3029	1.3056
3.7	1.3083	1.3110	1.3137	1.3164	1.3191	1.3218	1.3244	1.3271	1.3297	1.3324
3.8	1.3350	1.3376	1.3403	1.3429	1.3455	1.3481	1.3507	1.3533	1.3558	1.3584
3.9	1.3610	1.3635	1.3661	1.3686	1.3712	1.3737	1.3762	1.3788	1.3813	1.3838
4.0	1.3863	1.3888	1.3913	1.3938	1.3962	1.3987	1.4012	1.4036	1.4061	1.4085
4.1	1.4110	1.4134	1.4159	1.4183	1.4207	1.4231	1.4255	1.4279	1.4303	1.4327
4.2	1.4351	1.4375	1.4398	1.4422	1.4446	1.4469	1.4493	1.4516	1.4540	1.4563
4.3	1.4586	1.4609	1.4633	1.4656	1.4679	1.4702	1.4725	1.4748	1.4770	1.4793
4.4	1.4816	1.4839	1.4861	1.4884	1.4907	1.4929	1.4951	1.4974	1.4996	1.5019
4.5	1.5041	1.5063	1.5085	1.5107	1.5129	1.5151	1.5173	1.5195	1.5217	1.5239
4.6	1.5261	1.5282	1.5304	1.5326	1.5347	1.5369	1.5390	1.5412	1.5433	1.5454
4.7	1.5476	1.5497	1.5518	1.5539	1.5560	1.5581	1.5602	1.5623	1.5644	1.5665
4.8	1.5686	1.5707	1.5728	1.5748	1.5769	1.5790	1.5810	1.5831	1.5851	1.5872
4.9	1.5892	1.5913	1.5933	1.5953	1.5974	1.5994	1.6014	1.6034	1.6054	1.6074
5.0	1.6094	1.6114	1.6134	1.6154	1.6174	1.6194	1.6214	1.6233	1.6253	1.6273
5.1	1.6292	1.6312	1.6332	1.6351	1.6371	1.6390	1.6409	1.6429	1.6448	1.6467
5.2	1.6487	1.6506	1.6525	1.6544	1.6563	1.6582	1.6601	1.6620	1.6639	1.6658
5.3	1.6677	1.6696	1.6715	1.6734	1.6752	1.6771	1.6790	1.6808	1.6827	1.6845
5.4	1.6864	1.6882	1.6901	1.6919	1.6938	1.6956	1.6974	1.6993	1.7011	1.7029

TABLE 3 (continued)

N	.00	.01	.02	.03	.04	.05	.06	.07	.08	.09
5.5	1.7047	1.7066	1.7084	1.7102	1.7120	1.7138	1.7156	1.7174	1.7192	1.7210
5.6	1.7228	1.7246	1.7263	1.7281	1.7299	1.7317	1.7334	1.7352	1.7370	1.7387
5.7	1.7405	1.7422	1.7440	1.7457	1.7475	1.7492	1.7509	1.7527	1.7544	1.7561
5.8	1.7579	1.7596	1.7613	1.7630	1.7647	1.7664	1.7681	1.7699	1.7716	1.7733
5.9	1.7750	1.7766	1.7783	1.7800	1.7817	1.7834	1.7851	1.7867	1.7884	1.7901
6.0	1.7918	1.7934	1.7951	1.7967	1.7984	1.8001	1.8017	1.8034	1.8050	1.8066
6.1	1.8083	1.8099	1.8116	1.8132	1.8148	1.8165	1.8181	1.8197	1.8213	1.8229
6.2	1.8245	1.8262	1.8278	1.8294	1.8310	1.8326	1.8342	1.8358	1.8374	1.8390
6.3	1.8405	1.8421	1.8437	1.8453	1.8469	1.8485	1.8500	1.8516	1.8532	1.8547
6.4	1.8563	1.8579	1.8594	1.8610	1.8625	1.8641	1.8656	1.8672	1.8687	1.8703
6.5	1.8718	1.8733	1.8749	1.8764	1.8779	1.8795	1.8810	1.8825	1.8840	1.8856
6.6	1.8871	1.8886	1.8901	1.8916	1.8931	1.8946	1.8961	1.8976	1.8991	1.9006
6.7	1.9021	1.9036	1.9051	1.9066	1.9081	1.9095	1.9110	1.9125	1.9140	1.9155
6.8	1.9169	1.9184	1.9199	1.9213	1.9228	1.9242	1.9257	1.9272	1.9286	1.9301
6.9	1.9315	1.9330	1.9344	1.9359	1.9373	1.9337	1.9402	1.9416	1.9430	1.9445
7.0	1.9459	1.9473	1.9488	1.9502	1.9516	1.9530	1.9544	1.9559	1.9573	1.9587
7.1	1.9601	1.9615	1.9629	1.9643	1.9657	1.9671	1.9685	1.9699	1.9713	1.9727
7.2	1.9741	1.9755	1.9769	1.9782	1.9796	1.9810	1.9824	1.9838	1.9851	1.9865
7.3	1.9879	1.9892	1.9906	1.9920	1.9933	1.9947	1.9961	1.9974	1.9988	2.0001
7.4	2.0015	2.0028	2.0042	2.0055	2.0069	2.0082	2.0096	2.0109	2.0122	2.0136
7.5	2.0149	2.0162	2.0176	2.0189	2.0202	2.0215	2.0229	2.0242	2.0255	2.0268
7.6	2.0281	2.0295	2.0308	2.0321	2.0334	2.0347	2.0360	2.0373	2.0386	2.0399
7.7	2.0412	2.0425	2.0438	2.0451	2.0464	2.0477	2.0490	2.0503	2.0516	2.0528
7.8	2.0541	2.0554	2.0567	2.0580	2.0592	2.0605	2.0618	2.0631	2.0643	2.0656
7.9	2.0669	2.0681	2.0694	2.0707	2.0719	2.0732	2.0744	2.0757	2.0769	2.0782
8.0	2.0794	2.0807	2.0819	2.0832	2.0844	2.0857	2.0869	2.0882	2.0894	2.0906
8.1	2.0919	2.0931	2.0943	2.0956	2.0968	2.0980	2.0992	2.1005	2.1017	2.1029
8.2	2.1041	2.1054	2.1066	2.1078	2.1090	2.1102	2.1114	2.1126	2.1138	2.1150
8.3	2.1163	2.1175	2.1187	2.1199	2.1211	2.1223	2.1235	2.1247	2.1258	2.1270
8.4	2.1282	2.1294	2.1306	2.1318	2.1330	2.1342	2.1353	2.1365	2.1377	2.1389
8.5	2.1401	2.1412	2.1424	2.1436	2.1448	2.1459	2.1471	2.1483	2.1494	2.1506
8.6	2.1518	2.1529	2.1541	2.1552	2.1564	2.1576	2.1587	2.1599	2.1610	2.1622
8.7	2.1633	2.1645	2.1656	2.1668	2.1679	2.1691	2.1702	2.1713	2.1725	2.1736
8.8	2.1748	2.1759	2.1770	2.1782	2.1793	2.1804	2.1815	2.1827	2.1838	2.1849
8.9	2.1861	2.1872	2.1883	2.1894	2.1905	2.1917	2.1928	2.1939	2.1950	2.1961
9.0	2.1972	2.1983	2.1994	2.2006	2.2017	2.2028	2.2039	2.2050	2.2061	2.2072
9.1	2.2083	2.2094	2.2105	2.2116	2.2127	2.2138	2.2148	2.2159	2.2170	2.2181
9.2	2.2192	2.2203	2.2214	2.2225	2.2235	2.2246	2.2257	2.2268	2.2279	2.2289
9.3	2.2300	2.2311	2.2322	2.2332	2.2343	2.2354	2.2364	2.2375	2.2386	2.2396
9.4	2.2407	2.2418	2.2428	2.2439	2.2450	2.2460	2.2471	2.2481	2.2492	2.2502
9.5	2.2513	2.2523	2.2534	2.2544	2.2555	2.2565	2.2576	2.2586	2.2597	2.2607
9.6	2.2618	2.2628	2.2638	2.2649	2.2659	2.2670	2.2680	2.2690	2.2701	2.2711
9.7	2.2721	2.2732	2.2742	2.2752	2.2762	2.2773	2.2783	2.2793	2.2803	2.2814
9.8	2.2824	2.2834	2.2844	2.2854	2.2865	2.2875	2.2885	2.2895	2.2905	2.2915
9.9	2.2925	2.2935	2.2946	2.2956	2.2966	2.2976	2.2986	2.2996	2.3006	2.3016

TABLE 4 Values of Trigonometric Functions and Radians

Angle	Radians	Sin	Tan	Cot	Cos		
0° 00′	0.0000	0.0000	0.0000	....	1.0000	1.5708	**90° 00′**
10′	029	0.0029	0.0029	343.8	1.0000	679	**50′**
20′	058	0.0058	0.0058	171.9	1.0000	650	**40′**
30′	0.0087	0.0087	0.0087	114.6	1.0000	1.5621	**30′**
40′	116	0.0116	0.0116	85.94	0.9999	592	**20′**
50′	145	0.0145	0.0145	68.75	0.9999	563	**10′**
1° 00′	0.0175	0.0175	0.0175	57.29	0.9998	1.5533	**89° 00′**
10′	204	0.0204	0.0204	49.10	0.9998	504	**50′**
20′	233	0.0233	0.0233	42.96	0.9997	475	**40′**
30′	0.0262	0.0262	0.0262	38.19	0.9997	1.5446	**30′**
40′	291	0.0291	0.0291	34.37	0.9996	417	**20′**
50′	320	0.0320	0.0320	31.24	0.9995	388	**10′**
2° 00′	0.0349	0.0349	0.0349	28.64	0.9994	1.5359	**88° 00′**
10′	378	0.0378	0.0378	26.43	0.9993	330	**50′**
20′	407	0.0407	0.0407	24.54	0.9992	301	**40′**
30′	0.0436	0.0436	0.0437	22.90	0.9990	1.5272	**30′**
40′	465	0.0465	0.0466	21.47	0.9989	243	**20′**
50′	495	0.0494	0.0495	20.21	0.9988	213	**10′**
3° 00′	0.0524	0.0523	0.0524	19.08	0.9986	1.5184	**87° 00′**
10′	553	0.0552	0.0553	18.07	0.9985	155	**50′**
20′	582	0.0581	0.0582	17.17	0.9983	126	**40′**
30′	0.0611	0.0610	0.0612	16.35	0.9981	1.5097	**30′**
40′	640	0.0640	0.0641	15.60	0.9980	068	**20′**
50′	669	0.0669	0.0670	14.92	0.9978	039	**10′**
4° 00′	0.0698	0.0698	0.0699	14.30	0.9976	1.5010	**86° 00′**
10′	727	0.0727	0.0729	13.73	0.9974	1.4981	**50′**
20′	756	0.0756	0.0758	13.20	0.9971	952	**40′**
30′	0.0785	0.0785	0.0787	12.71	0.9969	1.4923	**30′**
40′	814	0.0814	0.0816	12.25	0.9967	893	**20′**
50′	844	0.0843	0.0846	11.83	0.9964	864	**10′**
5° 00′	0.0873	0.0872	0.0875	11.43	0.9962	1.4835	**85° 00′**
10′	902	0.0901	0.0904	11.06	0.9959	806	**50′**
20′	931	0.0929	0.0934	10.71	0.9957	777	**40′**
30′	0.0960	0.0958	0.0963	10.39	0.9954	1.4748	**30′**
40′	989	0.0987	0.0992	10.08	0.9951	719	**20′**
50′	0.1018	0.1016	0.1022	9.788	0.9948	690	**10′**
6° 00′	0.1047	0.1045	0.1051	9.514	0.9945	1.4661	**84° 00′**
10′	076	0.1074	0.1080	9.255	0.9942	632	**50′**
20′	105	0.1103	0.1110	9.010	0.9939	603	**40′**
30′	0.1134	0.1132	0.1139	8.777	0.9936	1.4573	**30′**
40′	164	0.1161	0.1169	8.556	0.9932	544	**20′**
50′	193	0.1190	0.1198	8.345	0.9929	515	**10′**
7° 00′	0.1222	0.1219	0.1228	8.144	0.9925	1.4486	**83° 00′**
10′	251	0.1248	0.1257	7.953	0.9922	457	**50′**
20′	280	0.1276	0.1287	7.770	0.9918	428	**40′**
30′	0.1309	0.1305	0.1317	7.596	0.9914	1.4399	**30′**
40′	338	0.1334	0.1346	7.429	0.9911	370	**20′**
50′	367	0.1363	0.1376	7.269	0.9907	341	**10′**
8° 00′	0.1396	0.1392	0.1405	7.115	0.9903	1.4312	**82° 00′**
10′	425	0.1421	0.1435	6.968	0.9899	283	**50′**
20′	454	0.1449	0.1465	6.827	0.9894	254	**40′**
30′	0.1484	0.1478	0.1495	6.691	0.9890	1.4224	**30′**
40′	513	0.1507	0.1524	6.561	0.9886	195	**20′**
50′	542	0.1536	0.1554	6.435	0.9881	166	**10′**
9° 00′	0.1571	0.1564	0.1584	6.314	0.9877	1.4137	**81° 00′**
		Cos	Cot	Tan	Sin	Radians	Angle

TABLE 4 (continued)

Angle	Radians	Sin	Tan	Cot	Cos		
9° 00′	0.1571	0.1564	0.1584	6.314	0.9877	1.4137	**81° 00′**
10′	600	0.1593	0.1614	6.197	0.9872	108	**50′**
20′	629	0.1622	0.1644	6.084	0.9868	079	**40′**
30′	0.1658	0.1650	0.1673	5.976	0.9863	1.4050	**30′**
40′	687	0.1679	0.1703	5.871	0.9858	1.4021	**20′**
50′	716	0.1708	0.1733	5.769	0.9853	1.3992	**10′**
10° 00′	0.1745	0.1736	0.1763	5.671	0.9848	1.3963	**80° 00′**
10′	774	0.1765	0.1793	5.576	0.9843	934	**50′**
20′	804	0.1794	0.1823	5.485	0.9838	904	**40′**
30′	0.1833	0.1822	0.1853	5.396	0.9833	1.3875	**30′**
40′	862	0.1851	0.1883	5.309	0.9827	846	**20′**
50′	891	0.1880	0.1914	5.226	0.9822	817	**10′**
11° 00′	0.1920	0.1908	0.1944	5.145	0.9816	1.3788	**79° 00′**
10′	949	0.1937	0.1974	5.066	0.9811	759	**50′**
20′	978	0.1965	0.2004	4.989	0.9805	730	**40′**
30′	0.2007	0.1994	0.2035	4.915	0.9799	1.3701	**30′**
40′	036	0.2022	0.2065	4.843	0.9793	672	**20′**
50′	065	0.2051	0.2095	4.773	0.9787	643	**10′**
12° 00′	0.2094	0.2079	0.2126	4.705	0.9781	1.3614	**78° 00′**
10′	123	0.2108	0.2156	4.638	0.9775	584	**50′**
20′	153	0.2136	0.2186	4.574	0.9769	555	**40′**
30′	0.2182	0.2164	0.2217	4.511	0.9763	1.3526	**30′**
40′	211	0.2193	0.2247	4.449	0.9757	497	**20′**
50′	240	0.2221	0.2278	4.390	0.9750	468	**10′**
13° 00′	0.2269	0.2250	0.2309	4.331	0.9744	1.3439	**77° 00′**
10′	298	0.2278	0.2339	4.275	0.9737	410	**50′**
20′	327	0.2306	0.2370	4.219	0.9730	381	**40′**
30′	0.2356	0.2334	0.2401	4.165	0.9724	1.3352	**30′**
40′	385	0.2363	0.2432	4.113	0.9717	323	**20′**
50′	414	0.2391	0.2462	4.061	0.9710	294	**10′**
14° 00′	0.2443	0.2419	0.2493	4.011	0.9703	1.3265	**76° 00′**
10′	473	0.2447	0.2524	3.962	0.9696	235	**50′**
20′	502	0.2476	0.2555	3.914	0.9689	206	**40′**
30′	0.2531	0.2504	0.2586	3.867	0.9681	1.3177	**30′**
40′	560	0.2532	0.2617	3.821	0.9674	148	**20′**
50′	589	0.2560	0.2648	3.776	0.9667	119	**10′**
15° 00′	0.2618	0.2588	0.2679	3.732	0.9659	1.3090	**75° 00′**
10′	647	0.2616	0.2711	3.689	0.9652	061	**50′**
20′	676	0.2644	0.2742	3.647	0.9644	032	**40′**
30′	0.2705	0.2672	0.2773	3.606	0.9636	1.3003	**30′**
40′	734	0.2700	0.2805	3.566	0.9628	1.2974	**20′**
50′	763	0.2728	0.2836	3.526	0.9621	945	**10′**
16° 00′	0.2793	0.2756	0.2867	3.487	0.9613	1.2915	**74° 00′**
10′	822	0.2784	0.2899	3.450	0.9605	886	**50′**
20′	851	0.2812	0.2931	3.412	0.9596	857	**40′**
30′	0.2880	0.2840	0.2962	3.376	0.9588	1.2828	**30′**
40′	909	0.2868	0.2994	3.340	0.9580	799	**20′**
50′	938	0.2896	0.3026	3.305	0.9572	770	**10′**
17° 00′	0.2967	0.2924	0.3057	3.271	0.9563	1.2741	**73° 00′**
10′	996	0.2952	0.3089	3.237	0.9555	712	**50′**
20′	0.3025	0.2979	0.3121	3.204	0.9546	683	**40′**
30′	0.3054	0.3007	0.3153	3.172	0.9537	1.2654	**30′**
40′	083	0.3035	0.3185	3.140	0.9528	625	**20′**
50′	113	0.3062	0.3217	3.108	0.9520	595	**10′**
18° 00′	0.3142	0.3090	0.3249	3.078	0.9511	1.2566	**72° 00′**
		Cos	Cot	Tan	Sin	Radians	Angle

TABLE 4 (continued)

Angle	Radians	Sin	Tan	Cot	Cos		
18° 00′	0.3142	0.3090	0.3249	3.078	0.9511	1.2566	**72° 00′**
10′	171	0.3118	0.3281	3.047	0.9502	537	50′
20′	200	0.3145	0.3314	3.018	0.9492	508	40′
30′	0.3229	0.3173	0.3346	2.989	0.9483	1.2479	30′
40′	258	0.3201	0.3378	2.960	0.9474	450	20′
50′	287	0.3228	0.3411	2.932	0.9465	421	10′
19° 00′	0.3316	0.3256	0.3443	2.904	0.9455	1.2392	**71° 00′**
10′	345	0.3283	0.3476	2.877	0.9446	363	50′
20′	374	0.3311	0.3508	2.850	0.9436	334	40′
30′	0.3403	0.3338	0.3541	2.824	0.9426	1.2305	30′
40′	432	0.3365	0.3574	2.798	0.9417	275	20′
50′	462	0.3393	0.3607	2.773	0.9407	246	10′
20° 00′	0.3491	0.3420	0.3640	2.747	0.9397	1.2217	**70° 00′**
10′	520	0.3448	0.3673	2.723	0.9387	188	50′
20′	549	0.3475	0.3706	2.699	0.9377	159	40′
30′	0.3578	0.3502	0.3739	2.675	0.9367	1.2130	30′
40′	607	0.3529	0.3772	2.651	0.9356	101	20′
50′	636	0.3557	0.3805	2.628	0.9346	072	10′
21° 00′	0.3665	0.3584	0.3839	2.605	0.9336	1.2043	**69° 00′**
10′	694	0.3611	0.3872	2.583	0.9325	1.2014	50′
20′	723	0.3638	0.3906	2.560	0.9315	1.1985	40′
30′	0.3752	0.3665	0.3939	2.539	0.9304	1.1956	30′
40′	782	0.3692	0.3973	2.517	0.9293	926	20′
50′	811	0.3719	0.4006	2.496	0.9283	897	10′
22° 00′	0.3840	0.3746	0.4040	2.475	0.9272	1.1868	**68° 00′**
10′	869	0.3773	0.4074	2.455	0.9261	839	50′
20′	898	0.3800	0.4108	2.434	0.9250	810	40′
30′	0.3927	0.3827	0.4142	2.414	0.9239	1.1781	30′
40′	956	0.3854	0.4176	2.394	0.9228	752	20′
50′	985	0.3881	0.4210	2.375	0.9216	723	10′
23° 00′	0.4014	0.3907	0.4245	2.356	0.9205	1.1694	**67° 00′**
10′	043	0.3934	0.4279	2.337	0.9194	665	50′
20′	072	0.3961	0.4314	2.318	0.9182	636	40′
30′	0.4102	0.3987	0.4348	2.300	0.9171	1.1606	30′
40′	131	0.4014	0.4383	2.282	0.9159	577	20′
50′	160	0.4041	0.4417	2.264	0.9147	548	10′
24° 00′	0.4189	0.4067	0.4452	2.246	0.9135	1.1519	**66° 00′**
10′	218	0.4094	0.4487	2.229	0.9124	490	50′
20′	247	0.4120	0.4522	2.211	0.9112	461	40′
30′	0.4276	0.4147	0.4557	2.194	0.9100	1.1432	30′
40′	305	0.4173	0.4592	2.177	0.9088	403	20′
50′	334	0.4200	0.4628	2.161	0.9075	374	10′
25° 00′	0.4363	0.4226	0.4663	2.145	0.9063	1.1345	**65° 00′**
10′	392	0.4253	0.4699	2.128	0.9051	316	50′
20′	422	0.4279	0.4734	2.112	0.9038	286	40′
30′	0.4451	0.4305	0.4770	2.097	0.9026	1.1257	30′
40′	480	0.4331	0.4806	2.081	0.9013	228	20′
50′	509	0.4358	0.4841	2.066	0.9001	199	10′
26° 00′	0.4538	0.4384	0.4877	2.050	0.8988	1.1170	**64° 00′**
10′	567	0.4410	0.4913	2.035	0.8975	141	50′
20′	596	0.4436	0.4950	2.020	0.8962	112	40′
30′	0.4625	0.4462	0.4986	2.006	0.8949	1.1083	30′
40′	654	0.4488	0.5022	1.991	0.8936	054	20′
50′	683	0.4514	0.5059	1.977	0.8923	1.1025	10′
27° 00′	0.4712	0.4540	0.5095	1.963	0.8910	1.0996	**63° 00′**
		Cos	Cot	Tan	Sin	Radians	Angle

TABLE 4 (continued)

Angle	Radians	Sin	Tan	Cot	Cos		
27° 00′	0.4712	0.4540	0.5095	1.963	0.8910	1.0996	**63° 00′**
10′	741	0.4566	0.5132	1.949	0.8897	966	**50′**
20′	771	0.4592	0.5169	1.935	0.8884	937	**40′**
30′	0.4800	0.4617	0.5206	1.921	0.8870	1.0908	**30′**
40′	829	0.4643	0.5243	1.907	0.8857	879	**20′**
50′	858	0.4669	0.5280	1.894	0.8843	850	**10′**
28° 00′	0.4887	0.4695	0.5317	1.881	0.8829	1.0821	**62° 00′**
10′	916	0.4720	0.5354	1.868	0.8816	792	**50′**
20′	945	0.4746	0.5392	1.855	0.8802	763	**40′**
30′	0.4974	0.4772	0.5430	1.842	0.8788	1.0734	**30′**
40′	0.5003	0.4797	0.5467	1.829	0.8774	705	**20′**
50′	032	0.4823	0.5505	1.816	0.8760	676	**10′**
29° 00′	0.5061	0.4848	0.5543	1.804	0.8746	1.0647	**61° 00′**
10′	091	0.4874	0.5581	1.792	0.8732	617	**50′**
20′	120	0.4899	0.5619	1.780	0.8718	588	**40′**
30′	0.5149	0.4924	0.5658	1.767	0.8704	1.0559	**30′**
40′	178	0.4950	0.5696	1.756	0.8689	530	**20′**
50′	207	0.4975	0.5735	1.744	0.8675	501	**10′**
30° 00′	0.5236	0.5000	0.5774	1.732	0.8660	1.0472	**60° 00′**
10′	265	0.5025	0.5812	1.720	0.8646	443	**50′**
20′	294	0.5050	0.5851	1.709	0.8631	414	**40′**
30′	0.5323	0.5075	0.5890	1.698	0.8616	1.0385	**30′**
40′	352	0.5100	0.5930	1.686	0.8601	356	**20′**
50′	381	0.5125	0.5969	1.675	0.8587	327	**10′**
31° 00′	0.5411	0.5150	0.6009	1.664	0.8572	1.0297	**59° 00′**
10′	440	0.5175	0.6048	1.653	0.8557	268	**50′**
20′	469	0.5200	0.6088	1.643	0.8542	239	**40′**
30′	0.5498	0.5225	0.6128	1.632	0.8526	1.0210	**30′**
40′	527	0.5250	0.6168	1.621	0.8511	181	**20′**
50′	556	0.5275	0.6208	1.611	0.8496	152	**10′**
32° 00′	0.5585	0.5299	0.6249	1.600	0.8480	1.0123	**58° 00′**
10′	614	0.5324	0.6289	1.590	0.8465	094	**50′**
20′	643	0.5348	0.6330	1.580	0.8450	065	**40′**
30′	0.5672	0.5373	0.6371	1.570	0.8434	1.0036	**30′**
40′	701	0.5398	0.6412	1.560	0.8418	1.0007	**20′**
50′	730	0.5422	0.6453	1.550	0.8403	0.9977	**10′**
33° 00′	0.5760	0.5446	0.6494	1.540	0.8387	0.9948	**57° 00′**
10′	789	0.5471	0.6536	1.530	0.8371	919	**50′**
20′	818	0.5495	0.6577	1.520	0.8355	890	**40′**
30′	0.5847	0.5519	0.6619	1.511	0.8339	0.9861	**30′**
40′	876	0.5544	0.6661	1.501	0.8323	832	**20′**
50′	905	0.5568	0.6703	1.492	0.8307	803	**10′**
34° 00′	0.5934	0.5592	0.6745	1.483	0.8290	0.9774	**56° 00′**
10′	963	0.5616	0.6787	1.473	0.8274	745	**50′**
20′	992	0.5640	0.6830	1.464	0.8258	716	**40′**
30′	0.6021	0.5664	0.6873	1.455	0.8241	0.9687	**30′**
40′	050	0.5688	0.6916	1.446	0.8225	657	**20′**
50′	080	0.5712	0.6959	1.437	0.8208	628	**10′**
35° 00′	0.6109	0.5736	0.7002	1.428	0.8192	0.9599	**55° 00′**
10′	138	0.5760	0.7046	1.419	0.8175	570	**50′**
20′	167	0.5783	0.7089	1.411	0.8158	541	**40′**
30′	0.6196	0.5807	0.7133	1.402	0.8141	0.9512	**30′**
40′	225	0.5831	0.7177	1.393	0.8124	483	**20′**
50′	254	0.5854	0.7221	1.385	0.8107	454	**10′**
36° 00′	0.6283	0.5878	0.7265	1.376	0.8090	0.9425	**54° 00′**
		Cos	Cot	Tan	Sin	Radians	Angle

TABLE 4 (continued)

Angle	Radians	Sin	Tan	Cot	Cos		
36° 00′	0.6283	0.5878	0.7265	1.376	0.8090	0.9425	**54° 00′**
10′	312	0.5901	0.7310	1.368	0.8073	396	50′
20′	341	0.5925	0.7355	1.360	0.8056	367	40′
30′	0.6370	0.5948	0.7400	1.351	0.8039	0.9338	30′
40′	400	0.5972	0.7445	1.343	0.8021	308	20′
50′	429	0.5995	0.7490	1.335	0.8004	279	10′
37° 00′	0.6458	0.6018	0.7536	1.327	0.7986	0.9250	**53° 00′**
10′	487	0.6041	0.7581	1.319	0.7969	221	50′
20′	516	0.6065	0.7627	1.311	0.7951	192	40′
30′	0.6545	0.6088	0.7673	1.303	0.7934	0.9163	30′
40′	574	0.6111	0.7720	1.295	0.7916	134	20′
50′	603	0.6134	0.7766	1.288	0.7898	105	10′
38° 00′	0.6632	0.6157	0.7813	1.280	0.7880	0.9076	**52° 00′**
10′	661	0.6180	0.7860	1.272	0.7862	047	50′
20′	690	0.6202	0.7907	1.265	0.7844	0.9018	40′
30′	0.6720	0.6225	0.7954	1.257	0.7826	0.8988	30′
40′	749	0.6248	0.8002	1.250	0.7808	959	20′
50′	778	0.6271	0.8050	1.242	0.7790	930	10′
39° 00′	0.6807	0.6293	0.8098	1.235	0.7771	0.8901	**51° 00′**
10′	836	0.6316	0.8146	1.228	0.7753	872	50′
20′	865	0.6338	0.8195	1.220	0.7735	843	40′
30′	0.6894	0.6361	0.8243	1.213	0.7716	0.8814	30′
40′	923	0.6383	0.8292	1.206	0.7698	785	20′
50′	952	0.6406	0.8342	1.199	0.7679	756	10′
40° 00′	0.6981	0.6428	0.8391	1.192	0.7660	0.8727	**50° 00′**
10′	0.7010	0.6450	0.8441	1.185	0.7642	698	50′
20′	039	0.6472	0.8491	1.178	0.7623	668	40′
30′	0.7069	0.6494	0.8541	1.171	0.7604	0.8639	30′
40′	098	0.6517	0.8591	1.164	0.7585	610	20′
50′	127	0.6539	0.8642	1.157	0.7566	581	10′
41° 00′	0.7156	0.6561	0.8693	1.150	0.7547	0.8552	**49° 00′**
10′	185	0.6583	0.8744	1.144	0.7528	523	50′
20′	214	0.6604	0.8796	1.137	0.7509	494	40′
30′	0.7243	0.6626	0.8847	1.130	0.7490	0.8465	30′
40′	272	0.6648	0.8899	1.124	0.7470	436	20′
50′	301	0.6670	0.8952	1.117	0.7451	407	10′
42° 00′	0.7330	0.6691	0.9004	1.111	0.7431	0.8378	**48° 00′**
10′	359	0.6713	0.9057	1.104	0.7412	348	50′
20′	389	0.6734	0.9110	1.098	0.7392	319	40′
30′	0.7418	0.6756	0.9163	1.091	0.7373	0.8290	30′
40′	447	0.6777	0.9217	1.085	0.7353	261	20′
50′	476	0.6799	0.9271	1.079	0.7333	232	10′
43° 00′	0.7505	0.6820	0.9325	1.072	0.7314	0.8203	**47° 00′**
10′	534	0.6841	0.9380	1.066	0.7294	174	50′
20′	563	0.6862	0.9435	1.060	0.7274	145	40′
30′	0.7592	0.6884	0.9490	1.054	0.7254	0.8116	30′
40′	621	0.6905	0.9545	1.048	0.7234	087	20′
50′	650	0.6926	0.9601	1.042	0.7214	058	10′
44° 00′	0.7679	0.6947	0.9657	1.036	0.7193	0.8029	**46° 00′**
10′	709	0.6967	0.9713	1.030	0.7173	0.7999	50′
20′	738	0.6988	0.9770	1.024	0.7153	970	40′
30′	0.7767	0.7009	0.9827	1.018	0.7133	0.7941	30′
40′	796	0.7030	0.9884	1.012	0.7112	912	20′
50′	825	0.7050	0.9942	1.006	0.7092	883	10′
45° 00′	0.7854	0.7071	1.000	1.000	0.7071	0.7854	**45° 00′**
		Cos	Cot	Tan	Sin	Radians	Angle

ANSWERS TO SELECTED EXERCISES

CHAPTER 1

Exercises (Pages 8–9)

1. true   **3.** false   **5.** true   **7.** true   **9.** false   **11.** false   **13.** false   **15.** true   **17.** $\frac{3}{8}$ **19.** $\frac{3}{5}$ **21.** $\frac{5}{6}$
23. $\frac{1}{12}$ **25.** $\frac{17}{12}$ **27.** $\frac{8}{9}$ **29.** $\frac{31}{30}$ **31.** $\frac{3}{10}$ **33.** $\frac{2}{75}$ **35.** $0.166666\ldots$ **37.** 0.0625 **39.** $0.090909\ldots$
41. $0.101010\ldots$

Exercises (Pages 18–19)

1. $5 > 2$ **3.** $-4 > -10$ **5.** $x > 0$ **7.** $x \nless 0$ **9.** $3x \le 18$ **11.** Property 1.24   **13.** Property 1.25
15. Property 1.25   **17.** Property 1.26

19.

21.

23.

25.

27.

29.

31.

33.

35. $(2, 6)$ **37.** $[-3, 5)$ **39.** $(-4, 6]$ **41.** $\left(-\frac{1}{2}, 1\right]$ **43.** $(-\infty, 2]$ **45.** $\left(-\infty, -\frac{5}{2}\right)$ **47.** 7   **49.** 2   **51.** 5

53. -5 **55.** -9 **57.** $-\dfrac{1}{4}$ **59.** Property 1.29   **61.** Property 1.29   **63.** Property 1.30   **65.** Property 1.31

Exercises (Pages 24–25)

1. 32   **3.** 64   **5.** 26, 873, 856   **7.** 16   **9.** 9   **11.** $\dfrac{3}{32}$ **13.** $\dfrac{5}{4}$ **15.** $\dfrac{1}{2916}$ **17.** 200   **19.** $\dfrac{9}{2}$ **21.** $6a^7$

23. $20a^5b^4$ **25.** $a^{14}b^6$ **27.** a^2b^5 **29.** $\dfrac{d^6}{c}$ **31.** $\dfrac{8a^2b^2}{3}$ **33.** $\dfrac{1}{9x^2}$ **35.** $\dfrac{x^3}{y^3z^6}$ **37.** $\dfrac{y^2z^4}{x^3}$ **39.** $\dfrac{y^3}{x^{10}z^9}$ **41.** $\dfrac{3y^6}{4x^6}$

43. $\dfrac{8y^{12}}{x^6}$ **45.** $\dfrac{4x^2y^{10}z^{22}}{9}$ **47.** 1   **49.** $\dfrac{y^{16}}{x^{14}z^2}$

Exercises (Pages 30–31)

1. 6   **3.** 9   **5.** $4\sqrt{2}$ **7.** 4   **9.** $3\sqrt[3]{2}$ **11.** $\dfrac{3}{5}$ **13.** $2\sqrt[4]{2}$ **15.** $\dfrac{2}{3}$ **17.** $2|x|$ **19.** $\dfrac{4x^2}{|y|}$ **21.** $\dfrac{3a^2}{c^3}\sqrt{2b}$ **23.** xy^2

25. $\dfrac{2x^2}{y}$ **27.** $2|xy|$ **29.** $-\dfrac{2x}{y}$ **31.** $|x|y^2$ **33.** $\dfrac{\sqrt{5}}{5}$ **35.** $\dfrac{\sqrt{y}}{y}$ **37.** $\dfrac{x\sqrt{y}}{y^2}$ **39.** $\dfrac{\sqrt[3]{2x^2y}}{x}$

Exercises (Page 35)

1. 27   **3.** 5   **5.** 16   **7.** 100   **9.** $\dfrac{1}{8}$ **11.** $\dfrac{1}{25}$ **13.** $\dfrac{8}{27}$ **15.** $\dfrac{25}{4}$ **17.** $\dfrac{27}{8}$ **19.** x **21.** $6y^{5/6}$ **23.** $y^{1/4}$

25. $64x^6$ **27.** $32x^{5/2}$ **29.** $\sqrt{2y}$ **31.** x^5y^2 **33.** x^4y^2 **35.** $4x^2y^4$ **37.** $\dfrac{y^2}{x^4}$ **39.** $\dfrac{x^2}{y}$ **41.** $\dfrac{y^4}{x^2}$ **43.** $\dfrac{1}{y\sqrt{x}}$

45. $\dfrac{1}{xy^2}$ **47.** $\dfrac{c^2}{a^2b^4}$ **49.** $\dfrac{8y^3}{27x^3}$ **51.** $\dfrac{z^9}{8x^3y^6}$ **53.** $\dfrac{y}{2x}$ **55.** $x + x^{3/2} + x^{1/2}$ **57.** $x^{5/6} + x^{2/3}$

59. $x^{5/3} + x^{7/6} + x^{2/3}$

Exercises (Pages 39–40)

1. (a) 9   **(b)** 23   **(c)** $\dfrac{27}{4}$ **3. (a)** -5 **(b)** -20 **(c)** -51 **(d)** 1   **(e)** 15   **(f)** 40   **5. (a)** 8   **(b)** 9

(c) 26   **(d)** 33   **(e)** 29   **(f)** no   **7.** $3x^2 - 2x + 4$ **9.** $x^3 + 4x^2 + 5x + 2$ **11.** $2x^3 - 8x^2 + 6x + 4$

13. $m^3 + 6m^2 + 11m - 14$ **15.** $x^3 + 5x^2y + 5xy^2 + y^3$ **17.** $x^4 + 2x^3y + 13x^2y^2 + 3xy^3 - 2y^4$

19. $-2x^4 - 2x^2y^2 + 3xy^3 - y^4$

Exercises (Pages 46–47)

1. $6x - 9y + 3$ **3.** $2x^2 - 4xy + 10x$ **5.** $3x^3y + 4x^2y^2$ **7.** $x^5y^2 + 3x^3y^3 + x^2y^4$ **9.** $a^3bc + ab^3c + abc^3$

11. $6x^2 + x - 1$ **13.** $28x^2 - 23x - 15$ **15.** $18x^2 - 27x - 56$ **17.** $4x^2 - 5xy - 6y^2$ **19.** $2x^2 + 7xy - 15y^2$

21. $a^2 + ab - 6b^2$ **23.** $4m^2 + 16mn + 15n^2$ **25.** $x^2 + 7x + 12$ **27.** $x^2 + 2x - 15$ **29.** $x^2 + x - 72$

31. $x^2 - 17x + 66$ **33.** $x^2 + 6xy + 9y^2$ **35.** $16x^2 - 40xy + 25y^2$ **37.** $4x^2 + 28x + 49$ **39.** $9x^2 - 30x + 25$

41. $x^3 - 27$ **43.** $y^3 + 8$ **45.** $27x^3 - 64y^3$ **47.** $1 - x^4$ **49.** $1 - x^4$ **51.** $x^2 + y^2 - z^2 + 2xy$

53. $x^{2m} - y^{2m}$ **55.** $x^3 + x^2(a + b + c) + x(ab + ac + bc) + abc$ **57.** $x^3 + 3a^2x + 3ax^2 + a^3$ **59.** $2a^2 + 2b^2$

61. $24xy$ **63.** $4ac + 4bc$ **65.** $2x^3 + 6xy^2$ **67.** $6xy + 5y^2$ **69.** $24xy$ **71.** $\sqrt{3} - \sqrt{2}$ **73.** $\sqrt{6} + \sqrt{3}$

75. $\dfrac{\sqrt{x} + \sqrt{y}}{x - y}$ **77.** $\dfrac{\sqrt{x + 4y} + \sqrt{x}}{4y}$ **79.** $\dfrac{\sqrt{x + 2y} - \sqrt{x - 2y}}{2y}$ **81.** $\dfrac{1}{2(\sqrt{6} + \sqrt{5})}$ **83.** $\dfrac{4}{3(\sqrt{7} - \sqrt{3})}$

85. $\dfrac{1}{4\sqrt{3} - 3\sqrt{5}}$ **87.** $\dfrac{y}{2(\sqrt{x + y} - \sqrt{x})}$ **89.** $\dfrac{1}{\sqrt{x + h} + \sqrt{x}}$

Exercises (Pages 55–56)

1. $x(x + 1)$ **3.** $x^2(2x^2 + 3)$ **5.** $x(x^2 + x + 2)$ **7.** $xy(x + y)$ **9.** $4m(m + 2n)$ **11.** $(a + b)(x + y)$
13. $(x + y)(x + 2y)$ **15.** $(x - 2)(x + 2)$ **17.** $(2x - 3y)(2x + 3y)$ **19.** $(4x - 7y)(4x + 7y)$
21. $9(m - 3n)(m + 3n)$ **23.** $(xy - 3z)(xy + 3z)$ **25.** $4(a + b)c$ **27.** $(x + 2)(x + 3)$ **29.** $(x - 7)(x - 2)$
31. $(y - 7)(y + 5)$ **33.** $(y + 7)(y - 3)$ **35.** $(m + 12)(m - 4)$ **37.** $(m + 9)(m - 4)$ **39.** $(p + 2)^2$ **41.** $(p - 7)^2$
43. $(2x + 3)(x + 4)$ **45.** $(3x + 1)(x + 1)$ **47.** $(3x + 2)(x + 1)$ **49.** $(2x + 1)(2x + 3)$ **51.** $(2x + 7)(2x - 1)$
53. $(3m + 4)(2m - 3)$ **55.** $(2m + n)(3m - 5n)$ **57.** $(5p - q)(2p - 3q)$ **59.** $(5p + 16q)(2p - 3q)$ **61.** $(2x - 3y)^2$
63. $(2x + y)^2$ **65.** $(a + 1)(a^2 - a + 1)$ **67.** $(a - 2b)(a^2 + 2ab + 4b^2)$ **69.** $(2x + 3y)(4x^2 - 6xy + 9y^2)$
71. $(6x - 5y)(36x^2 + 30xy + 25y^2)$ **73.** $(x^2 + y^2)(x^4 - x^2y^2 + y^4)$
75. $(2x - 3y)(2x + 3y)(4x^2 + 6xy + 9y^2)(4x^2 - 6xy + 9y^2)$ **77.** $(x^2 + y)(x + 1)$ **79.** $(a - b)(x + y)$
81. $(a + 2)(a + b)$ **83.** $(x^2 + 1)(x + 1)$ **85.** $(a + 1)(x + y)(x - y)$ **87.** $(x + y + 1)(x - y + 1)$
89. $(2x - 3y + 4z)(2x - 3y - 4z)$

Exercises (Pages 59–60)

1. $x - 4$; 2   **3.** $x^2 + 2x + 1$; 4   **5.** $x^2 - 2x - 1$; -5 **7.** $3x^2 + 4x$; 3   **9.** $2x^3 - x^2 + 3x - 11$; 25
11. $x^3 + x^2 + x + 1$; 0   **13.** $2x^4 + x^3 - 2x^2 + 9x - 21$; 22   **15.** $x^4 - 2x^3 + 4x^2 - 8x + 16$; 0
17. $x^5 + x^4y + x^3y^2 + x^2y^3 + xy^4 + y^5$; 0

Exercises (Pages 66–68)

1. x **3.** $x + 1$ **5.** $\dfrac{x + 2}{x - 2}$ **7.** $x + 3$ **9.** $\dfrac{x + 3}{x}$ **11.** $\dfrac{x - 2}{x - 5}$ **13.** $\dfrac{x - 5}{x + 1}$ **15.** $x + y$ **17.** $\dfrac{x^2 + xy + y^2}{x + y}$

19. $\dfrac{x + 4y}{x + y}$ **21.** $\dfrac{x}{x + 2y}$ **23.** $\dfrac{x + 4y}{x - 5y}$ **25.** $\dfrac{4x + 1}{4}$ **27.** $\dfrac{3x^2 + x + 1}{y}$ **29.** $\dfrac{2 - x - 4y}{y}$ **31.** $\dfrac{5x + y}{3y}$

33. $\dfrac{x^2 + 6y^2}{3xy}$ **35.** $\dfrac{2x^2 + 2y^2}{x^2 - y^2}$ **37.** $\dfrac{4x^2 - 4xy + y^2 - 2x^2y - xy^2}{4x^2 - y^2}$ **39.** $\dfrac{2x^2 + 6xy + 5y^2}{(x + y)(x + 2y)}$ **41.** $x - 1$

43. $\dfrac{4}{y}$ **45.** $\dfrac{x + 1}{5x}$ **47.** $\dfrac{x - 2}{x + 2}$ **49.** $\dfrac{m + 8}{n + 9}$ **51.** $\dfrac{m}{m + 4}$ **53.** $\dfrac{x - 5y}{x - 7y}$ **55.** $\dfrac{(3x - 1)(3x - 2)}{3(2x - 1)}$

57. $\dfrac{m^2 - mn + n^2}{m^2 + mn + n^2}$ **59.** $\dfrac{m + n}{m - n}$ **61.** $\dfrac{x - 1}{2x + 1}$ **63.** $x - y$ **65.** $\dfrac{1}{x}$ **67.** $\dfrac{3x + 5}{(x + 1)^2}$ **69.** $\dfrac{2x^2 + 3x + 3}{3x^2 + 2x - 5}$

Exercises (Pages 71–72)

1. $\dfrac{1}{x} - \dfrac{1}{x + 2}$ **3.** $\dfrac{2}{x + 1} + \dfrac{1}{x + 2}$ **5.** $\dfrac{1}{3(x - 1)} + \dfrac{8}{3(x + 2)}$ **7.** $\dfrac{1}{2(x - 1)} - \dfrac{1}{2(x + 1)}$ **9.** $\dfrac{4}{x - 2} - \dfrac{1}{x - 1}$

11. $\dfrac{4}{3x + 1} + \dfrac{1}{x + 1}$ **13.** $\dfrac{3}{2(x - 1)} - \dfrac{4}{x - 2} + \dfrac{5}{2(x - 3)}$ **15.** $\dfrac{1}{x} + \dfrac{1}{x^2} + \dfrac{1}{1 - x}$ **17.** $\dfrac{1}{x + 1} - \dfrac{1}{x - 2} + \dfrac{2}{(x - 2)^2}$

19. $\dfrac{-3}{25(x + 2)} + \dfrac{3}{25(x - 3)} + \dfrac{7}{5(x - 3)^2}$

Exercises (Pages 76–77)

1. $9 + 6i$ **3.** $6 - i$ **5.** $-1 - i$ **7.** $3 - 7i$ **9.** $1 + 27i$ **11.** $5 + 3i$ **13.** $3 + 11i$ **15.** $24 - 2i$ **17.** 61

19. $-4 + \frac{3}{2}i$ **21.** $-4 - 3i$ **23.** $4 - 7i$ **25.** $\frac{1}{2} - \frac{1}{3}i$ **27.** $0 + i$ **29.** $\dfrac{23}{25} + \dfrac{14}{25}i$ **31.** $-\dfrac{7}{25} - \dfrac{24}{25}i$

33. $\dfrac{1}{3} - \dfrac{2}{3}i$ **35.** $-\dfrac{3}{13} - \dfrac{2}{13}i$ **37.** $\dfrac{28}{65} + \dfrac{16}{65}i$ **39.** $\dfrac{\sqrt{2}}{3} + \dfrac{2}{3}i$

41. Let $z_1 = a + bi$ and $z_2 = c + di$, where a, b, c, and d are real numbers. Then

$$z_1 - z_2 = (a + bi) - (c + di) = (a - c) + (b - d)i$$

and

$$\overline{z_1 - z_2} = (a - c) - (b - d)i$$
$$= (a - bi) - (c - di)$$
$$= \bar{z}_1 - \bar{z}_2.$$

Exercises for Review (Pages 78–80)

1. false **2.** true **3.** true **4.** true **5.** true **6.** false **7.** true **8.** true **9.** true **10.** true

11. $\{-4, -2, 0, 1, 5\}$ **12.** $\{-4, -2, 0, 1, \frac{2}{3}, 5\}$ **13.** $\left\{\sqrt{2}, \pi, \dfrac{1}{\sqrt{3}}, \dfrac{2\pi}{3}\right\}$ **14.** all members of set A

15. $x > 0$ **16.** $4x < 12$ **17.** $2x \geq 13$ **18.** $5 \leq x \leq 7$ **19.** $[1, 3]$ **20.** $(-1, 4]$ **21.** $[-2, 5)$ **22.** $[1, \infty)$

23. 4 **24.** -4 **25.** -4 **26.** $(-2, 2)$ **27.** $\dfrac{9}{64}$ **28.** $\dfrac{25}{81}$ **29.** $\dfrac{9}{4}$ **30.** $-\dfrac{9}{2}$ **31.** $36x^6 y^{10} z^8$ **32.** $\dfrac{4a^2}{3b}$

33. $\dfrac{16y^5}{27x^9}$ **34.** $\dfrac{x^{10} y^{15}}{z^{20}}$ **35.** $\dfrac{16y^4 z^8}{81x^4}$ **36.** $\dfrac{20}{3} a^5 b^5$ **37.** $7\sqrt{2}$ **38.** $3(\sqrt[4]{2})$ **39.** xy^2 **40.** $\dfrac{2xy^2}{z^3}$ **41.** $\sqrt{3} - \sqrt{2}$

42. $\dfrac{\sqrt{x + 2y} + \sqrt{x}}{2y}$ **43.** (a) $P(-1) = 1$ (b) $P(2) = 7$ (c) $Q(-2) = -4$ (d) $Q(\frac{1}{2}) = \frac{7}{2}$ (e) $P(Q(3)) = 133$

(f) $Q(P(2)) = 23$ **44.** $6x^3 + 2x^2 + 2x + 6$ **45.** $12x^3 y^2 + 8x^2 y^3 - 4xy^4$ **46.** $8x^2 + 2x - 15$ **47.** $8x^3 - 27$
48. $x^2 + 4xy + 4y^2 - 9z^2$ **49.** $24xy$ **50.** $4x^2 + 12xz + 9z^2 - y^2$ **51.** $(x + 9)(x - 4)$ **52.** $(2y + 3)(y + 4)$
53. $(3x - 4y)(9x^2 + 12xy + 16y^2)$ **54.** $(2x - 3y)(2x + 3y)(4x^2 + 9y^2)$ **55.** $(a + 1)(b + 1)$

56. $(3x - 4y - 5z)(3x - 4y + 5z)$ **57.** $x^2 - 5x + 2$; 0 **58.** $3x^2 + 6x + 8$; 15 **59.** $\dfrac{z}{2y + 3z}$

60. $\dfrac{1}{a^2 - ab + b^2}$ **61.** $\dfrac{x + 1}{x + 4}$ **62.** $\dfrac{x}{x + y}$ **63.** $\dfrac{x^2 + 3xy + y^2}{(x + 2y)(x + 3y)}$ **64.** $\dfrac{3x^2 - xy - 3y^2}{(3x + 2y)(3x + y)}$ **65.** $\dfrac{xy}{x + y}$

66. $\dfrac{x}{2 + x}$ **67.** $\dfrac{1}{x} - \dfrac{1}{x + 1}$ **68.** $\dfrac{2}{x + 5} - \dfrac{1}{x - 3}$ **69.** $\dfrac{3}{2x - 1} - \dfrac{4}{3x - 2}$ **70.** $\dfrac{1}{x} - \dfrac{1}{x - 1} + \dfrac{1}{(x - 1)^2}$ **71.** $4 + 3i$

72. $1 + 4i$ **73.** $23 - 14i$ **74.** $\dfrac{41}{10} - \dfrac{3}{5}i$ **75.** $\dfrac{1}{2} + \dfrac{1}{2}i$ **76.** $\dfrac{3}{5} + \dfrac{4}{5}i$ **77.** $\dfrac{3}{13} + \dfrac{11}{13}i$ **78.** $\dfrac{47}{58} - \dfrac{13}{58}i$ **79.** $1 - 2i$

80. Let $z_1 = a + bi$, $z_2 = c + di$, and $z_3 = e + fi$, where a, b, c, d, e, and f are real numbers. Then

$$z_1 + z_2 + z_3 = (a + c + e) + (b + d + f)i$$

and

$$\overline{z_1 + z_2 + z_3} = (a + c + e) - (b + d + f)i$$
$$= (a - bi) + (c - di) + (e - fi)$$
$$= \bar{z}_1 + \bar{z}_2 + \bar{z}_3.$$

Chapter Test (Pages 80–81)

1. $\dfrac{20y^8}{x}$ **2.** $\dfrac{2x^5}{5y^2z^5}$ **3.** $\dfrac{5x^3y}{3}$ **4.** $\dfrac{81}{4}a^2b$ **5.** $\dfrac{1}{\sqrt{x+h}+\sqrt{x}}$ **6. (a)** 2 **(b)** 9 **(c)** 146 **(d)** 41

7. (a) $2x^3 + 9x^2 + 21x + 18$ **(b)** $x^4 - 16y^4$ **8. (a)** $(2x - 5y)(2x + 5y)$ **(b)** $(2x - 3)(3x + 4)$

9. $3x^2 + 8x + 6$; remainder $= 16$ **10. (a)** $\dfrac{x-1}{x+2}$ **(b)** $\dfrac{5x+2}{(x+2)(x-2)}$ **11.** $\dfrac{1}{2x} - \dfrac{1}{2(x+2)}$

12. (a) $18 + i$ **(b)** $-\frac{6}{13} + \frac{17}{13}i$

CHAPTER 2

Exercises (Pages 88–89)

1. equivalent **3.** not equivalent **5.** not equivalent **7.** not equivalent **9.** $x = 3$ **11.** $x = \dfrac{7}{2}$

13. $x = 14$ **15.** $x = 13$ **17.** $x = -0.4$ **19.** $x = 1.5$ **21.** $x = 12$ **23.** $x = \dfrac{10}{11}$ **25.** $x = 3$ **27.** $x = 12$

29. $x = \dfrac{7}{2}$ **31.** no solution **33.** $x = 2$ **35.** $x = 5$ **37.** $x = 3$ **39.** $x = \frac{7}{6}$ **41.** no solution

43. no solution **45.** no solution **47.** $x = 5$ **49.** $x = -\dfrac{1}{2}$ **51.** $x = \frac{5}{22}$ **53.** $x = \dfrac{1}{16}$ **55.** $x = -\frac{12}{5}$

57. $x = 2.5$ **59.** $x = 6.0022$

Exercises (Pages 97–99)

1. $a = \dfrac{F}{m}$ **3.** $r = \dfrac{C}{2\pi}$ **5.** $u = v - gt$ **7.** $m_1 = \dfrac{Fd^2}{km_2}$ **9.** $b = \dfrac{2A}{h} - c$ **11.** $x = \dfrac{ab - by}{a}$ **13.** $s = \dfrac{A}{r} - r$

15. $s = a + (n-1)d$ **17.** $u = \dfrac{2s - gt^2}{2t}$ **19.** $r_1 = \dfrac{rr_2}{r_2 - r}$ **21.** 63; 102 **23.** 84

25. 1500 at 5.5 percent; 2000 at 6 percent **27.** 14 nickels; 35 dimes **29.** 200 adults; 75 children
31. 25 kilometers per hour **33.** At 6.30 P.M.; Inez: 2 miles, Mike: 3 miles **35.** 360 miles

37. 360 miles **39.** 1 hour 12 minutes **41.** $\dfrac{40}{11}$ hours **43.** 32 liters **45.** 15 pounds

47. Solution A: 60 liters, solution B: 20 liters

Exercises (Pages 106–108)

1. $\{1, 3\}$ **3.** $\left\{-\dfrac{1}{2}, -4\right\}$ **5.** $\{-1, -2\}$ **7.** $\{-3, -4\}$ **9.** $\{2, 3\}$ **11.** $\{-5, 8\}$ **13.** $\{3, -2\}$

15. $\left\{-1, \dfrac{1}{2}\right\}$ **17.** $\left\{1, \dfrac{1}{2}\right\}$ **19.** $\left\{-\dfrac{1}{2}, \dfrac{4}{3}\right\}$ **21.** complex **23.** rational and unequal

25. rational and unequal **27.** real and unequal **29.** double root **31.** $\{-4, 3\}$

33. $\left\{\dfrac{-5 + \sqrt{13}}{2}, \dfrac{-5 - \sqrt{13}}{2}\right\}$ **35.** $\left\{3, -\dfrac{3}{2}\right\}$ **37.** $\left\{-4, \dfrac{2}{3}\right\}$ **39.** $\left\{-2, \dfrac{3}{5}\right\}$ **41.** $\left\{\dfrac{-1 + \sqrt{3}i}{2}, \dfrac{-1 - \sqrt{3}i}{2}\right\}$

43. $\{2 + i, 2 - i\}$ **45.** $\left\{\dfrac{1}{2} \pm \dfrac{3}{4}i\right\}$ **47.** $\{7, -5\}$ **49.** $\left\{2, \dfrac{1}{2}\right\}$ **51.** $\{4, -1\}$ **53.** $\{12, -1\}$ **57.** $k = 2$

59. $t = \dfrac{1}{4}\sqrt{s}$ **61.** $\dfrac{-u \pm \sqrt{u^2 + 2gs}}{g}$ or $t = [-u \pm \sqrt{u^2 + 2gs}]/g$ **63.** 15; 12   **65.** $t = 2$, 4 seconds

67. 10 and 14 inches   **69.** 36 and 48 miles per hour   **71.** 2 miles per hour   **73.** 30 miles per hour
75. Jeff: 30 days; Bonnie: 45 days

Exercises (Pages 112–113)

1. $\{9\}$ **3.** $\{16\}$ **5.** $\{15\}$ **7.** $\{12\}$ **9.** $\left\{\dfrac{1}{2}\right\}$ **11.** $\{7\}$ **13.** $\{-1, 3\}$ **15.** $\{0\}$ **17.** $\{-3, 3\}$

19. $\{2, -1\}$ **21.** $\{5\}$ **23.** $\{8\}$ **25.** $\{0, 5\}$ **27.** $\left\{-\dfrac{1}{2}, -\dfrac{2}{3}\right\}$ **29.** $\{4\}$ **31.** $\{27, 64\}$ **33.** $\{64, -8\}$

35. $\left\{8, -\dfrac{27}{8}\right\}$ **37.** $\{\pm\sqrt{3}, \pm\sqrt{2}\}$ **39.** $\left\{\pm 3, \pm\dfrac{\sqrt{3}}{2}\right\}$ **41.** $x = \pm\dfrac{\sqrt{3}}{3}, \pm\dfrac{\sqrt{2}}{2}$ **43.** $x = 1$

Exercises (Page 115)

1. $(-\infty, 2)$ **3.** $(3, \infty)$ **5.** $(-\infty, 5)$ **7.** $(-\infty, 4]$ **9.** $(3, \infty)$ **11.** $(-\infty, 9]$ **13.** $\left(-\infty, \dfrac{9}{4}\right]$

15. $\left[-\dfrac{7}{8}, \infty\right)$ **17.** $\left(-\infty, \dfrac{1}{4}\right]$ **19.** $(-\infty, 12)$ **21.** $-2 < x < 5$

23. $1 \leq x \leq 3$ **25.** $-1 \leq x \leq 2$

Exercises (Page 120)

1. $(-\infty, 0) \cup (1, \infty)$ **3.** $(0, 4)$ **5.** $(-\infty, 1) \cup (2, \infty)$ **7.** $(2, 3)$ **9.** $(-\infty, -1] \cup [2, \infty)$ **11.** all real numbers
13. $\left[-\dfrac{5}{2}, 1\right]$ **15.** $[-2, 3]$ **17.** $(-\infty, 0) \cup \left(\dfrac{9}{4}, \infty\right)$ **19.** $(-2, 3)$ **21.** $\left(-\dfrac{1}{2}, 2\right)$ **23.** $\left(-\dfrac{5}{2}, -2\right)$
25. $\left(-\infty, \dfrac{2}{3}\right) \cup \left(\dfrac{5}{3}, \infty\right)$ **27.** $\left(2, \dfrac{7}{2}\right)$ **29.** $(0, 2) \cup (-\infty, -1)$ **31.** $(0, 1) \cup (4, \infty)$ **33.** $1 < t < 5$

Exercises (Pages 124–125)

1. $\{5, -1\}$ **3.** $\left\{-1, \dfrac{7}{3}\right\}$ **5.** $\left\{\dfrac{4}{3}, 4\right\}$ **7.** $\{-5, 11\}$ **9.** $\left\{-\dfrac{7}{3}, 3\right\}$ **11.** $\{-3, -1\}$ **13.** $\left\{\dfrac{1}{5}, 3\right\}$ **15.** $\left\{-1; \dfrac{11}{13}\right\}$
17. $\left\{-\dfrac{4}{7}, -16\right\}$ **19.** $\left\{-\dfrac{1}{7}, 1\right\}$ **21.** $(-1, 1)$ **23.** $\left(-\infty, -\dfrac{1}{3}\right)$ **25.** $[-1, 5]$ **27.** $(-\infty, 1) \cup (5, \infty)$
29. $(-\infty, -1) \cup (5, \infty)$ **31.** $[-8, 4]$ **33.** $[0, 12]$

Exercises for Review (Pages 125–126)

1. -1 **2.** $\dfrac{7}{8}$ **3.** 3   **4.** $\dfrac{6}{5}$ **5.** $\dfrac{3}{5}$ **6.** 5   **7.** 4, 5   **8.** $\dfrac{1}{2}, \dfrac{11}{2}$ **9.** $2 + \sqrt{3}, 2 - \sqrt{3}$ **10.** $\dfrac{-1 - \sqrt{3}i}{2}, \dfrac{-1 + \sqrt{3}i}{2}$

11. $1 - i, -1 - i$ **12.** $-3, 10$ **13.** $\dfrac{2 - \sqrt{39}}{5}, \dfrac{2 + \sqrt{39}}{5}$ **14.** $\dfrac{6 + \sqrt{3}}{3}, \dfrac{6 - \sqrt{3}}{3}$ **15.** 4   **16.** 3   **17.** 60

18. $-1, 2$ **19.** $\pm\sqrt{2}, \pm\sqrt{3}i$ **20.** 16   **21.** $-\frac{5}{2}, \frac{7}{2}$ **22.** $-\frac{11}{2}, \frac{3}{4}$ **23.** 5   **24.** 6   **25.** 2   **26.** 4
27. $x \le -5$ **28.** $x < \frac{19}{6}$ **29.** $-4 < x < 5$ **30.** $-1 \le x \le 9$ **31.** $-3 \le x \le 3$ **32.** $-5 < x < 2$

33. $x \le 1$ or $x \ge 4$ **34.** $-1 < x < 0$ or $x < -2$ **35.** $x < 0$ or $x > \frac{3}{8}$ **36.** $1 < x < 2$ or $x > 3$

37. $\frac{1}{2} \le x \le \frac{5}{2}$ **38.** $x < -\frac{4}{3}$ or $x > \frac{8}{3}$ **39.** \$1500 at 6 percent; \$6000 at 9 percent   **40.** 50 miles; 45 miles

41. 1 hour 30 minutes   **42.** solution A: 6 liters; solution B: 4 liters   **43.** 6 centimeters   **44.** 3 feet
45. 175 miles per hour   **46.** pipe A: 6 hours; pipe B: 4 hours

Chapter Test (Page 127)

1. $x = 13$ **2.** $x = \frac{7}{8}$ **3.** $x = -2$ **4.** $x = \frac{3}{5}$ **5.** $x = \frac{b}{a}(a - y)$ **6.** $x = \frac{zy}{y - z}$ **7.** $x = \frac{-7 \pm \sqrt{61}}{6}$ **8.** $x = 10$

9. $x = -\frac{27}{8}, 8$ **10.** $x = \pm 3$

11. **(a)** $x < 0$ or $x > \frac{9}{4}$ **(b)** $(-\infty, 0) \cup \left(\frac{9}{4}, \infty\right)$

12. **(a)**   **(b)** $(-\infty, 4)$

13. $x = -\frac{5}{2}, \frac{11}{10}$ **14.** $-1 \le x \le 4$ **15.** 6 days   **16.** $\frac{4}{5}$ mile   **17.** solution A, 6 liters; solution B, 4 liters

18. 30 miles per hour

CHAPTER 3

Exercises (Pages 137–138)

9. $(3, 2)$ **11.** $(-3, 6)$ **13.** $(-4, -6)$ **15.** 6   **17.** 6   **19.** 20   **21.** 5   **23.** 10   **25.** 13   **27.** 6
29. shipping by van lines is cheaper
31. Let $(12, 8)$, $(-2, 6)$, and $(6, 0)$ be the coordinates of points A, B, and C, respectively. Using distance formula, we see that
$$d_{AB} = \sqrt{200}, \qquad d_{BC} = 10, \qquad \text{and} \qquad d_{AC} = 10.$$
Since $d_{AB}^2 = d_{BC}^2 + d_{AC}^2$, we conclude that triangle ABC is a right triangle.
33. $(0, 7)$ **35.** $(-6, -8)$ **37.** $(5, -7)$

Exercises (Pages 148–151)

1. a function   **3.** not a function   **5.** a function   **7.** a function   **9.** 1, 4, and 7 are one-to-one functions
11. a function   **13.** not a function   **15.** a function   **17.** a function   **19.** a function   **21.** not a function

23. **(a)** 3   **(b)** 5   **(c)** 5   **(d)** 11   **(e)** 21   **(f)** 35   **25.** **(a)** 1   **(b)** $\frac{7}{8}$ **(c)** $\frac{1}{2}$ **(d)** $\frac{2a + 1}{3a - 1}$ **(e)** $\frac{2(a + h) + 1}{3(a + h) - 1}$

   (f) $\frac{-5h}{(3a - 1)(3a + 3h - 1)}$ **27.** **(a)** $a^2 - a + 2$ **(b)** $a + \sqrt{a} + 2$ **(c)** $a^2 + a + 2$ **(d)** $(a + h)^2 + a + h + 2$

(e) $a^2 + h^2 + a + h + 4$ (f) $2a + h + 1$ **29.** (a) 4000   (b) 3200   (c) 2000   (d) 1200   (e) 400   (f) 0
31. $\{-3, -1, 1, 3, 5\}$ **33.** $\{0, 1, 2\}$ **35.** $\{1, \frac{1}{2}, \frac{1}{3}, \frac{1}{4}, \frac{1}{5}\}$ **37.** set of real numbers   **39.** set of real numbers
41. set of real numbers except $x = -3, 3$
43. The domain consists of two-decimal-place numbers between 0 and 25, inclusive; the range consists also of two-decimal-place numbers between 0 and 100, inclusive.

45. $R(x) = (100 + x)(40 - 0.20x)$ **47.** $P(x) = (90 + x)(12.00 - 0.10x)$ **49.** $A(x) = \dfrac{x(3000 - 3x)}{2}$

Exercises (Pages 158–160)

1.

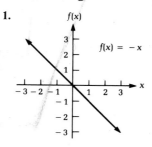

3.

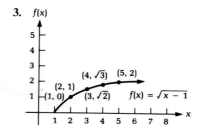

5.

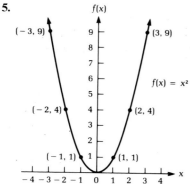

7.

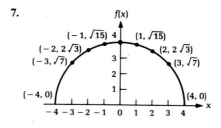

9.

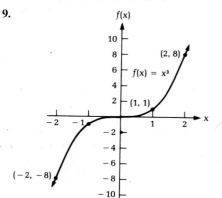

11.

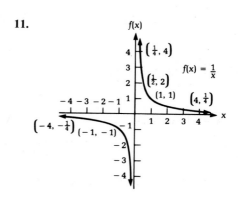

13. (a) $D(1) = 440$; $D(2) = 400$; $D(3) = 360$; $D(4) = 320$; $D(5) = 280$; $D(6) = 240$; $D(8) = 160$; $D(10) = 80$; $D(12) = 0$

(b)

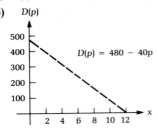

$D(p) = 480 - 40p$

15. (a) $g(2) = 50$; $g(5) = 20$; $g(10) = 10$; $g(20) = 5$; $g(50) = 2$

(b)

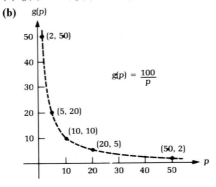

$$g(p) = \frac{100}{p}$$

17. (a) $s(3) = 0$; $s(4) = 10$; $s(5) = 15$; $s(6) = 18$; $s(7) = 20$; $s(11) = 24$; $s(16) = 26$; $s(21) = 27$; $s(31) = 28$

(b)

$$s(p) = 30 - \frac{60}{p - 1}$$

19. (a) $f(0) = 30{,}000$; $f(5) = 31{,}500$; $f(10) = 32{,}000$; $f(15) = 31{,}500$; $f(20) = 30{,}000$; $f(25) = 27{,}500$; $f(30) = 24{,}000$; $f(35) = 19{,}500$; $f(40) = 14{,}000$; $f(45) = 7500$; $f(50) = 0$

(b)

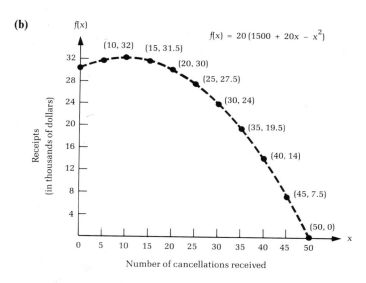

$f(x) = 20(1500 + 20x - x^2)$

21. (a) $4.60 (b) $13.40 (c) $20.00
23. (a) $0.70 (b) $1.10 (c) $1.50
(d) $1.90 (e) $2.30 (f) $2.70

Exercises (Pages 171–172)

1. -1 3. $\frac{2}{5}$ 5. 2 7. 0 9. $y = 5$ 11. $x = 2$ 13. undefined 15. $\frac{1}{2}$ 17. 3
19. slope $= \frac{2}{3}$, y intercept $= 2$ 21. slope $= -\frac{5}{6}$, y intercept $= 5$ 23. slope $= -4$, y intercept $= 0$
25. parallel 27. perpendicular 29. perpendicular 31. neither parallel nor perpendicular
33. $y = 4x + 3$ 35. $y = 4$ 37. $y = -4x + 5$ 39. $y = 2x - 4$ 41. $y = x + 1$ 43. $y = \frac{2}{3}x + \frac{7}{3}$
45. $y = x - 5$ 47. $y = \frac{9}{2}x - 22$ 49. (1) $y = -x + 7$ (2) $y = \frac{3}{7}x + \frac{20}{7}$ (3) $y = \frac{2}{5}x + 4$ (4) $x = 6$
(5) $y = 2x - 3$ (6) $y = 2x - 4$ (7) $y = 4$ (8) $y = -\frac{4}{3}x + 13$ 51. $4x + 3y - 12 = 0$
53. $2x - y + 2 = 0$ 55. $2x + 3y - 1 = 0$ 57. $x + y - 5 = 0$

Exercises (Pages 174–175)

1. (a) $y = 900 + 20x$ (b) $1400 3. (a) $y = 32.95 + 0.20x$ (b) $41.95 5. (a) $y = 400,000 - 10,000x$
(b) $280,000 7. (a) $y = 75,000 - 1200x$ (b) $43,800 9. (a) $y = 4000 + 125x$ (b) 5625
11. (a) $y = 25,000 - 500x$ (b) 13,500 gallons

Exercises (Pages 186–189)

1.

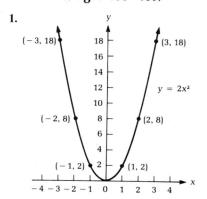

3.

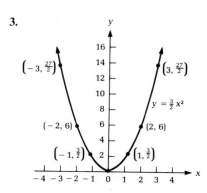

5.

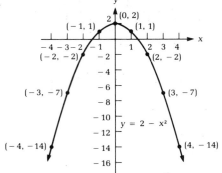

7.

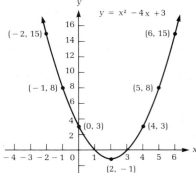

9.

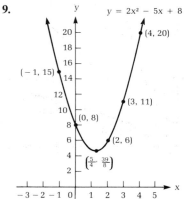

11.

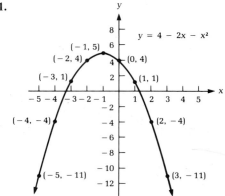

13.

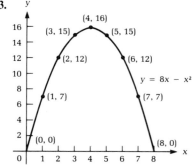

15. $(-1, 2)$; $x = -1$ **17.** $(-2, -8)$; $x = -2$ **19.** $(1, 9)$; $x = 1$
21. $(-1, 12)$; $x = -1$ **23.** 200   **25.** 4 million barrels   **27.** 10
29. 16 feet   **31.** $t = 5$ **33.** 7 hours; 11 miles
35. 2400 feet × 1200 feet   **37.** each side 10 inches long
39. 150 couples   **41.** 75 couples; \$562.50
43. 5 days; earnings, \$156.25

Exercises (Pages 198–199)

1. $g(f(1)) = 13$; $g(f(2)) = 19$; $g(f(3)) = 25$; $g(f(4)) = 31$; $g(f(5)) = 37$; $g(f(6)) = 43$ **3.** $g(f(1)) = 31$;
$g(f(2)) = 43$; $g(f(3)) = 55$; $g(f(4)) = 67$; $g(f(5)) = 79$; $g(f(6)) = 91$; $f(g(1)) = 25$; $f(g(2)) = 37$; $f(g(3)) = 49$;
$f(g(4)) = 61$; $f(g(5)) = 73$; $f(g(6)) = 85$ **5.** $g(f(x)) = 3x$; $f(g(x)) = 3x - 10$ **7.** $g(f(x)) = x + \frac{5}{3}$; $f(g(x)) = x + 5$
9. $g(f(x)) = 3x^2 + 3$; $f(g(x)) = 9x^2 + 1$ **11.** $g(f(x)) = 6x^2 + 14$; $f(g(x)) = 2(3x - 1)^2 + 5$
13. $g(f(x)) = 1 - (x^2 + 2x + 3)^3$; $f(g(x)) = (1 - x^3)^2 + 2(1 - x^3) + 3$ **15.** $g(f(x)) = \sqrt{x^2 + 1}$; $f(g(x)) = x + 1$

17. $g(f(x)) = x$; $f(g(x)) = x$ **19.** $g(f(x)) = \dfrac{1}{x^3 + 2x^2 + x + 2}$; $f(g(x)) = \dfrac{1}{(x-2)^3} + \dfrac{2}{(x-2)^2} + \dfrac{1}{x-2} + 4$

21. $g(f(x)) = x$; $f(g(x)) = x$ **23.** $g(f(x)) = x$; $f(g(x)) = x$ **25.** f is one-to-one; $f^{-1}(x) = \dfrac{x+3}{4}$

27. f is one-to-one; $f^{-1}(x) = x^2 + 5$, $x \geq 0$ **29.** f is one-to-one; $f^{-1}(x) = (x-1)^{1/3}$ **31.** $f^{-1}(x) = \dfrac{x}{2}$

33. $f^{-1}(x) = \dfrac{3-x}{4}$ **35.** $f^{-1}(x) = x^2 + 3$, $x \geq 0$ **37.** $f^{-1}(x) = \sqrt{4 - x^2}$, $-2 \leq x \leq 2$

39. $f^{-1}(x) = \dfrac{2x+1}{x}$, $x \neq 0$ **41.** $f^{-1}(x) = \dfrac{2}{x-1}$, $x \neq 1$ **43.** $f^{-1}(x) = \dfrac{1+x}{1-x}$, $x \neq 1$

45.

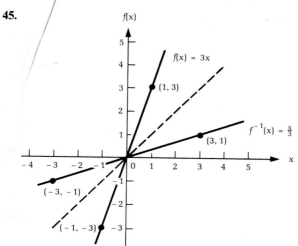

47.

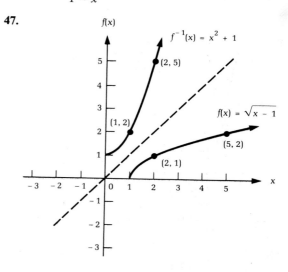

49.

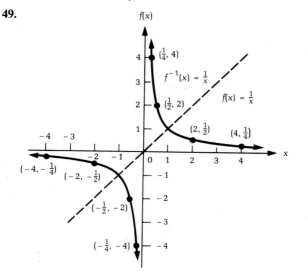

51. To show that $f^{-1}(x)$ exists, it is necessary to demonstrate that $f(x)$ is one-to-one. Suppose that a and b are two distinct elements in the domain of f and $f(a) = f(b)$. Then

$$|a| = |b|,$$

which implies that $a = b$ or $a = -b$. Since one element a in D_f has two images, b and $-b$, we conclude that $f(x)$ is not one-to-one. Therefore, $f^{-1}(x)$ does not exist.

Exercises (Pages 206–207)

1. 2 **3.** 3 **5.** 40 **7.** 25 **9.** π **11.** 80 **13.** 60 **15.** 128 feet **17.** 288π cubic inches **19.** 246 ohms
21. 15 feet **23.** 225 foot-pounds

Exercises for Review (Pages 208–210)

1. Using the distance formula, we see that

$$d_{AB} = \sqrt{116}, \qquad d_{BC} = \sqrt{29}, \qquad d_{AC} = \sqrt{145}.$$

Since

$$d_{AC}^2 = d_{AB}^2 + d_{BC}^2,$$

we conclude that triangle ABC is a right-angled triangle.

2. (a) 13 **(b)** 34 **(c)** 19 **(d)** $2(a + h)^2 + 3(a + h) - 1$ **(e)** $2a + 3\sqrt{a} - 1$ **(f)** $\dfrac{2}{a^2} + \dfrac{3}{a} - 1$

3. $6a + 5 + 3h$ **4. (a)** not a function **(b)** a function **5.** set of real numbers
6. set of real numbers greater than or equal to $\frac{1}{2}$
7. set of real numbers less than or equal to -1 or greater than or equal to 1
8. set of real numbers **9.** set of nonzero real numbers **10.** set of real numbers between -1 and 1, inclusive

11.

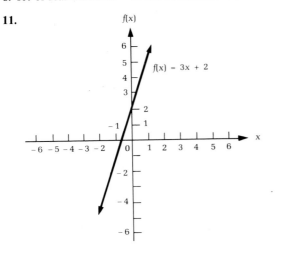

12.

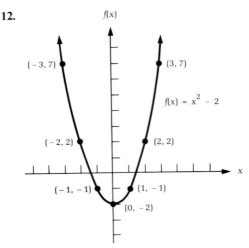

13.

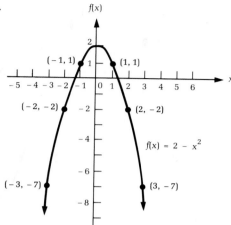

$f(x) = 2 - x^2$

14.

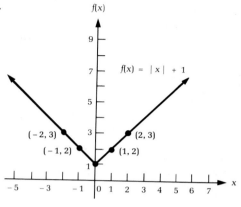

$f(x) = |x| + 1$

15.

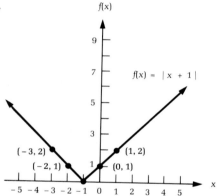

$f(x) = |x + 1|$

16.

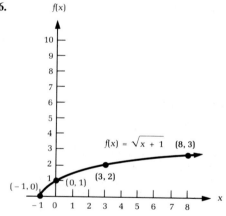

$f(x) = \sqrt{x + 1}$

17.

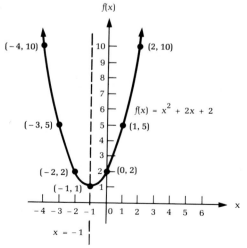

$f(x) = x^2 + 2x + 2$

$x = -1$

18.

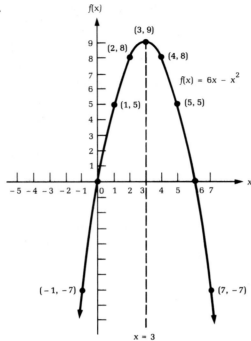

$f(x) = 6x - x^2$

19.

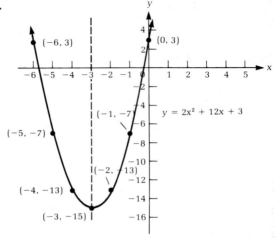

$y = 2x^2 + 12x + 3$

20.

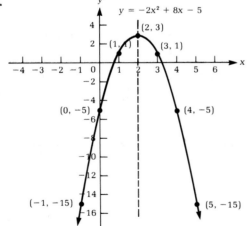

$y = -2x^2 + 8x - 5$

21. $m = 3, b = 5$ **22.** $m = \frac{2}{3}, b = 2$ **23.** neither **24.** parallel **25.** neither **26.** perpendicular
27. $y = 4x - 15$ **28.** $y = x + 2$ **29.** $x = 2$ **30.** $2x + 3y - 10 = 0$ **31.** $2x + y - 13 = 0$ **32.** $(1, 9); x = 1$
33. $g(f(x)) = 6x^2 + 3x + 16; f(g(x)) = 18x^2 + 15x + 8$ **34.** $g(f(x)) = x^2 + 2x + 1; f(g(x)) = x^2 + 1$

35. $g(f(x)) = x, f(g(x)) = |x|$ **36.** $g(f(x)) = \sqrt{x^2 + 1}, f(g(x)) = x + 1$ **37.** does not exist **38.** $f^{-1}(x) = \sqrt{\dfrac{x}{2}}$

39. $f^{-1}(x) = (x + 1)^{1/3}$ **40.** $f^{-1}(x) = x^2 + 1, x \geq 0$ **41.** **(a)** $C(x) = 0.25x + 32.95$ **(b)** $\$45.45$

42. $y = 5x - 190$ **43.** $h = 2500$ feet **44.** $t = 2.5$ hours; $N = 75$ **45.** 960 feet × 480 feet **46.** $r = \dfrac{128}{243}$

47. 6 men **48.** 1 ampere **49.** 15.811 centimeters

Chapter Test (Pages 210–211)

1. 13 **2.** **(a)** not a function **(b)** a function
3. **(a)** nonnegative set of real numbers **(b)** set of real numbers **(c)** nonzero set of real numbers
4. $h + 7$ **5.** $y = 2x - 2$ **6.** $y = x - 1$ **7.** $y = 6$ **8.** $5y - 6x - 7 = 0$

9. $g(f(x)) = 6x^2 + 5; f(g(x)) = 18x^2 + 5$ **10.** $y = \left(\dfrac{x}{2}\right)^{1/3}$ **11.** $\left(-\dfrac{5}{6}, \dfrac{119}{12}\right); x = -\dfrac{5}{6}$ **12.** $x = 10$

13. $s = 16t^2$ **14.** 39; 39
15. **(a)**

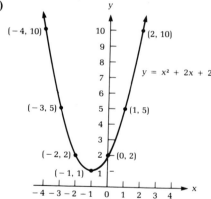

(b)

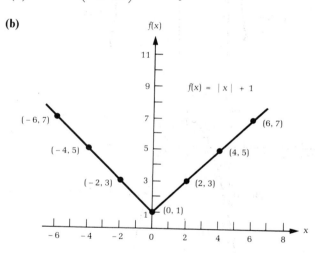

CHAPTER 4

Exercises (Pages 219–220)

1. 8 **3.** 6 **5.** 10 **7.** 0 **9.** $1 - 11i$ **11.** $1 + i$ **13.** 3 **15.** 0 **17.** $6 + 2i$ **19.** $4 - 3i$ **21.** $-2 - 6i$
23. factor **25.** factor **27.** factor **29.** factor **31.** factor **33.** -4 **35.** 1 **37.** $\frac{1}{5}$ **39.** $-2, -3, 1$
41. **(a)** 5.433 **(b)** 5.433 **43.** **(a)** 23.1045 **(b)** 23.1045

Exercises (Page 226)

1. 2, 3   **3.** $-3 \pm \sqrt{10}$ **5.** 0, i, $-i$ **7.** 0 and 1 are both of multiplicity 2   **9.** -2, 1; 1 is a root of multiplicity 2
11. $-\frac{1}{2}$, 1; 1 is a root of multiplicity 3   **13.** $x^3 + 2x^2 - 5x - 6$ **15.** $x^3 + 2x^2 + x + 2$
17. $x^4 - 2x^3 - 7x^2 + 8x + 12$ **19.** $(x-1)^3(x-2)$ **21.** $(x^2 - 4x + 5)(x-3)^2$ **23.** $x^3 + 2x^2 + x + 2$
25. $x^3 - 4x^2 + 6x - 4$ **27.** $x^4 - 6x^3 - 3x^2 - 24x - 28$ **29.** 2, -1, -3 **31.** -2, i, $-i$
33. -1, $3 - 2i$, $3 + 2i$ **35.** -1, -2, 2, 3   **37.** 1.1, 1.2, 1.3

Exercises (Page 230)

1. -1 **3.** $-\frac{3}{2}$ **5.** -1, 2   **7.** 3   **9.** 1, -1, -3 **11.** none   **13.** $-\frac{5}{2}$ **15.** $-\frac{1}{3}$ **17.** -1, 2, -2, 3
19. -2 **21.** $-\frac{1}{2}$, $\frac{1}{2}$, $\frac{1}{3}$, $\frac{3}{4}$ **23.** $\frac{3}{4}$
25. The set of rational zeros of $P(x)$ must be a subset of

$$\{1, -1, 3, -3, \tfrac{1}{2}, -\tfrac{1}{2}, \tfrac{3}{2}, -\tfrac{3}{2}\}$$

Substituting these numbers in $P(x)$, we find that none are zeros of $P(x)$. Hence $P(x)$ has no rational zeros.

Exercises (Pages 234–235)

1. upper bound: 2; lower bound: -2 **3.** upper bound: 3; lower bound: -1 **5.** upper bound: 2; lower bound: -3
7. upper bound: 5; lower bound: -5 **9.** upper bound: 1; lower bound: -2
11. one positive and two or no negative   **13.** one positive and two or no negative
15. three or one positive and one negative   **17.** five, three, or one positive and no negative
19. $P(2) = -2$, $P(3) = 11$; hence $P(x)$ has a real zero between 2 and 3.
21. $P(1) = -2$, $P(2) = 1$; hence $P(x)$ has a real zero between 1 and 2.
23. $P(1) = -1$, $P(2) = 21$; hence $P(x)$ has a real zero between 1 and 2.
25. $P(2.2) = -0.392$, $P(2.3) = 0.577$; hence $P(x)$ has a real zero between 2.2 and 2.3.   **27.** $\frac{5}{2}$; $1 \pm \sqrt{5}$
29. $\frac{1}{2}$, $2 \pm \sqrt{3}$ **31.** 1, -3, $1 \pm \sqrt{3}$ **33.** 1, -2, $\frac{2}{3}$, $-\frac{4}{3}$ **35.** 1.9   **37.** 0.2, 1.8   **39.** 0.3, 1.6, -1.9

41. $x^2 = \dfrac{-1 \pm \sqrt{3}i}{3}$, which is not a real number; hence $P(x)$ has no real zeros.

Exercises (Page 243)

1.

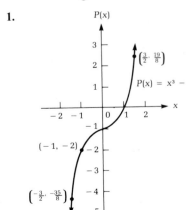

3.

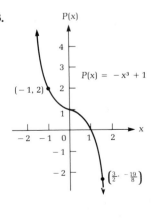

5.

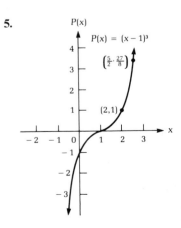

7.

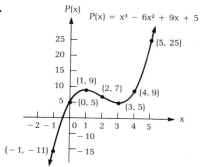

$P(x) = x^3 - 6x^2 + 9x + 5$

(5, 25)
(1, 9) (2, 7) (4, 9)
(0, 5) (3, 5)
(-1, -11)

9.

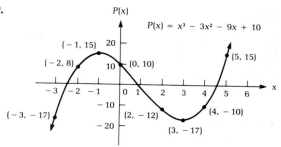

$P(x) = x^3 - 3x^2 - 9x + 10$

(-1, 15)
(-2, 8) (0, 10) (5, 15)
(-3, -17)
(2, -12) (4, -10)
(3, -17)

11.

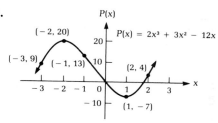

$P(x) = 2x^3 + 3x^2 - 12x$

(-2, 20)
(-3, 9) (-1, 13) (2, 4)
(1, -7)

13.

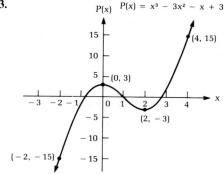

$P(x) = x^3 - 3x^2 - x + 3$

(4, 15)
(0, 3)
(2, -3)
(-2, -15)

15.

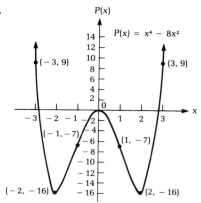

$P(x) = x^4 - 8x^2$

(-3, 9) (3, 9)
(-1, -7) (1, -7)
(-2, -16) (2, -16)

17.

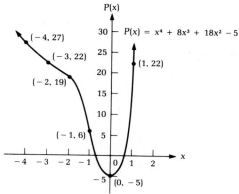

$P(x) = x^4 + 8x^3 + 18x^2 - 5$

(-4, 27)
(-3, 22)
(-2, 19) (1, 22)
(-1, 6)
(0, -5)

Exercises (Pages 252–253)

1. $x = 2$ **3.** $x = -2, x = 2$ **5.** $x = -2, x = 3$ **7.** horizontal asymptote: $y = 2$; vertical asymptote: $x = 1$
9. horizontal asymptote: $y = 0$; vertical asymptotes: $x = -2, x = -1$

11. horizontal asymptote: $y = 2$; vertical asymptotes: $x = -1$, $x = \dfrac{3}{2}$

13. vertical asymptote: $x = 1$; oblique asymptote: $y = x + 1$

15.

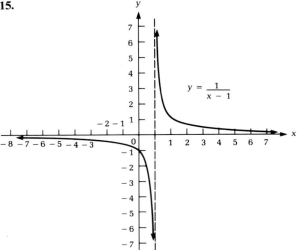

17.

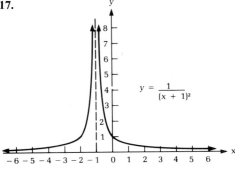

19.

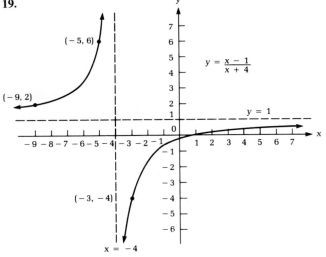

21.

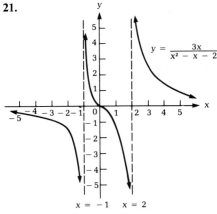

23.

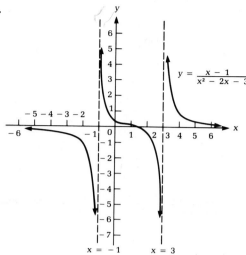

$$y = \frac{x-1}{x^2 - 2x - 3}$$

$x = -1$ $x = 3$

25.

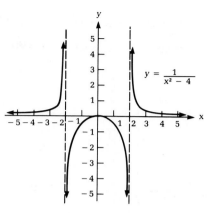

$$y = \frac{1}{x^2 - 4}$$

27.

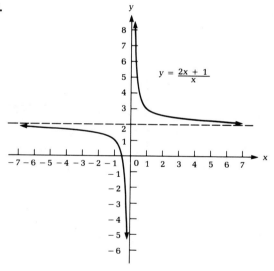

$$y = \frac{2x+1}{x}$$

29.

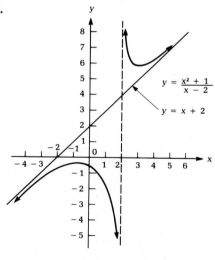

$$y = \frac{x^2+1}{x-2}$$

$$y = x + 2$$

Exercises for Review (Pages 254–255)

1. $2; 7$ **2.** $3; -7$ **3.** $-\frac{7}{2}; \frac{1}{2}$ **4.** $\frac{1}{5}; \frac{3}{2}$ **5.** $0, i, -i$ **6.** $0, 1, -2$ **7.** 4 **8.** -7 **9.** 10 **10.** $4 + i$
11. 34 **12.** 4 **13.** $6 + 2i$ **14.** $-2 - 6i$ **15.** factor   **16.** not a factor   **17.** factor   **18.** factor   **19.** $k = 1$
20. $k = 1$ **21.** $x^3 - 8x^2 + 22x - 20$ **22.** $x^4 - 7x^3 + 27x^2 - 47x + 26$

23. $x^4 - 5x^3 + x^2 + 21x - 18$ **24.** $x^4 - 10x^3 + 37x^2 - 72x + 80$ **25.** $1, -1, -2$ **26.** $-1, 3 \pm 2i$

27. $-1, 2, \pm 2i$ **28.** $-1, 7, \pm 2i$ **29.** $-1, 2, 3$ **30.** $-\frac{3}{2}$ **31.** $-3, \pm\sqrt{2}$ **32.** $\frac{5}{2}; 1 \pm \sqrt{5}$

33. $-3, 1, 1 \pm \sqrt{3}$ **34.** $\frac{1}{3}, \frac{3}{2}, 1 \pm \sqrt{2}$ **35.** $-0.9, 1.3, 2.5$ **36.** $\pm 1.7, \pm 1.4$

37.

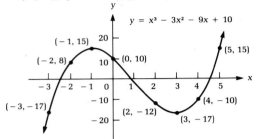

$y = x^3 - 3x^2 - 9x + 10$

$(-1, 15)$ $(-2, 8)$ $(0, 10)$ $(5, 15)$ $(-3, -17)$ $(2, -12)$ $(3, -17)$ $(4, -10)$

38.

$y = x^3 - 9x^2 + 24x$

$(2, 20)$ $(3, 18)$ $(5, 20)$ $(1, 16)$ $(4, 16)$

39.

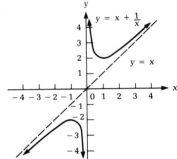

$y = x + \dfrac{1}{x}$

$y = x$

40.

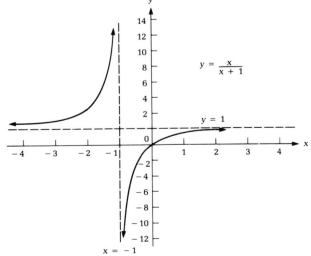

$y = \dfrac{x}{x + 1}$

$y = 1$

$x = -1$

Chapter Test (Page 255)

1. (a) $0, \pm i$ **(b)** $0, \frac{1}{2}, -\frac{7}{2}$ **2.** -17 **3.** is a factor   **4.** $x^4 - 5x^3 + 9x^2 - 7x + 2$ **5.** $-2, 2, 3$
6. $x^3 + 2x^2 + x + 2$ **7.** upper bound, 3; lower bound, -1 **8.** one positive and two or no negative
9. horizontal asymptote, $y = 2$; vertical asymptotes, $x = 2, x = 4$
10. (a)

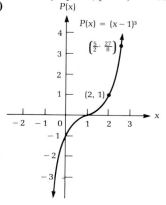

(b)

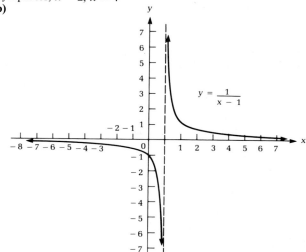

CHAPTER 5

Exercises (Pages 264–265)

1.

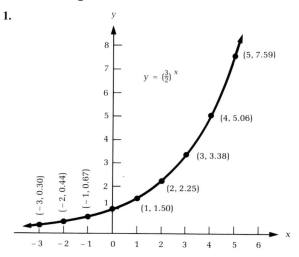

3.

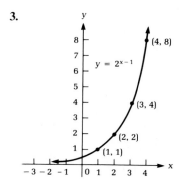

5.

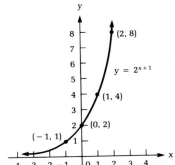

$y = 2^{x+1}$

(2, 8), (1, 4), (0, 2), (−1, 1)

7.

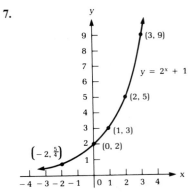

$y = 2^x + 1$

(3, 9), (2, 5), (1, 3), (0, 2), $\left(-2, \frac{5}{4}\right)$

9.

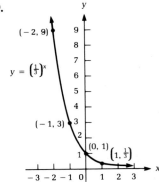

$y = \left(\frac{1}{3}\right)^x$

(−2, 9), (−1, 3), (0, 1), $\left(1, \frac{1}{3}\right)$

11.

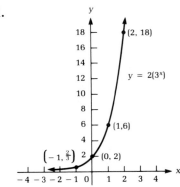

$y = 2(3^x)$

(2, 18), (1, 6), (0, 2), $\left(-1, \frac{2}{3}\right)$

13.

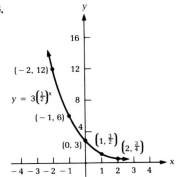

$y = 3\left(\frac{1}{2}\right)^x$

(−2, 12), (−1, 6), (0, 3), $\left(1, \frac{3}{2}\right)$, $\left(2, \frac{3}{4}\right)$

15.

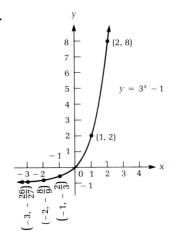

$y = 3^x - 1$

(2, 8), (1, 2), $\left(-1, -\frac{2}{3}\right)$, $\left(-2, -\frac{8}{9}\right)$, $\left(-3, -\frac{26}{27}\right)$

17.

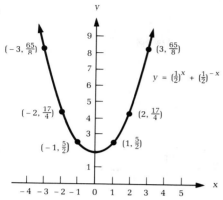

$y = \left(\frac{1}{2}\right)^x + \left(\frac{1}{2}\right)^{-x}$

$\left(-3, \frac{65}{8}\right)$ $\left(3, \frac{65}{8}\right)$

$\left(-2, \frac{17}{4}\right)$ $\left(2, \frac{17}{4}\right)$

$\left(-1, \frac{5}{2}\right)$ $\left(1, \frac{5}{2}\right)$

19.

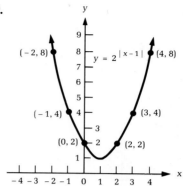

$y = 2^{|x-1|}$

$(-2, 8)$ $(4, 8)$

$(-1, 4)$ $(3, 4)$

$(0, 2)$ $(2, 2)$

21.

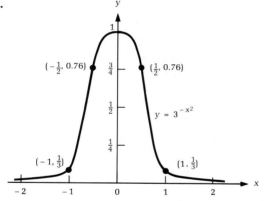

$\left(-\frac{1}{2}, 0.76\right)$ $\left(\frac{1}{2}, 0.76\right)$

$y = 3^{-x^2}$

$\left(-1, \frac{1}{3}\right)$ $\left(1, \frac{1}{3}\right)$

23.

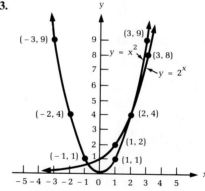

$(-3, 9)$ $(3, 9)$

$y = x^2$ $(3, 8)$

$y = 2^x$

$(-2, 4)$ $(2, 4)$

$(1, 2)$

$(-1, 1)$ $(1, 1)$

25.

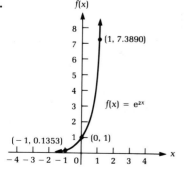

$(1, 7.3890)$

$f(x) = e^{2x}$

$(-1, 0.1353)$ $(0, 1)$

27.

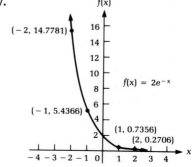

$(-2, 14.7781)$

$f(x) = 2e^{-x}$

$(-1, 5.4366)$

$(1, 0.7356)$

$(2, 0.2706)$

29.

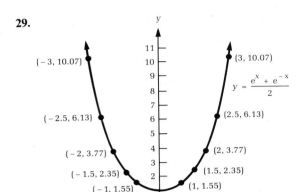

$$y = \frac{e^x + e^{-x}}{2}$$

Points labeled: $(-3, 10.07)$, $(3, 10.07)$, $(-2.5, 6.13)$, $(2.5, 6.13)$, $(-2, 3.77)$, $(2, 3.77)$, $(-1.5, 2.35)$, $(1.5, 2.35)$, $(-1, 1.55)$, $(1, 1.55)$

Exercises (Page 266)

1. $x = 2$ **3.** $x = -3$ **5.** $x = 2$ **7.** $x = -4$ **9.** $x = 4$ **11.** $x = 3$ **13.** $x = \frac{1}{2}$ **15.** $x = -2$
17. $x = -2$ **19.** $x = 1$ **21.** $x = \frac{2}{3}$ **23.** $x = -3$ **25.** $x = -3$ **27.** $x = -3$ **29.** $x = \frac{3}{2}$ **31.** $x = 5$

Exercises (Pages 273–274)

1. $929.12 **3.** $1425.76 **5.** $2174.92 **7.** $2577.68 **9.** $4294.22 **11.** $2579.88 **13.** $6560.57
15. $4098.62 **17.** $761.03 **19.** $1627.34 **21.** $1419.29 **23.** $3030.47 **25.** $1852.09 **27.** $2893.33
29. $2203.73 **31.** 16.0754 percent **33.** 19.5618 percent **35.** 21.5558 percent **37.** $3333.68
39. (a) $8929.13 **(b)** $8903.08 **(c)** $8890.26 **(d)** $8889.82

Exercises (Pages 282–283)

1. (a) 18,321 **(b)** 22,377 **(c)** 27,331
3. (a) 370.4091 **(b)** 274.4058 **(c)** 203.2848
5. $Q(1) = 3.70$; $Q(2) = 2.74$; $Q(3) = 2.03$; $Q(4) = 1.51$;
$Q(5) = 1.12$; $Q(6) = 0.83$; $Q(8) = 0.45$; $Q(10) = 0.25$
7. (a) $S(1) = \$6023.88$; $S(2) = \$1814.36$; $S(3) = \$546.47$;
$S(4) = \$164.60$; $S(5) = \$49.58$

7. (b)

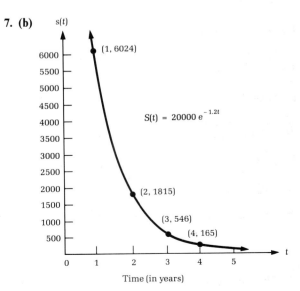

$S(t) = 20000 \, e^{-1.2t}$

Points labeled: $(1, 6024)$, $(2, 1815)$, $(3, 546)$, $(4, 165)$

Time (in years)

9. $N(1) = 41$; $N(2) = 71$; $N(3) = 93$; $N(4) = 108$; $N(5) = 120$; $N(6) = 128$; $N(8) = 138$; $N(10) = 144$; $N(12) = 147$; $N(14) = 148$; $N(16) = 149$

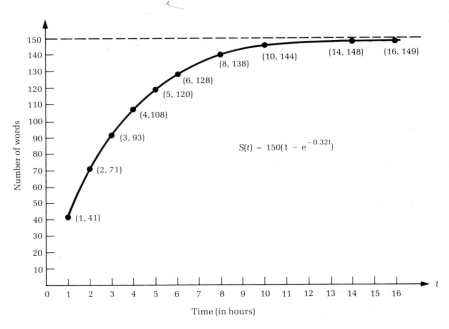

11. $P(1) = \$34{,}816$; $P(2) = \$30{,}976$; $P(3) = \$28{,}131$; $P(4) = \$26{,}024$; $P(5) = \$24{,}463$; $P(6) = \$23{,}306$; $P(8) = \$21{,}814$; $P(10) = \$20{,}996$

Exercises (Page 285)

1. $\log_2 16 = 4$ **3.** $\log_4 1024 = 5$ **5.** $\log_{36} 6 = \frac{1}{2}$ **7.** $\log_{216} 36 = \frac{2}{3}$ **9.** $3^3 = 27$ **11.** $2^6 = 64$ **13.** $4^4 = 256$
15. $6^2 = 36$ **17.** $x = 81$ **19.** $x = 3$ **21.** $x = 3$ **23.** $x = \frac{1}{25}$ **25.** $x = 3$

Exercises (Page 289)

1. -1.26 **3.** 3.56 **5.** 4.83 **7.** 4.66 **9.** 3.74 **11.** 5.51 **13.** 1.0212 **15.** 0.1505 **17.** 0.3891 **19.** 0.4407
21. 1.7713 **23.** 2.8076 **25.** $\log_b x + 2 \log_b y$ **27.** $3 \log_b x + 3 \log_b y + 3 \log_b z$ **29.** $2 \log_b x - \log_b y$
31. $\frac{1}{2} \log_b x + \frac{1}{2} \log_b y$ **33.** $\frac{2}{3} \log_b x + \frac{1}{3} \log_b y - \frac{1}{3} \log_b z$ **35.** $\log_b (xy)$ **37.** $\log_b\left(\dfrac{x^2}{y}\right)$ **39.** $\log_b\left(\dfrac{x^3 y^4}{z}\right)$
41. $\log_b (x^{1/2} y^{1/3})$ **43.** $\log_b\left(\dfrac{x^{1/4} y}{z^3}\right)$

45. Let $\log_a b = x$ and $\log_b a = y$. Then $b = a^x$ and $a = b^y$. We see that
$$b^y = (a^x)^y = a$$
$$a^{xy} = a$$
$$xy = 1.$$
Thus $(\log_a b)(\log_b a) = 1$.

Exercises (Page 300)

1. 2.3692   **3.** 1.3692   **5.** −0.6308   **7.** −2.2343   **9.** −4.1688   **11.** 7.8823   **13.** 3.2771   **15.** −5.9332
17. 5.9189   **19.** −2.3613   **21.** −7.9866   **23.** 2.5398   **25.** 1450.1068   **27.** 0.0045   **29.** 27.4979   **31.** 0.0047
33. 10.0001   **35.** 4.2905   **37.** 9.2804   **39.** 0.2367   **41.** 0.0257   **43.** 1.5682   **45.** 1.9395   **47.** 1.5391
49. −0.5595   **51.** 3.9069   **53.** 1.2619   **55.** 0.8614   **57.** 0.1475   **59.** 3.25   **61.** 13.2572   **63.** 2.3026
65. 4.1744   **67.** 2.3979   **69.** 1.0722   **71.** $x = 3.5$ **73.** $x = 3$ **75.** $x = \frac{19}{7}$ **77.** $x = \frac{500}{99}$ **79.** $x = 10$

Exercises (Pages 309–310)

1. 11.5525 years   **3.** 4.9765 centuries   **5.** 0.0462   **7.** 0.0001216   **9.** 19,188 years
11. (a) 32,000   **(b)** 512,000   **13.** 8.66 weeks   **15. (a)** 4   **(b)** 19   **(c)** 224   **17.** 796

Exercises for Review (Pages 311–313)

1. $x = 2$ **2.** $x = \frac{4}{3}$ **3.** $x = -3$ **4.** $x = 4$ **5.** $x = 81$ **6.** $x = 4$ **7.** $x = 5$ **8.** $x = \frac{2}{3}$ **9.** 0.7781
10. 1.4771   **11.** 0.8751   **12.** 0.23855   **13.** 0.6825   **14.** 1.4651   **15.** 3.7110   **16.** 1.3856   **17.** −1.6144
18. −2.6144   **19.** 6.3404   **20.** 3.4595   **21.** −1.4524   **22.** −6.0576   **23.** 70.4044   **24.** 3.5703   **25.** 0.0538
26. 5.2899   **27.** 45.5996   **28.** 0.5670   **29.** 1.4314   **30.** 0.5682   **31.** 0.2178   **32.** 3.8076   **33.** 1.2292
34. 2.5722   **35.** −0.0125   **36.** 1.0581   **37.** 0.6015   **38.** 1.7952   **39.** $x = 2$
40. $x = \frac{21}{8}$ **41. (a)** 20,000   **(b)** 18,096   **(c)** 16,374   **42.** 10,000
43. $P(1) = \$10,000$; $P(2) = \$16,487$; $P(3) = \$27,183$; $P(4) = \$44,817$; $P(5) = \$73,891$; $P(6) = \$121,820$

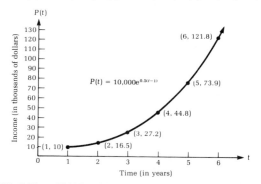

44. $P(1) = \$15,000$; $P(2) = \$11,703$; $P(3) = \$9493$; $P(4) = \$8012$; $P(5) = \$7019$; $P(6) = \$6353$; $P(10) = \$5273$

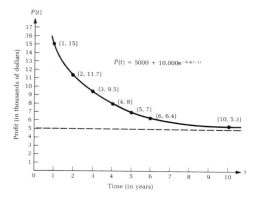

45. $N(5) = 26.37$; $N(10) = 44.05$; $N(15) = 55.90$; $N(20) = 63.84$; $N(25) = 69.17$; $N(30) = 72.74$

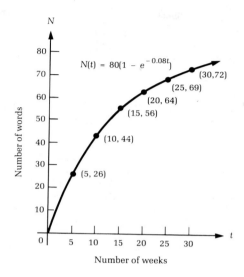

46. $D(1) = 1$; $D(2) = 4.855$; $D(3) = 6.587$; $D(4) = 7.365$; $D(5) = 7.715$; $D(6) = 7.872$; $D(8) = 7.974$; $D(10) = 7.995$; $D(12) = 7.999$

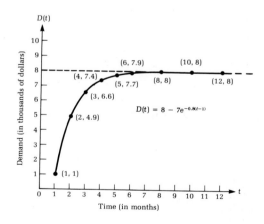

47. (a) $4990.16   **(b)** $5007.69   **(c)** $5016.65   **48.** $11,169.62   **49.** 7.30 years   **50.** 168.75 million
51. $k = 0.020433024$; 64.80 grams   **52.** 32.89 years   **53.** 2.45 hours   **54.** 8 hours   **55.** 634   **56.** 558

57.

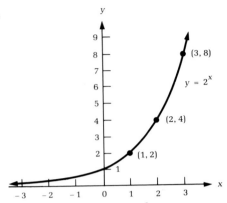

58.

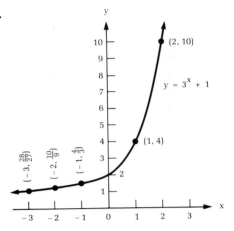

59.

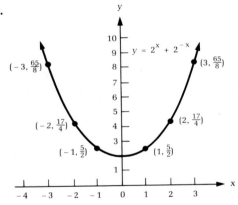

60.

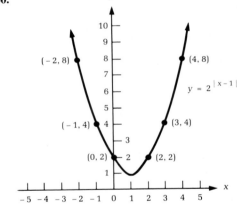

Chapter Test (Page 313)

1. (a) -5 **(b)** 5 **(c)** 3.5 **(d)** 125 **(e)** 2 **(f)** 4 **2. (a)** 2.71 **(b)** 3.91 **(c)** 1.4636 **3.** 2.4650
4. $x = 2$ **5. (a)** 8023.53 **(b)** 8061.13 **(c)** 8080.37 **6.** $k = 0.024755256$ **7.** 6.93 years
8. $k = 0.0071096$; 547,579

9.

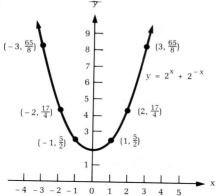

10.

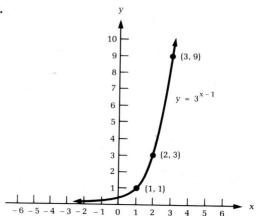

CHAPTER 6

Exercises (Pages 325–328)

1. 315° **3.** −40° **5.** 665° **7.** 207°50′ **9.** −202°15′ **11.** 485°42′ **13.** 180°18′56″ **15.** 45°44′40″
17. 280° **19.** 170° **21.** 100° **23.** $4\pi/3$ **25.** $4\pi/3$ **27.** $7\pi/4$ **29.** III **31.** II **33.** III **35.** $49\pi/36$
37. $5\pi/9$ **39.** $-5\pi/4$ **41.** $-3\pi/4$ **43.** $\pi/10$ **45.** $11\pi/6$ **47.** 300° **49.** −225° **51.** 10°50′ **53.** 135°
55. 1.6 radians **57.** 9.42 centimeters **59.** 14.32 meters **61.** 4.71 meters **63.** 4874 radians **65.** 7.64
67. 0.23 feet **69.** 3455.75 miles

Exercises (Pages 338–340)

1. 52° **3.** 68° **5.** 41° **7.** 8.5° **9.** $\sin A = \frac{12}{13} = \cos B$ $\quad \csc A = \frac{13}{12} = \sec B$
$\cos A = \frac{5}{13} = \sin B$ $\quad \sec A = \frac{13}{5} = \csc B$
$\tan A = \frac{12}{5} = \cot B$ $\quad \cot A = \frac{5}{12} = \tan B$

11. $\sin A = \frac{3}{5} = \cos B$ $\quad \csc A = \frac{5}{3} = \sec B$ **13.** $\sin A = \dfrac{2\sqrt{5}}{5} = \cos B$ $\quad \csc A = \dfrac{\sqrt{5}}{2} = \sec B$
$\cos A = \frac{4}{5} = \sin B$ $\quad \sec A = \frac{5}{4} = \csc B$
$\tan A = \frac{3}{4} = \cot B$ $\quad \cot A = \frac{4}{3} = \tan B$ $\qquad \cos A = \dfrac{\sqrt{5}}{5} = \sin B$ $\quad \sec A = \sqrt{5} = \csc B$
$\tan A = 2 = \cot B$ $\quad \cot A = \frac{1}{2} = \tan B$

	sin	cos	tan	csc	sec	cot
15.	$7\sqrt{85}/85$	$6\sqrt{85}/85$	$7/6$	$\sqrt{85}/7$	$\sqrt{85}/6$	$6/7$
17.	$5/17$	$2\sqrt{66}/17$	$5\sqrt{66}/132$	$17/5$	$17\sqrt{66}/132$	$2\sqrt{66}/5$
19.	$3/5$	$4/5$	$3/4$	$5/3$	$5/4$	$4/3$
21.	$\sqrt{21}/5$	$2/5$	$\sqrt{21}/2$	$5\sqrt{21}/21$	$5/2$	$2\sqrt{21}/21$
23.	$3\sqrt{13}/13$	$2\sqrt{13}/13$	$3/2$	$\sqrt{13}/3$	$\sqrt{13}/2$	$2/3$
25.	$4/5$	$3/5$	$4/3$	$5/4$	$5/3$	$3/4$

	sin	cos	tan	csc	sec	cot
27.	1/5	$2\sqrt{6}/5$	$\sqrt{6}/12$	5	$5\sqrt{6}/12$	$2\sqrt{6}$
29.	5/13	12/13	5/12	13/5	13/12	12/5
31.	1/2	$\sqrt{3}/2$	$\sqrt{3}/3$	2	$2\sqrt{3}/3$	$\sqrt{3}$
	$\sqrt{2}/2$	$\sqrt{2}/2$	1	$\sqrt{2}$	$\sqrt{2}$	1
	$\sqrt{3}/2$	1/2	$\sqrt{3}$	$2\sqrt{3}/3$	2	$\sqrt{3}/3$

33. $\frac{1}{2}=\frac{1}{2}$ **35.** $\frac{4}{3}=\frac{4}{3}$ **37.** $\frac{1}{2}=\frac{1}{2}$ **39.** 2 **41.** 1 **43.** $\sqrt{3}/2$ **45.** $(1+\sqrt{3})/2$ **47.** 1/2 **49.** undefined
51. 0 **53.** $\frac{1}{2}$ **55.** 1 **57.** $\frac{1}{2}$ **59.** 0

Exercises (Pages 344–345)

1. 0.2306 **3.** 0.5519 **5.** 0.0029 **7.** 0.3607 **9.** 0.3118 **11.** 0.23061587 **13.** 0.55193699 **15.** 0.00290889
17. 0.36067948 **19.** 0.31178219 **21.** 65°30′ **23.** 27° **25.** 72°10′ **27.** 3°10′ **29.** 46°20′ **31.** 65.50°
33. 27.00° **35.** 72.17° **37.** 3.17° **39.** 46.33° **41.** 0.8646 **43.** 0.4176 **45.** 2.356 **47.** 0.864573
49. 0.4176163 **51.** 2.3560209 **53.** 0.2311819 **55.** 0.5502379 **57.** 0.0023271 **59.** 0.1781521
61. 0.3050398 **63.** 0.2482418 **65.** 0.2101608 **67.** 0.2830790 **69.** 0.1987414

Exercises (Pages 353–355)

1. $B = 65°10′, a = 6.9, b = 14.9$ **3.** $A = 58°40′, B = 31°20′, c = 47.3$ **5.** $B = 51°20′, b = 31.6, c = 40.5$
7. $A = 77°20′, a = 436, c = 447$ **9.** $B = 58°, b = 3.2, c = 3.8$ **11.** $A = 54°, B = 36°, c = 15.5$
13. $B = 27°20′, a = 1.23, c = 1.39$ **15.** $B = 79°40′, a = 1.13, b = 6.19$ **17.** $A = 50°40′, B = 39°20′, b = 103.8$
19. $A = 45°, B = 45°, b = 27.44$ **21.** $B = 24°30′, b = 6.81, c = 16.40$ **23.** $A = 42°34′, a = 2853, b = 3106$
25. $A = 48°10′, B = 41°50′, c = 19.73$ **27.** $B = 76°30′, a = 146, b = 610$ **29.** $A = 53°10′, B = 35°50′, a = 0.0471$
31. 4.8 meters; 66°20′ **33.** 35°30′ **35.** $h = 19.29; d = 17.65$ **37.** 2767 meters **39.** 1.73 meters
41. 26°11′; 6.51 centimeters **43.** 50.2 meters **45.** 161 miles per hour **47.** 62°22′ **49.** 21.13; 45.32 **51.** 39.8°
53. 527 meters

Exercises (Pages 361–362)

1.

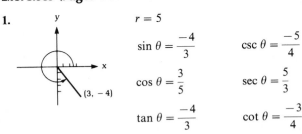

$r = 5$

$\sin \theta = \dfrac{-4}{3}$ $\csc \theta = \dfrac{-5}{4}$

$\cos \theta = \dfrac{3}{5}$ $\sec \theta = \dfrac{5}{3}$

$\tan \theta = \dfrac{-4}{3}$ $\cot \theta = \dfrac{-3}{4}$

3.

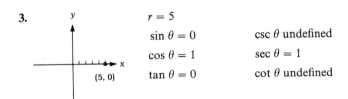

$r = 5$

$\sin \theta = 0$ $\csc \theta$ undefined

$\cos \theta = 1$ $\sec \theta = 1$

$\tan \theta = 0$ $\cot \theta$ undefined

5.

$r = 3\sqrt{2}$

$\sin \theta = \dfrac{-\sqrt{2}}{2}$ $\csc \theta = -\sqrt{2}$

$\cos \theta = \dfrac{-\sqrt{2}}{2}$ $\sec \theta = -\sqrt{2}$

$\tan \theta = 1$ $\cot \theta = 1$

7.

$r = \sqrt{2}$

$\sin \theta = \dfrac{\sqrt{2}}{2}$ $\csc \theta = \sqrt{2}$

$\cos \theta = \dfrac{\sqrt{2}}{2}$ $\sec \theta = \sqrt{2}$

$\tan \theta = 1$ $\cot \theta = 1$

9.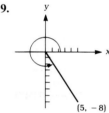

$r = \sqrt{89}$

$\sin \theta = \dfrac{-8\sqrt{89}}{89}$ $\csc \theta = \dfrac{-\sqrt{89}}{8}$

$\cos \theta = \dfrac{5\sqrt{89}}{89}$ $\sec \theta = \dfrac{\sqrt{89}}{5}$

$\tan \theta = \dfrac{-8}{5}$ $\cot \theta = \dfrac{-5}{8}$

11.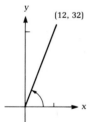

$r = 4\sqrt{73}$

$\sin \theta = \dfrac{8\sqrt{73}}{73}$ $\csc \theta = \dfrac{\sqrt{73}}{8}$

$\cos \theta = \dfrac{3\sqrt{73}}{73}$ $\sec \theta = \dfrac{\sqrt{73}}{3}$

$\tan \theta = \dfrac{8}{3}$ $\cot \theta = \dfrac{3}{8}$

13.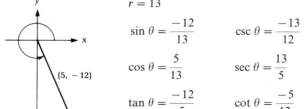

$r = 13$

$\sin \theta = \dfrac{-12}{13}$ $\csc \theta = \dfrac{-13}{12}$

$\cos \theta = \dfrac{5}{13}$ $\sec \theta = \dfrac{13}{5}$

$\tan \theta = \dfrac{-12}{5}$ $\cot \theta = \dfrac{-5}{12}$

15.

$r = 4$

$\sin \theta = \dfrac{\sqrt{3}}{2}$ $\csc \theta = \dfrac{2\sqrt{3}}{3}$

$\cos \theta = \dfrac{-1}{2}$ $\sec \theta = -2$

$\tan \theta = -\sqrt{3}$ $\cot \theta = \dfrac{-\sqrt{3}}{3}$

17.

	a	b	c	d	e	f	g	h	i	j	k
Quadrant	I	III	III	II	IV	I	IV	I	I	II	II
$\sin \theta$	+	−	−	+	−	+	−	+	+	+	+
$\cos \theta$	+	−	−	−	+	+	+	+	+	−	−
$\tan \theta$	+	+	+	−	−	+	−	+	+	−	−
$\csc \theta$	+	−	−	+	−	+	−	+	+	+	+
$\sec \theta$	+	−	−	−	+	+	+	+	+	−	−
$\cot \theta$	+	+	+	−	−	+	−	+	+	−	−

	sin	cos	tan	csc	sec	cot
19.	$3/5$	$-4/5$	$-3/4$	$5/3$	$-5/4$	$-4/3$
21.	$1/4$	$\sqrt{15}/4$	$\sqrt{15}/15$	4	$4\sqrt{15}/15$	$\sqrt{15}$
23.	$-\sqrt{3}/2$	$1/2$	$-\sqrt{3}$	$-2\sqrt{3}/3$	2	$-\sqrt{3}/3$
25.	$-2\sqrt{5}/5$	$-\sqrt{5}/5$	2	$-\sqrt{5}/2$	$-\sqrt{5}$	$1/2$
27.	$-8/17$	$-15/17$	$8/15$	$-17/8$	$-17/15$	$15/8$
29.	$-4/5$	$-3/5$	$4/3$	$-5/4$	$-5/3$	$3/4$
31.	$\sqrt{2}/2$	$-\sqrt{2}/2$	-1	$\sqrt{2}$	$-\sqrt{2}$	-1

Exercises (Pages 370–371)

1. $1 = 1$ **3.** $4 = 4$ **5.** $\sqrt{3}/2 = \sqrt{3}/2$ **7.** $\tfrac{1}{2} = \tfrac{1}{2}$ **9.** $0 = 0$ **11.** $\tfrac{1}{2} = \tfrac{1}{2}$ **13.** $-1 = -1$ **15.** $4 = 4$
17. $\sqrt{2}/2 = \sqrt{2}/2$ **19.** $\sqrt{3}/2 = \sqrt{3}/2$ **21.** $-\tfrac{1}{2} = -\tfrac{1}{2}$ **23.** $\sqrt{3}/3 = \sqrt{3}/3$ **25.** $1 = 1$ **27.** $-\sqrt{3} = -\sqrt{3}$
29. $\tfrac{4}{3} = \tfrac{4}{3}$

	α	$\sin \theta$	$\cos \theta$	$\tan \theta$	$\csc \theta$	$\sec \theta$	$\cot \theta$
31.	60°	$\sqrt{3}/2$	$-1/2$	$-\sqrt{3}$	$2\sqrt{3}/3$	-2	$-\sqrt{3}/3$
33.	60°	$-\sqrt{3}/2$	$1/2$	$-\sqrt{3}$	$-2\sqrt{3}/3$	2	$-\sqrt{3}/3$
35.	60°	$-\sqrt{3}/2$	$-1/2$	$\sqrt{3}$	$-2\sqrt{3}/3$	-2	$\sqrt{3}/3$
37.	60°	$\sqrt{3}/2$	$-1/2$	$-\sqrt{3}$	$2\sqrt{3}/3$	-2	$-\sqrt{3}/3$
39.	60°	$-\sqrt{3}/2$	$-1/2$	$\sqrt{3}$	$-2\sqrt{3}/3$	-2	$\sqrt{3}/3$
41.	30°	$-1/2$	$-\sqrt{3}/2$	$\sqrt{3}/3$	-2	$-2\sqrt{3}/3$	$\sqrt{3}$

Exercises (Pages 380–381)

1.

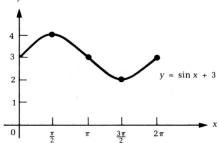

$y = \sin x + 3$

3.

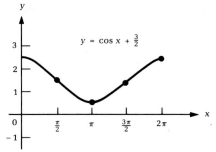

$y = \cos x + \frac{3}{2}$

5.

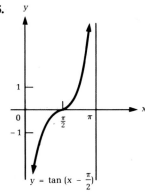

$y = \tan \left(x - \frac{\pi}{2}\right)$

7.

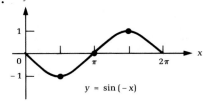

$y = \sin (-x)$

9. $y = \cot x$

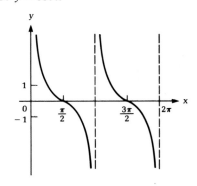

11.

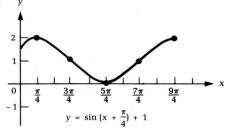

$y = \sin \left(x + \frac{\pi}{4}\right) + 1$

13.

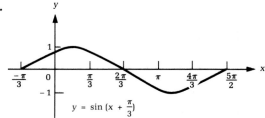

$$y = \sin\left(x + \frac{\pi}{3}\right)$$

15. $y = \cos\left(x - \dfrac{\pi}{2}\right)$

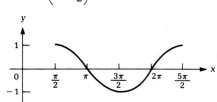

17.

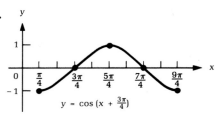

$$y = \cos\left(x + \frac{3\pi}{4}\right)$$

19.

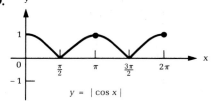

$$y = |\cos x|$$

21.

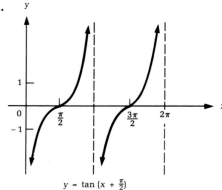

$$y = \tan\left(x + \frac{\pi}{2}\right)$$

23.

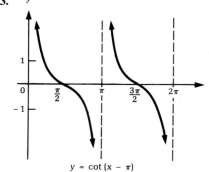

$$y = \cot(x - \pi)$$

25. $y = \sec x$

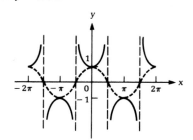

The secant function has a period of 2π; that is, $\sec(x + 2\pi) = \sec x$. Its domain is the set of all real numbers except odd multiples of $\pi/2$. The range consists of all real numbers greater than or equal to 1 and less than or equal to -1. The graph is symmetric with respect to the y axis (even symmetry), thus $\sec(-x) = \sec x$. Asymptotes occur at odd multiples of $\pi/2$.

27.

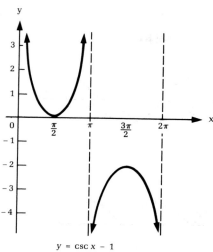

$y = \csc x - 1$

29.

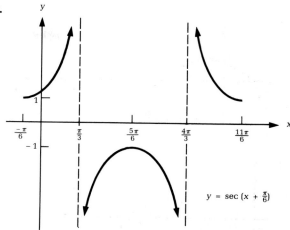

$y = \sec \left(x + \frac{\pi}{6}\right)$

Exercises (Page 386)

1.

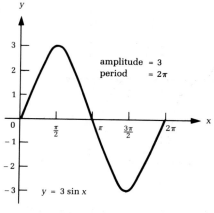

amplitude = 3
period = 2π

$y = 3 \sin x$

3.

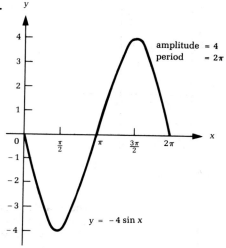

amplitude = 4
period = 2π

$y = -4 \sin x$

5.

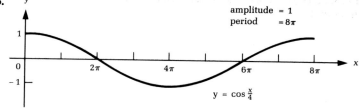

amplitude = 1
period = 8π

$y = \cos \frac{x}{4}$

7. $y = -2 \cos x$

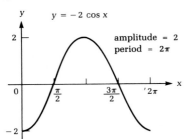

9. $y = \sin \dfrac{3x}{2}$

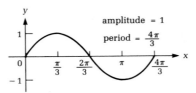

11. $y = 4 \cos (\pi/2)x$

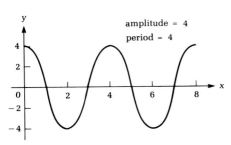

13. $y = \dfrac{1}{3} \sin 2\beta$

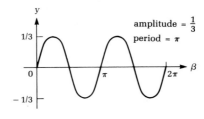

15.

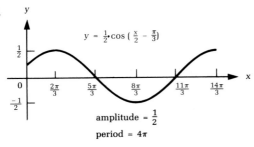

17.

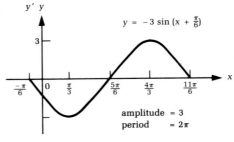

19. $y = 2 \sin x + \sin 2x$

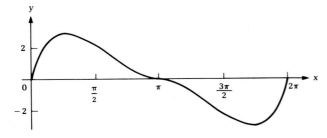

21.

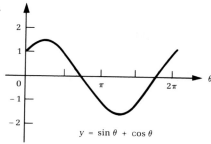

$y = \sin \theta + \cos \theta$

23. $y = 3 \cos x - \sin 2x$

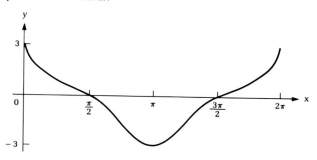

Exercises for Review (Pages 388–389)

1.

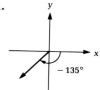

$-135°$

2. $45°$ **3.** $(-\sqrt{2}/2, -\sqrt{2}/2)$

4.

sin	cos	tan	csc	sec	cot
$-\sqrt{2}/2$	$-\sqrt{2}/2$	1	$-\sqrt{2}$	$-\sqrt{2}$	1

5.

$330°$

6. $30°$ **7.** $(\sqrt{3}/2, -1/2)$

8.

sin	cos	tan	csc	sec	cot
$-1/2$	$\sqrt{3}/2$	$-\sqrt{3}/3$	-2	$2\sqrt{3}/3$	$-\sqrt{3}$

9. $245°23'43''$ **10.** $313°38'$

	r	sin	cos	tan	csc	sec	cot
11.	$\sqrt{74}$	$-7\sqrt{74}/74$	$5\sqrt{74}/74$	$-7/5$	$-\sqrt{74}/7$	$\sqrt{74}/5$	$-5/7$
12.	$2\sqrt{5}$	$-2\sqrt{5}/5$	$-\sqrt{5}/5$	2	$-\sqrt{5}/2$	$-\sqrt{5}$	$1/2$
13.			$-\sqrt{7}/4$	$-3\sqrt{7}/7$	$4/3$	$-4\sqrt{7}/7$	$-\sqrt{7}/3$
14.		$-\sqrt{3}/2$	$-1/2$		$-2\sqrt{3}/3$	-2	$\sqrt{3}/3$

15.

θ	$0°$	$90°$	$150°$	$180°$	$240°$	$270°$	$315°$
$\sin \theta$	0	1	$1/2$	0	$-\sqrt{3}/2$	-1	$-\sqrt{2}/2$
$\cos \theta$	1	0	$-\sqrt{3}/2$	-1	$-1/2$	0	$\sqrt{2}/2$
$\tan \theta$	0	undefined	$-\sqrt{3}/3$	0	$\sqrt{3}$	undefined	-1

16. $1 = 1$ **17.** $1 = 1$ **18.** $\sqrt{3}/2$ **19.** 0 **20.** $4 = 4$

21.

	α	sin	cos	tan	csc	sec	cot
(a) $225°$	$45°$	$-\sqrt{2}/2$	$-\sqrt{2}/2$	1	$-\sqrt{2}$	$-\sqrt{2}$	1
(b) $-135°$	$45°$	$-\sqrt{2}/2$	$-\sqrt{2}/2$	1	$-\sqrt{2}$	$-\sqrt{2}$	1
(c) $600°$	$60°$	$-\sqrt{3}/2$	$-1/2$	$\sqrt{3}$	$-2\sqrt{3}/3$	-2	$\sqrt{3}/3$

22. $\pi/2$ **23.** $3\pi/4$ **24.** $5\pi/3$ **25.** $75°$ **26.** $-45°$ **27.** $24°$ **28.** 0.8124 **29.** 8.38 **30.** 35.81
31. $B = 40°30', a = 323, b = 276$ **32.** $A = 68°50', b = 116, c = 322$ **33.** $A = 41°20', B = 48°40', c = 10{,}000$
34. $A = 52°10', B = 37°50', a = 0.055$ **35.** $B = 24°40', b = 0.325, c = 0.778$ **36.** $A = 33°40'$ **37.** 418.6 meters
38. 300 feet   **39.** 62 feet, 370 feet, 308 feet

40. $y = -2 \sin 2\theta$

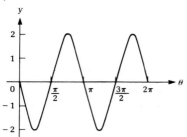

41. $y = -3 \sin 3x$

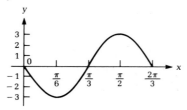

42. $y = \frac{1}{2} \cos \frac{\theta}{2}$

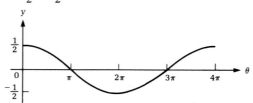

43. $y = \frac{1}{3} \cos \frac{x}{3}$

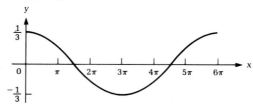

44. $y = 3 \sin \frac{\theta}{3}$

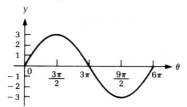

45. $y = 4 \tan 2x$

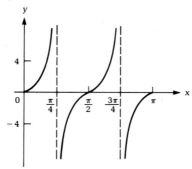

46. $y = \cot 3\theta$

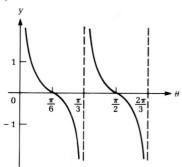

47. $y = 2 \sin \left(x + \frac{\pi}{4} \right)$

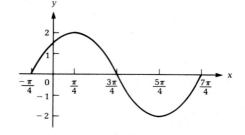

48. $y = -\cos(x - \pi/4)$

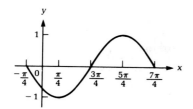

49. $y = \sin x + \cos x$

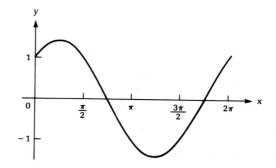

50. $y = \csc 2x$

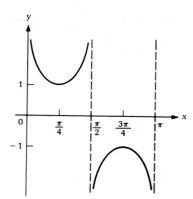

51. $y = 3 \sec 2x$

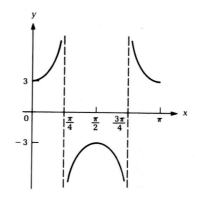

Chapter Test (Page 389)

1. $(\frac{1}{2}, -\sqrt{3}/2)$ **2.** $\sin(-30°) = -1/2$ $\csc(-30°) = -2$
$\cos(-30°) = \sqrt{3}/2$ $\sec(-30°) = 2\sqrt{3}/3$
$\tan(-30°) = -\sqrt{3}/3$ $\cot(-30°) = -\sqrt{3}$

3. $r = \sqrt{29}$ **4.** $\sin\theta = -\sqrt{5}/3$ $\csc\theta = -3\sqrt{5}/5$
$\sin\theta = 5\sqrt{29}/29$ $\csc\theta = \sqrt{29}/5$ $\cos\theta = 2/3$ $\sec\theta = 3/2$
$\cos\theta = -2\sqrt{29}/29$ $\sec\theta = -\sqrt{29}/2$ $\tan\theta = -\sqrt{5}/2$ $\cot\theta = -2\sqrt{5}/5$
$\tan\theta = -5/2$ $\cot\theta = -2/5$

5. $\sqrt{3}/2$ **6.** $\sin 630° = -1$ $\csc 630° = -1$
$\cos 630° = 0$ $\sec 630°$ undefined
$\tan 630°$ undefined $\cot 630° = 0$

7. (a) $35\pi/36$ **(b)** $17\pi/9$ **8. (a)** $50°$ **(b)** $-15°$ **9.** 1.62 radians

10. (a)

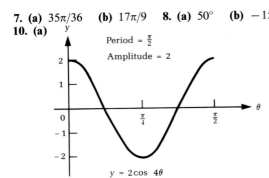

$y = 2\cos 4\theta$

(b)

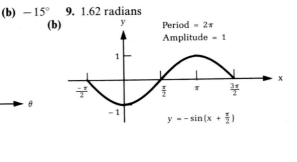

$y = -\sin(x + \frac{\pi}{2})$

11. $B = 48.4°$; $a = 106$; $b = 120$ **12.** $\theta = 31°$

CHAPTER 7

Exercises (Pages 404–405)

1. $\tan x \cot x = 1$

$$\tan x \cot x = \frac{\sin x}{\cos x} \frac{\cos x}{\sin x}$$
$$= 1$$

3. $\tan x = \dfrac{\sec x}{\csc x}$

$$\frac{\sec x}{\csc x} = \frac{1/\cos x}{1/\sin x}$$
$$= \frac{\sin x}{\cos x}$$
$$= \tan x$$

5. $\cos x = \dfrac{\sin x}{\tan x}$

$$\frac{\sin x}{\tan x} = \frac{\sin x}{\frac{\sin x}{\cos x}}$$
$$= \cos x$$

7. $\cos x = \sin x \cot x$

$$\sin x \cot x = \sin x \frac{\cos x}{\sin x}$$
$$= \cos x$$

9. -1 **11.** $-2\sqrt{2}$

13. $(1 + \cos \theta)(1 - \cos \theta) = \sin^2 \theta$

$$(1 + \cos \theta)(1 - \cos \theta) = 1 - \cos^2 \theta$$
$$= \sin^2 \theta$$

15. $\dfrac{\sec^2 x}{\sin^2 x} = \csc^2 x + \sec^2 x$

$$\csc^2 x + \sec^2 x = \frac{1}{\sin^2 x} + \frac{1}{\cos^2 x}$$
$$= \frac{\cos^2 x + \sin^2 x}{\sin^2 x \cos^2 x}$$
$$= \frac{1}{\sin^2 x \cos^2 x}$$
$$= \frac{\sec^2 x}{\sin^2 x}$$

17. $\csc^2 x(1 - \cos^2 x) = 1$

$$\csc^2 x(1 - \cos^2 x) = \frac{1}{\sin^2 x} \cdot \sin^2 x$$
$$= 1$$

21. $\dfrac{1 + \tan^2 x}{1 - \tan^2 x} = \sec 2x$

$$\dfrac{1 + \tan^2 x}{1 - \tan^2 x} = \dfrac{1 + \dfrac{\sin^2 x}{\cos^2 x}}{1 - \dfrac{\sin^2 x}{\cos^2 x}}$$

$$= \dfrac{\cos^2 x + \sin^2 x}{\cos^2 x - \sin^2 x}$$

$$= \dfrac{1}{\cos^2 x - \sin^2 x}$$

$$= \dfrac{1}{\cos 2x}$$

$$= \sec 2x$$

23. $1 - \sin 2x = (\sin x - \cos x)^2$

$(\sin x - \cos x)^2 = \sin^2 x - 2 \sin x \cos x + \cos^2 x$

$$= 1 - 2 \sin x \cos x$$

$$= 1 - \sin 2x$$

25. $2 \csc 2x = \sec x \csc x$

$$2 \csc 2x = \dfrac{2}{\sin 2x}$$

$$= \dfrac{2}{2 \sin x \cos x}$$

$$= \dfrac{1}{\sin x \cos x}$$

$$= \dfrac{1}{\sin x} \cdot \dfrac{1}{\cos x}$$

$$= \csc x \sec x$$

$$= \sec x \csc x$$

27. $\sec 2x = \dfrac{\sec^2 x}{2 - \sec^2 x}$

$$\dfrac{\sec^2 x}{2 - \sec^2 x} = \dfrac{\dfrac{1}{\cos^2 x}}{2 - \dfrac{1}{\cos^2 x}}$$

$$= \dfrac{1}{2 \cos^2 x - 1}$$

$$= \dfrac{1}{\cos^2 x + 1 - \sin^2 x - 1}$$

$$= \dfrac{1}{\cos^2 x - \sin^2 x}$$

$$= \dfrac{1}{\cos 2x} = \sec 2x$$

29. $\cos^2 3x - \sin^2 3x = \cos 6x$

$$\cos 6x = \cos (3x + 3x)$$

$$= \cos 3x \cos 3x - \sin 3x \sin 3x$$

$$= \cos^2 3x - \sin^2 3x$$

31. $\cos 4\theta = 8 \cos^4 \theta - 8 \cos^2 \theta + 1$

$$\cos 4\theta = \cos (2\theta + 2\theta)$$

$$= \cos 2\theta \cos 2\theta - \sin 2\theta \sin 2\theta$$

$$= (2 \cos^2 \theta - 1)(2 \cos^2 \theta - 1) - (2 \sin \theta \cos \theta)(2 \sin \theta \cos \theta)$$

$$= 4 \cos^4 \theta - 4 \cos^2 \theta + 1 - 4 \sin^2 \theta \cos^2 \theta$$

$$= 4 \cos^4 \theta - 4 \cos^2 \theta + 1 - 4 \cos^2 \theta (1 - \cos^2 \theta)$$

$$= 4 \cos^4 \theta - 4 \cos^2 \theta + 1 - 4 \cos^2 \theta + 4 \cos^4 \theta$$

$$= 8 \cos^4 \theta - 8 \cos^2 \theta + 1$$

33. $\tan \theta + \cot \theta = 2 \csc 2\theta$

$$\tan \theta + \cot \theta = \dfrac{\sin \theta}{\cos \theta} + \dfrac{\cos \theta}{\sin \theta}$$

$$= \dfrac{\sin^2 \theta + \cos^2 \theta}{\sin \theta \cos \theta}$$

$$= \dfrac{1}{\sin \theta \cos \theta}$$

$$= \dfrac{2}{2 \sin \theta \cos \theta}$$

$$= \dfrac{2}{\sin 2\theta} = 2 \csc 2\theta$$

35. $4 \sin \dfrac{\theta}{2} \cos \dfrac{\theta}{2} = 2 \sin \theta$

$$4 \sin \dfrac{\theta}{2} \cos \dfrac{\theta}{2} = 4 \sqrt{\dfrac{1 - \cos \theta}{2}} \sqrt{\dfrac{1 + \cos \theta}{2}}$$

$$= 4 \sqrt{\dfrac{(1 - \cos \theta)(1 + \cos \theta)}{4}}$$

$$= 2 \sqrt{1 - \cos^2 \theta}$$

$$= 2 \sqrt{\sin^2 \theta}$$

$$= 2 \sin \theta$$

35. $\dfrac{1 + \tan x}{1 + \cot x} = \dfrac{\sec x}{\csc x}$

$$\dfrac{1 + \tan x}{1 + \cot x} = \dfrac{1 + \dfrac{\sin x}{\cos x}}{1 + \dfrac{\cos x}{\sin x}} = \dfrac{\dfrac{\cos x + \sin x}{\cos x}}{\dfrac{\sin x + \cos x}{\sin x}}$$

$$= \dfrac{\cos x + \sin x}{\cos x} \cdot \dfrac{\sin x}{\sin x + \cos x}$$

$$= \dfrac{\sin x}{\cos x} = \dfrac{\sec x}{\csc x}$$

37. $\dfrac{1 + \tan x}{1 - \tan x} + \dfrac{1 + \cot x}{1 - \cot x} = 0$

$$\dfrac{1 + \tan x}{1 - \tan x} + \dfrac{1 + \cot x}{1 - \cot x} = \dfrac{(1 + \tan x)(1 - \cot x) + (1 - \tan x)(1 + \cot x)}{(1 - \tan x)(1 - \cot x)}$$

$$= \dfrac{1 - \cot x + \tan x - 1 + 1 + \cot x - \tan x - 1}{(1 - \tan x)(1 - \cot x)}$$

$$= 0$$

39. $\dfrac{1 - \cos x}{\sin x} = \dfrac{\sin x}{1 + \cos x}$

$$\dfrac{\sin x}{1 + \cos x} = \dfrac{\sin x}{1 + \cos x} \cdot \dfrac{1 - \cos x}{1 - \cos x}$$

$$= \dfrac{(\sin x)(1 - \cos x)}{1 - \cos^2 x}$$

$$= \dfrac{(\sin x)(1 - \cos x)}{\sin^2 x}$$

$$= \dfrac{1 - \cos x}{\sin x}$$

41. $\dfrac{\sin x - \cos x}{\sec x - \csc x} = \dfrac{\cos x}{\csc x}$

$$\dfrac{\sin x - \cos x}{\sec x - \csc x} = \dfrac{\sin x - \cos x}{\dfrac{1}{\cos x} - \dfrac{1}{\sin x}}$$

$$= \dfrac{\sin x - \cos x}{\dfrac{\sin x - \cos x}{\cos x \sin x}}$$

$$= \cos x \sin x$$

$$= \dfrac{\cos x}{\csc x}$$

43. $\sec(\theta + 90°) = \dfrac{1}{\cos(\theta + 90°)} = \dfrac{1}{-\sin \theta} = -\csc \theta$ **45.** $\cot(\theta + 90°) = \dfrac{\cos(\theta + 90°)}{\sin(\theta + 90°)} = \dfrac{-\sin \theta}{\cos \theta} = -\tan \theta$

47. $\csc(\theta - 90°) = \dfrac{1}{\sin(\theta - 90°)} = \dfrac{1}{-\cos \theta} = -\sec \theta$ **49.** $-\cot 12° + \sec 12°$ **51.** $[\cos(\pi/12)][-\tan(\pi/12)]$

53. $-\csc^2 2 + \sec^2 2$ **55.** $-(3 + 2\sqrt{3})/6$ **57.** $-(1 + \sqrt{3})/2$ **59.** -2 **61.** $-1 \le \sin \theta \le 1$
63. A squared real number cannot be negative. **65.** $\csc \theta \le -1$ or $\csc \theta \ge 1$, that is, $\csc^2 \theta \ge 1$

Exercises (Pages 409–410)

1. 30°, 150°, 210°, 330° **3.** 60°, 270°, 300° **5.** 0°, 45°, 135°, 180° **7.** 90° **9.** 30°, 150°, 270°
11. 60°, 180° **13.** 120°, 240° **15.** 43°40′, 316°20′, 74°, 286° **17.** 60°, 300° **19.** 60°, 104°30′, 255°30′, 300°
21. 90°, 120°, 240°, 270° **23.** 30°, 150°, 210°, 330° **25.** 0°, 30°, 150°, 180°, 210°, 330° **27.** 120°, 180°, 240°

29. 30°, 90°, 270°, 330°   **31.** 45°, 210°, 225°, 330°   **33.** 45°, 90°, 135°, 225°, 270°, 315°   **35.** 0°, 120°, 240°
37. 135°, 315°   **39.** 63°30′, 101°20′, 243°30′, 281°20′   **41.** 315°   **43.** 45°, 116°30′, 225°, 296°30′
45. 30°, 60°, 150°, 300°   **47.** 0°, 120°, 240°   **49.** 30°, 90°, 150°, 210°, 270°, 330°   **51.** 109°30′, 250°30′
53. 60°, 300°   **55.** 45°, 135°, 225°, 315°   **57.** 60°, 300°   **59.** 30°, 150°, 270°

Exercises (Pages 419–421)

1. $\dfrac{\sqrt{6}+\sqrt{2}}{4}, \dfrac{\sqrt{6}-\sqrt{2}}{4}, 2+\sqrt{3}$ **3.** $\dfrac{\sqrt{6}+\sqrt{2}}{4}, \dfrac{\sqrt{2}-\sqrt{6}}{4}, -\sqrt{3}-2$ **5.** $\dfrac{-\sqrt{6}-\sqrt{2}}{4}, \dfrac{\sqrt{6}-\sqrt{2}}{4}, -\sqrt{3}-2$

7. $\cos(90° + \theta) = -\sin\theta$

$\cos(90° + \theta) = \cos 90° \cos\theta - \sin 90° \sin\theta$

$\qquad = 0\cos\theta - 1\sin\theta = -\sin\theta$

9. $\cos(180° - \theta) = -\cos\theta$

$\cos(180° - \theta) = \cos 180° \cos\theta + \sin 180° \sin\theta$

$\qquad = (-1)\cos\theta + 0\sin\theta = -\cos\theta$

11. $\cos(270° - \theta) = -\sin\theta$

$\cos(270° - \theta) = \cos 270° \cos\theta + \sin 270° \sin\theta$

$\qquad = 0\cos\theta + (-1)\sin\theta = -\sin\theta$

13. $\sin(180° + A) = -\sin A$

$\sin(180° + A) = \sin 180° \cos A + \cos 180° \sin A$

$\qquad = 0\cos A + (-1)\sin A = -\sin A$

15. $\sin(90° + A) = \cos A$

$\sin(90° + A) = \sin 90° \cos A + \cos 90° \sin A$

$\qquad = 1\cos A + 0\sin A = \cos A$

17. $\sin(270° - A) = -\cos A$

$\sin(270° - A) = \sin 270° \cos A - \cos 270° \sin A$

$\qquad = (-1)\cos A - 0\sin A = -\cos A$

19. $\tan(180° - A) = -\tan A$

$\tan(180° - A) = \dfrac{\tan 180° - \tan A}{1 + \tan 180° \tan A} = \dfrac{0 - \tan A}{1 + 0} = -\tan A$

21. $\tan(90° + A) = -\cot A$

$\tan(90° + A) = \dfrac{\sin(90° + A)}{\cos(90° + A)} = \dfrac{\cos A}{-\sin A} = -\cot A$

23. $\tan(270° - A) = \cot A$

$\tan(270° - A) = \dfrac{\sin(270° - A)}{\cos(270° - A)} = \dfrac{-\cos A}{-\sin A} = \cot A$

25. 1   **27.** $\frac{24}{25}$ **29.** undefined   **31.** $\frac{33}{65}; \frac{63}{65}$ **33.** $\frac{77}{85}; -\frac{13}{85}$ **35.** $-1; -\frac{119}{120}$ **37.** $\frac{1}{2}\sin\theta + \sqrt{3}/2 \cos\theta$

39. $\dfrac{3\tan\theta + \sqrt{3}}{3 - \sqrt{3}\tan\theta}$ **41.** $\sin 58°$ **43.** $\tan 35°$ **45.** $\sin A$ **47.** $\cos(A + B)$ **49.** $2\cos 2\theta \cos\theta$

51. $12\cos\frac{5}{12}x \sin\frac{1}{12}x$ **53.** $\cos 4A$ **55.** $\cot(A + B) = \dfrac{\cot A \cot B - 1}{\cot B + \cot A}$ **57.** $2\sin\theta\cos\theta$

59. $\sin\left(\theta - \dfrac{\pi}{6}\right) + \cos\left(\theta - \dfrac{\pi}{3}\right) = \sqrt{3}\,\sin\theta$

$$\sin\left(\theta - \frac{\pi}{6}\right) + \cos\left(\theta - \frac{\pi}{3}\right) = \sin\theta\cos\frac{\pi}{6} - \cos\theta\sin\frac{\pi}{6} + \cos\theta\cos\frac{\pi}{3} + \sin\theta\,\sin\frac{\pi}{3}$$

$$= \frac{\sqrt{3}}{2}\sin\theta - \frac{1}{2}\cos\theta + \frac{1}{2}\cos\theta + \frac{\sqrt{3}}{2}\sin\theta$$

$$= 2\left(\frac{\sqrt{3}}{2}\sin\theta\right) = \sqrt{3}\,\sin\theta$$

61. $\cos(45° + x) - \cos(45° - x) = -\sqrt{2}\,\sin x$

$\cos(45° + x) - \cos(45° - x) = \cos 45°\cos x - \sin 45°\sin x$
$\qquad\qquad\qquad\qquad\qquad - (\cos 45°\cos x + \sin 45°\sin x)$
$\qquad\qquad\qquad = \cos 45°\cos x - \sin 45°\sin x$
$\qquad\qquad\qquad\quad - \cos 45°\cos x - \sin 45°\sin x$
$\qquad\qquad\qquad = -2\sin 45°\sin x$
$\qquad\qquad\qquad = -2\left(\dfrac{1}{\sqrt{2}}\right)\sin x = -\sqrt{2}\,\sin x$

63. $\tan(x + 45°) = \dfrac{\cos x + \sin x}{\cos x - \sin x}$

$\tan(x + 45°) = \dfrac{\tan x + \tan 45°}{1 - \tan x\tan 45°}$

$\qquad\qquad = \dfrac{\tan x + 1}{1 - \tan x}$

$\qquad\qquad = \dfrac{\dfrac{\sin x}{\cos x} + 1}{1 - \dfrac{\sin x}{\cos x}}$

$\qquad\qquad = \dfrac{\dfrac{\sin x}{\cos x} + 1}{1 - \dfrac{\sin x}{\cos x}} \cdot \dfrac{\cos x}{\cos x}$

$\qquad\qquad = \dfrac{\sin x + \cos x}{\cos x - \sin x}$

65. $\cot(x + y) = \dfrac{1 - \tan x\tan y}{\tan x + \tan y}$

$\cot(x + y) = \dfrac{1}{\tan(x + y)}$

$\qquad\qquad = \dfrac{1}{\dfrac{\tan x + \tan y}{1 - \tan x\tan y}}$

$\qquad\qquad = \dfrac{1 - \tan x\tan y}{\tan x + \tan y}$

67. $\tan 2x = \dfrac{2\tan x}{1 - \tan^2 x}$

$\tan 2x = \tan(x + x)$

$\qquad = \dfrac{\tan x + \tan x}{1 - \tan x\tan x}$

$\qquad = \dfrac{2\tan x}{1 - \tan^2 x}$

Exercises (Pages 430–431)

1. $\frac{1}{2}\sqrt{2 + \sqrt{3}}$ **3.** $\frac{1}{2}\sqrt{2 - \sqrt{2}}$ **5.** $\frac{1}{2}\sqrt{2 + \sqrt{3}}$ **7.** $2 + \sqrt{3}$ **9.** $-\frac{1}{2}\sqrt{2 - \sqrt{2}}$ **11.** $1 - \sqrt{2}$

13. $\frac{1}{10}\sqrt{2}; \frac{7}{10}\sqrt{2}; \frac{1}{7}$ **15.** $\dfrac{3\sqrt{13}}{13}; \dfrac{2\sqrt{13}}{13}; \dfrac{3}{2}$ **17.** $\dfrac{\sqrt{10}}{10}; \dfrac{3\sqrt{10}}{10}; \dfrac{1}{3}$

19. $\cos 4x = 1 - 8\sin^2 x\cos^2 x$

$\cos 4x = \cos(2x + 2x)$

$\qquad = \cos 2x\cos 2x - \sin 2x\sin 2x$

$\qquad = (\cos^2 x - \sin^2 x)(\cos^2 x - \sin^2 x) - (2\sin x\cos x)(2\sin x\cos x)$

$\qquad = \cos^4 x - 2\sin^2 x\cos^2 x + \sin^4 x - 4\sin^2 x\cos^2 x$

$\qquad = (\cos^2 x)(1 - \sin^2 x) + (\sin^2 x)(1 - \cos^2 x) - 6\sin^2 x\cos^2 x$

$\qquad = \cos^2 x - \sin^2 x\cos^2 x + \sin^2 x - \sin^2 x\cos^2 x - 6\sin^2 x\cos^2 x$

$\qquad = 1 - 8\sin^2 x\cos^2 x$

19.
$$\sin^3 x + \cos^3 x = (\sin x + \cos x)(1 - \sin x \cos x)$$
$$(\sin x + \cos x)(1 - \sin x \cos x) = \sin x - \sin^2 x \cos x + \cos x - \sin x \cos^2 x$$
$$= \sin x - \sin x \cos^2 x + \cos x - \sin^2 x \cos x$$
$$= (\sin x)(1 - \cos^2 x) + (\cos x)(1 - \sin^2 x)$$
$$= \sin x \sin^2 x + \cos x \cos^2 x$$
$$= \sin^3 x + \cos^3 x$$

21. $\dfrac{\cot B}{\sec B - \tan B} - \dfrac{\cos B}{\sec B + \tan B} = \sin B + \csc B$

$$\dfrac{\cot B}{\sec B - \tan B} - \dfrac{\cos B}{\sec B + \tan B} = \dfrac{(\cot B)(\sec B + \tan B) - (\cos B)(\sec B - \tan B)}{\sec^2 B - \tan^2 B}$$
$$= \dfrac{\csc B + 1 - 1 + \sin B}{1}$$
$$= \sin B + \csc B$$

23. $\dfrac{\cos \theta}{1 + \sin \theta} = \sec \theta - \tan \theta$

$$\sec \theta - \tan \theta = \dfrac{1}{\cos \theta} - \dfrac{\sin \theta}{\cos \theta}$$
$$= \dfrac{1 - \sin \theta}{\cos \theta} \cdot \dfrac{1 + \sin \theta}{1 + \sin \theta}$$
$$= \dfrac{1 - \sin^2 \theta}{(\cos \theta)(1 + \sin \theta)}$$
$$= \dfrac{\cos^2 \theta}{(\cos \theta)(1 + \sin \theta)}$$
$$= \dfrac{\cos \theta}{1 + \sin \theta}$$

25.
$$\sec^4 x = \sec^2 x + \tan^2 x \sec^2 x$$
$$\sec^2 x + \tan^2 x \sec^2 x = (\sec^2 x)(1 + \tan^2 x)$$
$$= \sec^2 x \sec^2 x$$
$$= \sec^4 x$$

27.
$$\cos^4 x = \cos^2 x - \cos^2 x \sin^2 x$$
$$\cos^2 x - \cos^2 x \sin^2 x = (\cos^2 x)(1 - \sin^2 x)$$
$$= \cos^2 x \cos^2 x$$
$$= \cos^4 x$$

29. $\dfrac{\sin x}{1 - \cos x} = \dfrac{1 + \cos x}{\sin x}$

$$\dfrac{1 + \cos x}{\sin x} = \dfrac{1 + \cos x}{\sin x} \cdot \dfrac{1 - \cos x}{1 - \cos x}$$
$$= \dfrac{1 - \cos^2 x}{(\sin x)(1 - \cos x)}$$
$$= \dfrac{\sin^2 x}{(\sin x)(1 - \cos x)}$$
$$= \dfrac{\sin x}{1 - \cos x}$$

31. $\dfrac{1}{\sec x - \tan x} = \tan x + \sec x$

$$\dfrac{1}{\sec x - \tan x} = \dfrac{1}{\sec x - \tan x} \cdot \dfrac{\sec x + \tan x}{\sec x + \tan x}$$
$$= \dfrac{\sec x + \tan x}{\sec^2 x - \tan^2 x}$$
$$= \dfrac{\sec x + \tan x}{1}$$

33. $\dfrac{1 + \cos x}{\sin x} + \dfrac{\sin x}{\cos x} = \dfrac{\cos x + 1}{\sin x \cos x}$

$$\dfrac{1 + \cos x}{\sin x} + \dfrac{\sin x}{\cos x} = \dfrac{(\cos x)(1 + \cos x) + \sin^2 x}{\sin x \cos x}$$
$$= \dfrac{\cos x + \cos^2 x + \sin^2 x}{\sin x \cos x}$$
$$= \dfrac{\cos x + 1}{\sin x \cos x}$$

37. $\cot 2\theta = \dfrac{\cot^2 \theta - 1}{2 \cot \theta}$

$\cot 2\theta = \dfrac{\cos 2\theta}{\sin 2\theta} = \dfrac{\cos^2 \theta - \sin^2 \theta}{2 \sin \theta \cos \theta}$

$= \dfrac{\dfrac{\cos^2 \theta}{\sin^2 \theta} - \dfrac{\sin^2 \theta}{\sin^2 \theta}}{\dfrac{2 \sin \theta \cos \theta}{\sin^2 \theta}} = \dfrac{\cot^2 \theta - 1}{2 \cot \theta}$

39. $0°, 180°, 120°, 240°; 0, \pi, \dfrac{2\pi}{3}, \dfrac{4\pi}{3}$ **41.** $30°, 90°, 150°, 270°; \dfrac{\pi}{6}, \dfrac{\pi}{2}, \dfrac{5\pi}{6}, \dfrac{3\pi}{2}$ **43.** $0°, 180°; 0, \pi$

45. $30°, 150°; \dfrac{\pi}{6}, \dfrac{5\pi}{6}$ **47.** $90°, 270°; \dfrac{\pi}{2}, \dfrac{3\pi}{2}$ **49.** $0°; 0$ **51.** $0°, 30°, 90°, 150°, 180°, 270°; 0, \dfrac{\pi}{6}, \dfrac{\pi}{2}, \dfrac{5\pi}{6}, \pi, \dfrac{3\pi}{2}$

53. $\cos^2 x - \sin^2 x = \cos 2x$

55. $\dfrac{\sin x}{1 + \cos x} = \tan \dfrac{x}{2}$

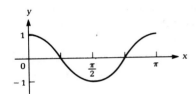

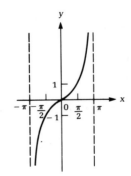

Exercises (Page 438)

1. $\pi/2$ **3.** $-\pi/2$ **5.** $\pi/6$ **7.** $-\pi/6$ **9.** $-\pi/4$ **11.** $\pi/2$ **13.** $-\pi/2$ **15.** 89°30′ or 1.5621 radians
17. 31°30′ or 0.5498 radian **19.** $-59°$ or -1.0297 radians **21.** $-44°$ or -0.7679 radian
23. 70°50′ or 1.2363 radians **25.** -1 **27.** $\sqrt{3}/2$ **29.** $-\sqrt{3}$ **31.** $-\sqrt{3}$ **33.** $\sqrt{2}$
35. $-1/2$ **37.** 45° **39.** 150° **41.** 0° **43.** 112° **45.** $\frac{5}{13}$ **47.** 0.9588 **49.** 0.5878
51. Arcsin $[(-3 \pm \sqrt{21})/6]$ **53.** Arccos $\pm\sqrt{21}/7$; Arccos $\pm\sqrt{5}/5$ **55.** Arctan $(\pm\sqrt{6}/6)$; Arctan $(\pm\sqrt{5})$

Exercises for Review

1. $1 = 1; \frac{4}{3} = \frac{4}{3}; 4 = 4$ **2.** $1 = 1; 4 = 4; \frac{4}{3} = \frac{4}{3}$ **3.** $\cos x = \pm\sqrt{1 - \sin^2 x}$ or $\cos x = \sin(x + 90°)$

4. $\tan x = \dfrac{\sin x \sqrt{1 - \sin^2 x}}{1 - \sin^2 x}$ **5.** $\begin{aligned} \csc x &= \pm\sqrt{\cot^2 x + 1} \\ \pm\sqrt{\cot^2 x + 1} &= \pm\sqrt{\csc^2 x} \\ &= \csc x \end{aligned}$

6. $$\cos x = \frac{\cot x}{\pm\sqrt{1 + \cot^2 x}}$$

$$\frac{\cot x}{\pm\sqrt{1 + \cot^2 x}} = \frac{\cot x}{\csc x}$$

$$= \frac{\cos x/\sin x}{1/\sin x}$$

$$= \cos x$$

7. $$\sin x = \frac{\tan x}{\pm\sqrt{1 + \tan^2 x}}$$

$$\frac{\tan x}{\pm\sqrt{1 + \tan^2 x}} = \frac{\tan x}{\sec x}$$

$$= \frac{\sin x/\cos x}{1/\cos x}$$

$$= \sin x$$

8. $(\sec x + 1)(\sec x - 1) = \tan^2 x$

$(\sec x + 1)(\sec x - 1) = \sec^2 x - 1$

$$= \tan^2 x$$

9. $(1 + \sin x)(1 - \sin x) = \cos^2 x$

$(1 + \sin x)(1 - \sin x) = 1 - \sin^2 x$

$$= \cos^2 x$$

10. $\csc x - \cos x \cot x = \sin x$

$$\csc x - \cos x \cot x = \frac{1}{\sin x} - \cos x \frac{\cos x}{\sin x}$$

$$= \frac{1 - \cos^2 x}{\sin x}$$

$$= \frac{\sin^2 x}{\sin x}$$

$$= \sin x$$

11. $-3\dfrac{\sqrt{5}}{4}$ **12.** $48/125$

13. $\sin^2 \theta - \cos^2 \theta = 1 - 2 \cos^2 \theta$

$\sin^2 \theta - \cos^2 \theta = 1 - \cos^2 \theta - \cos^2 \theta$

$$= 1 - 2 \cos^2 \theta$$

14. $$\tan x = \frac{\pm\sqrt{1 - \cos^2 x}}{\cos x}$$

$$\frac{\pm\sqrt{1 - \cos^2 x}}{\cos x} = \frac{\sqrt{\sin^2 x}}{\cos x}$$

$$= \frac{\sin x}{\cos x}$$

$$= \tan x$$

15. $2 \sin x + \tan x = \dfrac{2 + \sec x}{\csc x}$

$$2 \sin x + \tan x = 2 \sin x + \frac{\sin x}{\cos x}$$

$$= (\sin x)\left(2 + \frac{1}{\cos x}\right)$$

$$= (\sin x)(2 + \sec x)$$

$$= \frac{2 + \sec x}{\csc x}$$

16. $(1 + \cos x)(\csc x - \cot x) = \sin x$

$$(1 + \cos x)(\csc x - \cot x) = (1 + \cos x)\left(\frac{1}{\sin x} - \frac{\cos x}{\sin x}\right)$$

$$= (1 + \cos x)\frac{1 - \cos x}{\sin x}$$

$$= \frac{1 - \cos^2 x}{\sin x}$$

$$= \frac{\sin^2 x}{\sin x}$$

$$= \sin x$$

17. $(\sin x + \cos x)(\sec x - \csc x) = \tan x - \cot x$

$(\sin x + \cos x)(\sec x - \csc x) = \sin x \sec x - \sin x \csc x + \cos x \sec x - \cos x \csc x$

$$= \frac{\sin x}{\cos x} - 1 + 1 - \frac{\cos x}{\sin x}$$

$$= \tan x - \cot x$$

18. $\dfrac{\tan^2 x}{\sin^2 x} - \dfrac{\sin^2 x}{\cos^2 x} = 1$

$$\frac{\tan^2 x}{\sin^2 x} - \frac{\sin^2 x}{\cos^2 x} = \frac{\sin^2 x}{\cos^2 x} \cdot \frac{1}{\sin^2 x} - \frac{\sin^2 x}{\cos^2 x}$$

$$= \frac{1 - \sin^2 x}{\cos^2 x}$$

$$= \frac{\cos^2 x}{\cos^2 x} = 1$$

19. $(1 - \sin x)(\sec x + \tan x) = \cos x$

$(1 - \sin x)(\sec x + \tan x) = \sec x + \tan x - \dfrac{\sin x}{\cos x} - \sin x \cdot \dfrac{\sin x}{\cos x}$

$$= \frac{1}{\cos x} - \frac{\sin^2 x}{\cos x}$$

$$= \frac{\cos^2 x}{\cos x} = \cos x$$

20. $\dfrac{1}{\tan x + \cot x} = \sin x \cos x$

$$\frac{1}{\tan x + \cot x} = \frac{1}{\dfrac{\sin x}{\cos x} + \dfrac{\cos x}{\sin x}}$$

$$= \frac{1}{\dfrac{\sin^2 x + \cos^2 x}{\sin x \cos x}}$$

$$= \frac{1}{\dfrac{1}{\sin x \cos x}}$$

$$= \sin x \cos x$$

21. $\dfrac{1}{1 + \sin x} + \dfrac{1}{1 - \sin x} = 2 \sec^2 x$

$$\frac{1}{1 + \sin x} + \frac{1}{1 - \sin x} = \frac{1 - \sin x + 1 + \sin x}{(1 + \sin x)(1 - \sin x)}$$

$$= \frac{2}{1 - \sin^2 x}$$

$$= \frac{2}{\cos^2 x}$$

$$= 2 \sec^2 x$$

22. $4 \sin^2 x \cos^2 x = 1 - \cos^2 2x$

$$1 - \cos^2 2x = \sin^2 2x$$

$$= (2 \sin x \cos x)^2$$

$$= 4 \sin^2 x \cos^2 x$$

23. $\cos 4x = 1 - 8 \sin^2 x \cos^2 x$

$$\cos 4x = \cos (2x + 2x)$$

$$= \cos^2 2x - \sin^2 2x$$

$$= (\cos^2 x - \sin^2 x)^2 - (2 \sin x \cos x)^2$$

$$= \cos^4 x - 2 \sin^2 x \cos^2 x + \sin^4 x - 4 \sin^2 x \cos^2 x$$

$$= (\cos^2 x)(1 - \sin^2 x) + (\sin^2 x)(1 - \cos^2 x) - 6 \sin^2 x \cos^2 x$$

$$= \cos^2 x - \cos^2 x \sin^2 x + \sin^2 x - \sin^2 x \cos^2 x - 6 \sin^2 x \cos^2 x$$

$$= 1 - 8 \sin^2 x \cos^2 x$$

24. $\sec 2x = \dfrac{\sec^2 x}{2 - \sec^2 x}$

$$\frac{\sec^2 x}{2 - \sec^2 x} = \frac{1}{2 \cos^2 x - 1} \quad \begin{array}{l}\text{(dividing numerator} \\ \text{and denominator} \\ \text{by } \sec^2 x)\end{array}$$

$$= \frac{1}{\cos 2x}$$

$$= \sec 2x$$

25. $2 \csc 2x = \sec x \csc x$

$$2 \csc 2x = \frac{2}{\sin 2x}$$

$$= \frac{2}{2 \sin x \cos x}$$

$$= \sec x \csc x$$

26. $\cos^4 x - \sin^4 x = \cos 2x$

$\cos^4 x - \sin^4 x = (\cos^2 x + \sin^2 x)(\cos^2 x - \sin^2 x) = 1(\cos 2x) = \cos 2x$

27. $\tan 2x = (\tan x)(1 + \sec 2x)$

$(\tan x)(1 + \sec 2x) = (\tan x)\left(1 + \dfrac{1}{\cos 2x}\right)$

$= (\tan x)\left(1 + \dfrac{1}{2\cos^2 x - 1}\right)$

$= (\tan x)\left(\dfrac{2\cos^2 x - 1 + 1}{2\cos^2 x - 1}\right)$

$= (\tan x)\left(\dfrac{2\cos^2 x}{\cos^2 x - \sin^2 x}\right)$

$= (\tan x)\left(\dfrac{2}{1 - \tan^2 x}\right)$

$= \dfrac{2\tan x}{1 - \tan^2 x} = \tan 2x$

28. $2\cot 2x = \cot x - \tan x$

$2\cot 2x = 2\,\dfrac{\cos 2x}{\sin 2x}$

$= \dfrac{2(\cos^2 x - \sin^2 x)}{2\sin x \cos x}$

$= \dfrac{\cos^2 x}{\sin x \cos x} - \dfrac{\sin^2 x}{\sin x \cos x}$

$= \dfrac{\cos x}{\sin x} - \dfrac{\sin x}{\cos x}$

$= \cot x - \tan x$

29. $\sec^2 \dfrac{x}{2} = \dfrac{2\tan x}{\tan x + \sin x}$

$\dfrac{2\tan x}{\tan x + \sin x} = \dfrac{2}{1 + \dfrac{\sin x}{\tan x}}$

$= \dfrac{2}{1 + \cos x}$

$= \dfrac{1}{\dfrac{1 + \cos x}{2}}$

$= \dfrac{1}{\cos^2 \dfrac{x}{2}}$

$= \sec^2 \dfrac{x}{2}$

30. $\csc^2 \dfrac{x}{2} - 1 = \dfrac{\sec x + 1}{\sec x - 1}$

$\csc^2 \dfrac{x}{2} - 1 = \dfrac{1}{\sin^2 \dfrac{x}{2}} - 1$

$= \dfrac{1}{\dfrac{1 - \cos x}{2}} - 1$

$= \dfrac{2}{1 - \cos x} - 1$

$= \dfrac{2 - 1 + \cos x}{1 - \cos x}$

$= \dfrac{1 + \cos x}{1 - \cos x}$ (dividing numerator and denominator by $\cos x$)

$= \dfrac{\sec x + 1}{\sec x - 1}$

31. $\tan \dfrac{x}{2} = \csc x - \cot x$

$\csc x - \cot x = \dfrac{1}{\sin x} - \dfrac{\cos x}{\sin x}$

$= \dfrac{1 - \cos x}{\sin x}$

$= \tan \dfrac{x}{2}$

32. $\tan x \sin 2x = 2\sin^2 x$

$\tan x \sin 2x = \left(\dfrac{\sin x}{\cos x}\right)(2\sin x \cos x)$

$= 2\sin^2 x$

33. $\dfrac{1 - \cos 2x}{\sin 2x} = \tan x$

$\dfrac{1 - \cos 2x}{\sin 2x} = \dfrac{1 - (1 - 2\sin^2 x)}{2 \sin x \cos x}$

$= \dfrac{\sin x}{\cos x}$

$= \tan x$

34. $\sin^2 x = \dfrac{1 - \cos 2x}{2}$

$\dfrac{1 - \cos 2x}{2} = \dfrac{1 - (\cos^2 x - \sin^2 x)}{2}$

$= \dfrac{1 - \cos^2 x + \sin^2 x}{2}$

$= \dfrac{\sin^2 x + \sin^2 x}{2}$

$= \dfrac{2 \sin^2 x}{2}$

$= \sin^2 x$

35. $30°, 150°, 210°, 330°$ **36.** $60°, 120°, 240°, 300°$ **37.** $45°, 135°, 225°, 315°$ **38.** $0°, 90°, 180°$ **39.** $0°$

40. $30°, 120°, 210°, 240°$ **41.** $\dfrac{-\sqrt{2} - \sqrt{6}}{4}$ **42.** $\dfrac{\sqrt{2 + \sqrt{2}}}{2}$ **43.** $2 - \sqrt{3}$ **44.** $2 + \sqrt{3}$ **45.** $\dfrac{\sqrt{2 + \sqrt{2}}}{2}$

46. $\dfrac{\sqrt{2} - \sqrt{6}}{4}$ **47.** $\dfrac{6 - 4\sqrt{5}}{15}$ **48.** $\dfrac{6 + 4\sqrt{5}}{15}$ **49.** $\dfrac{-3\sqrt{5} + 8}{15}$ **50.** $\dfrac{-3\sqrt{5} - 8}{15}$ **51.** $\dfrac{108 + 50\sqrt{5}}{19}$

52. $\dfrac{-108 + 50\sqrt{5}}{19}$ **53.** $\sec^2 x$ **54.** $\cos^2 x$ **55.** $\dfrac{5\sqrt{26}}{26}, \dfrac{-\sqrt{26}}{6}, -5$ **56.** 1 **57.** $-1 = -1$ **58.** $-\frac{1}{2} = -\frac{1}{2}$

59. $\pi/6$ **60.** $0°$ **61.** $3\pi/4$ **62.** $-60°$ **63.** $2\pi/3$ **64.** x **65.** $\sqrt{1 - x^2}$ **66.** 0 **67.** $9°$ **68.** $43°40'$
69. $66°30'$ **70.** $72°$ **71.** $0°, 120°, 240°$

Chapter Test (Page 441)

1. $\frac{35}{36}$ **2.** $\dfrac{(1 + \cos x)^2 + \sin^2 x}{(\sin x)(1 + \cos x)} = 2 \csc x$

$\dfrac{(1 + \cos x)^2 + \sin^2 x}{(\sin x)(1 + \cos x)} = \dfrac{1 + 2\cos x + \cos^2 x + \sin^2 x}{(\sin x)(1 + \cos x)}$

$= \dfrac{2 + 2\cos x}{(\sin x)(1 + \cos x)}$

$= \dfrac{2(1 + \cos x)}{(\sin x)(1 + \cos x)}$

$= 2 \csc x$

3. $4\sqrt{\dfrac{1 - \cos x}{2}}\sqrt{\dfrac{1 + \cos x}{2}} = 2 \sin x$

$4\sqrt{\dfrac{1 - \cos x}{2}}\sqrt{\dfrac{1 + \cos x}{2}} = 4\sqrt{\dfrac{1 - \cos^2 x}{4}}$

$= 4\sqrt{\dfrac{\sin^2 x}{4}}$

$= 4\left(\dfrac{\sin x}{2}\right)$

$= 2 \sin x$

4. (a) $60°, 120°, 240°, 300°$ **(b)** $30°, 150°, 210°, 330°$ **(c)** $120°, 180°, 240°$

5. $\dfrac{1 + \sqrt{3}}{1 - \sqrt{3}}$ **6. (a)** $\frac{36}{85}$ **(b)** $\frac{84}{13}$ **7. (a)** $\dfrac{5\sqrt{26}}{26}$ **(b)** -5 **(c)** $\dfrac{119}{169}$ **8.** $\dfrac{-2\sqrt{2}}{3}$

CHAPTER 8

Exercises (Page 449)

1. $A = 75°, b = 7.32, c = 8.97$ **3.** $A = 35°16', C = 84°44', c = 6.9$ **5.** $C = 24°, A = 132°, a = 16.4$
7. impossible **9.** $A = 47°50', C = 72°40', a = 6093; A' = 13°10', C' = 107°20', a' = 1870$
11. $A = 123°18', a = 55.58, c = 36.63$ **13.** $A = 82°11', a = 28.89, b = 27.11$

Exercises (Pages 451–452)

1. $b = 3.58$ **3.** $a = 8.36$ **5.** $C = 93°49'$ **7.** $A = 70°55'$, $B = 49°5'$, $c = 13.75$
9. $d^2 = a^2 + b^2 - 2ab \cos(180° - \theta)$
$$= a^2 + b^2 - 2ab(\cos 180° \cos \theta + \sin 180° \sin \theta)$$
$$= a^2 + b^2 - 2ab[(-1)\cos \theta + 0]$$
$$= a^2 + b^2 + 2ab \cos \theta$$

11. $\left. \begin{matrix} a^2 = b^2 + c^2 - 2bc \cos A \\ b^2 = a^2 + c^2 - 2ac \cos B \\ c^2 = a^2 + b^2 - 2ab \cos C \end{matrix} \right\}$ law of cosines

Adding these three equations, we get

$$a^2 + b^2 + c^2 = 2a^2 + 2b^2 + 2c^2 - 2bc \cos A - 2ac \cos B - 2ab \cos C$$

$$\frac{a^2 + b^2 + c^2}{2abc} = \frac{a^2 + b^2 + c^2}{abc} - \frac{\cos A}{a} - \frac{\cos B}{b} - \frac{\cos C}{c} \qquad \text{(dividing by } 2abc)$$

$$\frac{-(a^2 + b^2 + c^2)}{2abc} = \frac{-\cos A}{a} - \frac{\cos B}{b} - \frac{\cos C}{c} \qquad \text{(subtraction)}$$

$$\frac{a^2 + b^2 + c^2}{2abc} = \frac{\cos A}{a} + \frac{\cos B}{b} + \frac{\cos C}{c} \qquad \text{(multiplying by } -1)$$

13.
$$a^2 = b^2 + c^2 - 2bc \cos A$$
$$\cos A = \frac{b^2 + c^2 - a^2}{2bc}$$
$$1 + \cos A = \frac{b^2 + c^2 - a^2}{2bc} + \frac{2bc}{2bc}$$
$$= \frac{b^2 + 2bc + c^2 - a^2}{2bc}$$
$$2 \cos^2 \frac{A}{2} = \frac{b^2 + 2bc + c^2 - a^2}{2bc}$$
$$= \frac{(b + c)^2 - a^2}{2bc}$$
$$= \frac{(b + c + a)(b + c - a)}{2bc}$$

Exercises (Pages 455–456)

1. 436   **3.** 30°16′; 149°44′   **5.** 17°50′   **7.** 4°52′; 42.7 nautical miles   **9.** 19.4 meters
11. 9.7 kilometers from A, 7.2 kilometers from B **13.** 38°25′
15. 99°22′   **17.** 0.9 and 1.8 hours later; 12:54 P.M. and 1:48 P.M.   **19.** 537 meters
21. $BD = 16.4$ kilometers; $CD = 7.5$ kilometers   **23.** 93 meters   **25.** 5.2 miles

Exercises (Pages 460–461)

1. 10.5 **3.** 73.2 **5.** 934.0 **7.** 53,839.2 **9.** 81.38 **11.** 370 **13.** 91,241 **15.** \$246.88
17. 1383 square meters

Exercises (Pages 472–473)

1. 41.6; 65°50′ **3.** 938; 281 **5.** 20.4; 11°20′ **7.** 1411 **9.** 84.9 **11.** 12°50′; yes **13.** 643; 26°40′
15. 68; 37.7 **17.** 4.12; 76° **19.** 5; −126°50′ **21.** 5; 180° **23.** 5.6; 45° **25.** 2; 30° **27.** (2.95, −0.521)
29. (3.46, 2) **31.** (−1.41, 1.41) **33.** (2.5, 4.33) **35.** (0, 0) **47.** (4.5, 277°) **49.** (3.5, 260°) **51.** (5.2, 74°)
53. (3.0, 230°) **55.** (1.7, 202°) **57.** (4.2, 88°) **59.** (3, 185°) **61.** (2, 152°) **63.** (4.5, 125°) **65.** (2, 330°)
67. 5.39 **69.** 1 **71.** 5 **73.** $\langle 8, 16 \rangle$; $\langle -2, -8 \rangle$ **75.** $\langle 4, 6 \rangle$; $\langle -4, -2 \rangle$ **77.** $\langle 4, -4 \rangle$; $\langle -2, 2 \rangle$

79. $\left\langle \dfrac{5}{13}, \dfrac{-12}{13} \right\rangle$ **81.** $\left\langle \dfrac{\sqrt{2}}{2}, \dfrac{\sqrt{2}}{2} \right\rangle$

Exercises (Pages 485–486)

1. $2(\cos 315° + i \sin 315°)$ **3.** $1(\cos 180° + i \sin 180°)$ **5.** $6(\cos 90° + i \sin 90°)$ **7.** $2\sqrt{2}(\cos 45° + i \sin 45°)$
9. $15(\cos 0° + i \sin 0°)$ **11.** $2(\cos 30° + i \sin 30°)$ **13.** $20(\cos 270° + i \sin 270°)$
15. $13(\cos 292.6° + i \sin 292.6°)$ **17.** $2(\cos 240° + i \sin 240°)$ **19.** $10(\cos 240° + i \sin 240°)$ **21.** $-1.732 + i$

23. $3.53 + 3.53i$ **25.** $2.29 + 3.28i$ **27.** $2 + 0i; 0 + i$ **29.** $-4 + 0i; \dfrac{-1}{2} - \dfrac{\sqrt{3}}{2}i$ **31.** $6 + 3i; 0.67 + 0.33i$

33. $12 + 0i; -\frac{4}{3} + 0i$ **35.** $16 + 0i$ **37.** $4096 + 0i$ **39.** $0.707 - 0.707i$ **41.** $-0.025 + 0.0238i$

43. $0.782 + 0.730i$ **45.** $\sqrt{3} + i; -\sqrt{3} - i$ **47.** $\dfrac{\sqrt{2}}{2} + \dfrac{\sqrt{2}}{2}i; \dfrac{-\sqrt{2}}{2} - \dfrac{\sqrt{2}}{2}i$ **49.** $0 + i; -\sqrt{3}/2 - \frac{1}{2}i; \sqrt{3}/2 - \frac{1}{2}i$

51. $1.148 + 0.004i, 1.148 + 0.029i, 1.147 + 0.055i, 1.146 + 0.080i, 1.144 + 0.105i$ **53.** $1; i; -1; -i$

Exercises for Review (Pages 487–488)

1. $A = 94°, b = 4.01, c = 6.65$ **2.** $B = 88°, a = 8.67, b = 13.76$ **3.** $B = 34°, a = 23, C = 57°$
4. $A = 64°50′, B = 66°10′, b = 12.12$ **5.** $C = 35°, B = 110°, b = 34.4$ **6.** $A = 41°10′, C = 106°30′, c = 1337$
7. $c = 3.66$ **8.** $B = 30°$ **9.** $b = 8.3$ **10.** $A = 78°10′$ **11.** $C_n = 336.6°; 52.5$ nautical miles
12. 37°41′ or 37.68° **13.** 41.568 **14.** 81.384 **15.** 24.249 **16.** 387 **17.** 83,891 **18.** 93,270
19. 45.69 lbs.; 66°48′ **20.** 25.88; 96.59 **21.** 258.82 **22.** 361 **23.** 17.68; 17.68 **24.** $\theta = 41.8°$ or 41°49′
25. $(3\sqrt{2}, 45°)$ **26.** $(2, 63°26′)$ **27.** $(\sqrt{5}, 330°)$ **28.** $(2\sqrt{7}, -19°6′)$ **29.** $\langle 0.866, 0.5 \rangle$ **30.** $\langle 1, \sqrt{3} \rangle$
31. $\langle -2.28, 1.928 \rangle$ **32.** $\langle 0, 4 \rangle$ **37.** $\sqrt{17}$ **38.** $\sqrt{13}$ **39.** $\langle 10, -1 \rangle, \langle -6, 15 \rangle$ **40.** $\langle -1, -5 \rangle, \langle 1, -1 \rangle$
41. $\langle 6\sqrt{61}/61, 5\sqrt{61}/61 \rangle$ **42.** $\langle -7\sqrt{170}/170, 11\sqrt{170}/170 \rangle$ **43.** $\sqrt{5}(\cos 333°30′ + i \sin 333°30′)$
44. $3\sqrt{2}(\cos 45° + i \sin 45°)$ **45.** $2(\cos 330° + i \sin 330°)$ **46.** $5(\cos 36°52′ + i \sin 36°52′)$
47. $2.0704 + 7.7272i$ **48.** $2.1213 - 2.1213i$ **49.** $-1.414 + 1.414i$ **50.** $-4.330 - 2.5i$
51. $5\sqrt{2}(\cos 315° + i \sin 315°); (\sqrt{2}/5)(\cos 315° + i \sin 315°)$
52. $\sqrt{145}(\cos 4°46′ + i \sin 4°46′); (\sqrt{145}/29)(\cos 48°22′ + i \sin 48°22′)$ **53.** $-8i$ **54.** $-46 - 9i$
55. $-\sqrt{3} + i$ and $\sqrt{3} - i$ **56.** $2.427 + 1.763i; -0.927 + 2.853i; -3 + 0i; -0.927 - 2.853i; 2.427 - 1.763i$

Chapter Test (Pages 488–489)

1. (a) $a = 24.8; A = 62°15′; C = 76°9′$ **(b)** $c = 23; A = 29°22′; B = 40°38′$
2. (a) $z = \sqrt{5}[\cos(\pi - 1.107) + i \sin(\pi - 1.107)]$ **(b)** $z = \sqrt{7}[\cos(2\pi - 0.857) + i \sin(2\pi - 0.857)]$

3. $-0.5176 + 1.9318i$ **4. (a)** $11 + 2i$ **(b)** $\frac{1}{5} - 2i/3$ **5.** $-3.538 + 0.6923i$
6. $\sqrt[4]{18}(-0.3827 + 0.9239i)$ and $\sqrt[4]{18}(0.3827 - 0.9239i)$ **7.** 1949 square meters **8.** 209 pounds
9. 1372 meters **10.** A is 61.5 yards; B is 67.1 yards; A reaches swimmer first.

CHAPTER 9

Exercises (Pages 500–501)

1. $x = 1, y = 0$ **3.** $x = 2, y = -3$ **5.** $x = 3, y = 2$ **7.** $x = \frac{14}{5}, y = \frac{6}{5}$ **9.** inconsistent **11.** independent
13. dependent **15.** $x = 3, y = 1$ **17.** $x = 2, y = 2$ **19.** $x = 3, y = 2$ **21.** 45, 33
23. adults: 1850; children: 1250 **25.** $7000 at 7 percent; $5000 at 5.5 percent **27.** 10 kilograms
29. solution A: 90 quarts; solution B: 30 quarts **31.** 52.5 gallons

Exercises (Page 505)

1. $x = 7, y = 3, z = -2$ **3.** $x = 2, y = 3, z = 1$ **5.** $x = 3 - 2c, y = c, z = 1$ **7.** $x = 2, y = -1, z = 2$
9. no solution **11.** $x = 2, y = 4, z = 3$

Exercises (Pages 510–512)

1. 2×3 **3.** 3×3 **5.** 1×3 **7.** $a_{12} = -4, a_{21} = 1, a_{32} = -2, a_{23} = 2$ **9.** $x = 3, y = 1, z = 4, u = 3$
11. $x = 3, y = 3, z = 7$ **13.** $x = \frac{5}{2}, y = \frac{-5}{2}, z = 1$ **15.** $A + B = [0 \quad 2 \quad 4 \quad 4 \quad 5], A - B = [4 \quad 4 \quad 4 \quad -2 \quad 9]$

17. $A + B = \begin{bmatrix} 1 \\ 4 \\ 5 \\ 1 \end{bmatrix}, A - B = \begin{bmatrix} 3 \\ 2 \\ -3 \\ -5 \end{bmatrix}$ **19.** $A + B = \begin{bmatrix} 4 \\ 4 \\ 4 \end{bmatrix}, A - B = \begin{bmatrix} -4 \\ 6 \\ 10 \end{bmatrix}$

21. $A + B = \begin{bmatrix} 3 & 1 \\ 2 & 7 \end{bmatrix}, A - B = \begin{bmatrix} -1 & 3 \\ 4 & 1 \end{bmatrix}$ **23.** not possible to compute **25.** not possible to compute

27. $\begin{bmatrix} 13 & 22 \\ 23 & 30 \end{bmatrix}$ **29.** $\begin{bmatrix} 28 & 27 \\ 31 & 44 \end{bmatrix}$ **31.** $\begin{bmatrix} 13 & 15 \\ 16 & 22 \end{bmatrix}$

Exercises (Pages 517–518)

1. 3×2 **3.** 2×4 **5.** 3×6 **7.** both **9.** neither **11.** both **13.** $AB = \begin{bmatrix} 19 & 11 \\ 13 & 8 \end{bmatrix}, BA = \begin{bmatrix} 17 & 23 \\ 7 & 10 \end{bmatrix}$

15. $AB = \begin{bmatrix} 7 & 15 \\ 22 & 39 \end{bmatrix}, BA = \begin{bmatrix} 7 & 11 \\ 30 & 39 \end{bmatrix}$ **17.** $AB = \begin{bmatrix} 14 & 43 & 48 \\ 20 & 71 & 74 \end{bmatrix}$ **19.** $BC = \begin{bmatrix} 17 & 26 & 33 \\ 28 & 44 & 56 \\ 8 & 28 & 28 \end{bmatrix}$ **21.** $\begin{bmatrix} 8 & 17 \\ 22 & 55 \end{bmatrix}$

23. $\begin{bmatrix} 9 & 6 \\ 33 & 6 \end{bmatrix}$ **25.** $\begin{bmatrix} -5 & 2 \\ 22 & 20 \end{bmatrix}$ **27.** $\begin{bmatrix} 74 & 52 \\ 220 & 176 \end{bmatrix}$ **29.** $\begin{bmatrix} 17 & 23 \\ 55 & 61 \end{bmatrix}$ **31.** $\begin{bmatrix} 61 & 56 \\ 53 & 40 \end{bmatrix}$

33. $A(B + C) = \begin{bmatrix} 21 & 43 \\ 42 & 67 \end{bmatrix}$, $AB = \begin{bmatrix} 13 & 20 \\ 24 & 29 \end{bmatrix}$, $AC = \begin{bmatrix} 8 & 23 \\ 18 & 38 \end{bmatrix}$, $AB + AC = \begin{bmatrix} 21 & 43 \\ 42 & 67 \end{bmatrix}$; $A(B + C) = AB + AC$

35. AB is a zero matrix, but this does not imply that either A or B is a zero matrix.

Exercises (Pages 526–527)

1. $x = 1, y = 2$ **3.** $x = 2, y = 3$ **5.** $x = 2, y = 3$ **7.** $x = 1, y = 2, z = -1$ **9.** $x = 2, y = -3, z = 3$
11. $x = 3, y = -1, z = 4$ **13.** $x = 1, y = 2, z = 1$ **15.** $x = 1, y = -1, z = 2$ **17.** $x = 1, y = 1, z = 2$
19. no solution   **21.** $x = 3c + 6, y = c$ **23.** $x = 2, y = 3, z = 1$ **25.** $x = 2 + c, y = -3 + 2c, z = c$
27. $x = \frac{24}{7} - \frac{3}{7}c, y = \frac{9}{7} + \frac{5}{7}c, z = c$ **29.** $x = \frac{4}{5} - \frac{2}{5}c, y = \frac{3}{5}c - \frac{11}{5}, z = c$

Exercises (Pages 534–535)

1. $\begin{bmatrix} -3 & 2 \\ 5 & -3 \end{bmatrix}$ **3.** $\begin{bmatrix} 12 & -11 \\ -13 & 12 \end{bmatrix}$ **5.** $-\dfrac{1}{30}\begin{bmatrix} 3 & -1 & -4 \\ 12 & -14 & 4 \\ -12 & -6 & 6 \end{bmatrix}$ **7.** $\dfrac{1}{26}\begin{bmatrix} -6 & -2 & 12 \\ -7 & 2 & 1 \\ 11 & 8 & -9 \end{bmatrix}$

9. $\dfrac{1}{10}\begin{bmatrix} -5 & 5 & 5 \\ 5 & -3 & -1 \\ 5 & 1 & -3 \end{bmatrix}$ **11.** $\dfrac{1}{27}\begin{bmatrix} 8 & 5 & -7 \\ -1 & -4 & 11 \\ -7 & -1 & 23 \end{bmatrix}$ **13.** $\dfrac{1}{6}\begin{bmatrix} 1 & 1 \\ -2 & 4 \end{bmatrix}$; $x = 4, y = -1$

15. $\dfrac{1}{5}\begin{bmatrix} -1 & 2 \\ 4 & -3 \end{bmatrix}$; **(a)** $x = 2, y = 3$ **(b)** $x = 4, y = -3$ **(c)** $x = -3, y = 10$ **(d)** $x = 1, y = -1$

17. $\begin{bmatrix} -7 & -5 & 6 \\ 13 & 10 & -11 \\ -1 & -1 & 1 \end{bmatrix}$; **(a)** $x = 2, y = 3, z = -1$ **(b)** $x = 3, y = -2, z = 5$ **(c)** $x = 2, y = 4, z = 7$

19. $\dfrac{1}{64}\begin{bmatrix} 2 & 18 & -14 \\ -7 & 1 & 17 \\ 9 & -15 & 1 \end{bmatrix}$; three planes of type 1, two planes of type 2, and one plane of type 3

Exercises (Page 542)

1.

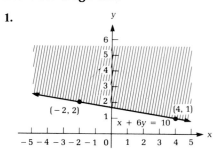

3.

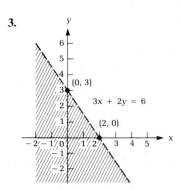

5.

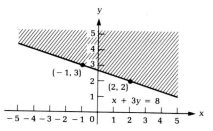

7.

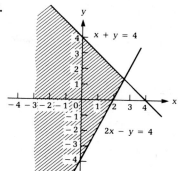

9.

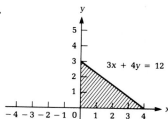

11.

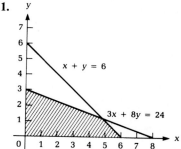

13.

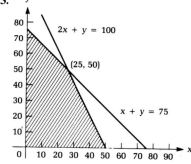

15.

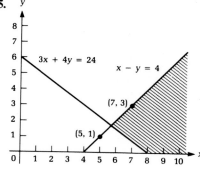

Exercises (Pages 548–549)

1. $x = 6$, $y = 4$ **3.** $x = 20$, $y = 10$ **5.** $x = \frac{5}{3}$, $y = \frac{28}{3}$ **7.** $x = 3$, $y = 4$ **9.** $x = 40$, $y = 30$
11. product 1: 20 units; product 2: 10 units   **13.** model 1: 15 units; model 2: 20 units
15. food 1: 5 units; food 2: 1 unit

Exercises for Review (Pages 549–552)

1. $x = 3$, $y = 5$ **2.** $x = 2$, $y = 0$ **3.** $x = 4$, $y = -3$ **4.** $x = \frac{19}{23}$, $y = -\frac{18}{23}$ **5.** 83, 68
6. adults: 1500; children: 700   **7.** 16 pounds   **8.** antifreeze A: 12.5 liters; antifreeze B: 7.5 liters

9. $x = 1$, $y = 2$, $z = 1$ **10.** $x = 2$, $y = -1$, $z = 2$ **11.** $x = 2$, $y = 6$, $z = 5$ **12.** $\begin{bmatrix} 6 & 5 & -1 \\ 8 & 1 & 1 \end{bmatrix}$

13. $\begin{bmatrix} -2 & 3 & 3 \\ 2 & -1 & -7 \end{bmatrix}$ **14.** $\begin{bmatrix} 14 & 14 & -1 \\ 21 & 2 & -1 \end{bmatrix}$ **15.** $\begin{bmatrix} -1 & 1 & 8 \\ 30 & 25 & 30 \end{bmatrix}$ **16.** $\begin{bmatrix} -8 & -4 & 4 \\ 59 & 24 & 31 \end{bmatrix}$

17. $\begin{bmatrix} 47 & 19 & 26 \\ 44 & 18 & 22 \\ 28 & 11 & 19 \end{bmatrix}$ **18.** $\begin{bmatrix} 27 & 23 & 29 \\ 22 & 18 & 19 \\ 32 & 28 & 39 \end{bmatrix}$ **19.** $A(BC) = (AB)C = \begin{bmatrix} 55 & 59 \\ 137 & 141 \end{bmatrix}$

20. $A(B + C) = \begin{bmatrix} 18 & 23 \\ 27 & 59 \end{bmatrix}$, $AB = \begin{bmatrix} 6 & 17 \\ 2 & 36 \end{bmatrix}$, $AC = \begin{bmatrix} 12 & 6 \\ 25 & 23 \end{bmatrix}$, $AB + AC = \begin{bmatrix} 18 & 23 \\ 27 & 59 \end{bmatrix}$; $A(B + C) = AB + AC$

21. $x = 2, y = 1$ **22.** $x = 2, y = 1$ **23.** $x = 1, y = -1, z = 2$ **24.** $x = 1, y = -1, z = 5$

25. $A^{-1} = \begin{bmatrix} 13 & -5 \\ -5 & 2 \end{bmatrix}$; **(a)** $x = 2, y = 3$ **(b)** $x = -3, y = 2$

26. $A^{-1} = \begin{bmatrix} -7 & -5 & 6 \\ 13 & 10 & -11 \\ -1 & -1 & 1 \end{bmatrix}$; $x = 0, y = 2, z = 1$

27.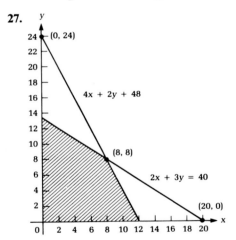

28.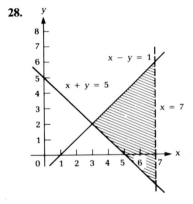

29.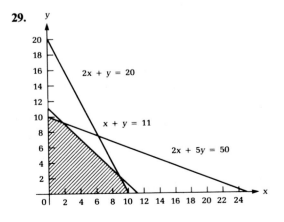

30.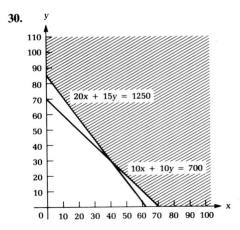

31. $x = 100$, $y = 0$; $x = 30$, $y = 105$ **32.** $x = 6$, $y = 4$ **33.** $x = 4$, $y = 2$ **34.** $x = 6$, $y = 2$
35. electric clocks: 400; battery-operated clocks: 200   **36.** product 1: 20 units; product 2: 20 units
37. food 1: 48 units; food 2: 0 unit; or food 1: 20 units; food 2: 35 units   **38.** regular: 100 gallons; low sulfur: 0 gallon

Chapter Test (Pages 552–553)

1. $\begin{bmatrix} 2 & 9 & 14 \\ 1 & -1 & 18 \end{bmatrix}$ **2.** $\begin{bmatrix} 4 & -1 & -2 \\ -3 & 5 & -2 \end{bmatrix}$ **3.** $\begin{bmatrix} 4 & 37 & 58 \\ 7 & -9 & 74 \end{bmatrix}$ **4.** $\begin{bmatrix} 34 & 23 \\ 8 & -12 \end{bmatrix}$ **5.** $\begin{bmatrix} 14 & 64 \\ 11 & 8 \end{bmatrix}$ **6.** $\begin{bmatrix} 46 & 23 \\ 13 & 25 \end{bmatrix}$

7. $\begin{bmatrix} 80 & 46 \\ 21 & 13 \end{bmatrix}$ **8.** $\begin{bmatrix} 80 & 46 \\ 21 & 13 \end{bmatrix}$ **9.** $\begin{bmatrix} 38 & 20 \\ 7 & -6 \end{bmatrix}$ **10.** $\begin{bmatrix} 239 & 58 \\ 4 & -56 \end{bmatrix}$ **11.** $\begin{bmatrix} 239 & 58 \\ 4 & -56 \end{bmatrix}$ **12.** $\begin{bmatrix} 52 & 84 \\ 18 & 2 \end{bmatrix}$

13. $x = 4$, $y = -3$ **14.** $x = 1$, $y = 2$, $z = 1$ **15.** $x = 2$, $y = 3$ **16.** $A^{-1} = \begin{bmatrix} 2 & -3 \\ -3 & 5 \end{bmatrix}$; $x = 1$, $y = 1$

17.

18. $x = 100$, $y = 0$ *or* $x = 30$, $y = 105$ **19.** $x = 4$, $y = 2$
20. fish, 2 pounds; milk, 3 quarts

CHAPTER 10

Exercises (Pages 560–561)

1. 2, 4, 6, 8, 10   **3.** 4, 7, 10, 13, 16   **5.** -1, 3, 7, 11, 15   **7.** 1, 4, 9, 16, 25   **9.** 0, 3, 8, 15, 24   **11.** 2, 4, 8, 16, 32

13. 3, 9, 27, 81, 243   **15.** $1, \dfrac{1}{2}, \dfrac{1}{4}, \dfrac{1}{8}, \dfrac{1}{16}$ **17.** $-1, 0, -1, 0, -1$ **19.** 1, 3, 6, 10, 15   **21.** $\dfrac{1}{2}, \dfrac{2}{5}, \dfrac{3}{10}, \dfrac{4}{17}, \dfrac{5}{26}$

23. 1, 5, 14, 30, 55   **25.** 2, 0, 2, 0, 2   **27.** $3n - 1$ **29.** $2n + 1$ **31.** 2^n **33.** $(-1)^{n+1}$ **35.** $\left(\dfrac{1}{2}\right)^{n-1}$ **37.** $\dfrac{n+1}{n}$

39. $\left(-\dfrac{1}{2}\right)^n$ **41.** 10   **43.** 35   **45.** 91   **47.** 50   **49.** 400   **51.** 3, 8, 15, 24, 35   **53.** 7, 18, 33, 52, 75

55. 1, 5, 14, 30, 55   **57.** 0, 3, 11, 26, 50   **59.** 1, 0, -1, 0, -1 **61.** 1, 4, 13, 40, 121   **63.** $0, \dfrac{1}{2}, \dfrac{7}{6}, \dfrac{23}{12}, \dfrac{163}{60}$
65. $\dfrac{1}{2}, \dfrac{3}{4}, \dfrac{7}{8}, \dfrac{15}{16}, \dfrac{31}{32}$

Exercises (Pages 566–567)

1. 31 **3.** 39 **5.** 10 **7.** −19 **9.** −18 **11.** 50; 435 **13.** 61; 495 **15.** −27; −90 **17.** −41; 195
19. 5050 **21.** 2550 **23.** 5, 7, 9, 11, 13 **25.** $a_1 = 3$, $d = 5$, $a_{20} = 98$, $S_{20} = 1010$ **27.** 6, 8, 10 *or* 10, 8, 6
29. 1.4 miles; 25.5 miles **31.** $9300 **33.** $10,200 **35.** $22,850 **37.** 168 **39. (a)** $70 **(b)** $2835 **(c)** 0
41. The first n odd positive integers are 1, 3, 5, . . . , $2n − 1$. Thus $a_1 = 1$, $d = 2$, and $a_n = 2n − 1$. Using (10.5), we

obtain $S_n = \dfrac{n}{2}(a_1 + a_n) = \dfrac{n}{2}(1 + 2n − 1) = \dfrac{n}{2}(2n) = n^2.$

Exercises (Pages 570–571)

1. 4374 **3.** 2187 **5.** 16,384 **7.** $\dfrac{1}{128}$ **9.** $\dfrac{128}{2187}$ **11.** 96 **13.** 972 **15.** $\dfrac{1}{9}$ **17.** 1022 **19.** 976,562

21. $\dfrac{87,381}{1024}$ **23.** $\dfrac{1533}{128}$ **25.** 1536; 3069 **27.** $\dfrac{1}{729}, \dfrac{29,524}{729}$ **29.** 3, 6, 12, 24, . . . **30.** $27,420.59

33. $2531.25 **35.** $\dfrac{1365}{16}$ feet **37.** $22,050; $110,512.63

Exercises (Pages 579–580)

1. 2 **3.** $\frac{2}{3}$ **5.** 2 **7.** 0 **9.** 0 **11.** limit does not exist **13.** 0 **15.** limit does not exist
17. limit does not exist **19.** $\frac{1}{2}$ **21.** limit does not exist **23.** 2 **25.** 1

27. We observe that $\displaystyle\lim_{n \to \infty} a_n = \lim_{n \to \infty} \frac{2n + 1}{3n + 2} = \lim_{n \to \infty} \frac{2 + 1/n}{3 + 2/n} = \frac{\displaystyle\lim_{n \to \infty}(2 + 1/n)}{\displaystyle\lim_{n \to \infty}(3 + 2/n)} = \frac{2}{3}$

and $\displaystyle\lim_{n \to \infty} b_n = \lim_{n \to \infty} \frac{n + 1}{2n − 1} = \lim_{n \to \infty} \frac{1 + 1/n}{2 − 1/n} = \frac{\displaystyle\lim_{n \to \infty}(1 − 1/n)}{\displaystyle\lim_{n \to \infty}(2 − 1/n)} = \frac{1}{2}$

(a) $\quad a_n b_n = \left(\dfrac{2n + 1}{3n + 2}\right)\left(\dfrac{n + 1}{2n − 1}\right) = \dfrac{2n^2 + 3n + 1}{6n^2 + n − 2} = \dfrac{2 + 3/n + 1/n^2}{6 + 1/n − 2/n^2}$

$\displaystyle\lim_{n \to \infty} a_n b_n = \frac{\displaystyle\lim_{n \to \infty}(2 + 3/n + 1/n^2)}{\displaystyle\lim_{n \to \infty}(6 + 1/n − 2/n^2)} = \frac{2 + 0 + 0}{6 + 0 − 0} = \frac{2}{6} = \frac{1}{3}$

$\left(\displaystyle\lim_{n \to \infty} a_n\right)\left(\displaystyle\lim_{n \to \infty} b_n\right) = \left(\dfrac{2}{3}\right)\left(\dfrac{1}{2}\right) = \dfrac{1}{3}$ Thus $\displaystyle\lim_{n \to \infty} a_n b_n = \left(\displaystyle\lim_{n \to \infty} a_n\right)\left(\displaystyle\lim_{n \to \infty} b_n\right).$

(b) $\quad \dfrac{a_n}{b_n} = \left(\dfrac{2n + 1}{3n + 2}\right)\left(\dfrac{2n − 1}{n + 1}\right) = \dfrac{4n^2 − 1}{3n^2 + 5n + 2} = \dfrac{4 − 1/n^2}{3 + 5/n + 2/n^2}$

$\displaystyle\lim_{n \to \infty} \frac{a_n}{b_n} = \frac{\displaystyle\lim_{n \to \infty}(4 − 1/n^2)}{\displaystyle\lim_{n \to \infty}(3 + 5/n + 2/n^2)} = \frac{4 − 0}{3 + 0 + 0} = \frac{4}{3}$ and

$\dfrac{\displaystyle\lim_{n \to \infty} a_n}{\displaystyle\lim_{n \to \infty} b_n} = \dfrac{2/3}{1/2} = \dfrac{4}{3}$ Thus $\displaystyle\lim_{n \to \infty} \frac{a_n}{b_n} = \frac{\displaystyle\lim_{n \to \infty} a_n}{\displaystyle\lim_{n \to \infty} b_n}.$

(c)

$$a_n + b_n = \frac{2n + 1}{3n + 2} + \frac{n + 1}{2n - 1} = \frac{4n^2 - 1 + 3n^2 + 5n + 2}{(3n + 2)(2n - 1)} = \frac{7n^2 + 5n + 1}{6n^2 + n - 2}$$

$$= \frac{7 + 5/n + 1/n^2}{6 + 1/n - 2/n^2}$$

$$\lim_{n \to \infty} (a_n + b_n) = \frac{\lim_{n \to \infty} (7 + 5/n + 1/n^2)}{\lim_{n \to \infty} (6 + 1/n - 2/n^2)} = \frac{7 + 0 + 0}{6 + 0 + 0} = \frac{7}{6}$$

$$\lim_{n \to \infty} a_n + \lim_{n \to \infty} b_n = \frac{2}{3} + \frac{1}{2} = \frac{7}{6} \qquad \text{Thus}$$

$$\lim_{n \to \infty} (a_n + b_n) = \lim_{n \to \infty} a_n + \lim_{n \to \infty} b_n.$$

Exercises (Page 583)

1. $\frac{4}{3}$ **3.** $\frac{2}{3}$ **5.** 6 **7.** $\frac{1}{5}$ **9.** $\frac{10}{9}$ **11.** $\frac{1}{9}$ **13.** $\frac{7}{9}$ **15.** $\frac{4}{33}$ **17.** $\frac{101}{999}$ **19.** $\frac{212}{999}$ **21.** $\frac{323}{999}$ **23.** $\frac{271}{999}$ **25.** $\frac{1001}{333}$
27. $\frac{100}{33}$ **29.** $\frac{485}{333}$

Exercises (Pages 586–587)

1. We wish to show that the statement $p(n)$ is true for all positive integers where $p(n)$: $2 + 4 + 6 + \cdots + 2n = n(n + 1)$.
(i) If we substitute $n = 1$ in $p(n)$, then the left-hand side equals 2 and the right-hand side is $1(1 + 1) = 2$. This shows that $p(1)$ is true.
(ii) Assume that for a positive integer k, $p(k)$ is true. That is,

$$p(k): 2 + 4 + 6 + \cdots + 2k = k(k + 1).$$

We must show that $p(k)$ implies the truth of $p(k + 1)$, where

$$p(k + 1): 2 + 4 + 6 + \cdots + 2k + 2(k + 1) = (k + 1)[(k + 1) + 1].$$

The induction hypothesis is

$$2 + 4 + 6 + \cdots + 2k = k(k + 1).$$

By adding $2(k + 1)$ to both sides of this equation, we get

$$2 + 4 + 6 + \cdots + 2k + 2(k + 1) = k(k + 1) + 2(k + 1)$$
$$= (k + 1)(k + 2).$$

This is precisely what we wanted to establish. Thus we have proved that if $p(k)$ is true, then $p(k + 1)$ is also true. This means that $p(n)$ is true for every positive integer n.
3. Let $p(n)$ denote the statement that

$$p(n): 2 + 6 + 10 + \cdots + (4n - 2) = 2n^2.$$

If we substitute $n = 1$ in $p(n)$, then the left-hand side equals 2 and the right-hand side is $2(1^2) = 2$. This shows that $p(1)$ is true.
Next assume that for a positive integer k, $p(k)$ is true. That is,

$$p(k): 2 + 6 + 10 + \cdots + (4k - 2) = 2k^2.$$

We must show that $p(k)$ implies the truth of $p(k + 1)$, where

$$p(k + 1): 2 + 6 + 10 + \cdots + (4k - 2) + [4(k + 1) - 2] = 2 + 6 + 10 + \cdots + (4k + 2)$$
$$= 2(k + 1)^2.$$

The induction hypothesis is

$$2 + 6 + 10 + \cdots + (4k - 2) = 2k^2.$$

By adding $4(k + 1) - 2 = 4k + 2$ to both sides of this equation, we get

$$2 + 6 + 10 + \cdots + (4k - 2) + (4k + 2) = 2k^2 + 4k + 2$$
$$= 2(k^2 + 2k + 1)$$
$$= 2(k + 1)^2.$$

This is what we wanted to establish. Thus we have proved that if $p(k)$ is true, then $p(k + 1)$ is also true. This means that $p(n)$ is true for every positive integer n.

5. Let $p(n)$ denote the statement that

$$p(n): 1 + 2 + 2^2 + \cdots + 2^{n-1} = 2^n - 1.$$

(i) If we substitute $n = 1$ in $p(n)$, then the left-hand side equals 1 and the right-hand side is $2^1 - 1 = 2 - 1 = 1$. This shows that $p(1)$ is true.

(ii) Assume that for a positive integer k, $p(k)$ is true. That is,

$$p(k): 1 + 2 + 2^2 + \cdots + 2^{k-1} = 2^k - 1.$$

We must show that $p(k)$ implies the truth of $p(k + 1)$, where

$$p(k + 1): 1 + 2 + 2^2 + \cdots + 2^k = 2^{k+1} - 1.$$

The induction hypothesis is

$$1 + 2 + 2^2 + \cdots + 2^{k-1} = 2^k - 1.$$

By adding 2^k to both sides of this equation, we get

$$1 + 2 + 2^2 + \cdots + 2^{k-1} + 2^k = 2^k - 1 + 2^k$$
$$= 2(2^k) - 1$$
$$= 2^{k+1} - 1.$$

This is precisely what we wanted to prove. Thus we have shown that if $p(k)$ is true, then $p(k + 1)$ is also true. This means that $p(n)$ is true for every positive integer n.

7. Let $p(n)$ denote the statement that

$$p(n): 1^2 + 2^2 + 3^2 + \cdots + n^2 = \frac{n(n + 1)(2n + 1)}{6}.$$

(i) If we substitute $n = 1$ in $p(n)$, then the left-hand side equals 1 and the right-hand side is $1(1 + 1)(2 + 1)/6 = 1$. This shows that $p(1)$ is true.

(ii) Assume that for a positive integer k, $p(k)$ is true. That is,

$$p(k): 1^2 + 2^2 + 3^2 + \cdots + k^2 = \frac{k(k + 1)(2k + 1)}{6}.$$

We must show that $p(k)$ implies the truth of $p(k + 1)$, where

$$p(k + 1):\ 1^2 + 2^2 + 3^2 + \cdots + (k + 1)^2 = \frac{(k + 1)[(k + 1) + 1][2(k + 1) + 1]}{6}$$

$$= \frac{(k + 1)(k + 2)(2k + 3)}{6}.$$

Our induction hypothesis is

$$1^2 + 2^2 + 3^2 + \cdots + k^2 = \frac{k(k + 1)(2k + 1)}{6}.$$

By adding $(k + 1)^2$ to both sides of this equation, we get

$$1^2 + 2^2 + 3^2 + \cdots + k^2 + (k + 1)^2 = \frac{k(k + 1)(2k + 1)}{6} + (k + 1)^2$$

$$= (k + 1)\left[\frac{k(2k + 1)}{6} + (k + 1)\right]$$

$$= (k + 1)\left[\frac{k(2k + 1) + 6(k + 1)}{6}\right]$$

$$= (k + 1)\left(\frac{2k^2 + k + 6k + 6}{6}\right)$$

$$= (k + 1)\left(\frac{2k^2 + 7k + 6}{6}\right)$$

$$= \frac{(k + 1)(k + 2)(2k + 3)}{6}.$$

9. Let $p(n)$ denote the statement that

$$p(n) = 1^3 + 2^3 + 3^3 + \cdots + n^3 = \left[\frac{n(n + 1)}{2}\right]^2.$$

(i) If we substitute $n = 1$ in $p(n)$, then the left-hand side equals $1^3 = 1$ and the right-hand side is $\left[\dfrac{1(1 + 1)}{2}\right]^2 = 1$.
This shows that $p(1)$ is true.
(ii) Assume that for a positive integer k, $p(k)$ is true. That is,

$$p(k) = 1^3 + 2^3 + 3^3 + \cdots + k^3 = \left[\frac{k(k + 1)}{2}\right]^2.$$

We need to show that $p(k)$ implies the truth of $p(k + 1)$, where

$$p(k + 1):\ 1^3 + 2^3 + 3^3 + \cdots + (k + 1)^3 = \left[\frac{(k + 1)(k + 2)}{2}\right]^2.$$

The induction hypothesis is

$$1^3 + 2^3 + 3^3 + \cdots + k^3 = \left[\frac{k(k + 1)}{2}\right]^2.$$

By adding $(k + 1)^3$ to both sides of this equation, we get

$$1^3 + 2^3 + 3^3 + \cdots + k^3 + (k + 1)^3 = \left[\frac{k(k + 1)}{2}\right]^2 + (k + 1)^3$$

$$= \frac{k^2(k + 1)^2}{4} + (k + 1)^3$$

$$= (k + 1)^2 \left[\frac{k^2}{4} + (k + 1)\right]$$

$$= (k + 1)^2 \left(\frac{k^2 + 4k + 4}{4}\right)$$

$$= \frac{(k + 1)^2(k + 2)^2}{4}$$

$$= \left[\frac{(k + 1)(k + 2)}{2}\right]^2.$$

This is precisely what we wanted to establish. Thus we have shown that if $p(k)$ is true, then $p(k + 1)$ is also true. This means that $p(n)$ is true for every positive integer n.

11. Let $p(n)$ denote the statement that

$$p(n): 3 + 3^2 + 3^3 + \cdots + 3^n = \frac{3(3^n - 1)}{2}.$$

(i) If we substitute $n = 1$ in $p(n)$, then the left-hand side equals 3 and the right-hand side equals $3(3 - 1)/2 = 3$. This shows that $p(1)$ is true.

(ii) Assume that for a positive integer k, $p(k)$ is true. That is,

$$p(k): 3 + 3^2 + 3^3 + \cdots + 3^k = \frac{3(3^k - 1)}{2}.$$

We must show that $p(k)$ implies the truth of $p(k + 1)$, where

$$p(k + 1): 3 + 3^2 + 3^3 + \cdots + 3^{k+1} = \frac{3(3^{k+1} - 1)}{2}.$$

The induction hypothesis is

$$3 + 3^2 + 3^3 + \cdots + 3^k = \frac{3(3^k - 1)}{2}.$$

By adding 3^{k+1} to both sides of this equation, we get

$$3 + 3^2 + 3^3 + \cdots + 3^k + 3^{k+1} = \frac{3(3^k - 1)}{2} + 3^{k+1}$$

$$= 3\left(\frac{3^k - 1}{2} + 3^k\right)$$

$$= 3\left[\frac{3^k - 1 + 2(3^k)}{2}\right]$$

$$= 3\left[\frac{3^k(1 + 2) - 1}{2}\right]$$

$$= 3\left(\frac{3^{k+1} - 1}{2}\right),$$

and this is what we wanted to establish. Thus we have shown that if $p(k)$ is true, so is $p(k + 1)$. This means that $p(n)$ is true for every positive integer n.

13. Let $p(n)$ denote the statement that

$$p(n): (1)(2) + (2)(3) + \cdots + n(n + 1) = \frac{n(n + 1)(n + 2)}{3}.$$

(i) If we substitute $n = 1$ in $p(n)$, then the left-hand side equals $(1)(2) = 2$ and the right-hand side is $1(1 + 1)(1 + 2)/3 = 2$. This shows that $p(1)$ is true.
(ii) Assume that for a positive integer k, $p(k)$ is true. That is,

$$p(k): (1)(2) + (2)(3) + \cdots + k(k + 1) = \frac{k(k + 1)(k + 2)}{3}.$$

We must show that $p(k)$ implies the truth of $p(k + 1)$, where

$$p(k + 1): (1)(2) + (2)(3) + \cdots + (k + 1)(k + 2) = \frac{(k + 1)(k + 2)(k + 3)}{3}.$$

The induction hypothesis is

$$(1)(2) + (2)(3) + \cdots + k(k + 1) = \frac{k(k + 1)(k + 2)}{3}.$$

By adding $(k + 1)(k + 2)$ to both sides of this equation, we get

$$(1)(2) + (2)(3) + \cdots + k(k + 1) + (k + 1)(k + 2) = \frac{k(k + 1)(k + 2)}{3} + (k + 1)(k + 2)$$

$$= (k + 1)(k + 2)\left(\frac{k}{3} + 1\right)$$

$$= \frac{(k + 1)(k + 2)(k + 3)}{3},$$

and this is what we wanted to prove. Thus we have shown that if $p(k)$ is true, so is $p(k + 1)$. This means that $p(n)$ is true for all positive integers n.

15. Let $p(n)$ denote the statement that

$$p(n): \frac{1}{(1)(2)} + \frac{1}{(2)(3)} + \cdots + \frac{1}{n(n + 1)} = \frac{n}{n + 1}.$$

(i) If we substitute $n = 1$ in $p(n)$, then the left-hand side equals $1/[(1)(2)] = \frac{1}{2}$, and the right-hand side is $1/(1 + 1) = \frac{1}{2}$. This shows that $p(1)$ is true.
(ii) Assume that for a positive integer k, $p(k)$ is true. That is,

$$p(k): \frac{1}{(1)(2)} + \frac{1}{(2)(3)} + \cdots + \frac{1}{k(k + 1)} = \frac{k}{k + 1}.$$

We must show that $p(k)$ implies the truth of $p(k + 1)$, where

$$p(k + 1): \frac{1}{(1)(2)} + \frac{1}{(2)(3)} + \cdots + \frac{1}{(k + 1)(k + 2)} = \frac{k + 1}{k + 2}.$$

The induction hypothesis is

$$\frac{1}{(1)(2)} + \frac{1}{(2)(3)} + \cdots + \frac{1}{k(k+1)} = \frac{k}{k+1}.$$

By adding $1/[(k+1)(k+2)]$ to both sides of this equation, we get

$$\frac{1}{(1)(2)} + \frac{1}{(2)(3)} + \cdots + \frac{1}{k(k+1)} + \frac{1}{(k+1)(k+2)} = \frac{k}{k+1} + \frac{1}{(k+1)(k+2)}$$
$$= \frac{k(k+2)+1}{(k+1)(k+2)}$$
$$= \frac{k^2 + 2k + 1}{(k+1)(k+2)}$$
$$= \frac{(k+1)^2}{(k+1)(k+2)}$$
$$= \frac{k+1}{k+2},$$

and this is what we wanted to prove. Thus we have shown that if $p(k)$ is true, so is $p(k+1)$. This means that $p(n)$ is true for every positive integer n.

17. Let $p(n)$ denote the statement that $n(n^2 + 2)$ is divisible by 3.
(i) Let $n = 1$. Then

$$n(n^2 + 2) = 1(1 + 2) = 3,$$

which is certainly divisible by 3. This means that $p(1)$ is true.
(ii) Assume the validity of the statement that

$$p(k): k(k^2 + 2) \text{ is divisible by 3.}$$

We must show that

$$p(k+1): (k+1)[(k+1)^2 + 2] = (k+1)(k^2 + 2k + 1 + 2)$$
$$= (k+1)(k^2 + 2k + 3)$$

is also divisible by 3. Our hypothesis that $p(k)$ is true implies the existence of a positive integer m such that $k(k^2 + 2) = 3m$. Consider $p(k+1)$.

$$p(k+1): (k+1)(k^2 + 2k + 3) = (k+1)(k^2 + 2k) + 3(k+1)$$
$$= (k+1)k(k+2) + 3(k+1)$$
$$= k(k+1)(k+2) + 3(k+1)$$
$$= k(k^2 + 3k + 2) + 3(k+1)$$
$$= k(k^2 + 2) + 3k^2 + 3(k+1)$$
$$= k(k^2 + 2) + 3(k^2 + k + 1).$$

We have assumed that $p(k): k(k^2 + 2)$ is divisible by 3. The second term, $3(k^2 + k + 1)$, is certainly divisible by 3. Thus $p(k+1)$ holds. This means that $p(n)$ holds for all integers n.

19. Let $p(n)$ denote the statement that 2^{4n} is divisible by 15.
(i) Let $n = 1$. Then

$$2^{4n} - 1 = 2^4 - 1 = 16 - 1 = 15,$$

which is certainly divisible by 15. This means that $p(1)$ is true.

(ii) Assume the validity of the statement that

$$p(k): 2^{4k} - 1 \text{ is divisible by 15.}$$

We must show that

$$p(k + 1): 2^{4(k+1)} - 1 = 2^{4k+4} - 1 \text{ is also divisible by 15.}$$

Our hypothesis that $p(k)$ is true implies the existence of a positive integer m such that

$$2^{4k} - 1 = 15m$$
$$2^{4k} = 1 + 15m.$$

Multiplying both sides of this equation by 2^4, we have

$$2^{4k+4} = 2^4(1 + 15m)$$
$$= 16(1 + 15m)$$
$$= 16 + 240m$$
$$= 15(1 + 16m) + 1;$$

that is,

$$2^{4k+4} - 1 = 15(1 + 16m),$$

which implies that $2^{4k+4} - 1$ is divisible by 15. This statement implies that $p(k + 1)$ is true. Thus we have shown that if $p(k)$ is true, so is $p(k + 1)$. This means that $p(n)$ is true for all positive integers n.

21. Let $p(n)$ denote the statement that $2^n \geq 2n$.
(i) Let $n = 1$. Then the left-hand side equals $2^1 = 2$, and the right-hand side also equals 2. This means that $p(1)$ is true.
(ii) Assume the validity of the statement that

$$p(k): 2^k \geq 2k.$$

We must show that

$$p(k + 1): 2^{k+1} \geq 2(k + 1).$$

The induction hypothesis is

$$2^k \geq 2k.$$

Multiplying both sides of this equation by 2, we obtain

$$2^{k+1} \geq 2(2k) = 4k.$$

Since $4k \geq 2k + 2$ for all positive integers k,

$$2^{k+1} \geq 4k \geq 2k + 2.$$

This means that $p(k + 1)$ is true. By the principal of mathematical induction, the assertion is true for all positive integers n.

Exercises for Review (Pages 588–589)

1. 5, 8, 11, 14, 17   **2.** 3, 7, 13, 21, 31   **3.** 3, 5, 9, 17, 33   **4.** $\dfrac{1}{3}, \dfrac{1}{9}, \dfrac{1}{27}, \dfrac{1}{81}, \dfrac{1}{243}$ **5.** $3n$ **6.** 2^{n-1}

7. $5n - 3$ **8.** $4n - 1$ **9.** 3^n **10.** $(-1)^{n+1}$ **11.** 20 **12.** 30 **13.** 34 **14.** 120 **15.** 80 **16.** 305

17. 7, 18, 33, 52 **18.** 2, 8, 20, 40 **19.** 2, 6, 14, 30 **20.** $\dfrac{1}{2}, \dfrac{3}{4}, \dfrac{7}{8}, \dfrac{15}{16}$ **21.** 17; 80 **22.** 23; 100

23. 36; 148 **24.** -15; -8 **25.** $3^{20} - 1$ **26.** $2^{20} - 1$ **27.** $\dfrac{5^{20} - 1}{4}$ **28.** $32\left[1 - \left(\dfrac{1}{2}\right)^{20}\right]$

29. 354,294; 531,440 **30.** 10,240; 20,475 **31.** 6, 7, 8, *or* 8, 7, 6 **32.** 9 years **33.** 5, 10, 20, 40, 80

34. \$506.25 **35.** 3 **36.** 0 **37.** $\dfrac{2}{3}$ **38.** 0 **39.** 1 **40.** limit does not exist **41.** 6 **42.** $\dfrac{10}{9}$ **43.** $\dfrac{21}{99}$

44. $\dfrac{105}{999}$ **45.** $\dfrac{1270}{999}$ **46.** $\dfrac{261}{99}$

47. Let $p(n)$ denote the given statement for all $n \in N$.
(i) If we substitute $n = 1$ in $p(n)$, then the left-hand side equals 1 and the right-hand side is $1(1 + 1)/2$, which also equals 1. This shows that $p(1)$ is true.
(ii) Assume that for a positive integer k, $p(k)$ is true. That is,

$$p(k): 1 + 2 + 3 + \cdots + k = \frac{k(k + 1)}{2}.$$

We must show that $p(k)$ implies that $p(k + 1)$ is true, where

$$p(k + 1): 1 + 2 + 3 + \cdots + (k + 1) = \frac{(k + 1)[(k + 1) + 1]}{2}.$$

The induction hypothesis is

$$1 + 2 + 3 + \cdots + k = \frac{k(k + 1)}{2}.$$

By adding $k + 1$ to both sides of this equation, we have

$$1 + 2 + 3 + \cdots + k + (k + 1) = \frac{k(k + 1)}{2} + (k + 1)$$

$$= (k + 1)\left(\frac{k}{2} + 1\right)$$

$$= (k + 1)\left(\frac{k + 2}{2}\right)$$

$$= \frac{(k + 1)[(k + 1) + 1]}{2}.$$

This is precisely the statement we wanted to establish. Thus we have proved that if $p(k)$ is true, then $p(k + 1)$ is also true. This means that $p(n)$ is true for every positive integer n.
48. see the solution on pages A-61 and A-62 to problem 7 (Exercise on page 586)

Chapter Test (Page 589)

1. 4, 7, 10, 13, 16 **2.** 224 **3.** 5, 6, 7; 7, 6, 5 **4.** 2025 **5.** 3, 6, 12, 24, . . . **6.** 1536; 3069

7. $\dfrac{2}{3}$ **8.** $\dfrac{7}{33}$ **9.** 120 **10.** \$857.50 **11.** does not exist

12. Let $p(n)$ denote the statement that

$$p(n): 2 + 4 + 6 + \cdots + 2n = n(n + 1).$$

(i) If we substitute $n = 1$ in $p(n)$, then the left-hand side equals 2, and the right-hand side is $1(1 + 1) = 1(2) = 2$. This shows that $p(1)$ is true.
(ii) Assume that for a positive integer k, $p(k)$ is true. That is,

$$p(k): 2 + 4 + 6 + \cdots + 2k = k(k + 1).$$

We must show that $p(k)$ implies the truth of $p(k + 1)$, where

$$p(k + 1): 2 + 4 + 6 + \cdots + 2k + 2(k + 1) = (k + 1)(k + 2).$$

The induction hypothesis is

$$2 + 4 + 6 + \cdots + 2k = k(k + 1).$$

By adding $2(k + 1)$ to both sides of this equation, we get

$$2 + 4 + 6 + \cdots + 2k + 2(k + 1) = k(k + 1) + 2(k + 1) = (k + 1)(k + 2).$$

This is precisely what we wanted to establish. Thus we have proved that if $p(k)$ is true, then $p(k + 1)$ is also true. This means that $p(n)$ is true for every positive integer n.

CHAPTER 11

Exercises (Page 596)

1. $x^2 + y^2 = 16$ **3.** $x^2 + y^2 - 2x - 4y - 20 = 0$ **5.** $x^2 + y^2 + 8x - 2y - 47 = 0$
7. $x^2 + y^2 + 6x + 8y - 24 = 0$ **9.** $x^2 + y^2 + 10y - 119 = 0$ **11.** $C(0, 0); r = 8$ **13.** $C(1, -1); r = 5$
15. $C(3, 4); r = 9$ **17.** $C(-3, 6); r = 5$ **19.** $C(-7, -4); r = 6$ **21.** $x^2 + y^2 - 10x - 6y + 29 = 0$
23.

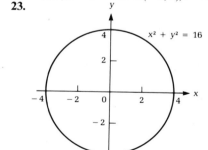

25.

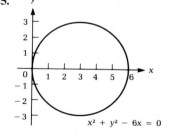

27.

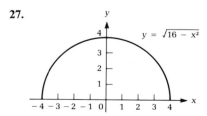

Exercises (Pages 608–609)

1.
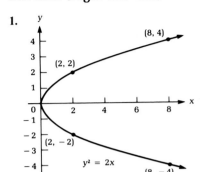

3.
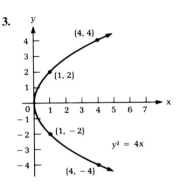

5.
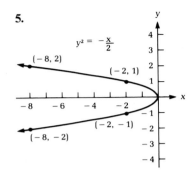

7.
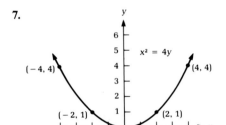

9.
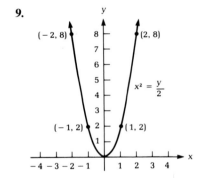

11.
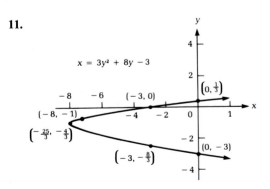

13.
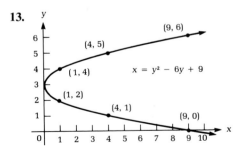

15. $V(0, 0)$, $F(10, 0)$; $x = -10$ **17.** $V(0, 0)$; $F(-\frac{7}{12}, 0)$; $x = \frac{7}{12}$ **19.** $V(0, 0)$; $F(0, \frac{5}{2})$; $y = -\frac{5}{2}$ **21.** $y^2 = 24x$
23. $x^2 = -12y$ **25.** $y^2 = 16x$

27.

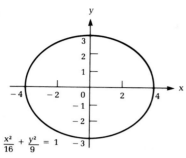

$$\frac{x^2}{16} + \frac{y^2}{9} = 1$$

29.

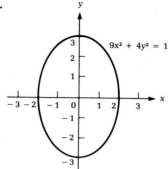

$9x^2 + 4y^2 = 1$

31.

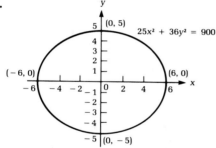

$25x^2 + 36y^2 = 900$

33. $\dfrac{x^2}{16} + \dfrac{y^2}{12} = 1$

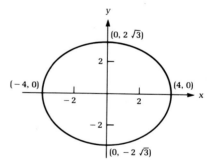

35. $\dfrac{x^2}{25} + \dfrac{y^2}{16} = 1$

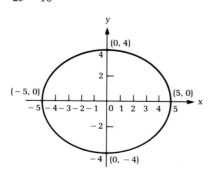

37.

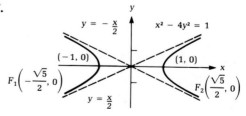

$y = -\dfrac{x}{2}$ $x^2 - 4y^2 = 1$

$F_1\left(-\dfrac{\sqrt{5}}{2}, 0\right)$ $(-1, 0)$ $(1, 0)$ $F_2\left(\dfrac{\sqrt{5}}{2}, 0\right)$

$y = \dfrac{x}{2}$

39.

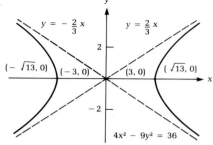

41.

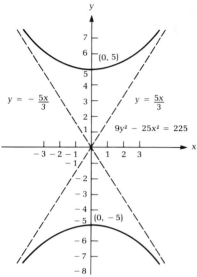

43. $\dfrac{x^2}{4} - \dfrac{y^2}{5} = 1$ **45.** $\dfrac{y^2}{9} - \dfrac{x^2}{16} = 1$ **47.** $\dfrac{x^2}{9} - \dfrac{y^2}{36} = 1$ **49.** $\dfrac{x^2}{20} - \dfrac{y^2}{5} = 1$

Exercises (Pages 612–613)

1. 48   **3.** 120   **5.** 1200   **7.** 20,160   **9.** 60   **11.** 720   **13.** 120; 72; 24   **15.** 40,320   **17.** 8,000,000

Exercises (Pages 619–620)

1. 696   **3.** 40,200   **5.** 720   **7.** 151,200   **9.** 10,080   **11.** 56   **13.** 330   **15.** 630,630   **17.** 8   **19.** 336
21. 504   **23.** 5040   **25.** 11,880   **27.** 2184   **29.** 1560   **31.** 132,600   **33.** n **35.** $n(n-1)(n-2)$ **37.** 24
39. 720   **41.** 720   **43.** 5040   **45. (a)** 210   **(b)** 840   **(c)** 2520   **(d)** 5040   **47.** 360   **49.** 30,240
51. 10,080   **53.** 720; 120; 24; 240   **55.** 720   **57.** 5040   **59.** 30,240

Exercises (Pages 625–627)

1. 56   **3.** 125,970   **5.** 4950   **7.** 35   **9.** $\dfrac{25!}{9!16!}$ **11.** 80   **13. (a)** 252   **(b)** 100   **15. (a)** 364   **(b)** 91

17. (a) 3003   **(b)** 1001   **(c)** 2002   **19. (a)** 3744   **(b)** 54,912   **(c)** 624   **(d)** 4   **(e)** 36   **21.** 27,720
23. 495; 165   **25.** 7000   **27.** 1, 7, 21, 35, 35, 21, 7, 1

Exercises (Pages 633–634)

1. $x^7 + 7x^6y + 21x^5y^2 + 35x^4y^3 + 35x^3y^4 + 21x^2y^5 + 7xy^6 + y^7$ **3.** $x^3 + 6x^2y + 12xy^2 + 8y^3$
5. $32x^5 + 240x^4y + 720x^3y^2 + 1080x^2y^3 + 810xy^4 + 243y^5$

7. $1 - 8x + 28x^2 - 56x^3 + 70x^4 - 56x^5 + 28x^6 - 8x^7 + x^8$ **9.** $64x^6 - 192x^4 + 240x^2 - 160 + \dfrac{60}{x^2} - \dfrac{12}{x^4} + \dfrac{1}{x^6}$

11. 16.   **13.** 64.   **15.** 2^n **17.** 1.3108   **19.** 0.9510   **21.** 1.0242   **23.** 2.8127   **25. (a)** 45   **(b)** 680   **(c)** $\dfrac{2n!}{n!n!}$
27. \$15,529.24

Exercises for Review (Pages 635–637)

1.
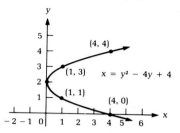

2.
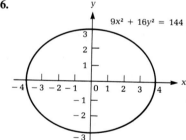

3.
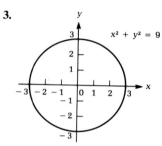

4.
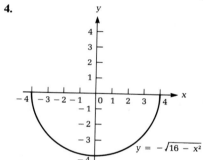

5.
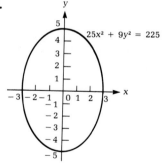

6.
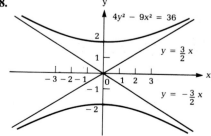

7.
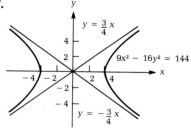

8.
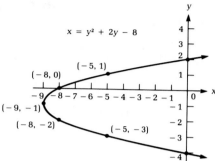

9. $C(3, 4), r = 6$ **10.** $C(-2, -3), r = 4$ **11.** $y^2 = 16x$ **12.** $x^2 = 8y$ **13.** $y^2 = 8x$ **14.** $x^2 = y$
15. $8x^2 + 9y^2 = 72$ **16.** $100x^2 + 9y^2 = 225$ **17.** $9y^2 - 16x^2 = 144$ **18.** $28x^2 - 36y^2 = 63$ **19.** 32 **20.** 32
21. 30,240 **22. (a)** 7,893,600 **(b)** 11,881,376 **23.** 720 **24.** 40,320 **25.** 1,048,576 **26.** 6,760,000
27. (a) 720 **(b)** 7776 **28.** 120 **29.** 126 **30.** 792 **31. (a)** 1680 **(b)** 70 **32.** 1400 **33.** 350
34. (a) 1,098,240 **(b)** 123,552 **35. (a)** 6188 **(b)** 2040 **(c)** 136
36. $81x^4 + 216x^3y + 216x^2y^2 + 96xy^3 + 16y^4$
37. $1024x^5 - 3840x^4y + 5760x^3y^2 - 4320x^2y^3 + 1620xy^4 - 243y^5$
38. $1 - 12x + 60x^2 - 160x^3 + 240x^4 - 192x^5 + 64x^6$
39. $1 + 18x + 135x^2 + 540x^3 + 1215x^4 + 1458x^5 + 729x^6$ **40.** $108,864x^3y^5$ **41.** -252 **42.** 1.4693
43. 0.9415 **44.** 1.6771 **45.** 0.9801

Chapter Test (Page 637)

1. (a)

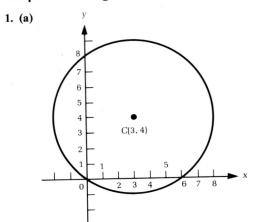

(b)

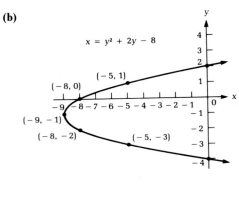

2. $C(3, 4), r = 6$ **3.** $y^2 = 16x$ **4.** $8x^2 + 9y^2 = 72$ **5.** Vertices: $(-3, 0), (3, 0)$ **6.** 8,000,000 **7.** 120; 32
Foci: $(-5, 0), (5, 0)$
Asymptotes: $y = \pm\frac{4}{3}x$

8. 1050 **9.** $\dfrac{15!}{(4!)(5!)(6!)} = 630,630$ **10.** 5148 **11.** $256x^4 - 1280x^3y + 2400x^2y^2 - 2000xy^3 + 625y^4$
12. $108,864x^3y^5$

INDEX